Jie Lu, Da Ruan, Guangquan Zhang (Eds.)

E-Service Intelligence

Studies in Computational Intelligence, Volume 37

Editor-in-chief
Prof. Janusz Kacprzyk
Systems Research Institute
Polish Academy of Sciences
ul. Newelska 6
01-447 Warsaw
Poland
E-mail: kacprzyk@ibspan.waw.pl

Further volumes of this series can be found on our homepage: springer.com

Vol. 20. Wojciech Penczek, Agata Półrola
Advances in Verification of Time Petri Nets and Timed Automata, 2006
ISBN 3-540-32869-6

Vol. 21. Cândida Ferreira
Gene Expression on Programming: Mathematical Modeling by an Artificial Intelligence, 2006
ISBN 3-540-32796-7

Vol. 22. N. Nedjah, E. Alba, L. de Macedo Mourelle (Eds.)
Parallel Evolutionary Computations, 2006
ISBN 3-540-32837-8

Vol. 23. M. Last, Z. Volkovich, A. Kandel (Eds.)
Algorithmic Techniques for Data Mining, 2006
ISBN 3-540-33880-2

Vol. 24. Alakananda Bhattacharya, Amit Konar, Ajit K. Mandal
Parallel and Distributed Logic Programming, 2006
ISBN 3-540-33458-0

Vol. 25. Zoltán Ésik, Carlos Martín-Vide, Victor Mitrana (Eds.)
Recent Advances in Formal Languages and Applications, 2006
ISBN 3-540-33460-2

Vol. 26. Nadia Nedjah, Luiza de Macedo Mourelle (Eds.)
Swarm Intelligent Systems, 2006
ISBN 3-540-33868-3

Vol. 27. Vassilis G. Kaburlasos
Towards a Unified Modeling and Knowledge-Representation based on Lattice Theory, 2006
ISBN 3-540-34169-2

Vol. 28. Brahim Chaib-draa, Jörg P. Müller (Eds.)
Multiagent based Supply Chain Management, 2006
ISBN 3-540-33875-6

Vol. 29. Sai Sumathi, S.N. Sivanandam
Introduction to Data Mining and its Applications, 2006
ISBN 3-540-34689-9

Vol. 30. Yukio Ohsawa, Shusaku Tsumoto (Eds.)
Chance Discoveries in Real World Decision Making, 2006
ISBN 3-540-34352-0

Vol. 31. Ajith Abraham, Crina Grosan, Vitorino Ramos (Eds.)
Stigmergic Optimization, 2006
ISBN 3-540-34689-9

Vol. 32. Akira Hirose
Complex-Valued Neural Networks, 2006
ISBN 3-540-33456-4

Vol. 33. Martin Pelikan, Kumara Sastry, Erick Cantú-Paz (Eds.)
Scalable Optimization via Probabilistic Modeling, 2006
ISBN 3-540-34953-7

Vol. 34. Ajith Abraham, Crina Grosan, Vitorino Ramos (Eds.)
Swarm Intelligence in Data Mining, 2006
ISBN 3-540-34955-3

Vol. 35. Ke Chen, Lipo Wang (Eds.)
Trends in Neural Computation, 2007
ISBN 3-540-36121-9

Vol. 36. Ildar Batyrshin, Janusz Kacprzyk, Leonid Sheremetov, Lotfi A. Zadeh (Eds.)
Perception-based Data Mining and Decision Making in Economics and Finance, 2006
ISBN 3-540-36244-4

Vol. 37. Jie Lu, Da Ruan, Guangquan Zhang (Eds.)
E-Service Intelligence, 2007
ISBN 3-540-37015-3

Jie Lu
Da Ruan
Guangquan Zhang (Eds.)

E-Service Intelligence

Methodologies, Technologies and Applications

With 190 Figures and 69 Tables

Springer

Prof. Dr. Jie Lu
Faculty of Information Technology
University of Technology
Sydney (UTS), PO Box 123
Broadway, NSW 2007
Australia
E-mail: jielu@it.uts.edu.au

Prof. Dr. Guangquan Zhang
Faculty of Information Technology
University of Technology
Sydney (UTS), PO Box 123
Broadway, NSW 2007
Australia
E-mail: zhangg@it.uts.edu.au

Prof. Dr. Da Ruan
The Belgian Nuclear Research Centre (SCK·CEN)
Boeretang 200, 2400 Mol
Belgium
E-mail: druan@sckcen.be

and

Department of Applied Mathematics
and Computer Science
Ghent University
Krijgslaan 281 (S9), 9000 Gent
Belgium
E-mail: Da.Ruan@UGent.be

Library of Congress Control Number: 2006930406

ISSN print edition: 1860-949X
ISSN electronic edition: 1860-9503
ISBN-10 3-540-37015-3 Springer Berlin Heidelberg New York
ISBN-13 978-3-540-37015-4 Springer Berlin Heidelberg New York

Springer is a part of Springer Science+Business Media
springer.com

Cover design: deblik, Berlin
Typesetting by the editors and SPi
Printed on acid-free paper SPIN: 11430056 89/SPi 5 4 3 2 1 0

PREFACE

Many business organizations and government departments are nowadays developing and providing Internet based electronic services (e-services) featuring various intelligent functions. This form of e-services is commonly called *e-service intelligence* (ESI). ESI integrates intelligent technologies and methodologies into e-service systems for realizing intelligent Internet information searching, presentation, provision, recommendation, online system design, implementation, and assessment to Internet users. These intelligent technologies include machine learning, soft computing, intelligent languages, and data mining etc. ESI has been recently identified as a new direction for the future development stage of e-services.

E-services offer great opportunities and challenges for many areas of services, such as government, education, tourism, commerce, marketing, finance, and logistics. They thus involve various online service providers, delivery systems and applications including e-government, e-learning, e-shopping, e-marketing, e-banking, and e-logistics. ESI is providing with a much higher quality presentation of web information, personalized online recommendation, customaries online decision support, direct user participation in organizational planning, and more integrated seamless link online services. It also has e-services evolved into online knowledge discovery and user analysis, and becomes adaptive, proactive and accessible from a broader variety of devices. We have begun to see more and more successful developments in building intelligent functions and systems of e-services such as web search by fuzzy matching; web usage mining by fuzzy sequential pattern; Internet shopping systems using multi-agents; product recommender systems supported by genetic algorithms; e-logistics systems using optimization models; online customer segments using data mining; rough set based ontology of mapping in online service integration; online question/answer service systems using fuzzy logic; visualized web information presentation; game theory based e-negotiation; inference approach in e-service cost benefic analysis; and knowledge discovery through using case-based reasoning in e-learning systems. It is thus instructive and vital to gather current trends and provide a high quality forum for theoretical research results of ESI and practical developments of intelligent e-service applications for various government and business organizations.

This book aims at offering a thorough introduction and systematic overview of the new field. It consists of 32 chapters invited and selected from more than 60 submissions distributed in about 15 countries and regions and covers the state-of-the-art of the research and development in various aspects including both theorems and applications of ESI in five parts: (1) E-Services and Intelligent Techniques; (2) Web Information Presentation, Search, and Mining; (3) Personalization, Privacy, and Trust in E-Services; (4) E-Service Evaluation, Optimization and Knowledge Discovery; and (5) Intelligent E-Service System Developments.

The intelligent techniques applied in e-services, reported in this book, include fuzzy logic, expert systems, case based reasoning, artificial neural networks, Bayesian network, game theory, multi-criteria decision analysis, rough sets, data mining, linguistic techniques, multi-agents, ontology, sensory model, Chaos theory, genetic algorithms, and many of their combinations. The detailed application fields of ESI, presented in this book, involve personal e-banking, e-negotiators, e-map, one-stop e-shopping, secure e-transactions, integrated e-supply chain, learner-oriented e-learning, e-government service integration, online auctions, online payments, online sports services, online human resource management, online customer experience management, online user behaviors analysis, and online user trust evaluation. The research methodologies shown in these chapters include theoretical investigations, framework development, model establishment, approach proposing, case based study, survey data analysis, hypothesis testing, software implementation, and experimental assessment.

There are more than 20 national research grants to have supported the completeness of the researches presented in this book. Special thanks are due to all the authors of all chapters for their timely cooperation. Each chapter of the book is self-contained and we hope this volume will benefit many readers around the world.

June 2006

Jie Lu, University of Technology, Sydney (UTS), Australia

Da Ruan, Belgian Nuclear Research Centre (SCK•CEN) & Ghent University, Belgium

Guangquan Zhang, University of Technology, Sydney (UTS), Australia

CONTENTS

E-Service Intelligence: An Introduction

Jie Lu[#], Da Ruan*, and Guangquan Zhang[#]

[#]Faculty of Information Technology, University of Technology, Sydney (UTS) PO Box 123, Broadway, NSW 2007, Australia. Email: {jielu, zhangg}@it.uts.edu.au

*Belgian Nuclear Research Centre (SCK•CEN), Boeretang 200, 2400 Mol, Belgium. Email: druan@sckcen.be

Abstract. *E-service intelligence* is a new research field that deals with fundamental roles, social impacts and practical applications of various intelligent technologies on the Internet based e-service applications that are provided by e-government, e-business, e-commerce, e-market, e-finance, and e-learning systems, to name a few. This chapter offers a thorough introduction and systematic overview of the new field e-service intelligence mainly based on computational intelligence techniques. It covers the state-of-the-art of the research and development in various aspects including both theorems and applications of e-service intelligence. Moreover, it demonstrates how adaptations of existing computational intelligent technologies benefit from the development of e-service applications in online customer decision, personalized services, web mining, online searching/data retrieval, and various web-based support systems.

1 E-Services

Electronic-services (e-services) involve various types, delivery systems, advanced information technologies, methodologies and applications of online services that are provided by e-government, e-business, e-commerce, e-market, e-finance, and e-learning systems, to name a few. The term e-services is typically used to describe a variety of internet based electronic interactions ranging from basic services, such as the delivery of news and stock quotes, to smart services, such as the delivery of context-aware emergency services (Chidambaram, 2001). In a fully web-enabled smart e-service environment, the services are likely to be composed of many interacting components and have the potential for "combinatorial

J. Lu et al.: *E-Service Intelligence: An Introduction*, Studies in Computational Intelligence (SCI) **37**, 1–33 (2007)
www.springerlink.com

explosion" described in cybernetics and systems theory. E-services are likely to push the limits of software engineering in terms of analysis, design, security, and testing. Moreover, it will have and is conducting long-term impacts of e-services on individuals, institutions, and society.

Over the last decade, many government and business online services have mainly gone through three stages in most industrialized countries: (a) online information presentation, (b) online transaction, and (c) online information integration. In the third stage, all possible related services that might be provided by the same agency, different agencies, agencies in other jurisdictions, and private partners have been integrated in an either vertical or horizontal way. Businesses and citizens can deal with online services as a single cohesive entity and for services to be delivered in a seamless manner-'one-stop shop.' Individuals wish to be able to get information and complete services without worrying whether it involves different agencies or layers of the business and government, and also wish to receive personalized services to avoid information overload problems. Clearly, the keyword *intelligence* will be the next paradigm shift in the e-services thanks to Internet technological advances (Lu et al., 2006). To provide intelligence for e-services, various intelligent technologies including fuzzy logic, expert systems, machine learning, neural networks, Bayesian network, game theory, optimization, rough sets, data mining, multi-agents and evolutionary algorithms etc. are being applied in various e-service approaches, systems, and applications. In the framework of intelligent technologies, government and business e-services will provide with a much higher quality for online information presentation, online information searching, personalized recommendation, website evaluation, and web based support systems. Some successful developments have appeared recently in applying various intelligent techniques to build intelligent e-service support systems, such as intelligent e-negotiation systems, intelligent e-shopping systems, intelligent online customer management systems, and intelligent online decision support systems. Literature has also showed some successful investigations based on intelligent approaches to evaluate e-service systems, conduct web user classification, help users' online trading and support users' online decision making. In the following we describe the application of intelligent techniques in the Internet based e-service development, implementation, and management. The chapter is organized as follows. Section 2 summarizes the role of intelligent techniques in e-service applications. Section 3 highlights applications of intelligent techniques in the web information presentation and online search. Section 4 presents how intelligent techniques are integrated with web mining. Section 5 discusses the implementation of e-service personalization supported by intelligent approaches. Section 6 presents how intelligent techniques can help our e-service evaluation. Section 7 analyses several typical intelligent e-service systems with various intelligent techniques. Section 8 displays most available intelligent e-service models. Finally, Section 9 concludes the chapter and its related future research direction.

2 The Role of Intelligent Techniques in E-Services

Artifitial intelligent techniques incluing conventional intelligence, such as expert systems, machine learning, case based reasong, Bayesian network, and computational intelligence, such as artificial neural networks, fuzzy systems, evolutionary computation are playing important roles in e-service applications. The power of each technique or methodology as a design tool is limited only by the designer's imagination. Two features, in particular, stand out: (1) many of them are biologically inspired, and (2) they are all capable of solving non-linear problems (Ruan, 1997). The techniques and methodologies are for the most part complementary and synergistic rather than competitive. Intelligent techniques have already enjoyed considerable success in e-services, which have proven to be instructive and vital (Lu et al., 2006).

Computational intelligence (CI) research, closely assosiated with soft computing (SC), aims to use learning, adaptive, or evolutionary computation to create programs that are, in some sense, intelligent. Fuzzy logic (FL) (Zadeh, 1965) is designed to handle imprecise linguistic concepts such as *small, big, low, high, young*, or *old.* Systems based on FL exhibit an inherent flexibility and have proven to be successful in a variety of industrial control and pattern-recognition tasks ranging from handwriting recognition to traffic control. Central to the flexibility that FL provides is the notion of fuzzy sets. Fuzzy sets are the basic concept supporting fuzzy theory. The main research fields in fuzzy theory are fuzzy sets, fuzzy logic, and fuzzy measure. Fuzzy reasoning or approximate reasoning is an application of fuzzy logic to knowledge processing. Fuzzy control is an application of fuzzy reasoning to control. One of the main strengths of FL compared with other schemes to deal with imprecise data is that their knowledge bases, which are in a rule format, are easy to examine and understand. This rule format also makes it easy to update and maintain the knowledge base. Experts think in imprecise terms, such as *very often* and *almost never, usually* and *hardly ever, frequently* and *occasionally*, and use linguistic variables such as the above-mentioned *small, big, low* and *high* etc. FL provides a means to compute with words. In concentrates on the use of fuzzy values that capture the meaning of words, human reasoning and decision making, and provides a way of breaking through the computational burden of traditional expert systems. As for the limitations of FL, the main shortcoming is that the membership functions and rules have to be specified manually. Determining membership functions can be a time-consuming, trial-and-error process. Moreover the elicitation of rules from human experts can be an expensive, error-prone procedure. As a new progress, fuzzy logic is moving its applications from computing with numbers to computing with words; and from manipulation of measurements to manipulation of perceptions (Zadeh, 1999). In particular, Fuzzy Logic and the Internet (FLINT) as an important topic has been proposed by Zadeh and has attracted many researchers to work on it (Loia et al., 2004). Zadeh (2003) also indicated that existing web search engines would need evolving into question-answering systems. Achievement of this goal requires a quantum jump in the web IQ of existing search engines. A view is that bivalent-logic-based methods have intrinsically limited capability to address complex problems which arise in deduction from information which is pervasively ill-structured, uncertain and imprecise. Web information is world knowledge that humans acquire through experience and education. Imprecision of perception-based information is a major obstacle to dealing with world knowledge through the use of methods based on bivalent logic and bivalent-logic-based probability theory. What is needed for this purpose is a collection of tools drawn from fuzzy logic-- a logic in which everything is, or is allowed to be, a matter of degree.

A neural network (NN), also called an artificial neural network (ANN) in computational intelligence, is an interconnected group of artificial neurons that uses a mathematical or computational model for information processing based on a connectionist approach to computation. In most cases an NN is an adaptive system that changes its structure based on external or internal information that flows through the network. Two of its main areas of application are classification and decision problems (Anagnostopoulos et al., 2004). Neural networks are chosen mainly for computational reasons since, once trained, they operate very fast and the creation of thesauri and indices is avoided. Many experimental investigations on the use of NNs for implementing relevance feedback in an interactive information retrieval system have been proposed. In these investigations, the anticipated outcome was to compare relevance feedback mechanisms with NNs based techniques on the basis of relevant and non-relevant document segmentation (Crestani, 1994, Anagnostopoulos et al., 2004).

Evolutionary Computation (EC) is the general term for several computational techniques that are based to some degree on the evolution of biological life in the natural world. It is a subfield of CI involving combinatorial optimization problems. It mostly involves metaheuristic optimization algorithms such as genetic algorithms, evolutionary programming, evolution strategy, learning classifier systems, ant colony optimization and particle swarm optimization and so on (http://en.wikipedia.org/wiki/Computational_intelligence). As a particular class of EC, a genetic algorithm (GA) is a search technique used to find approximate solutions to optimization and search problems. GAs are typically implemented as a computer simulation in which a population of abstract representations of candidate solutions to an optimization problem evolves toward better solutions.

As a broad subfield ofAI, machine learning is concerned with the development of algorithms and techniques, which allow computers to "learn" including inductive learning, and deductive learning. Extracting rules and patterns out from massive data sets is one of important tasks and has been widely used in the field of data mining. Machine learning research has also developed a set of useful inference algorithms. A wide spectrum of applications of machine learning such as search engines, stock market analysis, and object recognition have been well developed.

Data mining, also known as Knowledge-Discovery in Databases, is the process of automatically searching large volumes of data for patterns. Data mining is a fairly recent and contemporary topic in computing. It mainly applies many computational techniques from statistics, machine learning and pattern recognition and can be seen as a kind of intelligent techniques. Web mining refers to the use of data mining techniques to automatically retrieve, extract and evaluate information for knowledge discovery from web documents and services. Web mining can be divided into three categories: structure mining, usage mining, and content mining. Web structure mining is a research field focused on using the analysis of the link structure of the web, and one of its purposes is to identify more preferable documents. Web usage mining, also known as Web Log mining, is the process of extracting interesting patterns in web access logs. Analyzing the web access logs of different web sites can help understand the user behavior and the web structure, thereby improving the design of this colossal collection of resources. Web content mining is an automatic process that extracts patterns from on-line information, such as the HTML files, images, or E-mails, and it already goes beyond only keyword extraction or some simple statistics of words and phrases in documents. With more researchers continue to develop data mining techniques with intelligent functions, web mining intelligence is playing an increasingly important role in meeting the challenges of developing the intelligent e-services.

Expert systems apply reasoning capabilities to reach a conclusion. An expert system can process large amounts of known information and provide conclusions based on them. Case based reasong and Bayesian network techniques are two main techniques used in expert systems.

With more and more applications of intelligent techniques in e-services, the integration between intelligence and web-based technology has been appeared. Some hybrid technologies of CI and web technology are dedicating to the improvement of e-service intelligence. Figure 1 shows main intelligent techniques and their applications in popular e-service fields with some typical examples.

As shown in Figure 1, intelligent techniques and methodologies as additional useful tools have nevertheless been successfully applied to some of the most interesting e-service areas. Typical e-service applications with the support of intelligent techniques will be briefly outlined over the rest of the chapter and will be detailed given by the rest chapters of this book.

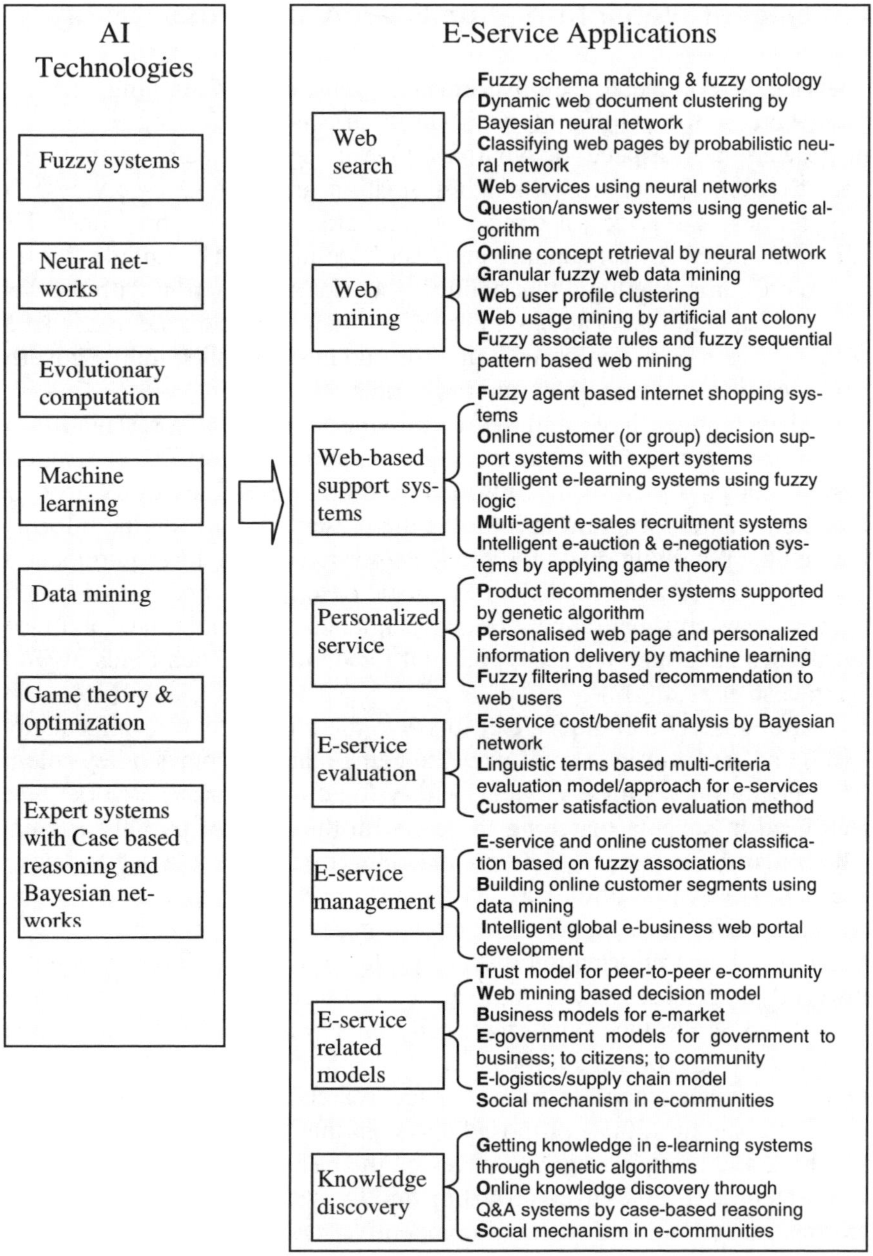

Figure 1 A list of intelligent techniques and web based applications

3 Intelligent Web Information Search and Presentation

E-service, e-commerce, e-government, e-business, e-learning, and e-science etc. -- the coining of such words reflects the growing importance of the web in all aspects of our lives. Consumers are spending more and more time on the web to search information and receive e-services. To provide these e-services, a typical web program involves three tiers. The front-end is the browser running on a user's computer. The middle-tier is a server, executing the logic that controls the user's interaction with the web site. The back-end is a database, providing the information the user wishes to access, such as a catalog of items for purchase, a collection of government records, or a store of learning materials.

AI techniques have abilities to provide a natural representation of human conceptualization and offer computational theory of perceptions (Zadeh, 2001). They are making many more contributions to each of the three tiers of web program to improve the presentation for online information and the searching on online products/services. With the requirement of semantic web from the facility of linguistic descriptions and partial matching, fuzzy systems, neural networks, machine learning and other intelligent approaches have shown a great potential for supporting such kinds of work (Nikravesh et al., 2002).

Applying AI techniques can improve the quality of web information search and retrieval in several ways. Schema matching plays a key role in data integration within e-services. Fuzzy logic and neural networks have been used in schema matching to deal with the issue of partial matching and similarity. In general, schema matching methods use a set of known rules. However, it is often hard to get such a set of rules. This is because the relationships of schema elements are usually uncertain and the availability of schema information may vary. Neural networks have emerged as powerful pattern recognition and can learn the similarities among data directly from its instances and empirically infer solutions from data without prior knowledge of regularities (Li et al., 2005). Neural network techniques have been applied in web page searching. For example, Anagnostopoulos et al. (2004) proposed a system that is capable of identifying, searching and categorizing web pages on the basis of information filtering. The system is a three-layer probabilistic NN with biases and radial basis neurons in the middle layer and competitive neurons in the output layer. The probabilistic NN scopes to identify e-commerce web pages and classify them to the respective type according to a framework, which describes the fundamental phases of commercial transactions in the web. Nikravesh and Azvin (2002) proposed a fuzzy conceptual-based search engine using conceptual semantic indexing, and discussed fuzzy queries, search, and decision support systems. Fuzzy set techniques based conceptual search models and methods can be used for intelligent online information retrieval through conceptual matching of both text and images. Under the support of intelligent techniques, the selected query does not need to match the decision criteria exactly, which gives the system a more human-like behavior.

Existing web search engines have many remarkable capabilities. However, what is not among them is the deduction capability-the capability to answer a query by a synthesis of information which resides in various parts of the knowledge base. A question-answering (Q&A) system is by definition a system which has this capability. At current stage of e-service development, Q&A systems are important aiding tools for helping users effectively obtain information from the web. Such a system often finds answers by matching the features of previous questions to current ones, and considering the weights of the features. To improve the prediction accuracy, case-based reasoning approach, genetic algorithms, fuzzy logic, neural networks, and their combinations all have been applied in practice. Literature shows that case-based reasoning shows significant promise for improving the effectiveness of complex and unstructured decision-making. In a Q&A system, it can use a user's feed back gradually and make inferences using analogy to obtain similar experience for solving problems. It often combines with other intelligent methods such as fuzzy logic and neural networks to improve its performance. For example, Zhang et al. (2004b) used a genetic algorithm and case-based reasoning into a Q&A system and developed an interactive Q&A engine.

The semantic web (Berners-Lee et al., 2001) aims at creating a platform where information has its semantics and can be understood and processed by computers themselves with minimum human interference. Ontology theory and its related technology have been developed to help construct such a platform because ontology promises to encode certain levels of semantics for information and orders a set of common vocabulary for people or computers to communicate with. Fuzzy ontology has been developed to support schema and data integration for web-based applications (Parry, 2004, Widyantoro and Yen, 2001). For example, Widyantoro and Yen (2001) implemented a web-based, domain-specific search engine for searching abstracts of research papers. The system uses fuzzy ontology of term associations to support the feature. The fuzzy ontology is automatically generated from the information contained in its collection. The FOGA (Fuzzy Ontology Generation Framework) for automatic generation of fuzzy ontology on uncertainty information is developed by Quan et al. (2004a, b, c). The FOGA framework comprises fuzzy formal concept analysis, fuzzy conceptual clustering and fuzzy ontology generation. First, fuzzy formal concept analysis incorporates fuzzy logic into formal concept analysis to form a fuzzy concept lattice. Fuzzy conceptual clustering then constructs the concept hierarchy from the fuzzy concept lattice. Finally, the fuzzy ontology generation generates the fuzzy ontology from the concept hierarchy.

The real way to get individualized interaction between a user and a website is to present the user with a variety of options and to let the user choose what is of interest at that specific time. If the information space is designed well, then this choice is easy, and the user achieves optimal information through the use of *natural intelligence,* that is, the choices are easy to understand so that users *know* what they will see if they click a link, and what they de-select by not following other links (Bonett, 2006). However, it is not easy to design a well information space. The **AI** approach is therefore used to help computers *guess* the user's needs, and find information and services the user needs to the user. This requirement needs more applications of intelligent techniques in web searching and information presentation.

4 Intelligent Web Mining

Classification and association-rule discovery are two of the most important tasks addressed in the web mining literature. The two tasks involve mining web data, web customers, web documents, web-based transactions and various web based applications and each has developed a set of approaches. Also, extensive research has been carried out to integrate approaches from both sides of the tasks to conduct high performance web mining. In the meantime, some intelligence techniques including neural networks, fuzzy logic (also its application in fuzzy associate rules and fuzzy sequential pattern), machine learning, genetic algorithms, Bayesian network, and case-based reasoning have also been applied into web mining. Some successful examples include web user profile clustering, web document classification, web document clustering, online concept retrieval, and web key information finding (Wang et al., 2006).

One of the main aims of web mining is to find out web consumers' customer behavior patterns. By these patterns, businesses and governments can target marketing, e.g. input patterns to a web server combining with web pages so that when pattern-related customers access the website corresponding ways of marketing can be created (Chen et al., 2004b). In general, web mining can help establish marketing patterns and customize marketing to bring right products and services to right customers. It can also establish potential customers' list to help make decisions in customer relationship management. Intelligent techniques can be combined with traditional web mining approaches to improve the quality of mining. For example, by applying fuzzy set techniques to set up corresponding marketing patterns, businesses can better predict which kind of customers are loyal for a particular time being, and therefore can help businesses hold best customers and filter potential customers precisely. Fuzzy Adaptive Resonance (ART) has also been used for clustering customers in groups for targeting (Jain and Krishnapuram, 2001).

A related important aspect of web mining is the clustering of customer similar profiles to create customer “segments” (Mobasher et al., 2000). Clustered user profiles are a good option when there exist insufficient data to build individual profiles. Most customer attributes such as “quality-conscious” and “calorie-conscious” are inherently uncertain or fuzzy. Also, customer segments are not crisp. Thus, fuzzy set theory can play a major role in customer profile representations and clustering (Jain and Krishnapuram, 2001). Bautista et al. (2000) used a genetic algorithm to build an adaptive consumer profile based on documents retrieved by users. A fuzzy classification and a genetic term-selection process together provide a better utilization of valuable knowledge to learn the current and future interests of users. Granular computing techniques have been used in web mining applications to enhance the intelligent functionality of mining systems. For example, Zhang et al. (2003) used both fuzzy computing and interval computing techniques to design a fuzzy-interval data mining system for credit card companies with actual large data sets. Park (2000) also introduced a neural network-based data mining method to a company's internal customer data for target marketing. A fuzzy ART NN proposed in this study takes customer's purchasing history as input values and cluster similar customers into groups.

Web document and web pages classification as another important web mining aspect has used NN architectures (Kohonen et al., 2000). In the meantime, the use of evolution-based genetic algorithms, and the utilization of fuzzy function approximation, have also been presented as possible solutions for the classification problems (Rialle et al., 1997, Petridis and Kaburlasos, 2001, Haruechaiyasak et al., 2002). Also, Anagnostopoulos et al. (2004) described a probabilistic NN that classifies web pages under the concepts of business media framework. The classification is performed by estimating the likelihood of an input feature vector according to Bayes posterior probabilities.

Web usage mining has become very critical for effective website management, creating adaptive websites, business and support services, personalization, network traffic flow analysis and so on (Abraham and Ramos, 2003). Krishnapuram et al. (2001) introduced the notion of uncertainty in web usage mining, discovering clusters of user session profiles using robust fuzzy algorithms. In the approach, a user or a page can be assigned to more than one cluster. A dissimilarity matrix is created that is used by fuzzy algorithms presented in order to cluster typical user sessions. The study of ant colonies behavior and their self-organizing capabilities is of interest to information/knowledge retrieval, decision support systems sciences and also web mining, because it provides models of distributed adaptive organization, which are useful to solve difficult optimization, classification, and distributed control problems. Abraham and Ramos (2003) proposed an ant clustering algorithm to discover web usage patterns (data clusters) and a linear genetic programming approach to analyze the visitor trends.

5 E-Service Personalizations

E-service personalization (ESP) is the process of getting web users' information online, and using the information to tailoring web pages to individual users' preferences and deliver services to the user's needs. It is as an act of response according to the individual web user's characteristics, interest and preference. Personalised e-service is a means of meeting the web user's needs more effectively and efficiently; making online interactions with user faster and easier; and increasing online user satisfaction and repeat visits (Bonett, 2006, Riecken, 2000). It aims at delivering right information to right user at right time, so that to enhance e-service quality. In a marketing environment, the purposes of applying information technology to provide personalization are expressed as to (1) Better serve the customer by anticipating needs; (2) Make the interaction efficient and satisfying for both parties; and (3) and Build a relationship that encourages the customer to return for subsequent purchases including products and services (The PersonalizationConsortium,
http://www.personalization.org/personalization.html Chen et al., 2004a).

E-service personalization is experiencing widespread adoption in the application areas such as e-commerce interaction, e-learning, online booking, and customer relationship management (Lu et al., 2006, Adomavicius and Tuzhilin, 2005). For example, given a customer, how to pick the right advertisement to target him/her? How to determine which product should be recommended to him/her? How to determine the content of a web page that he/she views? Therefore, personalization on the e-services is also referred as being about "building customer loyalty by building a meaningful one-to-one relationship." While creating user one-to-one relationships is to have its reflections on the web user satisfactions. User satisfaction is therefore the ultimate aim of personalization.

Many researchers have recently endeavored to provide personalization mechanisms, technologies and approaches for e-services (Adomavicius and Tuzhilin, 2005, Eirinaki and Vazirgiannis, 2003). Results involve five main aspects: (1) Profile based e-service personalization, (2) Link based e-service personalization, (3) Content based e-service personalization, (4) Structure based e-service personalization and (5) Recommendation based e-service personalization.

E-services will benefit from web personalization techniques in tailoring the interaction with their users according to an evolving customer model in both link based and content based personalization. In this context, relationship-value market segmentation becomes a central customer modeling activity. But value segmentation categories are inherently vague due to the use of imprecise linguistic categories, combined with a degree of uncertainty about customer behavior, and the difficulty inherent to estimating intangible variables. A fuzzy set approach to value segmentation is therefore developed, allowing more flexible customer segments. Fuzzy models of value estimations are represented by fuzzy triangular numbers for *directed* and *discovery-oriented* segmentation approaches. The usefulness of the approach has been illustrated through concrete personalization techniques based on those fuzzy categories (Sicilia and Garcia, 2003).

Recommender systems as the most typical personalization techniques have gained much attention in the past 10 years (Adomavicius and Tuzhilin, 2005). Recommender systems aim at filtering out the uninterested items (or predicting the interested ones) automatically on behalf of the users according to their personal preferences. A recommendation system considers user preferences, interests, or browsing behaviors when analyzing user behaviors for personalized services (Amoroso and Reinig, 2004). It therefore can either predict whether a particular user will like a particular item, or to identify a set of items that will be of interest to a certain user (Karypis, 2001).

Various approaches for recommender systems have been developed (Breese et al., 1998, Burke, 2000, Zeng et al., 2004). The main types of these approaches adopted in recommender systems are the content-based (CB) approach, the collaborative filtering (CF) approach, and the knowledge-based (KB) approach. The CB approach mainly relies on the content and relevant profiles to generate personalized recommendations. Using the approach, a recommender system recommends some web objects, to a user, which are similar to what the user has been interested in the past. The

CF approach offers recommendations based on the similarity of a group of users (Mobasher et al., 2000). The CF approach has been known to be the most popular recommendation approach. It has been used in various e-service applications such as recommending web pages, movies, articles and products. The CF approach can be divided into two types: user-based CF and item-based CF (Karypis, 2001). The user-based CF approach is implemented in two main steps: (1) a set of k-nearest neighbors of a target user is selected. This is performed by computing correlations or similarities between user profiles and a target user; (2) producing a prediction value for the target user on unrated (or unvisited) items, and generating recommendations to the target user. The item-based CF approach first considers the relationships among items. Rather than finding user neighbors, the system attempts to find k similar items that are rated (or visited) by different users in some similar ways. Then, for a target item, related predictions can be generated. For example, we can take a weighted average of a target user's item ratings (or weights) on these neighbor items. The third type is the KB approach. A knowledge-based recommender system attempts to suggest items based on inferences about a user's preferences. Such systems use knowledge in relevant to users and items to generate recommendations. In some sense, all recommendation techniques could be described as doing some kinds of inference. A knowledge-based recommender system avoids gather information about a particular user because its judgments are independent of individual taste (Burke, 2000).

There are still some spaces with current recommendation approaches to improve including the lack of scalability and sparsity, and the lack of the ability and accuracy to provide recommendations or predictions for new users and new items (Guo and Lu, 2006). Some intelligent techniques such as fuzzy approximate reasoning (Klir and Yuan, 1995), fuzzy matching, and fuzzy similarity, and case-based reasoning are being used in recommendation approaches to overcome these existing problems. Nasraoui and Petenes (2003) investigated the framework and provided a dynamic prediction in the web navigation space. Yager (2003) described a reclusive approach in which fuzzy set methods are used for the representation and subsequent construction of justifications and recommendation rules. Differing from CF, it is based solely on preferences of the single individuals for whom we provide the recommendation, without using preferences of other collaborators. It makes extensively use of an internal description of the items, and relies solely on the preferences of the target user. Carbo and Molina (2004) developed a CF-based algorithm in which ratings and recommendations can be linguistic labels represented by fuzzy sets. Perny and Zucker (1999, 2001) proposed a recommender system from a decision support perspective, noting that such applications position themselves between the archetypical problems of individual and group decision making. In that light, they pursued a hybrid approach that involves a number of fuzzy relations. Using appropriate fuzzy similarity measures, for each item *i*, and each user *u*, a neighborhood of *k* most similar elements is constructed. Also, based on a fuzzy similarity measure, a hybrid recommendation algorithm with fuzzy set theory was proposed. It is being used in a one-and-only item recommendation system in government e-services (Cornelis et al., 2005). Some of KB recommender systems employ the techniques of case-based reasoning for knowledge-based recommendation, such as Wasabi Personal Shopper (Burke, 1999) and a restaurant recommender system.

Personalization approaches have also been applied e-learning environments, which are mainly based on a range of delivery and interactive services. A personalized courseware recommendation system (PCRS) is proposed by Chen et al. (2004a). This system is developed based on the fuzzy item response theory to provide web-based learning services. In the proposed fuzzy item response theory, fuzzy set theory is combined with the original item response theory (Baker and Frank, 1992) to model uncertainly learning response. The PCRS can dynamically estimate learner ability based on the proposed fuzzy item response theory by collecting learner feedback information after studying the recommended courseware. Another example is a web-based personalized learning recommender system (Lu, 2004). This research aims to help students find learning materials they would need to read and therefore support students learn more effectively. Two related technologies are developed under the framework: one is a multi-attribute evaluation method to justify a student's need, and another is a fuzzy matching method to find suitable learning materials to best meet each student's needs.

The applications of intelligent techniques in e-service personalization also conduct some personalized e-service models. For example, Yang et al. (2004) combined genetic algorithms and k-nearest neighbor technology to model and reason a customer's personal preferences from a higher profile and then provide the most appropriate products to meet the user's higher needs. Genetic algorithms can help to obtain information more efficiently from the customers so that to help personalized products selection. Viswanathan and Childers (1999) considered online product categories as fuzzy sets. Products are said to have degrees of memberships in specific attributes. The memberships at the attribute level are then combined to obtain an overall degree of memberships of a product in a category. Fuzzy-set-based measures enable fine distinction among products and assist in the new product development, brand extension, and brand positioning. Fuzzy methods are also used in modeling market structure for e-business since they can handle the uncertainty associated with consumers' choice and their next purchase (Nishio and Shiizuka, 1995).

6 Intelligent E-Service Evaluations

Since the mid-1990s, businesses have spent quite a bit of time, money and effort developing web-based e-service applications. These applications are assisting businesses in building more effective customer relationships and gaining competitive advantage through providing interactive, personalized, faster e-services to fulfill customer demands (Chidambaram, 2001). Businesses in the earlier stages of employing web-based applications had little data, knowledge and experience for assessing and evaluating the potential of e-services for organizational impacts and benefits. Organisational efforts were largely geared toward customer service provision with little to no thought identifying and measuring the costs involved in moving services online against the benefits received by adopting e-services. After several years experience of e-service provision, businesses now urgently need to plan their further development in e-services (Lu and Zhang, 2003). Importantly, businesses have obtained related e-service running data and knowledge, which makes it possible to identify in what items of investment for an e-service application effectively contribute to what benefit aspects of business objectives.

Recent reports concerning the success, quality, usability and benefit of e-services have led researchers to express increasing interest in evaluating and measuring the development of e-service applications (Wade and Nevo, 2005, Schubert and Dettling, 2002). Much research has been conducted to evaluate e-services from various views and using various methods. In general, the research in e-service evaluation can be classified under four major categories.

The first one is the evaluation for the features, functions or usability of e-service systems. It is often combined with the evaluation of the use of related websites. Typical approaches used in this category of research are testing, inspection and inquiry (Hahn and Kauffman, 2002). These approaches are often used together in analyzing a web search or a desk survey. For example, Ng et al. (1998) reported a desk survey of business websites and discussed the features and benefits of web-based applications. Smith (2001) proposed a set of evaluation criteria to New Zealand government websites. Lu et al. (2001) showed their assessment results for e-commerce development in the businesses of New Zealand. The quality of websites needs to be measured using criteria focused on the effective website design (e.g., clear ordering of information, consistent navigation structure). However, from the information consumer's perspective the quality of a website may not be assessed independently of the quality of the information content that provides. Based on the information quality framework for the design of information systems defined in (Zadeh, 1975, Lee et al., 2002), Enrique et al. (2003) presented a computing-with-words based fuzzy-set method to measure the informative quality of Web sites used to publish information stored in XML documents.

The second category is the customer satisfactory evaluation. Meeting a web user's needs successfully is likely to lead to a satisfying relationship. Various evaluation criteria and factors about meeting users' needs and assessing user satisfactory degrees have been identified (Schubert and Dettling, 2002, Awan and Singh, 2006). Some evaluation systems have been designed for obtaining customers' feedback and measuring the degree of their satisfaction to current e-services provided (Lu and Lu, 2004). Questionnaire-based survey and multi-criteria evaluation systems are mainly used to conduct this kind of research. For example, Lin (2003) examined some customer satisfaction for e-commerce and proposed three main scales that play a significant role in influencing customer satisfaction: customer need, customer value, and customer cost. In the meantime, a related topic, customer loyalty, such as the antecedents and consequences of customer loyalty in e-commerce have been explored (Srinivasan et al., 2002). During the user satisfactory evaluation, fuzzy set techniques have been extended to discovery of fuzzy association rules (Kuok et al., 1998) and their extension to fuzzy sequential patterns (Hong et al., 1999). Fuzzy set theory provides a host of parameterized operators that can be used to model various aggregation strategies in web-based knowledge discovery (Jain and Krishnapuram, 2001). Setnes and Kaymak (2001) described an application of a fuzzy clustering algorithm to extract fuzzy rules from consumer response data collected by a sampling procedure. Such results will help e-service providers clear understanding their customers' satisfactory degrees to their e-services.

The third category is e-service investment analysis that has been conducted for evaluating and justifying investment in an e-service application. For example, Giaglis et al. (1999) presented a case-study of e-commerce investment evaluation. Furthermore, Drinjak et al. (2001) investigated the perceived business benefits of investing in e-service applications. While Amir et al. (2000) created a cost-benefit framework for online system management and evaluation. In particular, Lu and Zhang (2003) proposed a cost-benefit factor analysis model in e-services and conducted analysis for e-service development of businesses in Australia based on a questionnaire survey. Following the results, Zhang et al. (2004a) applied Bayesian network techniques to analyze and verify the relationships among cost factors and benefit factors in the development of e-services. A cost-benefit factor-relation model is first proposed and considered as domain knowledge. Data collected through a questionnaire based survey is as evidence to the inference-based verification. This study first creates a graphical structure for cost-benefit factors. It then calculates conditional probability distributions among these factors. Based on the established Bayesian network, the Junction-tree algorithm is used to conduct inference. A set of useful findings have been obtained for the costs involved in moving services online against the benefits received by adopting e-service applications. For example, 'increased investment in maintaining e-services' would significantly contribute to 'enhancing perceived company image,' and 'increased investment in staff training' would significant contribute to 'realizing business strategies.' These findings have potential to improve the strategic planning of businesses by determining more effective investment items and adopting more suitable development activities in the development of e-services. Fuzzy set approaches have been used to summarize and analyze the survey results in e-service evaluation in the form of linguistic knowledge that can be understood by merchants easily. Fuzzy set techniques are known to be effective for analysis even with sparse data especially when application-specific knowledge is available in terms of fuzzy rules. Hsu et al. (2002) proposed a fuzzy clustering approach for segment structure analysis of customer response to surveys.

Significant results have also been reported in the fourth category, the establishment of evaluation models, frameworks and systems. For example, Lee et al. (1999) created a model for evaluating the business value of business-to-business e-service through five propositions. Zhang and Dran (2000) developed a two-factor model for the website design and evaluation. More generally, Hahn and Kauffman (2002) presented a value-driven framework for evaluating e-commerce websites. In general the quality of websites is often measured and evaluated using criteria focused on the effective website design such as clear ordering of information or consistent navigation structure. However, from the information consumer's perspective the quality of a website may not be assessed and evaluated independently of the quality of the information content that it provides. The evaluation of websites focusing on the quality of the information is a difficult task. One of the reasons is that users cannot express their judgments with an exact numerical value sometimes. Therefore, a more realistic approach may be to use linguistic assessments to express the evaluation judgments instead of numerical values. As mentioned above, the fuzzy linguistic approach is a tool to manage linguistic information; it is suitable to model qualitative values used in human communication for representing qualitative concepts. Based on the principle, Herrera-Viedma et al. (2003) presented an evaluation model of informative quality of the websites based on fuzzy linguistic techniques. This model can be used by evaluators to use linguistic terms and finally can generate linguistic recommendations on quality websites by using information stored in multiple kinds of documents structured in the XML-format. A web-based fuzzy multi-criteria decision support system has been developed by Lu et al. (2005) to use for the website evaluation. A group of users can use the online decision support system to input linguistic terms such as '*good,*' '*very good*' for a set of selected websites respectively. Finally, evaluation results will show the level of the user satisfactory for each attribute of each website.

7 Intelligent E-Service Support Systems

As e-services become common, many Internet-based support systems such as software agents have been developed to assist users to receive high quality services in different aspects of e-services. These systems mainly perform tasks of intermediation and communication between users and the web (Yager, 2000), and many of them are developed under an intelligent framework.

One possibility to facilitate the communication processes between users and the web consists in the application of the fuzzy linguistic approach (Zadeh, 1975), which provides a flexible representation model of information by means of linguistic labels. The application of fuzzy linguistic techniques enables e-service providers to handle information with several degrees of truth and solving the problem of quantifying qualitative concepts. Some examples of the use of fuzzy linguistic techniques in the design of intelligent e-service systems, in particular multi-agent systems, can be found in (Delgado et al., 2001, Delgado et al., 2002, Herrera-Viedma et al., 2004). These papers presented some new models of fuzzy linguistic intelligent systems that involve the use of fuzzy set techniques and other intelligent approaches to improve the information access on the web. For example, a fuzzy linguistic multi-agent system can gather information on the web with a hierarchical architecture of seven action levels (Herrera-Viedma et al., 2004). E-negotiation, e-learning, e-decision, and e-shopping are the main forms of e-service support systems with a strong combination to intelligent techniques, which are displayed as follows.

E-negotiation as a kind of e-service support systems has well integrated with intelligent techniques. Kowalczyk and Bui (2000) presented some aspects of a customable fuzzy e-negotiation agents (FeNAs) system for autonomous multi-issue negotiation in the presence of limited common knowledge and imprecise/soft constraints and preferences. The FeNAs use the principles of utility theory and fuzzy constraint-based reasoning in order to find a consensus that maximizes the agent's utility at the highest possible level of fuzzy constraint satisfaction subject to its acceptability by other agents (Dubois et al., 1994, Kowalczyk, 1999, Zadeh, 1973). Through applying the fuzzy set approaches, a variety of e-negotiation problems with incomplete common knowledge and imprecise/soft constraints can be handled. Genetic algorithms and Bayesian rule updating methods have also been used in e-negotiation systems. For example, Ouchiyama et al. (2003) proposed an experience based evolutionary negotiation agent that can conduct negotiation process in e-commerce on behalf of users it represents. By emulating human being, skills of an agent in negotiations can be improved with increasing knowledge and experience.

Another kind of e-service support systems is web-based e-learning systems. With the rapid growth of computers and Internet technologies, e-learning has currently become a major trend in the computer assisted teaching and learning field. Many researchers made efforts in developing web based e-learning systems to assist distance online learning. To promote learning efficiency and effectiveness, some systems have applied fuzzy sets and other intelligent approaches to fully consider learner's behaviors, interests, or habits, and also to assist learners in selecting subjects, topics, and materials through learners gives a 'linguistic' based response of understanding percentage for the learned courseware. Results show that applying the proposed fuzzy set approaches to e-learning can achieve personalized learning and help learners to learn more effectively and efficiently.

E-service activity in automated procurement will benefit from the kinds of web based decision support systems that can be constructed using intelligent technology. Ngai and Wat (2005) developed a fuzzy decision support system for risk analysis in e-commerce. Lu et al. (2005) developed a web-based fuzzy group decision support system (WFGDSS) based on the a fuzzy group decision-making method. This system first identifies three factors from web users that may influence the assessment of utility of alternatives and the deriving of the group satisfactory solution. The first one is an individual's role (weight) in the ranking and selection of the satisfactory solutions. The second factor is an individual's preference for alternatives. The third factor is criteria for assessing these alternatives. The above-mentioned three factors also derive a crucial requirement for linguistic information processing techniques in an online group decision-making practice. Any individual role in an online decision process, a preference for alternatives, and a judgment for assessment-criteria are often expressed by linguistic terms. For example, an individual role can be described by using linguistic terms *important person* or *general decision person*. While a preference for an alternative can be described using linguistic terms *strong like it, just like ti,* or *don't like it any more*. Since these linguistic terms reflect the uncertainty, inaccuracy and fuzziness of decision makers, fuzzy set theory (Zadeh, 1965) is directly applied to deal with them. The WFGDSS uses a web environment as a development and delivery platform, and therefore allows decision makers distributed in different locations to participate in a group decision-making activity through the web. It manages the group decision-making process through criteria generation, alternative evaluation, opinion interaction and decision aggregation with the use of linguistic terms. This WFGDSS has a convenient and graphical user interface with visualization possibilities, and therefore is automatically available to many decision makers. Another example is a web based multi-objective decision support systems developed by Zhang and Lu (2005) and Lu et al. (2003). This system has a set of multi-objective decision-making methods. Some methods are more suitable than others for particular practical decision problems and particular decision makers. To help users choose a most suitable one in a particular situation, a linguistic term based intelligent guide is included in this system for selecting a desired method.

Intelligent e-shopping systems have been widely developed and used, because the Internet has been reaching almost every family in the world and anyone can build his/her own e-shops and also can purchase goods from any e-shops. Companies expect taking more benefit from online buying, selling, trading, etc and therefore continuing improve the functionality of their e-shopping systems. Some e-shopping systems applied traditional intelligent techniques such as multi-agent systems, rule-based or case-based process flows to coordinate communications for system automation. Some e-shopping systems propose fuzzy logic and neural networks or their combinations based approaches to tackle the uncertainties in practical online shopping activities such as consumer preferences, product specification, product selection, price negotiation, purchase, delivery, after-sales service, and evaluation (Liu and You, 2003). The fuzzy neural network provides an automatic and autonomous product classification and selection scheme to support fuzzy decision making by integrating fuzzy logic technology and the back propagation feed forward neural network. Fuzzy decision strategies are also proposed to guide the retrieval and evaluation of similar products that are used online shopping. For example, Liu and You (2003) presented a visualization approach to provide intelligent web browsing support for e-commerce using data warehousing and data mining techniques. This approach can overcome the limitations of the current web browsers, which lack flexibility for customers to visualize products from different perspectives. By using fuzzy logic for a fuzzy neural network to support decision making, an agent-based fuzzy shopper system is developed to implement the strategy. Furthermore, Chau and Yeh (2005) presented how intelligent techniques facilitate the automatic development of multilingual web portal for e-business operating as a global enterprise. E-auction can be seen as a special kind of e-shopping. Byde (2003) applied evolutionary game theory to auction mechanism design. The main idea is using an evolution-based method for evaluating auction mechanisms and developing a multi-agent system to evolve good players for each mechanism.

Intelligent techniques have also been used in e-recruitment systems. For example, Khosla and Goonesekera (2003) reported an e-recruitment multi-agent application for recruitment and benchmarking of salespersons. This multi-agent e-sales recruitment system (e-SRS) integrates a selling behavioral model with expert systems and soft computing techniques like fuzzy- K-means for predicting the selling behavior profile of a sales candidate.

8 E-Service Models and Management

Management techniques and related models have long existed and are quite mature for every segment of service industries. In a growing competitive marketplace, e-services are under constant pressure to optimize their utilization of information resources. E-service administrators and planners have tried to use intelligent technologies and approaches to model and manage e-service applications. Some models developed under intelligent approaches include fuzzy association based e-service customer classification model, intelligent global e-business web portal development model, e-commence trust model, and web mining based decision model. These models involve both e-business and e-government. In the meantime, research on getting knowledge and online knowledge discovery in various e-service systems through genetic algorithms, case-based reasoning etc. has been shown in literature. The rest of the section will list some typical developments.

Goldszmidt et al. (2001) proposed an approach for defining and quantifying effective e-business capacity that allows to translate quality of service objectives into the number of users that a website can support. This approach is based on inducing online models using machine learning and statistical pattern recognition techniques. The concept of e-business capacity allows us to naturally answer planning and operational questions about the information system infrastructure needed to support the e-business. The questions range from indicating which performance measures in the system are "important" to simulating "if-then" scenarios.

E-services have managed to place themselves in the society. However, there are many hindrance factors that cause them to fail to reach their full potential, mainly on the dissatisfaction of customers, such as a low level of personal data security and mistrust of the technology (Manchala, 2000). This has affected consumers' trust towards online business. Since the concept of trust is subjective, it creates a number of unique problems that obviate any clear mathematical result. Hence, fuzzy logic is currently being investigated as a possible best fit approach as it takes into account the uncertainties within e-commerce data and like human relationships, trust is often expressed by linguistic terms rather than numerical values. Nefti et al. (2005) identified two advantages of using fuzzy logic to quantify trust in e-commerce applications. (1) Fuzzy inference is capable of quantifying imprecise data and quantifying uncertainty in measuring the trust index of the vendors. For example, in the trust model, the community comments variable in the fulfillment factor has a wide range of values as we may have a small or large number of customers providing positive or negative feedback to the vendor; the number of comments will affect the decision made by the associated evaluation module. (2) Fuzzy inference can deal with variable dependencies in the system by decoupling dependable variables. The membership functions can be used to generate membership degrees for each variable. Any defined fuzzy rule set will be applied to the output space (trust index) through fuzzy '*and*' and '*or*' operators. Such a fuzzy trust module can describe more effectively users' trust behavior in e-services. Another trust model is developed under Peer-to-Peer e-commerce communities. A main way to minimize threats in such an open community is to use community-based reputations to help evaluating the trustworthiness and predicting the future behavior of peers. Xiong and Liu (2003) presented PeerTrust, a coherent adaptive trust model for quantifying and comparing the trustworthiness of peers based on a transaction-based feedback system. This study introduces two adaptive trust factors, the transaction context factor and the community context factor, to allow the basic trust metric to incorporate different contexts (situations) and to address common problems encountered in a variety of online e-commerce communities. Some studies have applied fuzzy-logic-based models to evaluate trust in e-commerce by taking into account the uncertainties within e-commerce data. For example, Chang et al (2005) demonstrated the application of a fuzzy trust model in an ecommerce platform. It aims to measure trustworthiness, reputation or credibility of e-service consumers and e-service providers in loosely coupled, distributed e-commerce systems.

The potential for government using web to enhance services to its citizens, businesses and communities is now more evident than ever before. Thus, in the most of developed countries and some of developing countries, e-government applications are growing rapidly. E-government development promises to make governments more efficient, responsive, transparent, and legitimate. The challenge for governments is to continually embrace the opportunities that the web provides, and ensure that the needs and expectations of citizens, businesses and communities are met (Guo and Lu, 2004). As the amount of information available on the web is overwhelming, the users of e-government are constantly facing the problem of information overload. The increasing information overload would hinder government e-service effectiveness (Guo and Lu, 2005). In order to explore how e-government better face the challenge and developing next innovations, Hinnant and O'Looney (2003) conducted an exploratory study of e-service personalization in the public sector by examining pre-adoption interest of government in online innovations. This study proposes a revised model of technological innovation with an emphasis on socio-technical factors associated with electronic service delivery. This model focuses on three primary dimensions of online innovation: perceived need, technical capacity, and risk mitigation. This model is then used to examine a single online innovation, personalization of online government information and services.

9 Conclusions

E-service intelligence, like every new technology, provides solutions and challenges, along with a new set of research questions. The preliminary research seemed promising, but more research and developments should be followed soon. Intelligent technology plays an important role for dealing with e-services as already briefly outlined by many successful applications. We can speculate about the potential and problems of an integrated intelligent e-service economy, there is a lot we need to explore about the technological or organizational issues involved. We strongly believe the use of intelligent technologies in cope with web technologies will significantly enhance the current development of e-services in general and the future intelligent e-services in particular.

References

1. Abraham, A., Ramos, V. (2003) Web usage mining using artificial ant colony clustering and linear genetic programming. Proceedings of the 2003 Congress on Evolutionary Computation, 1384- 1391.
2. Adomavicius, G., Tuzhilin, A. (2005) Personalization technologies: a process-oriented perspective. Communications of the ACM, 48**:** 83-90.
3. Amir, Y., Awerbuch, B., Borgstrom, R.S. (2000) A Cost–Benefit framework for online management of a metacomputing system. Decision Support Systems, 28**:** 155-164.
4. Amoroso, D. L., Reinig, B.A. (2004) Personalization management systems: Mini-track introduction. Proceedings of the 37th Annual Hawaii International Conference on System Sciences (HICSS'04), Big Island, Hawaii, Track 7.
5. Anagnostopoulos, I., Anagnostopoulos, C., Loumos, V., Kayafas, E. (2004) Classifying Web pages employing a probabilistic neural network. Proceedings of IEEE Vol. 151, 139- 150.
6. Awan, I., Singh, S. (2006) Performance evaluation of e-commerce requests in wireless cellular networks. Information and Software Technology, Available online**:** 1-9.
7. Baker, Frank, B. (1992) Item Response Theory: Parameter Estimation Techniques, Marcel Dekker, New York.
8. Bautista, M.M., Vila, M.A., Larsen, H. (2000) Building adaptive user profiles by a genetic fuzzy classifier with feature selection. Proceedings of The Ninth IEEE International Conference on Fuzzy Systems, 308-312.
9. Berners-Lee, T., Hendler, J., Lassila, O. (2001) The semantic web. Scientific American, May 2001**:** 29--37.
10. Bonett, M. (2006) Personalization of Web Services: Opportunities and Challenges http://www.ariadne.ac.uk/issue28/personalization/.
11. Breese, J. S., Heckerman, D., Kadie, C. (1998) Empirical analysis of predictive algorithms for collaborative filtering. Proceedings of the 14th Conference on Uncertainty in Artificial Intelligence, Madison, Wisconsin, USA, 43-52.
12. Burke, R. (1999) The Wasabi Personal Shopper: a case-based recommender systems. Proceedings of the 11th National Conference on Innovative Applications of Artificial Intelligence, Menlo Park, CA, 844-849.
13. Burke, R. (2000) Knowledge-based recommender systems. *Encyclopedia of Library and Information Systems*. Kent, A., Ed., Marcel Dekker, New York

14. Byde, A. (2003) Applying evolutionary game theory to auction mechanism design. Proceedings of IEEE International Conference on E-Commerce, 347- 354.
15. Carbo, J., Molina, J. M. (2004) Agent-based collaborative filtering based on fuzzy recommendations. International Journal of Web Engineering and Technology, 1: 414 - 426.
16. Chang, E. Schmidt, S. and Steele, R. and Dillon, T. (2005), Applying a fuzzy trust model to e-commerce systems, in Jarvis, R. and Zhang, S. (ed), *18th Australian Joint Conference on Artificial Intelligence (AI)*, Sydney, Australia, Dec. 2005, 318-329.
17. Chau, R., Yeh, C. H. (2005) Intelligent techniques for global e-business Web portal development. Proceedings of the 2005 IEEE International Conference on e-Technology, e-Commerce and e-Service, 334- 340.
18. Chen, C.M., Duh, L.J., Liu, C.Y. (2004a) A personalized courseware recommendation system based on fuzzy item response theory. Proceedings of 2004 IEEE International Conference on e-Technology, e-Commerce and e-Service (EEE'04), Taipei, Taiwan, 305-308.
19. Chen, Q. Z., Mao, K. J., Zhang, Y., Lu, L. Y. (2004b) Catching potential customers: an example of web-mining-aided e-commerce decision-making. Proceedings of the Fifth World Congress on Intelligent Control and Automation, 3986- 3990.
20. Chidambaram, L. (2001) The editor's column: Why e-Service Journal. e-Service Journal, 1: 1-3.
21. Cornelis, C., Guo, X., Lu, J., Zhang, G. Q. (2005) A fuzzy relational approach to event recommendation. Proceedings of the Second Indian International Conference on Artificial Intelligence (IICAI-05), Pune, India, 2231-2242.
22. Crestani, F. (1994) Comparing neural and probabilistic relevance feedback in an interactive information retrieval system. Proceedings of IEEE Int. Conf. on Neural Networks, Orlando, FL, 3426–3430.
23. Delgado, M., Herrera, F., Herrera-Viedma, E., Martín-Bautista, M. J., Martínez, L., Vila, M. A. (2002) A communication model based on the 2-tuple fuzzy linguistic representation for a distributed intelligent agent system on Internet. Soft Computing, 6: 320-328.
24. Delgado, M., Herrera, F., Herrera-Viedma, E., Martín-Bautista, M. J., Vila, M. A. (2001) Combining linguistic information in a distributed intelligent agent model for information gathering on the Internet. *Computing with Words*. Wang, P.P., Ed., John Wiley & Son, 251-276.

25. Drinjak, J., Altmann, G., Joyce, P. (2001) Justifying investments in electronic commerce. Proceedings of the Twelfth Australia conference on Information Systems, Coffs Habour, Australia, 187-198.
26. Dubois, D., Fargier, H., Prade, H., (1994) Propagation and Satisfaction of Flexible Constraints. *Fuzzy Sets, Neural Networks arid Sott Computing*. Yager, R. R. and Zadeh, L. A., Eds., 166-187
27. Eirinaki, M., Vazirgiannis, M. (2003) Web mining for web personalization. ACM Transactions on Internet Technology, 3: 1-27.
28. Enrique, H., Eduardo, P., Mar´ıa, D. O., Juan, C. H., Yusef, H. M. (2003) Evaluating the Informative Quality of Web Sites by Fuzzy Computing with Words. Proceedings of Atlantic Web Intelligence Conference. Madrid, Spain, Lecture Notes in Artificial Intelligence 2663: Springer, 62-72
29. Giaglis, G. M., Paul, R. J., Doukidis, G. I. (1999) Dynamic modelling to assess the business value of electronic commerce. International Journal of Electronic Commerce, 3: 35-51.
30. Goldszmidt, M., Palma, D., Sabata, B. (2001) On the quantification of e-business capacity. Proceedings of the 3rd ACM Conference on Electronic Commerce, Tampa, Florida, USA, 235-244.
31. Guo, X., Lu, J. (2004) Effectiveness of E-government Online Ser-vices in Australia. *Digital Government: Strategies and Implementation from Developing and Developed Countries*. Huang, Siau and Wei, Eds., Idea Group, Inc. 214-241.
32. Guo, X., Lu, J. (2005) Applying web personalization techniques in E-government services. Proceedings of the 11th Australian World Wide Web Conference, Gold Coast, Australia, 233-238.
33. Guo, X., Lu, J. (2006) Intelligent e-government services with recommendation techniques. International Journal of Intelligent Systems, Special issue on E-service intelligence: Accepted.
34. Hahn, J., Kauffman, R. J. (2002) Evaluating selling web site performance from a business value perspective. Proceedings of International conference on e-Business, Beijing, China, 435-443.
35. Haruechaiyasak, C., Mei-Ling, S., Shu-Ching, C., Xiuqi, L. (2002) Web document classification based on fuzzy association. Proceedings of the 26th Annual Int. Computer Software and Applications Conf., Oxford, UK, 487–492.
36. Herrera-Viedma, E., Herrera, F., Martínez, L., Herrera, J. C., López, A. G. (2004) Incorporating filtering techniques in a fuzzy multi-agent model for gathering of information on the Web. Fuzzy Sets and Systems, 148: 61-83.
37. E. Herrera-Viedma, E. Peis, M.D. Olvera, Y.H. Montero, J.C. Herrera (2003) Evaluating the informative quality of Web sites by Fuzzy

Computing with Words. Atlantic Web Intelligence Conference, AWIC'03. Madrid (Spain), Lecture Notes in Artificial Intelligence 2663, pp.62-72.

38. Hinnant, C. C., O'looney, J. A. (2003) Examining pre-adoption interest in online innovations: an exploratory study of e-service personalization in the public sector. IEEE Transactions on Engineering Management: 436- 447.
39. Hong, T., Kuo, C., Chi, S. (1999) Mining fuzzy sequential patterns from quantitative data. Proceedings of IEEE International Conference on Systems, Man, and Cybernetics, 962-966.
40. Hsu, T., Chu, K.M., Chan, H.C. (2002) The fuzzy clustering on market segment. Proceedings of the Ninth IEEE International Conference on Fuzzy Systems, 62 1-626.
41. Jain, V., Krishnapuram, R. (2001) Applications of fuzzy sets in personalization for e-commerce. Proceedings of IFSA-NAFIPS 2001 Conference, 263-268.
42. Karypis, G. (2001) Evaluation of item-based top-N recommendation algorithms, Proceedings of the ACM 10th International Conference on Information and Knowledge Management, Atlanta, Georgia, 247-254.
43. Khosla, R., Goonesekera, T. (2003) An online multi-agent e-sales recruitment system. Proceedings of IEEE/WIC International Conference on Web Intelligence, 111- 117.
44. Klir, G. J., Yuan, B. (1995) Fuzzy Sets and Fuzzy Logic: Theory and Applications. Upper Saddle River, Prentice Hall, N.J.247-254.
45. Kohonen, T., Kaski, S., Lagus, K., Salojarvi, J., Honkela, J., Paatero, V., Saarela, A. (2000) Self organization of a massive document collection. IEEE Trans. Neural Network, 11: 574–585.
46. Kowalczyk, R. (1999) On Linguistic Fuzzy Constraint Satisfaction Problems *Computing with Words in Intelligent Information Systems.* Zadeh, L. A. and Kacprzyk, J., Eds., Kluwer, 166- 187.
47. Kowalczyk, R., Bui, V. (2000) FeNAs: A fuzzy e-negotiation agents system. Proceedings of the Conference on Computational Intelligence for Financial Engineering (CIFEr 2000), NY, 26-29.
48. Krishnapuram, R., Nasraoui, O., Joshi, A., L.M (2001) Low complexity fuzzy relational clustering algorithms for web mining. IEEE Transactions on Fuzzy Systems 9:595-607.
49. Kuok, C., Fu, A., Wong, M. H. (1998) Mining fizzy association rules in databases. SIGMOD Record, 27: 41-46.
50. Lee, C., Seddon, P., Corbitt, B. (1999) Evaluating business value of internet-based business-to-business electronic commerce. Proceedings of 10th Australasian Conference on Information Systems, Wellington, New Zealand, 2: 508-519.

51. Lee, Y. W., Strong, D. M., Kahn, B. K., Wang, R. Y. (2002) AIMQ: A methodology for information quality assessment. Inf.ormation & Managment, 40: 133-146.
52. Li, Y., Liu, D. B., Zhang, W. M. (2005) Schema matching using neural network. Proceedings of The 2005 IEEE/WIC/ACM International Conference on Web Intelligence, Paris, 743- 746.
53. Lin, C. (2003) A critical appraisal of customer satisfaction and e-commerce. Managerial Auditing Journal, 18: 202-212.
54. Liu, J., You, J. (2003) Smart shopper: an agent-based web-mining approach to Internet shopping. IEEE Transactions on Fuzzy Systems, 11: 226- 237.
55. Loia, V., Nikravesh, M., Zadeh, L. A. (2004) Fuzzy Logic and the Internet, Physica-Verlag, Springer.
56. Lu, J. (2004) Framework and approach for developing intelligent personalized learning recommender systems. Proceedings of the 2nd International Conference on Information Technology and Applications, Harbin, China, CDROM.
57. Lu, J., Lu, Z. (2004) Development, distribution and evaluation for online tourism services in China. Electronic Commerce Research Journal, 4: 221-239.
58. Lu, J., Ruan, D., Zhang, G. Q., Zimmermann, H. J. (2007) International Journal of Intelligent Systems, Special Issue on E-Service Intelligence.
59. Lu, J., Shi, C. G., Zhang, G. Q. (2003) Framework and implementation of a web-based WMODSS. Proceedings of Workshop on Applications, Products and Ser-vices of Web-based Support Systems, in conjunction with IEEE/WIC Inter-national Conference on Web Intelligence, Halifax, Canada, 7-11.
60. Lu, J., Tang, S., Mccullough, G. (2001) An assessment for internet-based electronic commerce development in businesses of New Zealand Electronic Markets: International Journal of Electronic Commerce and Business Media, 11: 107-115.
61. Lu, J., Zhang, G. Q. (2003) A model for evaluating E-commerce based on cost/benefit and customer satisfaction. Journal of Information Systems Frontiers, 5: 265-277.
62. Lu, J., Zhang, G. Q., Wu, F. (2005) Web-Based Multi-Criteria Group Decision Support System with Linguistic Term Processing Function. The IEEE Intelligent Informatics Bulletin, 5: 35-43.
63. Manchala, D. W. (2000) E-commerce trust metrics and models. IEEE Internet Comp. 4: 36-44.

64. Mobasher, B., Dai, H., Luo, M., Wiltshire, (2002) Discovery of evaluation of aggregate usage profiles for web personalization. Data mining and knowledge discovery, 6 (1): 61-82.
65. Nasraoui, O., Petenes, C. (2003) An intelligent web recommendation engine based on fuzzy approximate reasoning. Proceedings of the IEEE International Conference on Fuzzy Systems, St. Louis, MO, 1116-1121.
66. Nefti, S., Meziane, F., Kasiran, K. (2005) A fuzzy trust model for E-commerce. Proceedings of Seventh IEEE International Conference on E-Commerce Technology (CEC'05), 401-404.
67. Ng, H., Pan, Y. J., Wilson, T. D. (1998) Business use of the world wide web: A report on further investigations. International Journal of Management, 18: 291-314.
68. Ngai, E. W. T., Wat, F. K. T. (2005) Fuzzy decision support system for risk analysis in e-commerce development. Decision Support Systems, 40: 235-255.
69. Nikravesh, M., Azvin, B. (2002) Fuzzy queries, search, and decision support system. International Journal of Soft Computing-Special Issue on Fuzzy Logic and the Internet, 6:373-399.
70. Nikravesh, M., Loia, V., Azvine, B. (2002) Fuzzy logic and the Internet (FLINT), Internet, World Wide Web, and Search Engines. International Journal of Soft Computing-Special Issue in fuzzy logic and the Internet, 6:33-37.
71. Nishio, C., Shiizuka, H. (1995) Competitive market structures constructed from brand-switching data by fuzzy structural modeling. Proceedings of IEEE International Conference on Fuzzy Systems, 819-824.
72. Ouchiyama, H., Huang, R., Ma, J., Sim, K. M. (2003) An experience-based evolutionary negotiation model. Proceedings of the Fifth International Conference on Computational Intelligence and Multi-media Applications, 212- 217.
73. S. Park (2000), Neural Networks and Customer Grouping in E-Commerce: A Framework Using Fuzzy ART. AIWoRC 2000: 331-336
74. Parry, D. (2004) A fuzzy ontology for medical document retrieval. Proceedings of the Second Workshop on Australasian information security, Data Mining and Web Intelligence, and Software Internationalisation - Volume 32 CRPIT '04, 121-126.
75. Perny, P., Zucker, J. D. (1999) Collaborative filtering methods based on fuzzy preference relations. Proceedings of the EUROFUSE-SIC, 279 - 285.

76. Perny, P., Zucker, J. D. (2001) Preference-based search and machine learning for collaborative filtering: the 'Film-Conseil' movie recommender system. Revue I3, 1: 1-40.
77. Petridis, V., Kaburlasos, V. G. (2001) Clustering and classification in structured data domains using fuzzy lattice neurocomputing (FLN). IEEE Trans. Knowl. Data Eng., 13: 245-260.
78. Quan, T. T., Hui, S. C., Cao, T. H. (2004a) A Fuzzy FCA-based Approach to Conceptual Clustering for Automatic Generation of Concept Hierarchy on Uncertainty Data. CLA 2004. Proceedings of CLA 2004, 1-12.
79. Quan, T. T., Hui, S. C., Fong, A. C. M., Cao, T. H. (2004b) Automatic Generation of Ontology for Scholarly Semantic Web. Proceedings of International Semantic Web Conference 2004, 726-740.
80. Quan, T. T., Siu Cheung Hui, S. C., Cao, T. H. (2004c) FOGA: A fuzzy ontology generation framework for scholarly semantic Web. Proceedings of the Knowledge Discovery and Ontologies (2004), 37-48.
81. Rialle, V., Meunier, J., Oussedik, S., Nault, G. (1997) Semiotic and modeling computer classification of text with genetic algorithm: analysis and first results. Proceedings of Int. Conf. on Intelligent Systems and Semiotics (ISAS), Gaithersburg, 325–330.
82. Riecken, D. E. (2000) Personalized views of personalization. Communications of the ACM, 43: 27-28.
83. Ruan, D. (1997) Intelligent Hybrid Systems, Kluwer Academic Publishers, Boston.
84. Schubert, P., Dettling, W. (2002) Extended Web assessment method (EWAM) - evaluation of E-commerce applications from the customer's view-point, Proceedings of the 35th Annual Hawaii International Conference on System Sciences (HICSS'02)-Volume 7, 175-184.
85. Setnes, M., Kaymak, U. (2001) Fuzzy modeling of client preference from large data sets: an application to target selection in direct marketing. IEEE Trans. on Fuzzy Systems 9:153-163.
86. Sicilia, M., Garcia, E. (2003) On fuzziness in relationship value segmentation: applications to personalized e-commerce. SIGecom Exch., 4: 1-10.
87. Smith, A. G. (2001) Applying evaluation criteria to New Zealand government websites. International Journal of Information Management, 21: 137-149.
88. Srinivasan, S., Anderson, R., Ponnavolu, K. (2002) Customer loyalty in e-commerce: an exploration of its antecedents and consequences. Journal of Retailing, 78: 41-50.

89. Viswanathan, M., Childers, T. L. (1999) Understanding how product attributes inhence product categorization: development and validation of fuzzy set-based measures of gradedness in product categories. Journal of Marketing Research, XXXVI: 75-94.
90. Wade, R. M., Nevo, S. (2005) Development and validation of a perceptual instrument to measure E-commerce performance. International Journal of Electronic Commerce, 10: 123.
91. Wang, C., Lu, J., Zhang, G. Q. (2006) Mining key information of Web pages: a method and its application. Expert Systems with Applications: Accepted.
92. Widyantoro, D. H., Yen, J. (2001) Using fuzzy ontology for query refine-ment in a personalized abstract search engine. Proceedings of IFSA World Congress and 20th NAFIPS International Conference
93. Xiong, L., Liu, L. (2003) A reputation-based trust model for peer-to-peer e-commerce communities. Proceedings of IEEE International Conference on E-Commerce, 275- 284.
94. Yager, R. R. (2000) Targeted E-commerce marketing using fuzzy intelligent agents. IEEE Intelligent Systems, 15: 42-45.
95. Yager, R. R. (2003) Fuzzy logic methods in recommender systems. Fuzzy Sets and Systems, 136: 133-149.
96. Yang, H. W., Pan, Z.G., Wang, X.Z., Xu, B. (2004) A personalized products selection assistance based on e-commerce machine learning. Proceedings of the 2004 International Conference on Machine Learning and Cybernetics, 2629- 2633.
97. Zadeh, L. A. (1965) Fuzzy sets. Information and Control, 8: 338-353.
98. Zadeh, L. A. (1973) Outline of a new approach to the analysis of complex systems and decision processes. IEEE Trans. Man. and Cybernetics, 3: 28-44.
99. Zadeh, L. A. (1975) The concept of a linguistic variable and its applications to approximate reasoning. Part I, in Information Sciences 8, 199-249. Part II, in Information Sciences 8, 301-357. Part III, in Information Sciences 9, 43-80.
100. Zadeh, L. A. (1999) From computing with numbers to computing with words -- from manipulation of measurements to manipulation of perceptions. IEEE Transactions on Circuits and Systems, 45: 105-119.
101. Zadeh, A. L. (2003) Web Intelligence and Fuzzy Logic-The Concept of Web IQ (WIQ) keynote in Web Intelligence 2003.
102. Zadeh, L. A. (2001) A new direction in AI: towards a computational theory of perceptions. AI Magazine, 22: 73-84.

103. Zeng, C., Xing, C.X., Zhou, L.Z., Zheng, X.H. (2004) Similarity measure and instance selection for collaborative filtering. International Journal of Electronic Commerce, 8: 115 - 129.
104. Zhang, G. Q., Lu, J. (2005) A linguistic intelligent guide for method selection in multi-objective decision support systems. Special issue on Linguistic Decision Making: Tools and Applications, Information Sciences: Accepted.
105. Zhang, G. Q., Lu, J., Bai, C. G., Zhang, C. (2004a) Bayesian network based cost benefit factor inference in e-services. Proceedings of the 2nd International Conference on Information Technology and Applications, Harbin, China, CDROM.
106. Zhang, P., Dran, V. G. (2000) Satisfiers and dissatisfiers: a two-factor model for website design and evaluation Journal of American Association for Information Science (JASIS), 51: 1253-1268.
107. Zhang, T. Z., Fu, Y. G., Shen, R. M. (2004b) Improve question & answer system by applying genetic algorithm. Proceedings of the 2004 International Conference on Machine Learning and Cy-bernetics, 2317- 2321.
108. Zhang, Y. Q., Shteynberg, M., Prasad, S. K., Sunderraman, R. (2003) Granular fuzzy Web intelligence techniques for profitable data mining. Proceedings of the 12th IEEE International Conference on Fuzzy Systems, 1462- 1464.

Rough Sets and Conflict Analysis

Zdzisław Pawlak and Andrzej Skowron[1]

Institute of Mathematics, Warsaw University
Banacha 2, 02-097 Warsaw, Poland
skowron@mimuw.edu.pl

Commemorating the life and work of Zdzisław Pawlak[1].

Summary. E-service intelligence requires tools for approximate reasoning about vague concepts. The rough set based methods make it possible to deal with imperfect knowledge. In particular, approximate reasoning about vague concepts can be based on the rough set approach. This approach seems to be of fundamental importance to AI and cognitive sciences, especially in the areas of machine learning and pattern recognition, knowledge acquisition, decision analysis, data mining and knowledge discovery from databases, expert systems, and conflict analysis. We discuss the basic concepts of rough sets and we outline some current research directions on rough sets. Conflict analysis is one of the basic issues in e-service intelligence. The contribution of this article is an extension of the existing approach based on rough sets to conflict analysis.

Keywords: Information and decision systems, indiscernibility, approximation space, set approximations, rough set, rough membership function, reducts, decision rule, dependency of attributes, conflicts, classifier, information granulation, vague concept approximation, rough mereology, ontology approximation

1 Introduction

E-service intelligence has been identified as a new e-service direction with many potential applications in different areas such as governmental management, medicine, business, learning, banking, science (e.g., habitat monitoring with wireless sensor networks) or logistics. To realize intelligent presentation of web content, intelligent online services, personalized support or direct customer participation in organizational decision-making processes, it is neces-

[1] Professor Zdzisław Pawlak passed away on 7 April 2006.

Z. Pawlak and A. Skowron: *Rough Sets and Conflict Analysis*, Studies in Computational Intelligence (SCI) **37**, 35–74 (2007)
www.springerlink.com

sary to develop methods that will make it possible to understand vague concepts and reasoning schemes expressed in natural language by humans who will cooperate with e-services. Hence, methods for approximation of vague concepts as well as methods for approximate reasoning along with reasoning performed in natural language are needed. In this article, we discuss some basic concepts of the rough set approach created for dealing with vague concepts and for approximate reasoning about vague concepts. Among the most important issues of e-service intelligence are also conflict analysis and negotiations. We also outline an approach for conflict analysis based on the rough set approach.

2 Preliminaries of Rough Sets

This section briefly delineates basic concepts in rough set theory. Basic ideas of rough set theory and its extensions, as well as many interesting applications can be found in books (see, e.g., [14, 18, 21, 22, 30, 35, 39, 40, 50, 52, 57, 67, 70, 71, 74, 84, 87, 90, 91, 100, 129, 151]), issues of the Transactions on Rough Sets [79, 80, 81, 82], special issues of other journals (see, e.g., [13, 48, 69, 78, 110, 130, 154, 155]), proceedings of international conferences (see, e.g., [1, 34, 51, 89, 109, 120, 137, 138, 142, 153, 156, 126, 127, 139]), tutorials (see, e.g., [38]), and on the internet such as www.roughsets.org, logic.mimuw.edu.pl, rsds.wsiz.rzeszow.pl.

2.1 Rough Sets: An Introduction

Rough set theory, proposed by Pawlak in 1982 [74, 73] can be seen as a new mathematical approach to vagueness.

The rough set philosophy is founded on the assumption that with every object of the universe of discourse we associate some information (data, knowledge). For example, if objects are patients suffering from a certain disease, symptoms of the disease form information about patients. Objects characterized by the same information are indiscernible (similar) in view of the available information about them. The *indiscernibility relation* generated in this way is the mathematical basis of rough set theory. This understanding of indiscernibility is related to the idea of Gottfried Wilhelm Leibniz that objects are indiscernible if and only if all available functionals take on them identical values (Leibniz's Law of Indiscernibility: The Identity of Indiscernibles) [2, 44]. However, in the rough set approach indiscernibility is defined relative to a given set of functionals (attributes).

Any set of all indiscernible (similar) objects is called an elementary set, and forms a basic granule (atom) of knowledge about the universe. Any union of some elementary sets is referred to as *crisp* (precise) set. If a set is not crisp then it is called *rough* (imprecise, vague).

Consequently, each rough set has *borderline cases* (*boundary-line*), i.e., objects which cannot be classified with certainty as members of either the set or its complement. Obviously crisp sets have no borderline elements at all. This means that borderline cases cannot be properly classified by employing available knowledge.

Thus, the assumption that objects can be "seen" only through the information available about them leads to the view that knowledge has granular structure. Due to the granularity of knowledge, some objects of interest cannot be discerned and appear as the same (or similar). As a consequence, vague concepts in contrast to precise concepts, cannot be characterized in terms of information about their elements. Therefore, in the proposed approach, we assume that any vague concept is replaced by a pair of precise concepts – called the lower and the upper approximation of the vague concept. The lower approximation consists of all objects which definitely belong to the concept and the upper approximation contains all objects which possibly belong to the concept. The difference between the upper and the lower approximation constitutes the boundary region of the vague concept. Approximations are two basic operations in rough set theory.

Hence, rough set theory expresses vagueness not by means of membership, but by employing a boundary region of a set. If the boundary region of a set is empty it means that the set is crisp, otherwise the set is rough (inexact). A nonempty boundary region of a set means that our knowledge about the set is not sufficient to define the set precisely.

Rough set theory it is not an alternative to classical set theory but it is embedded in it. Rough set theory can be viewed as a specific implementation of Frege's idea of vagueness, i.e., imprecision in this approach is expressed by a boundary region of a set.

Rough set theory has attracted attention of many researchers and practitioners all over the world, who have contributed essentially to its development and applications. Rough set theory overlaps with many other theories. Despite this overlap, rough set theory may be considered as an independent discipline in its own right. The rough set approach seems to be of fundamental importance in artificial intelligence and cognitive sciences, especially in research areas such as machine learning, intelligent systems, inductive reasoning, pattern recognition, mereology, knowledge discovery, decision analysis, and expert systems. The main advantage of rough set theory in data analysis is that it does not need any preliminary or additional information about data like probability distributions in statistics, basic probability assignments in Dempster–Shafer theory, a grade of membership or the value of possibility in fuzzy set theory. One can observe the following about the rough set approach:

- introduction of efficient algorithms for finding hidden patterns in data,
- determination of optimal sets of data (data reduction),
- evaluation of the significance of data,
- generation of sets of decision rules from data,

- easy-to-understand formulation,
- straightforward interpretation of obtained results,
- suitability of many of its algorithms for parallel processing.

2.2 Indiscernibility and Approximation

The starting point of rough set theory is the indiscernibility relation, which is generated by information about objects of interest (see Sect. 2.1). The indiscernibility relation expresses the fact that due to a lack of information (or knowledge) we are unable to discern some objects employing available information (or knowledge). This means that, in general, we are unable to deal with each particular object but we have to consider granules (clusters) of indiscernible objects as a fundamental basis for our theory.

From a practical point of view, it is better to define basic concepts of this theory in terms of data. Therefore we will start our considerations from a data set called an *information system*. An information system is a data table containing rows labeled by objects of interest, columns labeled by attributes and entries of the table are attribute values. For example, a data table can describe a set of patients in a hospital. The patients can be characterized by some attributes, like *age, sex, blood pressure, body temperature*, etc. With every attribute a set of its values is associated, e.g., values of the attribute *age* can be *young, middle*, and *old*. Attribute values can be also numerical. In data analysis the basic problem we are interested in is to find patterns in data, i.e., to find a relationship between some set of attributes, e.g., we might be interested whether *blood pressure* depends on *age and sex*.

Suppose we are given a pair $\mathcal{A} = (U, A)$ of non-empty, finite sets U and A, where U is the *universe* of *objects*, and A – a set consisting of *attributes*, i.e. functions $a : U \longrightarrow V_a$, where V_a is the set of values of attribute a, called the *domain* of a. The pair $\mathcal{A} = (U, A)$ is called an *information system* (see, e.g., [72]). Any information system can be represented by a data table with rows labeled by objects and columns labeled by attributes. Any pair (x, a), where $x \in U$ and $a \in A$ defines the table entry consisting of the value $a(x)$[2].

Any subset B of A determines a binary relation $I(B)$ on U, called an *indiscernibility relation*, defined by

$$xI(B)y \text{ if and only if } a(x) = a(y) \text{ for every } a \in B, \tag{1}$$

where $a(x)$ denotes the value of attribute a for object x.

Obviously, $I(B)$ is an equivalence relation. The family of all equivalence classes of $I(B)$, i.e., the partition determined by B, will be denoted by $U/I(B)$, or simply U/B; an equivalence class of $I(B)$, i.e., the block of the partition U/B, containing x will be denoted by $B(x)$ (other notation used: $[x]_B$ or

[2] Note, that in statistics or machine learning such a data table is called a sample [25].

$[x]_{I(B)}$). Thus in view of the data we are unable, in general, to observe individual objects but we are forced to reason only about the accessible granules of knowledge (see, e.g., [70, 74, 94]).

If $(x, y) \in I(B)$ we will say that x and y are *B-indiscernible*. Equivalence classes of the relation $I(B)$ (or blocks of the partition U/B) are referred to as *B-elementary sets* or *B-elementary granules*. In the rough set approach the elementary sets are the basic building blocks (concepts) of our knowledge about reality. The unions of *B-elementary sets* are called *B-definable sets*[3].

For $B \subseteq A$ we denote by $Inf_B(x)$ the B-signature of $x \in U$, i.e., the set $\{(a, a(s)) : a \in A\}$. Let $INF(B) = \{Inf_B(s) : s \in U\}$. Then for any objects $x, y \in U$ the following equivalence holds: $xI(B)y$ if and only if $Inf_B(x) = Inf_B(y)$.

The indiscernibility relation will be further used to define basic concepts of rough set theory. Let us define now the following two operations on sets $X \subseteq U$

$$B_*(X) = \{x \in U : B(x) \subseteq X\}, \tag{2}$$

$$B^*(X) = \{x \in U : B(x) \cap X \neq \emptyset\}, \tag{3}$$

assigning to every subset X of the universe U two sets $B_*(X)$ and $B^*(X)$ called the *B-lower* and the *B-upper approximation* of X, respectively. The set

$$BN_B(X) = B^*(X) - B_*(X), \tag{4}$$

will be referred to as the *B-boundary region* of X.

From the definition we obtain the following interpretation:

- The *lower approximation* of a set X with respect to B is the set of all objects, which can be for *certain* classified as X using B (are *certainly* X in view of B).
- The *upper approximation* of a set X with respect to B is the set of all objects which can be *possibly* classified as X using B (are *possibly* X in view of B).
- The *boundary region* of a set X with respect to B is the set of all objects, which can be classified neither as X nor as not-X using B.

In other words, due to the granularity of knowledge, rough sets cannot be characterized by using available knowledge. Therefore with every rough set we associate two *crisp* sets, called *lower* and *upper approximation*. Intuitively, the lower approximation of a set consists of all elements that *definitely* belong to the set, whereas the upper approximation of the set constitutes of all elements that *possibly* belong to the set, and the *boundary region* of the set consists of all elements that cannot be classified uniquely to the set or its complement, by employing available knowledge. The approximation definition is clearly depicted in Figure 1.

---

[3] One can compare data tables corresponding to information systems with relations in relational databases [26].

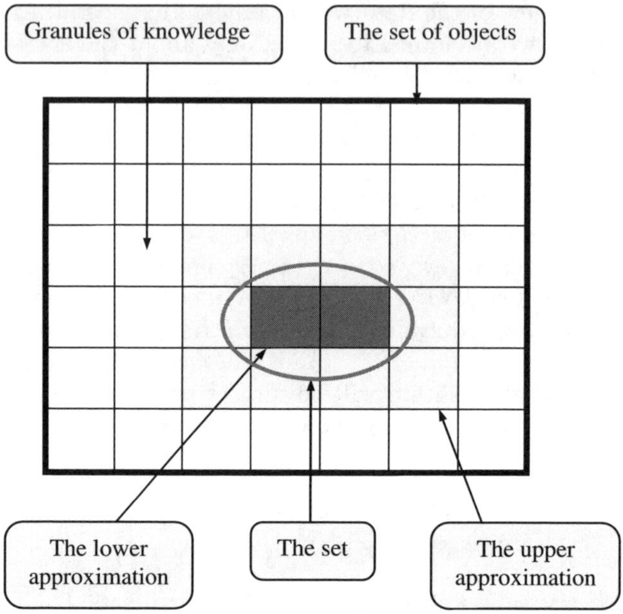

Fig. 1. A rough set

The approximations have the following properties:

$$
\begin{aligned}
&B_*(X) \subseteq X \subseteq B^*(X), \\
&B_*(\emptyset) = B^*(\emptyset) = \emptyset, B_*(U) = B^*(U) = U, \\
&B^*(X \cup Y) = B^*(X) \cup B^*(Y), \\
&B_*(X \cap Y) = B_*(X) \cap B_*(Y), \\
&X \subseteq Y \text{ implies } B_*(X) \subseteq B_*(Y) \text{ and } B^*(X) \subseteq B^*(Y), \\
&B_*(X \cup Y) \supseteq B_*(X) \cup B_*(Y), \\
&B^*(X \cap Y) \subseteq B^*(X) \cap B^*(Y), \\
&B_*(-X) = -B^*(X), \\
&B^*(-X) = -B_*(X), \\
&B_*(B_*(X)) = B^*(B_*(X)) = B_*(X), \\
&B^*(B^*(X)) = B_*(B^*(X)) = B^*(X).
\end{aligned} \tag{5}
$$

Let us note that the inclusions in (5) cannot be in general substituted by the equalities. This has some important algorithmic and logical consequences.

Now we are ready to give the definition of rough sets.

If the boundary region of X is the empty set, i.e., $BN_B(X) = \emptyset$, then the set X is *crisp* (*exact*) with respect to B; in the opposite case, i.e., if $BN_B(X) \neq \emptyset$, the set X is referred to as *rough* (*inexact*) with respect to B.

Thus any rough set, in contrast to a crisp set, has a non-empty boundary region.

One can define the following four basic classes of rough sets, i.e., four categories of vagueness:

$$\begin{aligned} &B_*(X) \neq \emptyset \text{ and } B^*(X) \neq U, \text{ iff } X \text{ is } \textit{roughly B-definable}, \\ &B_*(X) = \emptyset \text{ and } B^*(X) \neq U, \text{ iff } X \text{ is } \textit{internally B-indefinable}, \\ &B_*(X) \neq \emptyset \text{ and } B^*(X) = U, \text{ iff } X \text{ is } \textit{externally B-indefinable}, \\ &B_*(X) = \emptyset \text{ and } B^*(X) = U, \text{ iff } X \text{ is } \textit{totally B-indefinable}. \end{aligned} \tag{6}$$

The intuitive meaning of this classification is the following.

If X is roughly *B-definable*, this means that we are able to decide for some elements of U that they belong to X and for some elements of U we are able to decide that they belong to $-X$, using B.

If X is internally B-indefinable, this means that we are able to decide about some elements of U that they belong to $-X$, but we are unable to decide for any element of U that it belongs to X, using B.

If X is externally B-indefinable, this means that we are able to decide for some elements of U that they belong to X, but we are unable to decide, for any element of U that it belongs to $-X$, using B.

If X is totally B-indefinable, we are unable to decide for any element of U whether it belongs to X or $-X$, using B.

Thus a set is *rough* (imprecise) if it has nonempty boundary region; otherwise the set is crisp (precise). This is exactly the idea of vagueness proposed by Frege.

Let us observe that the definition of rough sets refers to data (knowledge), and is *subjective*, in contrast to the definition of classical sets, which is in some sense an *objective* one.

A rough set can also be characterized numerically by the following coefficient

$$\alpha_B(X) = \frac{card(B_*(X))}{card(B^*(X))}, \tag{7}$$

called the *accuracy of approximation*, where $card(X)$ denotes the cardinality of $X \neq \emptyset$. Obviously $0 \leq \alpha_B(X) \leq 1$. If $\alpha_B(X) = 1$ then X is *crisp* with respect to B (X is *precise* with respect to B), and otherwise, if $\alpha_B(X) < 1$ then X is *rough* with respect to B (X is *vague* with respect to B). The accuracy of approximation can be used to measure the quality of approximation of decision classes on the universe U. One can use another measure of accuracy defined by $1 - \alpha_B(X)$ or by $1 - \frac{card(BN_B(X))}{card(U)}$. Some other measures of approximation accuracy are also used, e.g., based on entropy or some more specific properties of boundary regions (see, e.g., [108, 122, 27]). The choice of a relevant accuracy of approximation depends on a particular data set. Observe that the accuracy of approximation of X can be tuned by B. Another approach to accuracy of approximation can be based on the Variable Precision Rough Set Model (VPRSM) [152] (see Section 3.1).

In the next section, we discuss decision rules (constructed over a selected set B of features or a family of sets of features) which are used in inducing classification algorithms (classifiers) making it possible to classify to decision classes unseen objects. Parameters which are tuned in searching for a classifier with the high quality are its description size (defined using decision rules) and its quality of classification (measured by the number of misclassified objects on a given set of objects). By selecting a proper balance between the accuracy of classification and the description size we expect to find the classifier with the high quality of classification also on unseen objects. This approach is based on the minimal description length principle [97, 98, 124].

2.3 Decision Systems and Decision Rules

Sometimes we distinguish in an information system $\mathcal{A} = (U, A)$ a partition of A into two classes $C, D \subseteq A$ of attributes, called *condition* and *decision* (*action*) attributes, respectively. The tuple $\mathcal{A} = (U, C, D)$ is called a *decision system.*

Let $V = \bigcup\{V_a | a \in C\} \bigcup\{V_d | d \in D\}$. Atomic formulae over $B \subseteq C \cup D$ and V are expressions $a = v$ called *descriptors* (*selectors*) over B and V, where $a \in B$ and $v \in V_a$. The set $\mathcal{F}(B, V)$ of formulae over B and V is the least set containing all atomic formulae over B and V and closed with respect to the propositional connectives $\wedge$ (conjunction), $\vee$ (disjunction) and $\neg$ (negation).

By $\|\varphi\|_{\mathcal{A}}$ we denote the meaning of $\varphi \in \mathcal{F}(B, V)$ in the decision table $\mathcal{A}$ which is the set of all objects in U with the property φ. These sets are defined by $\|a = v\|_{\mathcal{A}} = \{x \in U \mid a(x) = v\}$, $\|\varphi \wedge \varphi'\|_{\mathcal{A}} = \|\varphi\|_{\mathcal{A}} \cap \|\varphi'\|_{\mathcal{A}}$; $\|\varphi \vee \varphi'\|_{\mathcal{A}} = \|\varphi\|_{\mathcal{A}} \cup \|\varphi'\|_{\mathcal{A}}$; $\|\neg\varphi\|_{\mathcal{A}} = U - \|\varphi\|_{\mathcal{A}}$ The formulae from $\mathcal{F}(C, V)$, $\mathcal{F}(D, V)$ are called *condition formulae of* $\mathcal{A}$ and *decision formulae of* $\mathcal{A}$, respectively.

Any object $x \in U$ belongs to the *decision class* $\| \bigwedge_{d \in D} d = d(x)\|_{\mathcal{A}}$ of $\mathcal{A}$. All decision classes of $\mathcal{A}$ create a partition U/D of the universe U.

A *decision rule* for $\mathcal{A}$ is any expression of the form $\varphi \Rightarrow \psi$, where $\varphi \in \mathcal{F}(C, V)$, $\psi \in \mathcal{F}(D, V)$, and $\|\varphi\|_{\mathcal{A}} \neq \emptyset$. Formulae φ and ψ are referred to as the *predecessor* and the *successor* of decision rule $\varphi \Rightarrow \psi$. Decision rules are often called "$IF \ldots THEN \ldots$" rules. Such rules are used in machine learning (see, e.g., [25]).

Decision rule $\varphi \Rightarrow \psi$ is *true* in $\mathcal{A}$ if and only if $\|\varphi\|_{\mathcal{A}} \subseteq \|\psi\|_{\mathcal{A}}$. Otherwise, one can measure its *truth degree* by introducing some inclusion measure of $\|\varphi\|_{\mathcal{A}}$ in $\|\psi\|_{\mathcal{A}}$.

Given two unary predicate formulae $\alpha(x), \beta(x)$ where x runs over a finite set U, Łukasiewicz [53] proposes to assign to $\alpha(x)$ the value $\frac{card(\|\alpha(x)\|)}{card(U)}$, where $\|\alpha(x)\| = \{x \in U : x \text{ satisfies } \alpha\}$. The fractional value assigned to the implication $\alpha(x) \Rightarrow \beta(x)$ is then $\frac{card(\|\alpha(x) \wedge \beta(x)\|)}{card(\|\alpha(x)\|)}$ under the assumption that $\|\alpha(x)\| \neq \emptyset$. Proposed by Łukasiewicz, that fractional part was much later adapted by machine learning and data mining literature.

Each object x of a decision system determines a *decision rule*

$$\bigwedge_{a \in C} a = a(x) \Rightarrow \bigwedge_{d \in D} d = d(x). \quad (8)$$

For any decision table $\mathcal{A} = (U, C, d)$ one can consider a *generalized decision* function $\partial_A : U \longrightarrow \mathrm{Pow}(\times_{d \in D} V_d)$ defined by

$$\partial_A(x) = \left\{ i \ : \ \exists x' \in U \, [\, (x', x) \in I(A) \text{ and } d(x') = i] \right\}, \quad (9)$$

where $\mathrm{Pow}(V_d)$ is the powerset of the Cartesian product $\times_{d \in D} V_d$ of the family $\{V_d\}_{d \in D}$.

$\mathcal{A}$ is called *consistent (deterministic)*, if $card(\partial_A(x)) = 1$, for any $x \in U$. Otherwise $\mathcal{A}$ is said to be *inconsistent (non-deterministic)*. Hence, a decision table is inconsistent if it consists of some objects with different decisions but indiscernible with respect to condition attributes. Any set consisting of all objects with the same generalized decision value is called a *generalized decision class*. Now, one can consider certain (possible) rules (see, e.g. [31, 33]) for decision classes defined by the lower (upper) approximations of such generalized decision classes of $\mathcal{A}$. This approach can be extend, using the relationships of rough sets with the Dempster-Shafer theory (see, e.g., [108, 101]), by considering rules relative to decision classes defined by the lower approximations of unions of decision classes of $\mathcal{A}$.

Numerous methods have been developed for different decision rule generation that the reader can find in the literature on rough sets (see also Section 3.2). Usually, one is searching for decision rules (semi) optimal with respect to some optimization criteria describing quality of decision rules in concept approximations.

In the case of searching for concept approximation in an extension of a given universe of objects (sample), the following steps are typical. When a set of rules has been induced from a decision table containing a set of training examples, they can be inspected to see if they reveal any novel relationships between attributes that are worth pursuing for further research. Furthermore, the rules can be applied to a set of unseen cases in order to estimate their classificatory power. For a systematic overview of rule application methods the reader is referred to the literature (see, e.g., [56, 3] and also Section 3.2).

2.4 Dependency of Attributes

Another important issue in data analysis is discovering dependencies between attributes in a given decision system $\mathcal{A} = (U, C, D)$. Intuitively, a set of attributes D depends totally on a set of attributes C, denoted $C \Rightarrow D$, if the values of attributes from C uniquely determine the values of attributes from D. In other words, D depends totally on C, if there exists a functional dependency between values of C and D. Hence, $C \Rightarrow D$ if and only if the rule (8) is true on $\mathcal{A}$ for any $x \in U$. D can depend partially on C. Formally such a dependency can be defined in the following way.

We will say that D *depends on* C to a *degree* k $(0 \leq k \leq 1)$, denoted $C \Rightarrow_k D$, if

$$k = \gamma(C, D) = \frac{card(POS_C(D))}{card(U)} \tag{10}$$

where

$$POS_C(D) = \bigcup_{X \in U/D} C_*(X), \tag{11}$$

called a *positive region* of the partition U/D with respect to C, is the set of all elements of U that can be uniquely classified to blocks of the partition U/D, by means of C.

If $k = 1$ we say that D *depends totally* on C, and if $k < 1$, we say that D *depends partially* (to *degree* k) on C. If $k = 0$ then the *positive region* of the partition U/D with respect to C is empty.

The coefficient k expresses the ratio of all elements of the universe, which can be properly classified to blocks of the partition U/D, employing attributes C and will be called the *degree of the dependency.*

It can be easily seen that if D depends totally on C then $I(C) \subseteq I(D)$. It means that the partition generated by C is finer than the partition generated by D. Notice, that the concept of dependency discussed above corresponds to that considered in relational databases.

Summing up: D is *totally* (*partially*) dependent on C, if *all* (*some*) elements of the universe U can be uniquely classified to blocks of the partition U/D, employing C.

Observer, that (10) defines only one of possible measures of dependency between attributes (see, e.g., [122]). One also can compare the dependency discussed in this section with dependencies considered in databases [26].

2.5 Reduction of Attributes

We often face a question whether we can remove some data from a data-table preserving its basic properties, that is – whether a table contains some superfluous data.

Let us express this idea more precisely.

Let $C, D \subseteq A$, be sets of condition and decision attributes respectively. We will say that $C' \subseteq C$ is a *D-reduct* (reduct with *respect* to D) of C, if C' is a minimal subset of C such that

$$\gamma(C, D) = \gamma(C', D). \tag{12}$$

The intersection of all D-reducts is called a *D-core* (core with *respect* to D). Because the core is the intersection of all reducts, it is included in every reduct, i.e., each element of the core belongs to some reduct. Thus, in a sense, the core is the most important subset of attributes, since none of its elements can be removed without affecting the classification power of attributes. Certainly, the geometry of reducts can be more compound. For example, the core can be empty but there can exist a partition of reducts into a few sets with non empty intersection.

Many other kinds of reducts and their approximations are discussed in the literature (see, e.g., [5, 59, 60, 102, 121, 123, 124]). It turns out that they can be efficiently computed using heuristics based, e.g., on the Boolean reasoning approach.

2.6 Discernibility and Boolean Reasoning

Methodologies devoted to data mining, knowledge discovery, decision support, pattern classification, approximate reasoning require tools aimed at discovering *templates (patterns)* in data and classifying them into certain *decision classes*. Templates are in many cases most frequent sequences of events, most probable events, regular configurations of objects, the decision rules of high quality, standard reasoning schemes. Tools for discovering and classifying of templates are based on *reasoning schemes* rooted in various paradigms [20]. Such patterns can be extracted from data by means of methods based, e.g., on Boolean reasoning and discernibility.

The discernibility relations are closely related to indiscernibility and belong to the most important relations considered in rough set theory.

The ability to discern between perceived objects is important for constructing many entities like reducts, decision rules or decision algorithms. In the classical rough set approach the *discernibility relation* $DIS(B) \subseteq U \times U$ is defined by $xDIS(B)y$ if and only if $non(xI(B)y)$. However, this is in general not the case for the generalized approximation spaces (one can define indiscernibility by $x \in I(y)$ and discernibility by $I(x) \cap I(y) = \emptyset$ for any objects x, y where $I(x) = B(x), I(y) = B(y)$ in the case of the indiscernibility relation defined in Section 2.2 and in more general case (see Section 3) $I(x), I(y)$ are neighborhoods of objects not necessarily defined by the equivalence relation.

The idea of Boolean reasoning is based on construction for a given problem P of a corresponding Boolean function f_P with the following property: the solutions for the problem P can be decoded from prime implicants of the Boolean function f_P. Let us mention that to solve real-life problems it is necessary to deal with Boolean functions having large number of variables.

A successful methodology based on the discernibility of objects and Boolean reasoning has been developed for computing of many important for applications. These include reducts and their approximations, decision rules, association rules, discretization of real value attributes, symbolic value grouping, searching for new features defined by oblique hyperplanes or higher order surfaces, pattern extraction from data as well as conflict resolution or negotiation.

Most of the problems related to generation of the above mentioned entities are NP-complete or NP-hard. However, it is possible to develop efficient heuristics returning suboptimal solutions of the problems. The results of experiments on many data sets are very promising. They show very good quality of solutions generated by the heuristics in comparison with other methods reported in literature (e.g., with respect to the classification quality of unseen objects). Moreover, they are very efficient from the point of view of time necessary for computing of the solution. Many of these methods are based on discernibility matrices. Note, that it is possible to compute the necessary information about these matrices using directly[4] information or decision systems (e.g., sorted in preprocessing [3, 62, 144]) which significantly improves the efficiency of algorithms.

It is important to note that the methodology makes it possible to construct heuristics having a very important *approximation property* which can be formulated as follows: expressions generated by heuristics (i.e., implicants) *close* to prime implicants define approximate solutions for the problem.

2.7 Rough Membership

Let us observe that rough sets can be also defined employing the rough membership function (see Eq. 13) instead of approximation [77]. That is, consider

$$\mu_X^B : U \to < 0, 1 >,$$

defined by

$$\mu_X^B(x) = \frac{card(B(x) \cap X)}{card(X)}, \tag{13}$$

where $x \in X \subseteq U$.

The value $\mu_X^B(x)$ can be interpreted as the degree that x belongs to X in view of knowledge about x expressed by B or the degree to which the elementary granule $B(x)$ is included in the set X. This means that the definition reflects a subjective knowledge about elements of the universe, in contrast to the classical definition of a set.

The rough membership function can also be interpreted as the conditional probability that x belongs to X given B. This interpretation was used by several researchers in the rough set community (see, e.g., [4, 32, 123, 140, 143]). Note also that the ratio on the right hand side of the equation (13) is known as the confidence coefficient in data mining [25, 37]. It is worthwhile to mention that set inclusion to a degree has been considered by Łukasiewicz [53] in studies on assigning fractional truth values to logical formulas.

It can be shown that the rough membership function has the following properties [77]:

1) $\mu_X^B(x) = 1$ *iff* $x \in B_*(X)$;
2) $\mu_X^B(x) = 0$ *iff* $x \in U - B^*(X)$;
3) $0 < \mu_X^B(x) < 1$ *iff* $x \in BN_B(X)$;

[4] i.e., without the necessity of generation and storing of the discernibility matrices

4) $\mu^B_{U-X}(x) = 1 - \mu^B_X(x)$ *for any* $x \in U$;
5) $\mu^B_{X \cup Y}(x) \geq max(\mu^B_X(x),\ \mu^B_Y(x))$ *for any* $x \in U$;
6) $\mu^B_{X \cap Y}(x) \leq min(\mu^B_X(x),\ \mu^B_Y(x))$ *for any* $x \in U$.

From the properties it follows that the rough membership differs essentially from the fuzzy membership [149], for properties 5) and 6) show that the membership for union and intersection of sets, in general, cannot be computed – as in the case of fuzzy sets – from their constituents membership. Thus formally the rough membership is more general from fuzzy membership. Moreover, the rough membership function depends on an available knowledge (represented by attributes from B). Besides, the rough membership function, in contrast to fuzzy membership function, has a probabilistic flavor.

Let us also mention that rough set theory, in contrast to fuzzy set theory, clearly distinguishes two very important concepts, vagueness and uncertainty, very often confused in the AI literature. Vagueness is the property of sets and can be described by approximations, whereas uncertainty is the property of objects considered as elements of a set and can be expressed by the rough membership function.

Both fuzzy and rough set theory represent two different approaches to vagueness. Fuzzy set theory addresses *gradualness* of knowledge, expressed by the fuzzy membership, whereas rough set theory addresses *granularity* of knowledge, expressed by the indiscernibility relation. A nice illustration of this difference has been given by Dider Dubois and Henri Prade [19] in the following example. In image processing fuzzy set theory refers to gradualness of gray level, whereas rough set theory is about the size of pixels.

Consequently, both theories are not competing but are rather complementary. In particular, the rough set approach provides tools for approximate construction of fuzzy membership functions. The rough-fuzzy hybridization approach proved to be successful in many applications (see, e.g., [68, 71]).

Interesting discussion of fuzzy and rough set theory in the approach to vagueness can be found in [96]. Let us also observe that fuzzy set and rough set theory are not a remedy for classical set theory difficulties.

One of the consequences of perceiving objects by information about them is that for some objects one cannot decide if they belong to a given set or not. However, one can estimate the degree to which objects belong to sets. This is a crucial observation in building foundations for approximate reasoning. Dealing with imperfect knowledge implies that one can only characterize satisfiability of relations between objects to a degree, not precisely. One of the fundamental relations on objects is a rough inclusion relation describing that objects are parts of other objects to a degree. The rough mereological approach [70, 87, 88, 90] based on such a relation is an extension of the Leśniewski mereology [45].

3 Extensions

The rough set concept can be defined quite generally by means of topological operations, *interior* and *closure*, called *approximations* [84]. It was observed in [73] that the key to the presented approach is provided by the exact mathematical formulation of the concept of approximative (rough) equality of sets in a given approximation space. In [74], an approximation space is represented by the pair (U, R), where U is a universe of objects, and $R \subseteq U \times U$ is an indiscernibility relation defined by an attribute set (i.e., $R = I(A)$ for some attribute set A). In this case R is an equivalence relation. Let $[x]_R$ denote an equivalence class of an element $x \in U$ under the indiscernibility relation R, where $[x]_R = \{y \in U : xRy\}$.

In this context, R-approximations of any set $X \subseteq U$ are based on the exact (crisp) containment of sets. Then set approximations are defined as follows:

- $x \in U$ belongs with certainty to the R-lower approximation of $X \subseteq U$, if $[x]_R \subseteq X$.
- $x \in U$ belongs with certainty to the complement set of $X \subseteq U$, if $[x]_R \subseteq U - X$.
- $x \in U$ belongs with certainty to the R-boundary region of $X \subseteq U$, if $[x]_R \cap X \neq \oslash$ and $[x]_R \cap (U - X) \neq \oslash$.

Several generalizations of the above approach have been proposed in the literature (see, e.g., [29, 70, 113, 118, 131, 152]). In particular, in some of these approaches, set inclusion to a degree is used instead of the exact inclusion.

Different aspects of vagueness in the rough set framework are discussed, e.g., in [55, 65, 66, 96, 107].

Our knowledge about the approximated concepts is often partial and uncertain [30]. For example, concept approximation should be constructed from examples and counter examples of objects for the concepts [25]. Hence, the concept approximations constructed from a given sample of objects is extended, using inductive reasoning, on unseen so far objects. The rough set approach for dealing with concept approximation under such partial knowledge is presented, e.g., in [118]. Moreover, the concept approximations should be constructed under dynamically changing environments [107]. This leads to a more complex situation when the boundary regions are not crisp sets what is consistent with the postulate of the higher order vagueness, considered by philosophers (see, e.g., [36]). It is worthwhile to mention that a rough set approach to the approximation of compound concepts has been developed and at this time no traditional method is able directly to approximate compound concepts [11, 141]. The approach is based on hierarchical learning and ontology approximation [8, 61, 70, 111]). Approximation of concepts in distributed environments is discussed in [105]. A survey of algorithmic methods for concept approximation based on rough sets and Boolean reasoning in presented, e.g., in [103].

3.1 Generalizations of Approximation Spaces

Several generalizations of the classical rough set approach based on approximation spaces defined as pairs of the form (U, R), where R is the equivalence relation (called indiscernibility relation) on the set U, have been reported in the literature. Let us mention two of them.

A generalized approximation space[5] can be defined by a tuple $AS = (U, I, \nu)$ where I is the *uncertainty function* defined on U with values in the powerset $\textsc{Pow}(U)$ of U ($I(x)$ is the *neighboorhood* of x) and ν is the *inclusion function* defined on the Cartesian product $\textsc{Pow}(U) \times \textsc{Pow}(U)$ with values in the interval $[0, 1]$ measuring the degree of inclusion of sets. The lower AS_* and upper AS^* approximation operations can be defined in AS by

$$AS_*(X) = \{x \in U : \nu(I(x), X) = 1\}, \tag{14}$$
$$AS^*(X) = \{x \in U : \nu(I(x), X) > 0\}. \tag{15}$$

In the standard case $I(x)$ is equal to the equivalence class $B(x)$ of the indiscernibility relation $I(B)$; in case of tolerance (similarity) relation $\tau \subseteq U \times U$ [95] we take $I(x) = \{y \in U : x\tau y\}$, i.e., $I(x)$ is equal to the tolerance class of τ defined by x. The standard inclusion relation ν_{SRI} is defined for $X, Y \subseteq U$ by

$$\nu_{SRI}(X, Y) = \begin{cases} \frac{card(X \cap Y)}{card(X)}, & \text{if } X \text{ is non} - \text{empty}, \\ 1, & \text{otherwise}. \end{cases} \tag{16}$$

For applications it is important to have some constructive definitions of I and ν.

One can consider another way to define $I(x)$. Usually together with AS we consider some set F of formulae describing sets of objects in the universe U of AS defined by semantics $\| \cdot \|_{AS}$, i.e., $\|\alpha\|_{AS} \subseteq U$ for any $\alpha \in F$. Now, one can take the set

$$N_F(x) = \{\alpha \in F : x \in \|\alpha\|_{AS}\}, \tag{17}$$

and $I(x) = \{\|\alpha\|_{AS} : \alpha \in N_F(x)\}$. Hence, more general uncertainty functions having values in $\textsc{Pow}(\textsc{Pow}(U))$ can be defined and in the consequence different definitions of approximations are considered. For example, one can consider the following definitions of approximation operations in AS:

$$AS_\circ(X) = \{x \in U : \nu(Y, X) = 1 \text{ for some } Y \in I(x)\}, \tag{18}$$
$$AS^\circ(X) = \{x \in U : \nu(Y, X) > 0 \text{ for any } Y \in I(x)\}. \tag{19}$$

There are also different forms of rough inclusion functions. Let us consider some examples.

---

[5] Some other generalizations of approximation spaces are also considered in the literature (see, e.g., [47, 49, 104, 146, 147, 145, 148]).

In the first example of rough inclusion function a a threshold $t \in (0, 0.5)$ is used to relax the degree of inclusion of sets. The rough inclusion function ν_t is defined by

$$\nu_t(X,Y) = \begin{cases} 1, & \text{if} \quad \nu_{SRI}(X,Y) \geq 1-t, \\ \frac{\nu_{SRI}(X,Y)-t}{1-2t}, & \text{if} \quad t < \nu_{SRI}(X,Y) < 1-t, \\ 0, & \text{if} \quad \nu_{SRI}(X,Y) \leq t. \end{cases} \tag{20}$$

One can obtain approximations considered in the variable precision rough set approach (VPRSM) [152] by substituting in (14)-(15) the rough inclusion function ν_t defined by (20) instead of ν, assuming that Y is a decision class and $N(x) = B(x)$ for any object x, where B is a given set of attributes.

Another example of application of the standard inclusion was developed by using probabilistic decision functions. For more detail the reader is referred to [122, 123].

The rough inclusion relation can be also used for function approximation [118] and relation approximation [133]. In the case of function approximation the inclusion function ν^* for subsets $X, Y \subseteq U \times U$, where $U \subseteq \mathcal{R}$ and $\mathcal{R}$ is the set of reals, is defined by

$$\nu^*(X,Y) = \begin{cases} \frac{card(\pi_1(X \cap Y))}{card(\pi_1(X))}, & \text{if } \pi_1(X) \neq \emptyset, \\ 1, & \text{if } \pi_1(X) = \emptyset, \end{cases} \tag{21}$$

where π_1 is the projection operation on the first coordinate. Assume now, that X is a cube and Y is the graph $G(f)$ of the function $f : \mathcal{R} \longrightarrow \mathcal{R}$. Then, e.g., X is in the lower approximation of f if the projection on the first coordinate of the intersection $X \cap G(f)$ is equal to the projection of X on the first coordinate. This means that the part of the graph $G(f)$ is "well" included in the box X, i.e., for all arguments that belong to the box projection on the first coordinate the value of f is included in the box X projection on the second coordinate.

The approach based on inclusion functions has been generalized to the *rough mereological approach* [70, 87, 88, 90] (see also Section 3.6). The inclusion relation $x\mu_r y$ with the intended meaning *x is a part of y to a degree at least r* has been taken as the basic notion of the rough mereology being a generalization of the Leśniewski mereology [45, 46]. Research on rough mereology has shown importance of another notion, namely *closeness* of complex objects (e.g., concepts). This can be defined by $xcl_{r,r'}y$ if and only if $x\mu_r y$ and $y\mu_{r'}x$.

Rough mereology offers a methodology for synthesis and analysis of objects in a distributed environment of intelligent agents, in particular, for synthesis of objects satisfying a given specification to a satisfactory degree or for control in such a complex environment. Moreover, rough mereology has been recently used for developing the foundations of the *information granule calculi*, aiming at formalization of the Computing with Words paradigm, recently formulated by Lotfi Zadeh [150]. More complex information granules are defined recursively using already defined information granules and their measures of inclusion and closeness. Information granules can have complex structures like classifiers or approximation spaces. Computations on information granules are performed to discover relevant information granules, e.g., patterns or approximation spaces for complex concept approximations.

Usually there are considered families of approximation spaces labeled by some parameters. By tuning such parameters according to chosen criteria (e.g., minimal description length) one can search for the optimal approximation space for concept description (see, e.g., [3]).

3.2 Concept Approximation

In this section, we consider the problem of approximation of concepts over a universe U^∞ (concepts that are subsets of U^∞). We assume that the concepts are perceived only through some subsets of U^∞, called samples. This is a typical situation in the machine learning, pattern recognition, or data mining approaches [25, 37]. We explain the rough set approach to induction of concept approximations using the generalized approximation spaces of the form $AS = (U, I, \nu)$ defined in Section 3.1.

Let $U \subseteq U^\infty$ be a finite sample. By Π_U we denote a perception function from $P(U^\infty)$ into $P(U)$ defined by $\Pi_U(C) = C \cap U$ for any concept $C \subseteq U^\infty$. Let $AS = (U, I, \nu)$ be an approximation space over the sample U.

The problem we consider is how to extend the approximations of $\Pi_U(C)$ defined by AS to approximation of C over U^∞. We show that the problem can be described as searching for an extension $AS_C = (U^\infty, I_C, \nu_C)$ of the approximation space AS, relevant for approximation of C. This requires to show how to extend the inclusion function ν from subsets of U to subsets of U^∞ that are relevant for the approximation of C. Observe that for the approximation of C it is enough to induce the necessary values of the inclusion function ν_C without knowing the exact value of $I_C(x) \subseteq U^\infty$ for $x \in U^\infty$.

Let AS be a given approximation space for $\Pi_U(C)$ and let us consider a language L in which the neighborhood $I(x) \subseteq U$ is expressible by a formula $pat(x)$, for any $x \in U$. It means that $I(x) = \|pat(x)\|_U \subseteq U$, where $\|pat(x)\|_U$ denotes the meaning of $pat(x)$ restricted to the sample U. In case of rule based classifiers patterns of the form $pat(x)$ are defined by feature value vectors.

We assume that for any new object $x \in U^\infty \setminus U$ we can obtain (e.g., as a result of sensor measurement) a pattern $pat(x) \in L$ with semantics $\|pat(x)\|_{U^\infty} \subseteq U^\infty$. However, the relationships between information granules over U^∞ such as sets $\|pat(x)\|_{U^\infty}$ and $\|pat(y)\|_{U^\infty}$, for different $x, y \in U^\infty$, are, in general, known only if they can be expressed by relationships between the restrictions of these sets to the sample U, i.e., between sets $\Pi_U(\|pat(x)\|_{U^\infty})$ and $\Pi_U(\|pat(y)\|_{U^\infty})$.

The set of patterns $\{pat(x) : x \in U\}$ is usually not relevant for approximation of the concept $C \subseteq U^\infty$. Such patterns are too specific or not enough general, and can directly be applied only to a very limited number of new objects. However, by using some generalization strategies, one can search, in a family of patterns definable from $\{pat(x) : x \in U\}$ in L, for such new patterns that are relevant for approximation of concepts over U^∞. Let us consider a subset $PATTERNS(AS, L, C) \subseteq L$ chosen as a set of pattern candidates for relevant approximation of a given concept C. For example, in case of rule based classifier one can search for such candidate patterns among sets definable by subsequences of feature value vectors corresponding to objects from the sample U. The set $PATTERNS(AS, L, C)$ can be selected by using some quality measures checked on meanings (semantics) of its elements restricted to the sample U (like the number of examples from the concept $\Pi_U(C)$ and its complement that support a given pattern). Then, on the basis of properties of sets definable by these patterns over U we induce approximate values of the inclusion function ν_C on subsets of U^∞ definable by any of such pattern and the concept C.

Next, we induce the value of ν_C on pairs (X, Y) where $X \subseteq U^\infty$ is definable by a pattern from $\{pat(x) : x \in U^\infty\}$ and $Y \subseteq U^\infty$ is definable by a pattern from $PATTERNS(AS, L, C)$.

Finally, for any object $x \in U^\infty \setminus U$ we induce the approximation of the degree $\nu_C(\|pat(x)\|_{U^\infty}, C)$ applying a conflict resolution strategy $Conflict\_res$ (a voting strategy, in case of rule based classifiers) to two families of degrees:

$$\{\nu_C(\|pat(x)\|_{U^\infty}, \|pat\|_{U^\infty}) : pat \in PATTERNS(AS, L, C)\}, \tag{22}$$

$$\{\nu_C(\|pat\|_{U^\infty}, C) : pat \in PATTERNS(AS, L, C)\}. \tag{23}$$

Values of the inclusion function for the remaining subsets of U^∞ can be chosen in any way – they do not have any impact on the approximations of C. Moreover, observe that for the approximation of C we do not need to know the exact values of uncertainty function I_C – it is enough to induce the values of the inclusion function ν_C. Observe that the defined extension ν_C of ν to some subsets of U^∞ makes it possible to define an approximation of the concept C in a new approximation space AS_C.

Observe that one can also follow principles of Bayesian reasoning and use degrees of ν_C to approximate C (see, e.g., [76, 125, 128]).

In this way, the rough set approach to induction of concept approximations can be explained as a process of inducing a relevant approximation space.

3.3 Higher Order Vagueness

In [36], it is stressed that vague concepts should have non-crisp boundaries. In the definition presented in Section 2.2, the notion of boundary region is defined as a crisp set $BN_B(X)$. However, let us observe that this definition is relative to the subjective knowledge expressed by attributes from B. Different sources of information may use different sets of attributes for concept approximation. Hence, the boundary region can change when we consider these different views. Another aspect is discussed in [107, 117] where it is assumed that information about concepts is incomplete, e.g., the concepts are given only on samples (see, e.g., [25, 37, 56]). From [107, 117] it follows that vague concepts cannot be approximated with satisfactory quality by *static* constructs such as induced membership inclusion functions, approximations or models derived, e.g., from a sample. Understanding of vague concepts can be only realized in a process in which the induced models are adaptively matching the concepts in a dynamically changing environment. This conclusion seems to have important consequences for further development of rough set theory in combination with fuzzy sets and other soft computing paradigms for adaptive approximate reasoning.

3.4 Information Granulation

Information granulation can be viewed as a human way of achieving data compression and it plays a key role in the implementation of the strategy of divide-and-conquer in human problem-solving [150]. Objects obtained as the result of granulation are information granules. Examples of elementary information granules are indiscernibility or tolerance (similarity) classes (see Section 2.2). In reasoning about data and knowledge under uncertainty and imprecision many other more compound information granules are used (see, e.g., [92, 94, 104, 114, 115]). Examples of such granules are decision rules, sets of decision rules or classifiers. More compound information granules are defined by means of less compound ones. Note that inclusion or closeness measures between information granules should be considered rather than their strict equality. Such measures are also defined recursively for information granules.

Let us discuss shortly an example of information granulation in the process of modeling patterns for compound concept approximation (see, e.g., [6, 7, 8, 9, 10, 61, 134]. We start from a generalization of information systems. For any attribute $a \in A$ of an information system (U, A) we consider together with the value set V_a of a a relational structure $\mathcal{R}_a$ over the universe V_a (see, e.g., [119]). We also consider a language $\mathcal{L}_a$ of formulas (of the same relational signature as $\mathcal{R}_a$). Such formulas interpreted over $\mathcal{R}_a$ define subsets of Cartesian products of V_a. For example, any formula α with one free variable defines a subset $\|\alpha\|_{\mathcal{R}_a}$ of V_a. Let us observe that the relational structure $\mathcal{R}_a$ induces a relational structure over U. Indeed, for any k-ary relation r from $\mathcal{R}_a$ one can define a k-ary relation $g_a \subseteq U^k$ by $(x_1, \dots, x_k) \in g_a$ if and only if $(a(x_1), \dots, a(x_k)) \in r$ for any $(x_1, \dots, x_k) \in U^k$. Hence, one can consider any formula from $\mathcal{L}_a$ as a constructive method of defining a subset of the universe U with a structure induced by $\mathcal{R}_a$. Any such a structure is a new information granule. On the next level of hierarchical modeling, i.e., in constructing new information systems we use such structures as objects and attributes are properties of such structures. Next, one can consider similarity between new constructed objects and then their similarity neighborhoods will correspond to clusters of relational structures. This process is usually more complex. This is because instead of relational structure $\mathcal{R}_a$ we usually consider a fusion of relational structures corresponding to some attributes from A. The fusion makes it possible to describe constraints that should hold between parts obtained by composition from less compound parts. Examples of relational structures can be defined by indiscernibility, similarity, intervals obtained in discretization or symbolic value grouping, preference or spatio-temporal relations (see, e.g., [29, 37, 113]). One can see that parameters to be tuned in searching for relevant[6] patterns over new information systems are, among others, relational structures over value sets, the language of formulas defining parts, and constraints.

3.5 Ontological Framework for Approximation

In a number of papers (see, e.g., [116]) the problem of ontology approximation has been discussed together with possible applications to approximation of compound concepts or to knowledge transfer (see, e.g., [63, 99, 116, 106]).

In the ontology [132] (vague) concepts and local dependencies between them are specified. Global dependencies can be derived from local dependencies. Such derivations can be used as hints in searching for relevant compound patterns (information granules) in approximation of more compound concepts from the ontology. The ontology approximation problem is one of the fundamental problems related to approximate reasoning in distributed environments. One should construct (in a given language that is different from the ontology specification language) not only approximations of concepts from ontology but also vague dependencies specified in the ontology. It is worthwhile to mention that an ontology approximation should be induced on the basis of incomplete information about concepts and dependencies specified in the ontology. Information granule calculi based on rough sets have been proposed as tools making it possible to solve this problem. Vague dependencies have vague concepts in premisses and conclusions. The approach to approximation of vague dependencies based only on degrees of closeness of concepts from dependencies and their approximations (classifiers) is not satisfactory for approximate reasoning. Hence, more advanced approach should be developed. Approximation of any vague dependency is a method which allows us for any object to compute the arguments "for" and "against" its membership to the dependency conclusion on the basis of the analogous arguments relative to the dependency premisses. Any argument is a compound information granule (compound pattern). Arguments are fused by local schemes (production rules) discovered from data. Further fusions are possible through composition of local schemes, called approximate reasoning schemes (AR schemes) (see, e.g., [9, 92, 70]). To estimate the degree to which (at least) an object belongs to concepts from ontology the arguments "for" and "against" those concepts are collected and next a conflict resolution strategy is applied to them to predict the degree.

---

[6] for target concept approximation

3.6 Mereology and Rough Mereology

This section introduces some basic concepts of rough mereology (see, e.g., [85, 86, 88, 92, 93, 94]).

Exact and rough concepts can be characterized by a new notion of an element, alien to naive set theory in which this theory has been coded until now. For an information system $\mathcal{A}=(U, A)$, and a set B of attributes, the mereological element $el^{\mathcal{A}_B}$ is defined by letting

$$x el_B^{\mathcal{A}} X \text{ if and only if } B(x) \subseteq X. \tag{24}$$

Then, a concept X is B-exact if and only if either $x el_B^{\mathcal{A}} X$ or $x el_B^{\mathcal{A}} U \setminus X$ for each $x \in U$, and the concept X is B–rough if and only if for some $x \in U$ neither $x el_B^{\mathcal{A}} X$ nor $x el_B^{\mathcal{A}} U \setminus X$.

Thus, the characterization of the dichotomy exact–rough cannot be done by means of the element notion of naive set theory, but requires the notion of containment ($\subseteq$), i.e., a notion of mereological element.

The Leśniewski Mereology (theory of parts) is based on the notion of a part [45, 46]. The relation π of part on the collection U of objects satisfies

$$1.\ \text{if } x\pi y \text{ then not } y\pi x, \tag{25}$$
$$2.\ \text{if } x\pi y \text{ and } y\pi z \text{ then } x\pi z. \tag{26}$$

The notion of mereological element el_π is introduced as

$$x el_\pi y \text{ if and only if } x\pi y \text{ or } x = y. \tag{27}$$

In particular, the relation of proper inclusion $\subset$ is a part relation π on any non–empty collection of sets, with the element relation $el_\pi = \subseteq$.

Formulas expressing, e.g., rough membership, quality of decision rule, quality of approximations can be traced back to a common root, i.e., $\nu(X, Y)$ defined by equation (16). The value $\nu(X, Y)$ defines the degree of *partial containment* of X into Y and naturally refers to the Leśniewski Mereology. An abstract formulation of this idea in [88] connects the mereological notion of element el_π with the partial inclusion by introducing a *rough inclusion* as a relation $\nu \subseteq U \times U \times [0, 1]$ on a collection of pairs of objects in U endowed with part π relation, and such that

$$1.\ \nu(x, y, 1) \text{ if and only if } x el_\pi y, \tag{28}$$
$$2.\ \text{if } \nu(x, y, 1) \text{ then (if } \nu(z, x, r) \text{ then } \nu(z, y, r)), \tag{29}$$
$$3.\ \text{if } \nu(z, x, r) \text{ and } s < r \text{ then } \nu(z, x, s). \tag{30}$$

Implementation of this idea in information systems can be based on *Archimedean* t–norms [88]; each such norm T is represented as $T(r, s) = g(f(r) + f(s))$ with f, g pseudo–inverses to each other, continuous and decreasing on $[0, 1]$. Letting for (U, A) and $x, y \in U$

$$DIS(x,y) = \{a \in A : a(x) \neq a(y)\} \tag{31}$$

and

$$\nu(x,y,r) \text{ if and only if } g(\frac{card(DIS(x,y))}{card(A)}) \geq r \tag{32}$$

defines a rough inclusion that satisfies additionally the transitivity rule

$$\frac{\nu(x,y,r), \nu(y,z,s)}{\nu(x,z,T(r,s))}. \tag{33}$$

Simple examples here are: the Menger rough inclusion in the case $f(r) = -\ln r$, $g(s) = e^{-s}$ yields $\nu(x,y,r)$ if and only if $e^{-\frac{card(DIS(x,y))}{card(A)}} \geq r$ and it satisfies the transitivity rule:

$$\frac{\nu(x,y,r), \nu(y,z,s)}{\nu(x,y,r \cdot s)}, \tag{34}$$

i.e., the t–norm T is the Menger (product) t–norm $r \cdot s$, and, the Łukasiewicz rough inclusion with $f(x) = 1 - x = g(x)$ yielding $\nu(x,y,r)$ if and only if $1 - \frac{card(DIS(x,y))}{card(A)} \geq r$ with the transitivity rule:

$$\frac{\nu(x,y,r), \nu(y,z,s)}{\nu(x,y,max\{0, r+s-1\})}, \tag{35}$$

i.e., with the Łukasiewicz t–norm.

Rough inclusions [88] can be used in *granulation of knowledge* [150]. Granules of knowledge are constructed as aggregates of indiscernibility classes close enough with respect to a chosen measure of closeness. In a nutshell, a granule $g_r(x)$ about x of radius r can be defined as the aggregate of all y with $\nu(y,x,r)$. The aggregating mechanism can be based on the class operator of mereology (cf. rough mereology [88]) or on set theoretic operations of union.

Rough mereology [88] combines rough inclusions with methods of mereology. It employs the operator of mereological class that makes collections of objects into objects. The class operator Cls satisfies the requirements, with any non–empty collection M of objects made into the object $Cls(M)$

$$\text{if } x \in M \text{ then } xel_\pi Cls(M), \tag{36}$$

$$\text{if } xel_\pi Cls(M) \text{ then there exist } y,z \text{ such that } yel_\pi x, yel_\pi z, z \in M. \tag{37}$$

In case of the part relation $\subset$ on a collection of sets, the class $Cls(M)$ of a non–empty collection M is the union $\bigcup M$.

Granulation by means of the class operator Cls consists in forming the granule $g_r(x)$ as the class $Cls(y : \nu(y,x,r))$. One obtains a granule family with regular properties (see [142]).

4 Conflicts

Knowledge discovery in databases considered in the previous sections reduces to searching for functional dependencies in the data set.

In this section, we will discuss another kind of relationship in the data - not dependencies, but conflicts.

Formally, the conflict relation can be seen as a negation (not necessarily, classical) of indiscernibility relation which was used as a basis of rough set theory. Thus indiscernibility and conflict are closely related from logical point of view.

It turns out that the conflict relation can be used to the conflict analysis study.

Conflict analysis and resolution play an important role in business, governmental, political and lawsuits disputes, labor-management negotiations, military operations and others. To this end many mathematical formal models of conflict situations have been proposed and studied, e.g., [12, 15, 16, 24, 41, 42, 43, 54, 58, 64, 75, 136].

Various mathematical tools, e.g., graph theory, topology, differential equations and others, have been used to that purpose.

Needless to say that game theory can be also considered as a mathematical model of conflict situations.

In fact there is no, as yet, "universal" theory of conflicts and mathematical models of conflict situations are strongly domain dependent.

We are going to present in this paper still another approach to conflict analysis, based on some ideas of rough set theory – along the lines proposed in [75] and extended in this paper.

The considered model is simple enough for easy computer implementation and seems adequate for many real life applications but to this end more research is needed.

4.1 Basic Concepts of Conflict Theory

In this section, we give after [75] definitions of basic concepts of the proposed approach.

Let us assume that we are given a finite, non-empty set Ag called the *universe*. Elements of Ag will be referred to as *agents*. Let a *voting function* $v : Ag \rightarrow \{-1, 0, 1\}$, or in short $\{-, 0, +\}$, be given assigning to every agent the number $-1, 0$ or 1, representing his opinion, view, voting result, etc. about some discussed issue, and meaning *against, neutral* and *favorable*, respectively.

Voting functions correspond to situations. Hence, let us assume there is given a set U of situations and a set $Voting\_Fun$ of voting functions as well as a *conflict function* $Conflict : U \longrightarrow Voting\_Fun$. Any pair $S = (s, v)$ where $s \in U$ and $v = Conflict(s)$ will be called a *conflict situation*.

In order to express relations between agents from Ag defined by a given voting function v we define three basic binary relations in Ag^2 : *conflict, neutrality,* and *alliance.*

To this end we first define the following auxiliary function:

$$\phi_v(ag, ag') = \begin{cases} 1, & \text{if } v(ag)v(ag') = 1 \text{ or } ag = ag' \\ 0, & \text{if } v(ag)v(ag') = 0 \text{ and } ag \neq ag' \\ -1, & \text{if } v(ag)v(ag') = -1. \end{cases} \tag{38}$$

This means that, if $\phi_v(ag, ag') = 1$, then agents ag and ag' have the same opinion about issue v (are *allied* on v); $\phi_v(ag, ag') = 0$ means that at least one agent ag or ag' has neutral approach to issue v (is *neutral* on v), and $\phi_v(ag, ag') = -1$, means that both agents have different opinions about issue v (are in *conflict* on v).

In what follows we will define three basic binary relations $R_v^+, R_v^0, R_v^- \subseteq Ag^2$ called *alliance, neutrality* and *conflict* relations respectively, and defined by

$$\begin{aligned} &R_v^+(ag, ag') \text{ iff } \phi_v(ag, ag') = 1, \\ &R_v^0(ag, ag') \text{ iff } \phi_v(ag, ag') = 0, \\ &R_v^-(ag, ag') \text{ iff } \phi_v(ag, ag') = -1. \end{aligned} \tag{39}$$

It is easily seen that the alliance relation has the following properties:

$$\begin{aligned} &R_v^+(ag, ag), \\ &R_v^+(ag, ag') \text{ implies } R_v^+(ag', ag), \\ &R_v^+(ag, ag') \text{ and } R_v^+(ag', ag'') \text{ implies } R_v^+(ag, ag''), \end{aligned} \tag{40}$$

i.e., R_v^+ is an *equivalence* relation. Each equivalence class of alliance relation will be called a *coalition* with respect to v. Let us note that the last condition in (40) can be expressed as "a friend of my friend is my friend".

For the conflict relation we have the following properties:

$$\begin{aligned} &\text{not } R_v^-(ag, ag), \\ &R_v^-(ag, ag') \text{ implies } R_v^-(ag', ag), \\ &R_v^-(ag, ag') \text{ and } R_v^-(ag', ag'') \text{ implies } R_v^+(ag, ag''), \\ &R_v^-(ag, ag') \text{ and } R_v^+(ag', ag'') \text{ implies } R_v^-(ag, ag''). \end{aligned} \tag{41}$$

The last two conditions in (41) refer to well known sayings "an enemy of my enemy is my friend" and "a friend of my enemy is my enemy".

For the neutrality relation we have:

$$\begin{aligned} &\text{not } R_v^0(ag, ag), \\ &R_v^0(ag, ag') = R_v^0(ag', ag). \end{aligned} \tag{42}$$

Let us observe that in the conflict and neutrality relations there are no coalitions.

We have $R_v^+ \cup R_v^0 \cup R_v^- = Ag^2$ because if $(ag, ag') \in Ag^2$ then $\Phi_v(ag, ag') = 1$ or $\Phi_v(ag, ag') = 0$ or $\Phi_v(ag, ag') = -1$ so $(ag, ag') \in R_v^+$ or $(ag, ag') \in R_v^-$ or $(ag, ag') \in R_v^-$. All the three relations R_v^+, R_v^0 , R_v^- are pairwise disjoint, i.e., every pair of objects (ag, ag') belongs to exactly one of the above defined relations (is in conflict, is allied or is neutral).

With every conflict situation $S = (s, v)$ we will associate a *conflict graph*

$$G_S = (R_v^+, R_v^0, R_v^-). \tag{43}$$

An example of a conflict graph is shown in Figure 2. Solid lines are denoting

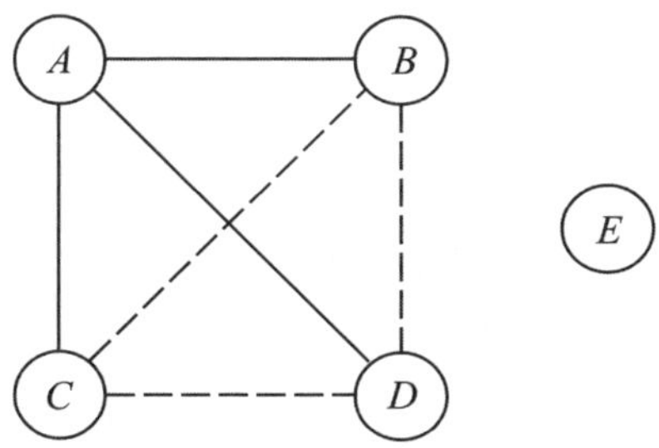

Fig. 2. Exemplary conflict graph

conflicts, doted line – alliance, and neutrality, for simplicity, is not shown explicitly in the graph. Of course, B, C, and D form a coalition.

A conflict degree $Con(S)$ of the conflict situation $S = (s, v)$ is defined by

$$Con(S) = \frac{\sum_{\{(ag,ag'):\ \phi_v(ag,ag')=-1\}} |\phi_v(ag, ag')|}{2\lceil \frac{n}{2} \rceil \times (n - \lceil \frac{n}{2} \rceil)} \tag{44}$$

where $n = Card(Ag)$.

One can consider a more general definition of conflict function $Conflict : U \longrightarrow Voting\_Fun^k$ where k is a positive integer. Then, a conflict situation is any pair $S = (s, (v_1, ..., v_k))$ where $(v_1, ..., v_k) = Conflict(s)$ and the conflict degree in S can be defined by

$$Con(S) = \frac{\sum_{i=1}^{k} Con(S_i)}{k} \tag{45}$$

where $S_i = (s, v_i)$ for $i = 1, \ldots, k$. Each function v_i is called a voting function on the $i - th$ issue in s.

4.2 An Example

In this section, we will illustrate the above presented ideas by means of a very simple tutorial example using concepts presented in the previous section. We consider a conflict situation $S = (s, v)$ where the domain ag of the voting function v is defined by $Ag = \{(1, A), \ldots, (240, A), (241, B), \ldots, (280, B), (281, C), \ldots, (340,\ C), (341, D), \ldots, (500, D)\}$ and $v(1, A) = \ldots = v(200, A) = +$, $v(201, A) = \ldots = v(230, A) = 0$, $v(231, A) = \ldots = v(240, A) = -$, $v(241, B) = \ldots = v(255, B) = +$, $v(256, B) = \ldots = v(280, B) = -$, $v(281, C) = \ldots = v(300, C) = 0$, $v(301, C) = \ldots = v(340, C) = -$, $v(341, D) = \ldots = v(365, D) = +$, $v(366, D) = \ldots = v(400, D) = 0$, $v(401, D) = \ldots = v(500, D) = -$.

This conflict situation is presented in Table 1. The maximal coalitions in this conflict situations are $v^{-1}(+)$ and $v^{-1}(-)$.

Table 1. Conflict situation with agents (*Member,Party*) and the voting function *Voting*

(*Member, Party*)	*Voting*
(1,A)	+
...	...
(200,A)	+
(201,A)	0
...	...
(230,A)	0
(231,A)	−
...	...
(240,A)	−
(241,B)	+
...	...
(255,B)	+
(256,B)	−
...	...
(280,B)	−
(281,C)	0
...	...
(300,C)	0
(301,C)	−
...	...
(340,C)	−
(341,D)	+
...	...
(365,D)	+
(366,D)	0
...	...
(400,D)	0
(401,D)	−
...	...
(500,D)	−

If one would like to keep only party name then Table 1 can be represented as it is shown in Table 2. This table presents a decision table in which the only condition attribute is *Party*, whereas the decision attribute is *Voting*. The table describes voting results in a parliament containing 500 members grouped in four political parties denoted A, B, C and D. Suppose the parliament discussed certain issue (e.g., membership of the country in European Union) and the voting result is presented in column *Voting*, where +, 0 and − denoted *yes*, *abstention* and *no* respectively. The column *support* contains the number of voters for each option.

The strength, certainty and the coverage factors for Table 2 are given in Table 3. The certainty and coverage factors have now a natural interpretation for the considered conflict situation.

From the certainty factors we can conclude, for example, that:

Table 2. Decision table with one condition attribute *Party* and the decision *Voting*

Fact	*Party*	*Voting*	*Support*
1	*A*	+	200
2	*A*	0	30
3	*A*	−	10
4	*B*	+	15
5	*B*	−	25
6	*C*	0	20
7	*C*	−	40
8	*D*	+	25
9	*D*	0	35
10	*D*	−	100

Table 3. Certainty and the coverage factors for Table 2

Fact	*Strength*	*Certainty*	*Coverage*
1	0.40	0.83	0.83
2	0.06	0.13	0.35
3	0.02	0.04	0.06
4	0.03	0.36	0.06
5	0.05	0.63	0.14
6	0.04	0.33	0.23
7	0.08	0.67	0.23
8	0.05	0.16	0.10
9	0.07	0.22	0.41
10	0.20	0.63	0.57

- 83.3% of party *A* voted *yes*,
- 12.5% of party *A* *abstained*,
- 4.2% of party *A* voted *no*.

From the coverage factors we can get, for example, the following explanation of voting results:

- 83.3% *yes* votes came from party *A*,
- 6.3% *yes* votes came from party *B*,
- 10.4% *yes* votes came from party *C*.

4.3 Conflicts and Rough Sets

There are strong relationships between the approach to conflicts and rough sets presented in Section 4.1. In this section, we discuss examples of such relationships. The presented in this section approach seems to be very promising for solving problems related to conflict resolution and negotiations (see, e.g., [41, 42, 43, 136]).

The application of rough sets can bring new results in the area related to conflict resolution and negotiations between agents because this make it possible to introduce approximate reasoning about vague concepts into the area.

Now, we would like to outline this possibility.

First, let us observe that any conflict situation $S = (s, V)$ where $V = (v_1, \ldots, v_k)$ and each v_i is defined on the set of agents $Ag = \{ag_1, \ldots, ag_n\}$ can be treated as an information system $\mathcal{A}(S)$ with the set of objects Ag and the set of attributes $\{v_1, \ldots, v_k\}$. The discernibility between agents ag and ag' in S can be defined by

$$disc_S(ag, ag') = \frac{\sum_{\{i:\ \phi_{v_i}(ag,ag')=-1\}} |\phi_{v_i}(ag, ag')|}{k}, \tag{46}$$

where $ag, ag' \in Ag$.

Now, one can consider reducts of $\mathcal{A}(S)$ relative to the discernibility defined by $disc_S$. For example, one can consider agents ag, ag' as discernible if

$$disc_S(ag, ag') \geq tr,$$

where tr a given threshold.[7] Any reduct $R \subseteq V$ of S is a minimal set of voting functions preserving all discernibility in voting between agents that are at least equal to tr. All voting functions from $V - R$ are dispensable with respect to preserving such discernibility between objects.

In an analogous way can be considered reducts of the information system $\mathcal{A}_T(S)$ with the universe of objects equal to $\{v_1, \ldots, v_k\}$ and attributes defined by agents and voting functions by $ag(v) = v(ag)$ for $ag \in Ag$ and $v \in V$. The discernibility between voting functions can be defined, e.g., by

$$disc(v, v') = |Con(S_v) - Con(S_{v'}|, \tag{47}$$

and makes it possible to measure the difference between voting functions v and v', respectively.

Any reduct R of $\mathcal{A}_T(S)$ is a minimal set of agents that preserves the differences between voting functions that are at least equal to a given threshold tr.

In our next example we extend a model of conflict by adding a set A of (condition) attributes making it possible to describe the situations in terms of values of attributes from A. The set of given situations is denoted by U. In this way we have defined an information system (U, A). Let us assume that there

[7] To compute such reducts one can follow a method presented in [112] assuming that any entry of the discernibility matrix corresponding to (ag, ag') with $disc(ag, ag') < tr$ is empty and the remaining entries are families of all subsets of V on which the discernibility between (ag, ag') is at least equal to tr [17].

is also given a set of agents Ag. Each agent $ag \in Ag$ has access to a subset $A_{ag} \subseteq A$ of condition attributes. Moreover we assume that $Ag = \bigcup_{ag \in AG} A_{ag}$. We also assume that there is also defined a decision attribute d on U such that $d(s)$ is a conflict situation $S = (s, V)$, where $V = (v_1, \ldots, v_k)$. Observe that $S = (s, V)$ can be represented by a matrix

$$[v_i(ag_j)]_{i=1,\ldots,n;j=1,\ldots,k}$$

where $v_i(ag_j)$ is the result of voting by jth agent on the ith issue. Such a matrix is a compound decision[8] corresponding to s. For the constructed decision system (U, A, d) one can use, e.g., the introduced above function (44) to measure the discernibility between compound decision values which correspond to conflict situations in the constructed decision table. The reducts of this decision table relative to decision have a natural interpretation with respect to conflicts.

The described decision table can also be used in conflict resolution. We would like to illustrate this possibility. First, let us recall some notation. For $B \subseteq A$ we denote by $Inf_B(s)$ the B-signature of the situation s, i.e., the set $\{(a, a(s)) : a \in A\}$. Let $INF(B) = \{Inf_B(s) : s \in U\}$. Let us also assume that for each agent $ag \in Ag$ there is given a similarity relation $\tau_{ag} \subseteq INF(A_{ag}) \times INF(A_{ag})$. In terms of these similarity relations one can consider a problem of conflict resolution relative to a given threshold tr in a given situation s described by $Inf_A(s)$. This is the searching problem for a situation s', if such a situation exists, satisfying the following conditions:

1. $Inf_A(s')|A_{ag} \in \tau_{ag}(Inf_{A_{ag}}(s))$, where $\tau_{ag}(Inf_{A_{ag}}(s))$ is the tolerance class of $Inf_{A_{ag}}(s)$ with respect to τ_{ag} and $Inf_A(s')|A_{ag}$ denotes the restriction of $Inf_A(s')$ to A_{ag}.
2. $Inf_A(s')$ satisfies given local constraints (e.g., specifying coexistence of local situations, see, e.g., [17, 83, 135]) and given global constraints (e.g., specifying quality of global situations, see, e.g., [17]).
3. The conflict degree in the conflict situation $d(s')$[9] measured by means of the chosen conflict measure[10] is at most tr.

In searching for conflict resolution one can apply methods based on Boolean reasoning (see,e.g., [17, 112]).

We have proposed that changes to the acceptability of an issue by agents can be expressed by similarity relations. Observe that in real-life applications these similarities can be more compound than it was suggested above, i.e., they are not defined directly by sensory concepts describing situations. However, they are often specified by high level concepts (see, e.g., [41, 116] and also

[8] For references to other papers on compound decision the reader is referred, e.g., to [4].

[9] Let us observe that s' is not necessarily in U. In such a case the value $d(s')$ should be predicted by the induced classifier from (U, A, d).

[10] For example, one can consider (45).

Section 3.5). These high level concepts can be vague and are linked with the sensory concepts describing situations by means of a hierarchy of other vague concepts. Approximation of vague concepts in the hierarchy and dependencies between them (see Section 3.5) makes it possible to approximate the similarity relations. This allows us to develop searching methods for acceptable value changes of sensory concepts preserving similarities (constraints) specified over high level vague concepts. One can also introduce some costs of changes of local situations into new ones by agents and search for new situations accessible with minimal or sub-minimal costs.

4.4 Negotiations for Conflict Resolution

In the previous section we have presented an outline to conflict resolution assuming that the acceptable changes of current situations of agents are known (in the considered example they were described by similarities). However, if the required solution does not exists in the current searching space then the negotiations between agents should start. Using the rough set approach to conflict resolution by negotiations between agents one can consider more advanced models in which actions and plans [28] performed by agents or their teams are involved in negotiations and conflict resolution.

We would like to outline such a model. Assume that each agent $ag \in Ag$ is able to perform actions from a set $Action_{ag}$. Each action $ac \in Action(ag)$ is specified by the input condition $in(ac)$ and the output condition $out(ac)$, representing conditions making it possible to perform the action and the result of action, respectively. We assume that $in(ac) \in INF(IN_{ag})$ and $out(ac) \in INF(OUT_{ag})$ for some sets $IN_{ag}, OUT_{ag} \subseteq A_{ag}$. The action ac can be applied to an object s if and only if $in(ac) \subseteq Inf_{A_{ag}}(s)$ and the result of performing of ac in s is described by $(Inf_{A_{ag}}(s) - Inf_{OUT_{ag}}(s)) \cup out(ac)$. Now, the conflict resolution task (see Section 4.3) in a given object (state) s can be formulated as searching for a plan, i.e., a sequence $ac_1, \ldots, ac_m$ of actions from $\bigcup_{ag \in Ag} Action(ag)$ transforming the objects s into object s' satisfying the requirements formulated in Section 4.3.[11]

In searching for the plan, one can use a Petri net constructed from the set $\bigcup_{ag \in Ag} Action(ag)$ of actions.[12] In this net places correspond to descriptors, i.e., pairs $(attribute, value)$ from $in(ac)$ and $out(ac)$ for $ac \in Action(ac)$ and $ag \in Ag$, transitions correspond to actions, and any transition is linked in a natural way with places corresponding to input and outputs conditions. Such a Petri net can be enriched by an additional control making it possible to preserve dependencies between local states (see, e.g., [83, 135]) or constraints

[11] To illustrate possible arising problems let us consider an example. Assume that two vague concepts are given with classifiers for them. For any state satisfying the first concept we would like to find a (sub-)optimal plan transforming the given state to the state satisfying the second concept.

[12] For applications of Petri nets in planning the reader is referred, e.g., to [23]

related to similarities. Next, a cooperation protocol between actions performed by different agents should be discovered and the Petri net should be extended by this protocol. Finally, markings corresponding to objects s' with the conflict degree at most tr, if such states exist, are reachable in the resulting Petri net from a given marking corresponding to the state s. Searching for such protocols is a challenge.

One of the very first possible approaches in searching for sequences of actions (plans) transforming a given situation into another one with the required decreasing level of conflict degree is to create a Petri net from specification of actions and perform experiments with such a net. The examples of markings reachable in the net are stored in an information system. This system is next extended to a decision system by adding a decision describing the conflict degree for each situation corresponding to the marking. From such a decision table one can induce a classifier for different levels of conflicts. Next, this classifiers can be used to create a new decision table. In this new decision table any object consists of a pair of situations together with a sequence of actions transforming the first situation to the second. The decision for a given object is equal to the difference in conflict degrees of situations from the object. Then, condition attributes which make it possible to induce rules for prediction differences in conflict degrees are discovered. These condition attributes express properties of the first situation in the pair and properties of the sequence of actions (of a given length) performed starting from this situation. Such rules specify the additional constraints for the net and they can be embedded into the net as an additional control. The resulting net makes it possible to select only such plans (i.e., sequences of actions) which decrease conflict degrees. Certainly, to make the task feasible one should consider a relevant length of the sequences of actions and next to develop a method for composing plans. In turn, this will require to use hierarchical modelling with of concept ontology and actions on different levels of hierarchy between them. In our current project we are developing the outlined above methods.

5 Summary

In this article, we have presented basic concepts of rough set theory and also some extensions of the basic approach. We have also discussed relationships of rough sets with conflict analysis which is of great importance for e-service intelligence. In our further study we would like to develop the approach based to conflict analysis outlined in the paper.

There are numerous active research directions on rough set theory, and applications of rough sets also in combination with other approaches to soft computing. For more details the reader is referred to the bibliography on rough sets and web pages cited in this paper.

Acknowledgements

The research of Andrzej Skowron has been supported by the grant 3 T11C 002 26 from Ministry of Scientific Research and Information Technology of the Republic of Poland.

Many thanks to Professors James Peters and Dominik Ślęzak for their incisive comments and for suggesting many helpful ways to improve this article.

References

1. J. J. Alpigini, J. F. Peters, A. Skowron, N. Zhong (Eds.). Third International Conference on Rough Sets and Current Trends in Computing (RSCTC'2002), Malvern, PA, October 14-16, 2002, Lecture Notes in Artificial Intelligence, vol. 2475. Springer-Verlag, Heidelberg, Germany, 2002.
2. R. Ariew, D. Garber (Eds.). Leibniz, G. W., Philosophical Essays. Hackett Publishing Company, Indianapolis, 1989.
3. J. Bazan, H. S. Nguyen, S. H. Nguyen, P. Synak, J. Wróblewski. Rough set algorithms in classification problems. In: Polkowski et al. [87], pp. 49–88.
4. J. Bazan, A. Osmólski, A. Skowron, D. Ślęzak, M. Szczuka, J. Wróblewski. Rough set approach to the survival analysis. In: Alpigini et al. [1], pp. 522–529.
5. J. G. Bazan. A comparison of dynamic and non-dynamic rough set methods for extracting laws from decision tables. In: Polkowski and Skowron [90], pp. 321–365.
6. J. G. Bazan, H. S. Nguyen, J. F. Peters, A. Skowron, M. Szczuka. Rough set approach to pattern extraction from classifiers. In: Skowron and Szczuka [120], pp. 20–29.
   URL `http://www.elsevier.nl/locate/entcs/volume82.html`
7. J. G. Bazan, H. S. Nguyen, A. Skowron, M. Szczuka. A view on rough set concept approximation. In: Wang et al. [142], pp. 181–188.
8. J. G. Bazan, J. F. Peters, A. Skowron. Behavioral pattern identification through rough set modelling. In: Ślęzak et al. [127], pp. 688–697.
9. J. G. Bazan, A. Skowron. Classifiers based on approximate reasoning schemes. In: Dunin-Kęplicz et al. [21], pp. 191–202.
10. S. Behnke. Hierarchical Neural Networks for Image Interpretation, Lecture Notes in Computer Science, vol. 2766. Springer, Heidelberg, Germany, 2003.
11. L. Breiman. Statistical modeling: The two cultures. Statistical Science 16(3) (2001) 199–231.
12. J. L. Casti. Alternate Realities? Mathematical Models of Nature and Man. John Wiley & Sons, New York, NY, 1989.
13. N. Cercone, A. Skowron, N. Zhong (Eds.). (Special issue), Computational Intelligence: An International Journal, vol. 17(3). 2001.
14. K. Cios, W. Pedrycz, R. Swiniarski. Data mining methods for knowledge discovery. Kluwer, Norwell, MA, 1998.
15. C. H. Coombs, G. S. Avruin. The Structure of Conflicts. Lawrence Erlbaum, London, 1988.
16. R. Deja. Conflict analysis, rough set methods and applications. In: Polkowski et al. [87], pp. 491–520.

17. R. Deja, A. Skowron. On some conflict models and conflict resolution. Romanian Journal of Information Science and Technology 5(1-2) (2002) 69–82.
18. S. Demri, E. Orłowska. Incomplete Information: Structure, Inference, Complexity. Monographs in Theoretical Cpmputer Sience, Springer-Verlag, Heidelberg, Germany, 2002.
19. D. Dubois, H. Prade. Foreword. In: Rough Sets: Theoretical Aspects of Reasoning about Data [74].
20. R. Duda, P. Hart, R. Stork. Pattern Classification. John Wiley & Sons, New York, NY, 2002.
21. B. Dunin-Kęplicz, A. Jankowski, A. Skowron, M. Szczuka (Eds.). Monitoring, Security, and Rescue Tasks in Multiagent Systems (MSRAS'2004). Advances in Soft Computing, Springer, Heidelberg, Germany, 2005.
22. I. Düntsch, G. Gediga. Rough set data analysis: A road to non-invasive knowledge discovery. Methodos Publishers, Bangor, UK, 2000.
23. A. E. Fallah-Seghrouchni. Multi-agent planning for autonomous agents coordination. In: Dunin-Kęplicz et al. [21], pp. 53–68.
24. M. Fedrizzi, J. Kacprzyk, H. Nurmi. How different are social choice functions: A rough sets approach. Quality and Quantity 30 (1996) 87–99.
25. J. H. Friedman, T. Hastie, R. Tibshirani. The Elements of Statistical Learning: Data Mining, Inference, and Prediction. Springer-Verlag, Heidelberg, Germany, 2001.
26. H. Garcia-Molina, J. Ullman, J. Widom. Database Systems: The Complete Book. Prentice Hall, Upper Saddle River, New Jersey, 2002.
27. G. Gediga, I. Düntsch. Rough approximation quality revisited. Artificial Intelligence 132 (2001) 219–234.
28. M. Ghallab, D. Nau, P. Traverso (Eds.). Automated Planning Theory and Practice. Morgan Kaufmann, San Francisco, 2004.
29. S. Greco, B. Matarazzo, R. Słowiński. Rough set theory for multicriteria decision analysis. European Journal of Operational Research 129(1) (2001) 1–47.
30. J. W. Grzymała-Busse. Managing Uncertainty in Expert Systems. Kluwer Academic Publishers, Norwell, MA, 1990.
31. J. W. Grzymała-Busse. LERS - A system for learning from examples based on rough sets. In: Słowiński [129], pp. 3–18.
32. J. W. Grzymala-Busse. A new version of the rule induction system LERS. Fundamenta Informaticae 31(1) (1997) 27–39.
33. J. W. Grzymala-Busse. LERS - A data mining system. In: The Data Mining and Knowledge Discovery Handbook. 2005, pp. 1347–1351.
34. S. Hirano, M. Inuiguchi, S. Tsumoto (Eds.). Proceedings of International Workshop on Rough Set Theory and Granular Computing (RSTGC'2001), Matsue, Shimane, Japan, May 20-22, 2001, Bulletin of the International Rough Set Society, vol. 5(1-2). International Rough Set Society, Matsue, Shimane, 2001.
35. M. Inuiguchi, S. Hirano, S. Tsumoto (Eds.). Rough Set Theory and Granular Computing, Studies in Fuzziness and Soft Computing, vol. 125. Springer-Verlag, Heidelberg, Germany, 2003.
36. R. Keefe. Theories of Vagueness. Cambridge Studies in Philosophy, Cambridge, UK, 2000.
37. W. Kloesgen, J. Żytkow (Eds.). Handbook of Knowledge Discovery and Data Mining. Oxford University Press, Oxford, 2002.
38. J. Komorowski, Z. Pawlak, L. Polkowski, A. Skowron. Rough sets: A tutorial. In: Pal and Skowron [71], pp. 3–98.

39. B. Kostek. Soft Computing in Acoustics, Applications of Neural Networks, Fuzzy Logic and Rough Sets to Physical Acoustics, Studies in Fuzziness and Soft Computing, vol. 31. Physica-Verlag, Heidelberg, Germany, 1999.
40. B. Kostek. Perception-Based Data Processing in Acoustics. Applications to Music Information Retrieval and Psychophysiology of Hearing, Studies in Computational Intelligence, vol. 3. Springer, Heidelberg, Germany, 2005.
41. R. Kowalski. A logic-based approach to conflict resolution. Report, Department of Computing, Imperial College (2003) 1–28.
URL `http://www.doc.ic.ac.uk/~rak/papers/conflictresolution.pdf`
42. S. Kraus. Strategic Negotiations in Multiagent Environments. The MIT Press, Cambridge, MA, 2001.
43. G. Lai, C. Li, K. Sycara, J. A. Giampapa. Literature review on multi-attribute negotiations. Technical Report CMU-RI-TR-04-66 (2004) 1–35.
44. G. W. Leibniz. Discourse on metaphysics. In: Ariew and Garber [2], pp. 35–68.
45. S. Leśniewski. Grungzüge eines neuen Systems der Grundlagen der Mathematik. Fundamenta Mathematicae 14 (1929) 1–81.
46. S. Leśniewski. On the foundations of mathematics. Topoi 2 (1982) 7–52.
47. T. Y. Lin. Neighborhood systems and approximation in database and knowledge base systems. In: M. L. Emrich, M. S. Phifer, M. Hadzikadic, Z. W. Ras (Eds.), Proceedings of the Fourth International Symposium on Methodologies of Intelligent Systems (Poster Session), October 12-15, 1989, Oak Ridge National Laboratory, Charlotte, NC. 1989, pp. 75–86.
48. T. Y. Lin (Ed.). Special issue, Journal of the Intelligent Automation and Soft Computing, vol. 2(2). 1996.
49. T. Y. Lin. The discovery, analysis and representation of data dependencies in databases. In: L. Polkowski, A. Skowron (Eds.), Rough Sets in Knowledge Discovery 1: Methodology and Applications, Physica-Verlag, Heidelberg, Germany, Studies in Fuzziness and Soft Computing, vol. 18. 1998, pp. 107–121.
50. T. Y. Lin, N. Cercone (Eds.). Rough Sets and Data Mining - Analysis of Imperfect Data. Kluwer Academic Publishers, Boston, USA, 1997.
51. T. Y. Lin, A. M. Wildberger (Eds.). Soft Computing: Rough Sets, Fuzzy Logic, Neural Networks, Uncertainty Management, Knowledge Discovery. Simulation Councils, Inc., San Diego, CA, USA, 1995.
52. T. Y. Lin, Y. Y. Yao, L. A. Zadeh (Eds.). Rough Sets, Granular Computing and Data Mining. Studies in Fuzziness and Soft Computing, Physica-Verlag, Heidelberg, Germany, 2001.
53. J. Łukasiewicz. Die logischen Grundlagen der Wahrscheinlichkeitsrechnung, 1913. In: L. Borkowski (Ed.), Jan Łukasiewicz - Selected Works, North Holland Publishing Company, Amsterdam, London, Polish Scientific Publishers, Warsaw. 1970, pp. 16–63.
54. Y. Maeda, K. Senoo, H. Tanaka. Interval density function in conflict analysis. In: Skowron et al. [109], pp. 382–389.
55. S. Marcus. The paradox of the heap of grains, in respect to roughness, fuzziness and negligibility. In: Polkowski and Skowron [89], pp. 19–23.
56. T. M. Mitchel. Machine Learning. McGraw-Hill Series in Computer Science, Boston, MA, 1999.
57. S. Mitra, T. Acharya. Data mining. Multimedia, Soft Computing, and Bioinformatics. John Wiley & Sons, New York, NY, 2003.

58. A. Nakamura. Conflict logic with degrees. In: Pal and Skowron [71], pp. 136–150.
URL `http://citeseer.nj.nec.com/komorowski98rough.html`
59. H. S. Nguyen, S. H. Nguyen. Rough sets and association rule generation. Fundamenta Informaticae 40(4) (1999) 383–405.
60. H. S. Nguyen, D. Ślęzak. Approximate reducts and association rules - correspondence and complexity results. In: Skowron et al. [109], pp. 137–145.
61. S. H. Nguyen, J. Bazan, A. Skowron, H. S. Nguyen. Layered learning for concept synthesis. In: Peters and Skowron [79], pp. 187–208.
62. S. H. Nguyen, H. S. Nguyen. Some efficient algorithms for rough set methods. In: Sixth International Conference on Information Processing and Management of Uncertainty on Knowledge Based Systems IPMU'1996. Granada, Spain, 1996, vol. III, pp. 1451–1456.
63. T. T. Nguyen, A. Skowron. Rough set approach to domain knowledge approximation. In: Wang et al. [142], pp. 221–228.
64. H. Nurmi, J. Kacprzyk, M. Fedrizzi. Theory and methodology: Probabilistic, fuzzy and rough concepts in social choice. European Journal of Operational Research 95 (1996) 264–277.
65. E. Orłowska. Semantics of vague conepts. In: G. Dorn, P. Weingartner (Eds.), Foundation of Logic and Linguistics, Plenum Press, New York. 1984, pp. 465–482.
66. E. Orłowska. Reasoning about vague concepts. Bulletin of the Polish Academy of Sciences, Mathematics 35 (1987) 643–652.
67. E. Orłowska (Ed.). Incomplete Information: Rough Set Analysis, Studies in Fuzziness and Soft Computing, vol. 13. Springer-Verlag/Physica-Verlag, Heidelberg, Germany, 1997.
68. S. K. Pal, P. Mitra. Pattern Recognition Algorithms for Data Mining. CRC Press, Boca Raton, Florida, 2004.
69. S. K. Pal, W. Pedrycz, A. Skowron, R. Swiniarski (Eds.). Special volume: Rough-neuro computing, Neurocomputing, vol. 36. 2001.
70. S. K. Pal, L. Polkowski, A. Skowron (Eds.). Rough-Neural Computing: Techniques for Computing with Words. Cognitive Technologies, Springer-Verlag, Heidelberg, Germany, 2004.
71. S. K. Pal, A. Skowron (Eds.). Rough Fuzzy Hybridization: A New Trend in Decision-Making. Springer-Verlag, Singapore, 1999.
72. Z. Pawlak. Information systems - theoretical foundations. Information Systems 6 (1981) 205–218.
73. Z. Pawlak. Rough sets. International Journal of Computer and Information Sciences 11 (1982) 341–356.
74. Z. Pawlak. Rough Sets: Theoretical Aspects of Reasoning about Data, System Theory, Knowledge Engineering and Problem Solving, vol. 9. Kluwer Academic Publishers, Dordrecht, The Netherlands, 1991.
75. Z. Pawlak. An inquiry into anatomy of conflicts. Journal of Information Sciences 109 (1998) 65–78.
76. Z. Pawlak. Decision rules, Bayes' rule and rough sets. In: Skowron et al. [109], pp. 1–9.
77. Z. Pawlak, A. Skowron. Rough membership functions. In: R. Yager, M. Fedrizzi, J. Kacprzyk (Eds.), Advances in the Dempster-Shafer Theory of Evidence. John Wiley & Sons, New York, NY, 1994, pp. 251–271.

78. J. Peters, A. Skowron (Eds.). Special issue on a rough set approach to reasoning about data, International Journal of Intelligent Systems, vol. 16(1). 2001.
79. J. F. Peters, A. Skowron (Eds.). Transactions on Rough Sets I: Journal Subline, Lecture Notes in Computer Science, vol. 3100. Springer, Heidelberg, Germany, 2004.
80. J. F. Peters, A. Skowron (Eds.). Transactions on Rough Sets III: Journal Subline, Lecture Notes in Computer Science, vol. 3400. Springer, Heidelberg, Germany, 2005.
81. J. F. Peters, A. Skowron (Eds.). Transactions on Rough Sets IV: Journal Subline, Lecture Notes in Computer Science, vol. 3700. Springer, Heidelberg, Germany, 2005.
82. J. F. Peters, A. Skowron, D. Dubois, J. W. Grzymała-Busse, M. Inuiguchi, L. Polkowski (Eds.). Transactions on Rough Sets II. Rough sets and fuzzy sets: Journal Subline, Lecture Notes in Computer Science, vol. 3135. Springer, Heidelberg, Germany, 2004.
83. J. F. Peters, A. Skowron, Z. Suraj. An application of rough set methods in control design. Fundamenta Informaticae 43(1-4) (2000) 269–290.
84. L. Polkowski. Rough Sets: Mathematical Foundations. Advances in Soft Computing, Physica-Verlag, Heidelberg, Germany, 2002.
85. L. Polkowski. Rough mereology: A rough set paradigm for unifying rough set theory and fuzzy set theory. Fundamenta Informaticae 54 (2003) 67–88.
86. L. Polkowski. Toward rough set foundations. mereological approach. In: Tsumoto et al. [139], pp. 8–25.
87. L. Polkowski, T. Y. Lin, S. Tsumoto (Eds.). Rough Set Methods and Applications: New Developments in Knowledge Discovery in Information Systems, Studies in Fuzziness and Soft Computing, vol. 56. Springer-Verlag/Physica-Verlag, Heidelberg, Germany, 2000.
88. L. Polkowski, A. Skowron. Rough mereology: A new paradigm for approximate reasoning. International Journal of Approximate Reasoning 15(4) (1996) 333–365.
89. L. Polkowski, A. Skowron (Eds.). First International Conference on Rough Sets and Soft Computing RSCTC'1998, Lecture Notes in Artificial Intelligence, vol. 1424. Springer-Verlag, Warsaw, Poland, 1998.
90. L. Polkowski, A. Skowron (Eds.). Rough Sets in Knowledge Discovery 1: Methodology and Applications, Studies in Fuzziness and Soft Computing, vol. 18. Physica-Verlag, Heidelberg, Germany, 1998.
91. L. Polkowski, A. Skowron (Eds.). Rough Sets in Knowledge Discovery 2: Applications, Case Studies and Software Systems, Studies in Fuzziness and Soft Computing, vol. 19. Physica-Verlag, Heidelberg, Germany, 1998.
92. L. Polkowski, A. Skowron. Towards adaptive calculus of granules. In: L. A. Zadeh, J. Kacprzyk (Eds.), Computing with Words in Information/Intelligent Systems. Physica-Verlag, Heidelberg, Germany, 1999, pp. 201–227.
93. L. Polkowski, A. Skowron. Rough mereology in information systems. a case study: Qualitative spatial reasoning. In: Polkowski et al. [87], pp. 89–135.
94. L. Polkowski, A. Skowron. Rough mereological calculi of granules: A rough set approach to computation. Computational Intelligence: An International Journal 17(3) (2001) 472–492.
95. L. Polkowski, A. Skowron, J. Zytkow. Rough foundations for rough sets. In: Lin and Wildberger [51], pp. 55–58.

96. S. Read. Thinking about Logic: An Introduction to the Philosophy of Logic. Oxford University Press, Oxford, New York, 1994.
97. J. Rissanen. Modeling by shortes data description. Automatica 14 (1978) 465–471.
98. J. Rissanen. Minimum-description-length principle. In: S. Kotz, N. Johnson (Eds.), Encyclopedia of Statistical Sciences, John Wiley & Sons, New York, NY. 1985, pp. 523–527.
99. A. Skowron. Rough sets in perception-based computing (keynote talk). In: First International Conference on Pattern Recognition and Machine Intelligence (PReMI'05) December 18-22, 2005, Indian Statistical Institute, Kolkata. pp. 21–29.
100. A. Skowron (Ed.). Proceedings of the 5th Symposium on Computation Theory, Zaborów, Poland, 1984, Lecture Notes in Computer Science, vol. 208. Springer-Verlag, Berlin, Germany, 1985.
101. A. Skowron. Boolean reasoning for decision rules generation. In: J. Komorowski, Z. W. Raś (Eds.), ISMIS'1993, Trondheim, Norway, June 15-18, 1993. Springer-Verlag, 1993, Lecture Notes in Artificial Intelligence, vol. 689, pp. 295–305.
102. A. Skowron. Extracting laws from decision tables. Computational Intelligence: An International Journal 11 (1995) 371–388.
103. A. Skowron. Rough sets in KDD - plenary talk. In: Z. Shi, B. Faltings, M. Musen (Eds.), 16-th World Computer Congress (IFIP'2000): Proceedings of Conference on Intelligent Information Processing (IIP'2000), Publishing House of Electronic Industry, Beijing. 2000, pp. 1–14.
104. A. Skowron. Toward intelligent systems: Calculi of information granules. Bulletin of the International Rough Set Society 5(1-2) (2001) 9–30.
105. A. Skowron. Approximate reasoning in distributed environments. In: Zhong and Liu [151], pp. 433–474.
106. A. Skowron. Perception logic in intelligent systems (keynote talk). In: S. Blair et al (Ed.), Proceedings of the 8th Joint Conference on Information Sciences (JCIS 2005), July 21-26, 2005, Salt Lake City, Utah, USA. X-CD Technologies: A Conference & Management Company, 15 Coldwater Road, Toronto, Ontario, M3B 1Y8, 2005, pp. 1–5.
107. A. Skowron. Rough sets and vague concepts. Fundamenta Informaticae 64(1-4) (2005) 417–431.
108. A. Skowron, J. W. Grzymala-Busse. From rough set theory to evidence theory. In: R. Yager, M. Fedrizzi, J. Kacprzyk (Eds.), Advances in the Dempster-Shafer Theory of Evidence, John Wiley & Sons, New York, NY. 1994, pp. 193–236.
109. A. Skowron, S. Ohsuga, N. Zhong (Eds.). Proceedings of the 7-th International Workshop on Rough Sets, Fuzzy Sets, Data Mining, and Granular-Soft Computing (RSFDGrC'99), Yamaguchi, November 9-11, 1999, Lecture Notes in Artificial Intelligence, vol. 1711. Springer-Verlag, Heidelberg, Germany, 1999.
110. A. Skowron, S. K. Pal (Eds.). Special volume: Rough sets, pattern recognition and data mining, Pattern Recognition Letters, vol. 24(6). 2003.
111. A. Skowron, J. Peters. Rough sets: Trends and challenges. In: Wang et al. [142], pp. 25–34. (plenary talk).
112. A. Skowron, C. Rauszer. The discernibility matrices and functions in information systems. In: Słowiński [129], pp. 331–362.
113. A. Skowron, J. Stepaniuk. Tolerance approximation spaces. Fundamenta Informaticae 27(2-3) (1996) 245–253.

114. A. Skowron, J. Stepaniuk. Information granules: Towards foundations of granular computing. International Journal of Intelligent Systems 16(1) (2001) 57–86.
115. A. Skowron, J. Stepaniuk. Information granules and rough-neural computing. In: Pal et al. [70], pp. 43–84.
116. A. Skowron, J. Stepaniuk. Ontological framework for approximation. In: Ślęzak et al. [126], pp. 718–727.
117. A. Skowron, R. Swiniarski. Rough sets and higher order vagueness. In: Ślęzak et al. [126], pp. 33–42.
118. A. Skowron, R. Swiniarski, P. Synak. Approximation spaces and information granulation. In: Peters and Skowron [80], pp. 175–189.
119. A. Skowron, P. Synak. Complex patterns. Fundamenta Informaticae 60(1-4) (2004) 351–366.
120. A. Skowron, M. Szczuka (Eds.). Proceedings of the Workshop on Rough Sets in Knowledge Discovery and Soft Computing at ETAPS 2003, April 12-13, 2003, Electronic Notes in Computer Science, vol. 82(4). Elsevier, Amsterdam, Netherlands, 2003.
URL `http://www.elsevier.nl/locate/entcs/volume82.html`
121. D. Ślęzak. Approximate reducts in decision tables. In: Sixth International Conference on Information Processing and Management of Uncertainty in Knowledge-Based Systems IPMU'1996. Granada, Spain, 1996, vol. III, pp. 1159–1164.
122. D. Ślęzak. Normalized decision functions and measures for inconsistent decision tables analysis. Fundamenta Informaticae 44 (2000) 291–319.
123. D. Ślęzak. Various approaches to reasoning with frequency-based decision reducts: a survey. In: Polkowski et al. [87], pp. 235–285.
124. D. Ślęzak. Approximate entropy reducts. Fundamenta Informaticae 53 (2002) 365–387.
125. D. Ślęzak. Rough sets and Bayes factor. In: Peters and Skowron [80], pp. 202–229.
126. D. Ślęzak, G. Wang, M. Szczuka, I. Düntsch, Y. Yao (Eds.). Proceedings of the 10th International Conference on Rough Sets, Fuzzy Sets, Data Mining, and Granular Computing (RSFDGrC'2005), Regina, Canada, August 31-September 3, 2005, Part I, Lecture Notes in Artificial Intelligence, vol. 3641. Springer-Verlag, Heidelberg, Germany, 2005.
127. D. Ślęzak, J. T. Yao, J. F. Peters, W. Ziarko, X. Hu (Eds.). Proceedings of the 10th International Conference on Rough Sets, Fuzzy Sets, Data Mining, and Granular Computing (RSFDGrC'2005), Regina, Canada, August 31-September 3, 2005, Part II, Lecture Notes in Artificial Intelligence, vol. 3642. Springer-Verlag, Heidelberg, Germany, 2005.
128. D. Ślęzak, W. Ziarko. The investigation of the Bayesian rough set model. International Journal of Approximate Reasoning 40 (2005) 81–91.
129. R. Słowiński (Ed.). Intelligent Decision Support - Handbook of Applications and Advances of the Rough Sets Theory, System Theory, Knowledge Engineering and Problem Solving, vol. 11. Kluwer Academic Publishers, Dordrecht, The Netherlands, 1992.
130. R. Słowiński, J. Stefanowski (Eds.). Special issue: Proceedings of the First International Workshop on Rough Sets: State of the Art and Perspectives, Kiekrz, Poznań, Poland, September 2–4 (1992), Foundations of Computing and Decision Sciences, vol. 18(3-4). 1993.

131. R. Słowiński, D. Vanderpooten. Similarity relation as a basis for rough approximations. In: P. Wang (Ed.), Advances in Machine Intelligence and Soft Computing Vol. 4, Duke University Press, Duke, NC. 1997, pp. 17–33.
132. S. Staab, R. Studer (Eds.). Handbook on Ontologies. International Handbooks on Information Systems, Springer, Heidelberg, Germany, 2004.
133. J. Stepaniuk. Approximation spaces, reducts and representatives. In: Polkowski and Skowron [91], pp. 109–126.
134. P. Stone. Layered Learning in Multi-Agent Systems: A Winning Approach to Robotic Soccer. The MIT Press, Cambridge, MA, 2000.
135. Z. Suraj. Rough set methods for the synthesis and analysis of concurrent processes. In: Polkowski et al. [87], pp. 379–488.
136. K. Sycara. Multiagent systems. AI Magazine (Summer 1998) 79–92.
137. T. Terano, T. Nishida, A. Namatame, S. Tsumoto, Y. Ohsawa, T. Washio (Eds.). New Frontiers in Artificial Intelligence, Joint JSAI'2001 Workshop Post-Proceedings, Lecture Notes in Artificial Intelligence, vol. 2253. Springer-Verlag, Heidelberg, Germany, 2001.
138. S. Tsumoto, S. Kobayashi, T. Yokomori, H. Tanaka, A. Nakamura (Eds.). Proceedings of the The Fourth Internal Workshop on Rough Sets, Fuzzy Sets and Machine Discovery, November 6-8, University of Tokyo , Japan. The University of Tokyo, Tokyo, 1996.
139. S. Tsumoto, R. Słowiński, J. Komorowski, J. G. Busse (Eds.). Proceedings of the 4th International Conference on Rough Sets and Current Trends in Computing (RSCTC'2004), Uppsala, Sweden, June 1-5, 2004, Lecture Notes in Artificial Intelligence, vol. 3066. Springer-Verlag, Heidelberg, Germany, 2004.
140. S. Tsumoto, H. Tanaka. PRIMEROSE: Probabilistic rule induction method based on rough sets and resampling methods. Computational Intelligence: An International Journal 11 (1995) 389–405.
141. V. Vapnik. Statistical Learning Theory. John Wiley & Sons, New York, NY, 1998.
142. G. Wang, Q. Liu, Y. Yao, A. Skowron (Eds.). Proceedings of the 9-th International Conference on Rough Sets, Fuzzy Sets, Data Mining, and Granular Computing (RSFDGrC'2003), Chongqing, China, May 26-29, 2003, Lecture Notes in Artificial Intelligence, vol. 2639. Springer-Verlag, Heidelberg, Germany, 2003.
143. S. K. M. Wong, W. Ziarko. Comparison of the probabilistic approximate classification and the fuzzy model. Fuzzy Sets and Systems 21 (1987) 357–362.
144. J. Wróblewski. Theoretical foundations of order-based genetic algorithms. Fundamenta Informaticae 28 (1996) 423–430.
145. T. Y. Yao. On generalizing rough set theory. In: Wang et al. [142], pp. 44–51.
146. Y. Y. Yao. Generalized rough set models. In: Polkowski and Skowron [90], pp. 286–318.
147. Y. Y. Yao. Information granulation and rough set approximation. International Journal of Intelligent Systems 16 (2001) 87–104.
148. Y. Y. Yao, S. K. M. Wong, T. Y. Lin. A review of rough set models. In: Lin and Cercone [50], pp. 47–75.
149. L. A. Zadeh. Fuzzy sets. Information and Control 8 (1965) 338–353.
150. L. A. Zadeh. A new direction in AI: Toward a computational theory of perceptions. AI Magazine 22(1) (2001) 73–84.
151. N. Zhong, J. Liu (Eds.). Intelligent Technologies for Information Analysis. Springer, Heidelberg, Germany, 2004.

152. W. Ziarko. Variable precision rough set model. Journal of Computer and System Sciences 46 (1993) 39–59.
153. W. Ziarko (Ed.). Rough Sets, Fuzzy Sets and Knowledge Discovery: Proceedings of the Second International Workshop on Rough Sets and Knowledge Discovery (RSKD'93), Banff, Alberta, Canada, October 12–15 (1993). Workshops in Computing, Springer–Verlag & British Computer Society, London, Berlin, 1994.
154. W. Ziarko (Ed.). Special issue, Computational Intelligence: An International Journal, vol. 11(2). 1995.
155. W. Ziarko (Ed.). Special issue, Fundamenta Informaticae, vol. 27(2-3). 1996.
156. W. Ziarko, Y. Yao (Eds.). Proceedings of the 2nd International Conference on Rough Sets and Current Trends in Computing (RSCTC'2000), Banff, Canada, October 16-19, 2000, Lecture Notes in Artificial Intelligence, vol. 2005. Springer-Verlag, Heidelberg, Germany, 2001.

Rough Ontology Mapping in E-Business Integration

Yi Zhao, Wolfgang Halang, and Xia Wang

Faculty of Electrical and Computer Engineering
FernUniversitaet Hagen, Hagen 58084, Germany
{yi.zhao, wolfgang.halang, xia.wang}@fernuni-hagen.de

Abstract

Ontology mapping represents a promising technique in E-business integration, agent communication and web services. Similarity measure plays an important role in ontology mapping. Investigations show that a multi-similarity measure usually gives a higher quality mapping results than that of single similarity value. In this chapter, we first give a brief introduction to the background knowledge of formal concept analysis, ontology mapping and similarity measure. A hybrid similarity measure based on rough formal concept analysis is then introduced. Based on Tversky's similarity model, the proposed measure combines rough set theory and formal concept analysis to calculate the similarity of the concept nodes from two ontologies. In this sense, the measure is both featural and structural. An example of the case study shows the applicability of our proposed similarity measure.

1. Introduction

Today, data or information can be retrieved from many different sources, such as databases, World Wide Web, knowledge bases, and other specific information systems. Integration of heterogeneous information sources is necessary in order to establish interoperation between agents or services. How to solve the interoperability problem is quite challenging due to the problems of structural and semantic heterogeneity: while structural heterogeneity concerns the different representations of information, semantic heterogeneity concerns the intended meaning of information described.

Y. Zhao et al.: *Rough Ontology Mapping in E-Business Integration*, Studies in Computational Intelligence (SCI) **37**, 75–93 (2007)
www.springerlink.com

The idea of the semantic web was proposed to add machine-processable information to web-based data in order to realize interoperability (Berners-Lee 2001). Ontologies play a prominent role in the concept of the semantic web (Doan 2002) to provide semantic information for assisting communication among heterogeneous information repositories. With the rapid development of the semantic web, it is likely that the number of ontologies will greatly boom over the next few years. Hence, the development of tools to assist in the ontology mapping process is crucial to the success of the semantic web. There are also a rapidly increasing number of web data sources, and E-business to be integrated which in turn shows the greatness of ontologies and data to be mapped. The use of ontologies and semantic mapping software can reduce the loss of semantics (meaning) in information exchange among heterogeneous applications like web services, E-Commerce, E-Business, Integrated intelligence analysis, and so on.

When comparing two ontologies, a similarity measure is quite important. There are different similarity measure approaches which can be broadly divided into two main groups (Tenenbaum 2001): continuous metric space models and set-theoretic matching models. An example of the former is the Shepard Model (Shepard 1980), which is based on probabilistic distributions; the latter consists of geometric, transformational, featural and alignment-based models, etc. Geometric models are based on distances, such as edit distance (Levenshtein 1966) measure. Transformational models are based on the number of transformations required to make two entities equal. Featural models consider the sets of common as opposed to distinctive features. One example is Tversky's ratio model (Tversky 1977). The similarity model can be single and multiple (Zhao 2006). The single model includes the Jaccard coefficient (Doan 2002), cosine measure (Lacher 2001), etc. The multi-similarity model is determined as a result of aggregating two or more similarity measures, which include the multi-weighted measure (Ehrig 2004), similarity flooding (Melnik 2002), and hybrid measure, etc. The hybrid measure can be the combination of linguistic and structural similarity measure (Madhavan 2001), or the combination of featural and structural similarity measure (de Souza 2004).

In this chapter, a novel similarity model for ontology mapping based on rough formal concept analysis (RFCA) (Kent 1993, Yao 2004, Zhao 2001) is proposed. RFCA is the combination of the rough set theory (Pawlak 1982) and formal concept analysis (FCA) (Ganter 1999, Wille 1982). The FCA theory is used here as a kind of tool to represent ontologies. Rough set theory (RST) is then applied on the basis of the FCA. With the proposed rough lower extent approximation as the similarity measure, the mapping between different ontology sources can be realized.

The organization of this chapter is as follows. In the next section, the background of ontologies, ontology mapping, and FCA are presented. A novel similarity measure method based on rough formal concept analysis is proposed in Section 3 to realize ontology mapping tasks. An example of case study is provided in the same section. Finally, conclusions are drawn in Section 4.

2. Background

2.1 Ontologies and Ontology Mapping

The term ontology is borrowed from philosophy, where it refers to a systematic account of what can exist or 'be' in the world. In the fields of artificial intelligence and knowledge representation the term ontology refers to the construction of knowledge models that specify a set of concepts, their attributes, and the relationships between them. Ontologies are defined as "explicit conceptualization(s) of a domain" (Gruber 1993), in which concepts, attributes and the relationships between them are defined as a set of representational terms, enabling knowledge to be shared and re-used. Ontologies are seen as the key to realize the vision of the semantic web, raising the level of specification of knowledge by incorporating semantics into the data, and promoting their exchange in an explicitly understandable form.

Many formal languages for representing ontologies in the semantic web have been proposed, such as RDF, RDFS, DAML-ONT (DAML Ontology)[1], OIL (the Ontology Inference Layer) (Fensel 2000), etc. RDFS in particular is recognised as an ontology representation language, talking about classes and properties, range and domain constraints, and subclass and subproperty (subsumption) relations. RDFS is a very primitive language, and more expressive power would clearly be necessary in order to describe resources in sufficient detail and to determine the semantic relationship between syntactically different terms.

Inspired by description logics, OWL (Dean 2002) and DAML+OIL (Horrocks 2002) are developed. DAML+OIL provides a richer set of vocabulary to describe the resources on web. Its semantic is a variation of

[1] www.daml.org

description logic with datatypes, which makes efficient ontology reasoning possible.

Given the decentralized nature of the semantic web's development, it is likely that the number of ontologies will greatly increase over the next few years (Doan 2002), and that many of them will describe similar or overlapping domains, providing a rich source of material. However, ontologies do not overcome per se any interoperability problems, since it is hardly conceivable that a single ontology is applied in all kinds of domains and applications. The ability to automatically and effectively perform mappings between these ontologies is becoming more and more important.

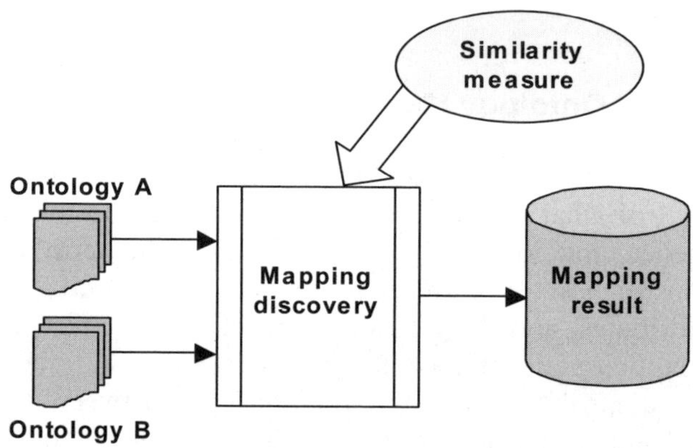

Fig. 1. Principle of ontology mapping

Ontology mapping, by specifying a set of ordered pairs or ontological correspondences, is a mechanism of communication between ontologies. An ontological correspondence specifies the term or expression in the target ontology that represents as closely as possible the meaning of the source ontology term or expression. The differences between ontologies comprise mismatch at the syntactic level, for instance, using the different representation languages, and mismatch at the semantic level, which includes using the same linguistic terms to describe different concepts or using different terms to describe the same concept, and so on. As shown in Fig. 1, source ontology *A* and target ontology *B* are put in correspondence in such a way as to preserve both the mathematical structure and ontological axioms. The mapping between *A* instances and *B* entities are discovered according to similarity measure. Finally, the mapping results can be expressed by mapping rules.

Ontology mapping does not intend to unify ontologies and their data, but to establish correspondences among the source ontologies. A good mapping tool should find the maximum number of potential mapping pairs. Of the methods of ontology mapping, we categorize them as follows (as to the details, please refer to (Zhao 2006)):

- **Manual ontology mapping.** The manual ontology mapping is the kind of mappings implemented by human experts. Manually specifying ontology mapping is quite accurate, but time consuming and error-prone. Nevertheless, when we choose semi-automatic ontology mapping algorithms, manual mapping is a good assistance to be considered.

- **Semi-automatic ontology mapping.** Most of the mapping discovery methods are semi-automatic, combining the knowledge of domain experts to get the mapping results through auxiliary information, machine learning techniques or soft computing techniques.

- **Automatic ontology mapping.** This kind of mapping produces the mappings without the participation of the human experts. Automatic mapping discovery is more desirable and practical in some cases considering the high heterogeneity and huge volume of information the web contains.

2.2 Formal Concept Analysis

Formal Concept Analysis (Wille 1982) is a mathematical approach to data analysis. It provides information with structure, and visualizes hierarchies by line diagrams. In artificial intelligence, FCA is used as a knowledge representation mechanism and as a conceptual clustering method. In database theory, FCA has extensively been used for design and management of class hierarchies. Concept lattices, or Galois lattices, constitute a mathematical framework which allows building embedded classes from a set of objects. They are the core of the mathematical theory of FCA. As a machine learning technique, concept lattice can be used to represent and discover knowledge. A concept lattice is a form of a hierarchy in which each node represents a subset of objects (extent) with their common attributes (intent). The characteristic of concept lattice theory lies in reasoning on the possible attributes of data sets. The Hasse diagram of a lattice represents a generalization / specification relationship between the concepts (nodes). Therefore, the concept lattice and its corresponding Hasse diagram with regard to a set of objects described by some attributes can be used as an effective tool for symbolic data analysis and knowledge acquisition.

We recall some necessary basic notions about FCA. FCA starts with a formal context, which is a triple $K := (O, A, R)$, where O (set of objects) and A (set of attributes) are two finite sets, and R is a binary relation between O and A. In order to express that an object o is in a relation R with an attribute a, we write oRa or $(o,a) \in R$, and read as "the object o has the attribute a", which is a unique ordered set which describes the inherent lattice structure defining natural groupings and relationships among the elements of O and A. This structure is known as concept lattice.

For a set $X \subseteq O$ of objects, X' can be defined as the set of attributes common to the objects in X, $X' := \{a \in A \mid oRa \text{ for all } o \in X\}$. Correspondingly, for a set Y of attributes, Y' is the set of objects which have all attributes in Y, $Y' := \{o \in O \mid oRa \text{ for all } a \in Y\}$.

A pair (X,Y), $X \subseteq O, Y \subseteq A$, is called a formal concept of the context K, with $X' = Y$, and $Y' = X$. Then, X is called the extent and Y the intent of concept (X,Y).

A partial order relation (<) can be defined on the concept sets of the context K, and it can be used to generate the lattice graph: given two nodes H_1 and H_2, if $H_1 < H_2$ and there is no other element H_3 in the lattice such that $H_1 < H_3 < H_2$, then H_1 is called parent of H_2, and H_2 child of H_1. The graph is usually called Hasse diagram. It is a valid visualization tool to data analysis and knowledge discovery, and reveals generalization / specification relationships between concept nodes. A Hasse diagram is a directed acyclic graph with an additional constraint: each pair of nodes in the graph has a unique nearest common descendant – or *meet* ($\wedge$) – and a unique nearest common ancestor – their *join* ($\vee$). These are referred to as greatest lower bounds, and least upper bounds, respectively. The meet and join of the lattice are characterized by:

$$\wedge_{t \in T}(X_t, Y_t) = \left(\bigcap_{t \in T} X_t, \bigcup_{t \in T} Y_t\right) \tag{1}$$

$$\vee_{t \in T}(X_t, Y_t) = \left(\bigcup_{t \in T} X_t, \bigcap_{t \in T} Y_t\right) \tag{2}$$

where T is an index set. For every $t \in T$, (X_t, Y_t) is a formal concept.

The set of all concepts forms a complete lattice, called concept lattice, and denoted by $L(O, A, R)$, or simply L.

Tab. 1 is an example formal context, where the column represents different attributes, and the rows are the objects. The concept lattice

$L(O,A,R)$ of the context given in Tab. 1 is shown in Fig. 2. As an example, Fig. 2 shows the corresponding Hasse diagram from the context of Tab. 1, where every node is represented with intent (*I*) and extent (*E*). Each pair of nodes in the concept lattice has a unique meet and join. From Fig.2, we can see that node 1 is the *join* of node 5 and node 6, and node 12 is their *meet*.

	a	b	c	d	e	f	g	h	i
1	1	1					1		
2	1	1					1	1	
3	1	1	1				1	1	
4	1		1				1	1	1
5	1	1		1		1			
6	1	1	1	1		1			
7	1		1	1	1				
8	1		1	1		1			

1: Leech; 2: Bream; 3: Frog; 4: Dog; 5: Spike-weed; 6: Reed; 7: Bean; 8: Maize; a: needs water to live; b: lives in water; c: lives on land; d: needs chlorophyll to produce food; e: two seed leaves; f: one seed leaf; g: can move around; h: has limbs; i: suckles its offspring.

Tab. 1. A formal context example (Ganter 1999).

As an example, Fig. 2 shows the corresponding Hasse diagram from the context of Tab. 1, where the column represents different attributes, and the rows are the objects. In Fig. 2, every node is represented with intent (*I*) and extent (*E*). Each pair of nodes in the concept lattice has a unique meet and join, for example, node 1 is the *join* of node 5 and node 6, and node 12 is their *meet*.

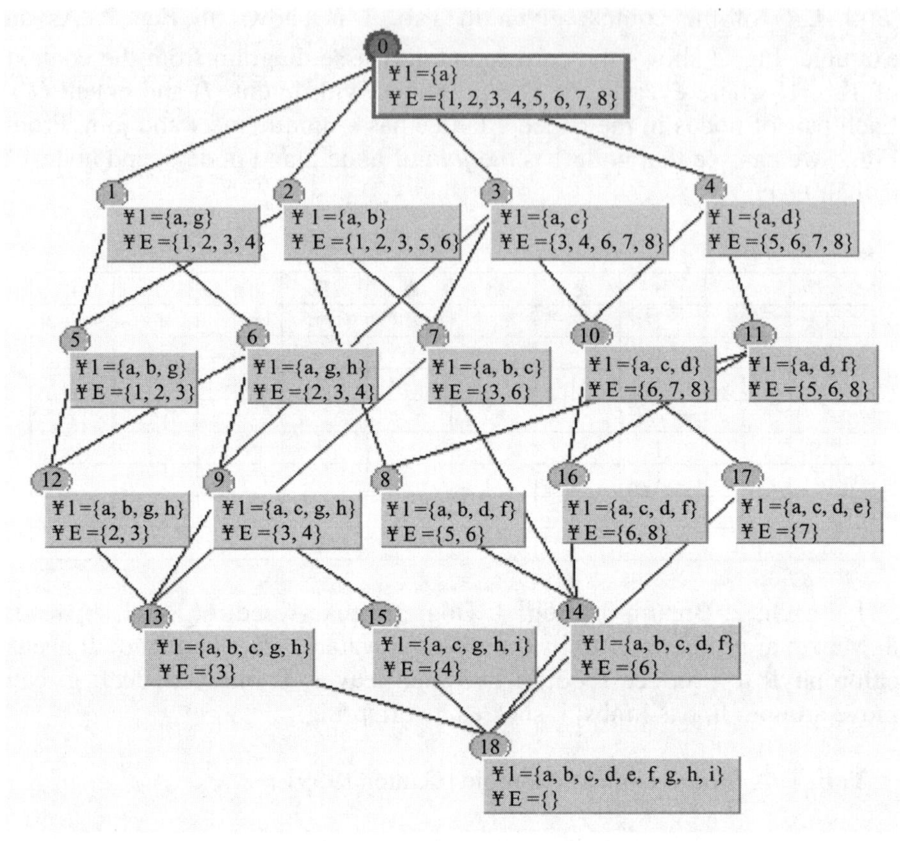

Fig. 2. Hasse diagram for the context of **Tab. 1,** with Galois Lattice Interactive Constructor (Galicia) (http://galicia.sourceforge.net)

3. Rough FCA based Similarity Model

In this part, the evolution of the similarity measure with the application of Tversky's model is first introduced, followed by the introduction of basics of rough set theory and rough formal concept analysis theory. The combination of RFCA theory with Tversky's model to produce our proposed similarity measure is then developed in the second part of the section.

3.1 Tversky's Similarity Model

A similarity measure has been proposed by Tversky in terms of a matching process. The measure produces a similarity value that is the result of common as well as different characteristics of objects. This approach is in agreement with an information-theoretic definition of similarity (Tversky 1977).

$$sim(m,n) = \frac{f(M \cap N)}{f(M \cap N) + \alpha \cdot f(M-N) + \beta \cdot f(N-M)} \tag{3}$$

where *M* and *N* are the feature sets of *m* and *n*, *f* denotes a measure over the feature sets, *f*(*M-N*) represents the sets of features present in *M* but not in *N* and *f*(*N-M*), those present in *N* but not in *M*. The parameters α and β are from Tversky's observation in psychological experimentation that the assessment of similarity is not symmetrical.

3.2 Rodriguez and Egenhofer's Similarity Model

In (Rodriguez 2003), Rodriguez and Egenhofer have proposed an assessment of semantic similarity among entity classes in different ontologies based on the normalization of Tversky's similarity model. The similarity model is a direct extension of Tversky's model, in which the function *f* is represented by the cardinality of a set, and the parameter ▯ is set to (1-▯).

$$sim(m,n) = \frac{|M \cap N|}{|M \cap N| + \alpha(m,n)|M-N| + (1-\alpha(m,n))|N-M|} \tag{4}$$

$$\alpha(M,N) = \begin{cases} \dfrac{depth(M)}{depth(M)+depth(N)} & if\ depth(M) \le depth(N) \\ 1 - \dfrac{depth(M)}{depth(M)+depth(N)} & if\ depth(M) > depth(N) \end{cases} \tag{5}$$

where *M* and *N* are the feature sets of the entity sets *m* and *n*. Here, the set theory functions of intersection (*M*▯*N*) and difference (*M-N*) are used in Eq. 4. "| |" represents the cardinality of the set. Values of α lie within the range [0, 0.5], as defined by Eq. 5. The common features are therefore considered to be more important than non-common features.

3.3 FCA based Similarity Model

Based on the theory of formal concept analysis, another extension of Tversky's model is introduced by de Souza and Davis (de Souza 2004). This model is designed with the set of common features and the set of common structural elements of the concept lattice, which are called meet-irreducible elements. An element is meet-irreducible if it can not be written as a meet of other elements in the concept lattice. With the typical lattice operations join (∨) and meet (∧), the set of common meet-irreducible elements is given by the meet-irreducible elements which are intent of $M \vee N$, i.e., $(M \vee N)^{\wedge}$. The construction of this measure is based on Tversky's similarity model. In this sense, this measure is both structural and featural at the same time. The similarity model is as follows

$$Sim(m,n) = \frac{|(M \vee N)^{\wedge}|}{|(M \vee N)^{\wedge}| + \alpha|(M-N)^{\wedge}| + (1-\alpha)|(N-M)^{\wedge}|} \qquad (6)$$

where $(M-N)^{\wedge}$ is the set of meet-irreducible elements which are in M but not in N, and $(N-M)^{\wedge}$ is the set of meet-irreducible elements which are in N but not in M. The value of is fixed to be 0.5, which means that the assessment of similarity is symmetrical, i.e. if node *c* is similar to node *d*, then node *d* is also similar to node *c* with the same similarity.

This kind of measure combines featural and structural information into decision, which is easy to understand and more reliable. Through the Hasse diagram it supports visualization easily. Therefore, FCA-based similarity measure is expected to give better mapping results than the other measures.

3.4 Rough FCA based Similarity Measure

Currently, more and more research interest has been focused on the incorporation of soft computing techniques in ontology mapping, either to apply them to existing ontology representation languages, or to specify the correspondence between source ontologies. As one of the soft computing techniques, rough set theory has provided the necessary formalism and ideas for the development of some propositional machine learning systems. It has also been used for knowledge representation and data mining. Here, we recall the basics of rough set theory and approximation operation in rough formal concept analysis. Then, a short overview of FCA based ontology interoperability approaches is given, followed by the basics of rough set theory and rough formal concept analysis. The proposed rough FCA based similarity measure model for ontology mapping is lastly presented.

3.4.1 Rough Set Theory

Rough set theory can be approached as an extension of the classical set theory, to be used when representing incomplete knowledge (Pawlak 1982). Rough sets can be considered as sets with fuzzy boundaries – sets that cannot be precisely characterized using the available set of attributes, and it is the approximation of a vague concept by a pair of precise concept, called lower and upper approximation, which are classification of domain of interest into disjoint categories.

Assume O is a non-empty set of objects, called universe, and E is an equivalence relation on O called an indiscernibility relation, $E \subseteq O \times O$. The equivalence relation divides the universe into a family of pairwise disjoint subsets. This ordered pair $C = (O, E)$ is called an *approximation space*. Each equivalence class induced by E, is called an *elementary set* in C. For an object $x \in O$, the equivalence class of E containing x is denoted by $[x]_E$. For each $X \subseteq O$, its *lower* and *upper approximation* in C are defined as:

$$\underline{X}_E = \{x \in O \mid [x]_E \subseteq X\} \tag{7}$$

$$\overline{X}_E = \{x \in O \mid [x]_E \cap X \neq \phi\} \tag{8}$$

The lower approximation of X is the greatest definable set contained in X, while its upper approximation is the least definable set containing X. If both approximations of X are exactly the same, then X is definable, otherwise it is a rough set.

3.4.2 Rough Formal Concept Analysis

A comparative examination of RST and FCA shows (Yao 2004) that each of them deals with a particular type of definability. The common notion of definability links the two theories together. The notions of formal concept and formal concept lattice can be introduced into RST by considering different types of formal concepts. Rough set approximation operators can be introduced into FCA by considering a different type of definability.

This has lead to the formulation of a new data analysis theory called Rough Formal Concept Analysis (Kent 1993).

Pioneered by R. E. Kent, RFCA is a synthesis of the theory of rough set and the theory of formal concept analysis. It studies the rough approximation of conceptual structures to provide an "approximation theory" for knowledge representation and knowledge discovery.

Given any approximation space (O, E) on objects and given any formal context (O, A, R), an attribute $m \in A$ is definable when its extent $Rm \subseteq O$ is a definable subset of objects with respect to indiscernibility relation *E*. The notions of rough lower and upper approximations were extended from subsets of objects to formal context to approximate the relation *R* in terms of definable contexts.

The upper *E*-approximation of *R*, denoted by $\overline{R}_E$, is defined element-wise: for each attribute $m \in A$, the extent of *m* in the upper approximation $\overline{R}_E$ is the upper approximation of its extent in *R*. The lower *E*-approximation of *R*, denoted by $\underline{R}_E$, is also defined element-wise: for each attribute $m \in A$, the extent of *m* in the lower approximation $\underline{R}_E$ is the lower approximation of its extent in *R*. A rough formal context in (*O*, *E*) is a collection of formal contexts of *O*-objects and *A*-attributes which have the same upper and lower approximation contexts.

Based on the two contextual approximations, the upper and lower conceptual approximations for concepts in $L(O, A, R)$ are defined in terms of *E*-definable concepts. Given any approximation space (*O*, *E*) on objects and given any formal context (O, A, R), a formal concept $(M, N) \in L(O, A, R)$ is a definable concept when its extent $M \subseteq O$ is a definable subset of objects with respect to indiscernibility relation *E*.

The upper *E*-approximation of a concept (M, N) is a monotonic function which assigns concepts in the upper approximation concept lattice to concepts in *L*; the lower *E*-approximation of a concept (M, N) is also a monotonic function which assigns concepts in the lower approximation concept lattice to concepts in *L*. Two concepts (M_1, N_1) and (M_2, N_2) of a formal context (O, A, R) are *E*-roughly equal, denoted by $(M_1, N_1) \equiv (M_2, N_2)$, when both $(M_1, N_1) \leq (M_2, N_2)$ and $(M_2, N_2) \leq (M_1, N_1)$. A formal rough concept of a formal context (O, A, R) with approximation space (O, E) is a collection of roughly equal concepts; or equivalently, a rough concept is a collection of concepts which have the same upper and lower conceptual approximations.

3.4.3 Rough FCA based Similarity Measure

To further improve the performance and the accuracy of the FCA-based similarity measure, a novel measure model based on the rough formal concept analysis theory is introduced in this section.

As in (Yao 2004), approximation operators have been defined based on lattice-theoretic operators. A formal concept consists of a definable set of concepts. The lattice is the family of all such definable concepts. Following the theory of rough sets, such a set of objects can be approximated by definable sets of objects, i.e., the extent of formal concepts. For a subset of objects $M \subseteq O$, we can approximate it by the extents of a pair of formal concepts in the concept lattice:

$$\underline{M} = \text{extent}(\vee\{(X,Y) \in L \mid X \subseteq M\}) \tag{9}$$

The extents of the resulting join of the two concepts are the lower approximation of *M*.

With this definition, the extent-based similarity measure between two concept sets *M* and *N* can be described as follows:

$$Sim(M,N) = \frac{|\underline{M} \vee \underline{N}|}{|\underline{M} \vee \underline{N}| + \alpha|\underline{M} - \underline{N}| + (1-\alpha)|\underline{N} - \underline{M}|} \tag{10}$$

In Eq. 10 we use the rough lower approximation to take the place of the meet-irreducible operation in Eq. 6. Different from the definition in Eq. 6, the function "| |" represents the number of different sets. Here, "$\underline{M} \vee \underline{N}$" means the rough lower approximation of the join of M and N; "$\underline{M} - \underline{N}$" represents the attribute sets in $\underline{M}$ but not in $\underline{N}$; and "$\underline{N} - \underline{M}$" presents those in $\underline{N}$ but not in $\underline{M}$. When α equals to 0.5, it means that $|\underline{M} - \underline{N}|$ and $|\underline{N} - \underline{M}|$ are equally important. With a specified similarity threshold σ, concepts whose similarity measure is greater than σ can be considered to be mapped. In the next section, we will show an example of case study with the application of the proposed ontology mapping method.

	Production	AnimalHusbandryMethods	AnimalFeeding	Growth	Sex	Poaceae	FeedingSystems	GrazingSystems	BeefCattle	DairyCattle	Postweaning	Preweaning	Pennisetum	DevelopmentalStages	Braquiaria
A production	1								1						
A processes	1								1						
A prod systems	1								1						
A intensive	1	1							1						
A fattening	1	1	1						1						
A growth	1	1		1					1					1	
A feeding system	1	1	1						1						
A males	1	1	1		1				1						
A brachiaria	1	1	1		1	1	1	1	1						1
A pasture usage	1	1		1			1	1	1					1	
A brachiaria2	1	1		1		1	1	1	1					1	1
A feeding syss	1	1		1			1		1					1	
B production	1									1					
B production sys	1									1					
B feeding	1						1			1					
B concentrate fd	1						1			1					
B calves	1						1			1				1	
B postweaning	1						1			1	1			1	
B preweaning	1						1			1		1		1	
B elephantGrass	1	1				1	1	1		1			1		
B intensive	1	1				1	1	1		1			1		

Tab. 2. Part of the contexts of ontologies Beef Cattle **A** and Dairy Cattle **B** (de Souza 2004)

3.5 Case Study

We use an example taken from (de Souza 2004) (see Tab. 2) to explain the mapping result with the above-mentioned similarity measure. Tab.2 shows a part of the formal contexts of the ontologies of Beef Cattle ***A*** and Dairy Cattle ***B***. The objects correspond to rows in the table, and attributes to columns. With the help of the software ConExp (Concept Explorer), we build the Hasse diagram as shown in Fig. 3. The concept nodes in white rectangle are objects, and the nodes in gray rectangle are attributes.

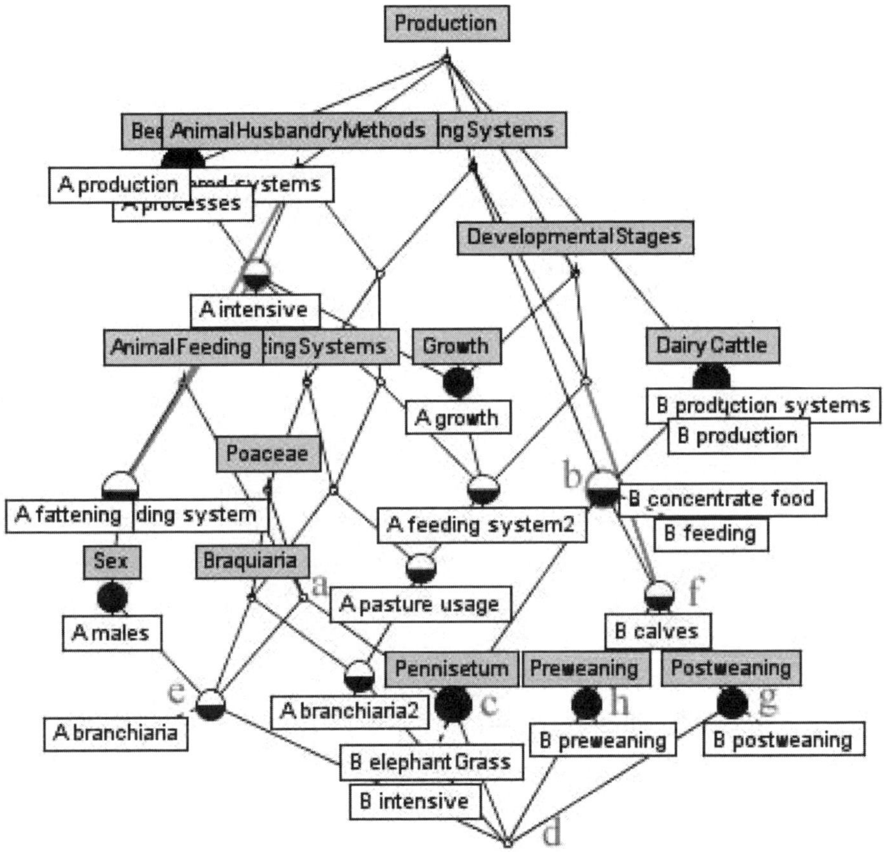

Fig. 3. Hasse diagram for the contexts in Table. 2

As an example, we calculate the similarity measure between concepts *a* and *b* (see Fig. 3). *c* is the join set of *a* and *b*, $c = a \vee b$. To simplify the expression the concept nodes, we use *a*, *b*, …, *g* to take the place of them. Then,

$$\underline{c} = \{\{c, d\}\}$$

$$\underline{a} = \{\{c, d\}, \{e, d\}\}$$

$\underline{b} = \{\{f, h, d\}, \{f, g, d\}, \{c, d\}\}$

Therefore,

$\underline{a} - \underline{b} = \{\{e, d\}\}$

$\underline{b} - \underline{a} = \{\{f, h, d\}, \{f, g, d\}\}$

Finally we have

$$Sim(a,b) = \frac{1}{1 + 0.5 \times 1 + 0.5 \times 2} = 0.4$$

Let the similarity measure threshold be 0.4, then *a* and *b* can be mapped onto each other. The mapping results with Eq. 10 to the different similarity thresholds are shown in Fig. 4.

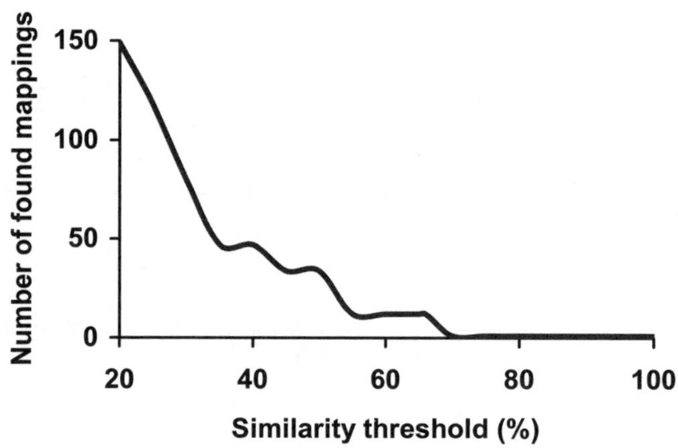

Fig. 4. The mapping results with different similarity thresholds

It is obvious from Fig. 4 that with the increase of the similarity threshold, the number of the found mappings decreases. The precision of the mapping can then be obtained by comparing the mapping results with that fulfilled by a human expert.

4. Conclusion

Ontology mapping represents a promising technique in E-business integration, agent communication and web services. Among others, similarity measure plays an important role in ontology mapping. Multi-similarity measure is proved to give higher quality mapping results than that of single measure. After a brief introduction to the background knowledge and an overview of different similarity measure, this chapter introduces a similarity measure based on rough set theory and formal concept analysis. The proposed approach combines RST and FCA to calculate the similarity measure of the concept nodes from two ontologies on the basis of Tversky's similarity model, under the condition that the ontologies are represented with concept lattices. The similarity measure is therefore both featural and structural in this sense. Our simulation results show the applicability of the proposed scheme. Further work is supposed to apply this mapping method to the real E-business integration tasks. In addition, the values of the parameter α and the similarity threshold σ are also of particular interest corresponding to specific applications.

References

Berners-Lee, T. (2001), The semantic web. Scientific American, Vol. 284, No. 5, 33-35.

Concept Explorer. http://sourceforge.net/projects/conexp.

Dean, M., Connolly D., et. al. (2002), OWL web ontology language 1.0 reference, W3C Working Draft 29 July 2002. http://www.w3.org/TR/owl-ref/.

de Souza, X.S. and Davis, J. (2004), Aligning ontologies and evaluating concept similarities. In Meersman, R. and Tari, Z. (Ed.): CoopIS/ DOA/ODBASE 2004, Lecture Notes in Computer Science, Vol. 3291, Springer-Verlag, 1012-1029.

Doan, A., Madhavan, J., Domingos, P. and Halevy, A. (2002), Learning to map between ontologies on the semantic web, In Proceedings of the 11th International WWW Conference, Hawaii, USA, 662 – 673.

Ehrig, M. and Sure, Y. (2004), Ontology mapping – an integrated approach, In Bussler, C., Davis, J., Fensel, D. and Studer, R. (Ed.), In Proceedings of the 1st ESWS, Lecture Notes in Computer Science, Vol. 3053, Springer-Verlag, 76-91.

Fensel, D., Horrocks, I., et al. (2000), OIL in a nutshell, In International Conference on Knowledge Engineering and Knowledge Management, 1–16.

Ganter, G. and Wille, R. (1999), Formal Concept Analysis: Mathematical Foundations, Springer-Verlag.

Gruber, T.R. (1993), A translation approach to portable ontology specification. Knowledge Acquisition, Vol. 5, No. 2, 199-220.

Horrocks, I. (2002), DAML+OIL: a description logic for the semantic web, Bulletin of the IEEE Computer Society Technical Committee on Data Engineering, Vol. 25, No. 1, 4-9.

Kent, R.E. (1993), Rough concept analysis. In Proceedings of the International Workshop on Rough Sets and Knowledge Discovery: rough sets, fuzzy sets and knowledge discovery, RSKD'93, Springer-Verlag, London, 248-255.

Lacher, S. and Groh, G. (2001), Facilitating the exchange of explicit knowledge through ontology mappings, In Proceedings of the 14th International Florida Artificial Intelligence Research Society Conference, 305 - 309.

Levenshtein, I.V. (1966), Binary codes capable of correcting deletions, insertions, and reversals. Cybernetics and Control Theory, Vol. 10, No. 8, 707-710.

Madhavan, J., Bernstein, P. and Rahm, E. (2001), Generic schema matching with cupid, In Proceedings of VLDB, 49-58.

Melnik, S., Garcia-Molina, H. and Rahm, E. (2002), Similarity flooding: a versatile graph matching algorithm and its application to schema matching, In Proceedings of the 18th International Conference on Data Engineering (ICDE'02), IEEE Computer Society, 117-128.

Pawlak, Z. (1982), Rough sets, International Journal of Information and Computer Science, 341-356.

Putnik, G.D. (Ed.), Adaptive Technologies and Business Integration: Social Managerial, and Organizational Dimensions, To be published.

Rodriguez, M.A. and Egenhofer, M.J. (2003), Determining semantic similarity among entity classes from different ontologies, IEEE Transactions on Knowledge and Data Engineering, Vol. 15, No. 2, 442-456.

Shepard, R. N. (1980), Multidimensional scaling, tree-fitting, and clustering, Science, Vol. 210, 390-398.

Tenenbaum, J.B. and Griffiths, T.L. (2001) Generalization, similarity, and bayesian inference, Behavioral and Brain Sciences 24, 629-640.

Tversky, A. (1977), Features of similarity, Psychological Review 84, 327-352.

Wille, R. (1982) Restructuring lattice theory: an approach based on hierarchies of concepts, In Rival, I. (Ed.), Ordered sets, Reidel, Dordrecht-Boston, 445-470.

Yao, Y.Y. and Chen, Y.H. (2004), Rough set approximations in formal concept analysis, In Proceedings of International Conference of the North American Fuzzy Information Processing Society (NAFIPS'04), 73-78.

Zhao, Y. and Shi, P.F. (2001), Restricted rough lattice-based implication rules discovery, The Journal of Shanghai Jiaotong University, Vol. 35, No. 2, 177 – 180.

Zhao, Y., Wang, X. and Halang, W.A. (2006), Ontology mapping techniques in information integration, in Cunha, M.M., Cortes, B. and Putnik, G.D. (Ed.), Adaptive Technologies and Business Integration: Social Managerial, and Organizational Dimensions, To be published.

Concept-based Semantic Web Search and Q&A

Masoud Nikravesh
BISC Program, Computer Sciences Division, EECS Department
and Imaging and Informatics Group-LBNL
University of California, Berkeley, CA 94720, USA
Email: nikravesh@cs.berkeley.edu,, URL: http://www-bisc.cs.berkeley.edu,
Tel: (510) 643-4522, Fax: (510) 642-5775

Abstract: *World Wide Web search engines including Google, Yahoo and MSN have become the most heavily-used online services (including the targeted advertising), with millions of searches performed each day on unstructured sites. In this presentation, we would like to go beyond the traditional web search engines that are based on keyword search and the* ***Semantic Web*** *which provides a common framework that allows* ***data*** *to be shared and reused across application,. For this reason, our view is that "Before one can use the power of web search the relevant information has to be mined through the concept-based search mechanism and logical reasoning with capability to Q&A representation rather than simple keyword search".*

In this paper, we will first present the state of the search engines. Then we will focus on development of a framework for reasoning and deduction in the web. A new web search model will be presented. One of the main core ideas that we will use to extend our technique is to change terms-documents-concepts (TDC) matrix into a rule-based and graph-based representation. This will allow us to evolve the traditional search engine (keyword-based search) into a concept-based search and then into Q&A model. Given TDC, we will transform each document into a rule-based model including it's equivalent graph model. Once the TDC matrix has been transformed into maximally compact concept based on graph representation and rules based on possibilistic relational universal fuzzy--type II (pertaining to composition), one can use Z(n)-compact algorithm and transform the TDC into a decision-tree and hierarchical graph that will represents a Q&A model. Finally, the concept of semantic equivalence and semantic entailment based on possibilistic relational universal fuzzy will be used as a basis for question-answering (Q&A) and inference from fuzzy premises. This will provide a foundation for approximate reasoning, language for representation of imprecise knowledge, a meaning representation language for natural languages, precisiation of fuzzy propositions expressed in a natural language, and as a tool for Precisiated Natural Language (PNL) and precisation of meaning. The maximally compact documents based on Z(n)-compact algorithm and possibilistic relational universal fuzzy--type II will be used to cluster the documents based on concept-based query-based search criteria.

Keywords: Semantic Web, Fuzzy Query, Fuzzy Search, PNL, NeuSearch, Z(n)-Compact, BISC-DSS

1. Introduction

What is Semantic Web? The Semantic Web is an extension of the current web in which information is given well-defined meaning, better enabling

This Paper is dedicated to Prof. Lotfi A. Zadeh, father of Fuzzy Logic "Zadeh Logic".

M. Nikravesh: *Concept-based Semantic Web Search and Q&A*, Studies in Computational Intelligence (SCI) **37**, 95–124 (2007)
www.springerlink.com

computers and people to work n cooperation." -- *Tim Berners-Lee, James Hendler, Ora Lassila, The Semantic Web, Scientific American, May 2001*

"The **Semantic Web** provides a common framework that allows **data** to be shared and reused across application, enterprise, and community boundaries. It is a collaborative effort led by W3C with participation from a large number of researchers and industrial partners. It is based on the Resource Description Framework (RDF), which integrates a variety of applications using XML for syntax and URIs for naming." – W3C organization (http://www.w3.org/2001/sw/)

"Facilities to put machine-understandable data on the Web are becoming a high priority for many communities. The Web can reach its full potential only if it becomes a place where data can be shared and processed by automated tools as well as by people. For the Web to scale, tomorrow's programs must be able to share and process data even when these programs have been designed totally independently. The Semantic Web is a vision: the idea of having data on the web defined and linked in a way that it can be used by machines not just for display purposes, but for automation, integration and reuse of data across various applications." (http://www.w3.org/2001/sw/).

Semantic Web is a mesh or network of information that are linked up in such a way that can be accessible and be processed by machines easily, on a global scale. One can think of Semantic Web as being an efficient way of representing and sharing data on the World Wide Web, or as a globally linked database. It is important to mention that Semantic Web technologies are still very much in their infancies and there seems to be little consensus about the likely characteristics of such system. It is also important to keep in mind that the data that is generally hidden in one way or other is often useful in some contexts, but not in others. It is also difficult to use on a large scale such information, because there is no global system for publishing data in such a way as it can be easily processed by anyone. For example, one can think of information about local hotels, sports events, car or home sales info, insurance data, weather information stock market data, subway or plane times, Major League Baseball or Football statistics, and television guides, etc..... All these information are presented by numerous web sites in HTML format. Therefore, it is difficult to use such data/information in a way that one might wanted to do so. To build any semantic web-based system, it will become necessary to construct a powerful logical language for making inferences and reasoning such that the system to become expressive enough to help users in a wide range of situations. This paper will try to develop a framework to address this issue. **Figure 1** shows Concept-Based Web-Based Databases for Intelligent Decision Analysis; a framework for next generation of Semantic Web. In the next sections, we will describe the components of such system.

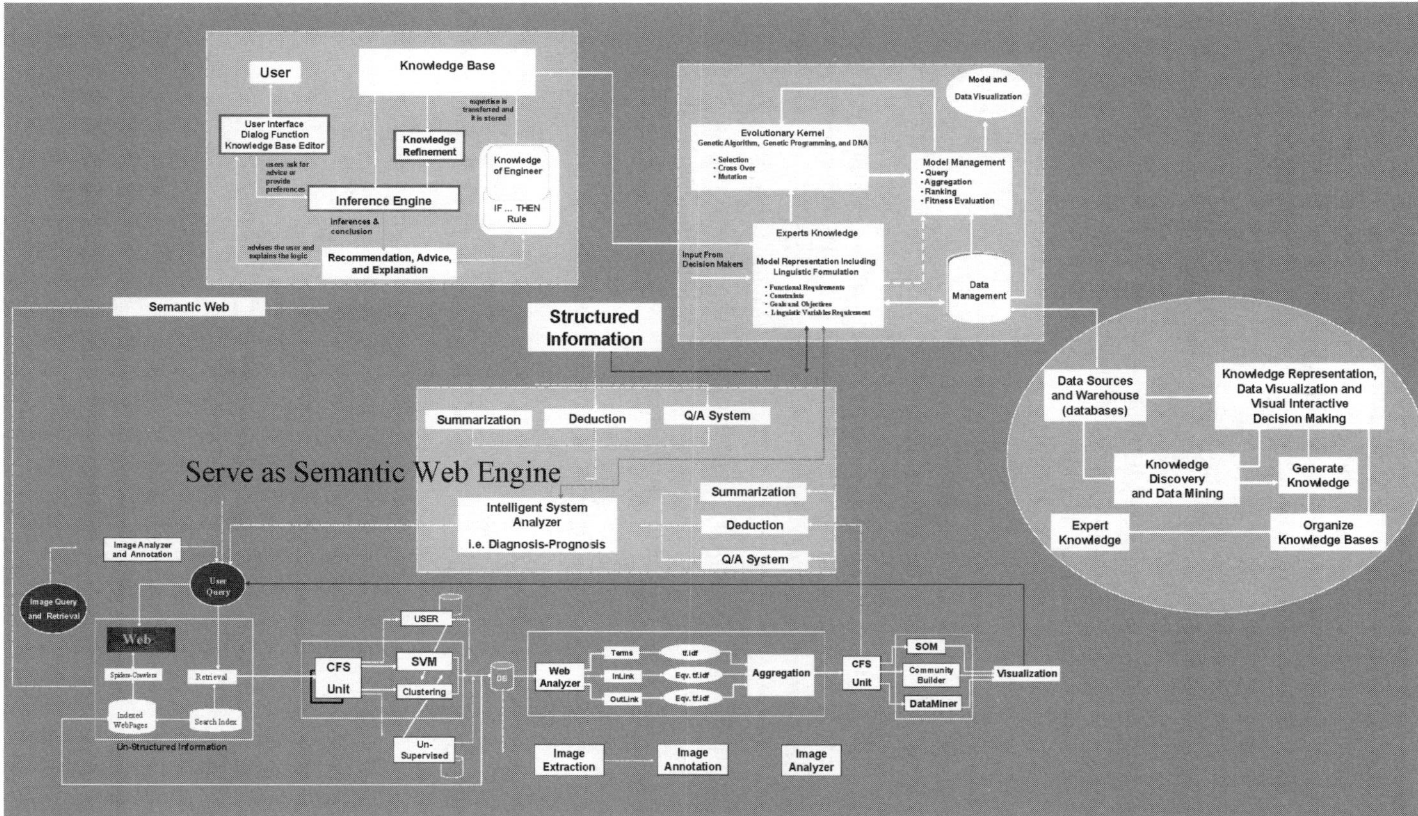

Figure 1. Concept-Based *Web-Based Databases for* Intelligent Decision Analysis; *a framework for next generation of Semantic Web*

2. Intelligent Search for Mining of Data and Textual Information

The central tasks for the most of the search engines can be summarize as 1) query or user information request- do what I mean and not what I say!, 2) model for the Internet, Web representation-web page collection, documents, text, images, music, etc, and 3) ranking or matching function-degree of relevance, recall, precision, similarity, etc. One can use clarification dialog, user profile, context, and ontology, into an integrated frame work to design a more intelligent search engine. The model will be used for intelligent information and knowledge retrieval through conceptual matching of text. The selected query doesn't need to match the decision criteria exactly, which gives the system a more human-like behavior. The model can also be used for constructing ontology or terms related to the context of search or query to resolve the ambiguity. The new model can execute conceptual matching dealing with context-dependent word ambiguity and produce results in a format that permits the user to interact dynamically to customize and personalized its search strategy. It is also possible to automate ontology generation and document indexing using the terms similarity based on Conceptual-Latent Semantic Indexing Technique (CLSI). Often time it is hard to find the "right" term and even in some cases the term does not exist. The ontology is automatically constructed from text document collection and can be used for query refinement. It is also possible to generate conceptual documents similarity map that can be used for intelligent search engine based on CLSI, personalization and user profiling.

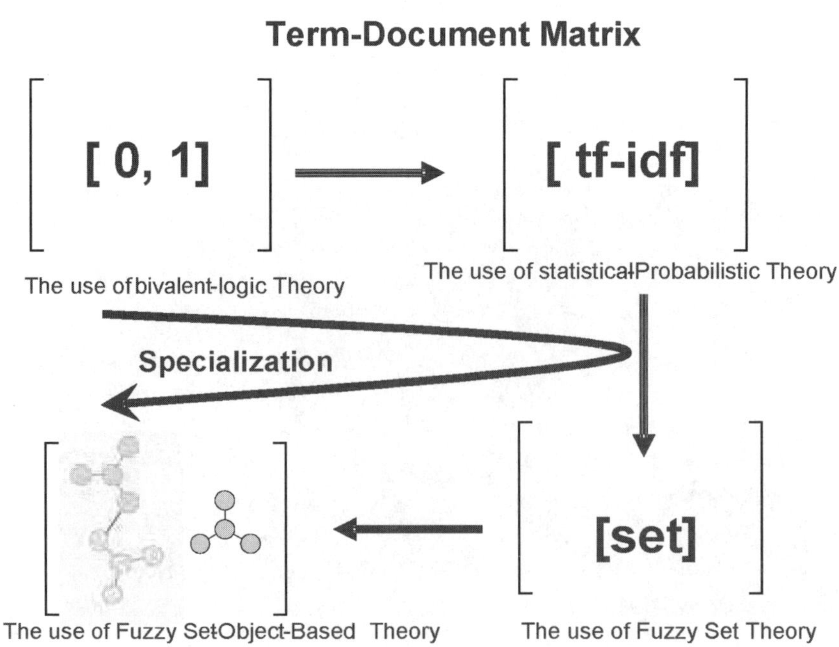

Figure 2. Evolution of Term-Document Matrix representation

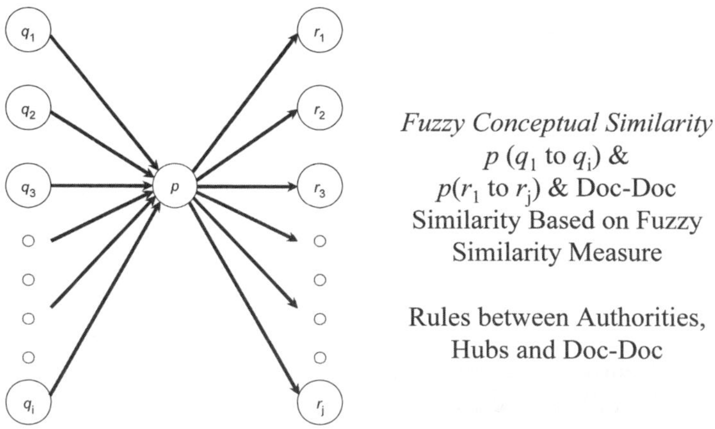

Figure 3. Fuzzy Conceptual Similarity.

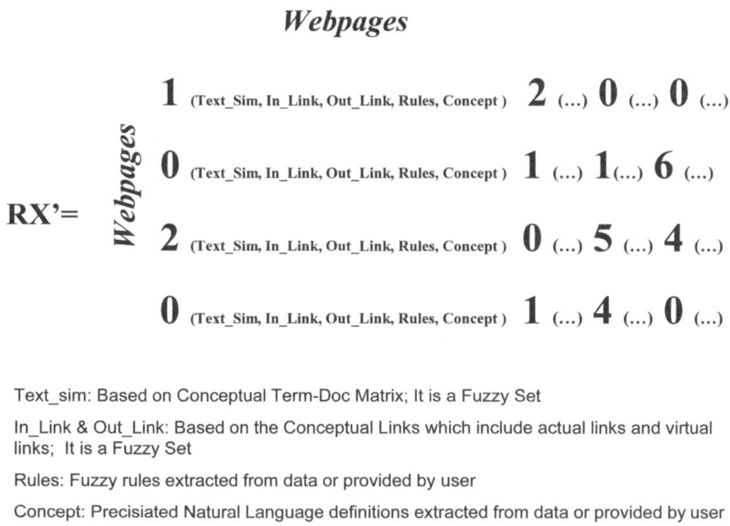

Figure 4. Matrix representation of Fuzzy Conceptual Similarity model.

The user profile is automatically constructed from text document collection and can be used for query refinement and provide suggestions and for ranking the information based on pre-existence user profile. Given the ambiguity and imprecision of the "concept" in the Internet, which may be described by both textual and image information, the use of Fuzzy Conceptual Matching (FCM) is a necessity for search engines.

In the FCM approach (**Figures 2 through 4**), the "concept" is defined by a series of keywords with different weights depending on the importance of each keyword. Ambiguity in concepts can be defined by a set of imprecise concepts. Each imprecise concept in fact can be defined by a set of fuzzy concepts. The fuzzy concepts can then be related to a set of imprecise words given the context. Imprecise words can then be translated into precise words given the ontology and ambiguity resolution through clarification dialog. By constructing the ontology and fine-tuning the strength of links (weights), we could construct a fuzzy set to integrate piecewise the imprecise concepts and precise words to define the ambiguous concept. To develop FCM, a series of new tools are needed. The new tools are based on fuzzy-logic-based method of computing with words and perceptions (CWP [1-4]), with the understanding that perceptions are described in a natural language [5-9] and state-of-the-art computational intelligence techniques [10-15]. **Figure 2** shows the evolution of Term-Document Matrix representation. The [0,1] representation of term-document matrix (or in general, storing the document based on the keywords) is the simplest representation. Most of the current search engines such as Google™, Teoma, Yahoo!, and MSN use this technique to store the term-document information. One can extend this model by the use of ontology and other similarity measures. This is the core idea that we will use to extend this technique to FCM. In this case, the existence of any keyword that does not exist directly in the document will be decided through the connection weight to other keywords that exist in this document. For example consider the followings:

- if the connection weight (based on the ontology) between term "i" (i.e. Automobile) and term "j" (i.e. Transportation) is 0.7; and the connection weight between term "j" (i.e. Transportation) and term "k" (i.e. City) is 0.3;
  - $w_{ij}=0.7$
  - $w_{jk}=0.3$
- and term "i" doesn't exist in document "I", term "j" exists in document "I", and term "k" does not exist in document I
  - $T_iD_I=0$
  - $T_jD_I=1$
  - $T_kD_I=0$
- Given the above observations and a threshold of 0.5, one can modify the term-document matrix as follows:
  - $T_iD_I'=1$
  - $T_jD_I'=1$
  - $T_kD_I'=0$
- In general one can use a simple statistical model such as the concurrence matrix to calculate w_{ij} [5, 12].

An alternative to the use of [1,0] representation is the use of the tf-idf (term frequency-inverse document frequency) model. In this case, each term gets a weight given its frequency in individual documents (tf, frequency of term in each document) and its frequency in all documents (idf, inverse document frequency). There are many ways to create a tf-idf matrix [7]. In FCM, our main focus is the use of fuzzy set theory to find the association between terms and documents. Given such association, one can represent each entry in the term-document matrix with a set rather then either [0,1] or tf-idf. The use of fuzzy-tf-idf is an alternative to the use of the conventional tf-idf. In this case, the original tf-idf weighting values will be replaced by a fuzzy set representing the original crisp value of that specific term. To construct such value, both ontology and similarity measure can be used. To develop ontology and similarity, one can use the conventional Latent Semantic Indexing (LSI) or Fuzzy-LSI [7]. Given this concept (FCM), one can also modify the link analysis (**Figure 3**) and in general Webpage-Webpage similarly (**Figure 4**). More information about this project and also a Java version of Fuzzy Search Tool (FST) that uses the FCM model is available at http://www.cs.berkeley.edu/~nikraves/fst//SFT and a series of papers by the author at Nikravesh, Zadeh and Kacprzyk [12]. Currently, we are extending our work to use the graph theory to represent the term-document matrix instead of the use of fuzzy set. While, each step increases the complexity and the cost to develop the FCM model, we believe this will increase the performance of the model given our understanding based on the results that we have analyzed so far. Therefore, our target is to develop a more specialized and personalized model with better performance rather than a general, less personalized model with less accuracy. In this case, the cost and complexity will be justified.

One of the main core ideas that we will use to extend our technique is to change terms-documents-concepts (TDC) matrix into a rule and graph. In the following section, we will illustrate how one can build such a model.

Consider a terms-documents-concepts (TDC) matrix presented as in **Table 1** where (Note that TDC entries ($k_{ii'}^{jj}$) could be crisp number, tf-idf values, set or fuzzy-objects (including the linguistic labels of fuzzy granular) as shown in **Figure 2**) .

D_i : Documents; where i = 1… m (in this example 12)
Key_j : Terms/Keywords in documents; where j = 1… n (in this example 3)
C_{ij} : Concepts; where ij = 1…l (in this example 2)

One can use Z(n)-compact algorithm (section 3.1.7) to represent the TDC matrix with rule-base model. **Table 2** shows the intermediate results based on Z(n)-compact algorithm for concept 1. **Table 3** shows how the original TDC (**Table 1**) matrix is represented in final pass with a maximally compact representation. Once the TDC matrix represented by a maximally compact representation (**Table 3**), one can translate this compact representation with rules as presented in **Tables 4** and **5**. **Table 6** shows the Z(n)-compact algorithm. Z(n) Compact is the basis to create web-web similarly as shown in **Figure 4.**

Table 1. Terms-Documents-Concepts (TDC) Matrix

Terms-Documents-Concepts					
Documents	key_1	key_2	key_3		Concepts
D_1	k_1^1	k_1^2	k_1^3		c_1
D_2	k_1^1	k_2^2	k_1^3		c_1
D_3	k_2^1	k_2^2	k_1^3		c_1
D_4	k_3^1	k_2^2	k_1^3		c_1
D_5	k_3^1	k_1^2	k_2^3		c_1
D_6	k_1^1	k_2^2	k_2^3		c_1
D_7	k_2^1	k_2^2	k_2^3		c_1
D_8	k_3^1	k_2^2	k_2^3		c_1
D_9	k_3^1	k_1^2	k_1^3		c_2
D_{10}	k_1^1	k_1^2	k_2^3		c_2
D_{11}	k_2^1	k_1^2	k_1^3		c_2
D_{12}	k_2^1	k_1^2	k_2^3		c_2

Table 2. Intermediate results for Z(n)-compact algorithm

The Intermediate Results/ Iterations					
Documents	key_1	key_2	key_3		C
D_1	k_1^1	k_1^2	k_1^3		c_1
D_2	k_1^1	k_2^2	k_1^3		c_1
D_3	k_2^1	k_2^2	k_1^3		c_1
D_4	k_3^1	k_2^2	k_1^3		c_1
D_5	k_3^1	k_1^2	k_2^3		c_1
D_6	k_1^1	k_2^2	k_2^3		c_1
D_7	k_2^1	k_2^2	k_2^3		c_1
D_8	k_3^1	k_2^2	k_2^3		c_1
D_2,D_3, D_4	*	k_2^2	k_1^3		c_1
D_6,D_7, D_8	*	k_2^2	k_2^3		c_1
D_5,D_8	k_3^1	*	k_2^3		c_1
D_2,D_6	k_1^1	k_2^2	*		c_1
D_3,D_7	k_2^1	k_2^2	*		c_1
D_4,D_8	k_3^1	k_2^2	*		c_1
D_2,D_6	*	k_2^2	*		c_1
$D_2,D_3, D_4,D_6,D_7, D_8$					
D_1	k_1^1	k_1^2	k_1^3		c_1
D_5,D_8	k_3^1	*	k_2^3		c_1
$D_2,D_3, D_4,D_6,D_7, D_8$	*	k_2^2	*		c_1

Table 3. Maximally Z(n)-compact representation of TDC matrix

Documents	key_1	key_2		key_n	Concepts
D_1	k_1^1	k_1^2		k_1^3	c_1
D_5,D_8	k_3^1	*		k_2^3	c_1
$D_2,D_3, D_4,D_6,D_7, D_8$	*	k_2^2		*	c_1
D_9	k_3^1	k_1^2		k_1^3	c_2
D_{10}	k_1^1	k_1^2		k_2^3	c_2
D_{11}, D_{12}	k_2^1	k_1^2		*	c_2

Table 4. Rule-based representation of Z(n)-compact of TDC matrix

Documents	Rules
D_1	If key_1 is k_1^1 and key_2 is k_1^2 and key_3 is k_1^3 THEN Concept is c_1
D_5, D_8	If key_1 is k_3^1 and key_3 is k_2^3 THEN Concept is c_1
$D_2, D_3, D_4, D_6, D_7, D_8$	If key_2 is k_2^2 THEN Concept is c_1
D_9	If key_1 is k_3^1 and key_2 is k_1^2 and key_3 is k_1^3 THEN Concept is c_2
D_{10}	If key_1 is k_1^1 and key_2 is k_1^2 and key_3 is k_2^3 THEN Concept is c_2
D_{11}, D_{12}	If key_1 is k_2^1 and key_2 is k_1^2 THEN Concept is c_2

Table 5. Rule-based representation of Maximally Z(n)-compact of TDC matrix (Alternative representation for **Table 4**)

Document	Rules
D_1 D_5, D_8 $D_2, D_3, D_4, D_6, D_7, D_8$	If key_1 is k_1^1 and key_2 is k_1^2 and key_3 is k_1^3 OR If key_1 is k_3^1 and key_3 is k_2^3 OR If key_2 is k_2^2 THEN Concept is c_1
D_9 D_{10} D_{11}, D_{12}	If key_1 is k_3^1 and key_2 is k_1^2 and key_3 is k_1^3 OR If key_1 is k_1^1 and key_2 is k_1^2 and key_3 is k_2^3 OR If key_1 is k_2^1 and key_2 is k_1^2 THEN Concept is c_2

Table 6. Z(n)-Compactification Algorithm

Z(n)-Compact Algorithm:
The following steps are performed successively for each column j; j=1 … n

1. Starting with k_{ii}^{jj} (ii=1, jj=1) check if for any k_{ii}^1 (ii=1, ..,3 in this case) all the columns are the same, then k_{ii}^1 can be replaced by *
   - For example, we can replace k_{ii}^1 (ii=1, …,3) with * in rows 2, 3, and 4 . One can also replace k_{ii}^1 with * in rows 6, 7, and 8. (**Table 1,** first pass).
2. Starting with k_{ii}^{jj} (ii=1, jj=2) check if for any k_{ii}^2 (ii=1, ..,3 in this case) all the columns are the same, then k_{ii}^2 can be replaced by *
   - For example, we can replace k_{ii}^2 (ii=1, …,3) with * in rows 5 and 8. (**Table 1,** first pass).
3. Repeat steps one and 2 for all jj.
4. Repeat steps 1 through 3 on new rows created Row* (Pass 1 to Pass nn, in this case, Pass 1 to Pass 3).
   - For example, on Rows* 2,3, 4 (Pass 1), check if any of the rows given columns jj can be replaced by *. In this case, k_{ii}^3 can be replaced by *. This will gives: * k_2^2 *.
   - For example, on Pass 3, check if any of the rows given columns jj can be replaced by *. In this case, k_{ii}^1 can be replaced by *. This will gives: * k_2^2 *
5. Repeat steps 1 through 4 until no compactification would be possible

As it has been proposed, the TDC entries could not be crisp numbers. The following cases would be possible:

A. The basis for the k_{ii}^{jj}s are [0 and 1]. This is the simplest case and Z(n)-compact will work as presented

B-- The basis for the k_{ii}^{jj}s are tf-idf or any similar statistical based values or GA-GP context-based tf-idf, ranked tf-idf or fuzzy-tf-idf. In this case, we use fuzzy granulation to granulate tf-idf into series of granular, two ([0 or 1] or [high and low]), three (i.e. low, medium, and high), etc. Then the Z(n)-compact will work as it is presented.

C-- The basis for the k_{ii}^{jj}s are set value created based on ontology which can be created based on traditional statistical based methods, human made, or fuzzy-ontology. In this case, the first step is to find the similarities between set values using statistical or fuzzy similarly measures. BISC-DSS software has a set of similarity measures, T-norm and T-conorm, and aggregator operators for this purpose. The second step is to use fuzzy granulation to granulate the similarities values into series of granular, two ([0 or 1] or [high and low]), three (i.e. low, medium, and high), etc. Then the Z(n)-compact will work as it is presented.

D --It is also important to note that concepts may also not be crisp. Therefore, steps B and C could also be used to granulate concepts as it is used to granulate the keyword entries values (k_{ii}^{jj}s).

E – Another important case is ho to select the right keywords in first place. One can use traditional statistical or probabilistic techniques which are based on tf-idf techniques, non traditional techniques such as GA-GP context-based tf-idf, ranked tf-idf or fuzzy-tf-idf, or clustering techniques. These techniques will be used as first pass to select the first set of initial keywords. The second step will be based on feature selection technique based on to maximally separating the concepts. This techniques are currently part of the BISC-DSS toolbox, which includes the Z(n)-Compact-Feature-Selection technique (Z(n)-CFS)).

F – Other possible cases: These include when the keywords are represented based on a set of concepts (such as Chinese-DNA model) or concepts are presented as a set of keywords (traditional techniques). In the following sections, we will use Neu-FCS model to create concepts automatically and relate the keywords to the concepts through a mesh of networks of neurons.

G-- Once the TDC matrix has been transformed into maximally compact concept based on graph representation and rules based on possibilistic relational universal fuzzy--type II (pertaining to composition), one can use Z(n)-compact algorithm and transform the TDC into a decision-tree and hierarchical graph that will represents a Q&A model. Finally, the concept of semantic equivalence and semantic entailment based on possibilistic relational universal fuzzy will be used as a basis for question-answering (Q&A) and inference from fuzzy premises. This will provide a foundation for approximate reasoning, language for representation of imprecise knowledge, a meaning representation language for natural languages, precisiation of fuzzy propositions expressed in a natural language, and as a tool for Precisiated Natural Language (PNL) and precisation of meaning. The maximally compact documents based on Z(n)-compact algorithm and possibilistic relational universal fuzzy--type II will be used to cluster the documents based on concept-based query-based search criteria. **Tables 7** through **9** show the technique based on possibilistic relational universal fuzzy--type II (pertaining to composition) [3-4] and series of examples to clarify the techniques. Types I through Types IV of possibilistic relational universal fuzzy in connection with Z(n)-Compact can be used as a basis for precisation of meaning. The focus of this paper is Type II and discussion of Type I, III, and IV are beyond the scope of this paper.

Table 7. Possibilistic Relational Universal Fuzzy--Type II -- operation of composition -- to be used to compact rules presented in **Table 5.**

R	X_1	X_2	...	X_n
	F_{11}	F_{12}	...	F_{1n}
	...	...	...	...
	F_{m1}	...	...	F_{mn}

$$\begin{aligned} R = {} & X_1 \text{ is } F_{11} \text{ and } X_2 \text{ is } F_{12} \text{ and } \ldots X_n \text{ is } F_{1n} \quad \text{OR} \\ & X_1 \text{ is } F_{21} \text{ and } X_2 \text{ is } F_{22} \text{ and } \ldots X_n \text{ is } F_{2n} \quad \text{OR} \\ & \quad \ldots \qquad \ldots \qquad \ldots \\ & X_1 \text{ is } F_{m1} \text{ and } X_2 \text{ is } F_{m2} \text{ and } \ldots X_n \text{ is } F_{mn} \end{aligned}$$

$$R \rightarrow (F_{11} \times \cdots \times F_{1n}) + \cdots + (F_{m1} \times \cdots \times F_{mn})$$

Table 8. Possibilistic Relational Universal Fuzzy--Type II -- pertaining to composition -- main operators

$p = q * r$

$q : M \text{ is } F$

$r : N \text{ is } G$

$M \text{ is } F \text{ and } N \text{ is } G : \overline{F} \cap \overline{G} = F \times G$

$M \text{ is } F \text{ or } N \text{ is } G : \overline{F} + \overline{G}$

$\text{if } M \text{ is } F \text{ then } N \text{ is } G = \overline{F}' \oplus \overline{G}$

or

$\text{If } M \text{ is } F \text{ then } N \text{ is } G = F \times G + F' \times V$

$\overline{F}' = F' \times V$

$\overline{G} = U \times G$

$\mu_{F \times G}(u,v) = \mu_F(u) \wedge \mu_G(v)$

$\mu_{\overline{F}' \oplus \overline{G}}(u,v) = 1 \wedge (1 - \mu_F(u) + \mu_G(v))$

$\wedge$: min, $+$ arithmetic sum,

$-$ arithmetic difference

If M is F then N is G else N is H

$$\Pi_{(X_1,\ldots,X_m,Y_1,\ldots,Y_n)} = \left(\overline{F}' \oplus \overline{G}\right) \cap \left(\overline{F} \oplus \overline{H}\right)$$

or

If M is F then N is G else N is H $\leftrightarrow$

$$\begin{pmatrix} (\text{If } M \text{ is } F \text{ then } N \text{ is } G) \text{ and} \\ (\text{If } M \text{ is not } F \text{ then } N \text{ is } H) \end{pmatrix}$$

Table 9 Possibilistic Relational Universal Fuzzy--Type II -- Examples

Example :

$U = V = 1 + 2 + 3,$

$M : X, N : Y$

$F : SMALL : 1/1 + 0.6/2 + 0.1/3$

$G : LARGE : 0.1/1 + 0.6/2 + 1/3.$

X is small and Y is large:

$$\Pi_{(X,Y)} = \begin{bmatrix} 0.1 & 0.6 & 1 \\ 0.1 & 0.6 & 0.6 \\ 0.1 & 0.1 & 0.1 \end{bmatrix}$$

X is small or Y is large:

$$\Pi_{(X,Y)} = \begin{bmatrix} 1 & 1 & 1 \\ 0.6 & 0.6 & 1 \\ 0.1 & 0.6 & 1 \end{bmatrix}$$

if X is small then Y is large:

$$\Pi_{(X,Y)} = \begin{bmatrix} 0.1 & 0.6 & 1 \\ 0.5 & 1 & 1 \\ 1 & 1 & 1 \end{bmatrix}$$

if X is small then Y is large:

$$\Pi_{(X,Y)} = \begin{bmatrix} 0.1 & 0.6 & 1 \\ 0.4 & 0.6 & 0.6 \\ 0.9 & 0.9 & 0.9 \end{bmatrix}$$

2.1 Fuzzy Conceptual Match and Search Engine

The main problem with conventional information retrieval and search such as vector space representation of term-document vectors are that 1) there is no real theoretical basis for the assumption of a term and document space and 2) terms and documents are not really orthogonal dimensions. These techniques are used more for visualization and most similarity measures work about the same regardless of model. In addition, terms are not independent of all other terms. With regards to probabilistic models, important indicators of relevance may not be term -- though terms only are usually used. Regarding Boolean model, complex query syntax is often misunderstood and problems of null output and Information overload exist. One solution to these problems is to use extended Boolean model or fuzzy logic. In this case, one can add a fuzzy quantifier to each term or concept. In addition, one can interpret the AND as fuzzy-MIN and OR as fuzzy-MAX functions. Alternatively, one can add agents in the user interface and assign certain tasks to them or use machine learning to learn user behavior or preferences to improve performance. This technique is useful when past behavior is a useful predictor of the future and wide variety of behaviors amongst users exist. In our perspective, we define this framework as ***Fuzzy Conceptual Matching based on Human Mental Model***. The Conceptual Fuzzy Search (CFS) model will be used for intelligent information and knowledge retrieval through conceptual matching of both text and images (here defined as "Concept"). The selected query doesn't need to match the decision criteria exactly, which gives the system a more human-like behavior. The CFS can also be used for constructing fuzzy ontology or terms related to the context of search or query to resolve the ambiguity. It is intended to combine the expert knowledge with soft computing tool. Expert knowledge needs to be partially converted into artificial intelligence that can better handle the huge information stream. In addition, sophisticated management work-flow need to be designed to make optimal use of this information. The new model can execute conceptual matching dealing with context-dependent word ambiguity and produce results in a format that permits the user to interact dynamically to customize and personalized its search strategy.

2.2 From Search Engine to Q/A Systems: The Need for New Tools

(Extracted text from Prof. Zadeh's presentation and abstracts; Nikravesh et al., Web Intelligence: Conceptual-Based Model, Memorandum No. UCB/ERL M03/19, 5 June 2003): "*Construction of Q/A systems has a long history in AI. Interest in Q/A systems peaked in the seventies and eighties, and began to decline when it became obvious that the available tools were not adequate for construction of systems having significant question-answering capabilities. However, Q/A systems in the form of domain-restricted expert systems have proved to be of value, and are growing in versatility, visibility and importance. Upgrading a search engine to a Q/A system is a complex, effort-*

intensive, open-ended problem. Semantic Web and related systems for upgrading quality of search may be viewed as steps in this direction. But what may be argued, as is done in the following, is that existing tools, based as they are on bivalent logic and probability theory, have intrinsic limitations. The principal obstacle is the nature of world knowledge. Dealing with world knowledge needs new tools. A new tool which is suggested for this purpose is the fuzzy-logic-based method of computing with words and perceptions (CWP [1-4]), with the understanding that perceptions are described in a natural language [5-9]. A concept which plays a key role in CWP is that of Precisiated Natural Language (PNL). It is this language that is the centerpiece of our approach to reasoning and decision-making with world knowledge. The main thrust of the fuzzy-logic-based approach to question-answering which is outlined here, is that to achieve significant question-answering capability it is necessary to develop methods of dealing with the reality that much of world knowledge—and especially knowledge about underlying probabilities is perception-based. Dealing with perception-based information is more complex and more effort-intensive than dealing with measurement-based information. In this instance, as in many others, complexity is the price that has to be paid to achieve superior performance."

Once the TDC matrix (**Table 1**) has been transformed into maximally Z(n)-compact representation (**Table 3**) and Rules (**Tables 4** and **5**), one can use Z(n)-compact algorithm and transform the TDC into a decision-tree/hierarchical graph which represent a Q&A model as shown in **Figure 5.**

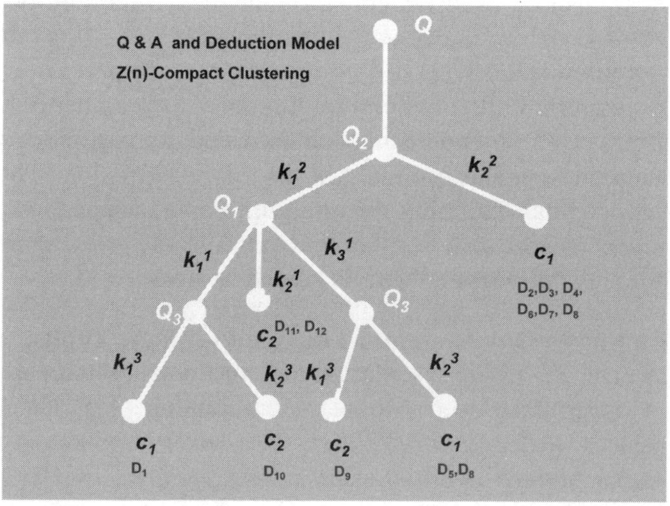

Figure 5. Q & A model of TDC matrix and maximally Z(n)-compact rules and concept-based query-based clustering

2.3 NeuSearch™

There are two types of search engine that we are interested and are dominating the Internet. First, the most popular search engines that are mainly for unstructured data such as Google ™, Yahoo, MSN, and Teoma which are based on the concept of Authorities and Hubs. Second, search engines that are task spcifics such as 1) Yahoo!: manually-pre-classified, 2) NorthernLight: Classification, 3) Vivisimo: Clustering, 4) Self-organizing Map: Clustering + Visualization and 5) AskJeeves: Natural Languages-Based Search; Human Expert. Google uses the PageRank and Teoma uses HITS for the Ranking. To develop such models, state-of-the-art computational intelligence techniques are needed [10-15]. **Figures 5** through **8** show, how neuro-science and PNL can be used to develop the next generation of the search engine.

Figure 6 shows a unified framework for the development of a search engine based on conceptual semantic indexing. This model will be used to develop the NeuSearch model based on the Neuroscience and PNL approach. As explained previously and represented by **Figures 2** through **4** with respect to the development of FCM, the first step will be to represent the term-document matrix. Tf-idf is the starting point for our model. As explained earlier, there are several ways to create the tf-idf matrix [7]. One such attractive idea is to use the term ranking algorithm based on evolutionary computing, as it is presented in **Figure 6**, GA-GP context-based tf-idf or ranked tf-idf.

Once fuzzy tf-idf (term-document matrix) is created, the next step will be to use such indexing to develop the search and information retrieval mechanism. As shown in **Figure 6,** there are two alternatives 1) classical search models such as LSI-based approach and 2) NeuSearch model. The LSI based models include 1) Probability based-LSI, Bayesian-LSI, Fuzzy-LSI, and NNNet-based-LSI models. It is interesting that one can find an alternative to each of such LSI-based models using Radial Basis Function (RBF). For example, one can use Probabilistic RBF equivalent to Probability based-LSI, Generalized RBF as an equivalent to Bayesian-LSI, ANFIS as an equivalent to Fuzzy-LSI, and RBF function neural network (RBFNN) as an equivalent to NNnet-LSI. RBF model is the basis of the NeuSearch model (Neuro-Fuzzy Conceptual Search-- NeuFCS). Given the NeuSearch model, one needs to calculate the w(i,j) which will be defined in next section (the network weights). Depends on the model and framework used (Probability, Bayesian, Fuzzy, or NNnet model), the interpretation of the weights will be different. **Figure 7** show the typical Neuro-Fuzzy Conceptual Search (NeuFCS) model input-output. The main idea with respect to NeuSearch is concept-context-based search. For example, one can search for word "Java" in the context of "Computer" or in the context of "Coffee". One can also search for "apple" in the context of "Fruit" or in the context of "Computer". Therefore, before we relate the terms to the document, we first extend the keyword, given the existing concepts-contexts and then we relate that term to the documents. Therefore, there will be always a two-step process, based on NeuSearch model as shown below:

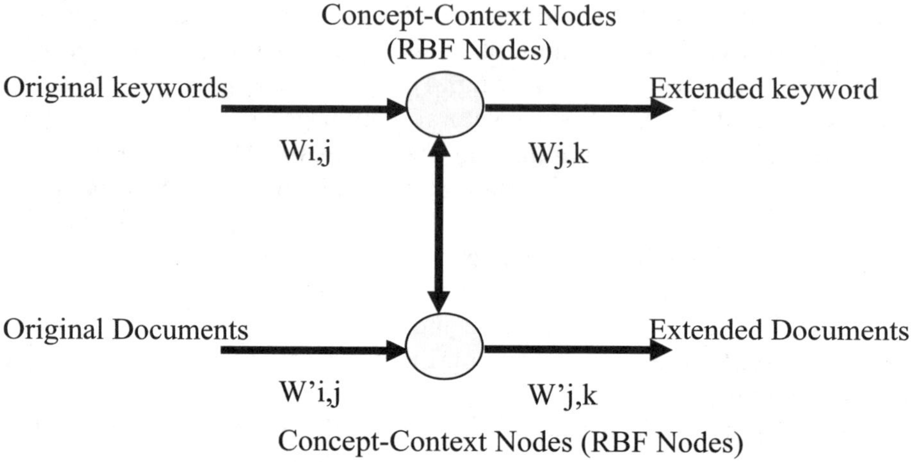

In general, w(i,j) is a function of p(i,j), p(i), and p(j), where p represent the probability. Therefore, one can use the probabilistic-LSI or PRBF given the NeuSearch framework. If the probabilities are not known, which is often times is the case, one can use Fuzzy-LSI model or ANFIS model or NNnet-LSI or RBFNN model using the NeuSearch model. In general, the PNL model can be used as unified framework as shown in **Figure 8**. **Figure 8** shows PNL-Based Conceptual Fuzzy Search Using Brain Science model and concept presented based on **Figures 2** through **7**.

Based on PNL approach, w(i,j) is defined based on ri,j as follows:

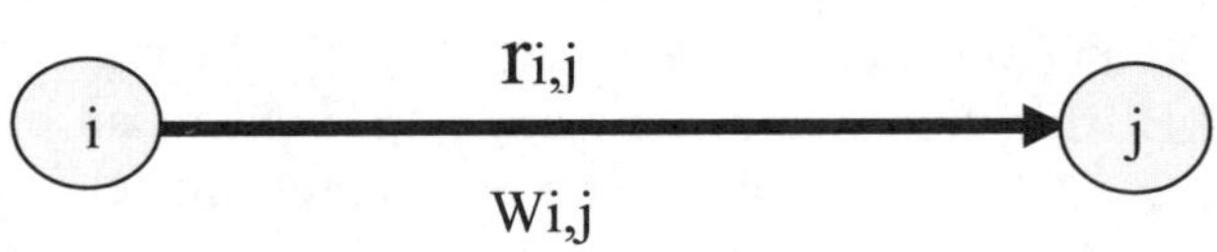

Where wi,j is granular strength of association between i and j, ri,j is epistemic lexicon, wi,j <== ri,j, and ri,j is defined as follows:

rij:	***i is an instance of j***	***(is or isu)***
	i is a subset of j	***(is or isu)***
	i is a superset of j	***(is or isu)***
	j is an attribute of i	

i causes j ***(or usually)***
i and j are related

Figure 9 shows a typical example of Neu-FCS result. Both term-document matrix and term-term document matrix/document-document matrix are reconstructed. For example, the original document corpus is very sparse as one can see in the figure; however, after processing through concept node, the document corpus is less spare and smoothens. Also, the original keywords are expanded given the concept-context nodes. While the nodes in the hidden space (Concept-Context nodes) are based on RBF (NueSearch model) and it is one dimensional nodes, one can use SOM to rearrange the nodes given the similarity of the concept (in this case the RBF in each node) and present the 1-D nodes into 2-D model as shown in **Figure 9.**

3. Intelligent Information Systems in Enterprise

To build any semantic web-based system, it will become necessary to construct a powerful logical language for making inferences and reasoning such that the system to become expressive enough to help users in a wide range of situations. This section will try to develop a framework to address this issue specially when we are dealing with structured data using web-based interface. Most of the existing search systems' 'software' are modeled using crisp logic and queries. We introduce fuzzy querying and ranking as a flexible tool allowing approximation where the selected objects do not need to match exactly the decision criteria resembling natural human behavior. The model consists of five major modules: the Fuzzy Search Engine (FSE), the Application Templates (AT), the User Interface (UI), the Database (DB) and the Evolutionary Computing (EC). We developed the software with many essential features. It is built as a web-based software system that users can access and use over the Internet. The system is designed to be generic so that it can run different application domains. To this end, the Application Template module provides information of a specific application as attributes and properties, and serves as a guideline structure for building a new application. The Fuzzy Search Engine (FSE) is the core module of the system. It has been developed to be generic so that it would fit any application. The main FSE component is the query structure, which utilizes membership functions, similarity functions and aggregators. Through the user interface a user can enter and save his profile, input criteria for a new query, run different queries and display results. The user can manually eliminate the results he disapproves of or change the ranking according to his preferences. The Evolutionary Computing (EC) module monitors ranking preferences of the user's queries. It learns to adjust to the intended meaning of the users' preferences.

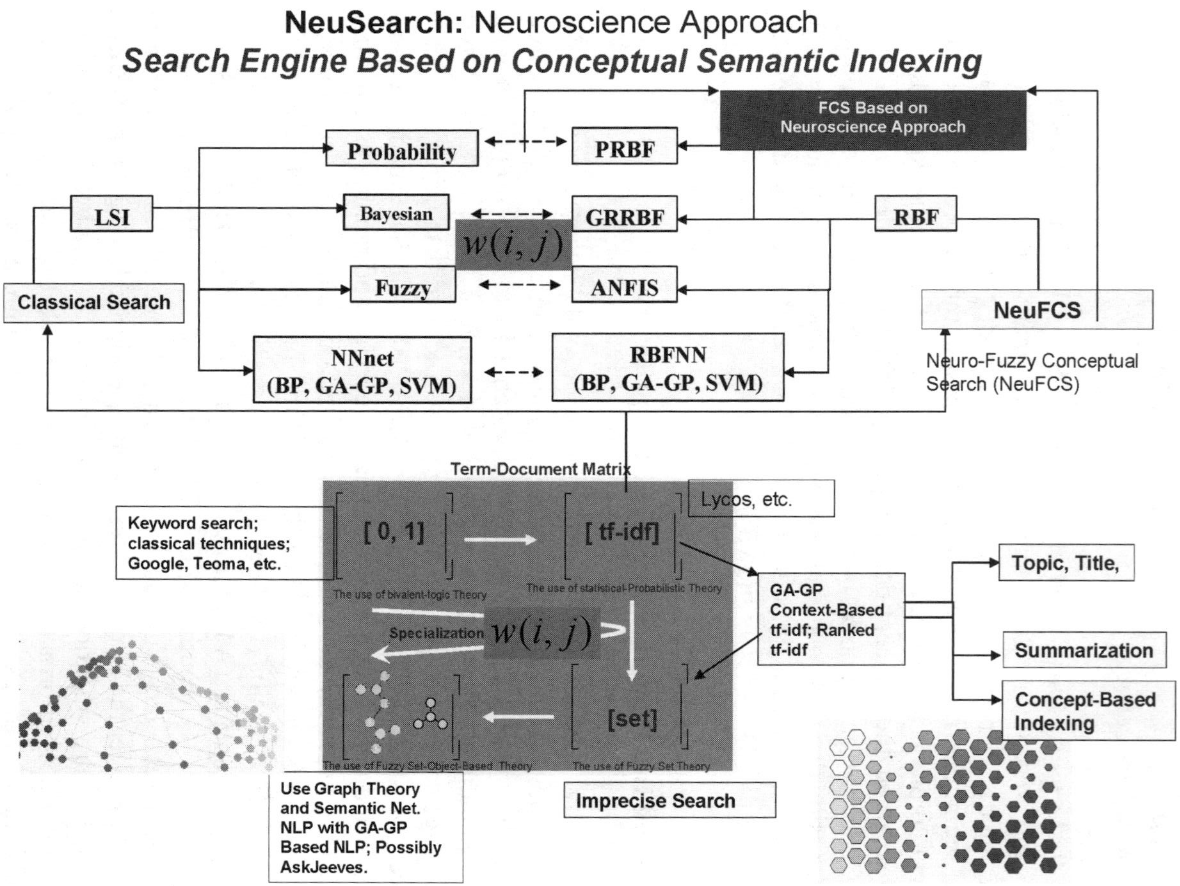

Figure 6. Search engine based on conceptual semantic indexing

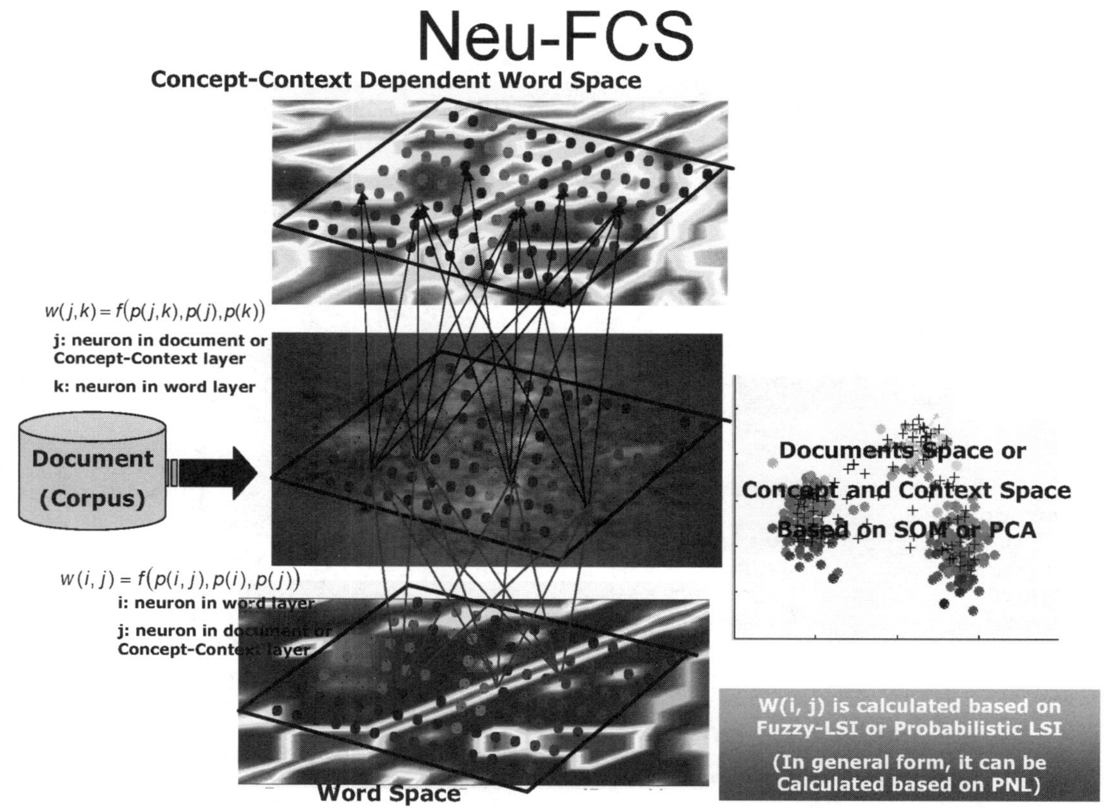

Figure 7. Neuro-Fuzzy Conceptual Search (NeuFCS)

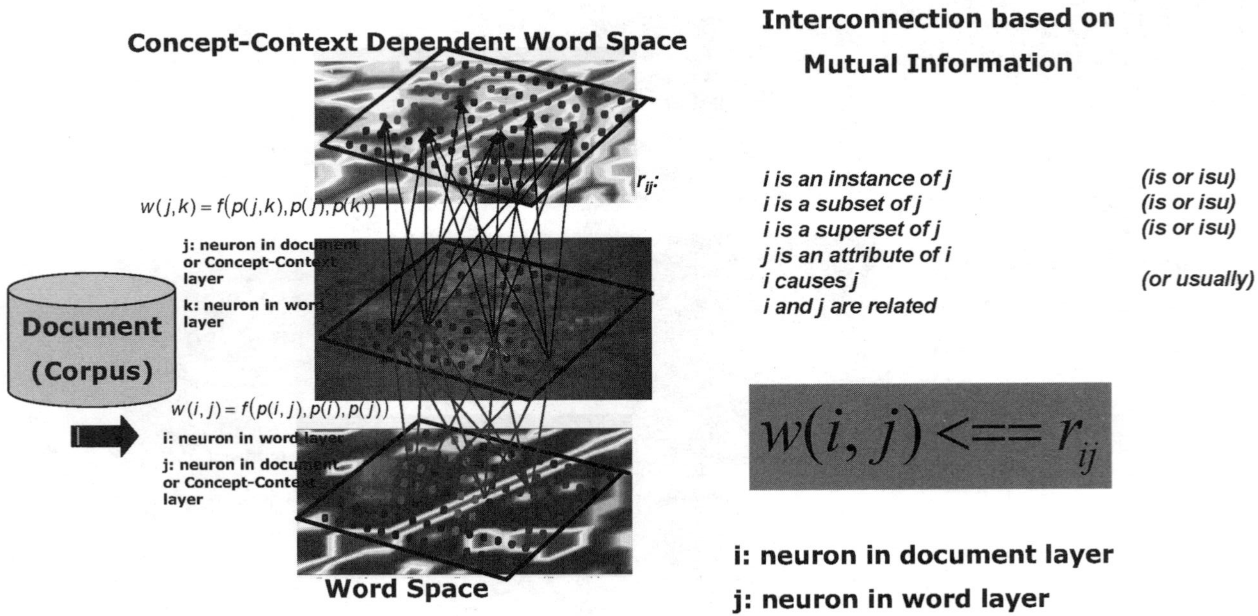

Figure 8. PNL-Based Conceptual Fuzzy Search Using Brain Science

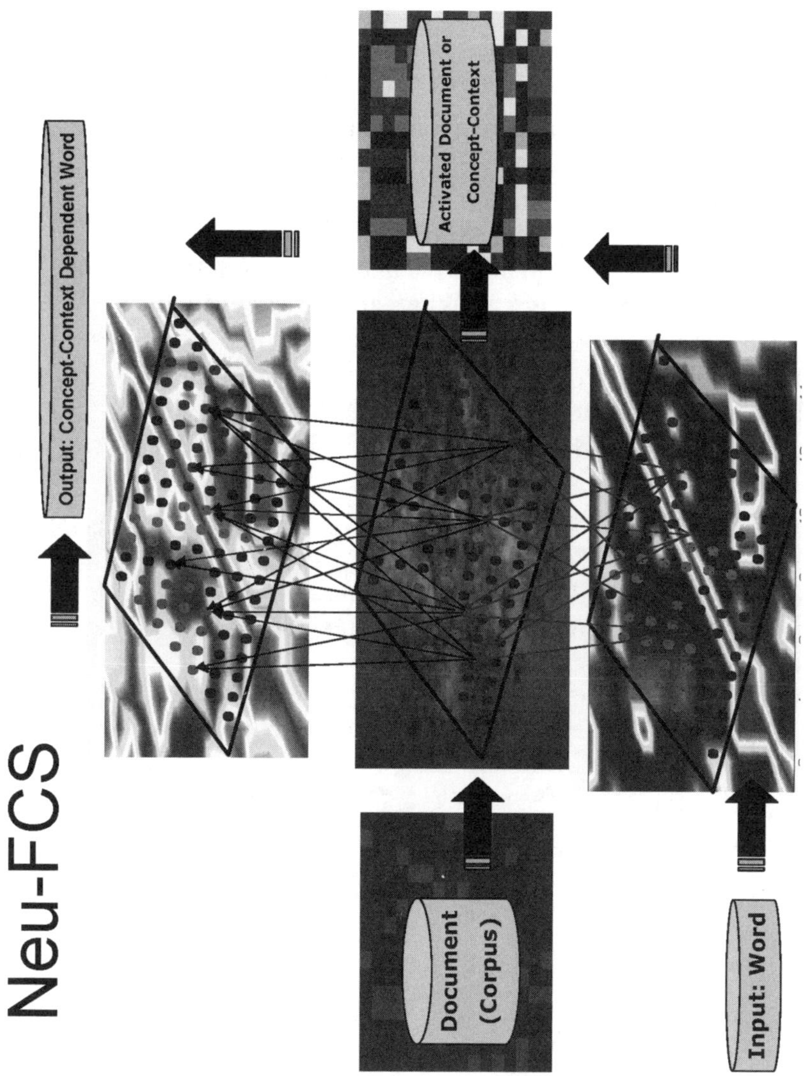

Figure 9. Typical example of Neu-FCS

3.1 BISC Decision Support System

The BISC (Berkeley Initiative in Soft Computing) Decision Support System Components include:

- Data Management
  - database(s) which contains relevant data for the decision process
- User Interface
  - users and DSS communication
- Model Management and Data Mining
  - includes software with quantitative and fuzzy_models including aggregation process, query, ranking, and fitness evaluation
- Knowledge Management and Expert System
  - model representation including linguistic formulation
- Evolutionary Kernel and Learning Process
- Data Visualization and Visual Interactive Decision Making

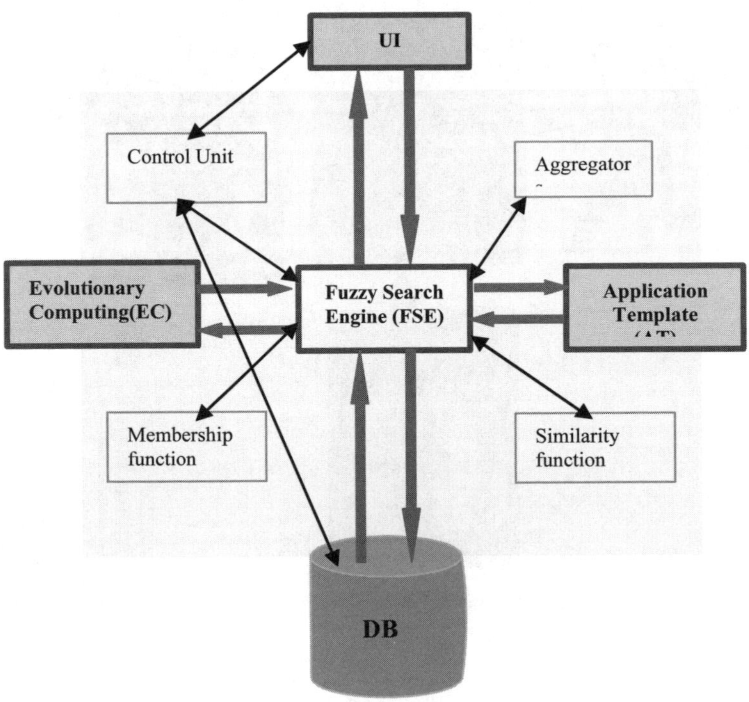

Figure 10. The BISC-DSS general framework

3.1.1 Model framework

The DSS system starts by loading the application template, which consists of various configuration files for a specific application and initializing the database for the application, before handling a user's requests, see **Figure 10.** Once the DSS system is initialized, users can enter their own profiles in the user interface or make a search with their preferences. These requests are handled by the control unit of the system. The control unit converts user input into data objects that are recognized by the DSS system. .Based on the request types, it forwards them to the appropriate modules. If the user wants to create a profile, the control unit will send the profile data directly to the database module which stores the data in the database for the application. If the user wants to query the system, the control unit will direct the user's preferences to the Fuzzy Search Engine which queries the database. The query results will be sent back to the control unit and displayed to the users.

3.1.2 Fuzzy Engine

Fuzzy Query, search and Ranking: To support generic queries, the fuzzy engine has been designed to have a tree structure. There are two types of nodes in the tree, category nodes and attribute nodes. While multiple category levels are not necessary, they are designed to allow various refinements of the query through the use of the type of aggregation of the children. Categories only act to aggregate the lower levels. The attribute nodes contain all the important information about a query. They contain the membership functions for the fuzzy comparison as well as the use of the various aggregation methods to compare two values. The attribute nodes handle the compare method slightly differently than the category nodes. There are two different ways attributes may be compared. The attribute nodes contain a list of membership functions comprising the fuzzy set. The degrees of membership for this set are passed to the similarity comparator object, which currently has a variety of different methods to calculate the similarity between the two membership vectors. In the other method, the membership vector created by having full membership to a single membership function specified in the fuzzy data object, but no membership value for the other functions. Membership function: Currently there are three membership functions implemented for the Fuzzy Engine. A generic interface has been created to allow several different types of membership functions to be added to the system. The three types of membership functions in the system are: Gaussian, Triangular and Trapezoidal. These functions have three main points, for the lower bound, upper bound and the point of maximum membership. For other functions, optional extra points may be used to define the shape (an extra point is required for the trapezoidal form).

3.1.3 Application Template

The DSS system is designed to work with different application domains. The application template is a format for any new application we build; it contains data of different categories, attributes and membership functions of that application. The application template module consists of two parts: 1) the application template data file specifies all the membership functions, attributes and categories of an application. We can consider it as a configuration data file for an application. It contains the definition of membership functions, attributes and the relationship between them; 2) The application template logic parses and caches data from the data file so that other modules in the system can have faster access to definitions of membership functions, attributes and categories. It also creates a tree data structure for the fuzzy search engine to transverse.

3.1.4 User Interface

It is difficult to design a generic user interface that suits different kind of applications for all the fields. For example, we may want to have different layouts for user interfaces for different applications. To make the DSS system generic while preserving the user friendliness of the interfaces for different applications, we developed the user interfaces into two parts.
First, we designed a specific HTML interface for each application we developed. Users can input their own profiles; make queries by specifying preferences for different attributes. Details for the DSS system are encapsulated from the HTML interface so that the HTML interface design would not be constrained by the DSS system. The second part of our user interface module is a mapping between the parameters in the HTML files and the attributes in the application template module for the application. The input mapping specifies the attribute names to which each parameter in the HTML interface corresponds.. With this input mapping, a user interface designer can use input methods and parameter names freely.

3.1.5 Database (DB)

The database module is responsible for all the transactions between the DSS system and the database. This module handles all queries or user profile creations from the Fuzzy Engine and the Control Unit respectively. For queries from the Fuzzy Search Engine, it retrieves data from the database and returns it in a data object form. Usually queries are sets of attribute values and their associated weights. The database module returns the matching records in a format that can be manipulated by the user, such as eliminating one or more record or changing their order. To create a user profile, it takes data objects from the Control Unit and stores it in the database. There are three components in the DB module.

The DB Manager is accountable for two things: 1) setting up database connections and allocating database connections to DB Accessor objects when needed. It also supplies information to the database for authentication purposes (e.g. username, password, path to the database etc); 2) The DB Accessor Factory creates DB Accessor objects for a specific application. For example, if the system is running the date matching application, DB Accessor Factory will create DB Accessor objects for the date matching application. The existence of this class serves the purpose of using a generic Fuzzy Search Engine; 3) the DB Accessor is responsible for storing and getting user profiles to and from the database. It is the component that queries the database and wrap result from the database into data objects that are recognized by our application framework.

3.1.6 Measure of Association and Fuzzy Similarity

As in crisp query and ranking, an important concept in fuzzy query and ranking applications is the measure of association or similarity between two objects in consideration. For example, in a fuzzy query application, a measure of similarity between two queries and a document, or between two documents, provides a basis for determining the optimal response from the system. In fuzzy ranking applications, a measure of similarity between a new object and a known preferred (or non-preferred) object can be used to define the relative goodness of the new object. Most of the measures of fuzzy association and similarity are simply extensions from their crisp counterparts. However, because of the use of perception based and fuzzy information, the computation in the fuzzy domain can be more powerful and more complex.

Various definitions of similarity exist in the classical, crisp domain, and many of them can be easily extended to the fuzzy domain. However, unlike in the crisp case, in the fuzzy case the similarity is defined on two fuzzy sets. Suppose we have two fuzzy sets *A* and *B* with membership functions $\mu_A(x)$ and $\mu_B(x)$, respectively. The arithmetic operators involved in the fuzzy similarity measures can be treated using their usual definitions while the union and the intersection operators need to be treated specially. It is important for these operator pairs to have the following properties: (1) conservation, (2) monotonicity, (3) commutativity, and (4) associativity. It can be verified that the triangular norm (T-norm) and triangular co-norm (T-conorm) [16-19] conform to these properties and can be applied here. A detailed survey of some commonly used T-norm and T-conorm pairs along with other aggregation operators can be find at Nikravesh and Azvine [20].

In many situations, the controlling parameters, including the similarity metric, the type of T-norm/conorm, the type of aggregation operator and associated weights, can all be specified based on the domain knowledge of a particular application. However, in some other cases, it may be difficult to specify a priori an optimal set of parameters. In those cases, various machine learning methods can be employed to automatically "discover" a suitable set of parameters using a supervised or unsupervised approach. For example, the Genetic Algorithm (GA) and DNA-based computing, as described in later sections, can be quite effective. Another important and unique component of our system is compactification algorithm or Z(n)-Compact [4, 21, 22]. Z(n)-Compact algorithm developed by Lotfi A. Zadeh [22] and it has been implemented for the first time as part of BISC-DSS for automatons multi-agents modeling as part of ONR project [4]. The algorithm has been extended and currently is part of the BISC-DSS software and it has been applied in several applications [4 and 21]. **Table 1** through **5** showed step by step how this algorithm works and how can be implemented [4, 21, 22].

3.1.7 Evolutionary Computing

In the Evolutionary Computing (EC) module of the BISC Decision Support System, our purpose is to use an evolutionary method to allow automatic adjusting of the user's preferences. These preferences can be seen as parameters of the fuzzy logic model in form of weighting of the used variables. These preferences are then represented by a weight vector and genetic algorithms will be used to fix them. In the Evolutionary Computation approach, Genetic Programming, which is an extension of Genetic Algorithms, is the closest technique to our purpose.

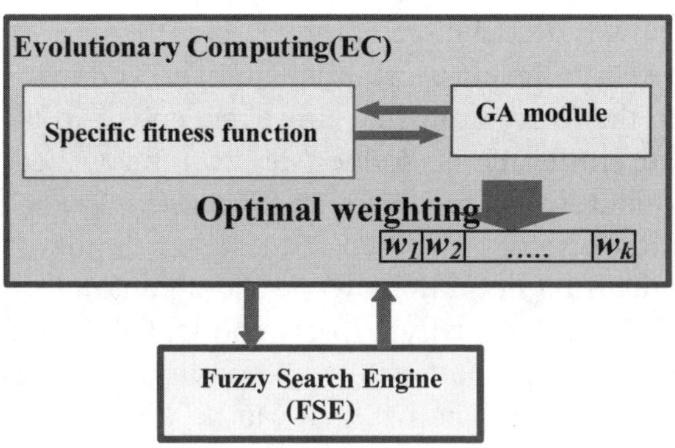

It allows us to learn a tree structure, which represents the combination of aggregators. The selection of these aggregators is included to the learning process using the Genetic Programming. In this section, we describe the GA [23] and GP [24] application to our problem [23-25].

Figure 11. Evolutionary Computing Module: preferences learning.

Our aim is learning fuzzy-DSS parameters which are the weight vectors representing the user preferences associated to the variables that have to be aggregated on the one hand, and the adequate decision tree representing the combination of the aggregation operators that have to be used on the other hand. Weight vector being a linear structure, can be represented by a binary string in which weight values are converted to binary numbers. This binary string corresponds to the individual's DNA in the GA learning process. The goal is to find the optimal weighting of the variables. A general GA module can be used by defining a specific fitness function for each application as shown in **Figure 11.**

Aggregators can be combined in the form of a tree structure which can be built using a Genetic Programming learning module (**Figure 12**). It consists in evolving a population of individuals represented by tree structures. The evolution principle remains the same as in a conventional GP module but the DNA encoding needs to be defined according to the considered problem. We propose to define an encoding for aggregation trees which is more complex than for classical trees and which is common to all considered applications. As shown in **Figure 11**, we need to define a specific encoding, in addition to the fitness function specification. Tree structures are generated randomly as in the conventional GP. But, since these trees are augmented according the properties defined above, the generation process has to be updated. So, we decided to randomly generate the number of arguments when choosing an aggregator as a node in the tree structure. And for the weights, we chose to generate them randomly for each node during its creation. Concerning the fitness function, it is based on performing the aggregation operation and the root node of the tree that has to be evaluated.

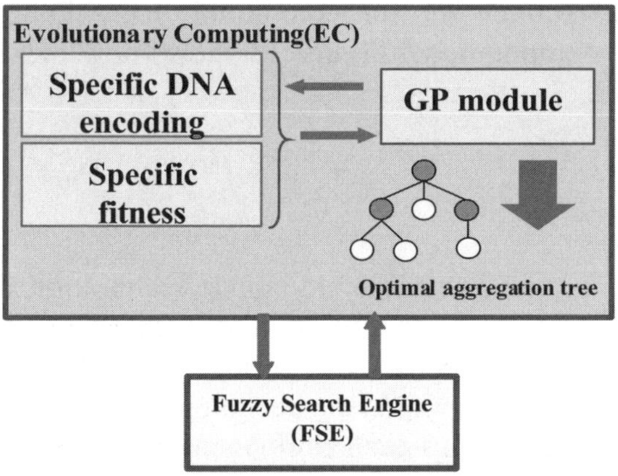

Figure 12. Evolutionary Computing Module: aggregation tree learning

For the university admissions application, the result of the root execution corresponds to the score that has to be computed for each value vector in the training data set. The fitness function, as in the GA learning of the user preferences, consists in simple or combined similarity measures. In addition, we can include to the fitness function a complementary measure that represents the individual's size which has to be minimized in order to avoid over-sized trees. We have described the use of evolutionary computation

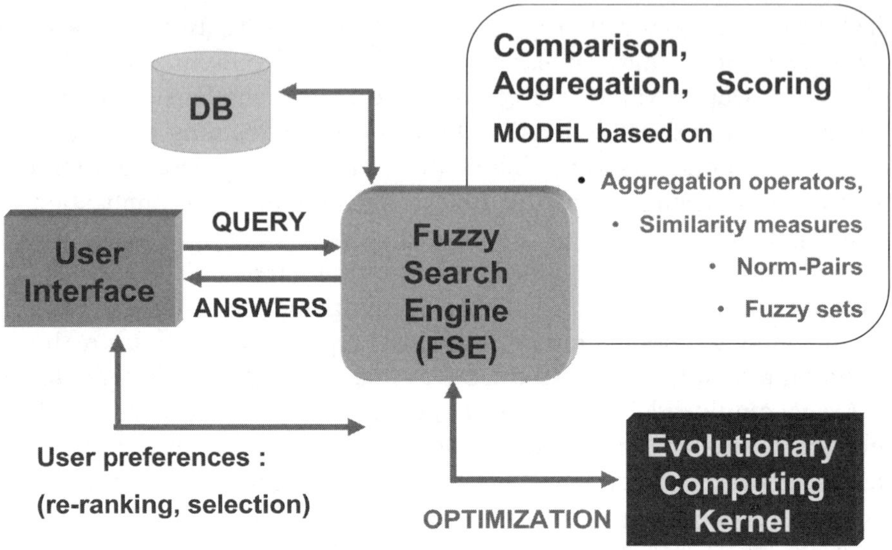

Figure 13. Interaction and Optimization

methods for optimization problems in the BISC decision support system. BISC-DSS is also a strategy and framework for combining fuzzy logic, machine learning and evolutionary computation. **Figure 13** shows how these components interact with each other.

4. Conclusions

Intelligent search engines with growing complexity and technological challenges are currently being developed. This requires new technology in terms of understanding, development, engineering design and visualization. While the technological expertise of each component becomes increasingly complex, there is a need for better integration of each component into a global model adequately capturing the imprecision and deduction capabilities. In addition, intelligent models can mine the Internet to conceptually match and rank homepages based on predefined linguistic formulations and rules defined by experts or based on a set of known homepages. The FCM model can be used as a framework for intelligent information and knowledge retrieval through conceptual matching of both text and images (here defined as "Concept"). The FCM can also be used for constructing fuzzy ontology or terms related to the context of the query and search to resolve the ambiguity. This model can be used to calculate conceptually the degree of match to the object or query.

Acknowledgements

Funding for this research was provided by the British Telecommunication (BT), the BISC Program of UC Berkeley, and Imaging and Informatics group at Lawrence Berkeley National Lab. The author would like to thanks Prof. Zadeh for his feedback, comments and allowing the authors to use the published and his unpublished documents, papers, and presentations to prepare this paper.

References

[1] L. A. Zadeh, From Computing with Numbers to Computing with Words -- From Manipulation of Measurements to Manipulation of Perceptions, IEEE Transactions on Circuits and Systems, 45, 105-119, 1999.

[2] L. A. Zadeh, "A new direction in AI: Towards a Computational Theory of Perceptions," *AI magazine*, vol. 22, pp. 73--84, 2001.

[3] L.A. Zadeh, Toward a Perception-based Theory of Probabilistic Reasoning with Imprecise Probabilities, *Journal of Statistical Planning and Inference*, 105 233–264, 2002.

[4] L. A. Zadeh and M. Nikravesh, Perception-Based Intelligent Decision Systems; Office of Naval Research, Summer 2002 Program Review, Covel Commons, University of California, Los Angeles, July 30th-August 1st, 2002. (L.A. PRUF-a meaning representation language for natural languages, Int. J. Man-Machine Studies 10, 395-460, 1978.)

[5] M. Nikravesh and B. Azvine; New Directions in Enhancing the Power of the Internet, Proc. Of the 2001 BISC Int. Workshop, University of California, Berkeley, Report: UCB/ERL M01/28, August 2001.

[6] V. Loia , M. Nikravesh, L. A. Zadeh, *Journal of Soft Computing*, Special Issue; fuzzy Logic and the Internet, Springer Verlag, Vol. 6, No. 5; August 2002.

[7] M. Nikravesh, et. al, "Enhancing the Power of the Internet", Volume 139, published in the Series Studies in Fuzziness and Soft Computing, Physica-Verlag, Springer (2004).

[8] M. Nikravesh, Fuzzy Logic and Internet: Perception Based Information Processing and Retrieval, Berkeley Initiative in Soft Computing, Report No. 2001-2-SI-BT, September 2001a.

[9] M. Nikravesh, BISC and The New Millennium, Perception-based Information Processing, Berkeley Initiative in Soft Computing, Report No. 2001-1-SI, September 2001b.

[10] M. Nikravesh, V. Loia,, and B. Azvine, Fuzzy logic and the Internet (FLINT), Internet, World Wide Web, and Search Engines, International Journal of Soft Computing-Special Issue in fuzzy logic and the Internet , 2002

[11] M. Nikravesh, Fuzzy Conceptual-Based Search Engine using Conceptual Semantic Indexing, NAFIPS-FLINT 2002, June 27-29, New Orleans, LA, USA

[12] M. Nikravesh and B. Azvin, Fuzzy Queries, Search, and Decision Support System, International Journal of Soft Computing-Special Issue in fuzzy logic and the Internet , 2002

[13] M. Nikravesh, V. Loia, and B. Azvine, Fuzzy logic and the Internet (FLINT), Internet, World Wide Web, and Search Engines, International Journal of Soft Computing-Special Issue in fuzzy logic and the Internet, 2002
[14] M. Nikravesh, Fuzzy Conceptual-Based Search Engine using Conceptual Semantic Indexing, NAFIPS-FLINT 2002, June 27-29, New Orleans, LA, USA
[15] V. Loia, M. Nikravesh and Lotfi A. Zadeh, "Fuzzy Logic and the Internet", Volume 137, published in the Series Studies in Fuzziness and Soft Computing, Physica-Verlag, Springer (2004)
[16] R. Fagin (1998) Fuzzy Queries in Multimedia Database Systems, Proc. ACM Symposium on Principles of Database Systems, pp. 1-10.
[17] R. Fagin (1999) Combining fuzzy information from multiple systems. J. Computer and System Sciences 58, pp 83-99.
[18] M. Mizumoto (1989) Pictorial Representations of Fuzzy Connectives, Part I: Cases of T-norms, T-conorms and Averaging Operators, Fuzzy Sets and Systems 31, pp. 217-242.
[19] M. Nikravesh (2001a) Perception-based information processing and re-trieval: application to user profiling, 2001 research summary, EECS, ERL, University of California, Berkeley, BT-BISC Project. (http://zadeh.cs.berkeley.edu/ & http://www.cs.berkeley.edu/~nikraves/ & http://www-bisc.cs.berkeley.edu/).
[20] M. Nikravesh and Ben Azvine (2002), Fuzzy Queries, Search, and Decision Support System, Journal of Soft Computing, Volum 6, # 5, August 2002.
[21] M. Nikravesh (2005), Evolutionary-Based Intelligent Information and Decision Systems, FuzzIEEE, May 22-25, Reno-Nevada, 2005 (Invited Talk)
[22] L.A. Zadeh, (19976), A fuzzy-algorithmic approach to the definition of complex or imprecise concepts, Int. Jour. Man-Machine Studies 8, 249-291, 1976.
[23] John H. Holland. Adaptation in Natural and Artificial Systems: An Introductory Analysis with Applications to Biology, Control and Artificial Intelligence. MIT Press, 1992. First Published by University of Michigan Press 1975.
[24] J. R. Koza, Genetic Programming : On the Programming of Computers by Means of Natural Selection, Cambridge, Mass. : MIT Press, USA 1992, 819 pages.
[25] M. Nikravesh (2001b) Credit Scoring for Billions of Financing Deci-sions, Joint 9th IFSA World Congress and 20th NAFIPS International Conference. IFSA/NAFIPS 2001 "Fuzziness and Soft Computing in the New Millenium", Vancouver, Canada, July 25-28, 2001.

E-Service Composition Tools from a Lifecycle Perspective

Wei Liu[1], Husniza Husni[1]*, and Lin Padgham[2]

[1] School of Computer Science & Software Engineering, the University of Western Australia, Australia wei@csse.uwa.edu.au

[2] School of Computer Science & Information Technology, Royal Melbourne Institute of Technology (RMIT), Australia linpa@cs.rmit.edu.au

Abstract. This chapter provides not only an overview of the current standard development in service composition, but also practical guidance on selecting and developing service composition tools. *Service-oriented programming* has come to be perceived as the key technology to leverage enterprise collaboration, moving from data and information sharing to business process integration. Most tools available today support *semi-automatic service composition*, while some are working towards fully *automated service composition*. This chapter takes a unique lifecycle view on evaluating Web service composition tools. It first identifies the desired properties at each stage of a composite service's lifecyle, then apply the criteria on evaluating both semi-automatic and automatic service composition tools. Current work on semi-automatic service composition mostly focuses on usability and is often process-oriented. The main task in this paradigm is connecting the input and output of available service components. On the other hand, automated service composition tools concentrate on the state of the world and the state transitions in the domain providing a list of actions or operators. Composition planners are used to automatically generate a list of applicable composition plans for flexible service composition. Consequently, tools supporting both usability and flexibility are scarce. Also, the integration with various verification and monitoring mechanisms to ensure failure recovery with secure and robust execution is not realised in any of the reviewed service composition tools.

1 Introduction to Web Services

The enormous opportunities offered by the Web and Internet Technologies are fueling a vision of changing today's document centric Web into an interoperable service backbone. Since Berners Lee et. al. [3] unleashed this semantic Web vision in 2001, *service-oriented programming* has come to be perceived as the key technology to leverage enterprise collaboration, moving from data and information sharing to business process integration.

Web service technology is built on standards-based technologies enabled by eXtensible Markup Language (XML). It leverages existing business processes

W. Liu et al.: *E-Service Composition Tools from a Lifecycle Perspective*, Studies in Computational Intelligence (SCI) **37**, 125–146 (2007)
www.springerlink.com

creation from *tightly coupled component-based* models, to *loosely coupled self-describing service-oriented* architectures. The main benefit of mainstream adoption of Web services is that business processes are no longer limited to a particular trust boundary (i.e. within an enterprise) but can span *across* organisational boundaries [18]. In other words, Web services are distributed autonomous software components designed by different vendors to provide certain business functionalities to *other* applications through an Internet connection. In particular, Web services are designed to be used by other software programs. Unlike humans, programs do not have any cognitive power in understanding various descriptions of a programming interface such as Java API or C++ Foundation Library. To enable programs to automatically discover desired business process components, integrate them meaningfully, invoke and coordinate them sensibly to provide the user desired output is the ultimate challenge faced by Web service researchers. This process of integrating existing Web services to achieve higher-level business tasks that cannot be fulfilled by any individual service alone is often referred to as *Web service composition*. The business process so achieved can also be exposed as a Web service, which we will term a *composite service*. The constituent services are termed *component services*.

The research into services-oriented programming tends to be roughly divided into two camps. The *industry approach* is focused primarily on converting existing business applications into Web services. For example, large online companies such as Amazon and Google released the Web service versions of their process interface so that potential vendors or service distributors can easily integrate Amazon and Google web services with their own Web applications. On the other hand, *academic research* has focused on endowing meaning or semantics to service descriptions so that automatic *discovery*, *integration*, *reasoning* and *verification* become possible with little or no human intervention. *Semantic annotations* using ontologies (domain ontology as well as service ontology), *automatic composition plan generation* and *fault-tolerant service integration* are major research topics in the academic research community. Much of the research is concerned with transferring existing research advances in knowledge representation, theorem proving and planning in particular, as well as agent technologies, into a service-oriented environment. Hereafter, we will use Web services to refer to the services that follow industry standards, and Semantic Web services to refer to those that follow semantic Web standards.

Many available publications compare and contrast the two camps of work [14, 18, 21, 24], which we will not duplicate here. Rather we will try to provide a unified view on how to evaluate existing tools and technologies according a set of identified requirements at each stage of the service composition lifecycle.

This chapter will first briefly revisit the current development of Web services in terms of both standards and technologies. Then we will focus on investigating the requirements of service composition and proposing evaluation criteria at each stage during the lifecycle of a composite service. We then explore current solutions to service composition, with a focus on the theoretical aspect of how to automatically generate composition plans. The lifecycle based evaluation strategy we proposed in this chapter is then applied on evaluating two existing tools. A small case study is used to demonstrate how *semi-automatic* and AI planning supported *automatic* service composition are achieved. The chapter concludes with an outlook to future directions of service composition tools.

2 Web Services and Service Composition

Interoperability between Web services are achieved through a stack of standards, as shown in Figure 1. Above the protocol layer, typical for any Internet applications, are the XML enabled layers for standardising message exchange (Simple Object Access Protocol - SOAP [9]), service interface description (Web Service Description language - WSDL [6]), service discovery and binding (Unified Service Description, Discovery and Integration - UDDI [15]), and service composition or business process integration (BPEL4WS [1], XLANG or WSFL [11]).

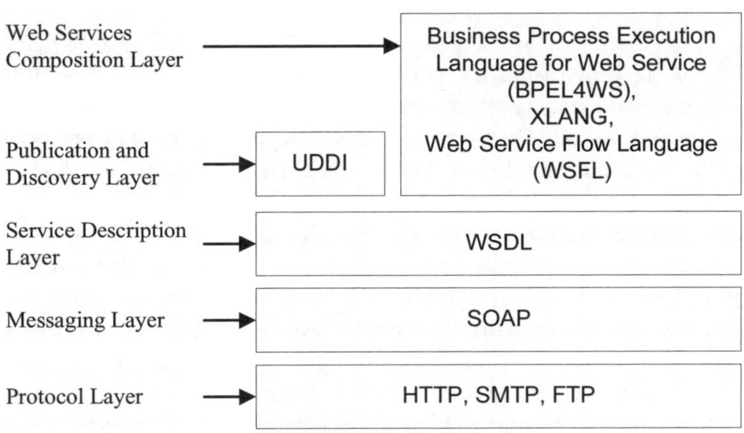

Fig. 1. Web Service Standards Stack

Current Web service architecture comprises three main entities - the *service registry*, the *service provider*, and the *service requester* as shown in Figure 2. Service requesters (also known as clients) are software programs or applications such as autonomous software agents that delegate tasks to others. A service provider is a program that can provide certain business functionality for the benefit of other software entities, either internal or external to the hosting environment. The service registry provides yellow pages, white pages and green pages facilities that enable the *binding* between providers and requesters possible. Binding is the key aspect that differentiates a service-oriented architecture from object-oriented or procedural-based systems [7]. Object-oriented or procedural-based systems rely upon type compatibility for matching and binding. Service compatibility, on the other hand, depends on a richer model that involves message structure (WSDL) and machine processable policy (WS-Policy) assertions for capabilities and requirement matching.

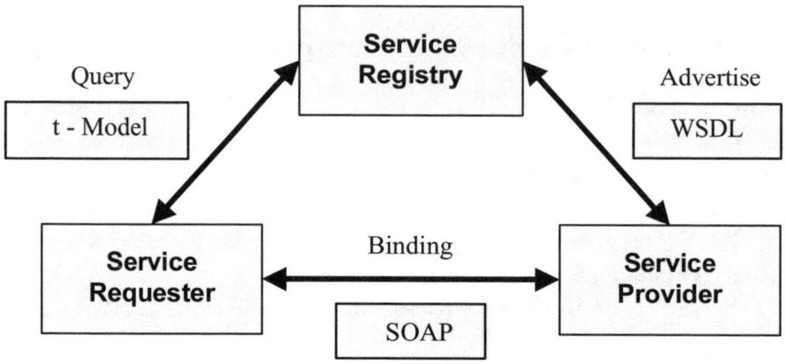

Fig. 2. Web Service Architecture [8]

By following the standards, computer programs (a service requester) can request the service registry for services that match its functional and more recently non-functional (Quality of Service) requirements. The service registry will return the binding address (also called the *end-point*) of the matching service program (a service provider). Then, the service requester can make the binding to the particular service provider and invoke the desired service. Moreover, the service requester may have a complex task to achieve that requires a team of service providers, each carrying out some subtasks. The subtasks are typically coordinated by the service requester, who can expose the coordinated activities as composite Web services. In other words, with standardised support infrastructure, Web services developed using different programming languages and deployed on different platforms are expected to be seamlessly inter-operable and composable.

3 The Lifecycle of a Composite Service

With the guarantee of interoperability, a composite service integrates and coordinates individual Web services in a similar way to any business process within a trusted organisational boundary. Therefore, the full lifecycle of a business process [18], including *modeling*, *executing*, *monitoring*, *management* and *optimisation*, can be used as the basis for the lifecycle of a composite service. Many emerging standards are available for supporting service compositions, including BPEL4WS and its predecessors (XLANG and WSFL), OWL-S and WSCI etc. In this section, we will compare these standards at various stages of a composite service's lifecycle and see how composition tools should support the standards in order to claim compliance. In the section that follows, we will investigate the service composition tools available and how they support the properties we discuss here.

3.1 Modeling and Design

Typical examples and case studies in the service composition domain revolve around simple examples, such as travel and tourism package assembly [16], or federated order processing and vendor management inventory systems for car dealers [7] or computer companies [18]. Given the problem of working out a tourism package, assembling a car or building a computer, the first task is to decompose the process into smaller subtasks. Process decomposition will require either a list of primitive tasks *known a priori* or support of *dynamic discovery* of primitive tasks. A process may be decomposed in many ways, for example, assembling a computer may require motherboards, memory chips, but may or may not require a sound card depending on whether there is an onboard sound card or not. Therefore, without sophisticated domain ontologies and reasoning support, such processes cannot be built automatically.

The goal of modeling and design in service composition is to deliver a correct model of composition and describe it in a machine processable language. We consider that modeling and design can be divided further into the *generation* and the *specification* stage. Depending on the assumption of the level of semantic and knowledge representation support, the model can be generated either *automatically* or *semi-automatically*. The specification of the generated composition plan can either follow certain standards such as BPEL4WS or OWL-S, or can be specified in the native syntax of declarative languages such as Prolog or LISP, which are used by many AI planners. In some other cases, formal specification in first-order logic and linear logic or protocol verification languages like Finite State Machine and Petri-Net are also used [14] for the verification of composition correctness.

Specifying the composition plan using standards will allow *recursive service composition*, where the composite service plan so defined or generated, can be reused by other service clients. In this way the specification language supports the reusability of the composite service. Web service composition is intended to be used for real-time business collaborations, which can be very complex. Therefore, the expressiveness of the specification language is of crucial importance. This should include essential control flow constructs: sequential, parallel, conditional branches, synchronisation points, nesting, recursion, iteration, dynamic late binding and reflection constructs [17]. Ideally, it should also be able to provide separate language support for *abstract* specification and *executable* description.

In addition to reusability, the correctness of the model generated also needs to be ensured during the modeling and design phase. Correctness and validation of the model are often achieved by AI knowledge representation techniques, for example, model checking, protocol verification using Petri-Nets or theorem proving (π-calculus and linear logic [20]).

The usability of the composition tool is another important dimension during evaluation. Managing XML based syntax can be quite distracting when trying to achieve the main goal of designing a composite service. Graphical support therefore is important to facilitate the design process and to allow the programmer to concentrate on the data and the control flow, rather than the syntactical details of a language. Also, a user friendly design interface will support users with less confidence in service-oriented programming and will maximise the accessibility of Web service technology. Intuitive graphical or visual tools are important in both automatic and semi-automatic generation of composite services.

3.2 Deployment and Mediation

Deployment is the process of mapping an abstract service composition plan into actual end-point bindings between the composite service and a set of selected service providers. Complexity increases as more Web services with similar functionalities become available. Instead of returning one functional compatible service, the service registry may return many. The composite service is then responsible for making selections among the applicable service providers. For example, composite service specification standards such as BPEL4WS support both abstract and executable business processes. Abstract processes are modeled as business protocols in BPEL4WS and are not executable [18]. Tools complying with BPEL4WS then should cater for the separation of abstract business protocols from executable bound instances of business interactions. During this phase, *service selection* becomes an important supplement to *service discovery*. Service discovery deals with finding the services which match the functional requirements. Service selection takes the functionally compatible providers and filters out one service provider that best matches other criteria such as Quality of Service (QoS) requirements as described in [26].

Different data representations or incompatible interaction styles are possibly to be expected in service composition. So during deployment, *mediation* services are also expected to resolve the data incompatibility and ensure effective and correct communication.

3.3 Execution

Executing the composite service involves generating client stubs that exchange messages with the remote service end points using an agreed protocol, e.g. SOAP, HTTP Get or HTTP Post. Interactions based on these protocols are simple and stateless. Therefore, a desired property during execution of a composed service is *persistence*. The runtime environment should support intermediate data storage either in the main memory (for fast retrieval of state data) or in persistent storage (for failure recovery in the *monitoring and management* stage).

For large complex composite services that may involve hundreds and thousands of tasks and sub-processes, *scalability* is another important factor for consideration during execution. Support for asynchronous communication should be provided so that sub-processes can be run in parallel, increasing the handling capacity of the composite service.

3.4 Monitoring and Management

Composite services are susceptible to failure because they depend on potentially unreliable services external to the enterprise boundary. Unlike traditional component-based services, composite services cannot assume robustness or failure handling of the components. Therefore, it is quite important for the runtime environment of a service composition tool to monitor the execution of component services, detect and report failures. In addition to failure monitoring and detection, it should also provide mechanisms to recover effectively and efficiently from the failed component service or sub-processes. Transactional attitudes (WSTx) [13] and exception handling in coordinated atomic actions (WSCA) [25] are two proposed standards for specifying expected behaviors for failure recovery.

Transactional attitudes build on the traditional distributed databases' ACID requirement. Because of the persistence between component service invocations, a composite service should return to its previous consistent state if any of the components fail. This will require intermediate states to be logged and resources to be locked until the component operations are committed. For certain actions, compensation operations should be provided to cancel the corresponding operation's effect. For example, in a hotel room booking scenario, to be able to roll back to its original state, a `bookRoom` action should be compensated by a `cancelBooking` operation. Therefore, the composite service tool needs to locate the compensation operation in face of failure.

Coordinated atomic action takes a forward recovery approach, viewing the composite service as the coordinator of the entire process. In the face of failure, the component service should notify the coordinator of failure by throwing exceptions. The coordinator then copes with the failure by coordinating the behaviors of all component services. Standards such as SOAP, WSDL and BPEL4WS all have language constructs for specifying faults and implementing fault handlers. So if a service composition tool claims compliance, it should also cater for exception handling.

Another way of recovering from failure is to take advantage of the intrinsic redundancy of a service-oriented environment. Often services of similar functionality will be available, so *dynamic role filling* at runtime by selecting another compatible service to substitute the failed service is another remedy in combating failure and building robust composite services.

3.5 Optimisation

The service composition process ideally should be built following a spiral model as an incremental process. It is not realistic to know a priori the invocation interface of all available services given the openness of a service oriented environment. Services can join and leave a service registry at any time. The dynamics of an incremental process cannot be optimised at compile time, but after the execution and in order to reuse the composite service, the process can be optimised with better control flow and data flow or reconsideration of the constituent service providers. To reconsider service providers, the composition tool should support tracking of the QoS parameters, such as response time, reliability and capacity of the service. A knowledge base on the QoS of the past services consumed can then be built or other service requesters can be consulted for recommendations.

3.6 Design Requirements of Composite Services

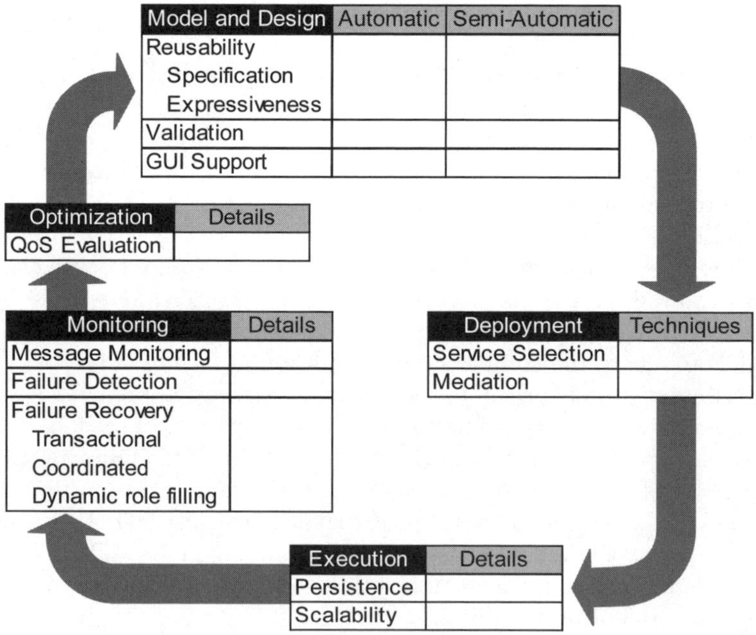

Fig. 3. Requirements at each stage of the composite service lifecycle

Figure 3 summarises the requirements as discussed above, this can be used to evaluate a service composition tool at each stage of the composite service development lifecycle. In the sections that follow, we will apply these evaluation criteria primarily on two representative composition tools, one for automatic composite service generation (SHOP2), the other for semi-automatic service generation (JOpera). Other composition tools will be discussed in lesser detail.

4 Evaluations of Service Composition Tools

4.1 Case Study - Planning a Holiday in Australia

Consider a scenario where Jimmy, a tourist from Malaysia, wanted to go on a holiday in a city or suburb in Australia that has some special attractions. He would like everything planned before he arrived at his destination. This can be done online by accessing Web services offered by tourism industries in Australia. First of all, Jimmy has to collect all the relevant information such as the accommodation, transportation, and places that he would like to visit. With a tight schedule, he wants to maximise his time by staying at a hotel close to the attractions and find a car rental service nearby. However, he has no time to collect such information from the Web manually and he knows he needs something to aid his trip planning. Hence, it is best to automate the search and information collection as well as planning of what must be done before others. As mentioned above, Jimmy needs more than one type of services– `accommodation`, `transportation` (hire), and `places of attractions`.

Here we will use this simple example to demonstrate how the two different composition tools support Jimmy to plan his holiday. The Web services used here are provided by the Australian Tourism Data Warehouse (ATDW)[3]. The ATDW web services provide generic access to accommodation, attractions, car rental services and information of tourism sites etc. To make service composition possible, as one of the service distributors of ATDW, we have fine tuned the generic WSDL file into many separate WSDL files according to the product category. For example, the accommodation, attraction and hire services connecting to ATDW's Web services are at:

```
http://perth-agencity.csse.uwa.edu.au/servlet/middleware/AttractionQuery?WSDL
http://perth-agencity.csse.uwa.edu.au/servlet/middleware/AccommodationQuery?WSDL
http://perth-agencity.csse.uwa.edu.au/servlet/middleware/HireQuery?WSDL
```

4.2 Semi-automatic Service Composition with JOpera

JOpera is a visual tool that supports specification of service composition using both data flow and control flow. Figure 4 and Figure 5 are screen shots of the

[3] `http://www.adtw.com.au/`

control flow and the data flow of Jimmy's composite service in JOpera as a plug-in for Eclipse[4], respectively.

In the Modeling and Design phase, JOpera for Eclipse provides a graphical interface that is easy to use yet powerful in defining and describing the composition process. Processes in JOpera interconnect many different types of software components, including Web Services, UNIX applications, Windows applications, Javascripts and others. In particular, Web services can be *imported* as external processes by simply providing the URLs that point to the WSDL files. Once imported, the services will be available in the JOpera process and program library. For example, as shown in Figure 4 and Figure 5, Jimmy can use the data flow and the control flow diagram to semi-automatically specify the connection among processes. The connection is done by linking the output of one process with the input of another process.

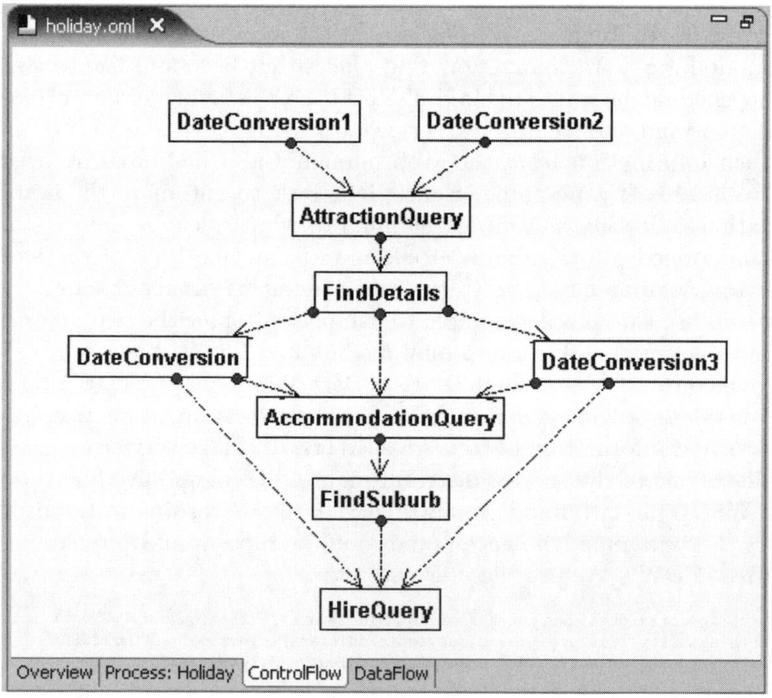

Fig. 4. Control Flow of the Sample Holiday Process

The visual flow diagrams for process specification are automatically translated into JOpera's visual composition language and stored in a file. In terms

[4] Eclipse is an open architecture platform that supports the integrations of web and application development tools as plug-ins. It is platform neutral and available for free download at `http://www.eclipse.org/`

of expressiveness, although JOpera's visual composition language is comparable to BPEL4WS, the process specified in JOpera is not exposed as WSDL for wider compatibility. The defined process can be re-used in the JOpera development environment, but with no direct support on exporting the process to comply with current Web service standards. Simple validation is supported through the user interface so that non-compatible data fields cannot be connected. Moreover, loops with race conditions are also considered and resolved with extra `wait` slots when race conditions occur.

In the Deployment phase, *mediation programs* are readily supported. The user is given a wide range of choice for integrating other software components such as Java, JavaScript code and etc. to enable the component services interconnection. In the example case, Jimmy may prefer to enter `date` in `MM-DD-YYYY` format as a string, Java code can be easily integrated for converting the user preferred date format into separate data fields (`day`, `month` and `year`) as required by the Web services. For example, the `DateConversion` program is to convert date format, and the `FindDetails` and `FindSuburb` programs are to parse the returned XML string for useful information. They are the mediators between component services. However, there is no direct support for dynamic service selection. Services and their URL addresses are known a priori.

In the Execution phase, JOpera compiles the process written in the JOpera visual composition language to executable Java code. Then the JOpera execution kernel runs the code automatically and communicates in SOAP with the external Web services. Pautasso and Alonso [17] detailed testing results on JOpera's scalability capability by replicating the JOpera execution kernel. JOpera also provides system variables to ensure the transactional aspect - the persistence of state variables in between component service invocation.

In the Monitoring phase, Jimmy can watch the process being executed and intercept the values of the state variables. However, there are no automated mechanisms for detecting possible failures or ensuring recovery from failure. Some code optimisation can be done during the Modeling and Design phase because JOpera supports refactoring. There are also flow patterns available as references to efficient and effective process design.

In summary, JOpera meets the design goal of being flexible and user friendly for rapid service composition. However, the composition flow is fixed at compile time. For example, the control flow in Figure 4 is only applicable when Jimmy wanted to find attractions first and use attraction details to narrow down the booking of accommodation and car hiring. However, if Jimmy is certain about the destination and would like to explore the attractions in the surrounding areas, then make a decision on car booking, a new control flow and data flow diagram need to be specified. Therefore, although semi-automatic generation of a composition plan is more practical as it incorporates human intelligence into service selection and plan generation, there is still a desire for fully automated composite plan generation.

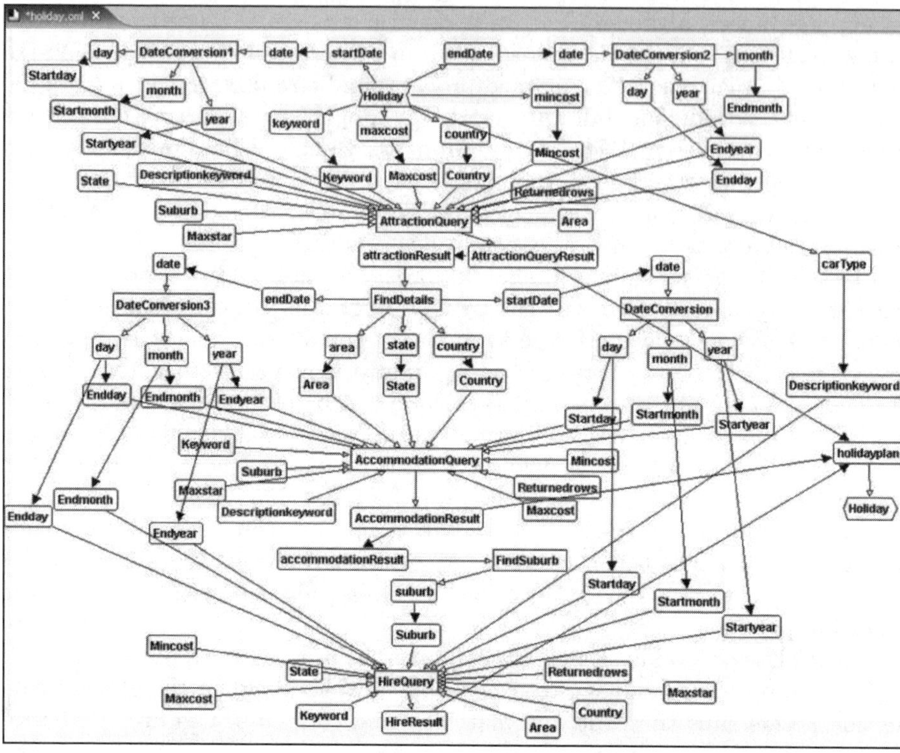

Fig. 5. Data Flow of the Sample Holiday Process

4.3 Automatic Service Composition using HTN Planning

An AI planning problem is viewed as generating a sequence of operators that will transform the current state of the environment into a goal state. Russel and Norvig [22] defined the problem of planning as one type of problem solving where an agent uses its beliefs about available actions and potential outcomes before it can identify a solution from an abstract view of possible plans. What makes AI Planning potentially possible for Web service composition is the description of Web services in Web Ontology Language for Services (OWL-S) [2], which describes the services in a machine readable manner. OWL-S is the descendant of DARPA Agent Markup Language for Services (DAML-S) [10]. It provides language constructs for describing service invocation interfaces as well as specifying the control flow between services. In other words, OWL-S fulfills a similar purpose to both WSDL and BPEL4WS (or WSFL) in normal Web service standards. Moreover, in addition to the functional input and output description, because of its formal syntax, OWL-S allows for specification of pre-conditions and post-conditions of services, and the relationship between services (e.g. service A *is equivalent to* service B). Therefore, with formal syntax, rich semantics and less ambiguity, OWL-S described services are more suitable for automatic composition using reasoning tools.

Hierarchical Task reduction Network (HTN) planners are a particular kind of AI planners, which use descriptions of abstract plans for tasks in a particular domain, to guide the search for a specific concrete plan. Work by Sirin et al [23] has explored using SHOP2, a popular HTN planner as a tool for web service composition, using semantically rich service descriptions in OWL-S.

HTN planning generates plans automatically by task decomposition (dividing problems into subtasks). Using schema reduction, a complex task can be reduced to a set of primitive operators. The domain knowledge provided in the abstract task descriptions is used to guide the search for the appropriate primitive operators. The primitive operators can be thought of as atomic services that cannot be decomposed any further. An operator is an action that describes how to achieve a primitive task. A *method* contains ways to decompose some tasks into partially ordered subtasks.

Given domain knowledge about the set of available *primitive operators* and the *reduction schema* (i.e. methods), plans are generated based on the order of task execution. Keeping track of the add list and delete list after each action, the planner is always aware of the expected state of the world during the planning process. SHOP2's knowledge base consists of operators, methods, and non-action items such as related facts and axioms. A planning problem for SHOP2 is given by a triple `(S, T, D)` where `S` is the initial state, `T` is a task list, and `D` is the domain description. SHOP2 takes this as its input and returns a plan, `P = (`p_1 p_2 p_3 `...` p_n`)`. `P` consists of a sequence of operators that will take the initial state `S` and perform the task list `T` in domain `D`.

The process of service composition using SHOP2 involves translating the OWL-S Web services descriptions into the SHOP2 planning domain (`D`) form. This requires a number of processes to be carried out as explained in detail in [23]. The SHOP2 planning process and the OWL-S description parsing and generation process are not tightly coupled. SHOP2 as a planner takes the input operators generated by parsing OWL-S service descriptions and produces composition plans that are in turn translated back into OWL-S descriptions. However, OWL-S specified services are not as widely available as those described in WSDL. As shown in Figure 6, SHOP2 can be used as a plan generator independent of the service description language. It takes a domain and a problem definition file and generates a set of plans potentially ordered according to plan costs. Therefore, rather than relying upon service descriptions in OWL-S like [23] did, we built necessary parsers as shown in Figure 7 to integrate the SHOP2 planing process with services described in WSDL. Now let's return to our tourist Jimmy's trip planning and consider composing the set of Web services provided by ATDW using SHOP2 as an independent plan generator.

In order to use SHOP2 to plan his holiday, Jimmy needs to manually create the domain and the problem definition file because of the lack of clear semantics in WSDL. In other words, the WSDL operations need to be trans-

Fig. 6. Planning Process in SHOP2

lated manually to SHOP2 operators with specified pre-conditions and post-conditions, as well as the add and the delete list to record state transitions before and after the action. This information is specified in the domain definition file. In the problem definition file, Jimmy should provide the initial state of the world, such as preferred city name, vehicle type, accommodation type etc.

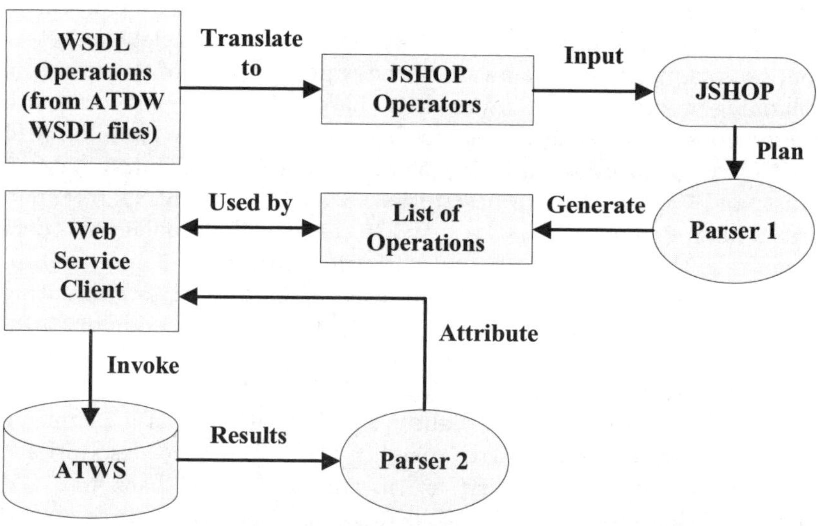

Fig. 7. Service Composition Framework using SHOP2

Figure 7 illustrates the composition process using an architectural overview of the conceptual components in the system. Figure 8 shows screenshots of the actual implementation of the Jimmy holiday case study. After feeding these two files into the SHOP2 planner (we used the Java version of SHOP2 - JSHOP), a set of applicable plans will be generated. Then a parser (`Parser 1` in Figure 7) converts the plan sets into a set of composition plans each corresponding to an ordered sequence of Web service invocations. The returned results from a service call may also need to be parsed (`Parser 2` in Figure 7) to get the result of a desired field. This is necessary when the Web service returns a single string as output containing structured information in XML format.

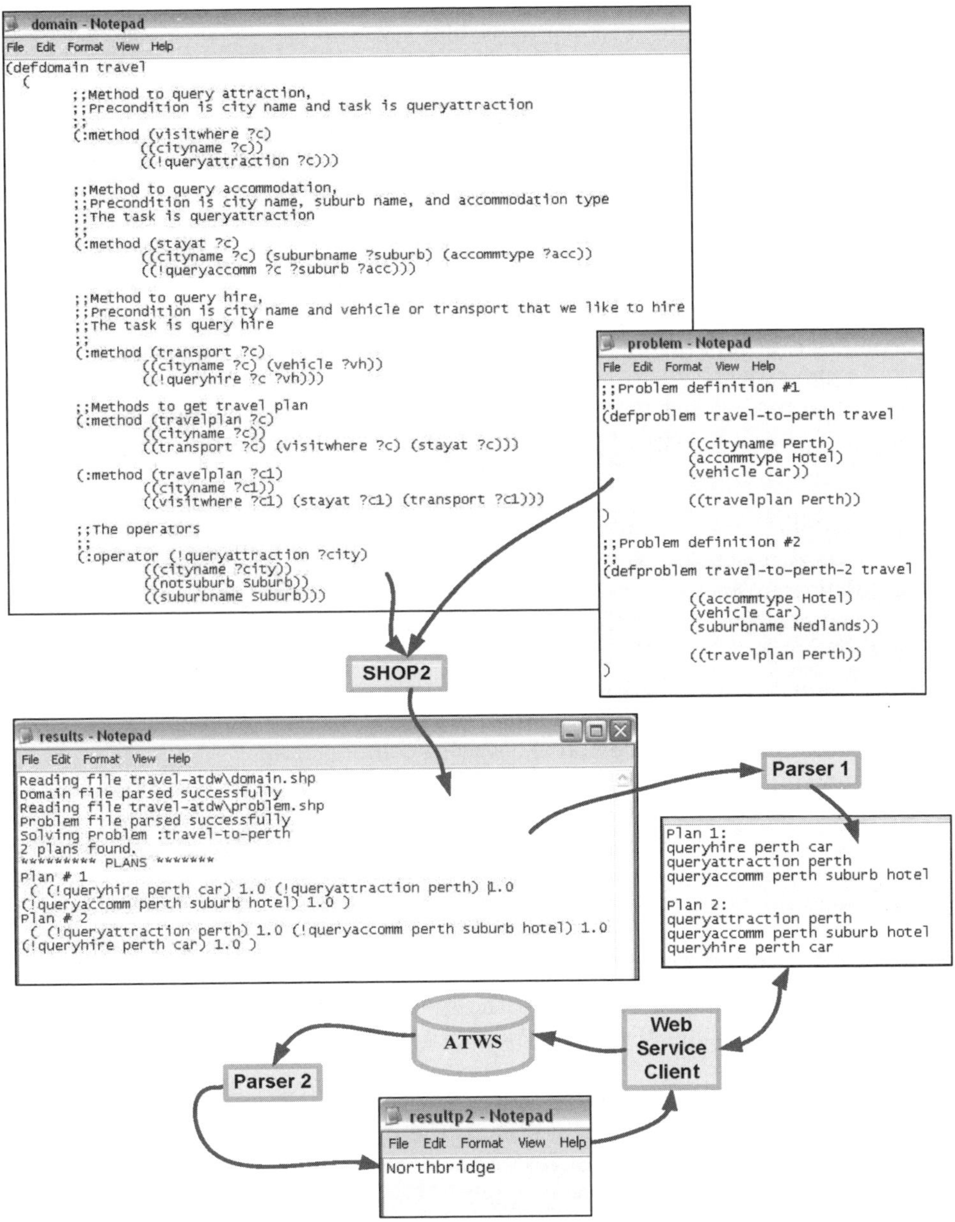

Fig. 8. Screenshots of an example service composition in SHOP2

One of the major benefits of using an AI planner is that alternative plans are automatically generated. The user can select the most suitable plan for execution. In case of failure, alternative composition plans can be tried out to recover from the failed process. Unlike JOpera, the composition process in SHOP2 is not confined to one particular scenario.

Taking a lifecycle view, an HTN planner such as SHOP2 only provides support in the design and modeling phase. Extra modules, such as mediation programs and Web service invocation need to be programmed to execute and monitor the composite service. Limited GUI support for editing the definition files is provided in the Modeling and Design phase. Consequently, SHOP2 (or JSHOP) is not a comprehensive tool that supports the entire lifecycle of service composition. However, through our experience with SHOP2, with sufficient ontology support to reduce the ambiguity in WSDL, it should be possible to incorporate SHOP2 as a plug-in of a comprehensive semi-automatic service composition tool (e.g. JOpera) for automated composition plan generation.

This exercise of using an HTN planner, SHOP2, for service composition also highlighted the following drawbacks of using AI planning for composition plan generation, which is in agreement with Srivastava and Koehler's observation in [24]:

- Inability to generate plans of complex control structure. Service composition involves complex control structures such as loops, choices, and non-determinism to cater for more broad and complex plans for Web service composition. So far, only planning as a model checking approach has been able to provide initial solutions to the generations of such complex plans. Planners like SHOP2 do not support the generation of parallel execution of service components, which requires asynchronous communication between the composite service and the service components.
- Inability to reuse the generated composite service. The service composition problem cannot be expected to take place at the level of primitive actions and control structures. It requires to take complex plans as building blocks and synthesise multi-partner interactions from them.
- Inability to support complex object representation. Web services rely on messaging to exchange objects, whereas AI planning concentrates on state transitions. Although some translation can be made, fully automated processing becomes difficult without extra-logical specification on how the objects exchanged between messages can affect the current state of the world.
- Inability to deal with incremental or incomplete domain information. In a dynamic open environment such as the Internet, services are autonomous. They can be available or disappear without notification. Therefore, it is impossible to create a list of available operators to facilitate planning regardless of semantic expressiveness of the service description language.

Composite service development should be viewed as a continuous process of synthesising, executing, optimising, and maintaining complex workflows rather than seeing it as one shot plan synthesis problem given explicit goal states. Srivastava and Koehler [24] suggested areas that need to be tackled before AI planning can be successfully applied to Web service composition. These include storage and retrieval of plans, plan analysis and optimisation, as well as plan execution and monitoring.

5 Other Tools for Intelligent Service Composition

Many other service composition tools are available. Some are based on business workflow management, such as Hewlett Packard's eFlow [4] and the Service Composition and Execution Tool (SCET) [5]. Some are based on the concept of service components - ServicCom [16]. Some are based on various AI techniques, such as the rule-based composition tool SWORD [19]. There also exist various suggestions of techniques for automatic composition generation and verification with minimal prototypical implementation. Most of these tools are not available for download like JOpera and SHOP2. Therefore, we will just provide an overview based on available literature discussions.

5.1 Web Service Composition Tools using Workflow

According to Rao and Su [21], Web service composition as workflow can be either static or dynamic. Static Web service composition requires the service requester to build a process model consisting of tasks and data dependencies before executing the composite service. Hence, component services are chosen at design time. A process model can be specified in a flow diagram like what JOpera provides. Dynamic service composition, however, generates the process model and selects the services automatically at run-time. The service requester just specifies some constraints including the dependency of atomic services and the user preferences.

SCET [5] is a static composition tool that composes Web services and stores them as WSFL based specifications. SCET consists of four components - the process designer, simulation model generator, Perl execution code generator, and an execution monitor using Java RMI server. The process designer designs the layout of the process structure and provides information about activities and links used in the process. An activity node stores information about the Web services that implements it including the services' WSDL specifications. The links are divided into two types - the control links and the data links. The control links model the control flow whilst the data links model how the output of one activity is linked to the input of another activity. The Perl execution code is automatically generated from the WSFL based specifications for easier execution of composed Web services. The separation of data flow from control flow is very similar to JOpera and indeed SCET depends on a human designer for the composition process so it is a semi-automatic composition tool.

Another example tool based on workflow management techniques is eFlow. The focus of eFlow is on providing dynamic and adaptive composition in an open environment. A composite service is described as a process schema that consists of other service nodes. This composite service is modeled as a graph of flow structure specifying the execution order of the service nodes. A service node represents the invocation of a basic or composite service. A service node specification includes the definition of data that the node has access to (i.e. read or modify), and the description of the service to be invoked.

In Figure 9, the rounded boxes represent invocations of single or composite services. The black circle represent the starting and ending point of a process. The horizontal lines indicate parallel invocation of asynchronous Web services and ensure synchronisation after service executions. The definition of a service node contains a search information that is used to query the service. This gives the dynamic feature of eFlow which allows it to discover Web services dynamically through plug-in *service brokers*. In particular, it supports the concept of *multi-service* node to dynamically invoke a number of instances of the same service, and *general service* node to allow dynamic bindings from a list of different services. eFlow also allows two ad-hoc changes – process schema modifications and process instance state modifications. It supports the persistence and ACID properties by definining *transactional regions*. Those features give eFlow its dynamic and adaptive properties.

5.2 Web Service Composition using other AI Techniques

SWORD [19] is a composition tool that allows basic services to be composed quickly by using keyword-based searching. The plan generation for service composition is realised by a rule-based technique. It does not require deployment of Web service standards such as WSDL, SOAP, RDF, or DAML though it could benefit from them. In SWORD, services are defined in a *world model* based on their pre-conditions and post-conditions. The world model is an Entity Relationship (ER) model, which consists of entities and relationships among them. However, instead of using the ER model as in traditional data modeling, SWORD uses it to describe the inputs and outputs of the Web services. Given the inputs and outputs of the services, a rule-based system is then defined for indicating which inputs produce which outputs through a sequence of services.

Generally, SWORD cannot handle services with multiple side effects, i.e. credit/debit of a bank account [19]. Instead, it is only suitable for information providing Web services that do not alter the state of the world. A good plan generator should be general enough to cope with either information providing or world altering Web services regardless of the service specification language. Furthermore, SWORD may generate non-deterministic results if the pre-condition fails to uniquely determine a post-condition [21].

A couple of comprehensive reviews are available with a special focus on Web service composition techniques. Rao and Su [21] provide a review focusing

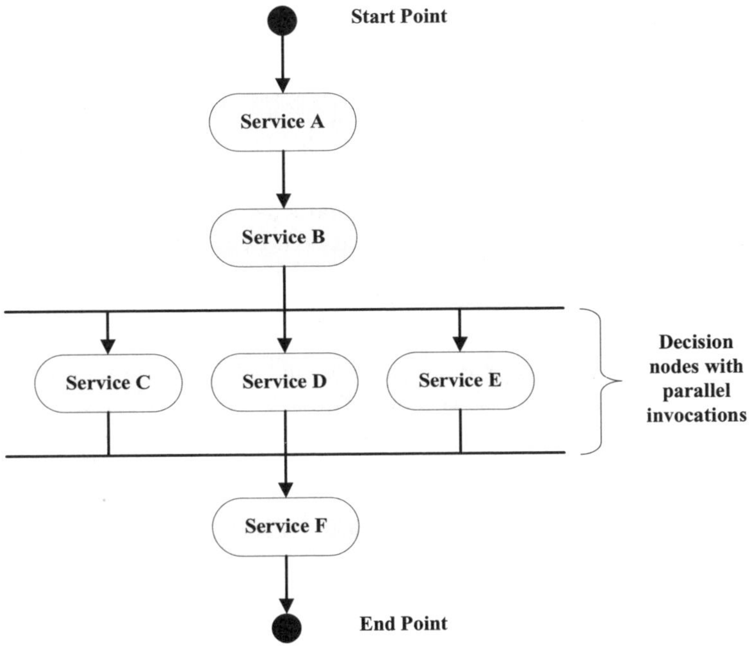

Fig. 9. An example of a composite service in eFlow's process schema [4]

on automatic generation of service composition using either workflow management or AI techniques. The AI techniques briefly reviewed include: theorem proving based on first-order logic, structural synthesis of programs, linear logic and π-calculus. The review suggested that linear logic can deal with both specification and semantic information. In particular, non-functional attributes can be directly considered in the theorem proving process. Milanovic and Malek's review on current solutions for service composition [14] concentrates on reviewing BPEL, OWL-S, Web Components, π-calculus, Petri nets and Model Checking/FSM against the following list of criteria:

- Service connectivity
- Ability to specify Nonfunctional Properties
- Composition correctness checking
- Automatic composition
- Composition scalability

The review shows all of these techniques are able to interconnect service components, most support composition correctness validation (except BPEL and OWL-S), only OWL-S is capable of specifying non-functional requirements, and only Model Checking/FSM supports automatic composition. They have varying degrees of scalability.

6 Conclusion and an Outlook for Open Problems

This chapter developed a structured process of evaluating Web service composition tools by first identifying the desired properties at each stage of a composite service's lifecyle. Most tools available today support semi-automatic service composition, while some are working towards fully automated service composition. Current work on semi-automatic service composition mostly focuses on usability and is often process-oriented. Like imperative programming languages, they rely on a human programmer to provide *how* a composite process can be created based on the available component steps. The main task in this paradigm is connecting the input and output of available service components. On the other hand, similar to declarative languages, the automated service composition tools concentrate on the state of the world and the state transitions in the domain giving a list of actions or operators. The planner generates a list of applicable composition plans for flexible service composition.

However, the support for both usability and flexibility is scarce. Also, the integration with various verification and monitoring mechanisms to ensure failure recovery with secure and robust execution is not realised in any of the reviewed service composition tools.

Ideally, with sufficient ontology support in both service description and domain specification, fully automated service composition can be achieved. However, to cope with an ever changing large distributed environment like the Internet, the `Plan` $\rightarrow$ `Compile` $\rightarrow$ `Execute` cycle is not going to be scalable enough to survive and thrive. In our recent work [12], a self-organising ecosystem view of a service-oriented environment is proposed. In this new paradigm, each Web service is viewed as a simple entity that has a self-organising tendency so organisations (potential candidates for a composite service) can emerge through interactions between services. The discovery of the services takes a less centralised architecture. It can be either through "word of mouth" (referral networks) or through service crawlers. We believe with this paradigm shift, incremental planning during service composition at run time will cope with the scalability issue that cannot be handled by techniques like classical AI planning.

*** About this author:**

Ms. Husniza Husni is a former honours student at the University of Western Australia under the supervision of Dr. Wei Liu. Ms. Husni is currently an academic at Universiti Utara, Malaysia. husniza@uum.edu.my

Acknowledgment:

The authors would like to thank the Australian Tourism Data Warehouse (ATDW) for the provision of live tourism data. The authors would also like to acknowledge the DEST IAP grant CG040014, and the European Union Project SATINE (IST-1-002104-STP) for funding support and international collaboration.

References

1. Tony Andrews, Francisco Curbera, Hitesh Dholakia, Yaron Goland, Johannes Klein, Frank Leymann, Kevin Liu, Dieter Roller, Doug Smith, Satish Thatte, Ivana Trickovic, and Sanjiva Weerawarana. Business process execution language for web services - language specification. May 2003.
2. Arhur Barstow, James Hendler, Mark Skall, Jeff Pollock, David Martin, Vincent Marcatte, Deborah L. McGuinness, Hideki Yoshida, and David De Roure. Owl web ontology language for services (owl-s). November 2004.
3. Tim Berners-Lee, James Hendler, and Ora Lassila. The semantic web. In *Scientific American*, 2001.
4. Fabio Casati, Ski Ilnicki, and Lijie Jin. Adaptive and dynamic service composition in eflow. In *Proceedings of 12th Intl. Conferences on Advanced Information Systems Engineering (CAiSE)*. Stockholm, Sweden, June 2000.
5. S. Chandrasekaran, J. A. Miller, G. S. Silver, B. Arpinar, and A. P. Sheth. Performance analysis and simulation of web services. *Electronic Markets*, 13(2):120–132, June 2003.
6. Erik Christensen, Francisco Curbera, Greg Meredith, and Sanjiva Weerawarana. Web services description language (wsdl) 1.1 language specification. March 2001.
7. Donald F. Ferguson, Tony Storey, Brad Lovering, and John Shewchuk. Secure, reliable, transacted web services: Architecture and composition. Technical report, MSDN Library, 2003.
8. K. Gottschalk, S. Graham, H. Kreger, and J. Snell. Introduction to web services architecture. *IBM Systems Journal*, 41(2), 2002.
9. Martin Gudgin, Marc Hadley, Noah Mendelsohn, Jean Jacques Moreau, and Henrik Frystyk Nielsen. Simple object access protocol (soap) version 1.2 specification. June 2003.
10. J. Hendler and D. McGuiness. The darpa agent markup language. *IEEE Intelligent Systems*, 15(6):72–73, November/December 2000.
11. Frank Leymann. Web services flow language (wsfl) specification version 1.0. May 2001.
12. Wei Liu. Trustworthy service selection and composition - reducing the entropy of service-oriented web. In *3rd International IEEE Conference on Industrial Informatics*, Perth, Australia, August 2005.
13. Thomas Mikalsen, Stefan Tai, and Isabelle Rouvellou. Transactional attitudes: Reliable composition of autonomous web services. In *Workshop on Dependable Middleware-based Systems (WDMS 2002), In the International Conference on Dependable Systems and Networks (DSN 2002)*. Washington D.C., June 2002.
14. Nikola Milanovic and Miroslaw Malek. Current solutions for web service composition. *IEEE Internet Computing*, 08(6):51–59, 2004.

15. Organisation for the Advancement of Structured Information Standards (OASIS). The uddi white paper - introduction to uddi: Important features and functional concepts. 2000.
16. Bart Orriens, Jian Yang, and Mike P. Papazoglou. Servicecom: A tool for service composition reuse and specialisation. *Proceedings of the Fourth International Conference on Web Information Systems Engineering (WISE03)*, 2003.
17. C. Pautasso and G. Alonso. Jopera: a toolkit for efficient visual composition of web services. *International Journal of Electronic Commerce*, 9(2):104–141, Winter 2004/2005.
18. Chris Peltz. Web services orchestration - a review of emerging technologies, tools and standards. Janurary 2003.
19. Shankar R. Ponnekanti and Armando Fox. Sword: A developer toolkit for web service composition. In *Eleventh World Wide Web Conference*. Honolulu, HI, USA, 2002.
20. Jinghai Rao, Peep Kungas, and Mihhail Matskin. Application of linear logic to web service composition. In *First International Conference on Web Services*. Las Vegas, USA, June 2003.
21. Jinghai Rao and Xiaomeng Su. A survey of automated web service composition methods. *Proceedings of the First International Workshop on Semantic Web Services and Web Process Composition, SWSWPC 2004, San Diego, California, USA*, July, 2004.
22. S. Russel and P. Norvig. *Artificial Intelligence: A Modern Approach (2nd Edition)*. Prentice-Hall Inc, 1995.
23. Evren Sirin, Bijan Parsia, Dan Wu, James Hendler, and Dana Nau. Htn planning for web service composition using shop2. *Journal of Web Semantics*, 1(4), 2004.
24. B. Srivastava and J. Koehler. Web service composition - current solutions and open problems. *ICAPS*, 2003.
25. F. Tartanoglu, V. Issarny, A. Romanovsky, and N. Levy. Coordinated forward error recovery for composite web services. In *The 22nd IEEE Symposium on Reliable Distributed Systems*. Florence, Italy, October 2003.
26. W3C. Qos for web services: Requirements and possible approaches. *W3C Working Group Note*, 2003 http://www.w3c.or.kr/kr-office/TR/2003/ws-qos/.

Web Information Representation, Extraction and Reasoning based on Existing Programming Technology

Fei Liu[1], Jidong Wang[2], and Tharam S. Dillon[3]

1. Department of Computer Science & Computer Engineering, La Trobe University, Melbourne Vic 3083 Australia Email liufei@cs.latrobe.edu.au

2. School of Electrical and Computer Engineering, Royal Melbourne Institute of Technology, Melbourne Vic 3004 Australia Email jidong.wang@rmit.edu.au

3. Faculty of Information Technology, University of Technology Sydney, Sydney NSW 2007 Australia Email Tharam.Dillon@uts.edu.au

Abstract

The chapter presents a framework on web information representation, extraction and reasoning utilising an object-oriented approach. A structure, which is similar to a class in object-oriented design and programming, is defined as an extended class (eClass). An eClass contains data attributes, member functions/methods, inference rules and presentations, and therefore, facilities both web presentation and information reasoning. As an object-oriented approach, the framework also supports encapsulation, polymorphism and inheritance. It can be implemented as an extension of an existing object-oriented programming language. An eClass can be extremely effective in describing content-dependent entities. It describes a value under its context, consequently, the search engine not only searches the value, but also understands the meaning of the value in the context. The chapter attempts to integrate GOPT-resolution, which is a reasoning mechanism based on partial intersection and truncation, into OOWIS. GOPT-resolution is ideal for reasoning in a web environment because, comparing with other approaches, it requires less search during its derivations.

F. Liu et al.: *Web Information Representation, Extraction and Reasoning based on Existing Programming Technology*, Studies in Computational Intelligence (SCI) **37**, 147–168 (2007)
www.springerlink.com

1. Introduction

With the fast expansion of the World Wide Web and rapid advances in Information Technology, our expectations of the search engine on the World Wide Web are constantly increasing. We are no longer satisfied with merely searching a phone number or checking a road map, we like our search engine to be intelligent. To be precise, we like the search engine to be able to reason. This requires an effective mechanism in information representation, extraction and reasoning.

The research on web information representation, extraction and derivation can be approximately divided into two categories: (1) information extraction by text learning and (2) web semantics. Information extraction by text learning is based on the research in natural language processing. It attempts to build the intelligent search engine in such a way that it is able to “understand” the content of a text file (such as a webpage), and consequently, can extract information from it [1]. Web Semantics, on the other hand, focuses mainly on defining the semantics of information representation. As pointed by Schwartz [24]: We can approach data representation in a Semantic Web environment as a series of layers, each of which adds a degree of semantics depth to a given data model. The four basic layers are Semantics, Ontology, Logic and Web trust. On Semantics and Ontology, a substantial amount of research has been conducted including XML, RDF and OWL.

XML is a markup language which contains a set of tags to structure its contents. There are significant differences between XML and HTML files. While the HTML file uses tags to specify its web presentation, the XML file uses tags to specify the structure of information. RDF [12] which utilises XML as its presenting language, provides interoperability between applications that exchange machine-understandable information on the Web. RDF emphasizes facilities to enable automated processing of Web resources. RDF with digital signatures can be utilized in building the "Web of Trust" for electronic commerce, collaboration, and other applications. The OWL [25] is a semantic markup language for publishing and sharing ontologies on the World Wide Web. OWL is developed as a vocabulary extension of RDF and is derived from the OIL [5] Web Ontology Language. Consequently, any RDF graph forms an OWL Full ontology. Furthermore, the meaning given to an RDF graph by OWL includes the meaning given to the graph by RDF. OWL Full ontologies can thus include arbitrary RDF content, which is treated in a manner consistent with its treatment by RDF. OWL assigns an additional meaning to certain RDF triples.

OWL has important features that capture many of significant relationship provided in object-oriented system, however several features of object oriented analysis (such as composition and encapsulation) which make the language extremely powerful in entity description are not fully supported. Since OWL, OIL and RDF are all based on XML, and require XML schema and parsing, the representation and reasoning, in some circumstances, can be tedious.

In this chapter, we propose the framework of an object-oriented approach in web information representation and reasoning, and we name it object-oriented web information specification (OOWIS)[21]. OOWIS can be easily planted into an existing object-oriented programming language, such as C++, JAVA or C#, and can be considered as an extension of the programming language.

2. Extended Class (EClass)

The eClass, which is an extension of a class in the object-oriented sense and which is the centre of OOWIS, will be defined in this section.

Definition 1

An extended class (eClass) is a structure that contains data (attributes), functions that are a group of operations associated with the data, inference rules that are associated with the data derivation, and a group of presentations each specifies a style that data (attributes) are presented on the web. The template of an eClass is as follows.

```
[Modifiers] EClass TitleOfClass{
//attributes
[Modifiers] dataDefinition1:datatype1 attribute1 [= value1]
[:unit1];
   ... ...
[Modifiers] dataDefinitionL:datatypeL attributeL [= valueL]
[:unitL];

//functions
[Modifiers] returnType1 operation1(listOfParamenters1);
   ... ...
[Modifiers] returnTypeM operationM(listOfParamentersM);
```

```
//inference rules
[Modifiers] inferenceRule1;
   ... ...
[Modifiers] inferenceRuleN;

//presentations
[Modifiers] presentation1 = "markupFile1";
   ... ...
[Modifiers] presentationP = "markupFileP";
}
```

An eClass contains 4 types of members: attributes, functions, rules and presentations. An attribute describes a particular value and it's meaning under certain context. From logic perspective, an attribute represents a triple which, as described in RDF, includes the subject (`attribute`), the object (`value`) and the predicate (`dataDefinition`). `dataDefinition` is defined in OOWIS. OOWIS contains a dictionary in which all data definitions are defined. The dictionary which is shared by all OOWIS users on the Internet, serves as a guide on common understanding of terms.

A function performs a certain action that is associated with the attributes. An inference rule (a horn clause) can be used in information derivation based on the attributes and functions of the eClass. A presentation is a specification in a markup language to specify the manner that its instance should be presented on the web. An eClass may have a number of presentations which allow its instances to be presented in different ways.

Example 1 demonstrates the construction of an eClass and its instantiation. As a theory, OOWIS is language independent, and consequently, an eClass can be implemented in any object-oriented programming language. In this particular example, however, we will use JAVA for the implementation.

Example 1

The example defines the eClass `Person` which describes certain characters of a person (an instance of the eClass), and his/her relationship with the other person which is, again, an instance of the eClass.

```
public eClass Person {
private Name:String name; //the person's name
private Mother:String mother; //name of the person's mother
private Father:String father; //name of the person's father
private Height:double height:cm; //the person's height (in
//cm)
private Quantity:int numberOfchildren; //number of children
```

```
//default constructor. Implementation details deleted
public  Person() {}

//constructor. Implementation details deleted
public Person(String _name, String _m, String _f, int _h, int
_noc) {}

//to change the height of this person.
private void changeHeight (int _change) {}

//a number of member functions acting as accessors have
//been omitted, however, are used later.

//to define the predicate mother(X, Y).
public static bool mother(Person X, Person Y) {
return X.getName() ==Y.getMother();
}

//to define the predicate father(X, Y)
public static bool father(Person X, Person Y) {
return X.getName() == Y.getFather();
}

//a set of inference rules
public static siblings(Person X, Person Y)
← mother(Person Z, Person X) ^ mother(Person Z, Person Y)
public static siblings(Person X, Person Y)
← father(Person Z, Person X) ^ father(Person Z, Person Y)
public static parent(Person X, Person Y)
← father(Person X, Person Y)
public static parent(Person X, Person Y)
← mother(Person X, Person Y)
public grandMother(Person X)
← parent(Person Y) ^ Y.mother(Person X)

//a set of presentation files
//two different presentation files which describe how
//instances of the eClass should be presented.
//"welcome.html" and "personalDetails.xml"
//are in two different markup languages.
public static presentation1 = "welcome.html";
private presentation2 = "personalDetails.xml";
}
```

Like object-oriented analysis and design, members of an eClass can be defined with certain modifiers such as public, protected, private and static etc. which indicate the member's accessibility and its accessing methods. An eClass consists of attributes and functions which allow it to behave as an ordinary class. Meanwhile it contains inference rules and presentations that make information derivation and web presentation easy to achieve.

An eClass is instantiated in exactly the same fashion as an ordinary class. The instance of an eClass is named as an object. Objects of an eClass share the same group of functions, inference rules and presentations, but have their own value for each attribute. Like an ordinary class, an eClass uses constructors to create objects. For example, the statements

```
Person p1 = new Person("Laura Wills", "Wendy Wills", "John
Wills", 135, 0);
Person p2 = new Person();
```

construct two objects of the `Person` eClass. One with default values, and the other have `"Laura Wills"` as its `name`, `"Wendy Wills"` as the value of its `mother`, `"John Wills"` as the value of its `father`, `135` as the value of its `height`, and `0` as the value of its `numberOfChildren`.

An eClass has a set of functions. For example

```
public Person()
public Person(String _name, String _m, String _f, int _h,
int _noc)
private void changeHeight (int _change)
public static bool father(Person X, Person Y)
public static bool mother(Person X, Person Y)
```

are functions of the eClass. Except for constructors, accessors and any other functions which can normally be found in an ordinary class, there is a group of special purposed functions in an eClass – predicator builders. The purpose of a predicate builder is to define the predicates in the eClass. Predicate builders normally have a Boolean return type. They act as a bridge linking attributes and inference rules (*ie.* they define the truth value of each predicate by using the values of attributes). For example

```
public static bool mother (Person X, Person Y)
public static bool father(Person X, Person Y)
```

are predicate builders. They define the truth value of the predicate based on attributes of the eClass. Obviously, predicate builders are extremely important in an eClass -- they make inference in an eClass possible.

An eClass contains a set of inference rules which makes information derivation possible. In the `Person` eClass, for example,

```
public static siblings(Person X, Person Y)
← mother(Person Z, Person X) ^ mother(Person Z, Person Y)
public static siblings(Person X, Person Y)
← father(Person Z, Person X) ^ father(Person Z, Person Y)
```

```
public static parent(Person X, Person Y)
← father(Person X, Person Y)
public static parent(Person X, Person Y)
← mother(Person X, Person Y)
public grandMother(Person X)
← parent(Person Y) ^ Y.mother(Person X)
```

make the set of rules. Including inference rules in the class structure represents the major difference between a class and an eClass.

Like functions, inference rules can be defined at the class level (*ie.* static) or object level (*ie.* non-static). A non-static rule is fired by being called by an object of the eClass. A static rule is fired by being called either using the eClass name or an object name. Details of the derivation based on inference rules will be discussed in Section 4.

A presentation is a markup language file which specifies the manner that attributes and predicates are displayed on the web. As previously indicated, an eClass may have a number of presentations that specify different ways that the attributes of an object are displayed on the web. In the Person eClass, for example, there are two presentations. They are

```
public static presentation1 = "welcome.html";
private presentation2 = "personalDetails.xml";
```

`presentation1` is defined as static while `presentation2` is non-static, and this indicates that `presentation1` is probably defined more generic and does not display any data attribute of a specific object, while `presentation2` is, possibly, more specifically related to an individual object (ie. to display the values of data attributes).

3. Object Oriented Features of OOWIS

In this section, we will view OOWIS from the object-oriented prospective to examine its object-oriented features. Theses include inheritance, encapsulation and polymorphism[22]. We will discuss the advantages and disadvantage of the features in the framework.

3.1 Inheritance

In object-oriented analysis and design, inheritance refers to the phenomenon that one structure inherits data attributes and functions from another structure, and therefore the construction of the structure is based on the other existing structure. Inheritance is a result of an important principle in Software Engineering – re-use. It saves the effort to re-define data attributes and functions have been defined previously. Inheritance is also important in batch data processing

In OOWIS inheritance refers to precisely the definition of an eClass (the derived eClass) is based on an existing eClass (the base eClass), and so that all the public members of the base eClass become public members of the derived eClass.

Definition 2

The eClass `A` is defined as follows

```
[modifier] EClass A {
//data attributes
public def_1:type_1 attr_1;
  ... ...
public def _ka:type_ka attr_ka;
private def_ka+1:type_ka+1 attr_ka+1;
  ... ...
private def_ka+na:type_ka+na attr_ka+na;

//functions
public type_1 function_1(X_1);
  ... ...
public type_kf function_kf(X_kf);
private type_kf+1 function_kf+1(X_kf+1);
  ... ...
private type_kf+nf function_kf+nf(X_kf+nf);

//inference rules
public rule_1(X_1);
  ... ...
public rule_kr(X_kr);
private rule_kr+1(X_kr+1);
  ... ...
private rule_kr+nr(X_kr+nr);

//presentations
public presention_1 = "file1.ext";
  ... ...
public presention_kp = "file_kp.ext";
private presention_kp+1 = "file_kp+1.ext";
  ... ...
private presention_kp+np = "file_kp+np.ext";
}
```

The eClass B is defined as a derived eClass of A

```
[modifiers] EClass B extends A {
//data attributes
[modifiers] defB_1:typeB_1 attrB_1;
  ......
[modifiers] defB_ma:typeB_ma attrB_ma;

//functions
[modifiers] typeB_1 functionB_1(XB_1);
  ......
[modifiers] typeB_mf functionB_mf(XB_mf);

//inference rules
[modifiers] ruleB_1(XB_1);
  ......
[modifiers] ruleB_mr(XB_mr);

//presentations
[modifiers] presentB_1 = "file_B_1.ext";
  ......
[modifiers] presentB_mr = "fileB_mr.ext";
}
```

The eClass B is effectively defined as

```
[modifiers] EClass B {
//data attributes
[modifiers] defB_1:typeB_1 attrB_1;
  ... ...
[modifiers] defB_ma:typeB_ma attrB_ma;
public def_1:type_1 attr_1;
  ... ...
public def _ka:type_ka attr_ka;

//functions
[modifiers] typeB_1 functionB_1(XB_1);
  ... ...
[modifiers] typeB_mf functionB_mf(XB_mf);
public type_1 function_1(X_1);
  ... ...
public type_kf function_kf(X_kf);

//inference rules
[modifiers] ruleB_1(XB_1);
 ... ...
[modifiers] ruleB_mr(XB_mr);
public rule_1(X_1);
  ... ...
public rule_kr(X_kr);

//presentations
```

```
[modifiers] presentB_1 = "file_B_1.ext";
  … …
[modifiers] presentB_mr = "fileB_mr.ext";
public presention_1 = "file1.ext";
  … …
public presention_kp = "file_kp.ext";
}
```

Not only `B` contains members defined in its own eClass, but also all the public members defined in `A`. As a consequence, the modification of `A` will affect the structure of `B`, and hence the definition of `A` and `B` are linked by this relation of inheritance. Below is an example demonstrating principles of inherence in OOWIS.

Example 2
Assume that `Organization` is an eClass being defined as follows

```
EClass Organization {
public name:String name;
private numberOfEmployees:int eNumber;
    … …
}
```

Now we can define various organizations such as `University, Faculty` and `Department` based on `Organization.`

```
EClass Department extends Organization {
private numberOfEnrolledStudents:int sNumber ;
public titleOfCourses:String[] courses;
   … …
}
```

This allows `Department` to inherit `numberOfEmployees` from `Organization`, furthermore when the definition of `numberOfEmployees` changes in `Organization`, it changes in `Department` as well.

Inheritance is one of the most important features in object-oriented design and programming and it makes re-use possible. In OOWIS, however the advantages are not limited to re-use. Another advantage of inheritance in OOWIS is that it allows information sharing. As indicated previously, in an eClass, the definition of a data attribute is not merely the definition of a value/term, but also the definition of the meaning of the value/term. Deriving an eClass from another actually passed the definition of attributes from the base eClass to the derived eClass, and hence the definition is shared. This is a significant advantage of utilizing inheritance in OOWIS, and this is regarded as the starting point of definition sharing.

3.2 Polymorphism

In his book "*Absolute JAVA*", Savitch defined polymorphism as "*the ability to associate many meanings to one method name by means of the late binding mechanism*".

In object-oriented design and programming, polymorphism allows one function to have a number of different implementations, and therefore the binding of the function call and the function will be determined at run-time. Hence Polymorphism is also referred as run-time binding (late binding). The benefit of having run-time binding is that it allows a number of behaviors to share the same name, and therefore provides flexibility in coding.

The concept of polymorphism in OOWIS is broader, because an eClass contains not only data attributes and functions, but also inference rules and web presentations. The idea of polymorphism is extended to the definition of inference rules and also web presentations. Precisely, it can happen that a number of inference rules may share the same name, however, have different contents. Meanwhile, there may be a number of presentations sharing the same name, but display data attributes in different ways.

Polymorphism not only provides flexibility in OOWIS, but also is essential in inference rule definitions. It facilitates multiple definitions of a predicate in an eClass which is essential in a rule-based system [6].

For example, polymorphism allows the definitions of `p(X, Y)` by using

Inference rule 1: `p(X, Y) ← q(X, Y)`

and

Inference rule 2: `p(X, Y) ← r(X) ^ s(Y)`

to exist concurrently in a single eClass. In this case, the signature of the rules (the heads of the rules) are exactly the same. Although this is not allowed in function definitions, it is allowed in inference rule definitions. An ordering rule that is defined in the inference engine will be used to determine which inference rule should be fired at a specific point of the derivation.

3.3 Encapsulation

Encapsulation is the ability to separate the detailed implementation from the general specification in a class, so that the implementation details can be hidden. In several major object-oriented languages, encapsulation is achieved by using the modifiers "public" and "private".

In OOWIS, encapsulation not only can achieve its original goal – data hiding, but can also be used in controlling data presentation. This means modifiers "public" and "private" control the visibility of data attributes on the website. Encapsulation is an important feature of OOWIS – to link the modifier of an attribute with its visibility on the web.

Example 3

The example partially presents the eClass hierarchy of an e-services system for information technology equipment. It demonstrates the principles of inheritance and encapsulation. The system contains eClasses such as `Merchandise`, `Hardware`, `Software`, `Computer` and `Printer` as follows.

```
EClass Merchandise {
protected ID:int iD; //the unique ID number
public productName:String name; //name of the item
public price:double price; //price of the item

public void discount (double percentage) {
  price = price * (1 - percentage);
}
  ... ...
}

EClass Hardware extends Merchandise {
public costOfDelivery:double deliveryCost;
//cost for home delivery
private recyclable:bool recyclable;
//if the item is recyclable

public addingDelCost() {
  price = price + deliveryCost;
}
  ... ...
}

EClass Computer extends Hardware {
public afterSaleServiceInformation:URI afterSaleServiceInfo;
//The website providing after sale services
public oSInstallationCost:double oSInstallationCost;
```

```
//charge for operating system installation

public addingOSInstCost() {
   price = price + oSInstallationCost;
}
       ... ...
}

EClass Printer extends Hardware {
public printingPaperSpec:String paperSpec;
//Specification of papers
public driverSite:URI driverDownload;
//Site for downloading drivers
  ... ...
}

EClass Software extends Merchandise {
private static final discount:double downloadDiscountRate =
0.15;
//rate of discount if the software is downloaded by the
//customer

public double downloadPrice() {
  price = price * (1 - donwloadDiscountRage);
}
       ... ...
}
```

The figure below describes eClass hierarchy of the system where `Hardware` and `Software` are derived from `Merchandise`, and `Computer` and `Printer` are subclasses of `Hardware`.

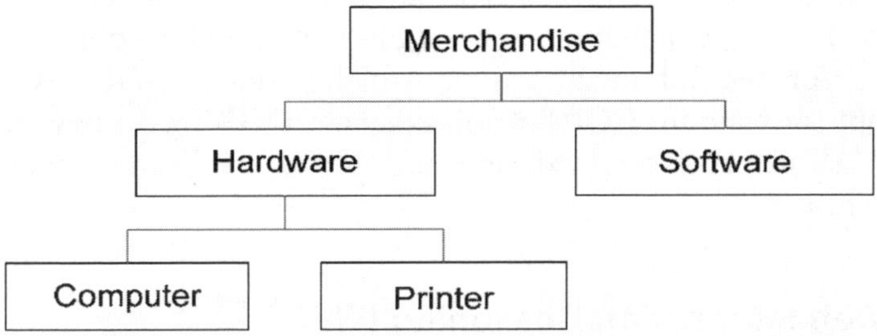

Fig. 1. IT equipment eClass Hierarchy

Based on principles of inheritance and encapsulation, `Hardware` and `Software` inherit all members of `Merchandise` (*ie.* `iD, name, price` and the member method `discount`). Note that this includes the protected member `iD`. Similarly, `Computer` and `Printer` inherit all the public members of `Hardware`, but not `recyclable` which is private.

4. Reasoning

The major difference between a class and an eClass lays on the fact that an eClass contains a set of inference rules, and hence supports reasoning. This section concentrates on the details of the inference engine of OOWIS.

Although automated reasoning has been a major research field in Artificial Intelligence for more than a half of a century, web information reasoning (WIR) [23] is still in its infancy. Theoretically, reasoning should be all based on the same principles regardless of the source of information; in practice, however, there are significant differences between the two types. WIR refers to the situation where reasoning is conducted based on information from more than one source. Precisely, the logic program is formed using the clauses (facts and rules) from different locations [7]. Apparently, WIR requires searching substantially. The search rule being utilized and the manner in which search is conducted will influence the performance of the derivation significantly.

In this section, we will examine various resolution strategies, and identify the type which fits web information derivations. We will, in particular, discuss goal-oriented resolution with partial intersection and truncation (GOPT-resolution) [15] in which the derivation is guided by the goal, and is based on set calculations. We argue that GOPT-resolution is a suitable choice for the inference engine which requires WIR. We will, then, attempt to integrate GOPT-resolution into OOWIS, so that derivations in OOWIS are goal-oriented and are based on partial intersection and truncation.

4.1 Web Information Reasoning (WIR)

Various resolution strategies have been developed for the last three decades [8], and those strategies can be approximately classified into three categories – top-down, bottom-up and combined top-down and bottom-up [11].

The first category [2, 3] starts the derivation from the top -- the goal, and backtrack down to the bottom -- the facts. The derivation is mainly based on pattern matching and backtracking. The second category [4, 13, 15] starts its derivation from the bottom -- a set of facts and a set of rules. It derives new facts from the existing facts and rules, updates the fact set, and again derives new facts based on the updated fact set till the stage that no new fact can be derived or the goal is proved. The third approach is a combined of the first and second approaches. In other words, the inference starts from both the top and the bottom concurrently till the sub-goals and the derived facts meet.

Although top-downs have always been the favorites in automated deduction, especially in logic programming, we argue that bottom-up approaches are probably more appropriate for WIR. The reasons are

1. WIR is reasoning based on an infinite (open) universe, and hence "closed world assumption" will on longer hold. Consequently, search is the major issue in relation to the performance of the inference engine. In general, a top-down approach which is based on backtracking and pattern-marching requires significantly higher amount of search than a bottom-up approach [8]. Therefore it is expected that an engine based on bottom-up principles performs better.

2. Unlike a bottom-up approach which can gather information by searching the web at the beginning of the derivation, a top-down approach requires search continuously during its derivation. This will, again, be an important factor that jeopardizes the performance of the engine.

3. Searching in an open domain is likely to lead to an infinite derivation, and therefore should be limited to the minimum.

At this point, we also like to point out that using mark-up languages in web information specification will not lead the research far. This is due to the fact that (1) a mark-up language is normally difficult to learn and complicated to use; (2) a single mark-up language file cannot normally accomplish its mission to properly define data, and therefore extra files will have to be created in the information specification [1, 9]; (3) comes to RDF in specific, it can be difficult to use triples to represent a complicated object. Besides, even it is managed to use multiple triples to represent the object, then those triples are individual items, and can be difficult to store and retrieve; (4) comparing with a compiled programming language, using a mark-up language will slow down the speed of the inference engine [8].

4.2 Goal-oriented Resolution with Partial Intersection and Truncation

GOPT-resolution stands for goal-oriented PT-resolution. PT-resolution stands for resolution with partial intersection and truncation.

GOPT-resolution is a deduction strategy that is based on set calculations: partial intersection and truncation. It belongs to the bottom-up family and it prevents derivations on logic programs which have a finite Herbrand universe from infinite recursion. Partial intersection is an intersection based on a certain variable order. The calculation involves comparing the tuples from each set according to a variable order, and concatenating the tuples. Truncation truncates a tuple or a set of tuples according to a variable order and appends free elements to the tuple if it is necessary.

Definition 3[15]
Let ***P*** be a logic program. A PT-calculation is a process that derives tuples from equations and restricted P-domains and adds those tuples to the corresponding P-domains. The process involves precisely two steps:

(1) Selecting each of the active equations of P according to a certain ordering rule and deriving tuples from the equation and the restricted P-domains of the equation;

(2) Adding the derived tuples to the corresponding P-domains.

A PT-derivation is simply a sequence of PT-calculations finite or infinite. Below is a precise definition of a PT-derivation.

Definition 4[15]
Let ***P*** be a logic program and

$$\boldsymbol{G} : \leftarrow \mathrm{p}_1(\mathrm{X}_1^1,\ldots,\mathrm{X}_{\mathrm{k}_1}^1)\wedge\ldots\wedge\mathrm{p}_\mathrm{m}(\mathrm{X}_1^\mathrm{m},\ldots,\mathrm{X}_{\mathrm{k}_\mathrm{m}}^\mathrm{m})$$

be a goal of ***P***. The PT-derivation on $\boldsymbol{P} \cup \{\boldsymbol{G}\}$ proceeds as follows.

Step 1. Mark all the equations as active.

Step 2. Iterate Steps 2 - 4 until at least one of the following termination conditions is satisfied.

Condition 1. The goal is proved.

Condition 2. No new tuple can be derived from the derivation.

If one of the conditions is satisfied, then go to Step 5.

Step 3. Apply the PT-calculation to the program.

Step 4. Mark all the saturated equations which have been selected in the previous calculation as inactive. Mark all new saturated P-domains and saturated equations which may arise.

Step 5. Prove the goal with the current P-domains. Any tuple which is in the P-domain, must satisfy the predicate symbol. Conversely, any tuple which is not in the P-domain, does not satisfy the predicate symbol.

Finally, GOPT-resolution is PT-resolution with a specifically selected rule set and fact set – goad-related rule set and fact set.

The soundness and completeness of GOPT-resolution were discussed in [20] and [19]. It was proved that GOPT-resolution is sound and complete if the derivation is finite. Since derivations on logic programs with a finite Herbrand universe are always finite, for those programs, GOPT-resolution is both sound and complete.

The control part of the resolution consists of two rules: selecting rule and ordering rule. A selecting rule determines the order in which partial intersections and truncations are conducted. An ordering rule decides the way that equations are selected. Both the selecting rule and ordering rule are independent of the GOPT-derivation.

A substantial amount of research has been conducted since the definition of GOPT-resolution. This includes the integration of PT-resolution and conditional proof [17]; introducing approximate reasoning into PT-resolution [18]; optimizing the selecting and ordering rules [16] etc.

4.3 GOPT-Resolution in OOWIS

Based on the discussion above, we now attempt to integrate GOPT-resolution which is from the bottom-up category, and in which the search and derivation are guided by the goal, into OOWIS.

Definition 5

In the OOWIS framework, a derivation based on GOPT-resolution is carried out as follows.

(1) When the eClass which contains data, functions, inference rules and web presentations, is defined, and instances of the eClass are created, the program execution environment creates the P-domain for each predicate symbol, whether it will be used in the derivation or not.

(2) For a given goal, the inference engine searches on the web for all goal-related rules, hence the set of goal-related rules is established.
(3) The inference engine, then, searches on the web for goal-related facts, and establishes the set of goal-related facts.
(4) The inference engine conducts the GOPT-derivation based on the set of goal-related rules and the set of goal-related facts.

Below is an example which illustrates the GOPT-derivation in the OOWIS framework.

Example 4
The class `Person` is defined and utilized to describe a person on the web.

```
EClass Person {
//data attributes
public name:string myName;
public workFor:Organisation empolyer;
private mother:string motherName;
private father:string fatherName;

//functions
public Person(string _my, string _m, string _f, string _e){
  myName = _my;
  motherName = _m;
  fatherName = _f;
  employer = _e;
}

public boolean mother(Person X) {
  if (X.myName == motherName)
       return true;
  else
       return false;
}

public boolean father(Person X) {
  //details omitted.
}

//rules
public grandMother(Person X)
← parent(Person Y) ^ Y.mother(Person X)
public parent(Person X) ← mother (Person X)
public parent(Person X) ← father(Person X)
public colleague(Person X)
← X.workFor = Y ^ workFor = Z ^ Y = Z

//presentation details omitted.
}
```

Assume the following objects of the class are defined on a number of different servers, and assume the goal is to search `grandmother` of `Verena`.

```
  Person a = new Person ("verena", " ", "michael", " ");
  Person b = new Person ("helen", "mary", "", "");
  Person c = new Person ("michael", "helen", "jack", "la Trobe University");
  Person d = new Person ("aimee", "helen", "", "la Trobe University);
```

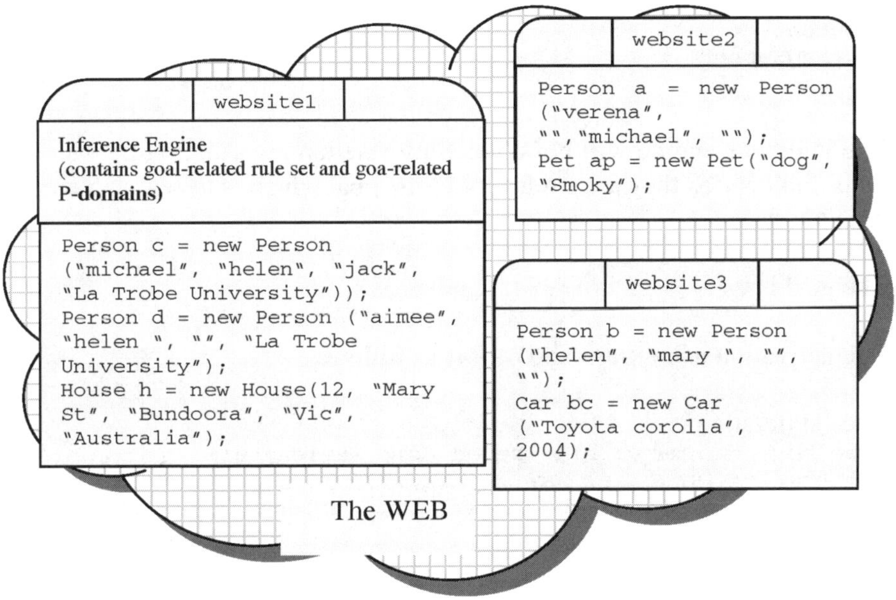

Fig. 2. Information on the web

When an object is created, the program execution environment calls the predicate definition functions automatically, so that the universe of the program and the P-domains of the predicate symbols can be established.

As the derivation is goal-oriented, it establishes p-domains for those goal-related predicate symbols only [15]. Hence, by the time when the objects are created, the universe, goal-related rule set and the goal-related P-domains are

```
U = {verena, michael, helen, mary, jack, laTrobeUniversity,
aimee, helen, Mary_St, Bundoora, Vic, Australia, dog, Smoky,
toyotaCorolla, 2004}
```

```
goal-related rule set GRR = {
public grandMother(Person X)← parent(Person Y) ^
Y.mother(Person X)
public parent(Person X) ← mother (Person X)
public parent(Person X) ← father(Person X)
}

goal-related P-domains GRP = {
D[mother] = {(mary, b), (helen, c), (helen, d)}
D[father] = {(michael, a), (jack, c}}
D[parent] = { }
D[grandMother] = { }
}
```

The inference engine will not collect information regarding `Pet`, `Car` and `House` objects, as they are irrelevant to the goal which is to search Verena's grandmother.

The goal is `G: ←grandMother(X, verena)`

The derivation using `GRR` and `GRP` is as follows.

<u>The 1st iteration</u>

```
D[mother] = {(mary, b), (helen, c), (helen, d)}
D[father] = {(michael, a), (jack, c}}
D[parent] = {(mary, b), (helen, c), (helen, d), (michael, a),
(jack, c}}
D[grandMother] = {}
```

<u>The 2nd iteration</u>

```
D[mother] = {(mary, b), (helen, c), (helen, d)}
D[father] = {(michael, a), (jack, c}}
D[parent] = {(mary, b), (helen, c), (helen, d), (michael, a),
(jack, c}}
D[grandMother] = {(mary, c), (mary, d), (helen, a)}
```

Therefore the answer to the goal is

`X= helen.`

5. Conclusions

The chapter established a framework for using the object-oriented approach in web information representation, extraction and reasoning. The framework can be implemented as an extension of an existing object-oriented language, and hence saves the effort to "reinvent the wheel". This also prevents the tedious work with writing up XML files, creating XML schemas and using XML parsing. Although in the circumstance when the quantity of information is relatively small, saving information into XML and HTML is probably simpler than coding it in the proposed framework, when quantity of information is relatively large, however, the effort is worthwhile.

The chapter also proposed the integration of GOPT-resolution into OOWIS. We discussed the advantages and disadvantages of top-down and bottom-up approaches, and concluded that bottom-up approaches would be more appropriate for reasoning in the web environment.

References

[1] Ankolekar, Seo and Sycara, Investigating Semantic Knowledge for Test Learning, Proceedings of Semantic Web Workshop in associated with 26th Annual International ACM SIGIR Conference, 2003.

[2] Apt K.R. and Doests K., *A New Definition of SLDNF-Resolution*, Journal of Logic Programming, Vol 18 1994, pp. 177 - 190.

[3] Bozzano M., Delzanno G. and Martelli M., *An Effective Fixpoint Semantics for Liner Logic Programs*, Journal of Theory and Practices of Logic Programming, Vol 2, Part 1, 2002, pp.85 – 122.

[4] Brewka G. and Eiter T., *Preferred Answer Sets for Extended Logic Programs*, Proceedings of Sixth International Conference on Principles of Knowledge Representation and Reasoning, 1998, pp.86-97.

[5] Connolly, D., Harmelen F., Horrocks, I., McGuinness D., Stein L., *DAML+OIL Reference Description,* http://www.w3.org/TR/daml+oil-reference, 2001.

[6] Dantsin E.et al., *Complexity and Expressive Power of Logic Programming*, ACM Computing Surveys, Vol 33, Num. 3, 2001, pp. 374 – 425.

[7] Fensel D., *The Semantic Web*, Tutorial notes, The 9th IFIP 2.6 Working Conference on Database Semantics, April 2001.

[8] Greiner R., *Efficient Reasoning*, ACM Computing Surveys, Vol. 33, Num. 2, 2001, pp. 1 – 30.

[9] Gruber T. R. *A translation approach to portable ontologies.* Knowledge Acquisition, 5(2):199-220, 1993.

[10] Kiryakov, Borislav and Manov, Semantic Indexing and Retrival, Proceedings of Semantic Web Workshop in associated with 26th Annual International ACM SIGIR Conference, 2003.

[11] Kondrak G. and van Beek P., *A theoretical Evaluation of Selected Backtracking Algorithms.* Artificial Intelligence Vol. 89, 1997, pp. 365 – 387.

[12] Lassila O. and Swick R. R., *Resource Description Framework (RDF) Model and Syntax Specification*, http://www.w3.org/TR/1999/REC-rdf-syntax-19990222/, 2001.
[13] Liu F. and Moore D.H., *An Implementation of Kripke-Kleene Semantics*, Journal of Information Sciences, Vol. 108, 1998, pp. 31 - 50.
[14] Liu F. and Moore D.H., *Double Defined Logic Programming*, Proceedings of the Sixth Australian Joint Conference on Artificial Intelligence, 1993, pp. 27 - 32.
[15] Liu F. and Moore D.H., *GOPT-Resolution and Its Applications*, Proceedings of the Eighth International Conference on Artificial Intelligence Applications, 1996, pp. 9 - 14.
[16] Liu F. and Moore D.H., *Independence of Selecting Rule and Ordering Rule in PT-Resolution*, Proceedings of IEEE International Conference on Intelligent Processing Systems, November 1997, pp. 1082 – 1086.
[17] Liu F. *PT(CP)-Resolution*, Proceedings of the 10th IEEE International Conference on Fuzzy Systems, December 2001, Vol. 2 pp 368
[18] Liu F., *Approximate Reasoning and PT-Resolution*, Proceedings of IEEE International Symposium on Computational Intelligence in Robotics and Automation, July 2003, pp. 609 - 613.
[19] Liu F., Moore D.H. and Wang J., *The Completeness of GOPT-Resolution*, Proceedings of the Second ECPD International Conference on Advanced Robotics, Intelligent Automation and Active Systems, 1996, pp. 92 - 98.
[20] Liu F., Moore D.H. and Wang J., *The Soundness of GOPT-Resolution*, Proceedings of the International Conference on Genetic Algorithms, 1996, pp. 66 - 70.
[21] Liu F., Wang J. and Dillon T., *An Object-Oriented Approach on Web Information Representation and Derivation,* Proceedings of the 2004 IEEE International Conference on e-Technology, e-Commerce and e-Service, April 2004, pp. 309 - 314.
[22] Liu F., Wang J. and Dillon T., GOPT-Resolution in Web Information Derivation, Proceedings of The Sixth International Conference on Information Integration and Web-based Applications and Services, September 2004, pp. 49 - 56.
[23] Liu F., Wang J. Wang H. and Dillon T., *The Object-Oriented Features of OOWIS*, Proceedings of International Conference on Internet Computing, to appear.
[24] Schwartz David G., From Open IS Semantics to the Semantic Web: The Road Ahead, IEEE Intelligent Systems, Volume 18, No.3, May/Jane 2003.
[25] van Harmelen F., Hendler J., Horrocks I., McGuinness D., Stein L., *OWL Web Ontology Language Reference*, http://www.w3.org/TR/2004/REC-owl-features-20040210/, 2004

Techniques and Technologies Behind Maps of Internet and Intranet Document Collections

Mieczysław A. Kłopotek, Sławomir T. Wierzchoń, Krzysztof Ciesielski, Michał Dramiński, and Dariusz Czerski

Institute of Computer Science, Polish Academy of Sciences,
ul. Ordona 21, 01-237 Warszawa,Poland
{klopotek,stw,kciesiel,mdramins,dcz}@ipipan.waw.pl

Summary. Visual presentation of document collections can provide with new insights into their content. A number of research projects has been launched to create document maps, like WebSOM, Themescape and other. The WebSOM like document map representation is regrettably time and space consuming, and rises also questions of scaling and updating of document maps. In this chapter we will describe some approaches we found useful in coping with these challenges, concentrating on growing neural gas. Solutions have been verified by creating a full-fledged visual search engine for collections of documents (up to a million). We extended WebSOM's goals by a multilingual approach, new forms of geometrical representation and we experimented also with various modifications to the clustering process itself. We also present some map quality evaluation techniques along with experimental results confirming the validity of our approach.

1 Introduction

Document maps become gradually more and more attractive as a way to visualize the contents of a large document collection. In section 2 we present a brief overview of the research in the area of visual Web mining.

The lesson, that may be learned from such important projects like WebSOM [28], is that the overall complexity of the map creation process, results n long run times for document collections of a couple of millions of documents. This imposes a heavy burden in cases when the document collection is dynamically growing or changing, so that recalculating the map from scretch is hardly acceptable. WebSOM has also the disadvantage that recomputation from sctrech generates different maps even for the same collection of documents due to random seeding of some processes. So, to reduce computational burden, as well as to avoid "revolutionary" changes of the image of the whole document collection, an incremental process of accommodation of new incoming documents into the collection needs to be designed.

M.A. Kłopotek et al.: *Techniques and Technologies Behind Maps of Internet and Intranet Document Collections*, Studies in Computational Intelligence (SCI) **37**, 169–190 (2007)
www.springerlink.com

Within our BEATCA project we have devoted much effort to enable such a gradual growth. To ensure intrinsic incremental formation of the map, all the computation-intense stages involved in the process of map formation (crawling, indexing, document and map cell clustering) need to be reformulated in terms of incremental growth. We provide with an overview of our approach in section 3. Section 4 concentrates on the toughest, most time consuming and most quality determining part of the process, that is on looking for general structure in the document collection.

In this study, we demonstrate also the effectiveness of our methodology both for vertical (new topics) and horizontal (new documents on current topics) growth of document collection and its effect on the map formation capability of the system (see section 5). To evaluate the effectiveness of the overall incremental map formation process, we compared it to the "from scratch" map formation in our experimental section. We also present topic-sensitive approach, which appears to be a robust solution to the problem of map generation process scalability (both in terms of time complexity and memory requirements). The conclusions from our research work can be found in section 6.

2 Related works

Modern man faces a rapid growth in the amount of written information. Therefore he needs a means of reducing the flow of information by concentrating on major topics in the document flow, including the one on the World Wide Web. Grouping documents based on similar contents may be helpful in this context as it provides the user with meaningful classes or clusters. Document clustering and classification techniques help significantly in organizing documents in this way.

For years now, therefore, we can observe a growing research effort around so-called "Web Mining". Web mining is generally understood as application area of data mining techniques to extraction of information from the World Wide Web. It is oriented either towards the improvement of site design and site structure, the generation of dynamic recommendations, and improving marketing. Inside web mining research, there exist essentially three main streams: web-content mining (around which the research presented in this paper concentrates), web-(link)structure mining and web usage mining.

Web content mining concentrates around the discovery of useful information from data contained in Web documents, including text, images, audio, video. The main difficulty is that on the one hand the Web content can be truly understood only by humans, and on the other hand the size of this content is immense. Two groups of methodologies are applicable here: Information retrieval, oriented towards unstructured (free text) data, and data based techniques, committed primarily to structured and semi-structured Web data. Information retrieval methods treat the documents as bugs of words, while database oriented methods exploit the semi-structured nature of HTML documents.

Database oriented methods attempt to find schemes (including type hierarchies) for semistructured data, to identify frequent substructures etc., by applying e.g. association rules or attribute oriented induction.

Information retrieval methods seek to find keywords and keyphrases, to discover grammatical rules, collocations, to predict word relations, to classify and/or cluster text/hypertext documents or named entities, to learn information extraction models (e.g. for event detection and tracking), to learn ontologies, and to find patterns. Methods used here are decision trees, kNN, rule learning methods, include inductive logic programming, reinforcement learning, support vector machines, self-organizing maps, but also Naive Bayes, Bayesian networks, logistic regression and other statistical methods.

In the recent time, the importance of visual presentation of Web mining results has been increasingly appreciated. For recent results concerning visual Web-structure mining the Reader may refer to [34], and with respect to visual Web-usage mining please consult e.g. [6].

As visual presentation of Web-content mining is concerned, there have been several important projects in the recent years. The early project SPIRE (Spatial Paradigm for Information Retrieval & Exploration) [33] represented documents as "stars in the skies", where the distance was reflecting word similarity and word pattern matching (with severe limitation of the document set related to screen resolution). The project *Themescape* (later also known as *Aureka!*), presents the document collection as a typical topological map, where "hills" correspond to frequent terms ("themes") [37]. While *Themescape* is "flat" (height presented with colors), the competing SpaceCast [38] project insists on intrinsic 3-D representation. Similar in spirit is the patented VxInsight [39] project.

Another project, worth mentioning, is the DocMiner [2]. It shows relationships between text documents, depicted by special marks, and groups of documents in terms of a document map. "Valleys" of similar documents are separated by "mountains" (dark colored, as on geographical maps).

Another interesting example is the project Island of Music [40], working with "music documents" (MP3 tunes), relying on similarity of compositions based on some psychoaccustic approximations.

A prominent position among the techniques of Visual Web-content mining is taken by the WebSOM (Self Organizing Maps) of Kohonnen and co-workers. However, the overwhelming majority of the existing document clustering and classification approaches rely on the assumption that the particular structure of the currently available static document collection will not change in the future. This seems to be highly unrealistic, because both the interests of the information consumer and of the information producers change over time.

A recent study described in [23] demonstrated deficiencies of various approaches to document organization under non-stationary environment conditions of growing document quantity. The mentioned paper pointed to weaknesses among others of the original SOM approach (which itself is adaptive to some extent) and proposed a novel dynamic self-organizing neural model, so-called Dynamic Adaptive Self-Organising Hybrid (DASH) model. This model is based on an adaptive hierarchical document organization, supported by human-created concept-organization hints available in terms of WordNet.

Other strategies like that of [30, 16], attempt to capture the move of topics, enlarge dynamically the document map (by adding new cells, not necessarily on a rectangle map).

We take a different perspective in this paper claiming that the adaptive and incremental nature of a document-map-based search engine cannot be confined to the map creation stage alone and in fact engages all the preceding stages of the whole document analysis process.

3 Overview of our appproach

One of main goals of the project is to create multidimensional document map in which geometrical vicinity reflects conceptual closeness of documents in a given document set. Additional navigational information (based on hyperlinks between documents) is introduced to visualize directions and strength of between-group topical connections.

The process of mapping a document collection to a two-dimensional map is a complex one and involves a number of steps which may be carried out in multiple variants. In our search engine BEATCA [11, 8, 9, 10, 12], the mapping process consists of the following stages: (1) document crawling (2) indexing (3) topic identification, (4) document grouping, (5) group-to-map transformation, (6) map region identification (7) group and region labeling (8) visualization. At each of theses stages various decisions can be made implying different views of the document map.

Our efforts started with creating a crawler, that can collect documents from the Internet devoted to a selected set of topics, based on ideas from [1]. The crawler learning process runs in a kind of horizontal growth loop while it improves its performance with increase of the amount of documents collected. It may also grow vertically, as the user can add new topics for search during its run time.

We needed also to construct an indexer that can achieve incremental growth and optimization of its dictionary with the growing collection of documents. The indexing process involves dictionary optimization, which may reduce the documents collection dimensionality and restrict the subspace in which the original documents are placed. Topics identification establishes basic dimensions for the final map and may involve such techniques as SVD analysis [3], fast Bayesian network learning (ETC [24]) and other. Document grouping may involve various variants of growing neural gas (GNG) techniques, [18]. The group-to-map transformation, used in BEATCA, is based on

SOM ideas, [28], but with variations concerning dynamic mixing of local and global search, based on diverse measures of local convergence.

With a strongly parameterized map creation process, the user of BEATCA can accommodate map generation to his particular needs, by selecting among diverse implemented technologies, or even generate multiple maps covering different aspects of document collection.

At the heart of the overall process is the issue of document clustering . It seems that clustering and content labeling is the crucial issue for understanding the two-dimensional map by the user. That is why we started our research with the WebSOM approach. It appeared however that both the speed and clustering stability were not very encouraging.

From our experience it follows that the basic problem with WebSOM lies in the initialization process of so-called reference vectors, being the centroids of the clusters to grow. In the original algorithm they are initialized randomly, and are gradually updated in the course of the training process. Such an initialization possibly leads to an instability during clustering, because the learning process of WebSOM possesses a "learning speed" parameter $\alpha(t)$, which may turn out to be too low to ensure convergence for a particular initialization.

Another problem lies in the general concept of clustering. In WebSOM, it is tightly coupled with a (non-linear) projection from a multidimensional to the two-dimensional space. Now, there may be infinitely many such projections with equal rights. So one needs really a sense of goal for selecting the appropriate one.

The first issue we tackled was dictionary optimization strategies and their speed-up effects to tackle the complexity issue. Another research direction was to obtain better clustering via fuzzy-set approach and immune-system-like clustering, [25]. In effect we proposed a multi-state approach to document clustering consisting of the following steps:

- clustering for identification of major topics (see [10], [9])
- cellular document clustering (see [11])
- cellular document clusters to WebSOM map projection (see [8])
- cell clusters extraction and cell labelling (see [25])

To obtain a stable map, one needs to fix the perspective from which he/she perceives the documents collection. This can be achieved by identifying major topics of the documents collection. This is done in the step "clustering for identification of major topics". In [11] we suggest a Bayesian approach, which is a result of our investigation of the behavior of the PLSA algorithm [22]. Alternatively, different initialization techniques could be used: in [10] we described an approach to major topic identification based on LSI/SVD, and in [13] we described usage of a version of Fuzzy-ISODATA algorithm for this purpose. Having identified the major topics, we can initialize the map in a more appropriate way [27].

After the topics have been identified, the documents need to be assigned to these and intermediate ones, and the relationships between the topics have to be identified. This process, termed "Cellular document clustering", leads to creation of a graph model of document collection. Three different techniques may be used at this point: the WebSOM (plain or hierarchical) approach [28], the GNG approach [18], or artificial immune systems (AIS) approach [27].

The graph model of document collection is visualized in a form of a document map. Therefore, a step of "cellular document clusters to WebSOM map projection" is applied. It is surplus in case, when WebSOM (plain or hierarchical) is used to cluster documents, but is necessary for more elastic topologies like GNG model and AIS model. This step is described in section 4.4. 4.

Finally, for purposes of better readability of the document map, cells need to be joined into larger, topically uniform areas which is done in the step "cell clusters extraction" [27]. Also cells have to be labeled, as described in [27].

4 An incremental model for Clustering of Text Documents

In this section we describe how we explore clustering properties of growing neural gas to accommodate new documents to the current clustering framework.

In our approach, objects (text documents as well as graph nodes, described below) are represented in the standard way, i.e. as vectors of the dimension equal to the number of distinct dictionary terms. A single element of so-called *referential vector* represents importance of a corresponding term and is calculated on the basis of the tfidf measure or our own context-sensitive w_{tdG} measure, which will be described later. Similarity measure between two objects is defined as the cosine of the angle between corresponding vectors.

4.1 GNG - pros and cons

It must be stressed that in our approach we heavily rely on Growing Neural Gas (GNG) networks proposed in [18]. Like Kohonen (SOM) networks, GNG can be viewed as topology learning algorithm, i.e. its aim can be summarized as follows: Given some collection of high-dimensional data, find a topological structure which closely reflects the topology of the collection. In typical SOM the number of units and topology of the map is predefined. As observed in [18], the choice of SOM structure is difficult, and the need to define a decay schedule for various features is problematic. On the contrary, GNG network does not require specification of the size of the network, and the resulting network adapts very well to a given data topology.

GNG starts with very few units and new units are inserted successively every each k iterations. To determine where to insert new units, local error measures are gathered during the adaptation process; new unit is inserted near the unit, which has accumulated maximal error. Interestingly, GNG cells of the GNG network are joined automatically by links, hence as a result a possibly disconnected graph is obtained, and its connected components can be treated as different data clusters.

The major drawback of the "plan" GNG is that it does not adopt well to changing structure of document collection. that is while it properly detects the shift of topic and creates new nodes to accomodate them, it leaves behind the old ones, useless for the representation of new collection structure. For this reason we turned to GNG with utility factor [20].

4.2 GNG Extension with Utility factor

Typical problem in web mining applications is that processed data is constantly changing - some documents disappear or become obsolete, while other enter analysis. All this requires models which are able to adapt its structure quickly in response to non-stationary distribution changes. Thus, we decided to adopt and implement GNG with utility factor model [20].

A crucial concept here is to identify the least useful nodes and remove them from GNG network, enabling further node insertions in regions where they would be more necessary. The utility factor of each node reflects its contribution to the total classification error reduction. In other words, node utility is proportional to expected error growth if the particular node would have been removed. There are many possible choices for the utility factor. In our implementation, utility update rule of a winning node has been simply defined as $U_s = U_s + error_t - error_s$, where s is the index of the winning node, and t is the index of the second-best node (the one which would become the winner if the actual winning node would be non-existent). Newly inserted node utility is arbitrarily initialized to the mean of two nodes which have accumulated most of the error: $U_r = (U_u + U_v)/2$.

After utility update phase, a node k with the smallest utility is removed if the fraction $error_j / U_k$ is greater then some predefined threshold; where j is the node with the greatest accumulated error.

4.3 Robust winner search in GNG network

Similarly to Kohonen algorithm [28], most computationally demanding part of GNG algorithm is the winner search phase. Especially, in application to web documents, where both the text corpus size and the number of GNG network nodes is huge, the cost of even a single global winner search phase is prohibitive.

Unfortunately, neither local-winner search method (i.e. searching through the graph edges from some staring node) nor joint-winner search method (our own approach devoted to SOM learning [25]) are directly applicable to GNG networks. The main reason for this is that a graph of GNG nodes can be unconnected. Thus, standard local-winner search approach would prevent document from shifting between separated components during the learning process.

A simple modification consist in remembering winning node for more than one connected component of the GNG graph and to conduct in parallel a single local-winner search thread for each component. Obviously, it requires periodical (precisely, once for an iteration) recalculation of connected components, but this is not very expensive (in order of $O(V + E)$, where V is the number of nodes and E is the number of connections (graph edges), as two winners are just sufficient to overcome the problem of components separation).

A special case is the possibility of a node removal. When the previous iteration's winning node for a particular document has been removed, we activate search processes (in parallel threads) from each of its direct neighbors in the graph.

We have implemented another method, a little more complex (both in terms of computation time and memory requirements) but, as the experiments show, more accurate. It exploits well-known Clustering Feature Tree [35] to group similar nodes in dense clusters. Node clusters are arranged in the hierarchy and stored in a balanced search tree. Thus, finding closest (most similar) node for a document requires $O(log_t V)$ comparisons, where V is the number of nodes and t is the tree branching factor (refer to [35]). Amortized tree structure maintenance cost (node insertion and removal) is also proportional to $O(log_t V)$.

4.4 Adaptive visualization of the model

Despite many advantages over SOM approach, GNG has one serious drawback: high-dimensional networks cannot be easily visualized. However, we can build Kohonen map on the referential vectors of GNG network, similarly to the case of single documents, i.e. treating each vector as a centroid representing a cluster of documents.

To obtain the visualization that singles out the main topics in the corpus and reflects the conceptual closeness between topics, the proper initialization of SOM cells is required. We have developed special initialization method, intended to identify broad topics in the text corpus. Briefly, in the first step we find the centroids of a few main clusters (via fast ETC Bayesian tree [24] and SVD eigenvectors decomposition [15]). Then, we select *fixpoint cells*, uniformly spread them on the map surface and initialize them with the centroid vectors. Finally, we initialize remaining map cells with *intermediate* topics, calculated as the weighted average of main topics, with the weight proportional to the Euclidean distance from the corresponding fixpoint cells.

After initialization, the map is learned with the standard Kohonen algorithm [28]. Finally, we adopt so-called *plastic clustering* algorithm [31] to adjust the position of GNG model nodes on the SOM projection map, so that the distance on the map reflects as close as possible the similarity of the adjacent nodes. The topical initialization of the map is crucial here to assure the stability of the final visualization [25].

The resulting map is a visualization of GNG network with the detail level depending on the SOM size (a single SOM cell can gather more than one GNG node). User can access document content via corresponding GNG node, which in turn can be accessed via SOM node - interface here is similar to the hierarchical SOM map case.

Exemplary map can be seen in Figure 1. The color brightness is related to the number of documents contained in the cell. Each cell which contains at least one document is labeled with a few descriptive terms (only one is visible here, the rest is available via BEATCA search engine). The black lines represents borders of topical areas. Clustering of map nodes, based on the combination of Fuzzy C-Means algorithm and minimal spanning tree technique is described [12]. It is important to stress that this planar representation is in fact a torus surface (which can also be visualized in 3D), so the cells on the map borders are adjacent.

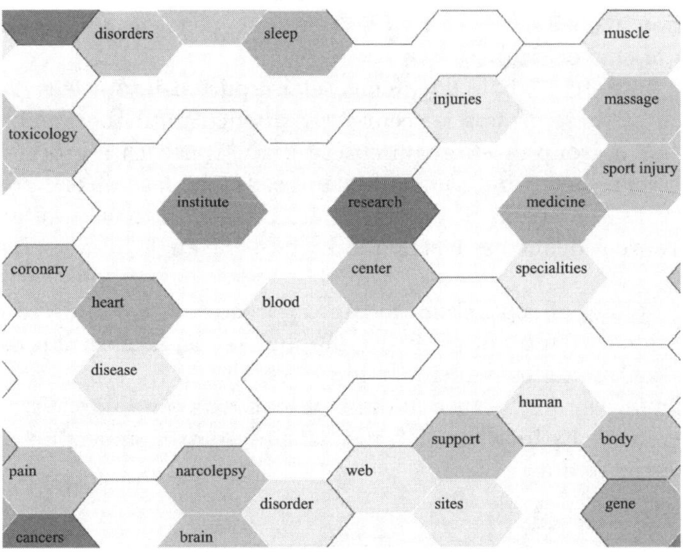

Fig. 1. Example of GNG model visualization

4.5 Contextual maps

In our work we use well known approach of representing documents as points in term vector space. It is a known phenomenon that text documents are not uniformly distributed over the space. Characteristics of frequency distributions of a particular term depend strongly on document location. On the basis of experimental results presented in the previous section, we suggest to identify automatically groups containing similar documents as the preprocessing step in document maps formation. We argue that after splitting documents in such groups, term frequency distributions within each group become much easier to analyze. In particular, it appears to be much easier to select significant and insignificant terms for efficient calculation of similarity measures during map formation step. Such document clusters we call *contextual* groups. For each contextual group, separate maps are generated. To obtain more informative maps there is a need to balance (during initial contextual clustering) size of each cluster. The number of documents presented on a map cannot be too high because the number of cells in the graph (and time required to create a map) would have to grow adequately. On the other hand, single map model should not hold only a few irrelevant documents.

Constraints on cluster size are obtained by recurrent divisions and merges of fuzzy document groups, created by a selected algorithm (e.g. PLSA [14] combined with ETC [24] or Chow-Liu [7] Bayesian net, SVD [15], Fuzzy C-Means [4]). In the case of Fuzzy C-means algorithm there is an additional modification in optimized quality criterion, that penalizes for inbalanced splits (in terms of cluster size).

In the first step, whole document set is splitted into a few (2-5) groups. Next, each of these groups is recursively divided until the number of documents inside a group meets required criteria. After such a process we obtain hierarchy, represented by a tree of clusters. In the last phase, groups which are smaller than predefined constraint, are merged to the closest group. Similarity measure is defined as a single-linkage cosine angle between both clusters centroids.

Crucial phase of contextual document processing is the division of terms space (dictionary) into - possibly overlapping - subspaces. In this case it is important to calculate fuzzy membership level, which will represent importance of a particular word or phrase in different contexts (and implicitly, ambiguity of its meaning). Estimation of fuzzy within-group membership of the term (m_{tG}) is estimated as:

$$m_{tG} = \frac{\sum_{d \in G} (f_{td} \cdot m_{dG})}{f_G \cdot \sum_{d \in G} m_{dG}} \tag{1}$$

where f_G is the number of documents in the cluster G, m_{dG} is the degree of document d membership level in group G, f_{td} is the number of occurrences of term t in document d.

Finally, vector-space representation of a document is modified to take into account document context. This representation increases weights of terms which are significant for a given contextual group and decrease weights of insignificant terms. In the boundary case, insignificant terms are ignored, what leads to the reduction of representation space dimensionality. To estimate the significance of term in a given context, the following measure is applied:

$$w_{tdG} = f_{td} \cdot m_{tG} \cdot log\left(\frac{f_G}{f_t \cdot m_{tG}}\right) \tag{2}$$

Main idea behind the proposed approach is to replace a single model (growing neural gas, or hierarchical SOM maps) by a set of independently created contextual models and to merge them together into a hierarchical model. Training data for each model is a single contextual group. Each document is represented as a standard referential vector in term-document space. However, tfidf measure of vector components is replaced by w_{tdG}.

To represent visually similarity relation between contexts (represented by a set of contextual models), additional "global" map is required. Such a model becomes a root of contextual maps hierarchy. Main map is created in a manner similar to previously created maps, with one distinction: an example in training data is a weighted centroid of referential vectors of the corresponding contextual model:

$$x_i = \sum_{c \in M_i} (d_c \cdot v_c) \tag{3}$$

Finally, cells and regions on the main map are labeled with keywords selected by the following contextual term quality measure:

$$Q_{tG} = ln(1 + f_{tG}) \cdot (1 - |EN_{tG} - 0.5|) \tag{4}$$

where EN_{tG} denotes normalized entropy of term frequency within the group.

Learning process of the contextual model is to some extent similar to the classic, non-contextual learning. However, it should be noted that each constituent model (and the corresponding contextual map) can be processed independently, in particular it can be distributed and calculated in parallel. Also a partial incremental update of such models appears to be much easier to perform, both in terms of model quality, stability and time complexity. The possibility of incremental learning stems from the fact that the very nature of the learning process is iterative. So if new documents come, we can consider the learning process as having been stopped at some stage and it is resumed now with all the documents. We claim that it is not necessary to start the learning process from scratch neither in the case that the new documents "fit" the distribution of the previous ones nor when their term distribution is significantly different. This claim is supported by experimental results presented in the section 5.2. In the section 5.3 we present some thoughts on scalability issues of contextual approach.

5 Experimental results

To evaluate the effectiveness of the overall incremental map formation process, we compared it to the "from scratch" map formation. In this section we describe the overall experimental design, quality measures used and the results obtained.

The architecture of our system supports comparative studies of clustering methods at the various stages of the process (i.e. initial document grouping, initial topic identification, incremental clustering, model projection and visualization, identification of topical areas on the map and its labeling). In particular, we conducted series of experiments to compare the quality and stability of GNG and SOM models for various model initialization methods, winner search methods and learning parameters [26]. In this paper we focus only on evaluation of the GNG winner search method and the quality of the resulting incremental clustering model with respect to the topic-sensitive learning approach.

5.1 Quality Measures for the Document Maps

Various measures of quality have been developed in the literature, covering diverse aspects of the clustering process. The clustering process is frequently referred as "learning without a teacher", or "unsupervised learning", and is driven by some kind of similarity measure. The term "unsupervised" is not completely reflecting the real nature of learning. In fact, the similarity measure used is not something "natural", but rather it reflects the intentions of the teacher. So we can say that clustering is a learning process with hidden learning criterion. The criterion is intended to reflect some esthetic preferences, like: uniform split into groups (topological continuity) or appropriate split of documents with known a priori categorization. As the criterion is somehow hidden, we need tests if the clustering process really fits the expectations. In particular, we have accommodated for our purposes and investigated the following well known quality measures of clustering [36, 5, 21]:

- **Average Map Quantization**: the average cosine distance between each pair of adjacent nodes. The goal is to measure topological continuity of the model (the lower this value is, the more "smooth" model is):

$$AvgMapQ = \frac{1}{|N|} \sum_{n \in N} \left(\frac{1}{|E(n)|} \sum_{m \in E(n)} c(n,m) \right) \tag{5}$$

  where N is the set of graph nodes, $E(n)$ is the set of nodes adjacent to the node n and $c(n,m)$ is the cosine distance between nodes n and m.

- **Average Document Quantization**: average distance (according to cosine measure) for the learning set between the document and the node it was classified into. The goal is to measure the quality of clustering at the level of a single node:

$$AvgDocQ = \frac{1}{|N|} \sum_{n \in N} \left(\frac{1}{|D(n)|} \sum_{d \in D(n)} c(d,n) \right) \tag{6}$$

where $D(n)$ is the set of documents assigned to the node n.

Both measures have values in the [0,1] interval, the lower values corresponds respectively to more "smooth" inter-cluster transitions and more "compact" clusters. To some extent, optimization of one of the measures entails increase of the other one. Still, experiments [26] show that the GNG models are much more smooth than SOM maps while the clusters are of similar quality.

The two subsequent measures evaluate the agreement between the clustering and the a priori categorization of documents (i.e. particular newsgroup in case of newsgroups messages).

- **Average Weighted Cluster Purity**: average "category purity" of a node (node weight is equal to its density, i.e. the number of assigned documents):

$$AvgPurity = \frac{1}{|D|} \sum_{n \in N} max_c\left(|D_c(n)|\right) \tag{7}$$

  where D is the set of all documents in the corpus and $D_c(n)$ is the set of documents from category c assigned to the node n.

- **Normalized Mutual Information**: the quotient of the total category and the total cluster entropy to the square root of the product of category and cluster entropies for individual clusters:

$$NMI = \frac{\sum_{n \in N} \sum_{c \in C} |D_c(n)| \; log\left(\frac{|D_c(n)| \; |D|}{|D(n)| \; |D_c|}\right)}{\sqrt{\left(\sum_{n \in N} |D(n)| \; log\left(\frac{|D(n)|}{|D|}\right)\right)\left(\sum_{c \in C} |D_c| \; log\left(\frac{|D_c|}{|D|}\right)\right)}} \tag{8}$$

  where N is the set of graph nodes, D is the set of all documents in the corpus, $D(n)$ is the set of documents assigned to the node n, D_c is the set of all documents from category c and $D_c(n)$ is the set of documents from category c assigned to the node n.

Again, both measures have values in the [0,1] interval. The higher the value is, the better agreement between clusters and a priori categories.

5.2 Incrementality study

Model evaluation were executed on 2054 of documents downloaded from 5 newsgroups with quite well separated main topics (antiques, computers, hockey, medicine and religion). Each GNG network has been trained for 100 iterations with the same set of learning parameters, using previously described winner search method.

In the main case (depicted with the black line), network has been trained on the whole set of documents. This case was the reference one for the quality measures of adaptation as well as comparison of the winner search methods.

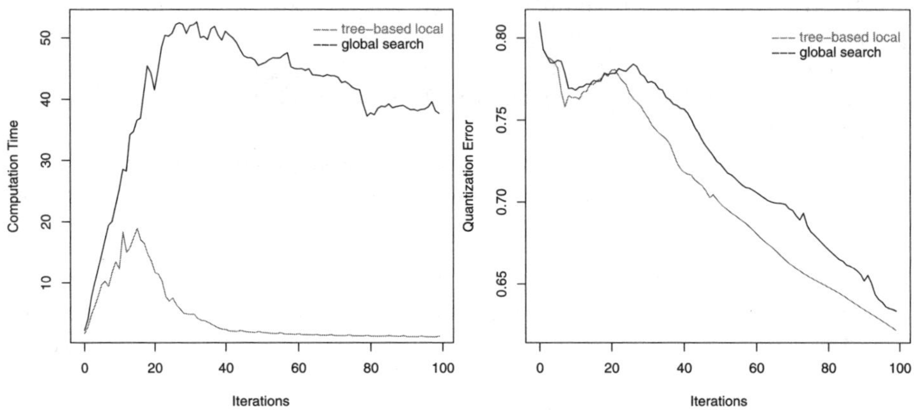

Fig. 2. Winner search methods (a) computation time (b) model quality

Figure 2 presents comparison of a standard global winner search method with our own CF-tree based approach. Local search method is not taken into consideration since, as it has already been mentioned, it is completely inappropriate in case of unconnected graphs. Obviously, tree-based local method is invincible in terms of computation time. The main drawback of the global method is that it is not scalable and depends on the total number of nodes in the GNG model.

The results seemed to be surprizing at first glance. On one hand, the quality was similar, on the other - global search appeared to be worse of the two! We have investigated it further and it turned out to be the aftermath of process divergence during the early iterations of the training process. We'll explain it later, on the example of another experiment.

In the next experiment, in addition to the main reference case, we had another two cases. During the first 30 iterations network has been trained on 700 documents only. In one of the cases (light grey line) documents were sampled uniformly from all five groups and in the 33rd iteration another 700 uniformly sampled were introduced to training. After the 66th iteration the model has been trained on the whole dataset.

In the last case (dark grey line) initial 700 documents were selected only from two groups. After the 33rd iteration of training, documents from the remaining newsgroups were gradually introduced in the order of their newsgroup membership. It should be noted here that in this case we had an a priori information on the categories of documents. In the general case, we are collecting fuzzy category membership information from Bayesian Net model.

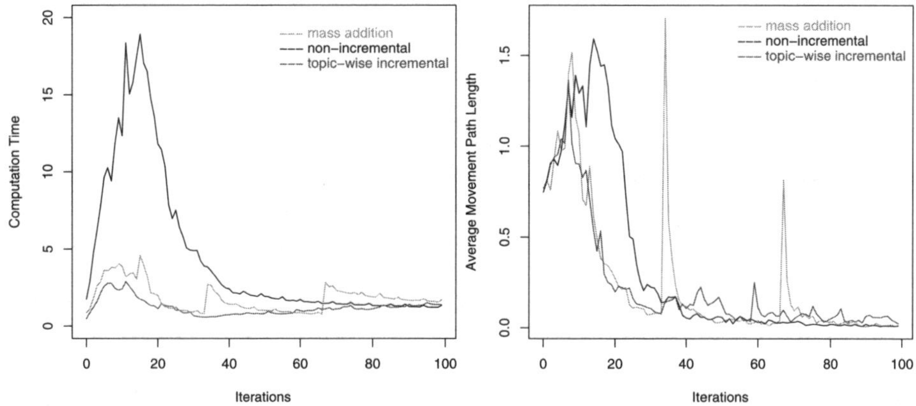

Fig. 3. Computation complexity (a) execution time of a single iteration (b) average path length of a document

As expected, in all cases GNG model adapts quite well to the topic drift. In the global and the topic-wise incremental case, the quality of the models were comparable, in terms of Average Document Quantization measure (see figure 4(a)), Average Weighted Cluster Purity, Average Cluster Entropy and Normalized Mutual Information (for the final values see table 1). Also the subjective criteria such as visualizations of both models and the identification of topical areas on the SOM projection map were similar.

Table 1. Final values of model quality measures

	Cluster Purity	*Cluster Entropy*	*NMI*
non-incremental	0.91387	0.00116	0.60560
topic-wise incremental	0.91825	0.00111	0.61336
massive addition	0.85596	0.00186	0.55306

The results were noticeably worse for the massive addition of documents, even though all covered topics were present in the training from the very beginning and should have occupied their own, specialized areas in the model. However, and it can be noticed on the same plot, a complex mixture of topics can pose a serious drawback, especially in the first training iterations. In the global reference case, the attempt to cover all topics at once leads learning process to a local minimum and to subsequent divergence (what, in fact, is quite time-consuming as one can notice on figure 3(a)).

As we have previously noticed, the above-mentioned difficulties apply also to the case of global winner search (figure 2(b)). In a matter of fact, the quality of the final models when we take advantage of the incremental approach is almost the same for global search and CF-tree based search (Cluster Purity: 0.92232 versus 0.91825, Normalized Mutual Information: 0.61923 versus 0.61336, Average Document Quantization: 0.64012 versus 0.64211).

The figure 3(b) presents average number of GNG graph edges traversed by a document during a single training iteration. It can be seen that a massive addition causes temporal instability of the model. Also, the above mentioned attempts to cover all topics at once in case of a global model caused much slower stabilization of the model and extremely high complexity of computations (figure 3(a)). The last reason for such slow computations is the representation of the GNG model nodes. The referential vector in such node is represented as a balanced red-black tree of term weights. If a single node tries to occupy too big portion of a document-term space, too many terms appear in such tree and it becomes less sparse and - simply - bigger. On the other hand, better separation of terms which are likely to appear in various newsgroups and increasing "crispness" of topical areas during model training leads to highly efficient computations and better models, both in terms of previously mentioned measures and subjective human reception of the results of search queries.

The last figure, 4(b), compares the change in the value of Average Map Quantization measure, reflecting "smoothness" of the model (i.e. continuous shift between related topics). In all three cases the results are almost identical. It should be noted that extremely low initial value of the Average Map Quantization is the result of the model initialization via broad topics method [25], shortly described in the section 4.4.

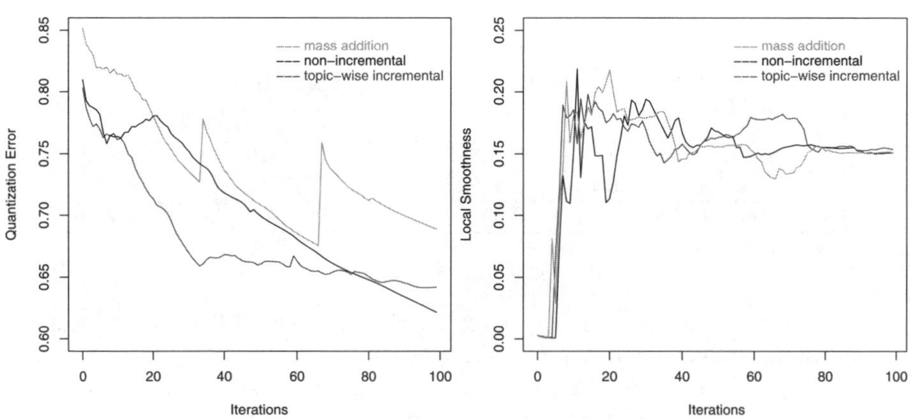

Fig. 4. Model quality (a) Average Document Quantization (b) Average Map Quantization

5.3 Scalability issues

To evaluate scalability of the proposed contextual approach (both in terms of space and time complexity), we built a model for a collection of million documents crawled by our topic-sensitive crawler starting from several Internet news sites (cnn, reuters, bbc).

Resulting model consisted of 412 contextual maps, which means that the average density of a single map was about 2500 documents. Experimental results in this section are presented in series of box-and-whisker plots, which allows to present a distribution of a given evaluation measure (e.g. time, model smoothness or quantization error) over all 412 models, measured after each iteration of the learning process (horizontal axis). Horizontal line represents median value, area inside the box represents 25% - 75% quantiles, whiskers represent extreme values and each dot represents outlier values.

Starting with initial document clustering/context initialization via hierarchical Fuzzy ISODATA (see section 4.5), followed by GNG model learning (see section 4.1) and GNG-SOM projection (see section 4.4), the whole cycle of map creation process took 2 days. It is impressing result, taking into account that Kohonen and his co-workers reported processing times in order of weeks [29]. It should also be noted that the model was built on a single personal computer (Pentium IV HT 3.2 GHz, 1 GB RAM). As it has been stated before, contextual model construction can be easily distributed and parallelized, what would lead to even shorter execution times.

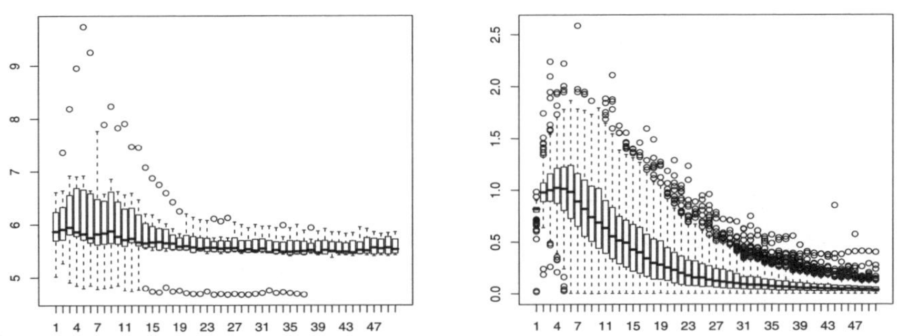

Fig. 5. Contextual model computation complexity (a) execution time of a single iteration (b) average path length of a document

The first observation is the complexity of a single iteration of GNG model learning (Figure 5(a)), which is almost constant, regardless of the increasing size of the model graph. It confirms the observations from section 5, concerning efficiency of the tree-based winner search methods. One can also observe the positive impact of homogeneity of the distribution of term frequencies in documents grouped to a single map cell. Such homogeneity is - to some extent - acquired by initial split of a document collection into contexts. Another way of the processing time reduction can be the contextual reduction of vector representation dimensionality, described in the previous section.

In the Figure 5(b), the dynamic of the learning process is presented. The average path length of a document is the number of shifts over graph edges when documents is moved to a new, optimal location. It can be seen that model stabilizes quite fast; actually, most models converged to final state in less than 30 iterations. The fast convergence is mainly due to topical initialization. It should also be noted here that the proper topical initialization can be obtained for well-defined topics, which is the case in contextual maps.

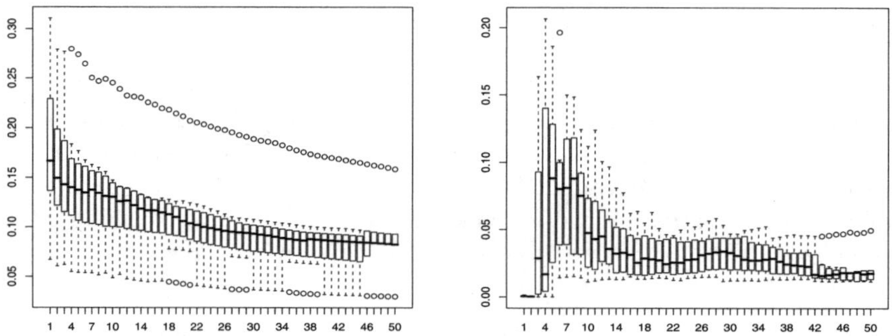

Fig. 6. Contextual model quality (a) Average Document Quantization (b) Average Map Quantization

The Figure 6 presents the quality of the contextual models. The final values of average document quantization (Figure 6(a)) and the map quantization (Figure 6(b)) are low, which means that the resulting maps are both "smooth" in terms of local similarity of adjacent cells and precisely represent documents grouped in a single node. Moreover, such low values of document quantization measure have been obtained for moderate size of GNG models (majority of the models consisted of only 20-25 nodes - due to their fast convergence - and represented about 2500 documents each).

6 Concluding remarks

As indicated e.g. in [23], most document clustering methods, including the original WebSOM, suffer from their inability to accommodate streams of new documents, especially such in which a drift, or even radical change of topic occurs.

Though one could imagine that such an accommodation could be achieved by "brute force" (learning from scratch whenever new documents arrive), but there exists a fundamental technical obstacle for a procedure like that: the processing time. But the problem is deeper and has a "second bottom": the clustering methods like those of WebSOM contain elements of randomness so that even re-clustering of the same document collection may lead to a radical change of the view of the documents.

The results of this research are concerned with both aspects of adaptive clustering of documents.

First of all, the whole process of document map formation has been designed in an incremental way, that is crawling, indexing and all the stages of map formation (document grouping, document group to WebSOM mapping and map region identification). So the Bayesian Network driven crawler is capable of collecting documents around an increasing number of distinct topics. Incremental structures of the indexer accommodate to changing dictionary. The query answering interface, in particular its query extension capability based on Bayesian network and GNG derived dynamic automated thesaurus, accommodates also to the growing document collection. Though the proper clustering algorithms used in our system, like GNG, SOM, or fuzzy-k-means, are by their nature adaptive, nonetheless their tuning and modification was not a trivial task, especially with respect to our goal to achieve quality of incremental map comparable to the non-incremental one.

Second, special algorithms for topical map initialization as well as for identification of document collection topics, based on GNG, SVD and/or Bayesian networks, lead to stabilization of the overall map. At the same time GNG detects the topic drift and so it may be appropriately visualized, due to plastic clustering approach, as new emerging map regions. It should be stressed at this point, that the map stabilization does not preclude obtaining different views of the same document collection. Our system permits to maintain several maps of the same document collection, obtained via different initializations of the map, and, what is more important, automatically tells the user which of the maps is most appropriate to view the results of his actual query.

The most important contribution of this paper is to demonstrate, that the whole incremental machinery not only works, but it works well. For the quality measures we investigated, we found that our incremental architecture compares well to non-incremental map learning both under scenario of "massive addition" of new documents (many new documents, not differing in their topical structure, presented in large portions) and of scenario of "topic-wise-increment" of the collection (small document groups added, but with new emerging topics). The latter seemed to be the most tough learning process for incremental learning, but apparently the GNG application prior to WebSOM allowed for cleaner separation of new topics from ones already discovered, so that the quality (e.g. in terms of cluster purity and entropy) was higher under incremental learning than under non-incremental learning.

The experimental results indicate, that the real hard task for an incremental map creation process is a learning scenario where the documents with new topical elements are presented in large portions. But also in this case the results proved to be satisfactory.

A separate issue is the learning speed in the context of crisp and fuzzy learning models. Apparently separable and topically "clean" models allow for faster learning as the referential vectors of SOM are smaller (contain fewer non-zero components).

From the point of view of incremental learning under SOM, a crucial factor for the processing time is the global winner search for assignment of documents to neurons. We were capable to elaborate a very effective method of mixing global with local winner search which does not deteriorate the overall quality of the final map and at the same time comes close to the speed of local search.

Our future research will concentrate on exploring further adaptive methods like artificial immunological systems for more reliable extraction of context-dependent thesauri and adaptive parameter tuning.

Acknowledgements

This research is partially supported under KBN research grant 4 T11C 026 25 "Maps and intelligent navigation in WWW using Bayesian networks and artificial immune systems". Authors would like to thank Dawid Weiss, who provided Polish and English "stop words" lists and Polish stemmer.

References

1. Aggarwal C, Al-Garawi F, Yu P (2001) Intelligent crawling on the World Wide Web with arbitrary predicates. In: Proc. 10th International World Wide Web Conference, 96–105.
2. Becks A (2001) Visual Knowledge Management with Adaptable Document Maps. GMD Research Series, Sank Augustin
3. Berry M, Drmac Z, IJessup E (1999) Matrices, vector spaces and information retrieval. SIAM Review 41:2:335–362
4. Bezdek J, Pal S (1992) Fuzzy Models for Pattern Recognition: Methods that Search for Structures in Data. IEEE, New York
5. Boulis C, Ostendorf M (2004) Combining multiple clustering systems. In: Proceedings of 8th European conference on Principles and Practice of Knowledge Discovery in Databases (PKDD-2004), LNAI 3202, Springer-Verlag
6. Chen J, Sun L, Zaiane O, Goebel R (2004) Visualizing and Discovering Web Navigational Patterns. `webdb2004.cs.columbia.edu/papers/1-3.pdf`
7. Chow C, Liu C (1968) Approximating discrete probability distributions with dependence trees. IEEE Transactions on IT, IT-14:3:462–467

8. Ciesielski K, Dramiński M,Kłopotek M, Kujawiak M, Wierzchoń S (2004) Architecture for graphical maps of Web contents. In: Proc. WISIS'2004,Warsaw
9. Ciesielski K, Dramiński M,Kłopotek M, Kujawiak M, Wierzchoń S (2004) Mapping document collections in non-standard geometries. In: De Beats B, De Caluwe R, de Tre G, Fodor J, Kacprzyk J, Zadrony S (eds): Current Issues in Data and Knowledge Engineering. Akademicka Oficyna Wydawnicza EXIT Publ., Warszawa, 122–132
10. Ciesielski K, Dramiński M,Kłopotek M, Kujawiak M, Wierzchoń S (2004) Clustering medical and biomedical texts - document map based approach. In: Proc. Sztuczna Inteligencja w Inynierii Biomedycznej SIIB'04, Kraków. ISBN-83-919051-5-2
11. Ciesielski K, Dramiński M,Kłopotek M, Kujawiak M, Wierzchoń S (2005) On some clustering algorithms for Document Maps Creation. In: Proceedings of the Intelligent Information Processing and Web Mining (IIS:IIPWM-2005), Gdansk, 2005
12. Ciesielski K, Dramiński M,Kłopotek M, Kujawiak M, Wierzchoń S (2005) Co-existence of crisp and fuzzy concepts in document maps. in: Duch w, Kacprzyk J, eds, Proc. ICAINN, LNCS vol. 3697/2005, Springer Verlag, part II pp. 859.
13. Ciesielski K, Dramiński M, Kłopotek M, Czerski D, Wierzchoń S (2006) Adaptive document maps. To appear in: Proceedings of the Intelligent Information Processing and Web Mining (IIS:IIPWM-2006), Ustroń
14. Cohn D, Hofmann T (2001) The missing link - a probabilistic model of document content and hypertext connectivity. In: Leen T, Dietterich T, Tresp V (eds): Advances in Neural Information Processing Systems, Vol. 10, `http://citeseer.nj.nec.com/cohn01missing.html`
15. Deerwester S, Dumais S, Landauer T, Furnas G, Harshman R (1990) Indexing by Latent Semantic Analysis. Journal of the American Society of Information Science, 41(1990)6:391–407 `citeseer.nj.nec.com/deerwester90indexing.html`
16. Dittenbach M, Rauber A, Merkl D (2002) Uncovering hierarchical structure in data using the Growing Hierarchical Self-Organizing Map. Neurocomputing 48 (1-4): 199–216.
17. Dubois D, Prade H (1980) Fuzzy Sets and Systems. Theory and Applications, Academic Press
18. Fritzke B (1995) A growing neural gas network learns topologies. In: Tesauro G, Touretzky D, Leen T (Eds.): Advances in Neural Information Processing Systems 7, MIT Press Cambridge, MA, 625–632.
19. Fritzke B (1996) Some competitive learning methods, draft available from `http://www.neuroinformatik.ruhr-uni-bochum.de/ini/VDM/research/gsn/JavaPaper`
20. Fritzke B (1997) A self-organizing network that can follow non-stationary distributions. In: Proceeding of the International Conference on Artificial Neural Networks '97, Springer, 613–618
21. Halkidi M, Batistakis Y, Vazirgiannis M (2001) On clustering validation techniques. Journal of Intelligent Information Systems, 17(2-3):107–145
22. Hoffmann T (1999) Probabilistic Latent Semantic Analysis. In: Proceedings of the 15th Conference on Uncertainty in AI.
23. Hung C, Wermter S (2005) A constructive and hierarchical self-organising model in a non-stationary environment. In: International Joint Conference in Neural Networks,

24. Kłopotek M (2002) A new Bayesian tree learning method with reduced time and space complexity. Fundamenta Informaticae, 49(4):349–367
25. Kłopotek M, Dramiński M, Ciesielski K, Kujawiak M, Wierzchoń S (2004) Mining document maps. In: Proceedings of Statistical Approaches to Web Mining Workshop (SAWM) at PKDD'04, M. Gori, M. Celi, M. Nanni (eds.), Pisa, 87–98
26. Kłopotek M, Wierzchoń S, Ciesielski K, Dramiński M, Czerski D, Kujawiak M (2005) Understanding nature of map representation of document collections map quality measurements. In: Proc. Int.Conf. Artificial Intelligence Siedlce,
27. Kłopotek M, Wierzchoń S, Ciesielski K, Dramiński M, Czerski D (2006) Conceptual maps and intelligent navigation in document space (in Polish). To appear in: Akademicka Oficyna Wydawnicza EXIT Publishing, Warszawa.
28. Kohonen T (2001) Self-Organizing Maps. Springer Series in Information Sciences, vol. 30, Springer, Berlin, Heidelberg, New York.
29. Kohonen T, Kaski S, Somervuo P, Lagus K, Oja M, Paatero V (2003) Self-organization of very large document collections. Helsinki University of Technology technical report. `http://www.cis.hut.fi/research/reports/biennial02-03`
30. Rauber A (1996) Cluster Visualization in Unsupervised Neural Networks. Diplomarbeit, Technische Universitt Wien, Austria
31. Timmis J (2001) aiVIS: Artificial Immune Network Visualization. In: Proceedings of EuroGraphics UK 2001 Conference, Univeristy College London, 61–69
32. Wierzchoń S (2001) Artificial immune systems. Theory and applications (in Polish), ICS PAS Publishing House.
33. Wise J, Thomas J, Pennock K, Lantrip D, Pottier M, Schur A, Crow V (1995) Visualizing the non-visual: Spatial analysis and interaction with information from text documents. In: IEEE Information Visualization. 51–58.
34. Youssefi A, Duke D, Zaki M (2004) Visual Web Mining. `http://www2004.org/proceedings/docs/2p394.pdf`, WWW2004, May 1722, 2004, New York, NY USA.
35. Zhang T, Ramakrishan R, Livny M (1997) BIRCH: Efficient data clustering method for large databases. In: Proceedings of ACM SIGMOD International Conference on Data Management.
36. Zhao Y, Karypis G (2005) Criterion functions for document clustering: Experiments and analysis, available at `http://www-users.cs.umn.edu/~karypis/publications/ir.html`
37. `http://www.micropatent.com/static/index.htm`
38. `http://www.geog.ucsb.edu/~sara/html/research/spacecast/spacecast.html`
39. `http://www.cs.sandia.gov/projects/VxInsight.html`
40. `http://www.oefai.at/~elias/music/index.html`

Exchange Rate Modelling for E-Negotiators Using Text Mining Techniques

Debbie Zhang, Simeon Simoff, and John Debenham

Faculty of Information Technology, University of Technology, Sydney
{debbiez, simeon, debenham}@it.uts.edu.au

Abstract

The Curious Negotiator project aims at the automation (to the extent possible) of the delivery and use of information by negotiation agents in electronic market environment. This chapter presents a framework for using text mining agents to provide processed online information to negotiation agents. It includes a news extraction algorithm, a quantitative process model based on the extracted news information, which is exemplified by an exchange rate prediction model, and a communication protocol between data mining agents and negotiation agents. This information is critical for the negotiation agents to form their negotiation strategies.

1 Introduction

Electronic negotiation is an area in intelligent technologies that has witnessed significant development over the last decade, aiming at increased opportunities for entrepreneurs for global trade. Successful negotiation relies on an understanding of how to 'play' the negotiation mechanism and how to utilise contextual information available at the time of negotiation (Simoff and Debenham 2002). Identifying, requesting and evaluating contextual information is part of the negotiation strategies as the negotiation proceeds. The significance of information to the negotiation process was formally analysed (Milgrom and Weber, 1982) in which the Linkage Principle, relating the revelation of contextual information to the price that a purchaser is prepared to pay, was introduced. Substantial effort has gone in the development of a variety of negotiation strategies, including offers with argumentation. However, very little work

D. Zhang et al.: *Exchange Rate Modelling for E-Negotiators Using Text Mining Techniques*, Studies in Computational Intelligence (SCI) **37**, 191–211 (2007)
www.springerlink.com

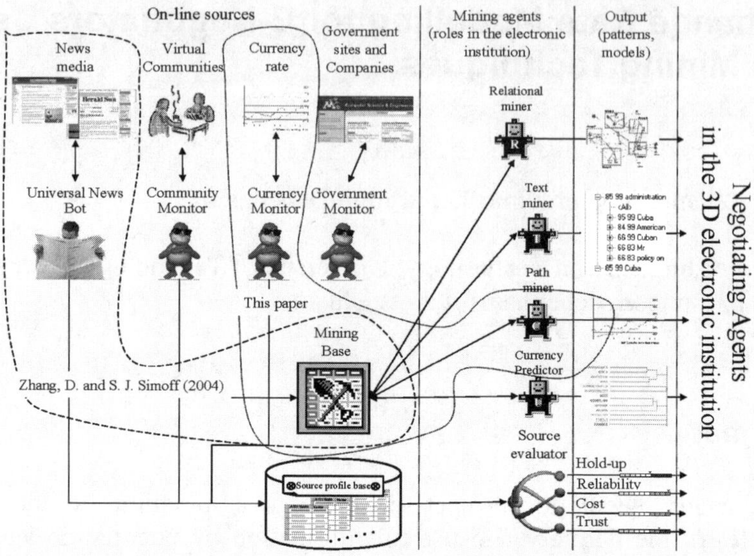

Fig. 1. The smart data mining system that provides information support to negotiation agents

has been done in automating the utilisation of contextual information and the development of mechanisms that allow incorporating this information in a computationally feasible (and executable) negotiation strategy.

This chapter focuses on the process of using intelligent agents for gathering, processing and presenting information to support negotiation agents. This work is part of the Curious Negotiator project. One of the main aims of the project is to automate (to the extent possible) the delivery and use of information by negotiation agents. The curious negotiator which was proposed by the authors is an internet-based multiagent system of competitive agents supporting multiple attributes, illustrated in Figure 1.

The Curious Negotiator is designed to incorporate data mining and existing information discovery methods that operate under time constraints, including methods from the area of topic detection and event tracking research. The intelligent data mining agents in the system operate in tandem with human and/or negotiation agents. Initially the information is extracted from various sources including on-line news media, virtual communities, and company and government on-line publications (including websites). Extracted information is converted into a structured representation. Both pre and post processed representations are stored in the mining database. They are used for further analysis by different data mining algorithms, including different text and network data mining agents. The output of the analysis results is used by the negotiation agents.

Illustrated by the example of currency exchange trading, this chapter describes the framework of online information extraction and data modelling for negotiation agents. The rest of the chapter is organised as follows: Section 2 presents an algorithm of online news extraction. An exchange rate model between currencies using news articles and economic data is provided in Section 3. Section 4 describes the communication protocol between data mining agents and negotiation agents based on ontology, followed by the conclusions in Section 5.

2 On Line News Extraction

Internet contains vast and timely information that can be used by negotiation agents. However, obtaining and verifying information from on-line sources takes time and resources. To reduce the impact of some delay factors on the net, the architecture of the data mining system in Figure 2 allows not only just-in-time operation, but also 'pre-fetching' some of the information that is expected to be necessary for a scheduled negotiation.

There are a number of challenges for online data extraction in real world that the smart data mining system needs to address, including (i) critical pieces of information being held in different repositories; (ii) non-standard formats; (iii) changes in formats at the same repository; (iv) possible duplicative, inconsistent and erroneous data. This section addresses the first three issues, and the fourth issue is partly addressed. Although there are many types of information available from the Internet, news is unarguable the most important information source that inferences market

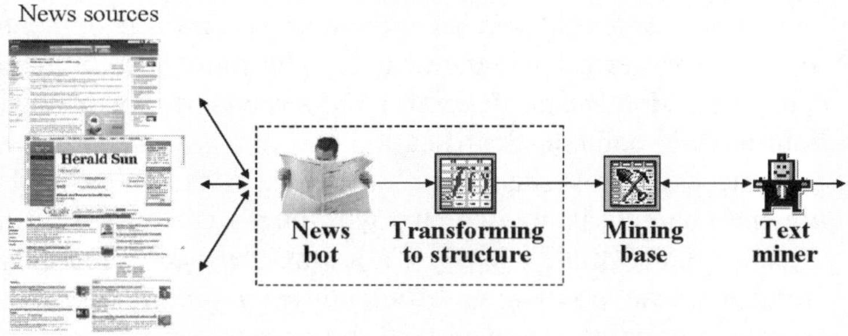

Fig. 2. The news mining portion of the system

dynamics. Therefore, a generic news bot is developed to monitor varied web sites of major news papers and obtain the latest news from news searching engines.

To retrieve related news URL is not a problem. Most news web sites nowadays provide RSS channels that provide news feed in XML format. However, they normally contain only news titles and URLs; the actual news data is not available to avoid readers to skip their advertisements. Another way for obtaining the URLs of related news is to use search engines such as Yahoo or Google. Since the news content can come from different web sites, developing a computer program for retrieving an individual news article from obtained URL could be a tedious job. Different news sources have different layout and format as illustrated by the news sources in Figure 2. The layout may vary from time to time even in the news coming from the same source. Hence when automating news retrieval, even for the same news site, it is impractical to develop a static template, as it will stop working when the layout is changed. It is even more impractical (if not impossible) to develop a predefined program (template) for each news web site in the whole Internet.

Data extraction from Web documents is usually performed by software modules called wrappers. To overcome the problems caused by using hard-coded wrappers, significant research has been done in the area of wrapper induction, which typically applies machine learning technology to generate wrappers automatically. WIEN is the first wrapper induction system that defined six wrapper classes (templates) to express the structures of web sites (Kushmerick, 2000). STALKER – a wrapper, more efficient than WIEN (Muslea et al, 1998), treats a web page as a tree-like structure and handles information extraction hierarchically. Significant work on knowledge-based information extraction from the Internet has also been conducted (Gao et al, 2000). However, most of the earlier wrapper techniques were tailored to particular types of documents and none are specific for news content retrieval. The more recent techniques aim on data extraction from general semi-structured documents. The application of general content identification and retrieval methods to news data brings unnecessary overhead in processing. A technique that takes into account the characteristics of news web pages was proposed by the authors (Zhang and Simoff, 2004). Without loss of generality, the approach improves the processing efficiency and requires neither user specified examples or priori knowledge of the pages.

The data extraction process is divided into three stages. The logical structure of the tagged file is firstly identified and the text, which is most likely to be the news article, is extracted. During the second stage a filter is dynamically built and some extra text is filtered out if multiple documents from the same web site are available. During the third stage extracted data is validated by the developed keyword based validation method. The details are presented below.

Stage 1: Identifying the logical structure of the tagged file

News pages normally not only contain the news article, but more often, also related news headings, the news category, advertisements, sometimes a search box, and other unrelated information. Although each web site may have a different format, web pages can always be broken down into content blocks. The layout in which these content blocks are arranged varies considerably across sites. The news article is expected to be the content block that is "central" to the content of the page. Therefore, it is reasonable to assume that the biggest block of text on the news web page is the news article, which is detected by counting the number of words in each block.

Most of web sites employ visible and invisible tables in conjunction with Cascading Style Sheets (CSS) to arrange their logical structures by using HTML table tags. Tables are designed to organize data into logical rows and columns. A table is enclosed within the <table></table> tag. Nested tables are normally used to form a complex layout structure. It is common for news web sites to display advertisements within news articles to attract reader's attention. This is normally done by inserting nested tables that contain advertisements and other contents in the table that contains the news article. The pseudo code of the process is presented in Figure 3.

Stage 2: Building adaptive internal filters dynamically

Although most of news web sites use tables for partitioning content blocks, there are some web sites that use other methods. Also, even for the web pages that use tables as the partition method, the table with the news article may contain a few extra lines of text at the beginning or the end of the article. Therefore, extraction accuracy can be improved by developing algorithms that do not rely on table tag information. Many web sites use templates to automatically generate pages and fill them with results of a database query, in particular, for news web sites. Hence, news under the same category from the same source is often with the same format. When two or more web pages from the same source become available, an adaptive filter can be constructed by comparing the extracted text from these pages. The filter contains the common header and tail of the text. The text is compared sentence by sentence from the beginning to the end between two files. Common sentences are regarded as part of web page template. Therefore, they should be removed from the file. The pseudo code of the process is shown in Figure 4. Once the filter is generated, text is refined by removing the common header and tail text in the filter. Since the filter is dynamically generated, it is adjusted automatically when the web site format is changed.

Input: HTML file
Output: The largest body of text contained in a table
Begin

1. Break down the HTML file into a one dimensional array, where each cell contains a line of text or an HTML tag
2. Remove the HTML tags except <table> and </table>
3. Set *table_counter* to 0
4. For each cell in the array:
    a. if <table> tag is encountered, increase *table_counter* by 1
    b. if <\table> tag is encountered, decrease *table_counter* by 1
    c. if it is a text element, append it to the end of container[*table_counter*]
5. Return container[*i*] that contains the largest body of text by counting the number of words.

End

Fig. 3. Pseudo code of the algorithm for identifying the largest text block in tagged text.

Stage 3: Keyword based validation

Incorrect and out of date URLs can cause errors in the results of data extraction. Such errors can not be identified by the data extracting methods described in the previous sections. A simple validation method based on keyword frequency is developed to validate the retrieved data.

The basic assumption is that a good news title should succinctly express the article's content. Therefore, the words contained in the news title are expected to be normally among the most frequent words appearing in a news article. Consequently, the words from the news title (except the stop words, which are filtered out) are considered as keywords. For situations when the news title is not available at the time of text extraction, the words in the first paragraph of the extracted data are considered as keywords, based on the assumption that title is always placed at the beginning of an article. The extracted text is regarded as the requested news article if it satisfies the following condition:

$$\min\left(w_1 \frac{l_t}{l_m}, w_2 \frac{n_k}{t_k}, w_3 k_f \right) > th \tag{1}$$

Where l_t and l_m represent total and minimum length respectively; n_k is the number of keyword that appears in the text at least once; t_k denotes the total number of keywords; k_f represents the average keyword frequency; w_1, w_2, w_3 are the weighting values; and th is the predefined threshold value.

Input: two text files from the same web site, each contains a news article
Output: a data structure contains:
String *URL*
String *Header*
String *Tail*

1. Remove all the html tags in the files.
2. Break down the files into one dimensional arrays (a and b), each cell contains a line of text.
3. For each cell of the array from beginning
    1. if **a**[*i*] == **b**[*i*], append **a**[*i*] at the end of *Header* string
    2. if **a**[*i*] != **b**[*i*], break;
4. For each cell of the array from the end
    1. if **a**[*i*] == **b**[*i*], insert **a**[*i*] at the beginning of *Tail* string
    2. if **a**[*i*] != **b**[*i*], break
5. Set the *URL* value to the common part of the URLs of two text files

Return the data structure that contains *URL*, *Header* and *Tail*.
End

Fig. 4. The pseudo code of dynamic filter generation.

The first term in equation 1 considers the total length of the extracted text. If the text length is unreasonably short, the text is unlikely to be a news article. The second term in the equation represents the percentage of the keywords that appeared in the text. The third term in the equation stands for the average frequency of the keywords that appeared in the text. The validation value takes the minimum value of these three and then compares with a predefined threshold to validate if the extracted text is the news article.

3 Exchange Rate Modelling Using Text Mining Techniques

To assist the negotiation agents to form the negotiation strategies, data mining agents need to process the extracted information further. The implementation of this step varies depending upon the actual application. In this chapter, a currency trading example (in particular, between the Euro and the US dollar) is used to illustrate the procedures of processing news data using text mining agents and incorporating the effect of the news stream into the currency rate.

Exchange rates prediction is one of the most challenging applications of modern time series forecasting. Until recently, most of the models are empirical models based on macro economic data. Among the enormous amount of empirical models, the sticky price monetary model of Dornbusch and Frankel remains the workhorse of policy-oriented analysis of exchange rate fluctuations (Dornbusch, 1976), which can be expressed as follows:

$$s_t = \beta_0 + \beta_1 \hat{m}_t + \beta_2 \hat{y}_t + \beta_3 \hat{i}_t + \beta_4 \hat{\pi}_t + \mu_t \quad (2)$$

where s is the changes of interest rate during each sampling period; m and y denote the logarithm of the money value and real GDP respectively; i and π are the interest and inflation rate, respectively; $\hat{\bullet}$ denotes the inter-country difference of the corresponding variable; μ is the error term. These models have performed reasonably well in explaining exchange rate development in the long term, but little success in predicting exchange rate in short and middle term movement. The general consensus of the poor performance of the traditional empirical models using economic fundamentals to account for exchange rate developments on short to medium term is caused by the irrationality of the market participants, bubbles, and herd behaviour, which are hard to be captured in econometric models.

Recent literature shows that news about fundamentals has played an important role in creating market dynamics. Prast and De Vor (2005) have studied the reaction of investors in foreign exchange markets to news information about the euro area and the United States on days of large changes in the euro-dollar exchange rates. Unlike the traditional models, daily changes in the euro/dollar rate on news about economic variables in the United States and the euro area, and the variables capturing news in the two economies were used in the regression model, which is:

$$E_t = \alpha + \sum_{i=1}^{8} \beta_i D_i + \varepsilon \tag{3}$$

where E_t is the percentage daily change in the euro-dollar exchange rate; D_{1-8} represent the following variables: 1 - real economy, euro area; 2 - inflation, euro area; 3 - change in official interest rate, ECB; 4 - statements/political events, euro area; 5 - real economy, United States; 6 - inflation, United States; 7 – change in official interest rate, United States; 8 - statements/political events, United States. It has been found that there is strong correlation between exchange rate daily movement and the market participants' responses to the daily economy news and political events.

More recent research has confirmed that news has statistically significant effects on daily exchange rate movement. Ehrmann and Fratzscher (2005) have evaluated the overall impact of macro news by analysing the daily exchange rate responses using similar regression models with news variables. Three key results were found. Firstly, the news about fundamentals can explain relatively well the direction, but only a much smaller extent to the magnitude of exchange rate development. Secondly, news about US economy has a larger impact on exchange rates than news about the euro area. Thirdly, higher degree of market uncertainty will lead to more significant effects of news releases on exchange rate movements.

The above findings motivated the research reported in this section. By using the text mining techniques, the manual process of identification and classification of positive and negative news can be automated. As the correlation between news and currency exchange rate has only been identified recently, there is not much work reported in this area. Eddelbuttel (1998) and Wong (2002) both tried to use the keywords in news headlines to forecast intraday currency exchange rate movements. Eddelbuttel used a set of keywords to identify the relevant news and sorted them into three groups: "All", "DEM" and "USD". Then the number of news pieces in three groups are calculated and used as the variable in the GARCH(1,1) model for prediction. The news analysis is restricted to the counting of the number of relevant news headlines to avoid qualitative judgement about "good" and "bad" news. Wong etc. proposed a prediction

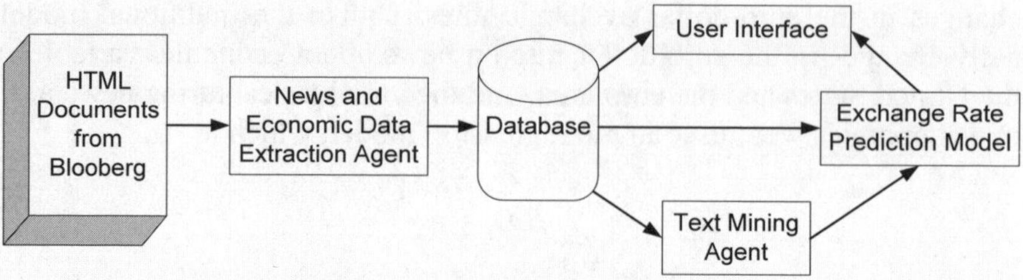

Fig. 5. Structure of the exchange rate model

model based on the occurrence of keywords in news titles. The keywords in the news title are identified by selecting the words with the highest weighting values. A set of rules for predicting the exchange rate movement direction from the keywords in the news titles are generated. These over simplified approaches only utilise news information to a very limited extent. A more sophisticated text mining approach for news filtering and classification was firstly proposed by the authors (Zhang et al, 2005) and is presented in this chapter. The design of this sub-system is shown in Figure 5.

3.1 Data Collection and Pre-processing

Before incorporating the news effect into an exchange rate model, it is important to identify the relevant news and classify them into "good" or "bad" news category, which would have opposite impact on the market behaviours. News articles are labelled "good" if they may cause an appreciation of one of the currencies, in this case, the US dollars; and the "bad" news have vice versa effect. This section describes the training process of news filtering and classification.

Prior to the processing, the news articles used for training are manually classified into two groups: news affect exchange rate (target corpus) and other news (generic corpus). Choosing the news articles in the target corpus is crucial for the process since the target corpus contains the underlying knowledge of what factors affect exchange rate movement. Much research has studied the factors that affect currency exchange rate, which can be macroeconomic data, statements by central bankers and politicians and political events that affect macroeconomics. Therefore, only the news that is relevant to these is chosen. To improve the process efficiency and avoid noise distraction, stop words in the target corpora are replaced by a stop word symbol but are not removed completely to avoid incorrect word co-occurrence. Porter stem algorithm is also applied to remove the common morphological and inflexional endings from words in the documents.

3.2 Automatically Keyword Extraction

Text mining operations are mainly based on the frequency of keywords. The goal of this step is to generate the best set of keywords that can distinguish news documents related to exchange rate from other news documents. To reduce the calculation complexity and increase the processing efficiency, the number of keywords is kept to the minimum amount but is still a good approximation of the original document set in its full space. There are two types of keyword frequencies used in this paper: term frequency and document frequency. The term frequency is calculated by the number of times a term appears in the corpora. The document frequency is the number of the documents that contain this term in the corpora.

Keywords are not restricted to single words, but can be phrases. Therefore, the first step is to identify phrases in the target corpus. The phrases are extracted based on the assumption that two constituent words form a collocation if they co-occur a lot (Manning et al, 1999).

Once the phrases have been extracted, the key terms are extracted amongst the single words except stop words and the phrases in the target corpus. The generic corpus is the background filter. The distribution of terms in the target corpus and the generic corpus are compared. The terms in the target corpus that stand out are considered as the features of the corpus, indicating that these terms are domain-specific terminology. The importance of each term is tested by the log likelihood ratio (LLR) Chi-square statistic test.

The likelihood ratio for a hypothesis is the ratio of the maximum value of the likelihood function over the subspace represented by the hypothesis to the maximum value of the likelihood function over the entire parameter space. In this case, the null hypothesis H_0 is formulated to test the distribution of a term is the same in the generic corpus and target corpus. H_a measures the actual distribution of the term in the whole data set. The log likelihood ratio for this test is:

$$-2\log\left(\frac{H_0(p;k_1,n_1,k_2,n_2)}{H_a(p_1,p_2;k_1,n_1,k_2,n_2)}\right) \tag{4}$$

The binomial distribution of the log likelihood statistic is given by:

$$\begin{aligned}-2\log\lambda &= 2\log L(p_1,k_1,n_1)+2\log L(p_2,k_2,n_2)\\ &\quad -2\log L(p,k_1,n_1)-2\log L(p,k_2,n_2)\end{aligned} \tag{5}$$

where $\log L(p,n,k)=k\log p+(n-k)\log(1-p)$; k_1 and k_2 denote the document frequency of a term in the target corpus and generic corpus respectively; n_1 and n_2 denote the size of the target corpus and generic corpus respectively; $p_1=\frac{k_1}{n_1}$, $p_2=\frac{k_2}{n_2}$, and $p=\frac{k_1+k_2}{n_1+n_2}$.

The method scores the terms based on the difference in the percentage of documents containing the term in the target and generic corpus. It does not distinguish whether the difference is caused by the term occurring more or less in the target corpus. As in this research only the terms significant in the target corpus are concerned, a simple condition $p_1/p_2\geq 1$ is added to the ranking equation so the terms that are significant in the generic corpus are filtered out.

3.3 News Relevance Classification

The news relevance classification is divided into two steps: the first step is to identify the news that has potential to cause movement in exchange rates; the second step is to identify the news that is Euro and/or US dollar related.

The exchange rate related news can be separated from other news based on the key terms extracted from the previous section, which often well represent the characteristics of the data set. In this case, a modified k-Means classification algorithm, which is particular suitable for this case, is chosen as being computationally simple and efficient. The centroid of the target corpus and the maximum Euclidean distance in the training data are calculated. The maximum distance is used as the threshold to determine if the data belongs to a target cluster.

News related to exchange rate may not be discussing Euro and US dollar currencies, which is further identified by using the frequency of the words of currency and country names it contains.

3.4 Positive and Negative News Classification

It is important to further classify the relevant news into "positive news" and "negative news" categories, as news in different groups have entirely different effects on the market behaviour.

Recent studies show that the effect of the news is the combined effect of market expectation and the news itself. A piece of news could have positive or negative impact to the market depending on the market expectation. Therefore, unlike some studies that define good and bad news by their immediately effect to the market, in this research, the news is defined to be good or bad according to its fundamental effect to the market. The market expectation is incorporated into the model in a later stage. For example, a news about US increased its interest rate is defined to be positive news to US dollars. The task of identifying "good" and "bad" news of exchange rate is not straight forward since both groups of news use similar set of keywords. For example, the following two pieces of news have exactly same set of words, but one is considered to be positive and the other is considered to be negative to the appreciation of US dollars:

1. The interest rate has gone up. The US dollar has gone down.
2. The interest rate has gone down. The US dollar has gone up.

The positive and negative news can use similar set of key terms, which causes great difficulties in the classification. However, the sequences of the key terms can represent the meaning of sentences better, which is well illustrated in the previous example. Therefore, a term is defined as the sequence of key terms in a sentence, which is used as the input features for the positive and negative news classification. The feature vectors of the above example can be represented as in table 1.

The idea of using keyword sequence as features is a step forward beyond words. It does not only consider the words contained in the sentences but also the word sequences in them. It compares the sentences by means of the keyword sequences they contain: the more keywords with the same sequence in common, the more similar they are.

Table 1. Example of feature vector representation for "good"/"bad" news classification

features	document 1	document 2
interest rate up	1	0
interest rate down	0	1
US dollar down	1	0
US dollar up	0	1

However, using key term sequence as classification features leads to a high dimensional vector space with sparsely distributed elements, which causes difficulty in separating instances into classes (subspaces). Therefore, the discriminant analysis is implemented to combine features of the original data in a way that most effectively discriminates between classes. The detailed algorithm can be found in (Berry, 2003).

3.5 The Econometric Model of Exchange Rate Responding to News and Economic Data

This section focuses on using text mining methods to incorporate the information in the news articles into a currency prediction model. As euro/dollar exchange rate will be used as the testing case, the empirical model presented by Galati and Ho to study the news effect on economic data particular for euro/dollar exchange rate is chosen (Galati et al, 2001). In this work, the above model is modified to incorporate a news index (I_{news}), which reflects the news effect on exchange rate. The regression equation has the following form:

$$\Delta \ln S(t) = \alpha_0 + \alpha_i x_i(t) + \beta I_{news} + \varepsilon \qquad (6)$$

where x_i represent the economic data variables which include: US non-farm payrolls, US unemployment rate, US employment cost index, US durable goods orders, NAPM manufacturing, NAPM non-manufacturing, US advance retail sales, US industrial production, US CPI, Ifo index, Germany unemployment rate, Germany industrial production, INSEE industrial trends, Germany CPI and EU 11 PPI.

The news index (I_{news}) in equation 6 corresponds to the news effects in the exchange rate model. The sampling period used in the regression is of a daily base. Historical data show that contradictory information seldom happens on daily bases since each day, there are mainly handful pieces of news affect the exchange rate market. However, when it does happen, the news index considers the effect of each piece of news and takes the total effects into account. In the future work, each piece of news will be validated by its source. When contradictory news items appear, only the news with higher reliability level is chosen. Also, the news effect will not only be measured by qualitatively but also quantitatively.

3.6 Information Summation

Ontology offers a way to share information between agents. A simple generic ontology using XML is built to integrate and present the data mining results of intelligent agents to the negotiation agents, which includes service ID, time stamp, estimate price, market price, and price explanation. In the future, a domain specific ontology will be built for each data mining agent to represent the key factors and their relationships with the estimate price. In this case, it is the relationships between economic data, the key factors discovered from the most relevant news and the estimate exchange rate. The ontology should be evolved when the factors that affect the exchange rate have been changed. With the domain specific ontology, the text mining agents are capable of answering a wide range of questions which can help the negotiation agents to generate the negotiation argumentation in a much flexible way.

4 Agent Communication Protocol

The Curious Negotiator as shown in Figure 1 is a multi-agent system. A simple agent communication protocol is developed for. It is based on the ontology negotiation protocol between information agents for scientific archives, which well suites this particular application (Bailin el at., 2002). This protocol can be implemented using KQML format.

Figure 6 shows the high level structure of message passing scheme between negotiation agents and text mining agents, where A is a negotiation agent and B is a text mining agent. Time flow is from top to bottom.

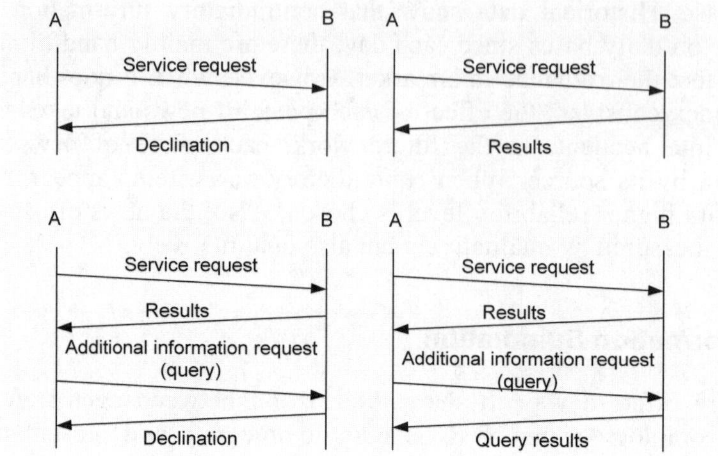

Fig. 6. High level structure of a message passing scheme.

Text mining agents and negotiation agents send and receive information through XML encoded messages. The ontology developed in section 3.6 provides a set of concepts, or meta-data, which can be queried and used to control the behaviour of agent cooperation. These concepts are marked using XML tags to underlie message interpretation. The structures and the semantics of the documents are represented by the corresponding DTDs and interpreters.

When a new text mining agent is added to the curious negotiator system, it advertises the service that it can provide. There are four scenarios at the current design. A negotiation agent initiates a service request to a text mining agent. The text mining agent can decline the request or fulfil the service. In the case that the negotiation agent receives the results of the requested service, it can request for additional information contained in the ontology. Query can only be handled after the service has been processed. The text mining agents keep the data of current service. The text mining agents record the query history. Therefore the information can only be query once.

With the domain specific ontology developed, the text mining agent can answer some sophisticated questions such as:

1. What is the most important factor affecting the current market?
2. How does the market change if a particular event happens i.e. interest rate goes up 0.5%?

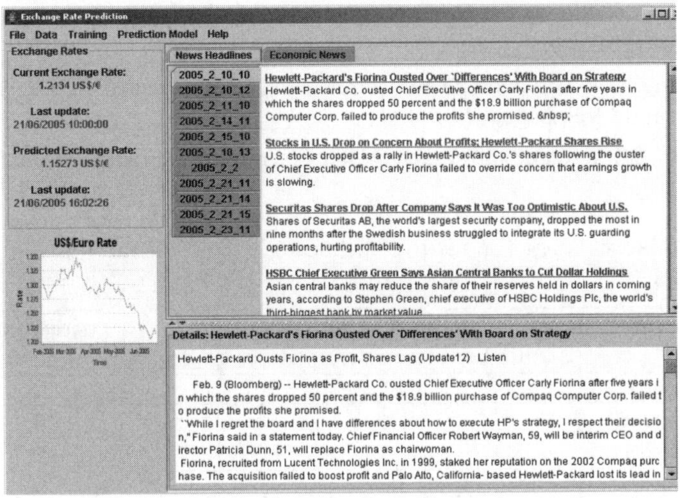

Fig. 7. User interface

These provide a broad knowledge base to negotiation agents and enable them to generate offers backed by arguments which summarise the reasons why the offers should be accepted. This type offer is more persuasive and thus places the negotiation agents in a better position to succeed.

5 Preliminary Results

The exchange rate prediction system is currently under construction. Figure 7 shows the user interface. Each part of the system has been implemented and tested individually. This section provides the experiment results of several parts of the system.

5.1 News Extraction

The proposed methods of extracting news articles were evaluated by the experiments using 19 most popular Australian and International news web sites. 200 pages from each URL location were tested. The average process time for each page was 436 milliseconds on a Pentium 4 1.60 GHz computer. The results are shown in **Table 2.** The notions used in the table are explained below:

- *Correct* – on average 0% error rate in the extracted text of a single web page;

- *Minor Error* – on average less than 5% error rate in the extracted text of a single web page;
- *Major Error* – on average between 5% to 30% error rates in the extracted text of a single web page;
- *Error* – on average more than 30% error rate in the extracted text of a single web page

Table 2. News sites for testing the news article extraction algorithm and the results

Accuracy	**without Filter** (number of news web sites)	**with Filter** (number of news web sites)
Correct	8	15
Minor Error	10	4
Major Error	1	0
Error	0	0

Experiment results show that news articles were mostly extracted properly except one web site, which is BBC News (UK). After analysing the web pages carefully, it was found that these web pages contained more than one content blocks in the table that also contains the news article, namely, the news article only occupies one of the table cell. Therefore, more experiments were conducted on this web site by using multiple documents. Experiment results show that the accuracy rate have been increased dramatically. It is because that although the content block is not correctly classified by the first step, other content blocks in the table are also extracted, but these extra content blocks in the extracted data are removed by the filtering process at the second step.

Experiments also show by using the dynamically generated filter, the extraction accuracy has been increased considerably. The experiment confirmed the approach, which assumes that the news article is contained in a table formatting structure, and the advertisements and other content block data are embedded in nested table structure within the news article table, works well. This layout method is commonly used in most of news web sites, which makes proposed algorithms and their implementation a practically valuable tools.

During the experiment, the threshold value for validation was set to 1. Different combinations of weighting values have been tested. Experiment results showed that the validation process is highly effective. Moreover, the experimental validation results are not sensitive to the choices of weighting values.

5.2 News Classification

The news classification was tested using the news collected from www.bloomberg.com between February 7, 2005 and July 5, 2005. In total, 2589 pieces of news have been collected. After the manual classification, each piece of news is classified into one of the categories shown in Figure 8. It was found that 1885 pieces of news are unrelated and 704 pieces are related. Among the related news, 200 pieces are good news and 113 pieces are bad news. The rest related news were identified to have no effect on the exchange rate or unrelated to either Euro or Dollars.

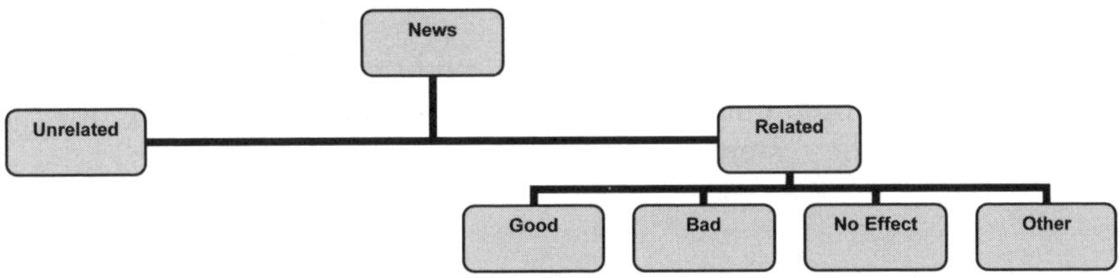

Fig. 8. News Categories

The average accuracy of related/unrelated news classification was 81.85%, while the average accuracy of good/bad news classification was 64.1%. These results show that the good/bad news classification is much more complicated than the related/unrelated news classification.

6 Conclusions

This chapter presents a framework for providing related information to the negotiation agents in the electronic market environment. It includes a news extraction algorithm, a quantitative process model based on the extracted news information, which is exemplified by an exchange rate prediction model, and a communication protocol between data mining agents and negotiation agents. The news extraction algorithm utilises data mining techniques for automatically collection of relevant set of news articles from varied news sources on the web, regardless of the format and structure of the sources. A novel approach to building an exchange rate prediction model using the extracted news articles and economical data are also presented. Recent research suggests that the market daily movement is the result of the market reaction to daily news. However, this type of news is not included in most of existing models due to its non-quantitative nature. This chapter is the first attempt to apply text mining methods to incorporate the daily economic news as well as economic and political events into prediction models with economic data. This approach leads to a more accurate exchange rate model with better prediction rate.

The Curious Negotiator defines the way the information is represented. The goal is to blend 'strategic negotiation sense' with 'strategic information sense' as the negotiation unfolds. In the case of the currency exchange, it can be a single rule, a set of values within a time interval, or a set of values with probabilities attached to them. The information is critical for the negotiation agents to form their negotiation strategies. The smart data mining system presented in this chapter supports the negotiation agents that operate under time-constraints and over dynamically changing corpus of information. Future developments include incorporating the trust information into the system.

References

1. Simoff, S. and J. K. Debenham (2002): Curious negotiator. *Proceedings 6th International Workshop Cooperative Information Agents VI CIA2002,* Madrid, Spain, Springer, 104-111.
2. Milgrom, P. and R. A. Weber (1982): "Theory of Auctions with Competitive Bidding." *Econometrica* **50**(5).
3. Kushmerick, N. (2000): "Wrapper induction: Efficiency and expressiveness." *Artificial Intelligence* **118**(1-2): 15-68.
4. Muslea, I., S. Minton, et al. (1998): STALKER: Learning extraction rules for semistructured, Web-based information sources. *Proceedings of AAAI-98 Workshop on AI and Information Integration*, Menlo Park, CA, AAAI Press.
5. Gao, X. and L. Sterling, (2000) Semi-structured Data Extraction from Heterogeneous Sources. In T. Bratjevik D. Schwartz, M. Divitini, editor, *Internet-based Knowledge Management and Organizational Memories*, pages 83--102. Idea Group Publishing.
6. Zhang, D. and S. J. Simoff (2004): Informing the Curious Negotiator: Automatic news extraction from the Internet. *Proceedings 3rd Australasian Data Mining Conference, 6 - 7th December, 2004*, Cairns, Australia, UTS, 55-72.
7. Dornbusch, R. (1976): Expectations and exchange rate dynamics. *Journal of Political Economy* 84, 1161-1176

8. Prast, H.M., de Vor, M.P.H. (2005): Investor reactions to news: a cognitive dissonance analysis of the euro-dollar exchange rate. *European Journal of Political Economy* 21, 115-141 TY - JOUR.
9. Ehrmann, M., Fratzscher, M. (2005): Exchange rates and fundamentals: new evidence from real-time data. *Journal of International Money and Finance* 24, 317-341 TY - JOUR.
10. Eddelbttel, D., McCurdy, T. (1998): The impact of news on foreign exchange rates: evidence from high frequency data. Technical report, University of Toronto
11. Peramunetilleke, D., Wong, R.K. (2002): Currency exchange rate forecasting from news headlines. *Aust. Comput. Sci. Commun.* 24, 131-139
12. Zhang, D., Simoff, S and Debenham, J (2005): Exchange Rate Modelling using News Articles and Economic Data. *The 18th Australian Joint Conference on Artificial Intelligence,* Sydney, Australia
13. Manning, C.D., Schutze, H. (1999): Foundations of statistical natural language processing. MIT Press, Cambridge, Mass. Christopher D. Manning, Hinrich Schutze.
14. Berry, M.W. (2003): Survey of text mining: clustering, classification, and retrieval. Springer, New York
15. Galati, G., Ho, C. (2001): Macroeconomic news and the euro/dollar exchange rate. Technical Report 105, Bank for International Settlements
16. Bailin, S. and Truszkowski, W. (2002): Ontology negotiation between intelligent information agents. *Knowledge Engineering Review*, Vol. 17, Issue 1

A Similarity-aware Web Content Caching Scheme and Agent-based Web Document Pre-fetching

Jitian Xiao

School of Computer and Information Science, Edith Cowan University, 2 Bradford Street, Mount Lawley, WA 6050, Australia, j.xiao@ecu.edu.au

Abstract

Web caching can be studied at different stages. At the first stage, web cache can support efficient implementation of sophisticated replacement policies. At the advanced stage, it is expected to implement cache based web information integration, in which cache is not just a passive tool in respect to changing of web access patterns, but it may become an active environment for integrating useful web information. This study focuses on the later to explore the architectural aspect of web caching. We propose a flexible and customizable web content management scheme for web caching that references the similarity information between cached documents, namely, *similarity-aware caching*. An agent-based Web document pre-fetching mechanism is developed to support the similarity-aware caching to further reduce the bandwidth consumption and network traffic latency, therefore to improve the Web access performance. Trace-driven simulations have been conducted to show the superiority of the proposed scheme.

1 Introduction

Restrictions inherent in the differences of bandwidth between remote and local access to web content impose additional costs when accessing remotely hosted web resources (Fan et al.1999). Content caching, in its various forms, is seen as a set of techniques based upon historical analysis and/or projection, to alleviate the effects of server bottlenecks and the vagaries of network traffic volume, thereby reducing latency experienced by a server, user or by client programs. Traditional caching, at its basic level, locally stores recently requested pages so they do not have to be retrieved

J. Xiao: *A Similarity-aware Web Content Caching Scheme and Agent-based Web Document Pre-fetching*, Studies in Computational Intelligence (SCI) **37**, 213–232 (2007)
www.springerlink.com

subsequently every time each is accessed. In brief, recently requested pages or files are held, or cached, on a local, or less remote, server in anticipation that they will be accessed again by clients. Such caching does much to reduce repeat network traffic.

Pre-fetching is an active technique that attempts to guess those documents that are likely to be requested when a page leading to them is accessed – success of this technique is measured as a "hit-ratio". However, in such guessing, there is a need for an effective balance to be achieved between user comfort and computational overheads – the extremes are: too little effort applied, resulting in too many on-demand-fetches, while too much effort results in too many pre-fetches. The consequence of either is that of slower response to a user.

Previous work by Xiao et al (2001) in developing pre-fetching predictions between caching proxies and browsing clients was based on measures of similarity between web users established that pre-fetching is capable of increasing the hit-ratio. Their work further established that organization of the cache affects opportunities for successful pre-fetching.

In this Chapter we describe a means of similarity based content management to improve the relative performance of pre-fetching techniques based upon document similarity detection. Pre-fetch caching in the context of this study will be based upon similarity detection and involve several phases. Similarities will be sought from previously cached documents by employing several concurrently applied, but differing, algorithms to detect equivalences of broad-content or keywords and links contained within pages under scrutiny. Similarities between web documents, having been detected, will then be ranked for candidature to be fetched in anticipation of a user's intentions. Following the ranking exercise, content settings may be realized for sub-caches and pre-fetching may then proceed.

The rest of the chapter is organized as follows. Section 2 defines the similarity measures. Section 3 describes a similarity-based web cache architecture. In Section 4, we propose an agent-based the similarity-aware web document pre-fetching scheme. Section 5 presents the simulation results, and Section 6 concludes the chapter.

2 Similarity Detection

There are generally two main streams in measuring similarities among documents: one uses a single relationship between documents1 or data objects while the other uses multiple relationships. Early research used a single relationship to measure the similarity of data objects. In the original vector space model (VSM) (Salton 1968), "terms" (e.g. key words or stems) were used to characterize queries and documents, yielding a document-term relationship matrix to compute similarities among terms and documents by taking the inner product of the two corresponding row or column vectors. Dice, Jaccard and Cosine measurements (Rasmussen 1992) are a few classical methods that used such document-term relationships to measure the similarity of documents for retrieval and clustering purposes. Deerwester (1994) saw that a document might not be well represented by its contained keywords and developed a Latent Semantic Index (LSI). In this, they apply a singular value decomposition (SVD) method to map the document-term matrix into some lower dimensional matrix where each dimension associates with a hidden "concept", where any similarity of text objects (documents and queries) is measured by relationships to those "concepts" rather than the keywords they contained.

With the advent of Word Wide Web, relationships with document objects, e.g., their hyperlink relationships, were used to derive similarity; a mechanism employed by Kleinberg (1998) to discover similar web pages. Further, Pitknow et al. (1997) applied co-citation to a hyperlink structure to measure any similarity of two web pages. Flesca and Masciari (2003) proposed a method to measure the similarity of two documents that represents the current and the previous version of monitored pages for effective web change detection.

The above approaches all relied upon a single relationship to measure any similarity of data objects. However, such approaches may run into serious problems when applications require accurate similarity e.g. where multiple types of data objects and relationships must be handled in an integrated manner. Accordingly, in the extended VSM (Fox 1983), feature vectors of data objects were augmented by adding attributes from objects of other related spaces. Similarity computation is then obtained from calculation on these enhanced feature vectors. The extended feature vectors were used for document search (Shaw & Fox 1994) or clustering purposes (Chakrabarti

[1] In this chapter, a *document* refers to a text document or a web page that may contain text, images and/or pictures.

et al. 1999). Racchio et al. (1997) expanded the query vector using those frequently-used terms appearing in the foremost documents retrieved by a query to improve search effectiveness. Recently, it has been tried to calculate the similarity of two data objects based upon any similarity of their related data objects. For example, it has been tried (Popescul et al. 2000) to measure the similarity of two queries by correspondences found in their respective search lists. Beeferman and Berger (2000) clustered queries using the similarity of both their selected web pages and cluster web pages based upon similarities of the queries that lead to the selection of those web pages. Works by Wen et al. (2002) calculated the query similarity based on both the query contents similarity and the similarity of the documents that were retrieved by the queries.

In this chapter, we define similarity measures of web documents for effective web document caching and pre-fetching. To pre-fetch documents that are of similar topic to the document a user is currently viewing, we need to derive the similarity of contents of web documents, ignoring any structural elements, e.g. HTML formatting. For efficacy of on-line pre-fetching, we propose different levels of similarity measures to capture levels of similarity between web documents. Consider a search of scientific papers over the web. A keyword based search usually returns a list of documents containing some or all of the given keywords. The matched keywords in the returned documents may appear in the title, keywords section, or other parts. Title/author-based searches follow similar principles. However, when a user is viewing a document and wishes to search for documents of similar topic, then the matching strategy may be quite different because the words to be matched may be related rather than explicitly stated. In our study, similarities between text documents are measured based on topics, page titles, keywords or page contents or combinations thereof. Compared with a keyword-based similarity measure, a content-based similarity is much complicated by the need for special techniques, e.g., from the area of information retrieval (Baeza-Yates et al. 1999). However, any computation of similarity still needs to be completed within a reasonable time limit.

2.1 Document Model

The similarities among web documents are computed based on a document model similar to that of Flesca's (2003), wherein structured web documents are represented as unordered labeled trees. We consider containment rather than order of appearance of words within a document. Our model differs from that of Flesca's in two ways: first, we don't consider the

HTML formatting elements and, second, we consider a document's structure to be based on sectional elements, e.g. *Abstract* and *subsections*, while their work specifies texts in terms of pairs of start and end tags, e.g., <table> … </table>, <ul. … </ul>.

In the resultant tree, each non-leaf node corresponds to a subsection of the document (e.g. characterizing the title of the subsection), except that the root-node might also contain a set of *keywords*, a list of *authors*, a string for *title*, or/and a set of words comprising the *abstract*. Each leaf node corresponds to the text of that (sub)section. Notably, such a structure allows us to determine sectional similarities between particular elements such as titles; between the various contents, and, implicitly, between the structures of compared documents. In brief, a document tree is an unordered tree wherein each node is characterized by an associated set of type-value pairs. Given a document tree *T*, of root *r*, with a node n_r we may represent a sub-tree of *T* rooted at n_r as $T(n_r)$. We define a set of functions, each characterizing some element, on the document tree: *keyword*(*r*), *title*(*r*), *authors*(*r*), *abstract*(*r*) and *text*(*r*). For a document tree rooted at *r*, *keyword*(*r*) = $\{s \mid s$ is a keyword contained in the keyword section of $r\}$. The *title*(*r*), *authors*(*r*) and *abstract*(*r*) can be defined similarly. If $n_1, n_2, \ldots, n_k$ are child nodes of *r*, then

$$text(r)=\begin{cases} title(r) \cup \cup_{i=1}^{k} \{s \mid s \in text(n_i)\} & \text{if } r \text{ is a non-leaf node, with children } n_1 \ldots, n_k \\ \{s \mid s \text{ is a word in } leaf(T(r))\} & \text{if } r \text{ is a leaf node of } T \end{cases}$$

Essentially *text*(*r*) is a set of words contained in the various strings associated with nodes of the (sub-)tree rooted at *r*. *text*(*r*) is defined recursively.

Our similarity computation algorithm works on this tree structure by exploiting the information contained in individual nodes and the whole tree. Observe that each node keeps track of its level in the tree, its content and the content of its child nodes.

2.2 Levels of Document Similarity Measures

When a user finds a web page that is on-topic, she or he may desire to find other documents of similar topic (this is one of the main driven points for similarity-aware pre-fetching). There are different parameters which determine the similarity of two documents: some are related according to their content and others by their in and out hyperlinks. In this section we consider only document content. There are three different models of document-to-document similarity measuring (Barfourosh et al. 2002), the *string distance* model which considers the distance as amount of difference between strings; the *statistics of words* which considers frequency of words in documents to judge on similarity (e.g., *term frequency* × *inverse document frequency* (TFIDF), see Salton 1973), and the *document components or structure* (DCS) model which considers structure of components of documents, for example references, abstract, title, keywords etc in research papers (e.g., Citation analysis, see Barfourosh et al. 2002). Of these, the last two models are closely related to the application scenario of this work.

TFIDF is based on word frequencies in documents. In fact it is suitable for sets of documents, especially as part of a large number of documents. In this approach, for each word in document, the weight of the word is calculated based on word frequency in a given document, the number of documents that include the word, the highest word frequency in a document, and the number of all documents in the document pool. Then the distance between two documents is calculated by dot product of the two word vectors for those documents.

The DCS model uses the knowledge about document components or structure to judge the similarity between two documents. This approach is well suited for situations in which documents are of a specific type, or have special components or structure. In the case of research papers, for instance, they have similar structure and components, such as title, abstract, keywords and references.

We combine the TFIDF and DCS models to forming our new document similarity measures in this work. Levels of document similarity measures are defined by making use of the text extracted from elements of document (sub-)trees. To compute the similarities efficiently, the measures must be normalized, allowing the comparison of pairs of documents and the selection of different levels of elements/components.

Given two document trees T_1 and T_2, and two nodes $r_1 \in T_1$ and $r_2 \in T_2$, we define an intersection function, which is based on Broder's resemblance function (Broder 1997), to assess the similarity between the document components represented by r_1 and r_2,

$$\mathit{intersect}\,(w(r_1), w(r_2)) = \frac{|\,w(r_1) \cap w(r_2)\,|}{|\,w(r_1) \cup w(r_2)\,|} \tag{1}$$

where $w(r)$ is a set of strings associated with nodes of the (sub-)tree rooted at r. The function $intersect(w(r_1),w(r_2))$ returns the percentage of the number of common words divided by the number of all words that appear in both $w(r_1)$ and $w(r_2)$. Clearly, $intersect(w(r_1),w(r_2)) \leq 1$, while equality exists when $w(r_1) = w(r_2)$.

For two trees rooted at r_1 and r_2, their similarities of keywords, titles and abstracts, respectively, may be defined by the formulae (2) through (4):

$$SIM_{KB}(r_1, r_2) = intersect(keyword(r_1), keyword(r_2)) \quad (2)$$

$$SIM_{TB}(r_1, r_2) = intersect(title(r_1), title(r_2)) \quad (3)$$

$$SIM_{AB}(r_1, r_2) = intersect(abstract(r_1), abstract(r_2)) \quad (4)$$

while the content-based similarity is defined as

$$SIM_{CB}(r_1, r_2) = intersect(w(r_1), w(r_2)) \quad (5)$$

where $\mathrm{w}(r_i) = \mathrm{text}(r_i) \cup \mathrm{keywords}(r_i) \cup \mathrm{abstract}(r_i)$, $i = 1, 2$.

Generally, the higher a word occurrence in a document, the closer that word relates to the theme of the document and this may be used to measure similarity between documents. To take the significance level of words into account, let $weight_r(s)$ be the number of occurrence of the word s in document represented by r, then the intersect function defined in (1) can be redefined as

$$intersect_{\mathrm{wt}}(w(r_1), w(r_2)) = \frac{\sum_{s \in w(r_1) \cap w(r_2)} \min\{weight_{r_1}(s), weight_{r_2}(s)\}}{\frac{1}{2}\sum_{s \in w(r_1) \cup w(r_2)} | weight_{r_1}(s) + weight_{r_2}(s) |} \quad (6)$$

Formula (6) is similar to the definition of TFIDF weight of documents, but it is used to calculating the similarity between two documents rather than evaluating how important a word is to a document. Based on this function, the *weighted* similarity measures $\mathrm{SIM_{KB}}()$, $\mathrm{SIM_{TB}}()$, $\mathrm{SIM_{AB}}()$ and $\mathrm{SIM_{CB}}()$ can all be re-defined by replacing *intersect*() in (2) to (5) with $intersect_{wt}()$ defined in (6). The weighted similarity measures will be used only when the frequency of word occurrences is considered important to the content of the documents.

2.3 Data Pre-Processing

To calculate similarities among documents, a text filter was developed to extract meaningful words from related sections of a document, and count them per section. The method is described briefly below:

In the text filter, raw text is first parsed into generalized words, called *tokens*. Tokens include meaningful strings, abbreviations, punctuation and other specialized symbols that have been derived from the document's sections. For example, while typical words such as "web" and "page" are tokens, the punctuation mark "$" and the URL "www.ecu.edu.au" are also tokens. However, digits and other insignificant words, e.g. pronouns and prepositions, are not treated as tokens.

For each section, the text filter produces a list of *(token, c(token))* pairs, where *c(token)* is the count of that token within the section – in effect, a *bag-of-words* basis for our representation. Note that for brevity of the token list and subsequent comparison, each word is reduced to its stem (e.g., *server* and *service* into *serve*). While the unordered *bag-of-words* model will not suffice for linguistic analysis, we assume it captures most of the information needed for calculating similarities using formula (2) to (6).

3. The Similarity-based Web Cache Scheme

The basic idea of web-caching is to reduce network traffic load and reduce retrieval latency by holding recent requested documents at the proxy caches so that they do not have to be fully retrieved upon identical request.

Document similarity information is fundamental to effective caching and pre-fetching, yet it has never been incorporated directly in cache replacement algorithms. Rather, other properties of the request stream (e.g., document size and access frequency etc.), being easier to capture on-line, are used to infer similarity, and hence drive cache replacement policies. In this section, we describe a similarity-based multi-cache web content management scheme and on-line algorithm to capture and maintain an apposite similarity profile of documents requested through a caching proxy.

3.1 Caching Architecture

We now present a similarity-based multi-cache web caching architecture (see Figure 1). There are four major components: *central router*, *similarity profiles*, *Cache Similarity Knowledge Base*, *sub-caches,* and *document allocator*. Of these, the central router is pivotal in controlling and coordinating other components.

To configure the multi-cache web caching architecture, based on particular caching similarity level, we first cluster documents in cache based on combination of the similarity measures (2), (3) and (4) introduced before (i.e., by taking a similarity measure SIM(r_1, r_2) = intersect ($w(r_1)$, $w(r_2)$) where $w(r_i)$ = title(r_i) ∪ keywords(r_i) ∪ abstract (r_i), i = 1, 2), and determine the number, N, of themes of the documents. For the initial cache content clustering purpose, we examined a number web content classification approaches (e.g., decision trees, k-nearest neighbor classifiers, neural networks and support vector machine (SVM) (Dumais & Chen 2000), etc.) and finally adopted SVM classification algorithm (with slightly modification on it) for this purpose. SVM classification algorithm is more suitable than others to this work because it can work with short summary descriptions of web pages (such as title, keywords and/or a small number of starting words of the document body), and it has been shown to be both very fast and effective for text classification problems (Dumais & Chen 2000).

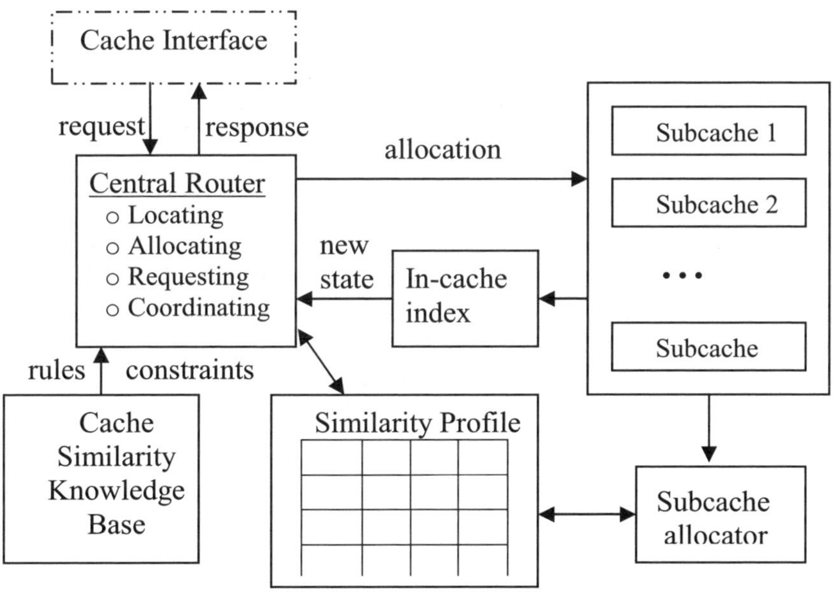

Figure 1. The similarity-based web cache scheme

For each theme (or cluster), a number of *stems* relating to it were chosen (e.g., by looking at all stems produced by the text filter when similarity profile vectors were computed). Then the cache is divided into *N*+1 subcaches. Each of the first *N* sub-caches stores documents of one particular theme, and the last sub-cache stores other documents not belonging to any of the *N* themes. In this way, we ensure that similarities among documents in any sub-cache are relatively higher, while relegating those among documents across sub-caches.

3.1.1 Similarity Profiles

The similarity profile (SP) comprises N two-dimensional arrays $A_i(*, *)$, $i=1, 2, ..., N$, of which each corresponds to one of the first N sub-caches. For each document j in sub-cache i, SP counts the number of occurrences of the stems that relate to the theme of the sub-cache, storing the numbers in vector $A_i(j,*)$. This information is useful when performing similarity-aware pre-fetching from the sub-cache to a client. For each theme, we limit the number of stems to be a specific number (e.g., 128).

3.1.2 Cache Similarity Knowledge Base

Cache Similarity Knowledge Base (CSKB) consists of a set of rules designed for classification of web contents. Semantics information can be used here to direct the cache. Based on these rules, more advanced caching management can be implemented. For example, the rule

*R1: allocate(X, y):- url(X, U), match(U, *.au), content(X, y(football)).*

directs a document in sub-cache *y* if it is about football and comes from Australia. The number of sub-cache *y* is determined by the theme most closely related stem “football”.

The CSKB not only directs similarity information in its rules but it may also impose various restrictions on sub-caches. For example, it is essential that dynamic documents, too large documents, or documents that are prohibited from caching will not be accepted in any sub-cache. These conditions can be combined into CSKB rules. As an example, the following rule R2 restricts that only those football related documents whose size is less than 2 megabytes could be cached.

R2: allocate(X, y):- content(X, y(football)), size(X, <2M).

3.1.3 Sub-caches

A sub-cache is an independent cache that has its own cache space, contents and replacement policy. Since documents in a same sub-cache are usually of similar theme, simpler replacement policies, such as LRU, LFU and FIFO, may be applied.

3.1.4 Sub-cache Document Allocator

The sub-cache allocator assesses comprehensively a candidate set of evictions selected by sub-caches, with possible results of: re-caching, eviction or probation. Of these, re-caching and eviction are instantaneous, while a probation document will be held by the allocator in its own space pending a final decision. A document to be re-cached will be cached at once in a certain sub-cache. An eviction document will be purged immediately.

3.2 Framework of Similarity-aware Caching Scheme

A request for a document *d* invokes the similarity-aware caching algorithm as follows: an instance of *d* is sought in an in-cache index; if *d* is already cached (i.e., cache hit) and still fresh its containing sub-cache is noted whereupon *d* will be returned to the requesting client. If the instance of *d* is not fresh, then re-cache from an origin server, updating related parameters such as SP vectors. For a cache miss, a request for *d* will be forwarded to the origin server and a resultant downloaded document d_{new} is returned to the client. Based on the content of d_{new}, a SP vector will be calculated to determine a sub-cache c_d in which d_{new} is to be cached. Where there is insufficient space for d_{new}, the sub cache c_d makes room according to its eviction (e.g. LRU, LFU) and/or space sharing policies. The document allocator of c_d will then assess and purge any eviction candidates.

The central router mediates between cooperating sub-caches. Although a document may be cached "conceptually" in several sub-caches in terms of sub-cache document allocator evaluation, only one actual copy will be maintained.

4 Web Document Pre-Fetching

In this section, we focus on web document pre-fetching between caching proxies and browsing clients. If the proxy can predict those cached documents a user might access next, the idle periods of network links may be used to push (or to have the browser/client pull) them to the user while the user is viewing the web document. Since the proxy only initiates pre-fetches for documents in its caches, there is no extra internet traffic increase.

4.1 Architecture of Agent-based Similarity-aware Pre-fetching

According to (Bradshaw 1997), an agent is a software entity that carries out some set of operations on behalf of a client/program with some degree of independence or autonomy. An agent employs some knowledge or representation of the client's goals or desires. Agents have the following properties: (1) agents operate without the direct intervention of humans or others, and have some kind of control over their actions and internal state. (2) agents interact with other agents (and possibly humans) via some kinds of agent-communication language. (3) agents perceive their environment, and respond in a timely fashion to changes that occur in it. (4) agents do not simply act in response to their environment; they are able to exhibit goal-directed behaviour by taking the initiative. (5) agents are able to transport themselves from one machine to another. (6) agents change their behavior based on their previous experience.

In this study, we employ both proxy-side and client-side agents that exchange messages using a predefined protocol for actualizing similarity detection, document prediction, network traffic monitoring and proxy-client coordination intentions during the process where they negotiate to reach the most probable solution. We are concerned with coordinating intelligent behaviour among these agents, i.e. how they coordinate their knowledge, goals, skills, and plans jointly to take action or to solve problems.

In the similarity-aware web document pre-fetching process, three activities are crucial and are the focuses in this section, including:

- identifying similarities between documents in the proxy cache and the document a user is viewing;
- predicting documents that a client is most likely to access next; and
- monitoring idle network periods to pre-fetch the documents.

Among these activities, the first one is similar to the similarity detection in caching new document (see Section 3). The third activity involves traffic handling and resource utilization, and is, thus, beyond the scope of this chapter. Therefore, we focus on the second activity. In this architecture, agents and other software components are described as follows:

Client Agent (CA): The agent plays the role of a client. It delivers a pre-fetching request to the *Coordination Agent (CoA)*. Upon receipt of an initial pre-fetching plan (i.e., a list of candidate documents to be pre-fetched) from the CoA, it modifies the plan by removing the candidates that were hit by its local cache, and then returns the modified plan to the CoA for final pre-fetching.

Coordination Agent (CoA): the agent is responsible for receiving the pre-fetching requests from clients, and coordinates among the similarity detection agent (SDA), access pattern matching agent (PMA), pre-fetching agent (PFA) and network traffic monitoring agent (TMA) for document pre-fetching. Through the interaction between the agents and the client in the architecture, the detailed job of the CoA involves following steps:

1) receive pre-fetching requests from CAs;
2) invoke SDA to identify a set of cached documents (in one or more sub-caches) whose similarities with the document the client is viewing surpass certain threshold;
3) invoke PMA to assess and identify a set of users' past (historical) access patterns that could be referenced for prediction of future access patterns of the client;
4) upon receipt of the responses from steps 2) and 3), assign a process that calls PFA to produce an initial pre-fetch plan (e.g., a list of candidate documents for pre-fetching);
5) send the initial pre-fetch plan to the CA to determine which in-list candidates should not be pre-fetched due to local cache hit; and
6) upon receipt of the modified pre-fetch plan from a CA, assign the plan to a TMA for document pre-fetching.

Similarity detection agent (SDA): The agent determines a set of documents whose similarity with the given document surpasses the given similarity threshold. These documents will be referenced when similarity-aware PFA performs document prediction.

Access pattern matching agent (PMA): The agent matches a number of other users whose past access patterns with the given user's is greater than or equal to a certain threshold. These access patterns will also be referenced when the PFA performs document prediction.

Pre-fetching agent (PFA): The agent is responsible for predicting a set of cached documents as candidates of an initial pre-fetching plan (see Section 4.2).

Network traffic monitoring agents (TMA): the agent is responsible for monitoring the network traffic between the proxy and a given client. Once a suitable idle period is identified, the agent sends (if a proxy-side agent) a candidate document of the pre-fetch plan from the proxy cache to the client within the idle period. This monitoring-identifying-sending process continues until all candidate documents were sent, or the pre-fetching time limit is reached.

Conversation Manager (CM): The CM coordinates the activities of agents in the documents pre-fetching circle. It is responsible for receiving events from an agent, and informing other agents of messages. For example, each agent routes all its outgoing messages through the CM, and all its incoming messages are received via the CM as well.

4.2 Pre-fetching Prediction

We propose two agent-based prediction algorithms to guide similarity-aware pre-fetching from proxy caches to clients. The first one is a pure similarity-based pre-fetching predictor which considers only those documents whose similarities with the document in viewing surpass a certain threshold. The second algorithm (i.e., similarity-aware pre-fetching) combines the *prediction by partial matching (PPM)* method (Palpanas 1998) and the pure similarity-based pre-fetching strategies. These algorithms are the main functionalities and behavior of PFAs.

4.2.1 Similarity-based Pre-fetching Predictor

The similarity-based pre-fetching agent predicts the next k documents in the proxy cache based on document similarities. With the support of the similarity-aware web cache architecture, the similarity-based document pre-fetching predictor works based on a very simple rule. Suppose a client is viewing a document, say d (at this time, a copy of d must be cached in a certain sub-cache, say c_i, or being held by the allocator). When a pre-fetching request is received, the CoA invokes an SDA which computes the similarities between d and those documents in sub-cache c_i by referencing the similarity information in i_{th} SP. No documents in other sub-caches are considered because of their low similarities with d. Then the predictor simply chooses k documents whose similarities with d are among the top k highest ones. These k documents, together with those cached pages to which hyperlinks exist from d, will form an initial pre-fetching plan and be returned to CoA for possible pre-fetching.

4.2.2 Similarity-aware Pre-fetching Predictor

Palpanas (1998) adopted the PPM to predict the next l requests based on the past m accesses of a user, limiting candidates by an access probability threshold t. The performance metrics of the algorithm depend on the (m, l, t) configurations (Fan et al. 1999). The algorithm uses patterns observed from all users' references to predict a particular user's behavior. Referencing too many contexts makes the prediction inaccurate, inefficient and unwieldy. Our previous work (Xiao et al. 2001) extended the PPM algorithm by referencing only those access patterns from a small group of other users exhibiting high similarities in their past access patterns to predict a current user's next accesses. The number of times the algorithm can make prediction is reduced because of the smaller sample size, but the hit ratio of the pre-fetching increases because more related access patterns are referenced. We call the method *pattern-similarity based PPM* (or *psPPM*).

To be more similarity-aware, we now modify *PPM* and *psPPM* by replacing the access threshold t with s, where s is the similarity threshold between the document to be pre-fetched and the document the client is viewing. Thus the new algorithm has the following parameters:

- r: the number of users whose access patterns are referenced to predict future accesses of the current user.
- m: the number of past accesses that are used to predict future ones. We call m the *prefix depth*.
- l: the number of steps that the algorithm tries to predict into the future.
- s: the similarity threshold used to weed out candidate document. Only those documents whose similarity with the viewing document is greater than s, where $0 \leq s \leq 1$, is considered for possible pre-fetching.

Suppose a user u is viewing a document d. When a pre-fetching request is received, the CoA invokes a PMA to assess and identify a set of r users' access patterns of relatively high similarities with u (sorted in descending order). For $l>1$, not only the immediate next request, but the next few requests after a URL are also considered for potential pre-fetching. For example, if $l = 2$, the PFA predicts both the immediate next and its successor for the user. If $m > 1$, more contexts of the r users' past accesses are referenced for the purpose of improving the accuracy of the prediction.

The PFA maintains a data structure that tracks the sequence of URLs for every user. For prediction, the past reference, the past two references, and up to the past m references are matched against the collection of succession to the users' past access patterns to produce a list of URLs for the next l steps. If a longer match sequence can be found from the other r users' patterns, the next URL to the longest match is also taken as a potential document to be accessed next by the user. The outcome of each prediction is a list of candidate documents, ordered by their similarities with d. For those candidate documents with the same similarity value, the URL matched with longer prefix is put first in the list.

5 Simulation

We conducted two series of simulations. The first series of simulations is to demonstrate the capability of our similarity measures for document comparison to determine the document themes (or clusters). Using the obtained similarity information, our second series of simulations demonstrates the improvement in prediction accuracy (and thus the pre-fetching hit rate) of the pre-fetching between caching proxies and browsing clients using the proposed similarity-based/-aware predictors. This section presents and analyse only the results of the second series of simulations.

5.1 The Data Sets Used

We recorded all HTTP URL requests made to our proxy servers for a period of four weeks (i.e., from 14 March 2005 to 11 April 2005) from groups of users. Two datasets are used: (1) UL: HTTP URL requests from workstations in a undergraduate computer science lab representing about 322 different user ids, containing 46,277 accesses to the caching proxy; and (2) GL: HTTP URL requests from a popular host used by graduate computer science students, containing 21,103 accesses. These data sets represent a mixture of Web usage for lab classes; for general lab or recreational web browsing. One distinction between UL and GL is that material for many units that the undergraduates take is provided through the web, in fact more than half of the units are wholly on-line (or "paperless"). UL also differs from GL in that the tutors often direct students to look at certain URLs or to search for certain topics, and hence there should be a definite correlation between the URLs requested by different clients.

5.2 Simulation on Pre-fetching Prediction

In this simulation, the agent-based similarity-aware web pre-fetching predictor is simulated against the PPM algorithm. We also tested the similarity measures for their suitability for prediction in pre-fetching. During the simulations, we first set up the Cache Similarity Knowledge Base (CSKB), and then cluster documents in cache using each of the four similarity measures given in (2)~(5), thus determine the number of sub-caches, N. Then, $N+1$ sub-caches are built; documents in cache are reallocated to the sub-caches; the stems of each sub-cache are determined; and SP vectors are produced. Afterward, the similarity information in SP is used to predict document pre-fetching for users' late requests. We call our predictor *cb*PPM (*kb*PPM, *tb*PPM, *ab*PPM), if the similarity measure used is content-based (keyword-based, title-based, abstract-based). As our simulations aim to evaluate mainly on the prediction accuracy (or hit ratio) of the proposed pre-fetching scheme, we fix $m=4$ and $l=2$, respectively, because these values are the best choice[2] in PPM. We test some combinations of s and r values (e.g., $s = 0.1, 0.2, 0.3, 0.4, 0.5$, and $r = 1, 2, \ldots, 8$).

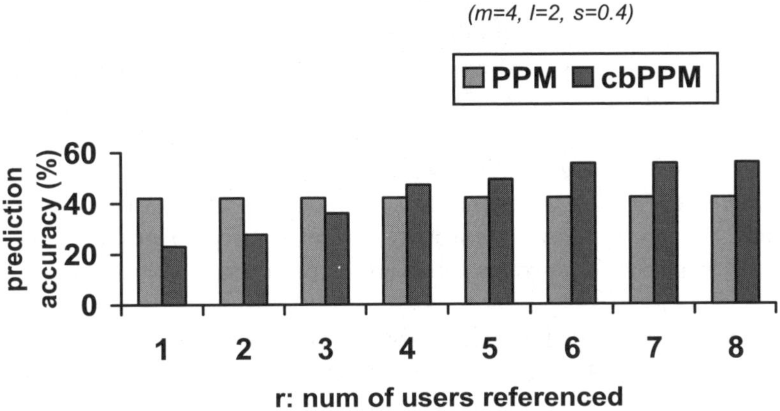

Figure 2. Simulation results: PPM against *cb*PPM

For our data sets, PPM holds on average a prediction accuracy of 42.2%. Among our predictors, *cb*PPM can predict more accurately than others do. This is because the content-based similarity measure takes more stems than the others thus reflect more precisely similarities between documents. On average, when the similarity threshold s is set to 0.4, about 24.2% accuracy can be achieved by only referencing one user's access pattern (i.e., $r=1$), and 27.1% achieves when $r = 2$. When the number of users' access patterns referenced increases to 4, *cb*PPM achieves an accuracy of 47.0%, indicating that *cb*PPM is a practical predictor when comparing with PPM. The prediction accuracy can achieve up to 55.7% when (r, m, l, s) was configured as (4, 4, 2, 0.4). The simulations have shown that *cb*PPM can predict at least as accurate in pre-fetching prediction as PPM does by only referencing up to 4 users' past access patterns (recall that PPM has to reference all other users' access patterns for prediction of document pre-fetching for individual users (Xiao et al. 2002)). In this sense, *cb*PPM has

---

2 The prediction accuracy is strongly related to the combination of parameters *(r, m, l, s)*. It is a challenge to find a systematic method for optimizing these parameters so that the prediction accuracy can be maximized, however it is beyond the scope of this chapter.

made an improvement to the PPM. Figure 2 shows the simulation results of our predictor against PPM (when $m=4$, $l=2$, and $s=0.4$).

6 Conclusions

Large scale web caches are localized sources of web contents rather than a heap of meaningless data as in traditional caching. Generally, replacement policies are required to decide the content (i.e., what to cache) to achieve high hit rates or other performance metrics. In order to make best use of such contents, it is important to develop suitable content management strategies, because semantics information related to a document are important indicators to its usage. Web pre-fetching builds on regular web caching and tries to predict the next set of documents that will be requested, and guide the pre-fetching accordingly. This greatly speeds up access to those documents, and improves the users' experience.

This chapter presented a similarity-based web content management and agent-based similarity-aware web document pre-fetching scheme. We described the underlying web-caching architecture and developed similarity-aware predictors for web document pre-fetching between proxy caches and browsing clients. Simulations have shown that our predictor is capable of practical prediction for web document pre-fetching in the sense that it may predict more accurately and effectively than the traditional PPM does by only referencing a reduced set of users' past access patterns.

References

Baeza-Yates, R., Ribeiro-Neto, B. (1999), Modern Information Retrieval, Addison Wesley Longman Publishing Co. Inc., ACM Press.

Barfourosh, A. A., Nezhad, H.R. M., Anderson, M. L. and Perlis, D. (2002), Information Retrieval on the World Wide Web and Active Logic: A Survey and Problem Definition, Technical report UMIACS-TR-2001-69, DRUM: Digital Repository at the University of Maryland.

Beefermand, D., Berger, A. (2000), Agglomerative clustering of a search engine query log. Proceedings of the sixth ACM SIGKDD International Conference on Knowledge Discovery and Data Mining, Boston, MA, 407-415.

Bradshaw, J. M. (1997), Software Agents. San Francisco, CA, USA: AAAI Press/MIT Press.

Broder, A.Z. (1997), On the Resemblance and Containment of Documents. Proceedings of Compression and Complexity of SEQUENCES 1997. Salerno, Italy, 21-29.

Chakrabarti, S., Dom, B.E., Kumar, S.R, Raghavan, P., Rajagopalan, S., Tomkins, A., Gibson, D. and Kleinberg, J.M. (1999), Mining the Web's Link Structure, IEEE Computer, 32 (8). 60-67.

Deerwester, S., Dumais, S.T., Landauer, T.K., Furnas, G.W., and Harshman., R.A. (1994), Indexing by Latent Semantics Analysis, Journal of the Society for Information Science, 41(6), 391-407.

Dumais, S.T., Furnas, G.W., Landauer, T.K., and Deerwester, S. (1988), Using Latent Semantic Analysis to Improve Information Retrieval, Proceedings of the CHI'88: Conference on Human Factors in Computing Systems, New York, ACM, 281-285.

Fan, L., Cao, P., Lin, W. and Jacobson, Q. (1999), Web Prefetching between Low-Bandwidth Client and Proxies: Potential and Performance, SIGMETRICS'99.

Flesca, S. and Masciari, E.(2003), Efficient and Effective Web Change Detection, Data & Knowledge Engineering, Elsevier.

Fox, E. (1983), Extending the Boolean and Vector Space Models on Information Retrieval with P-Norm Queries and Multiple Concepts Types. Cornell University Dissertation.

Kleinberg, J.M. (1998), Authoritative Sources in a Hyperlinked Environment, J. of the ACM (JACM), 46(5). 604-632.

Palpanas, T. (1998), Web Prefetching using Partial Matching Prediction, Technical report CSRG-376, University of Toronto.

Pitkow, J. and Pirolli, P. (1997), Life, Death, and Lawfulness on the Electronic Frontier. Proceedings of the Conference on Human Factors in Computing Systems, Atlanta, Georgia.

Popescul, A., Flake, G., Lawrence, S., Ungar, L.H., and Gile, C.L. (2000), Clustering and Identifying Temporal Trends in Document Database. Proceedings of the IEEE advances in Digital Libraries, Washington.

Rasmussen, E. (1992), Clustering algorithms. Information Retrieval: Data Structure and Algorithms. Prentice Hall, 419-442.

Rocchio, J. J. and McGill, M.J. (1997), Relevance Feedback in Information Retrieval. Prentice-Hall Inc., Englewood Cliff, NJ.

Salton, G. (1968), Automatic Information Organization and Retrieval. McGraw-Hill.

Salton, G. and Yang, C. (1973), On the specification of term values in automatic indexing, Journal of Documentation, Vol. 29, pp. 351–372.

Shaw, J.A., and Fox E.A. (1994), Combination of Multiple Searches. Proceedings of the 3rd Text Retrieval Conference (TREC-3), 105.

Wen, J.R., Nie, J.Y., and Zhang, H.J. (2002), Querying Clustering Using User Logs. ACM Transactions on Information Systems, 20(1), 59-81.

Xiao, J., Zhang, Y., Jia, X., and T. Li (2001), Measuring Similarity of Interests for Clustering Web-Users. Proceedings of the 12th Australian Database Conference 2001 (ADC'2001). Gold Coast, Australia, 107-114.

Representation and Discovery of Intelligent E-Services

Xia Wang, Bernd J. Krämer, Yi Zhao, and Wolfgang A. Halang

Faculty of Electrical and Computer Engineering
Fernuniversität in Hagen, 58084 Hagen, Germany
{xia.wang, bernd.kraemer, yi.zhao, wolfgang.halang}@fernuni-hagen.de

Abstract

The current electronic services (e-services) are considered as a kind of web-based applications, which accomplish a certain tasks, such as on-line shopping or flight booking. The next-generation service model is a trend towards intelligent e-services, named semantic web services, aiming to automatically and intelligently discover, customise, compose, execute, and provide personalised services to end user. This chapter is right to focus on the representation of e-service intelligence and its discovery machenism. Two main contributions are presented. First, after comparing the current work on representing semantic services, we take OWL-S as the service description language, and extend it by more features w.r.t. quality of service. Second, we simplify a matchmaking algorithm to quantitatively measure the similarity of services, in order to rank sets of matched services.

1 Introduction

Nowadays, the amount of data on the World Wide Web is exploding and the web's infrastructure is a huge passive storage of electronically linked documents. Moreover, one is dissatisfied with browsing such static information (e.g., hotel information), or manually filtering desired information out of thousands of results returned by search engines. It is desirable that the next generation of web should be intelligent, and should provide electronic services instead of static data to help users automatically accomplish certain complex tasks. Therefore, the field of e-service intelligence is emerging. Based on our understanding, there are four aspects of "intelligence" with respect to e-services:

a) E-services per se are intelligent, i.e., they are designed to be machine-understandable. Based on formally defined and sharable data, machines are able to understand themselves and other ones.

X. Wang et al.: *Representation and Discovery of Intelligent E-Services*, Studies in Computational Intelligence (SCI) **37**, 233–252 (2007)
www.springerlink.com

b) E-services can be intelligently discovered and automatically be composed according to some given requirements, and can flexibly be invoked and executed.
c) E-services intelligently adapt to user requirements by extracting implicit information to detect the highest potential of usefulness in applications.
d) E-services are intelligently customised according service requirements, e.g., with respect to input and output formats, in order to provide personalised services.

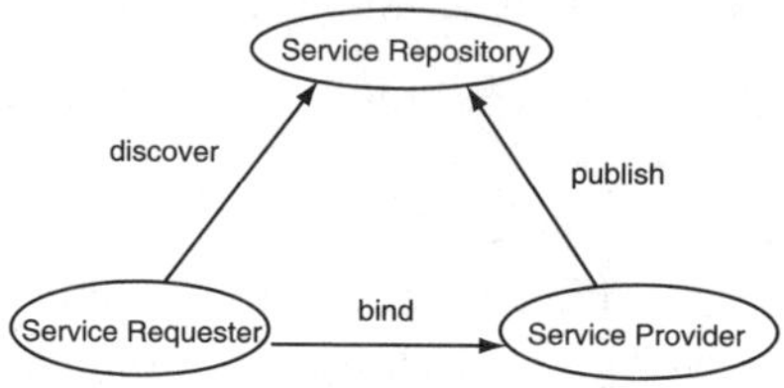

Fig. 1 Architecture of web services

Since 2001, web services are becoming an acknowledged field in research and industry. Web services are an effective way to provide for interoperation of heterogeneous applications and systems. It bases a stack of XML-based standard technologies, including the Web Service Description Language (WSDL), the Simple Object Access Protocol (SOAP), or Universal Description Discovery Integration (UDDI), to implement description, publication, and invocation of services. Fig.1 shows a simple architecture of web services with three roles: service requester, service provider, and service repository. The service repository is a place to store available services published by service providers, and a place for service requesters to find services matching given requirements, from where service requesters can obtain the Universal Reference Identifiers (URIs) of matched services to directly bind the corresponding services.

Meanwhile, the semantic web is predicted as the web's next generation by Berner-Lee [1]. It is not envisaged as a separate web, but as an extension to the current one, in which data are given well-defined meanings, facilitating machines and people to work on in co-operation. The semantic web is expected to bring structure to the meaningful content of web pages by using the concept of ontologies [4]. Data on the web, including meta-data, links and graphics, can be described as ontologies. Thus, data are not isolated anymore, but linked by their relationships, and can be interpreted automatically by machines using ontology reasoners.

Apparently, the future web is required to provide not only simple static information, but also complex dynamic services. The integration of semantic

web and web services is naturally an approach to intelligent e-services. However, to implement intelligent e-services for routine applications, there is still a long way to go and many challenges need to be overcome, including service representation, discovery, authentification, composition, invocation, execution, customisation, personalisation etc.
This chapter mainly focuses on the representation and discovery of semantic web services [2]. Two main contributions will be presented. First, after comparing the current work on representing semantic services, we select OWL-S as service description language and extend it by features oriented at providing better quality of service. Further a service selection model is formed, which effectively and compactly represents the capabilities of services. Second, we simplify a matchmaking algorithm to quantitatively measure the similarity of services, in order to rank matched services which all fulfil a service requestor's requirements.

2 Backgrounds

In this section, we present an overview on the state of the art in representing and discovering semantic web services and the challenges encountered. Also, we summarise and discuss related work.

2.1 Problem Statement and Current Situation

Web services uniformly and abstractly describe services using an abstract representation language like WSDL, in which the information consists of service name, operations, parameters, messages types, and optional binding protocols, aiming to advertise services and to let users know how to invoke them. Obviously, WSDL is very useful for service invocation, but not for discovery. When measuring the similarity of two services simply by literally matching their *.wsdl* documents, the result may be imprecise or even erroneous due to lacking semantic support [5,11].
There is a similar problem with UDDI [3,5]. It universally describes and discovers services, and ultimately returns service locations by a key defined in the *tModel*. However, lacking semantic description, UDDI only uses keywords to match services. Hence, the results may also be erroneous or just wrong.
Therefore, a semantic representation language for services is needed, which should have the capabilities to describe services with semantical information and to reason the explicit descriptions. Correspondingly, more powerful matching algorithms should be defined.

Currently, there are two mainstream web ontology languages used to describe services: the Web Ontology Language for Service (OWL-S) supported by W3C and DAML, and the Web Services Modeling Ontology (WSMO [7]) developed by DERI and SDK WSMX. Based on analysis and comparison of OWL-S and WSMO, we can briefly summarise as follows.
- OWL-S [6] per se is an OWL ontology used as semantic mark-up language for web services. It has three main parts: service profile, process model, and grounding (for details cp. [6]). Only the service profile aiming to advertise and discover services, presents the capabilities of services, including non-functional and functional properties as well as a few quality-of-service (QoS) attributes. However, its reasoning capabilities are somewhat weak.
- WSMO is an ontology and conceptual framework to completely describe web services and most related aspects in terms of the Web Service Modeling Framework (WSMF). Although WSMO describes capabilities of web services, it covers too wide a range of service scenarios to improve interoperability, and mainly focuses on solving the integration problem.
- OWL-S and WSMO adopt similar views on the service ontology, while relying on different logics, viz., OWL/SWRL and WSML, respectively.

From our point of view, OWL-S is better suited to represent and discover services. Here, we consider OWL-S as service ontology description language and use it to match services.

The current matchmaking algorithms of semantic services may be classified into four kinds according to the ways of their similarity measurements, namely, capability-based, description logic-based, QoS-based, and ontology-based (cp. the next section for details). In their construction, the following problems were encountered:
- Most algorithms [10,11] are based on the hypothesis that a single ontology is used. In fact, however, applications use their own ontology vocabularies, so that ontology similarities need to be considered.
- Most algorithms [10,11,16] yield approximate results when expressing the similarity of two services, only, e.g., [11] presents four matching degrees (exact, plug-in, subsume, and fail), thus failing to provide quantitative service measures.
- Reasoning capabilities based on service representations are weak [16,17,18], and several rule languages, such as SWRL by *daml.org* or RuleML by *ruleml.org*, are still in the development phase.

Based on the above analysis, this chapter does not aim to address all challenges (so, we talk little on reasoning languages, and address the similarity of multiple ontologies in [35]). Here, the only objective is to clearly present representation and discovery of service intelligence. Our positions on these two points are: to specify OWL-S and WSDL as the description languages to represent the intelligence of services; and to propose an effective matchmaking algorithm to discover services by measuring service similarity based on both capabilities and QoS properties of services.

2.2 Related Work

Discovery of services is done by measuring the similarity between service requirements (s_R) given by a service requester and service advertisements (s_A) of service providers. Service similarity is defined as $simService(s_A, s_R) = match(s_A, s_R) \in [0,1]$. If $simService(s_A, s_R) > \tau$ (with τ being a threshold), then s_A is assumed a matched service.

Matchmaking algorithm is one of the main challenges in the area of e-services. Generally, the current algorithms originate in component retrieval (signature [14] and specification matching [12,13]) and information retrieval relying on textual descriptions of artifacts [15]. We roughly distinguish them into four categories of similarity measurements, which are employed in the contributions on semantic service discovery:

- Capability-based algorithms (e.g., [10,11]) are based on a specific service description language to represent capabilities, which include non-functionalities and functionalities. They basically use subsumption reasoning [35] to measure service similarity. For instance, [10] defines a language called *Larks* for agent advertisements and requests. Its matchmaking process performs both syntactic and semantic matching. The process has five different filters (context matching, profile comparison, similarity matching, signature matching, and constraint matching), which can freely be combined to result in different degrees of matches [16,11]. In [11], similar methods are described, but based on the description language DAML-S (the previous version of OWL-S).
- QoS-based algorithms (cp. [19,20,23-25]) particularly stress the consideration of the service qualities rendered during service selection. For example, [19] enumerated objective (like reliability, availability, and request-to-response time) and subjective (focusing on user experience) QoS attributes, and presented an agent framework coupled with a QoS ontology to address the selection problem.
- Approaches based on Description Logic (DL) [31], as described in [10,11,18,22], use the language DAML-S and a Description Logic reasoner to compare service descriptions. Similarly, [22] is based on the WSMO concept model using the *F-Logic* [28] reasoning engine *Flora-2*. Generally, these approaches formulate service profiles in a logic format, and reason about their similarity based on the subsumption tree given by a reasoner (e.g., *Racer* [26] and *FaCT* [27]).

- Ontology-based algorithms, as in [8,9], consider service similarity from an ontology or concept similarity point of view. In [8], for instance, the similarity of concept terms is measured by a clustering algorithm according to their association degrees expressed in terms of their conditional probabilities of occurrence. In [9], the similarity of services is measured by taking into consideration all aspects of OWL object constructors present in the RDF format.

Any of the above algorithms has specific limitations. For instance, the capability-based ones are based on a latent and fatal hypothesis, viz., advertised and required services share a single ontology. In practice, however, applications use their own ontology vocabularies. QoS-based algorithms mainly emphasise metrics of QoS attributes. Description Logic-based approaches have somewhat weak reasoning capabilities, as they are mostly based on A-Box reasoning [31], only. Finally, ontology-based ones are essentially based on the concept level to consider the similarity of services, neglecting to match structural and semantic information.

Our position on service matching gives most attention to the advantages of above methods. We present a service model for selection, which compactly retrieves not more than the information really necessary to describe service capabilities. Our objective is to quantitatively analyse the semantic similarity of services, in contrast to former work just providing similarity degrees as defined in [11]. Correspondingly, in the sequel

- we discuss the representation of services by reviewing and specifying service description languages such as OWL-S and WSDL, and present the description capability of OWL-S with Description Logic;
- As the service profile of OWL-S contains too much information for effective matching, we propose a compact and effective service model whose content is extracted from *owl* and *wsdl* documents (the information in such models consists of non-functional and functional features and extended QoS attributes.); and
- we propose an effective matching algorithm of semantic services, finally returning a numerical value as relative and quantitative measure of service similarity, enabling to set up a ranked list of candidate services.

3 Representation of Intelligent E-Services

In this section, we review OWL-S together with WSDL with respect to describing e-service intelligence, and proposal a selection model of semantic services. Also a travel scenario is explained in our model.

3.1 Service Description Languages - WSDL and OWL-S

WSDL based on XML schema provides an abstract interface for services, which is especially helpful during service invocation. Shown in Fig.2 (a) is the structure of WSDL1.1, containing operations be provided (defined as *portType* in WSDL documents), data needed to be communicated (as *message*), and the way to bind services (as *binding*, which defines a concrete protocol and data format). Fig.2 (b) explains an example WSDL document of *BravoAirGrounding.wsdl* (BravoAir is a fictitious airline site developed by DAML Services). It describes a service named *BravoAir_Service*. One of its operations and binding protocols is designated by *GetDesiredFlightDetails_Port*, and the needed message is defined in the part of *GetDesiredFlightDetails_Input*.

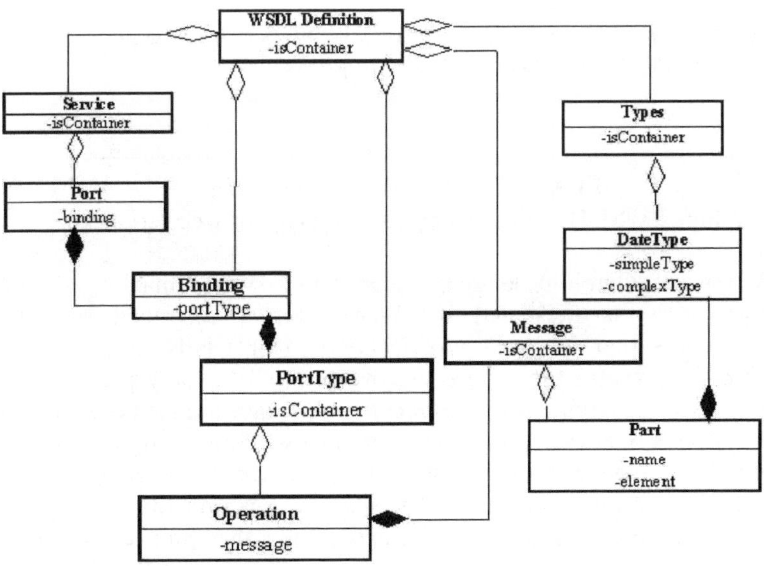

(a) Structure of WSDL Specification

```
<?xml version="1.0" ?>
-<definitions name="BravoAir_WSDL„
   targetNamespace="http://www.daml.org/services/owl-s/1.0/BravoAirGrounding.wsdl"
   xmlns:soap="http://schemas.xmlsoap.org/wsdl/soap/"
   ...
   xmlns:xsd="http://www.w3.org/2001/XMLSchema"
+<message name="GetDesiredFlightDetails_Input">
...
-<portType name="GetDesiredFlightDetails_PortType">
   <operation name="GetDesiredFlightDetails_operation"
     ...
     <input message="tns:GetDesiredFlightDetails_Input" />
   </operation>
 </portType>
...
+<binding name="GetDesiredFlightDetails_SoapBinding" type="tns:GetDesiredFlightDetails_PortType">
....
-<service name="BravoAir_Service">
 <documentation>...</documentation>
 -<port name="GetDesiredFlightDetails_Port" binding="tns:GetDesiredFlightDetails_SoapBinding">
     <soap:address location="http://www.BravoAir.com/GetDesiredFlightDetails" />
   </port>
   ....
 </service>
```

(b) Part of BravoAirGrounding.wsdl

Fig. 2 WSDL as the description language of web service

OWL-S is a web Ontology language supplying service semantics. Its top Ontology *service* is presented by *ServiceProfile*, described by *ServiceModel,* and supported by *ServiceGrounding* (for details see [6]). Only *ServiceProfile* is useful for service discovery. Fig.3 (a) only presents the ontology of *ServiceProfile*. *Advertisement* and *Request* are the potential actors who provide *ServiceProfile*, and *Actor* gives information on an actor. *ServiceProfile* defines all possible properties of a service, such as *serviceName*, *textDescription,* and *contactInformation*, functionality description (*inputs*, *outputs*, *preconditions,* and *effects*), and other profile attributes, which users should be aware of (such as *quality guarantee*, possible classification of the service, and additional parameters that the service may want to specify).

Fig.3 (b) shows only part of the profile extracted from *BravoAirProfile.owl*. For lack of space, we omit most of the concrete information, but still show that the service has the name *BravoAir_ReservationAgent*, a short *textDescription*, some *contactInformation*, and functionalities like-*hasInput* (e.g., codes of departure and arrival airports, dates etc.), *hasOut-*

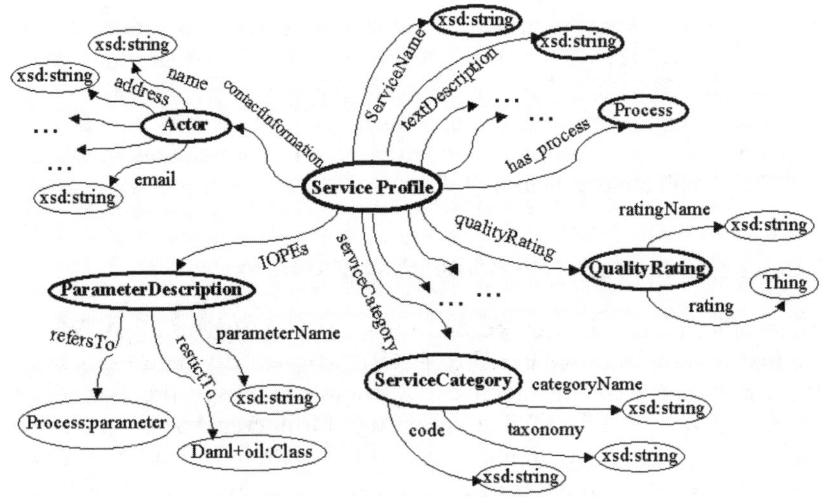

(a) Ontology of ServiceProfile

```
<?xml version="1.0" encoding="ISO-8859-1" ?>
 <rdf:RDF xmlns:rdf=http://www.w3.org/1999/02/22-rdf-syntax-ns#
  xmlns:rdfs=http://www.w3.org/2000/01/rdf-schema#
  ...
  xmlns:owl="http://www.w3.org/2002/07/owl#" >
- <owl:Ontology rdf:about="">
  ...
 <owl:imports rdf:resource="http://www.daml.org/services/owl-s/1.0/Service.owl" />
</owl:Ontology>
 ...
- <profileHierarchy:AirlineTicketing rdf:ID="Profile_BravoAir_ReservationAgent">
  ...
  <profile:serviceName>BravoAir_ReservationAgent</profile:serviceName>
  <profile:textDescription>...</profile:textDescription>
  <profile:contactInformation>...  </profile:contactInformation>
...
  <profile:qualityRating>
   <profile:QualityRating rdf:ID="BravoAir-goodRating">
      <profile:ratingName>SomeRating</profile:ratingName>
      <profile:rating rdf:resource="http://www.daml.org/services/owl-s/1.0/Concepts.owl#GoodRating" />
      </profile:QualityRating>
  </profile:qualityRating>
<profile:serviceCategory> ... </profile:serviceCategory>
 <profile:hasInput rdf:resource="..." />
  ...
 <profile:hasOutput rdf:resource="..." />
  ...
 <profile:hasEffect rdf:resource="..." />
</profileHierarchy:AirlineTicketing>
 </rdf:RDF>
```

(b) Part of *BravoAirGrounding.owl*

Fig. 3 ServiceProfile Ontology in OWL-S

put (like available flight itinerary), or *hasEffect* (like seats reserved), and some additional parameters about service quality and category.

Together, OWL-S and WSDL provide sufficient information for structural and semantic description of services. For selection of services, however, they require too much information to be considered. Therefore, a condensed model is necessary, which should be powerful enough to express service capabilities, and have a reasoning capability to extract implicit knowledge required for more accurate matchmaking. This is the objective for the construction of a service model based on specifications in OWL-S and WSDL as elaborated below.

3.2 Selection Model of Semantic Web Services

A conceptual model for service discovery is illustrated in Fig.4. We assume that a graphical user interface (GUI) is provided to end users to retrieve service requirements. Such a GUI is application specific. Service requirements gained via the GUI are re-written for internal use in our service model, yielding a formal requirement profile s_R; On the other side, all services published by service providers are pre-processed according to our model, yielding respective service profiles s_A, which are stored in a services repository. Then, a matching engine is invoked to select similar services, providing a set of matched services as output. Request re-writing [29] is beyond the scope of this chapter, as is GUI design.

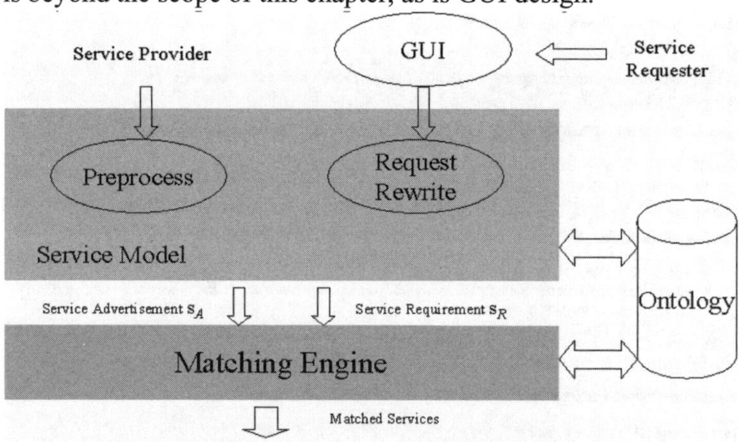

Fig. 4 Conceptual Model for Service Discovery

To describe a service, its non-functionality and functionality information should be defined. That is, static information (like service name, service categories, or short textual description) describes its non-functionality, whereas dynamic information like operations (which further have input and output parameters, pre-conditions, constraints and effects), stands for its functionalities. For instance, a flight-booking service is denoted by a ser-

vice name, sometimes with a short service description, and advertises its ticket booking function. If one would like to use this service, one must provide the input information, like flight dates, number of passengers and so on. Pre-condition is that one has a valid credit card for payment. The final effect of the service's execution is to draw money from the credit card account and to issue a flight ticket.
Based on the reference work [6,30], our service model is stated as service profile $s=(NF,F,Q,C)$, in which NF is a set defining the non-functionalities of a service, F is a set of service's functionalities, Q defines the service's quality attributes, and C is the overall cost of the service's invocation.

- Based on the OWL-S specification, we model *NF* as an array of *String*. For conciseness and efficiency it only includes:
  - *serviceName*, a *String* as the name of the service;
  - *serviceCategory*, a *String array* enumerating any possible categories of the service, classified according to its application fields, or some industry taxonomies such as North American Cartographic Information Society (NACIS) or United Nations Standard Products and Services Code (UN-SPSC);
  - *textDescription*, a short text readable by humans briefly describing the service.
- $F=\{f_1,f_2,\ldots,f_i\}$, $i \in N$, and $f_i=(op_i, \Sigma_{Ii}, \Sigma_{Oi}, Cons_i, P_i, E_i)$ is a set of functions published by the service. Each function f_i is described by:
  - op_i: *String*, the name of operation i;
  - Σ_{Ii}: an array of *String*, consisting of all the input parameters;
  - Σ_{Oi}: an array of *String*, consisting of all the output parameters;
  - $Cons_i$: an assertion expressed in DL syntax as constraint, under which the service is executed;
  - P_i : an assertion in DL syntax as pre-condition;
  - E_i : an assertion in DL syntax as effect;
  - $\delta: (op_i \times \Sigma_{Ii} \times Cons_i \times P_i) \rightarrow (\Sigma_{Oi} \times E_i)$: an implication from the set $(op_i \times \Sigma_{Ii} \times Cons_i \times P_i)$ to set $(\Sigma_{Oi} \times E_i)$ expressing the logic statement that "If $(op_i \times \Sigma_{Ii} \times Cons_i \times P_i)$ is true, then $(\Sigma_{Oi} \times E_i)$ is true ".
- Q denotes the quality attributes of the service.

In [32], all possible QoS attributes for a web service were identified, including performance, reliability, scalability, capacity, robustness, exception handling, accuracy, integrity, accessibility, availability, interoperability, security, and network-related QoS. For the purpose of optimum manipulability, our model categorises QoS only into two sets from the customer's perspective: a necessary and an optional set.

Intuitively, the necessary qualities comprise the attributes typically required by customers. For instance, a customer who wants to find an on-line flight booking service is more concerned about the response time to his order rather than about the scalability of the service itself. Therefore, response time will be emphasised in his service requirements and, thus, defined as necessary.

The necessary and optional QoS requirements are set in a dynamic way. Sometimes a customer may stress some specific qualities in his requirements, which are not belonging to the default setting of necessary attributes. In this case, the announced qualities will be moved from the category of optional ones to the necessary ones. This kind of design renders high extensibility and flexibility.

Adopting the contributions of [32] with respect to the quality attributes, we define them as the set Q_I, and let the necessary quality set be Q_n. Then, the optional quality set is $Q_o=Q_I-Q_n$. Based on cursory statistics, in our case we define the default set of necessary QoS attributes as Q_n={*qualityResponseTime,qualityExecutionTime,qualityReliability,quality ExceptionHandling,qualityAccuracy,qualitySecurity*}. Following the same syntax, we simply use Q_n to extend the quality part of the service profile in OWL-S. Thus, implicitly our model defines $Q=Q_n=\{q_1, q_2,\ldots,q_j\}$, $j\in N$, which is a set of attribute assertions expressed in DL syntax.

- *C* is the overall cost of consuming a service, expressed as an assertion in DL syntax.

Above service model will be applied to a realistic scenario in next section.

3.3 A Travel Scenario

Professor J. Heineken is planning to attend an international conference in Orlando, Florida, USA, during 14-17 July 2005. He lives in London, UK. At present he has to spend a lot of time manually arranging his trip with a web browser, including searching in on-line booking services, booking a return flight (based on his itinerary and locations), reserving a room in a convenient and low-cost hotel, and renting a car in Orlando.

Bob Heineken's requirements		
Flight Ticket Booking	Hotel Booking	Car Rental

Fig. 5 A travel scenario

Let us now consider how to help him by applying our service model. In the scenario, Prof. Heineken basically needs three kinds of services (cp. Fig.5): flight booking, hotel reservation, and car rental. Strongly supported by industry, there are many web services which can meet his requirements with similar functionalities. These services have been developed and deployed using specification standards such as WSDL and OWL-S.

Now, we apply our model to this scenario, and only consider the flight booking as the other operations are similar. The re-writen requirements of Prof. Heineken is as follows:

- ***NF*** = {*TicketBook*, *Travel*, "*TicketBookDescription*"}; Note that *TicketBookDescription* will be a short text about the anticipated service, such as "I would like to book a return ticket. I want the cheapest price and the best service…";
- For clarity we model $F=\{f_1\}$ (here, *F* has only one function) as:

op ={TicketBook};
Σ_I={LondonUK, OrlandoUSA, July1405, July1705, 1Person, 0Children, 0Infants, Return};
Σ_O ={TicketDepart, TicketReturn};
Cons={∩(≤800cost.Ticket)∩(≥8amtime.Ticket)∩(≤ 20pmtime.Ticket)∩(≥2class.Ticket)};
P={(=1*valid.PaymentCard*)};
E={(money.PaymentCard - cost.Ticket)};

Fig. 6 Description of F in the travel scenario

In Fig. 6, Prof. Heineken's requirements are mentioned: the price of the return flight should not exceed $800, the departure time should be between 8:00 and 20:00, and an economy class ticket is required.

- ***Q***={∩(≤15*time.qualityResponseTime*)∩(≤15*time.qualityExecuteTime*)∩(≥6*rank.Reliability*)∩(≥6*rank.qualityExceptionHandling*)∩(≥7*rank.qualityAccuracy*)∩(≥6*rank.qualitySecurity*)};

The quality requirements are that the service's response time should be less than 15s, the security of its execution better than rank 7, and so on.

- **C**={(≤20*cost.CostService*)}; The service charge should not exceed $20.

Fig. 7 gives two real examples of advertised services described in form of our service model.

NF={*CheapFlightSearch*, *Flight*, *"Search for a cheap flight ticket.UK citizens"* }; OP={*CheapFlightSearch*}; Σ_I={*LondonUK*, *OrlandoUSA*, *July1405*,*July1705*,*Adults*,*0Children*,*0Infants*,*ReturnTrip*}; Σ_O = {*TicketDepart*, *TicketReturn*}; Cons={∩(=645.70*cost.Ticket*)∩(=morningtime.*DepartTicket*)∩(=*eveningtime.ReturnTicket*)∩(=*economyclass.Ticket*)}; *P*={(=1*valid.PaymentCard*)}; *E*={(*money.InsurenceCard-cost.Ticket*)}; Q={∩(≤8*time.qualityResponseTime*)∩(≤8*time.qualtyExecuteTime*)∩(≥8*rank.Reliability*)∩(≥7*rank.qualityExceptionHandling*)∩(≥9*rank.qualityAccuracy*)∩(≥8*rank.qualitySecurity*)}; C= {(= 155.70*cost.CostService*)};	*NF*={*Flights*,*Flight*,*"travel expertise, unrivalled knowledge, excellent value..."*}; OP ={*FindFlight*}; Σ_I={*LondonCityAirport*, *OrlandoUSA*, *July1405*, *July1705*, *1Adults*, *0Children*, *0Infants*, *ReturnTrip*}; Σ_O = {*TicketDepart*, *TicketReturn*}; *Cons*={∩(=2280.35*cost.Ticket*)∩(=*morningtime.DepartTicket*)∩(=eveningtime.*ReturnTicket*)∩ (=2*class.Ticket*) }; *P* ={(=1*valid.PaymentCard*)}; *E* ={(*money.InsurenceCard-cost.Ticket*)}; Q={∩(≤12*time.qualityResponseTime*)∩(≤8*time.qualityExecuteTime*)∩(≥8*rank.Reliability*)∩(≥7*rank.qualityExceptionHandling*)∩(≥8*rank.qualityAccuracy*)∩(≥8*rank.qualitySecurity*)}; C = {(≤ 134.35*cost.CostService*)};

Advertisement1 *http://www.squareroutetravel.com/Flights/cheap-flight-ticket.htm*
Advertisement2 *http://www.travelbag.co.uk/flights/index.html*

Fig. 7 Two real examples of an e-service

4 Discovering Intelligent E-Services

Usually, there is always a service requirement s_R and a number of service advertisements s_A. A matchmaking algorithm filters available services based on the service requirements. During the process of matching, it will sequentially yield abstract values of the similarities of functional, non-functional, and QoS attributes as well as cost. Therefore, we take their weighted sum as an abstract measurement of the similarity of services:
$simService(s_R,s_A)=w_1 \bullet simNF(s_R,s_A)+w_2 \bullet simF(s_R,s_A)+w_3 \bullet simQ(s_R,s_A)+ w_4 \bullet simC(s_R,s_A)$, where $w_1+w_2+w_3+w_4=1$.
The weight of each attribute should be adjusted specific to the application domain. The value of $simService(s_R,s_A)$ is a real number in [0,1] used to rank the final selection set.
Refering to Fig. 4, we illustrate the principle of our matching engine in Fig. 8. A requester *R* provides a service named $s_R=(NF_R, F_R, Q_R, C_R)$. Similarly, a provider *A* publishes its service (by deploying its OWL-S and WSDL documents), from which the necessary information is retrieved to construct a service profile, named as $s_A=(NF_A, F_A, Q_A, C_A)$. In fact, the four constituents of s_A and s_R are independent of each other. A user may customise his matchmaking process by taking only unitary properties or combing multiform properties.
For example, if one would like to do a selection based on service name and text description, only, then the matching process in Fig. 8 follows the real

lines from *NF* to M_{NF}, and yields a result. Shown by the dashed lines is a matching process sequentially considering four properties, that is also taken in the remainder of this chapter as an example to present our matchmaking algorithm. The *analyser* is used to simply deal with the matching result, and the output is a ranked set of matched services.

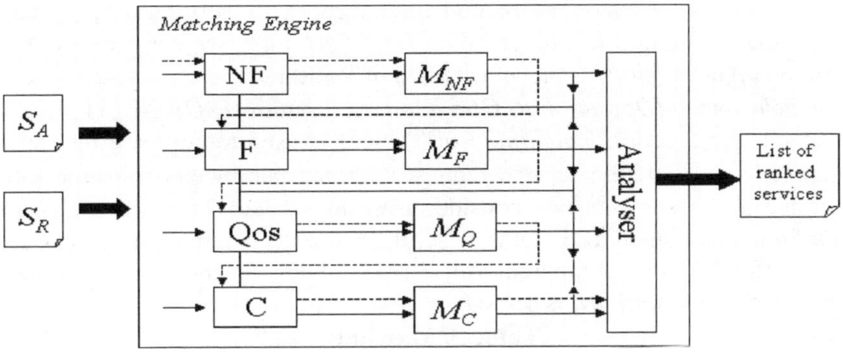

Fig. 8 Matching Engine

The matching process is carried out by three filters as follows:

1) ***NF*** – The first filter is based on service name, service category, and service description to filter all available services, and to obtain a collection of likely candidates, named *reultSet*$_1$.

The process starts with running the pseudo-code given in Table 1 both on the input parameter *serviceName* and on *serviceCategory*, respectively. In the algorithm, the function *distance* taken from [10] is used to measure the similarity of different concepts. The smaller the distance of two concepts is, the more similar they are. The corresponding outputs are *simName* and *simCategory* with their values lying between 0 and 1.

Table 1. Matching *serviceName* and *serviceCategory*

```
matchConcept (ConceptR, ConceptA){
  double simConcept;
  simConcept = distance(ConceptR, ConceptA);
  return simConcept;}// simConcept ∈ [0,1];
```

The theory of frequency-inverse document frequency (TF-IDF), which originates in the area of the information retrieval, is adopted here to compute the similarity of *serviceDescription*. In the algorithm of Table 2, *serviceDescription* is the input, and the similarity of descriptions, *simDes,* is the output in the range between 0 and 1.

Table 2. Matching *serviceDescription*

```
matchDescription(DescriptionR, DescriptionA){
  double simDes, DesR, DesA;
  for DesRij, DesAij computig weights:
      w = wf(w,d) · log(N / df(w));
  simDes= DesR· DesA / ( | DesR | · | DesA | );
  return simDes; } // simDes ∈ [0,1];
```

2) ***op***, Σ_I, Σ_O – The second filter works with *resultSet*$_1$ for further selection, yielding a result set called *resultSet*$_2$. First, we compare the operation ***op*** using *matchConcept* (defined in Table 1.):

matchConcept(Operation R, Operation A); // return *simOP*$\in$ [0,1]

Second, the function *matchInput* of Table 3 is used to compare the input parameters Σ_I. If the service requester's input parameters subsume [35] service advertisements, we consider them as similar. Then, the value of *simPara* is assigned as 1. Otherwise, the requester's input parameters are not sufficient to be compared with service advertisements. That is to say, the match fails, *simPara* is set to 0.

Table 3. Matching Σ_I

```
matchInput(Parameter PR, Parameter PA){
  int simPara = 0;
  if PR subsumes PA then simPara = 1;
  return simPara; }// simInput∈{0,1}
```

The function *matchOutput*, however, does the opposite. If the service requester's output parameters cannot be subsumed in the service advertisements, it means that the results of advertisement comparisons meet the service requester's requirement only partly. In this case the match failed, noted as 0. For example, if a flight service only provides one-way journeys, we consider the matchmaking of a return flight as failed.

Table 4. Matching Σ_O

```
matchOutput(parameter PR, parameter PA){
  if (simPara! = 0){
    if PA subsumes PR then simPara = 1;
       else simPara = 0;}
  return simPara;}// return simOutput∈{0,1}
```

3) ***Cons***, ***P***, ***E***, ***Q*** and ***C*** – Similarly, the third filter processes *resultSet*$_2$ to obtain the final *resultSet*. In this step, the matchmaking compares a set of expressions. Here, we cannot simply consider the subsumption between them, because we care more about the degree of similarity than their subsumption relationship. Taking *cost.Ticket,* for example, requester *R* states that the ticket price must be less than \$800, *Advertisement*$_1$ (noted as A_1) asks for \$645.70 and *Advertisement*$_2$ (A_2) for \$2280.35. In this case, we should scale the degree of proximity rather than consider the difference of the values:

Table 5. Proximity degree

closeExpCost (R, A_1)= (800− 645.70)/800 = 0.192875;
closeExpCost (R, A_2)= (800− 2280.35)/800= -1.8504375;

Obviously, the bigger a positive scaled value is, the better is the result. Hence, $Advertisement_1$ is a better choice than $Advertisement_2$.

However, this is only one expression under consideration. If extended to a set of expressions (in our case *Cons*, *P*, *E*, *Q* and *C*), the multi-attribute utility theory of [30,33] should be applied. That is, after computing the degree of proximity, we assign a *utility score* between 1 and 5, which strictly bases on its specific attribute definitions. For example, we assume that the definition of utility score for the flight ticket price could be 5 ($0.10 \leq closeExpCost \leq 1.00$), 4 ($0.05 \leq closeExpCost < 0.099$), 3 ($0 \leq closeExpCost < 0.05$), 2 ($-1 \leq closeExpCost < 0$), 1 ($-2.00 \leq closeExpCost < -1.00$), 0 ($closeExpCost < -2.00$).

Re-considering the above example, the values of proximity are $utilityScore(R,A_1)=5$ and $utilityScore(R,A_2)=1$. We assume expressions $E=\{e_1,e_2,\ldots,e_n\}$, $n \in N$, and a related utility score $U=\{u_1,u_2,\ldots,u_n\}$. A weight between 0 and 1 is assigned to each attribute. Then, $simExpression= (w_1 \cdot u_1 + w_2 \cdot u_2 + \ldots + w_n \cdot u_n)/(w_1 + w_2 + \ldots + w_n)$. Also, as *Cons*, *P*, *E*, *Q*, and *C* are such expressions, we could calculate their corresponding similarities as *simCons*, *simP*, *simE*, *simQoS* and *simC*.

Above, we have mentioned that utility score assignment must be based on attribute definitions. The reason is that different attributes have different scaling standards. When we consider *qualitySecurity*, in comparison with *cost.Ticket,* a requester may require the rank of a service's security to be more than degree 6, and $Advertisement_1$ is evaluated to have rank 8 in service quality, while $Advertisement_2$ has rank 10. In this case, the smaller the negative value (cp. Table 5) is, the better the match is and, accordingly, this attribute has a different definition of utility score. Fortunately, this part of matchmaking is trivial. We can finally compute the *simExpression* based on the specific definitions in a specific application.

Now, after the above matching has yielded a *resultSet*, and each step also has provided similarity values of different aspects, we use them to compute the final similarity *sumSimilarity*, and rank the *resultSet* based on this parameter.

In a similar way, we assign to each attribute involved in the matchmaking a weight between 0 and 1. Then, the final combined score becomes:

$$sumSimilarity=(w_1 \cdot simName+w_2 \cdot simCategory+ w_3 \cdot simDes + w_4 \cdot simOP+ w_5 \cdot simInput+w_6 \cdot simOutput+w_7 \cdot simCons+w_8 \cdot simP+w_9 \cdot simE+w_{10} \cdot simQoS+ w_{11} \cdot simC) / (w_1+ w_2+ w_3+ w_4+ w_5+ w_6+ w_7+ w_8+ w_9+ w_{10}+ w_{11});$$

This algorithm quantifies the similarity between service requirements and advertisements. On the basis of their similarity degrees it can be decided which service matches best.

5 Conclusions

The work reported in this chapter takes semantic web services as a model for intelligent e-services, aiming to show how to represent and to discover e-services. We discussed the current situation and challenges with respect to these two aspects, and presented our position. In short, we used OWL-S in combination with WSDL as the representation language for services, constructed a service selection model, and presented an original matchmaking algorithm, which quantitatively measures the similarity of services allowing ranking services matched.

Although many aspects of semantic web services have been studied thoroughly, such as service description, discovery, composition, QoS, or security, still many more contributions are needed to let intelligent e-services become a reality. These contributions are not only expected from the area of artificial intelligence, but also from other academic disciplines and from industry. With respect to service discovery or selection, two trends are interesting:

a) Uncertainty or fuzzy discovery. In fact, user requirements are invariably ambiguous. Especially in more professional domains, it is difficult to retrieve all trivial details to perform an exact match. Uncertainty discovery means to base on limited user-given information to extract the implicit information as far as possible, in order to achieve the highest potential for applications.

b) Service customisation and recommendation. The intelligence of e-services should have the capability for customisation according to user requirements, e.g., with respect to input and output formats, in order to provide personalised services. To recommend likely candidates, it is very useful to take analysis and elicitation of user interest as basis.

References

[1] Berners-Lee, T., Hendler, J., and Lassila O. (2001). The Semantic Web. Scientific American, 284(5), pp.34-43.

[2] McIlraith, S.A., Son, T.C., and Zeng, H. (2001). Semantic Web Services. IEEE Intelligent Systems 16(2), pp. 46-53.

[3] Verma, K., Sivashanmugam, K., Sheth, A., Patil, A., Oundhakar, S. and Miller, J. (2005): METEOR–S WSDI: A Scalable P2P Infrastructure of Registries for Semantic Publication and Discovery of Web Services. Journal of Information Technology and Management, 6(1):17-39.
[4] Maedche, A., and Staab, S. (2001). Ontology learning for the semantic web. IEEE Intelligent Systems, 16(2):72-79.
[5] Wang, Y., Stroulia, E. (2003). Semantic Structure Matching for Assessing Web-Service Similarity, First International Conference on Service Oriented Computing (ICSOC), pp.194-207.
[6] Martin, D., Burstein, M. et al (2004). OWL-S: Semantic Markup for Web Services, from http://www.w3.org/Submission/OWL-S/.
[7] Lausen, H., Roman, D., and Keller, U. (2004). Web Service Modeling Ontology (WSMO). Working draft, DERI, March 2004.
[8] Dong, X., Halevy, A.Y., Madhavan, J., Nemes, E., Zhang, J. (2004). Similarity Search for Web Services. VLDB 2004, pp. 372-383.
[9] Hau, J., Lee, W., Darlington, J. (2005). A Semantic Similarity Measure for Semantic Web Services. Conference WWW2005, Japan.
[10] Sycara, K., Widoff, S., Klusch, M., and Lu, J. (2002). LARKS: Dynamic Matchmaking Among Heterogeneous Software Agents in Cyberspace. Autonomous Agents and Multi-Agent Systems, 5 (2):173-203.
[11] Paolucci M., Kawamura T., Paye T. and Sycara K. (2002). Semantic matching of web services capabilities. Proc. of the 1st International Semantic Web Conference (ISWC), pp. 333-347.
[12] Cho, I., McGregor, J., and Krause, L.(1998). A protocol-based approach to specifying interoperability between objects. Proceeding. of the 26th Technology of Object-Oriented Languages and Systems (TOOLS'26), pp.84-96.
[13] Zaremski, A. M., and Wing, J. M.(1997). Specifications Matching of Software Components. ACM Transactions on Software Engineering and Methodology, 6(4):333-369.
[14] Zaremski, A.M., and Wing, J. M. (1995). Signature Matching: a Tool for Using SoftwareLibraries. ACM Transactions on Software Engineering and Methodology, 4(2):146-170.
[15] Faloutsos, C. and Oard, D. W. (1995). A survey of Information Retrieval and Filtering Methods. University of Maryland, Technical Report CS-TR-3514.
[16] Li,L. and Horrocks, I. (2004). A software framework for matchmaking based on semantic web technology. International Journal of Electronic Commerce, 8(4):39-60.
[17] Gonzalez-Castillo, J., Trastour, D., and Bartolini, C. (2001). Description Logics for Matchmaking of Services. In Proceedings of Workshop on Application of Description Logics. September 2001.
[18] Hacid, M., Leger, A., Rey, C., Toumani, F. (2002). Dynamic discovery of e-services: a Description Logics based approach. Proceeding of the 18th French conference on advanced databases (BDA), Paris, pp.21-25.
[19] Maximilien, E. M., and Singh, M. P. (2004). A Framework andOntology for Dynamic Web Services Selection. IEEE Internet Computing, 8(5):84–93.

[20] Ran, S. (2003). A model for web services discovery with QoS. ACMSIGecom Exchanges, 4(1):1–10.
[21] Horrocks, I., Patel-Schneider, P. F., and Harmelen, F. (2003). From SHIQ and RDF to OWL: The making of a web ontology language. Jouranl of Web Semantics, 1(1):7-26.
[22] Kifer, M., Lara, R., Polleres, A., Zhao, C., Keller, U., Lausen, H., Fensel, D. (2004). A Logical Framework for Web Service Discovery. In ISWC 2004 Workshop on Semantic Web Services: Preparing to Meet the World of Bussiness Applications, 2004.
[23] Zhou, C., Chia, L-T, Lee, B. S. (2005). Service discovery and measurement based on DAML-QoS ontology. WWW (Special interest tracks and posters) 2005, pp-1070-1071.
[24] Zhou, C., Chia, L-T., and Lee, B.B. (2004). DAML-Qos Ontology for Web Services. In International Conference on Web Services, pp.472-479.
[25] Vu, L-H., Hauswirth, M., Aberer, K. (2005). Towards P2P-based semantic web service discovery with QoS supports. Workshop on Business Processes and Services (BPS), Sept. 2005.
[26] Haarslev, V., and Möller, R. (2003). Racer: An OWL Reasoning Agent for the Semantic Web. Proceedings of the International Workshop on Applications, Products and Services of Web-based Support Systems, pp.91–95.
[27] Horrocks, I. (1998). The FaCT system. International Conference Tableaux'98, number 1397 in Lecture Notes in Artificial Intelligence, pp.307-312.
[28] Kifer, M., and Lausen, G. (1989). F-Logic: A Higher-Order language for Reasoning about Objects, Inheritance, and Scheme. ACM SIGMOD, pp.134-146.
[29] Benatallah, B., Hacid, M-S., Rey, C., and Toumani, F. (2003). Request Rewriting-Based Web Service Discovery, SemanticWeb ISWC2003, pp.242–257.
[30] Menasce, D. (2004). Composing Web Services: A QoS View, IEEE Internet Computing, 8(6): 88-90.
[31] Baader, F., Calvanese, D., McGuinness, D.L., Nardi, D., Patel-Schneider, P.F. (Eds.): The Description Logic Handbook: Theory, Implementation, and Applications. Cambridge University Press 2003.
[32] W3C, QoS for Web Services: Requirements and Possible Approaches, W3C Working Group Note, 25 Nov. 2003.
[33] Maximilien, E.M., and Singh, M.P. (2004). A Framework and Ontology for Dynamic Web Services Selection, IEEE Internet Computing, 8(5), pp.84-93.
[34] Zhao, Y., Halang, W. A., and Wang, X. (2006). Rough Ontology Mapping in E-Business Integration. This volume.
[35] Borgida, A., and Patel-Schneider, P. F. (1994). A semantics and complete algorithm for subsumption in the classic description logic. Journal of Artificial Intelligence Research, 1:277-308.

Learning the Nonlinear Dynamics of Cyberlearning

Giacomo Patrizi, Claudio Cifarelli, and Laura Di Giacomo

Dipartimento di Statistica, Probabilità e Statistiche Applicate
Università degli Studi "La Sapienza", Rome, Italy.

Summary. The aim of this paper is to present a method to model the dynamics of a distance interactive learning process, so that the navigation through a course is achieved by optimal control techniques and the learner becomes proficient in the material in the least time possible, given his attitude, learning style, basic knowledge and propensities for study. A nonlinear dynamical system representation of the learning process, permits to enact optimal adaptive controls of the process at the general level, by defining a system which is controllable, observable and reachable in a technical sense, determined through a simultaneous estimation and optimization algorithm, which assures a correct representation.

1 Introduction

Learning is a dynamic process dependent on the quality of the teaching , the method and the instruments available. In Ancient Times, teaching was principally argumentative, in the form of a discussion, known as the Socratic method . As writing instruments were costly and not easily accessible, learning relied on memorization and teaching on Oratory . The same method lasted well into the era of printing and cheap paper (for instance, it is not known whether Galileo Galilei used the Socratic style in his major works to make his presentation didactic or to offend Pope Bonifacius VIII [11]). Eventually, the process was integrated with formal lectures and a blackboard. Many innovations have been tried, such as the Harvard case method, overhead projectors, teaching machines and projective teaching methods. In the ensuing discussion, none of the alternatives have been shown to be superior [1], because essentially, the underlying dynamic learning process is unobservable, in a technical sense, (given the history of the process, one can not determine the initial state of the process).

The process of e-Learning, that is interactive distance learning will depend on the quality of the teaching process, the method and the instruments used [21]. As such it defines a dynamic process which is affected by synergy [4] .

G. Patrizi et al.: *Learning the Nonlinear Dynamics of Cyberlearning*, Studies in Computational Intelligence (SCI) **37**, 253–272 (2007)
www.springerlink.com

Evidence suggests that learning is highly nonlinear with substantial lags and the learning dynamics form irreversible processes. Thus nonlinear dynamical modelling seems to be an appropriate representation of e-Learning knowledge acquisition process, since a linear dynamical system would be a too simple process [5].

A number of difficulties preclude identification and control by traditional methods for general nonlinear systems [12], so to avoid biases and suboptimization a simultaneous estimation and optimization method must be applied [9] [10].

The aim of this paper is to formulate a nonlinear dynamic system implementation of distance interactive learning through computer techniques (e-Learning), such that the underlying process is observable, identifiable, controllable and reachable, in a well defined technical sense. Further the system must satisfy the statistical properties for a maximum likelihood estimate. Adaptive optimal control trajectories will be formulated to guide the succession of learning frames to be studied.

The outline of the paper is the following. In the next section, the experimental set up to define an e-Learning optimal control process is described and the various aspects are examined. In the third section, the mathematical dynamic learning system is formalized and relevant mathematical properties indicated. In section four the implementation aspects of two cybercourses are discussed present on the World-Wide Web.

2 Knowledge Acquisition by e-Learning Dynamics

Distance Education is the delivery of education courses from one location to students at other locations [14], while Cyberschools are the institutions that deliver these courses, Cyberlearning is the process by which content is designed, transmitted and acquired.

The effect of Distance Education, depends on the content of what is taught and method by which it is imparted. Students have differing learning styles to acquire Knowledge, which depend on [1]:

- the method of exposition [13] :
  - a formal axiomatic or deductive exposition of the material,
  - an informal presentation, followed by its formalization,
  - an intuitive and illustrative development of the material,
- the structure of the presentation [21]:
  - a linear structure, as in a book,
  - a guided tree structure with a limited capacity of selection,
  - an adaptive feedback mechanism in a tree like decision network.
- the interaction policy envisaged [6]:
  - no interaction allowed,
  - periodical question periods,

– full interaction.

Only with an adaptive feedback structure of the method of exposition can the content be varied and adapted in real time according to the needs of the student, so here, an adaptive dynamic interactive course, based on browser technology will be examined. This type of structure will be termed a cybercourse and the learning process will be indicated as the Cyberlearning process.

The learner can start in his own time at the initial page and navigate through the course on the web, add his own notes to the material presented, do the exercises, carry out the assignments and contact by e-mail other students, the instructor or Intelligent Agents, which are just pieces of software to analyze the output of the student, verify his progress while navigating, etc. Other intelligent Agents will monitor his performance and generally supervise on the work that is being done.

The system will monitor all his actions and advise on the next actions to be taken. The learner is of course absolutely free to chose other actions. The indication of the actions to take are based on the development by the system of an optimal trajectory to the completion of the course, based on the monitored proficiency shown. After any deviation on the part of the learner for whatever reason, the system recalculates the new optimal control path to completion based on the new position reached and capabilities shown by the learner.

A cybercourse will consist of a set of units each composed of a set of frames. Each unit is composed of multiple sets of similar frames to handle different methods of exposition, while the structure of the presentation and the interaction policy adopted will be handled by the underlying nonlinear dynamical process, based on the requirements of the student.

The utilization of a client computer, or a browser, allows the system server to record for each individual connection, the succession of frames traversed, the actions that have been performed on the client system and the length of time involved. Thus, at the server, all actions performed by the student can be monitored [6]. Through the state space formulation of the dynamic representation of the learning process, the input and output sequence will result well defined, so that the optimal control trajectory can be formulated.

3 The Mathematical Algorithm

The representation of phenomena by dynamic systems is more general than their representation by a static system, since the latter will always constitute a special case of the former.

Modelling a phenomenon by a dynamic system means imposing the structure of the phenomenon on the system variables. It also means that the dated variables, which represent the phenomenon, must agree with the dated estimated values of the system and the mathematical properties of the system variables must apply to the variables of the phenomenon. In dynamic modelling, this requirement is not just that the variables in the two systems be defined compatibly (both integer variables etc.), but their compatibility must extend to their dynamic structure . In short, both must posses compatible properties in their controllability, observability and stability: important properties which will be examined. If a stable phenomenon is modelled by an unstable system, the realisation may agree over certain limited intervals, but it is bound to diverge. Properties of the latter cannot be used to represent the former, since they are different.

The functional form of the dynamic system must be identified and the relevant parameters estimated. Since the system is nonlinear the value of the parameters will depend on the point considered. This precludes the application of standard system identification techniques for dynamic systems [22] and requires an algorithm that will solve simultaneously the estimation and the optimal control problem, the former in the space of parameters, the latter in the space of the decision variables.

To this end, the aim of this section is to present such an algorithm. First the dynamic system will be characterized, then it is examined how to ensure that the correct statistical properties of the estimates be obtained by solving an optimization problem which will also determine the optimal control strategy. Finally the solution of the optimization problem is discussed.

3.1 The Dynamical System Formulation

Mathematical System Theory deals essentially with the study of the dynamical relationships of systems under various conditions. A Dynamical System is a precise mathematical object [15]. Not every relationship can be modelled by mathematical system theory, since a representation which is non anticipatory is required [15].

Dynamical Systems have been defined at a high level of generality, to refine concepts and perceive unity in a diversity of applications and by appropriate modelling, whole hierarchies of phenomena can be represented as systems defined at different levels.

Definition 1. *[15]: A Dynamical System is a composite mathematical object defined by the following axioms:*

1. *There is a given time set T, a state set X, a set of input values U, a set of acceptable input functions $\Omega = \omega : \Omega \rightarrow U$, a set of output values Y and a set of output functions $\Gamma = \gamma : \Gamma \rightarrow Y$.*
2. *(Direction of time). T is an ordered subset of the reals.*
3. *The input space Ω satisfies the following conditions.*
   a) (Nontriviality). Ω is nonempty.

b) *(Concatenation of inputs) An input segment* $\omega_{(t_1,t_2]}$, $\omega \in \Omega$ *restricted to* $(t_1, t_2] \cap T$. *If* $\omega, \omega' \in \Omega$ *and* $t_1 < t_2 < t_3$ *there is an* $\omega'' \in \Omega$ *such that* $\omega''_{(t_1,t_2]} = \omega_{(t_1,t_2]}$ *and* $\omega''_{(t_2,t_3]} = \omega'_{(t_2,t_3]}$.

4. *There is a state transition function* $\varphi : T \times T \times X \times \Omega \to X$ *whose value is the state* $x(t) = \varphi(t; \tau, x, \omega) \in X$ *resulting at time* $t \in T$ *from the initial state* $x = x(\tau) \in X$ *at the initial time* $\tau \in T$ *under the action of the input* $\omega \in \Omega$. φ *has the following properties:*
   a) *(Direction of time).* φ *is defined for all* $t \geq \tau$, *but not necessarily for all* $t < \tau$.
   b) *(Consistency).* $\varphi(t; t, x, \omega) = x$ *for all* $t \in T$, *all* $x \in X$ *and all* $\omega \in \Omega$.
   c) *(Composition property). For any* $t_1 < t_2 < t_3$ *there results:*

$$\varphi(t_3; t_1, x, \omega) = \varphi(t_3; t_2, \varphi(t_2; t_1, x, \omega), \omega)$$

   for all $x \in X$ *and all* $\omega \in \Omega$.
   d) *(Causality). If* $\omega, \omega' \in \Omega$ *and* $\omega_{(\tau,t]} = \omega'_{(\tau,t]}$ *then* $\varphi(t; \tau, x, \omega) = \varphi(t; \tau, x, \omega')$.

5. *There is a given readout map* $\eta : T \times X \to Y$ *which defines the output* $y(t) = \eta(t, x(t))$. *The map* $(\tau, t] \to Y$ *given by* $\sigma \mapsto \eta(\sigma, \varphi(\sigma, \tau, x, \omega))$, $\sigma \in (\tau, t]$, *is an output segment, that is the restriction* $\gamma_{(\tau,t]}$ *of some* $\gamma \in \Gamma$ *to* $(\tau, t]$.

The following mathematical structures in definition 1 will be indicated by:

- the pair $(t, x), t \in T, x \in X \quad \forall t$ is called an event,
- the state transition function $\varphi(x_t, u_t)$ is called a trajectory.

Phenomena can be modelled by a dynamical systems in the input/output sense.

Definition 2. *A Dynamical System in an input/output sense is a composite mathematical object defined as follows:*

1. *There are given sets* T, U, Ω, Y *and* Γ *satisfying all the properties required by definition 1*
2. *There is a set* A *indexing a family of functions*

$$\mathcal{F} = \{f_\alpha : T \times \Omega \to Y, \alpha \in A\}$$

   each member of $\mathcal{F}$ *is written explicitly as* $f_\alpha(t, \omega) = y(t)$ *which is the output resulting at time* t *from the input* ω *under the experiment* α. *Each* f_α *is called an input/output function and has the following properties:*
   a) *(Direction of time). There is a map* $\iota : A \to T$ *such that* $f_\alpha(t, \omega)$ *is defined for all* $t \geq \iota(\alpha)$.
   b) *(Casuality) Let* $\tau, t \in T$ *and* $\tau < t$ *If* $\omega, \omega' \in \Omega$ *and* $\omega_{(\tau,t]} = \omega'_{(\tau,t]}$, *then* $f_\alpha(t, \omega) = f_\alpha(t, \omega')$ *for all* α *such that* $\tau = \iota(\alpha)$.

While the input/output approach may determine a family of functions, the state space approach represents the trajectories in the way indicated, through a unique function, so the latter approach is intuitively more appealing, especially in applications. However, both representations show the relationships of the time series of the single inputs on the state and the outputs. The first representation defines a unique mapping, while the second representation, restricted to a subspace, does not.

The representations are equivalent. It is easy to transform a given system from a state space formulation to a input/output formulation and vice versa [2] [15], so each may be used as convenience suggests.

A sufficiently general representation of a dynamic system may be formulated by applying definition 1, recalling the equivalence of an input-output system and a system in state form:

$$x_{t+1} = \varphi(x_t, u_t) \tag{1}$$
$$y_t = \eta(x_t) \tag{2}$$

where $x_t \in X \subseteq R^r$ may simply be taken as a r-dimensional vector in an Euclidean space X, indicating the state of the system at time t, $u_t \in U \subseteq R^q$ may be taken as a q-dimensional vector in an Euclidean subspace U of control variables and $y_t \in Y \subseteq R^p$ is a p-dimensional vector in an Euclidean space Y of output variables, in line with the definitions 1, 2.

The definition of a dynamical system is based on an intermediary set of states and a transition function or a family of functions. Neither of these constructions are unique, so if it is desired to represent a system by such structures, equivalence of the possible structures must be shown.

Definition 3. *Given two states x_{t_0} and $\hat{x}_{t_0}$ belonging to systems S and $\hat{S}$ which may not be identical, but have a common input space Ω and output space Y, the two states are said to be equivalent if and only if for all input segments $\omega_{[t_0,t)} \in \Omega$ the response segment of S starting in state x_{t_0} is identical with the response segment of $\hat{S}$ starting in state $\hat{x}_{t_0}$; that is*

$$x_{t_0} \cong \hat{x}_{t_0} \Leftrightarrow \eta(t, \varphi(x_{t_0}, \omega_{[t_0,t)})) = \hat{\eta}(t, \hat{\varphi}(\hat{x}_{t_0}, \omega_{[t_0,t)}))$$
$$\forall t \in T, t_0 \leq t, \forall \omega_{[t_0,t)} \in S, \hat{S} \tag{3}$$

Definition 4. *A system is in reduced form if there are no distinct states in its state space which are equivalent to each other.*

Definition 5. *Systems S and $\hat{S}$ are equivalent $S \equiv \hat{S}$ if and only if to every state in the state space of S there corresponds an equivalent state in the state space of $\hat{S}$ and vice versa.*

A number of important questions must be asked of the system description of the cyberlearning representation:

- can a certain state $x^* \in S$ be reached from the present state, or if the dynamical system attains a given state x_0 at time 0 can it also be made to reach a certain state x^*. Evidently it is required to determine the set of states reachable from a specific state x_t.
- can a dynamical system be driven to a given state by an input u. Thus controllability is concerned with the connectedness properties of the system representation.
- Reachability and controllability lead naturally to the determination of a dynamical system's observability, which provides the conditions to determine the given actual state uniquely.
- The stability of the system is important since it provides conditions on the way the trajectories will evolve, given a perturbation or an admissible control.

These conditions are very important, since they allow trajectories to be defined, the initial point of trajectories to be determined and their stability properties to be derived. Moreover they can be applied at any moment in time to determine if the goals of the cyberlearning system are still attainable.

Definition 6. *Given a state $x^* \in M \subseteq X$, it is reachable from the event (t_0, x_0) at time T if there exists a bounded measurable input $u_t \in \Omega$ such that the trajectory of the system satisfies:*

$$x_{t_0} = x_0 \tag{4}$$

$$x_T = x^* \qquad \forall x_{t_0} \in M, \quad 0 \leq t \leq T \tag{5}$$

The sets of states reachable from x_{t_0} is denoted by:

$$\Re(x_{t_0}) = \bigcup_{0 \leq T \leq \infty} \{x_T | x_T \quad reachable \quad at \quad time \quad T\} \tag{6}$$

the system is reachable at x_{t_0} if $\Re(x_{t_0}) = M$ and reachable if $\Re(x_{t_0}) = M \quad \forall x \in M$.

Definition 7. *A system is locally reachable at x_{t_0} if for every neighbourhood $N(x_{t_0}, h)$ of x_{t_0}, $\Re(x_{t_0}) \cap N_{x_0}$ is also a neighbourhood of x_{t_0} with the trajectory from the event (t_0, x_{t_0}) to $\Re(x_{t_0}) \cap N_{x_0}$ lying entirely within N_{x_0}. The system is locally reachable if it is locally reachable for each $x \in M$.*

These definitions lead to an important property for many systems, namely that reachability may not be symmetric, that is: if x_T is reachable from x_{t_0} the converse may not hold. Thus a weaker notion of reachability is opportune.

Definition 8. *Two states x^* and $\hat{x}$ are weakly reachable from each other if and only if there exist states $x^0, x^1, ..., x^k \in M$ such that $x^0 = x^*, x^k = \hat{x}$ and either x^i is reachable from x^{i-1} or x^{i-1} is reachable from x^i $(\forall 1 = 1, 2, ..., k)$. The system is weakly reachable if it is weakly reachable from every $x \in M$.*

Theorem 1. *The following implications apply:*

- *If the system is locally reachable then it is reachable,*
- *if the system is reachable then it is weakly reachable,*

Proof: Immediate from the definitions.

Definition 9. *State x_{t_0} of a system is controllable if and only if there exists a $u \in \Omega$ such that:*

$$\varphi(t; t_0, x_{t_0}, u) = \emptyset \tag{7}$$

The system is said to be controllable if and only if every state of the system is controllable.

Theorem 2. *A system which is controllable and in which every state is reachable from the zero state ($\emptyset$) is strongly connected*

Proof: Follows from definition 9 and 6, see [15].

Definition 10. *Simple and Multiple experiments:*

- *A simple experiment is an input/output pair $(u_{[t_0,t)}, y_{[t_0,t)})$ that is, given the system in an unknown state an input $u_{[t_0,t)}$ is applied over the interval of time (t, t_0) and the output $y_{[t_0,t)}$ is observed.*
- *A multiple experiment of size N consists of N input/output pairs $(u^i_{[t_0,t)}, y^i_{[t_0,t)}$ $i = 1, 2, ..., N$ where on applying on the i-th realization of the N systems the input $(u^i_{[t_0,t)})$ the i-th output $y^i_{[t_0,t)}$ is observed.*

Definition 11. *A system is simply (multiply) observable at state x_{t_0} if and only if a simple experiment (a multiple experiment) permits the determination of that state uniquely.*

Definition 12. *Equivalence of Systems:*

- *Two systems are simply equivalent if it is impossible to distinguish them by any simple experiment,*
- *Two systems are multiply equivalent if it is impossible to distinguish them by any multiple experiment.*

Theorem 3. *If two systems are simply equivalent and strongly connected, then they are multiply equivalent.*

Theorem 4. *If two systems are multiply equivalent then they are equivalent, (definition 5).*

Definition 13. *A system is initial-state determinable if the initial state x_0 can be determined from an experiment on the system started at x_0.*

Theorem 5. *A system is in reduced form if and only if it is initial-state determinable by an infinite multiple experiment.*

The definitions 10 - 13 and the theorems 3 - 5 formally justify the possibility of defining one or more representations of the dynamical system considered at a chosen level of detail. Notice however the distinction between systems that are simply equivalent and multiply equivalent. This distinction is crucial, if dynamical systems are considered, while with comparative static models, the distinction does not apply and the latter are consequently limited.

It is usual, since stability for nonlinear systems is an equilibrium concept, to examine autonomous systems in continuous time.

Thus consider:

$$\dot{x} = \varphi(x, t) \tag{8}$$

$$x(t_0) = x_0 \tag{9}$$

an autonomous nonlinear system, while x_0 is the initial state of the system.

Definition 14. *The equilibrium point $x = 0$ is called a stable equilibrium point of the system (8) if for all t_0, $\epsilon > 0$, there exists $\delta(t_0, \epsilon)$ such that:*

$$| x_0 |< \delta(t_0, \epsilon) \Rightarrow | x(t) |< \epsilon \quad \forall t \geq t_0 \tag{10}$$

The solution of the dynamic system given in equations (1) - (2) may be determined in a number of different ways, depending on the structure of the functions that are given [15].

3.2 Simultaneous Estimation and Optimization

A given finite dimensional estimation and optimization problem is considered, which is nonlinear and dynamic to determine simultaneously the maximum likelihood parameter estimates and the optimal control trajectory to the dynamic system.

It is important to apply a suitable data driven statistical method to determine the most appropriate statistical form and precise values of the parameters, which should have the following properties [16]:

1. the parameter estimates are unbiased, this means that:
   - as the size of the data set grows larger, the estimated parameters tend to their true values,
2. the parameter estimates are consistent, which require the following conditions to be satisfied:
   - the estimated parameters are asymptotically unbiased,
   - the variance of the parameter estimate must tend to zero as the data set tends to infinity.
3. the parameter estimates are asymptotically efficient,

- the estimated parameters are consistent,
- the estimated parameters have smaller asymptotic variance as compared to any other consistent estimator,

4. the residuals have minimum variance, which will require to ensure that this is so::
   - the variance of the residuals must be minimum,
   - the residuals must be homoscedastic,
   - the residuals must not be serially correlated.
5. the residuals are unbiased (have zero mean),
6. the residuals have a noninformative distribution (usually, the Gaussian distribution). If the distribution of the residuals is informative, the extra information can be used to reduce the variance of the residuals to yield better estimates.

In short, through correct implementation of statistical estimation techniques the estimates are as close as possible to their true values, all the information that is available is applied and the uncertainty surrounding the estimates and the data fit is reduced to the maximum extent possible. Thus the estimates of the parameters, which satisfy all these conditions, are the 'best' possible in a 'technical sense' [16].

By setting up the statistical properties, that a given estimate must fulfil, as constraints to the maximum likelihood problem to be solved, the parameters are defined implicitly by this optimization problem. The latter can be inserted into the optimal control system for the policy determination, so that statistically correct estimates will always result. The solution yielding the best policy can be chosen, where $N+1,...,T$ is the forecast period, by solving the optimization problem given below. By recursing on the specifications, better and better fits can be derived. At each iteration, the best combination of parameterization and policy is obtained.

The unknowns to be determined are the input and output variables considered and the parameters of the functional form specified in the current iteration.

The mathematical program is formulated with respect to the residual variables, but it is immediate that for a given functional form, the unknown parameters will be specified and thus the unknowns of the problem will also be defined and available. Hence the mathematical program is fully specified for each functional form to be considered.

Consider the data set of a phenomenon consisting of measurements (y_i, x_i, u_i) over $(i = 1, 2, .., N)$ periods, where it is assumed, that $y_i \in R^p$ is a p-dimensional vector, while $x_i \in R^r$ is a r-dimensional vector of explanatory or state variables of the dynamic process of dimension. Also, u_i is a q-dimensional vectors of control variables. Let $w_i \in R^r$, $v_i \in R^p$ be stochastic processes also to be determined.

The optimization problem to be solved is the following:

$$Min \quad J = \sum_{i=N+1}^{\mathcal{T}} c(x_i, u_i, y_i) \tag{11}$$

$$\varphi(x_i, u_i, y_i, w_i : \theta_1) = x_{i+1} \tag{12}$$

$$\eta(x_i, u_i, v_i : \theta_2) = y_{i+1} \tag{13}$$

$$\frac{1}{N}\sum_{i=1}^{N} w_i = 0 \tag{14}$$

$$\frac{1}{N}\sum_{i=1}^{N} v_i = 0 \tag{15}$$

$$\frac{1}{N}\sum_{i=1}^{N} w_i^2 \leq k_w \tag{16}$$

$$\frac{1}{N}\sum_{i=1}^{N} v_i^2 \leq k_v \tag{17}$$

$$-\epsilon_0 \leq \frac{1}{N}\sum_{i=1}^{N} v_i w_i \leq \epsilon_0 \tag{18}$$

$$-\epsilon_1 \leq \frac{1}{N}\sum_{i=1}^{N} w_i w_{i-1} \leq \epsilon_1 \tag{19}$$

$$-\epsilon_2 \leq \frac{1}{N}\sum_{i=1}^{N} v_i v_{i-1} \leq \epsilon_2 \tag{20}$$

$$-\epsilon_3 \leq \frac{1}{N}\sum_{i=1}^{N} v_i w_{i-1} \leq \epsilon_3 \tag{21}$$

$$-\epsilon_4 \leq \frac{1}{N}\sum_{i=1}^{N} w_i v_{i-1} \leq \epsilon_4 \tag{22}$$

$$\cdots\cdots\cdots\cdots\cdots\cdots$$

$$-\epsilon_{2s} \leq \frac{1}{N}\sum_{i=1}^{N} v_{i-s} w_{i-s} \leq \epsilon_{2s} \tag{23}$$

$$-\epsilon_{2s+1} \leq \frac{1}{N}\sum_{i=1}^{N} w_i w_{i-s} \leq \epsilon_{2s+1} \tag{24}$$

$$-\epsilon_{2s+2} \leq \frac{1}{N}\sum_{i=1}^{N} v_i v_{i-s} \leq \epsilon_{2s+2} \tag{25}$$

$$-\epsilon_{2s+3} \leq \frac{1}{N}\sum_{i=1}^{N} v_i w_{i-s} \leq \epsilon_{2s+3} \tag{26}$$

$$-\epsilon_{2s+4} \leq \frac{1}{N}\sum_{i=1}^{N} w_i v_{i-s} \leq \epsilon_{2s+4} \qquad (27)$$

$$\frac{1}{2} g_w^T \Psi(\Psi^T\Psi)^{-1}\Psi^T g_w - \frac{N}{2} \leq \chi^2_{1-\alpha:p-1} \qquad (28)$$

$$\frac{1}{2} g_v^T \Psi(\Psi^T\Psi)^{-1}\Psi^T g_v - \frac{N}{2} \leq \chi^2_{1-\alpha:p-1} \qquad (29)$$

$$-\epsilon_{2r+1} \leq \frac{1}{N}\sum_{i=1}^{N} w_i^{2r+1} \leq \epsilon_{2r+1}; r = 3, 4, ... \qquad (30)$$

$$\frac{1}{N}\sum_{i=1}^{N} w_i^{2r} \leq \frac{2r!}{r!2^r}\sigma_w^{2r}; r = 3, 4, ... \qquad (31)$$

$$-\epsilon_{2r+1} \leq \frac{1}{N}\sum_{i=1}^{N} v_i^{2r+1} \leq \epsilon_{2r+1}; r = 3, 4, ... \qquad (32)$$

$$\frac{1}{N}\sum_{i=1}^{N} v_i^{2r} \leq \frac{2r!}{r!2^r}\sigma_v^{2r}; r = 3, 4, ... \qquad (33)$$

$$x_i \in X, y_i \in Y, u_i \in U, w_i \in W, v_i \in V \qquad (34)$$

The conditions indicated above are met for an optimal solution of the program (11)-(34). The formal proof of these properties are presented in [10]. Here we shall show the connection between the constraints and the statistical properties indicated above which must be satisfied.

The abstract model of the dynamical system is to be optimized with regard to a given merit function(11) such that the sum of squares of the residuals to be less than a critical value k_w, k_v which can be decreased by dichotomous search at every iteration, until the problem does not yield a feasible solution.

The least values obtained for these parameters, while retaining a feasible solution to the whole problem, are equivalent to a minimization of the statistical estimation error and the maximum likelihood estimate of the parameters, under appropriate distributional assumptions concerning the residuals.

All the serial correlations between the residual are not significantly different from zero, as enforced by the constraints (18) - (27).

Moreover to ensure that these conditions hold throughout the possible variation of the independent variables, the residuals must be homoscedastic and thus satisfy (28) - (29). The homoscedasticity condition on the residuals is obtained by regressing the original variables of the problem, indicated by the data matrix Ψ, on the normalised square of the residuals, which are indicated respectively by: g_w, g_v. This leads to a set of nonlinear equations in the squared residuals. The χ^2 test is applied at a confidence level of $(1 - \alpha)$ with $m - 1$ degrees of freedom and a significance level of α, [3].

The conditions 4 and 5 hold at the solution of the optimization problem. Conditions 2 and 1 also hold because of the following consideration.

The constraints (14) - (27) as well as (30) - (33) are sample moments, so they will converge in probability to their population values. The ones indicated by (14) - (27) will converge to zero. For the second group, those representing the odd moments of the distribution, indicated by (30) and (32), will converge in probability to their population value of zero, while the even moments will converge in probability to their population values. These constraints enforce the residuals to have a noninformative distribution, here a Gaussian.

Thus the condition 6 is also met. Condition 3, which is also very important will hold in all cases that the constrained minimization problem (11) - (34) has a solution.

Finally it is easy to show that this constrained minimization problem (11) - (34) will dominate the solutions obtainable by the traditional three phase procedure, since whenever the latter has a statistically correct solution, the new procedure will also have such a solution, but not conversely.

3.3 Solving the Optimization Problem

An iterative procedure is here specified to minimize a given function subject to equality and inequality constraints, by solving a linear complementarity problem at each iteration, subject to a suitable trust region defined by a set of inequalities [18]. The detailed procedure and convergence results have been presented in [10] to which the reader is referred.

Consider the following optimization problem:

$$Min \quad Z = f(w) \qquad f : R^n \rightarrow R \tag{35}$$

$$g(w) \geq 0 \qquad g : R^n \rightarrow R^p \tag{36}$$

$$h(w) = 0 \qquad h : R^n \rightarrow R^q \tag{37}$$

The proposed algorithm consists in defining a quadratic approximation to the objective function, a linear approximation to the constraints and determining a critical point of the approximation by solving a linear complementarity problem (LCP), as given in [18].

Expanding the functions in a Taylor series, at the given iteration point w^k, the equality constraints may be eliminated simply by converting them into $p + 1$ inequality constraints.

Unconstrained variables can be transformed into nonnegative variables for the LCP algorithm, by defining a suitable offset.

A set of trust region constraints can be imposed on the problem as a system of linear inequalities centered around the iteration point, to limit the change in the possible solution. Thus an LCP results which, it can be shown, has a solution either on a trust region constraint or inside. If the solution point occurs on a trust region constraint and the solution is feasible while a reduction in the objective function has occurred, the solution point is taken as the new starting point and a new iteration is started. Otherwise, if the new point is infeasible, the trust region is reduced. Finally if there has been an increase in the objective function, the trust region is enlarged and the iteration is repeated, with suitable safeguards to force a reduction in the objective function. If instead the solution point occurs inside the trust region, it can be considered an approximate stationary point. If the objective function is bounded from below, for all values of the variables which satisfy the constraints, a local minimum point will be found. The minimum time free end optimal control problem will therefore be solved, which specifies the optimal learning trajectory. A great advantage of this algorithm is that variables restricted to binary values can be considered and the solutions determined will respect these conditions [7].

4 Implementation of Cybercourses

A practical application of the relevant methodology, succinct enough to fit the limited space available, would constitute a toy application, which would not reveal the possibility and the advantages of using Cyberlearning, as defined here, for actual courses. Thus the aim of this section is to describe briefly two cybercourses realized with the methodology **I**nteractive **D**istance **E**lectronic and **A**daptive **L**earning **S**ystem (**I.D.E.A.L.S.**) described in the previous sections, which the interested reader can find on the web, so that they can be analyzed in detail and examine the scientific methodology which lies behind the logical and numerical constructs, which ensures the correctness.

4.1 Cybercourses

The courses develop interactive adaptive tutorials, based a common set of general structural and organizational principles.

An array containing indications of the frames successively examined, the time spent on studying each, the collateral actions performed: e-mail sent and received, library information systems consulted, and how the tasks assigned were executed is generated frame by frame. Intelligent Agents can be queried and questions addressed to the instructor levied. On the output side, the next period's predicted state vector is determined, which contains the next frame to be studied, and an output array indicating the expected time taken to study the next frame, the tasks which are expected to be tackled and the performance which will result on the that frame. All this information as well as the results are saved in a database for each individual, which will be then used to update the system.

Similar proposals have been indicated [21] [6], but in these papers the next frame is either chosen by the learner or on the basis of a weighting function of the results. There is no adaptive optimal control exercised nor are special techniques used to avoid excessive oscillations between learning styles, (see section 4.2).

An initial estimate to the individual's learning style can be obtained from the characteristics indicated and from his performance over the first few exemplary frames. Here use is made of a pattern recognition technique called **T.R.A.C.E.** (**T**otal **R**ecognition by **A**daptive **C**lassification **E**xperiments) [17], which has given good results. Also initial representative trajectories can also be defined experimentally. As the individual moves through the frames, pertinent information is obtained, which is used to improve the predictive accuracy of the system representation. A cybercourse should be equipped with a cultural dimension to promote conceptual knowledge and purposeful reflection [13]. To this end full interaction between the learner, the instructor and the Cyberlearning process is envisaged.

Thus a number of instruments are incorporated in the frames structure so as to encourage this cultural extension of the material in many directions.

Any part of a frame may be highlighted and one of five action buttons can be clicked, leading to different actions. Clicking on:

- The demonstration button: a new window is presented with a detailed proof of the highlighted material,
- The reference button: a new window appears with a set of references on the matter highlighted. To this end the **S.I.B.I.L.L.A.** (**S**istema **I**nterattivo **BI**b **L**iografico con **L**iste **A**utomatiche), which is a bibliographic referential system [19], may be used.
- The example button: a new window appears and a worked example is shown step by step.
- The apropos button: a new window appears and a set of links appear with indications to more general information.
- The history button: a new window appears and a short history of the development of the concepts is given.

Each window opened also contains the five buttons and a label for that window so that the interaction process can continue. Other command buttons are also provided, such as repeat buttons, path visualization and e-mail connectivity.

The Cyberlearning Course

Cyberlearning (http://banach.sta.uniroma1.it/ideals/cyberlearning.html) is an application of the **I.D.E.A.L.S.** methodology which is explained in the tutorial in the form of an interactive distance learning course. The interactive course provides an adaptive and interactive tutorial on how it works and how to formulate the frames at the various levels of learning style adopted. As such it consists of three units:

1. Introduction and general principles of the **I.D.E.A.L.S.** methodology,
2. The framework for the dynamic adaptive control and the justification of its correctness and adequacy.

3. the rules and the method to construct the frames.

Because of the aim of this course, some additional instruments are provided, which are not usually part of the **I.D.E.A.L.S.** methodology, which may be characterised as follows:

- backward branching: at any frame the user may return to a previous frame and re-execute it in a debugging mode. Thus he may alter in some way his interaction and compare the optimal control that had been formulated before and the new formulation of the optimal control. This will be useful to study the consequences of different decisions on the optimal path formulated by the system,
- The user can force a choice of a frame, so as to inspect the frames that would have been provided with other outcomes. In short he can vary the state of the system and examine the effects of such a modification.
- The recursive estimation of the relationships, which define the nonlinear dynamic system can be checked at each frame, so that the precise mechanism that governs the transitions studied.

Thus the interested reader can study the pedagogical reasons behind this approach, experiment with the various alternatives, test himself for a better understanding and determine the time taken to accomplish a certain task under the different learning styles. The principles governing the nonlinear dynamical system can be studied in the most suitable way, based on his prior knowledge and preferred style and through the exercises he will obtain a deep knowledge of the subject matter. Finally the principles to construct frames are specified, again under different learning styles.

The Mathematical Programming Course

A cybercourse (http://banach.sta.uniroma1.it/proma06/initial.html) is a mathematical programming course with applications in decision making, imparted to senior undergraduates.

Different learning styles characterize the material presented: from an axiomatic approach distinguished by formal analysis and results to an approach built up by examples with development of intuitive explanations to justify rules, principles and algorithms and of course the tasks to be carried out in each frame will be very different.

The course itself is structured in 18 units (chapters) from traditional mathematical programming methods to more advanced methods such as nonlinear complementarity theory and variational inequalities, as well as the techniques to handle dynamic optimization problems and simultaneous estimation and optimization problem as indicated in section 3. Special evidence is given to interior point methods and a particular emphasis is given to modeling principles and the relevant methodology, since this aspect is considered an important element in a course on mathematical programming, while less so in an optimization course where the concern is only with the solution techniques of such problems.

4.2 Correctness and Adequacy

The justification why such a complex construction will work in reality, are of course implicit in the results presented in section 3. If the derivations are correct and if the concepts can be adequately interpreted by observable measures taken from the behaviour exhibited by the learner, given an acceptable indeterminacy level, then the results obtained by the learner will coincide, modulo the acceptable indeterminacy level, with the results indicated by the nonlinear dynamic system.

To see this, consider two nonlinear dynamical systems which interact as the course proceeds:

- The individual's learning procedure can be regarded as a nonlinear dynamic system, which is not observable as indicated in definition 11, since no single or set of experiments can determine uniquely what he has learnt.
- a representation of certain aspects of his behaviour may be realized by a specification of a nonlinear dynamical system which does not consider the unobservable aspects of the individual's learning process, but just the optimal sequence of frames and learning tasks to achieve the specific desired result.

Certainly, if there were a sufficient set of identical students learning the same subject matter, a design of experiments could be set up to choose for this group the optimal sequence of frames and tasks.

This would require many copies of that individual's learning procedure to determine the best set of tasks, since each task administered alters his learning capabilities. This is not possible, since the individual is unique.

Alternatively, a nonlinear dynamical system in the input/output sense, as indicated in definition 2, could be considered and a number of experiments conducted on the individuals' learning capabilities, so as to determine which sets of tasks are optimal to achieve the desired result.

Rather than experimenting on the individual, as in the traditional design of experiments, a representative nonlinear dynamical system can be used, defined just with the required properties to determine the sequence of tasks to administer. Multiple copies of this system can be generated, if simulations are to be performed, or just a single copy is required, if the optimal sequence is to be determined by an optimal adaptive control algorithm. In each period the optimal policy is determined to completion, the selected task is fed to the learner and the results noted, so that the nonlinear dynamical system can be updated if warranted.

If the results of the section 3 are implemented correctly, the simultaneous estimation and optimization algorithm will converge to a dynamical system representation, which cannot be distinguished by simple and multiple experiments, modulo the acceptable level of indeterminacy, thus ensuring that they are equivalent within the level of acceptable indeterminacy, as derived in theorem 3 and 4.

To verify the optimality of the policy enacted it is not possible to perform experiments, again because of the uniqueness of the individual, but optimality must follow through the properties of the results if these has been derived correctly [10].

The adequacy of the representation must also be evaluated. This means that the aspects of the learning behaviour of the individual, which are measured, satisfy a number of properties which ensures that at each period an appropriate frame is indicated and that the trajectory to completion of the course is well defined.

Reachability of the Cybercourse Representation

A set of states, say M, will lead to the final frame indicating satisfactory completion of the course. This set of states must be reachable from intermediate states as otherwise the optimal control problem will not have a feasible solution.

Once the dynamic system has been identified, System techniques can be applied to ensure that the system is reachable with respect $\Re(x_{t_0}) = M \quad \forall x \in M$, see definition 6 [20]. Also by theorem 1, the system will be weakly reachable. Thus the learner will not be left in the lurch, with no indication on how to progress.

Controllability of the Cybercourse Representation

The learner starts from a given frame, so that it must be ensured that in all circumstances a new state is formulated which leads to a new frame. Thus every state should be reachable from the initial state. Obviously, if the trajectory under given circumstances goes into a loop, the system is not strongly connected and it must be altered to render it controllable. Thus for **I.D.E.A.L.S.** representation, definition 9 and theorem 2 must apply and there are Dynamic system techniques to check and ensure that this is so [20].

Observability of the Cybercourse Representation

It must be ensured that the two systems: certain aspects of the learning process of the student and their dynamic system representation are equivalent, so that the optimal policy determined for the latter applies to those aspects of the learner and so ensure an optimal sequence of frames.

For this to hold, the dynamic system and the real system must be simply observable (definition 11) so that the two systems be simply equivalent (see definition 12). If the two systems are controllable and reachable, see above, then the systems will be, by theorem 2, strongly connected. By theorem 3 the two systems will be multiple equivalent and therefore equivalent by theorem 4.

Stability of the Cybercourse Representation

Any path through the cybercourse must be stable, which means that the path must not oscillate or cycle so that the sequence of frames will converge to the final frame of the course. Opportune constraints are added to problem specification to ensure the stability of every optimal trajectory [8].

5 Conclusion

Open loop policies can be specified for every student which desires to enroll in the course. Special recursive estimation and optimization techniques ensure the identification of the nonlinear dynamic system formulation of the required aspects of the learning process and its optimal control to completion. The properties of the results derived ensure the equivalence of the two systems, so that the model can be used to determine the optimal control.

A close scrutiny of the structure of this paper makes it evident that the method of exposition applied in each section reflects differences in learning styles, so as to appeal to a wide audience. Thus from an intuitive description in section 2, a formal and axiomatic exposition is used in section 3 and an informal but structured presentation in section 4 to allow interested readers to understand the material to completion.

The **I.D.E.A.L.S.** methodology improves the given structure by interactively adapting presentations to the desires of the reader and ensuring that the material is presented as efficiently as possible.

References

1. T. Anderson and F. Elloumi (eds.). *Theory and Practice of Online Learning.* Athabasca University, Athabasca, Canada, 2004.
2. M. Aoki. *Optimal Control and System Theory in Dynamic Economic Analysis.* North-Holland, New York, 1976.
3. T. S. Breusch and A. R. Pagan. A simple test for heteroschedasticity and random coefficient variation. *Econometrica*, 47:1287 – 1294, 1979.
4. R. R. Bush and F. Mosteller. *Stochastic Models of Learning.* Wiley, New York, 1955.
5. J. L. Casti. *Dynamical Systems and their Applications.* Academic Press, New York, 1977.
6. C.-M. Chen, H.-M. Lee, and Y.-H. Chen. Personalized e-learning system using item response theory. *Computers and Education*, 44:237–253, 2005.

7. L. Di Giacomo, E. Argento, and G. Patrizi. Linear complementarity methods for the solution of combinatorial problems. *to appear in Journal of Computing, copy at http://banach.sta.uniroma1.it/patrizi/*, 2005.
8. L. Di Giacomo and G. Patrizi. Distributed decision making in dynamic stabilized markets. *Paper presented at European Working Group on Distributed Decision Making, Louvain, April 16 -17, 2004, copy at http:/banach.sta.uniroma1.it/patrizi/*, pages 1 – 17, 2004.
9. L. Di Giacomo and G. Patrizi. Dynamic nonlinear modelization of operational supply chain systems. *to appear in the Journal of Global Optimization, copy at http://banach.sta.uniroma1.it/patrizi/*, 2005.
10. L. Di Giacomo and G. Patrizi. A general algorithm for simultaneous nonlinear estimation and optimization of constrained problems. *submited for publication, copy at: http://banach.sta.uniroma1.it/patrizi/*, pages 1–37, 2005.
11. A. Favaro. *Le Opere di Galileo Galilei.* Barbera, Padova, 1907.
12. A. R. Gallant and H. White. *A Unified Theory of Estimation and Inference for Nonlinear Statistical Models.* Basil Blackwell, Oxford, 1988.
13. H. Gardner. *Frames of Mind: The Theory of Multiple Intelligences.* Basic Books, New York, 1993.
14. G. R. Jones. *Cyberschools and Education Renaissance.* Jones Digital Century Inc., Englewood Co., 1997.
15. R. E. Kalman, P. L. Falb, and M. A. Arbib. *Topics in Mathematical System Theory.* McGraw-Hill, New York, 1969.
16. E. Mailnvaud. *Méthodes Statistiques de l' économétrie.* Dunod, Paris, 3eme ed., 1978.
17. L. Nieddu and G. Patrizi. Formal properties of pattern recognition algorithms: A review. *European Journal of Operational Research*, 120:459–495, 2000.
18. G. Patrizi. The equivalence of an lcp to a parametric linear program with a scalar paramter. *European Journal of Operational Research*, 51:367 – 386, 1991.
19. G. Patrizi. Sibilla: An implementation of an intelligent library system. *European Journal of Operational Research*, 64:12–37, 1993.
20. S. Sastry. *Nonlinear Systems: analysis, stability and Control.* Springer, Berlin, 1999.
21. K.-D. Schewe, B. Thalheim, A. Binemann-Zdanovicz, R. Kascheck, T. Kuss, and B. Tschiedel. A conceptual view of web-based e-learning systems. *Education and Information Technologies*, 10:p. 81–108, 2005.
22. T. Söderström and P. Stoica. *System Identification.* Prentice-Hall, Englewood Cliffs, N.J., 1989.

Personalizing E-Commerce with Data Mining

Matthew Smith[1], Brent Wenerstrom[1], Christophe Giraud-Carrier[1], Steve Lawyer[2], and Wendy Liu[3]

[1] Brigham Young University, Provo, UT 84602 (dml@byu.edu)
[2] BYU Bookstore, Provo, UT 84602
[3] Overstock.com, Salt Lake City, UT 84121

Summary. E-commerce greatly facilitates and enhances the level of interaction between a retailer and its customers, thus offering the potential for smarter marketing through thoughtful site design and analytics. This chapter presents and illustrates the three stages of knowledge discovery that move e-retailers from simple on-line catalog providers to finely-tuned, customer-centric service providers.

1 Introduction

As the Internet continues its expansion, most retailers have successfully added the Web to their other, more traditional distribution channels (e.g., stores, mailings). Unfortunately, for too many companies, the story ends there. That is, the Web channel is just that, another distribution channel, often consisting of little more than an on-line catalog tied to a secure electronic point of sale. Although valuable in its own right, such use of the Web falls far short of some of the unique possibilities it offers for intelligent marketing. Consider the following intrinsic differences between physical (i.e., brick-and-mortar) stores and on-line (i.e., Web-based) stores.

Physical stores are rather static and mostly customer-blind. In particular,

- The store's layout and content are the same for all customers.
- Changes to layout and/or content are generally costly.
- Visits are not traceable except in the case of sale's data, generally limited to what was bought, when it was bought and by what method of payment.

On-line stores or commercial Web sites, on the other hand, are naturally dynamic and customer-aware. Indeed,

- Layout and content can be modified easily and cheaply.
- Layout and content can be tailored to individual visitors.
- Every visit automatically generates a rich trail of information on the customer's experience (e.g., visit duration, pages viewed, items bought if any,

M. Smith et al.: *Personalizing E-Commerce with Data Mining*, Studies in Computational Intelligence (SCI) **37**, 273–286 (2007)
www.springerlink.com

etc.), and possibly on the customer's persona (e.g., demographics gathered through an online questionnaire at registration time).

With such flexibility and nearly everything traceable and measurable, the Web is a marketer's dream come true. Unfortunately, due to ignorance, Web capabilities and data are too often under-exploited in e-commerce, leading to sub-optimal performance both for the business and its customers.

The full benefit of the emerging and growing Web channel belongs to those who both gather and adequately leverage the rich information it provides [10, 14, 31]. In this chapter, we present an intuitive, staged approach, that illustrates what may be accomplished in e-commerce using data mining. Moving through the three proposed stages requires increasing sophistication, but also produces increasing return-on-investment. Our focus here is mainly on showing e-retailers (i.e., the knowledge consumers) what can be done with e-commerce data. An excellent companion paper, addressed more specifically to analysts and researchers (i.e., the knowledge producers) is found in [18], where lessons learned and typical challenges faced are thoroughly discussed, based on a number of implementations for a variety of customers.

2 First Stage: Clickstream Analysis

Web servers are typically configured so as to store Web usage data, also known as clickstream data, automatically in Web server log files. If not, they can be set up to do so easily and quickly. Web server logs grow as visitors interact with the Web site creating a clickstream. For each visitor, the log contains basic identifying information (e.g., originating IP address) and time-stamped entries for all pages visited, from visitor entry to exit. Figure 1 shows a small excerpt from a typical Web server log.

The amount of data found in Web server logs is enormous and clearly evades direct human interpretation. Yet, it is rich in potential, making Web server logs the readiest source of data to analyze on a Web site (e.g., see [24, 32, 12, 23]). Web log analysis tools, such as AWStats [2], The Webalizer [3], SiteCatalyst [26], ClickTracks [6], and NetTracker [30] have been designed specifically to summarize and interpret Web server log data, allowing marketers to gain basic knowledge about e-commerce customers' activities. Figure 2 shows a screen shot from the reporting tool used by one of our sponsors.

Log analysis tools generally report unique number of visitors and hits, visit duration, visitors' paths through the site (i.e., clickstream), visitors' host and domain, search engine referrals, robot or crawler visits, visitors' browser and operating system, search keywords used, and more. Clickstream analysis reports may provide insight into business questions such as:

- Where do most visitors come from?
- What proportion of visitors come from a direct link or bookmark, a search engine, or a partner Web site (if any)?

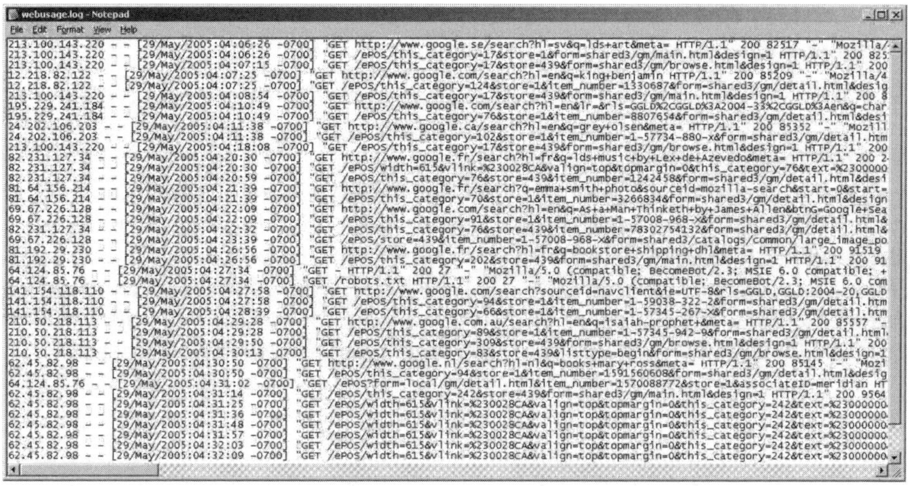

```
213.100.143.220 - - [29/May/2005:04:06:26 -0700] "GET http://www.google.se/search?hl=sv&q=lds+art&meta= HTTP/1.1" 200 82517 "-" "Mozilla/
213.100.143.220 - - [29/May/2005:04:06:26 -0700] "GET /ePOS/this_category=17&store=1&form=shared3/gm/main.html&design=1 HTTP/1.1" 200 825
213.100.143.220 - - [29/May/2005:04:07:15 -0700] "GET /ePOS/this_category=17&store=439&form=shared3/gm/browse.html&design=1 HTTP/1.1" 200
12.218.82.122 - - [29/May/2005:04:07:25 -0700] "GET http://www.google.com/search?hl=en&q=king+benjamin HTTP/1.1" 200 85209 "-" "Mozilla/4
12.218.82.122 - - [29/May/2005:04:07:25 -0700] "GET /ePOS/this_category=124&store=1&item_number=1330687&form=shared3/gm/detail.html&desig
213.100.143.220 - - [29/May/2005:04:08:54 -0700] "GET /ePOS/this_category=17&store=439&form=shared3/gm/main.html&design=1 HTTP/1.1" 200 8
195.229.241.184 - - [29/May/2005:04:10:49 -0700] "GET http://www.google.com/search?hl=en&lr=&rls=GGLD%2CGGLD%3A2004-33%2CGGLD%3Aen&q=char
195.229.241.184 - - [29/May/2005:04:10:49 -0700] "GET /ePOS/this_category=76&store=1&item_number=8807654&form=shared3/gm/detail.html&desi
24.202.106.203 - - [29/May/2005:04:11:38 -0700] "GET http://www.google.ca/search?hl=en&q=grey+olsen&meta= HTTP/1.1" 200 85352 "-" "Mozill
24.202.106.203 - - [29/May/2005:04:11:38 -0700] "GET /ePOS/this_category=102&store=1&item_number=1-57734-880-x&form=shared3/gm/detail.htm
213.100.143.220 - - [29/May/2005:04:18:08 -0700] "GET /ePOS/this_category=17&store=439&form=shared3/gm/browse.html&design=1 HTTP/1.1" 200
82.231.127.34 - - [29/May/2005:04:20:30 -0700] "GET http://www.google.fr/search?hl=fr&q=lds+music+by+Lex+de+Azevedo&meta= HTTP/1.1" 200 2
82.231.127.34 - - [29/May/2005:04:20:30 -0700] "GET /ePOS/width=615&vlink=%230028CA&valign=top&topmargin=0&this_category=76&text=%2300000
82.231.127.34 - - [29/May/2005:04:20:59 -0700] "GET /ePOS/this_category=76&store=439&item_number=1242458&form=shared3/gm/detail.html&desi
81.64.156.214 - - [29/May/2005:04:21:39 -0700] "GET http://www.google.fr/search?q=emma+smith+photo&sourceid=mozilla-search&start=0&start=
81.64.156.214 - - [29/May/2005:04:21:39 -0700] "GET /ePOS/this_category=70&store=1&item_number=3266834&form=shared3/gm/detail.html&design
69.67.226.128 - - [29/May/2005:04:22:09 -0700] "GET http://www.google.com/search?hl=en&q=As+a+Man+Thinketh+by+James+Allen&btnG=Google+Sea
69.67.226.128 - - [29/May/2005:04:22:09 -0700] "GET /ePOS/this_category=91&store=1&item_number=1-57008-968-X&form=shared3/gm/detail.html&
82.231.127.34 - - [29/May/2005:04:22:32 -0700] "GET /ePOS/this_category=76&store=439&item_number=78302754132&form=shared3/gm/detail.html&
69.67.226.128 - - [29/May/2005:04:23:39 -0700] "GET /ePOS/store=439&item_number=1-57008-968-X&form=shared3/catalogs/common/large_image_po
81.192.29.230 - - [29/May/2005:04:26:56 -0700] "GET http://www.google.fr/search?hl=fr&q=bookstore+shipping+dhl&meta= HTTP/1.1" 200 91519
81.192.29.230 - - [29/May/2005:04:26:56 -0700] "GET /ePOS/this_category=202&store=439&form=shared3/gm/main.html&design=1 HTTP/1.1" 200 91
64.124.85.76 - - [29/May/2005:04:27:34 -0700] "GET - HTTP/1.1" 200 27 "-" "Mozilla/5.0 (compatible; BecomeBot/2.3; MSIE 6.0 compatible; +
64.124.85.76 - - [29/May/2005:04:27:34 -0700] "GET /robots.txt HTTP/1.1" 200 27 "-" "Mozilla/5.0 (compatible; BecomeBot/2.3; MSIE 6.0 com
141.154.118.110 - - [29/May/2005:04:27:58 -0700] "GET http://www.google.com/search?sourceid=navclient&ie=UTF-8&rls=GGLD,GGLD:2004-20,GGLD
141.154.118.110 - - [29/May/2005:04:27:58 -0700] "GET /ePOS/this_category=94&store=1&item_number=1-59038-322-2&form=shared3/gm/detail.htm
141.154.118.110 - - [29/May/2005:04:28:39 -0700] "GET /ePOS/this_category=66&store=1&item_number=1-57345-267-X&form=shared3/gm/detail.htm
210.50.218.113 - - [29/May/2005:04:29:28 -0700] "GET http://www.google.com.au/search?hl=en&q=isaiah-prophet+&meta= HTTP/1.1" 200 85557 "-
210.50.218.113 - - [29/May/2005:04:29:28 -0700] "GET /ePOS/this_category=89&store=1&item_number=1-57345-942-9&form=shared3/gm/detail.html
210.50.218.113 - - [29/May/2005:04:29:50 -0700] "GET /ePOS/this_category=309&store=439&form=shared3/gm/browse.html&design=1 HTTP/1.1" 200
210.50.218.113 - - [29/May/2005:04:30:13 -0700] "GET /ePOS/this_category=83&store=439&listtype=begin&form=shared3/gm/browse.html&design=1
62.45.82.98 - - [29/May/2005:04:30:50 -0700] "GET http://www.google.nl/search?hl=nl&q=books+mary+ross&meta= HTTP/1.1" 200 85145 "-" "Mozi
62.45.82.98 - - [29/May/2005:04:30:50 -0700] "GET /ePOS/this_category=94&store=1&item_number=1591560608&form=shared3/gm/detail.html&desig
64.124.85.76 - - [29/May/2005:04:31:02 -0700] "GET /ePOS?form=local/gm/detail.html&item_number=1570088772&store=1&associateID=meridian HT
62.45.82.98 - - [29/May/2005:04:31:14 -0700] "GET /ePOS/this_category=242&store=439&form=shared3/gm/main.html&design=1 HTTP/1.1" 200 9564
62.45.82.98 - - [29/May/2005:04:31:25 -0700] "GET /ePOS/width=615&vlink=%230028CA&valign=top&topmargin=0&this_category=242&text=%23000000
62.45.82.98 - - [29/May/2005:04:31:36 -0700] "GET /ePOS/width=615&vlink=%230028CA&valign=top&topmargin=0&this_category=242&text=%23000000
62.45.82.98 - - [29/May/2005:04:31:48 -0700] "GET /ePOS/width=615&vlink=%230028CA&valign=top&topmargin=0&this_category=242&text=%23000000
62.45.82.98 - - [29/May/2005:04:31:57 -0700] "GET /ePOS/width=615&vlink=%230028CA&valign=top&topmargin=0&this_category=242&text=%23000000
62.45.82.98 - - [29/May/2005:04:32:03 -0700] "GET /ePOS/width=615&vlink=%230028CA&valign=top&topmargin=0&this_category=242&text=%23000000
62.45.82.98 - - [29/May/2005:04:32:09 -0700] "GET /ePOS/width=615&vlink=%230028CA&valign=top&topmargin=0&this_category=242&text=%23000000
```

Fig. 1. Excerpt from a Typical Web Log File

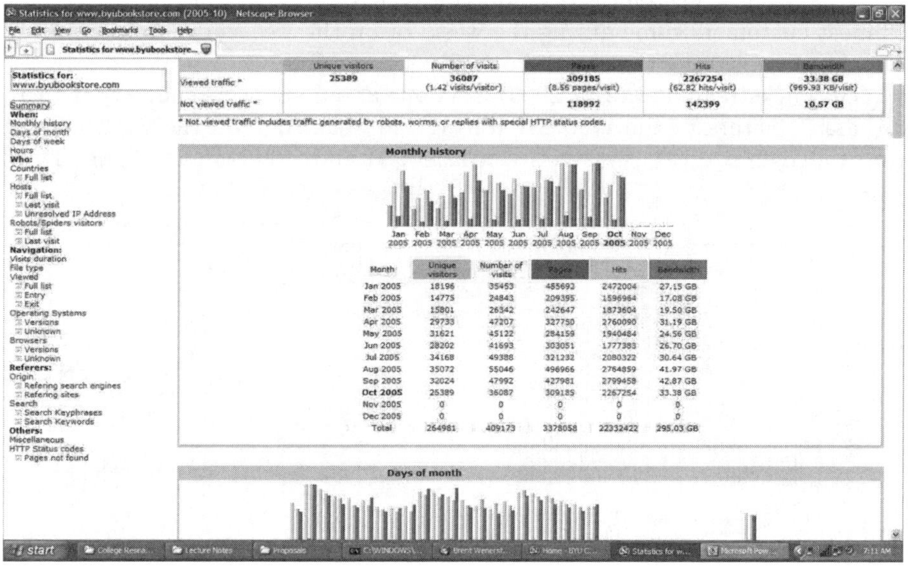

Fig. 2. Log Analysis Report Sample

- Which search engines (e.g., Google, Yahoo, MSN, etc.) and search terms are most frequently used?
- How long do visitors stay on the Web site?
- How many pages do visitors see on average?
- Which pages are most popular?

- How does Web site activity evolve throughout the day, week or month?
- From which pages do visitors commonly exit the Web site?

This information in turn helps e-retailers better understand the dynamics of customers' interactions with on-line offerings, and aids in such decisions as: which referrer to invest in, which pages to remove or replace, which pages to improve, etc. For instance, if clickstream analysis shows that a substantial number of visitors are accessing content several clicks deep into the Web site, then it might be valuable to make that content more accessible to visitors (e.g., maintaining and linking to "Top Sellers," "Most Wished for Items," or "Most Popular Items" pages on the home page). We illustrate the value of clickstream analysis further with two simple case studies from our sponsors.

2.1 Site Activity

Regularly analyzing the clickstream is useful for understanding some aspects of customer behavior. For example, as mentioned above, web usage data reveals when most customers come to the Web site. Interesting patterns, such as most customers come during the week, or on the weekends, or only during certain seasons, may thus be discovered.

The clickstream data (for 2005) from one of our sponsors shows that there are more visitors to the Web site during the week than on the weekend and that the Web site is busiest between 12:00PM and 2:00PM (see Figure 3).

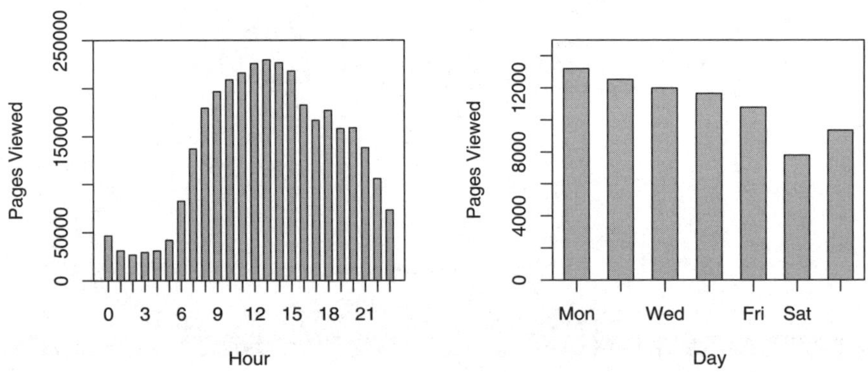

Fig. 3. Evolution of the Number of Pages Viewed

Such knowledge about customer behavior can be utilized to schedule special offers and promotions when the number of viewing visitors is maximized, as well as to ensure adequate back-up to the on-line experience (e.g., staffing customer call centers and server administration personnel). As customer behavior often fluctuates over time, it is important to continually analyze this type of information.

2.2 Search Engine Optimization

To increase Web site traffic, it is common to design/modify Web pages so that they rank higher on search engine results pages. These design activities, known as search engine optimization (SEO), include such things as ensuring Web pages have descriptive titles, unique textual content, and useful meta tags [17]. The impact of SEO changes (or the diagnostic of their absence) can be measured through clickstream analysis.

As we monitored one of our sponsors' Web sites, we noticed a significant surge in site activity in April 2005, which has continued since. An inspection of the clickstream data revealed that the search terms used to access the Web site prior to April contained very few unique keywords. Additionally, the most common keywords used were words in the actual domain name of the site, a common problem with unoptimized sites. Indeed, the former observation suggests that pages within the site were not found by search engines, while the latter observation suggests that visits were mostly restricted to those individuals already aware of the existence of the Web site. In April, the number of search keywords leading to the site soared (see Figure 4), accompanied by a diversification of the search terms, often corresponding to new products/offers.

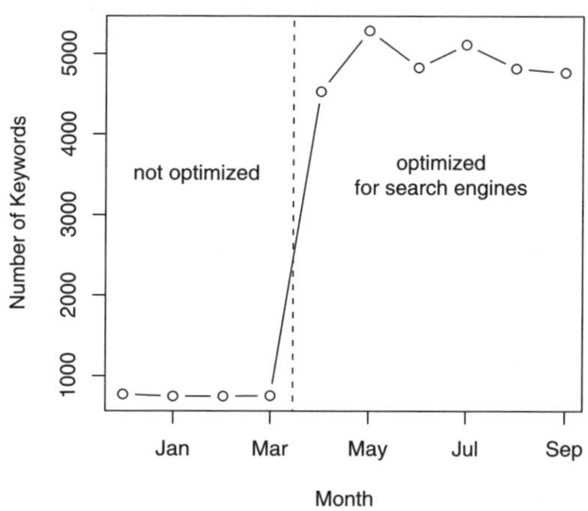

Fig. 4. Unique Search Keywords by Month

Upon consultation with our sponsor, we discovered that they had launched a SEO initiative at the beginning of April. Significantly for them, not only did the number of search terms increase, but the number of unique visitors doubled (and the trend seems to hold), confirming that many who missed the Web site before were now led to it by search engines. Although in this instance, clickstream was used a posteriori to evaluate the impact of the change, it could — and should — have been used a priori to diagnose and remedy the problem.

Although clearly useful, there is a limit to the knowledge that may be extracted from Web server logs. Much can be gleaned about people's attitudes, but nothing about their actual buying behavior, which is clearly the more relevant aspect for the e-retailer. Indeed, retailers generally wish to 1) understand attitudes to influence behavior, and 2) understand behavior to maximize both their customers' experience and their own revenue.

3 Second Stage: Behavior Analysis

In general, transactional data and order information are not stored directly in standard Web server logs. Yet, both are essential in discovering patterns of buying and non-buying customers. The next stage of knowledge discovery requires linking clickstream data to order information. Together with thoughtful design, this allows marketers to take further control of their e-commerce activity through more advanced data analysis (e.g., see [22, 9, 29]).

With adequate design, a host of new business-relevant questions may be answered when the front-end clickstream data is linked to the back-end transactional data. The following are a few classical examples.

- What is the conversion rate (i.e., how many Web site visitors become buying customers)?
- How many would-be customers begin shopping (i.e., partially filling up their shopping cart) but drop out before proceeding to or completing checkout?
- How well did special offer X do (i.e., how much revenue vs. interest did it generate)?
- Who buys product X?
- Who are the most profitable customers?
- What is being bought by whom?

Answers to these questions help e-retailers better understand their customers' buying behavior as well as the value of their offerings. This information, in turn, leads to considerations such as what products to focus on, how to turn browsers into customers, how to recognize customer types so as to respond to their needs/interests on-the-fly, etc.

It is important to recognize that behavior analysis cannot be an afterthought. Indeed, it must be carefully planned and reflected in the e-commerce site's design. For example, assume an e-retailer wishes to run a special offer and measure its impact. Then, a mechanism capable of capturing the supporting data must be set in place. In the on-line world, this may be accomplished simply and cost-effectively by adding a special ad somewhere on the Web site and tracking the visitors who click on it. Tracking is easily done using a distinguisher in the URL, yet this must be reflected in the implementation prior to launching the special offer. Once the mechanism is in place, one can easily measure the offer's success rate by computing the ratio of those who click and convert (i.e., buy the special offer) to those who only click.

Extending to multiple offers, this type of functionality provides a method to determine which offers are most successful, thus informing marketing decisions. Interestingly, referral programs can be implemented and evaluated using a similar approach. We again illustrate the value of behavior analysis with simple case studies from our sponsors.

3.1 Measuring the Impact of Shipping Policy Change

At the beginning of August 2005, one of our sponsors, fearing that their existing shipping policy might deter some people from completing their purchases, decided to revise the policy. In particular, they significantly lowered the shipping cost associated with small orders and implemented a free shipping policy when the total amount purchased exceeds a certain threshold. The announcement of the new policy was clearly displayed on the Web site's home page and accessible via a single click.[4]

We set out to examine the impact of the new policy on customers' behavior and on revenue. As a first step, we considered whether the new policy caused more customers to proceed to check-out and complete their purchase. To do so, we measured the percentage of dropped carts, i.e., carts that are created and (at least) partially filled but that do not make it to check-out. We assumed that prior to the policy change some people may have created a cart only to find out about shipping costs or may have been displeased with the late discovery of high shipping costs, thus potentially leading to high abandonment rate. The evolution of the rate of dropped carts from March 2005 to September 2005 is shown in Figure 5, where the vertical bar marks the introduction of the new shipping policy.

Surprisingly perhaps, there is no significant decrease in the percentage of dropped carts after August. Hence, we refined our analysis by looking at the number of purchases in each shipping category. Figure 6 shows the difference in average number of purchases between the two months prior to the policy change and the two months following.

There is a marked increase in the number of purchases in all categories, and most remarkably, there is an increase of 100% for the high-end category

[4] Information about shipping prior to this change was available to the customers in a remote part of the site, but more easily found *after* they had created a cart and begun the check-out process.

Fig. 5. Dropped Carts Rate Over Time

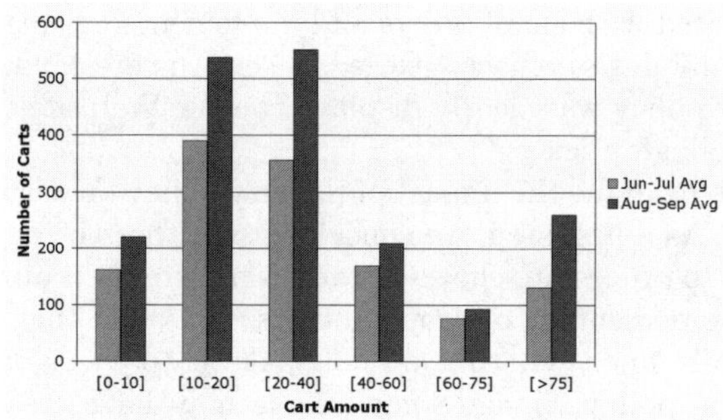

Fig. 6. Average Number of Purchases Pre and Post Policy Change

where shipping is now free. Although there is insufficient evidence to assign the observed increase in sales solely to the change in shipping policy, the graph does give some indication that the change in policy had an effect on customer purchasing behavior. Part of the difficulty is that the sponsor chose to make several changes at once, namely: 1) lower the cost of shipping for small orders, 2) provide free shipping for larger orders, and 3) make the shipping policy visible from the home page. It is clear that focusing on a single change at a time would allow better interpretation of the results.

3.2 Segmenting Customers

For another of our sponsors, we looked at clustering the customer base on the basis of customers' buying behavior. It is generally well accepted that Recency (i.e., the number of days since last purchase), Frequency (i.e., the total number of purchases) and Monetary value (i.e., the total amount spent) are relevant predictors of customer behavior. Although the analysis is typically performed statically (using quintiles), we chose to characterize each customer along these three dimensions and then run a clustering algorithm (Kohonen SOM from SPSS Clementine) to see what patterns might emerge. The results are summarized in Table 1.

Table 1. RFM Clustering

	R	F	M	% base
1	78	3.0	306	27
2	197	2.2	209	7
3	242	2.0	197	6
4	280	2.0	204	4
5	313	1.8	167	10
6	401	1.8	161	8
7	536	1.7	163	8
8	657	1.5	129	7
9	730	1.5	124	4
10	824	1.5	121	3
11	943	1.5	133	4
12	1440	1.4	146	13

These results highlight a large cluster (27% of the total number of customers) of very recent customers whose frequency and monetary value are well above average. These are clearly the customers most likely to generate revenue if they can be identified and catered for on the Web site. We also note that there are a few sizable clusters of relatively high monetary value whose customers have not interacted with the e-retailer for some time. It may be worthwhile to design ways to re-engage the corresponding customers. Finally, customers in the last clusters may not even be considered customers anymore.

Behavior analysis requires at least correlating both visitors' clickstreams and transaction information. However, in order to answer the more customer-related questions, one must also be able to associate specific behavior (i.e., Web usage and buying patterns) to specific customers. Again, this impacts site design.

The only way to accurately identify visitors is by having them sign in (e.g., log on to the Web site), thus creating a session. This is particularly important as visitors may be accessing the Web site from multiple computers in varied locations such as home, work, or grandma's house. Log-only information would in this case give the impression of three distinct visitors. When visitors sign in, however, they can be uniquely identified and associated with their information, such as interests, preferences, purchasing habits, clickstream history, and profile (e.g., gender, address, age, etc.). This type of session tracking is commonly implemented with "cookies" (a mechanism for storing data in the remote visitor's browser) or by propagating a unique session id in the URL. With this additional data, questions of segmentation and profitability can readily be addressed.

Note, however, that in many instances signing in to a Web site may be a deterrent, mostly due to concerns with privacy and spamming (e.g., see [5, 35]). A clear privacy statement and special privileges for signed-in users are often sufficient incentive. Experience suggests that both customers and e-retailers are learning that the benefit of collecting more information about visitors exceeds the cost.

With advanced behavior analysis, e-retailers take control of their marketing activities and learn what works well, which customers deserve extra attention, and what should be offered to whom. Although much of this additional knowledge may be leveraged within the context of a static Web site, its real benefit comes as e-retailers recognize that they can engage more fully with their customers by offering the right thing to the right customer at the right time.

4 Third Stage: Personalization

Every marketer's dream is to know enough about customers so as to tailor offers to each individually, in terms of both products and prices. Even when the needed knowledge is available, this is nearly impossible in a traditional store. Internet technology, on the other hand, makes it possible to adapt layout, contents and services offered "to the needs of a particular user or a set of users, taking advantage of the knowledge gained from the users' navigational behavior and individual interests, in combination with the content and the structure of the Web site" [11].

Furthermore, the Web channel taps into an unusually diverse and large customer pool. Web customers are not restricted by physical geography; they can come from all over the world and exhibit widely different demographic and socio-economic characteristics. This abundance of data offers a unique opportunity for personalization (e.g., see [27, 21, 1, 11]).

Unlike others, who distinguish several forms of personalization, such as content-based filtering or collaborative filtering, we explicitly view personalization as the final stage of knowledge discovery in e-commerce, where data, made available through careful site design, is analyzed and translated into finely-honed marketing actions, such as:

- Show product P to customer C.

- Offer a discounted price on bundle B to customer C.
- Suggest services and products that customer C is likely to be interested in.
- Provide timely chat or co-browsing to most valuable customers [15].

As a simple example, consider the prevalent notion of "gold" or high-value customers, consisting of that (generally small) group of customers who generate most of the revenue and profit [16]. Here, personalization makes it possible for both high-value and low-value customers to be precisely identified and treated accordingly during each visit. Customer appreciation programs and benefits can be used to retain high-value customers, while low-value customers can be enticed to become high-value customers through special offers.

Interestingly, with dynamic web design, one may use both information previously obtained and data provided in real-time (i.e., as the visitor interacts with the site) to tailor the exchange between the parties. Some of the largest e-businesses have had enormous success personalizing Web content. We mention a few here to illustrate.

Google presents relevant advertisements based on keywords in which a visitor is interested [13]. Overture, now Search Marketing [38], a Yahoo! company, provides a similar service for sites including AltaVista [33], MSN [20], and Yahoo! [36]. Amazon.com uses collaborative filtering to recommend products to users on-the-fly. In this case, order information is leveraged to identify clusters of users with similar purchasing habits, the underlying assumption being that people with similar buying behavior are very likely to have similar interests [11, 19]. Yahoo!'s LAUNCHcast also uses collaborative filtering to recommend music to listeners. In this scenario, however, music ratings are elicited from visitors and used to identify clusters of users with similar musical tastes [37].

Although quite useful in their own right, these implementations only scratch the surface of what is achievable with personalization in e-commerce. One can indeed gather and use richer sets of data to build more versatile models for personalization, to support activity such as cross-selling and up-selling. In particular, we have worked on uncovering the social networks [34] that often remain hidden within the mounds of customer usage data. As community members interact with one another through one or more Web sites, a social network emerges (see Figure 7 for an example).

Leveraging the social network of the customers presents additional opportunities for strategic personalized marketing. Identifying the group(s) to which a customer belongs is an effective method for personalizing content. Furthermore, since some customers have a disproportionate ability to shape public opinion [8], the social network can be utilized to identify such individuals as well as the customers they influence. After identifying these groups and the influential people within them, ads and products can be strategically targeted to effectively create "buzz" [28].

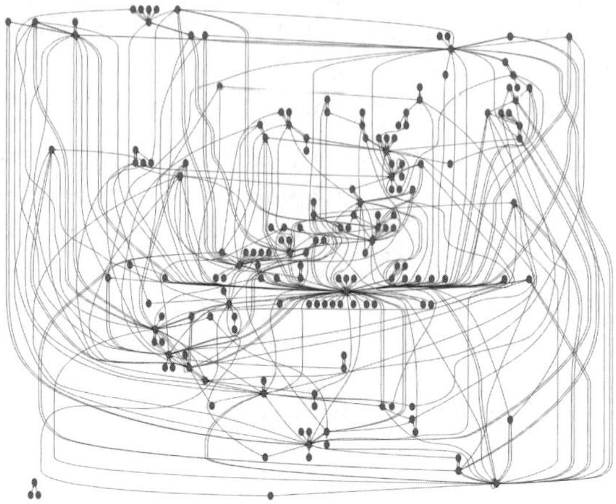

Fig. 7. A Sample Social Network

Ideally, personalization benefits both the customer and the e-retailer. Successful recommendations benefit customers by readily providing them with items most likely to be of interest, sometimes even introducing items that they were previously unaware of. In fact, not only are the most relevant products delivered but the transaction itself requires less time [7], thus improving overall customer satisfaction. In turn, more visitors will convert (i.e., buy what is suggested to them) and existing customers will return (i.e., build loyalty), so that the e-retailer sees increased revenue and profit at a relatively small cost. Thanks to the Internet's flexibility, e-commerce is probably the closest one can hope for to get back to Mr. Miller's successful one-to-one marketing [4].

5 Conclusion

E-commerce activities have richer data footprints than their physical counterparts, thus providing much greater opportunities for intelligent e-business. In this chapter, we have briefly reviewed the value of planning for knowledge discovery in the design of e-commerce applications. Three incremental stages of knowledge discovery have been presented and illustrated with simple case studies from our sponsors.

The sequence of stages has been designed to bring increasing return-on-investment at each stage and to help e-retailers get closer to an optimal use of the Web channel. Indeed, the unique nature of the on-line world makes achievable the double objective of maximizing the customer's experience and maximizing revenues.

We have focused here on enhancing e-commerce through personalization, in particular by matching offers to interests (i.e., supply and demand) in an efficient way through clickstream and behavior analysis. There are many other data-independent things e-retailers can do to enhance the user experience, such as offering social interactions (e.g., user forums) or providing virtual versions of physical stores (e.g., displays, lighting, music) [25].

References

1. C. Allen, D. Kania, and B. Yaeckel. *One-to-One Web Marketing: Build a Relationship Marketing Strategy One Customer at a Time, 2nd Edition.* John Wiley & Sons, 2001.
2. AWStats. Logfile Analyzer. http://awstats.sourceforge.net/docs/, 2005.
3. B. L. Barret. The Webalizer. http://www.mrunix.net/webalizer, 2005.
4. P. Cabena, P. Hadjinian, R. Stadler, J. Verhees, and A. Zanasi. *Discovering Data Mining: From Concept to Implementation.* Prentice Hall, 1997.
5. Center for Democracy and Technology. CDT's Guide to Online Privacy. http://www.cdt.org/privacy/guide/introduction, 1998.
6. ClickTracks. http://www.clicktracks.com, 2005.
7. Personalization Consortium. Personalization information. http://www.personalization.org/personalization.html, 2005.
8. R. Dye. The buzz on buzz. *Harvard Business Review*, November-December 2000.
9. E-Commerce Intelligence: Measuring, Analyzing, and Reporting on Merchandising Effectiveness of Online Stores. Stephen gomory and robert hoch and juhnyoung lee and mark podlaseck and edith schonberg, July 1999.
10. H. A. Edelstein. Pan For Gold In The Clickstream. *Information Week*, March 2005.
11. M. Eirinaki and M. Vazirgiannis. Web mining for web personalization. *ACM Transactions on Internet Technologies*, 3(1):1–27, 2003.
12. X. Fu, J. Budzik, and K.J. Hammond. Mining navigation history for recommendation. In *Proceedings of the International Conference on Intelligent User Interfaces*, pages 106–112, 2000.
13. Google. Advertising Programs. http://www.google.com/ads, 2005.
14. D. R. Greening. Data Mining on the Web. *Web Techniques*, January 2000.
15. Beagle Research Group. Turning browsers into buyers with value-based routing: Methodology enhanced e-commerce, white paper. http://www.beagleresearch.com/DownloadPDFfiles/ProficientFINAL.pdf, 2004.
16. A. Hughes. 25 Key Database Marketing Techniques. *DM News*, January 2005.
17. P. Kent. *Search Engine Optimization For Dummies.* Wiley Publishing, Inc., 2004.
18. R. Kohavi, L. Mason, R. Parekh, and Z. Zheng. Lessons and challenges from mining retail e-commerce data. *Machine Learning*, 57(1-2):83–113, 2004.
19. G. Linden, B. Smith, and J. York. Amazon.com Recommendations: Item-to-Item Collaborative Filtering. *IEEE Internet Computing, Industry Report*, January/February 2003.
20. Microsoft. MSN Search. http://search.msn.com, 2005.
21. B. Mobasher, R. Cooley, and J. Srivastava. Automatic personalization based on web usage mining. *Communications of the ACM*, 43(8):142–151, 2000.

22. B. Mobasher, N. Jain, E. Han, and J. Srivastava. Web mining: Pattern discovery from world wide web transactions. Technical Report TR-96050, Department of Computer Science, University of Minnesota, M inneapolis, 1996.
23. W. Moe and P. Fader. Capturing evolving visit behavior in clickstream data. *Journal of Interactive Marketing*, 18(1):5–19, 2004.
24. M. Monticino. Web-analysis: Stripping away the hype. *Computer*, 31(12):130–132, 1998.
25. S. Oberbeck. Internet shopping is big business, but more can be done by e-retailers. *The Salt Lake Tribune*, December 2004.
26. Omniture, Inc. SiteCatalyst 12. http://www.omniture.com/global/sitecatalyst, 2005.
27. M. Perkowitz and O. Etzioni. Towards adaptive Web sites: conceptual framework and case study. *Computer Networks*, 31(11–16):1245–1258, 1999.
28. E. Rosen. *The Anatomy of Buzz.* New York: Doubleday, 2000.
29. P. Rusmevichientong, S. Zhu, and D. Selinger. Identifying early buyers from purchase data. In *Proceedings of the Tenth ACM SIGKDD International Conference on Knowledge Discovery and Data Mining*, pages 671–677, 2004.
30. Sane Solutions LLC. NetTracker Web Analytics Solutions. http://www.sane.com, 2005.
31. M. Spiliopoulou and C. Pohle. Data mining for measuring and improving the success of web sites. *Data Mining and Knowledge Discovery*, 5(1–2):85–114, 2001.
32. J. Srivastava, R. Cooley, M. Desphande, and P-M. Tan. Web usage mining: Discovery and applications of usage patterns from web data. *SIGKDD Explorations*, 1(2), 2000.
33. Alta Vista. http://www.altavista.com, 2005.
34. S. Wasserman and K. Faust. *Social Network Analysis: Methods and Applications.* Cambridge University Press, 1994.
35. World Wide Web Consortium (W3C). Platform for Privacy Preferences (P3P). http://www.w3.org/P3P, 2004.
36. Yahoo! Inc. http://www.yahoo.com, 2005.
37. Yahoo! Inc. LAUNCHcast. http://launch.yahoo.com, 2005.
38. Yahoo! Inc. Search Marketing. http://searchmarketing.yahoo.com, 2005.

Personal eBanking Solutions based on Semantic Web Services

Oscar Corcho[1], Silvestre Losada[1], Richard Benjamins[1], José Luis Bas[2], and Sergio Bellido[2]

[1]Intelligent Software Components, S.A.
C/ Pedro de Valdivia, 10. 28006 Madrid (Spain). Tel: +34913349797, Fax: +34913349799
{ocorcho,slosada,rbenjamins}@isoco.com
[2]Fundación de la Innovación. Bankinter
Paseo Castellana, 29. 28046 Madrid (Spain). Tel: +34916234238
{jlbas,sbellidog}@bankinter.es

Abstract. We describe how Semantic Web Service technology can be used for the provision of personal e-banking online services. We describe two deployed applications: an overdraft notification service and a mortgage comparison service. The former accesses the bank accounts of a user as well as utility goods Web sites where invoicing information is stored and estimates whether the user will be in an overdraft situation in the near future, alerting him/her by e-mail or SMS. The latter accesses the mortgage information provided by the heterogeneous Web sites of different banks and allows users to compare them according to different types of criteria. The chapter not only focuses on the technological decisions followed to implement and deploy these services, but also on the added value of applying Semantic Web Services for them.

1 Introduction and motivation

The Internet has revolutionized our lives in areas such as communication (very low cost, large scope), online business (transactions), and access to content (millions of resources, irrespective of location and language). Also in our "financial lives", the Internet has provoked significant changes: instead of going to a physical bank branch, we can go to the bank's web site and make many different types of transactions.

There are important differences in the use of Internet banking between different countries [2], with a complex set of factors that influence adoption, such as access technology and infrastructure related factors, sector-specific Internet banking factors, and other socioeconomic factors. Most financial institutions allow their clients to access their accounts and consult their account information via Internet. Moreover, most of these institutions allow their clients to make transactions via Internet. In both cases, several options are made available for clients, ranging from regular bank services to associated services, such as information about grants, e-commerce services, mobility, shopping or on-line payments services, and so on.

O. Corcho et al.: *Personal eBanking Solutions based on Semantic Web Services*, Studies in Computational Intelligence (SCI) **37**, 287–305 (2007)
www.springerlink.com

In general, the main processes covered in banking operations can be classified into three categories:

- o ***Inter-banking processes***. They refer to the exchange of documents or account entries (cheques, receipts, international and national transfers). For example, when a Bank pays a cheque from another bank, a movement of funds is produced from the second Bank (the payer) to the first (the payee). This movement of funds is always accompanied by information on the operation (e.g., the cheque number) and, in some cases, the document itself.
- o ***Bank-provider processes***. This refers to basic supplies common to any industrial sector (paper, IT equipment, software, office furniture, etc.), with the sole exception of those information providers that are specific to banking (defaulter registry, real estate appraisal entities, etc.).
- o ***Bank-client processes***. This refers to product sale processes and service usage processes through the different channels made available by the bank. Here, we must include as well the internal operations of the Bank, since they make the relationship with the client possible.

From another perspective, the banking business can be also divided according to the following categories: products, services and channels.

- o ***Products***. They are contractual operations that involve the deposit of money (accounts, mortgages, deposits, investment and pension funds, etc.) or money loans (credit, loans, mortgages, guarantees, etc.).
- o ***Services***. They are operations that involve the entry or exit of funds from a banking product (i.e.: credit cards, cheques, promissory notes, receipts, transfers). In a broad sense this category would include every kind of service that is offered by the bank within its corporate purpose, even if it may not be purely related to banking, such as virtual stores, Internet services, etc.
- o ***Channels***. They are the means through which the bank reaches its clients: Branches, Telephone Banking Services, Internet, Agents, Commission Agents, etc. The most common situation to date is that the non-traditional channels offer only services. However, it is increasingly becoming more frequent that they also offer the possibility of contracting products.

Let us focus on the provision of products and services using the Internet channel, that is, on e-Banking. Internet banking services are considered as a cost-effective delivery channel, driven by cost reduction, market share increase and customer retention targets. However, profitability still remains a challenge. In line with this, there are still many fields subject to technological improvements. Within these, those most capable of being improved are the ones that fall in line with the requirements of the financial institutions and objectives, and usually are:

- o Those relating to task-performing by people. They consume valuable time and resources that may be used for marketing.
- o Those relating to cost reduction. One of major commitments of banks is efficiency in terms of competitiveness. The prices of the banking products and services are marked by the cost plus a differential (profit margin).

o Those relating to new products/services/quality. This is highly related to the strategic position based on the differentiation in the market.

In this situation, the eBanking sector provides good opportunities for the deployment of Semantic Web Service technology. These opportunities are mainly realted to those products and services that currently have a large complexity (involving the consumption of a large amount of time and human resources), and consequently are costly, and to those that have a large market potential (related to the provisiong of new added-value products and services that can differentiate the bank position in the market).

In this chapter we present two software prototypes that fall into the two previous categories: a mortgage comparison service (as an improvement of a service with a large complexity) and an overdraft notification service (as the provision of a new added-value service for bank clients. In both cases the services are deployed using Semantic Web Service technology. The application of such technology results in significant improvements and attractive return of investment (ROI).

The rest of this chapter is structured as follows: section 2 describes the mortgage comparison service, providing the service background as a set of general requirements, showing its architecture and describing in detail some of the aspects of its implementation and deployment. Section 3 describes the overdraft notification service, following the same structure as for the other one: general requirements, architecture and details of implementation and deployment. Section 4 provides some conclusions to this chapter, with a focus on the technological requirements that can be derived from the implemenation of both applications: data mediation, service discovery and composition, invocation, etc., which will have to be considered in the future by Semantic Web Service research. Finally, section 5 will describe our expected future trends in the application of this and other similar technology in the e-banking domain.

2 Mortgage Comparison Service

The mortgage application process requires a large amount of effort in time and resources, mainly due to the need for the compilation of information from external sources. Figure 1 graphically depicts this process, which can be described as follows:

1 The client looks for a property in accordance with his/her requirements and which a priori considers he/she can deal with. The search for information is outside the scope of the financial institution service offering and, consequently, it can help little or nothing.

2 Once a market search has been done, the client makes his/her calculations.
   a. Which type of mortgage can I deal with?
   b. What are my current resources?

c. How much of my monthly income can I set aside for the payment of the loan to acquire a property?

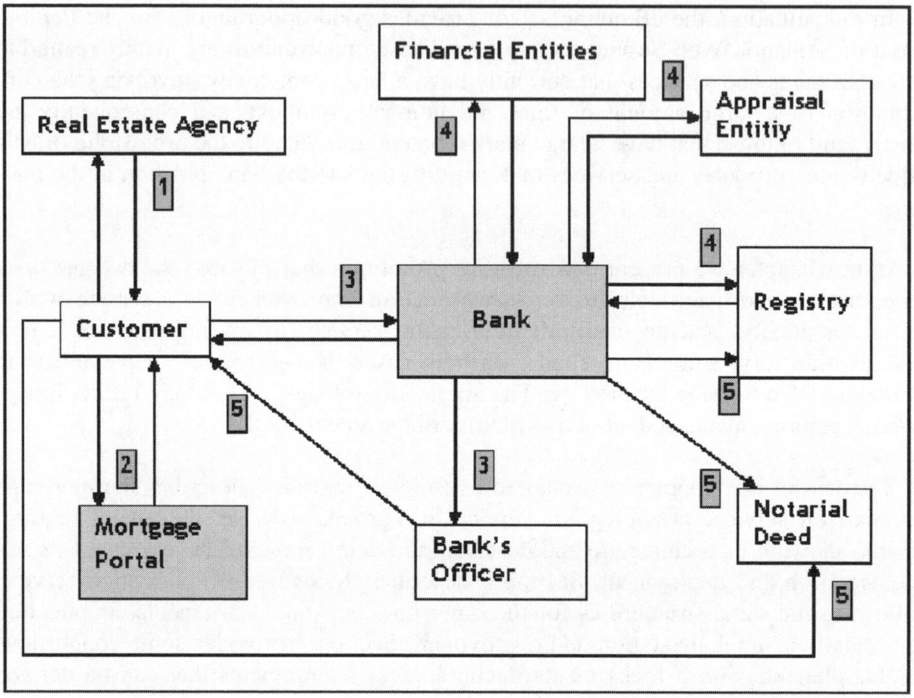

Figure 1. A simplified view of the mortgage application process.

This type of information can be provided by a mortgage simulator. Mortgage simulators follow a mechanism outside any financial institution (let's say that it follows the principle of an abstract mortgage). The simulator should be able to respond to the previous questions. Using a simple form that takes into account the client's income, it should be able to respond with the maximum amount that may be paid (taking into account a maximum depreciation time), or the number of instalments that should be paid if he/she wants to pay a fixed amount of money. A standard, fixed, reference can be used to make these calculations.
Once the reference information is obtained, the client knows the type of mortgage that can be contracted. Now, the client needs to know the mortgage offer which adapts best to his/her possibilities (or tastes). For this, information is obtained from different banks on the mortgage offers they provide. This data gathering can be done online and automatically, accessing a mortgage comparator. Given reference data, mortgage comparators calculate the amounts, the terms and conditions typified in the bank offers, and provide clients with an ordered list (in accordance with order parameters that the client can modify) of mortgage offers.

3 Once the mortgage offer to be contracted has been selected, the client contacts the bank providing the offer. The bank starts a client data gathering process, in which the client itself is involved.

4 During the client data gathering process, the bank obtains as much information as possible on the property to be acquired. This information gathering process is made by contacting different types of entities, such as appraisal entity agencies, public registries, notaries, etc. The bank also contacts other financial entities and its own records to retrieve information about the client.

5 Once all the necessary data has been gathered, it is necessary to make a decision on whether or not to provide the mortgage to the client. In case that the decision is positive, all the relevant information will be sent to the notary and the contract between the client and the financial entity will be signed.

The service described in this section is related to the second step in the mortgage application process: mortgage comparison and simulation.

The current situation is as follows: the client obtains information about the mortgages offered by different banks or building societies following a complex and tedious process: he/she is forced to access the Web sites of different organisations, find the relevant links inside them and perform a manual comparison taking into account many different factors, revealed by the heterogeneity of the existing mortgage offers (different vocabularies, different types of fees, etc.).

To alleviate the burden of performing such task, some financial institutions already offer comparison services[1] to their customers. The comparisons offered by these services are based on information that is input manually by persons (hence error prone and not always up to date) or using costly, and sometimes brittle, screen-scraping technologies. Furthermore, these services do not usually solve the heterogeneity problem highlighted above and hence are not so useful in the decision making process made by clients. Similarly, simulation services are currently accessible in many bank Web sites for the mortgages that they offer[2]. These services are more common than those aimed at comparing mortgages. Third parties, such as intermediary webs sites that recommend mortgages, normally perform the simulations and access the results using screen-scraping techniques.

The short-term benefit of applying Semantic Web Service technology in mortgage comparison and simulation is that human intervention is reduced in the process (this means a cost reduction in the maintenance of these services) and that data reliability increases.

Furthermore, as a middle-term result, new added value services and market opportunities may be generated with the optimum development of applications based on the use of semantics. For instance, new kinds of mortgages may appear, which are rarely taken into account nowadays due to high costs/human-task mediation required (e.g.,

[1] For instance, http://www.comparador.com/, http://www.secureapp.com/ome/qual/qualifier.asp
[2] For instance, https://www.ebankinter.com/www/es-es/cgi/ebk+hip+cuota_mes

small amount operations in which the costs are far higher than the incomes). Besides, other new services may appear, which make use of the new banking service and may mediate in the sale of mortgages. In this case, there would be a market opportunity for a bank, not only due to anticipating these intermediaries, but also its own competition. The bank could additionally become a wholesaler for these intermediaries.

2.1 Architecture of the Mortage Comparison Service

As aforementioned, this service aims at easing the task of mortgage comparison by providing a simple interface where users specify constraints about the type of mortgage that they want, and receive back the information in a homogeneous way.

The main difference between this service and others currently available (as the ones whose URLs have been provided above) is that each time a client makes a request to the service, the data used for the comparison will be obtained on-line from each bank or building society, so that the information is never outdated. We rely on the existence of Web services in each of the accessed bank Web sites that can provide the mortgage simulations with the data provided by the client (be it modified or not according to the Web service requirements).

Furthermore, the different types of results will be homogeneized, and filtered, in the user interface, so that the comparison can be performed according to the same set of parameters instead of those provided by each of the mortgage providers (since there is few standardization on the content of what a mortgage is).

Figure 2 summarises the architectural components needed to perform the operations of the mortgage comparison service. The architectural components in the middle box are based on the architecture defined in the context of the DIP project[3], which is the architecture on top of which the mortgage comparison service is deployed[4]. Though it is not the purpose of this description, we will briefly explain the role that those components play in the context of the DIP architecture and in the context of this service:

- o The *discovery component* is used to find the services suitable to perform an action. In DIP the service discovery process is focused on finding services that comply with an abstract service description (that is, a set of concrete service instances that share a well-defined set of characteristics). In the context of the mortgage comparison service, service discovery is used to determine which of the services published by different banks provide simulations of mortgages that comply with the constraints specified by the users.
- o The *mediation component* is used to overcome the communication problems that may arise between the service requestor and the service provider during the service invocation, due to the use of syntactically different data formats and different vocabularies to describe the data. In the context of the mortgage comparison service, service mediation will be needed to solve the problem of heterogeneity in

[3] http://dip.semanticweb.org/documents/D6.2-DIP-Architecture.pdf

[4] This architecture is the basis for the WSMO architecture (available at http://www.wsmo.org/).

the parameters used as inputs and outputs by each of the mortgage simulation services, which includes data formats (e.g., different data types for expressing dates, currency amounts, etc.) and vocabularies (e.g., different representations of what quota means, differences in types of fees applicable, etc.).

- The *invocation engine* is needed to provide a platform for the execution of the services. It makes it possible to execute the Web services to which the comparison service accesses and obtain the execution results from them.
- The *repositories* are used for different purposes: the goal repository is used by the comparison service in the construction of the goal expressed by the client when a mortgage request is done, that is, it contains goal templates that are filled-in by the service; the Web service repository is used as a registry of the Web services that are available, and is used during the discovery process; finally, the ontology repository maintains the ontologies (vocabularies) used by the different Web services and is used during the task of ontology mediation.

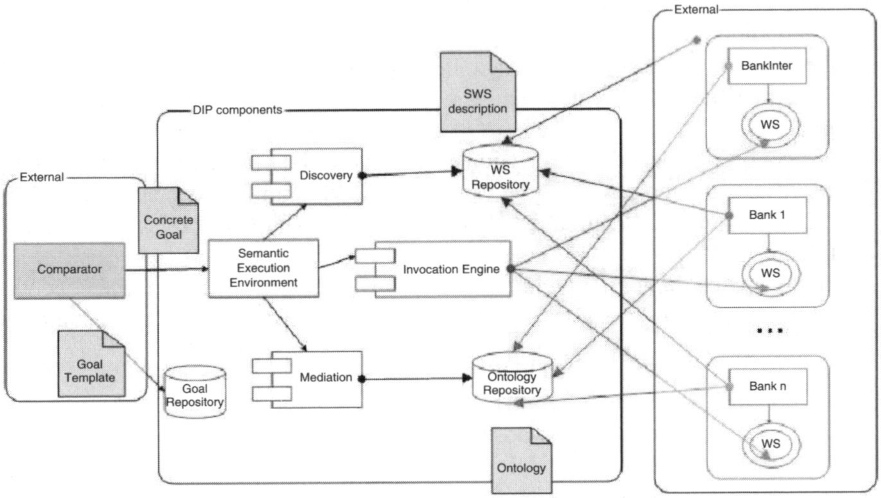

Figure 2. Architecture of the mortgage comparison service in relation to the Web services contacted and the components needed from a Semantic Web Service execution environment.

2.2 Implementation and Deployment of the Mortage Comparison Service

The service has been implemented[5] according to the architecture described in section 2.1. It is deployed on the WSMX execution environment[6], which is one of the two platforms that realise the WSMO architecture (the other one being IRS-III [4]).

[5] The service is available at http://comparador.isoco.com/.
[6] http://www.wsmx.org/

We will show now some examples of the goals and service descriptions that are used by this service. These descriptions are available in the WSML language[7], which is the one proposed by the WSMO framework and used by the WSMX platform.

Goals denote what a user wants as the result of the Web Service invocation. Figure 3 shows a goal expressed in WSML, more specifically the goal used to obtain the number of payments of a mortgage given the total amount to be lent and the foreseen quota to be paid. First of all, we can see the namespace declaration, which provides the namespaces used in the goal description. Then some non-functional properties of the goal are provided, such as title, type, date, language, etc. Then the ontologies used in the goal description are provided (in this case a financial ontology developed for Bankinter, which contains not only terms related to the mortgage application process, but also other aspects related to the financial domain).

Finally, the capability is expressed, describing the restrictions on the information elements that the user wants to get from the service (preconditions, posconditions) and the state of the world (assumptions, effects).

Preconditions describe restrictions that have to be true before invoking the service. In the case of this goal, the loan capital has to be known and less than 200000, the quota has to be less than 900, and the maximum interest rate has to be 0.5 over the reference rate.

Postconditions describe restrictions that have to be true after invoking the service. In the case of this goal, the mortgage requires having a life insurance contract, but not a home insurance contract, and the opening commission will be less than 0.7% of the total loan capital.

Finally, since there are no changes in the real world as a result of the execution of a mortgage simulation, the goal expressed in figure 3 does not contain assumptions nor effects.

[7] http://www.wsmo.org/wsml/

```
namespace {_"http://users.isoco.net/~slosada/ontologies/bankinter/GoalGetNumberOfPayments.wsml#",
    dc      _"http://purl.org/dc/elements/1.1#",
    foaf _"http://xmlns.com/foaf/0.1/",
    xsd     _"http://www.w3c.org/2001/XMLSchema#",
    wsml _"http://www.wsmo.org/2004/wsml#",
    fin     _"http://users.isoco.net/~slosada/ontologies/bankinter/FinancialOntology.wsml#"}

goal _"http://users.isoco.net/~slosada/ontologies/bankinter/GoalGetNumberOfPayments.wsml"

    nfp
        dc#title hasValue "Goal to find mortgage simulator with value restrictions"
        dc#type hasValue _"http://www.wsmo.org/2004/d2#goals"
        dc#description hasValue ""
        dc#subject hasValue { "Simulator", "Mortgage", "Financial", "Product"}
        dc#date hasValue _date("2005-03-07")
        dc#format hasValue "text/html"
        dc#language hasValue "en-US"
        dc#rights hasValue _"http://www.isoco.com/privacy.html"
        wsml#version hasValue "$Revision: 10 $"
    endnfp


    importsOntology { _"http://users.isoco.net/~slosada/ontologies/bankinter/FinancialOntology.wsml#"}

    capability
        sharedVariables ?mortgageLoan

          precondition
          nfp
                dc#description hasValue "The input has number of payments, type of interest and
                mortgage amount to simulate morgage result is a monthly payment."
          endnfp
          definedBy
          ?mortgageLoan memberOf fin#MortgageLoan
      [ loanCapital hasValue ?capital,
          intitalQuota hasValue ?quota,
          interestRateType hasValue ?interest]
          and ?capital < 200000
          and ?quota < 900
          and ?interest memberOf productRateApplicationVariable[interestRateValue hasValue 0.5,
          referenceType hasValue _# ].

          postcondition
          nfp
                dc#description hasValue "Result of webService is a list of mortgage. The person who wants
                contract this mortgage must have life insurance. "
          endnfp
          definedBy
                ?mortgageLoan memberOf fin#MortgageLoan
                     and ?mortgageLoan [
                     term hasValue _#,
                     openingCommission hasValue ?opCommission,
                     lifeInsurance hasValue true,
                     homeInsurance hasValue false ]
                     and ?opCommission < 0.7.
```

Figure 3. Goal to find a mortgage simulator, given a known loan capital, the interest rate required and the maximum monthly payment.

Similarly, **Web Service** descriptions also consist of restrictions about their preconditions, postconditions, assumptions and effects. The description presented in figure 4 provides information about a mortgage simulator Web service that is available from an external financial institution. It contains, as in the case of the goal description from figure 3, a set of preconditions and postconditions, but not assumptions or effects, since the mortgage simulation itself does not have any impact on the external world.

From this description we can conclude that the service is compliant with the goal presented in figure 3, and consequently would be discovered by a discovery service.

```
namespace {_"http://users.isoco.net/~slosada/ontologies/bankinter/WSGetMortgageCapital.wsml#",
    dc    _"http://purl.org/dc/elements/1.1#",
    foaf _"http://xmlns.com/foaf/0.1/",
    xsd    _"http://www.w3c.org/2001/XMLSchema#",
    wsml _"http://www.wsmo.org/2004/wsml#",
    fin    _"http://users.isoco.net/~slosada/ontologies/bankinter/FinancialOntology.wsml#"}

webService _"http://users.isoco.net/~slosada/ontologies/bankinter/WSGetMortgageCapital.wsml"

    nfp
        dc#title hasValue "Web Service that is a    mortgage simulator"
        dc#type hasValue {_"http://www.wsmo.org/2004/d2#webservice"    }
        dc#description hasValue ""
        dc#subject hasValue { "Simulator", "Mortgage", "Financial", "Product"}
        dc#date hasValue _date("2005-03-30")
        dc#format hasValue "text/html"
        dc#language hasValue "en-US"
        dc#rights hasValue {_"http://www.isoco.com/privacy.html" }
        wsml#version hasValue "$Revision: 1.0 $"
    endnfp

    importsOntology { _"http://users.isoco.net/~slosada/ontologies/bankinter/FinancialOntology.wsml#"}

    capability
        sharedVariables ?mortgageLoan
        precondition
         nfp
                dc#description hasValue "The input has number of payments, type of interest and
                mortgage amount to simulate morgage result is a monthly payment."
         endnfp
         definedBy
?mortgageLoan memberOf fin#MortgageLoan
    [ term hasValue ?term,
        intitalQuota hasValue ?quota,
        interestRateType hasValue ?interest]
                    and ?term[ totalTerm hasValue ?totalTerm,
                              typeTerm hasValue MONTH ]
                    and ?totalTerm < 300
        and ?quota > 600
        and ?interest memberOf productRateApplicationVariable
        [interestRateValue hasValue 0.5,
        referenceType hasValue _# ].

        postcondition
        nfp
                dc#description hasValue "Result of  webService is a list of mortgage. The person who wants
                contract this mortgage must have life insurance. "
        endnfp
        definedBy
                ?mortgageLoan memberOf fin#MortgageLoan
                    and ?mortgageLoan [
                    loanCapital hasValue _#,
                    openingCommission hasValue ?opCommission,
                    lifeInsurance hasValue true,
                    homeInsurance hasValue false ]
                    and ?opCommission < 0.7.
```

Figure 4. Web service description for one of the mortgage simulators.

3 Overdraft Notification Service

People normally have several bank accounts, with the same or different banks, where they have different amounts of money. Besides, they normally have contracts with consumer goods companies, such as telephone companies, gas and electricity providers, broadband providers, etc., whose bills are charged to any of the bank accounts that they have.

Account aggregation has been identified as one of the key services that will have to be provided by financial institutions in the future [2]. Currently, account aggregation tools exist, such as GetSee®[8], which perform this task, normally by means of screen-scraping techniques, and are available for customers in many bank Web sites.

The overdraft notification service described in this section goes a little bit further, and can be seen as one of the many added value services that can be created on top of account aggregation tools. This service detects whether any of the customer accounts is going to be in an overdraft situation, taking into account estimations of the next invoices that will be sent by consumer good companies to the customer's accounts. The service notifies the customer if the balance of the account is less than the expected invoice amount or is under a specific offset, using any of the notification channels available for the customer. Hence the customer can perform the corresponding transactions in order to avoid that situation[9].

Figure 5 describes the typical operational scenario for this service. In this scenario, the following actors are involved: the customer, the banks, and the consumer goods companies. And the following services are involved: customer notification agent (CNA), Sentinel and some estimation services. Finally, the iSOCO GETsee® application is at the core of this scenario, in charge of the aggregation of data from bank accounts and consumer goods companies.

[8] https://www.getsee.com/

[9] According to the current law in Spain the system cannot perform transactions like bank transfers automatically, without the explicit consent of the user. This functionality could be easily provided if this action was allowed.

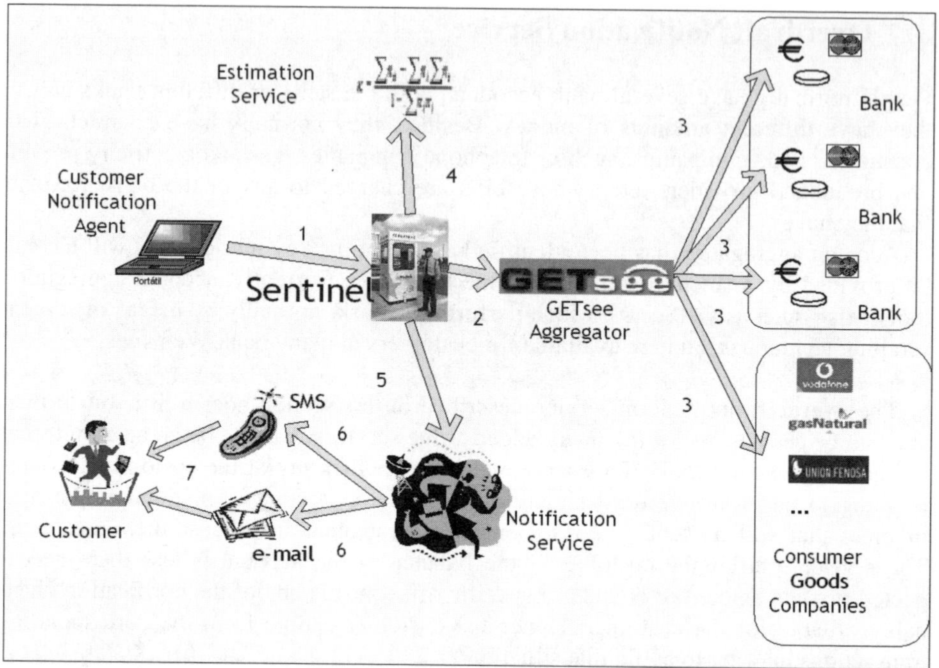

Figure 5. Sample scenario diagram for the Overdraft Notification Service.

The following steps will be normally done:

Step 1: Everyday, the Customer Notification Agent dynamically configures and invokes the Sentinel Service. This agent has the entire customer's information needed for invoking the composed service (online username, password and other data). The update frequency of this agent can be customized.

Step 2: The Sentinel Service uses iSOCO GETsee® for collecting information from the customer's accounts.

Step 3: iSOCO GETsee® collects the amount balance of all the customer's accounts (of banks B1, B2, …, Bn). In one (or more) of this accounts some consumer goods companies (E1, E2, …, En) can charge invoices. The invoices have their notification and value dates. The frequency of those invoices is always the same (weekly, monthly, bimonthly, annually).

Step 4: For each invoice of consumer goods companies (E1, E2, …, En) associated with the account, the Estimation Service estimates the probable amount at the end of the period, Ae (estimated amount) in terms of heuristics or mathematical models. Ae has a relationship with a consumer good company (Ee) and an account of a bank (ABe). If the Ae is less than the established threshold for the account, then an alert has to be raised.

Step 5: The Notification Service looks in a (de)centralized registry different ways to communicate with the user. It can find different services involving many different devices (phone calls using VoIP, SMS, electronic mail, telegram) and personal data (phone number, cell phone number, e-mail, postal address). The services discovered must have the ability to perform the action defined in the Notification Service.

Step 6: The invocation engine sorts in terms of cost, time to deliver, etc., the different possibilities and chooses the first service in this particular ranking. Some data mediation could be needed if terms of the ontology used differ from the one used by the Notification Service. If the service chosen has an irrecoverable mismatching of process or data, or some communication error occurs in the invocation, the service has to be able to choose another service and invoke it.

Step 7: The service chosen is invoked and the user is notified.

3.1 Architecture of the Overdraft Notification Service

The general architecture of this service is shown in figure 6, which resembles the top-level diagram already presented in figure 5. In this case we have opted for presenting the Semantic Web Services involved in the scenario instead of providing details about the components of the architecture that are used at each time, since this is quite similar to what was presented for the mortgage comparison service.

The figure shows that there are three main services that are executed at some point in time during the execution of the service. These are the GETsee service, in charge of the aggregation of accounts from different bank and consumer good companies, the estimation service, in charge of providing estimations of the amounts of the invoices that will be sent to each bank account by the consumer good companies, and the notification service, in charge of notifying customers about their possible overdraft situation.

Furthermore, the GETsee Service is decomposed into five atomic services (openSession, getAccounts, getInvoices, getBalance, closeSession). These five services are annotated using the same ontology as the GETsee service (although this is not mandatory in our approach). Those atomic services invoke other services, which are annotated according to other ontologies. In these cases, data mediation is needed for the exchange of messages. At last, the Notification Service looks for a service able to notify something to a person and finds at least two services (notification by SMS and notification by e-mail), which might be annotated according to other two more ontologies.

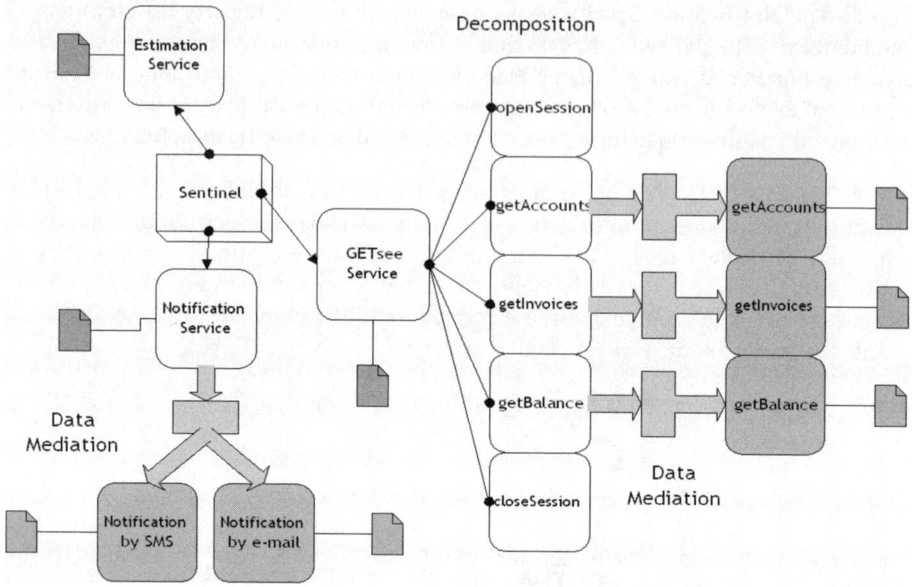

Figure 6. A diagram of the Semantic Web Services used for the notification scenario.

3.2 Implementation and Deployment of the Overdraft Notification Service

Unlike the mortgage comparison service presented in the previous section, this service has not been implemented on top of any existing Semantic Web Service execution platform, such as WSMX, IRS-III or the OWL-S [8] virtual machine, but using an ad-hoc approach. The reason for this decision is that this service was implemented and deployed when those platforms were still in an unstable stage of their development. However, it would be easy to adapt the service implementation to any of them since it follows a similar approach to that of those execution environments (using service discovery functions, data and ontology mediation functions, etc.), and the services are semantically described in a similar way to that required in those frameworks.

From all the processes involved in this service, the discovery process deserves special attention, since it is different to the one presented for the previous service on mortgage comparison. The discovery process works as a 2-step process, where the first stage is used to make a first filtering of the services that could provide the service requested, and the second stage is used to make a more fine-grained selection. This process is described in detail in [12].

For the first step the capabilities of services and the goals of service requestors are expressed using the description logic formalism [1] (more specifically using the OWL language [3]). As an example, below we have the description of the capabilities of two notification services (one for sending e-mail notifications and the another for sending SMS notifications) and a request to send a notification. With this approach we do not consider as well the difference between preconditions, postconditions, assumptions and effects, as we did in the previous service, but we just consider that all the descriptions are of preconditions.

Capabilities and a Request:

```
Cap_A ≡
EmailNotification  ⊓ ∃ from.User ⊓ ∃ to.User ⊓ ∀to.User ⊓
∃usedProvider.{Provider_A} ⊓ ∃sendingTime.Timestamp ⊓ ∃content.String ⊓
∀acknowledgement.=_F

Cap_B  ≡
SMSNotification ⊓ ∃from.User ⊓ ∃to.CellphoneUser ⊓ ∀to.CellphoneUser ⊓
∃ usedProvider.{Provider_B} ⊓ ∃ sendingTime.(Timestamp ⊓  ≤_currentTime+1week)
⊓ content.String

Req  ≡
Notification ⊓ ∃ from.{User_X} ⊓ ∃ to.{User_Y} ⊓ ∃ to.{User_Z} ⊓ =2to ⊓ ∀
usedProvider.Provider ⊓ ∃ sendTime ≤ _200406250900 ⊓ ∃ content.String ⊓ ∀
acknowledgment =_T ⊓ ∀ cost ≤_5
```

Besides, we have the following domain-level facts:

```
Notification ⊑ Action
EmailNotification ⊑ Notification
⊤  ⊑  =1 from
```

The basic idea under DL-based discovery matching is to check whether the conjunction of a request and a capability is satisfiable, i.e. they can have at least one instance in common. If $\texttt{Request} \sqcap \texttt{Capability}_X \sqsubseteq \bot$ holds true there is no such common instance and the request cannot be fulfilled by this capability

However, with this type of discovery we are just checking which are the classes of services that can fulfil a request, but we cannot determine exactly which of the instances of those classes of services can actually perform the operations. Furthermore, some specific constraints related to actual values of some properties cannot be used in the reasoning process, and hence a second step is needed, using that information in the process.

For this second step we use individual service descriptions in the F-Logic language [6] and a F-Logic enabled reasoner like Flora-2[10]. In this approach for discovery we check whether the capability entails the goal (capability ≤ goal). Current limitations with respect to available reasoners led to the current modeling, where the goal-postcondition is expressed as a fact (which may not be fully specified) and the capability-postcondition is expressed as a rule.

```
myGoal:goal[
  postCondition->myNotification].
myNotification:notification[
  ntf_userToBeNotified -> johndoe,
```

---

[10] http://flora.sourceforge.net/

```
      ntf_date -> d040606:date[dayOfMonth->5, monthOfYear->5,year->2004],
      paymentMethod -> creditCard,
      cost -> 0.2,
      ntf_body -> "Your Account Z will be in minus in 2 weeks",
      ntf_from -> sentinel].
    johndoe:user[
      nif -> 123,
      name -> "John Doe",
      password -> "p",
      login -> "l",
      firstPreference -> jdMobile,
      contacts ->>
       {jdEmail1:eml_account[eml_account->"jon@doe.com"],
        jdMobile:phone[phn_number->"0123456", phn_type->mobile],
        jdHome:phone[phn_number->"6543210", phn_type->home]}].
    sentinel:user[
      name -> "Sentinel System",
      contacts ->> {jdEmail1:eml_account[
         eml_account->"sentinel@isoco.com"]}].
```

The capability postcondition describes the state of the information space the service has after its execution. Here we use some prolog build in predicates, e.g. '//' which is an integer division, but that might also be replaced by more declarative predicate names like "integerDivision(X,Y,Z)".

```
smsProvider[postcondition] :-
 _AnyNotification:notificationSMS[
    phn_number -> _X:phone[phn_type->mobile],
    ntf_receiptAcknowledgement -> false,
    ntf_time -> Time:dateAndTime,
    content -> AnyMessage:message,
    payment -> Payment],
 is_charlist(
  AnyMessage.msg_body, AnyMessageLength)@prolog(),
  AnyMessageLength < 800,
  Tokens is '//'(AnyMessageLength,160)@prolog()+1,
  Cost is Tokens * 0.05,
  Payment.cost >= Cost,
  (Payment.paymentMode = creditCard; Payment.paymentMode = account),
  secondsBetween(currentDate,Time,X),   X < 5*60.
```

4 Conclusions

Internet technology is widely extended in the banking processes, especially in the context of bank-customer relationships, where Internet is used as another of the available channels that can be used by customers to access their accounts and perform transactions with them. Different banks and financial institutions adopt different strategies with respect to the types of services that they offer to their customers through this channel. Some of them only provide basic services to them and others are increasingly offering more ranges of products through their eBanking Websites.

However, new technologies are not only being applied to the bank-costumer relations, but much more to the bank-to-bank and the bank-providers relations. These relations are highly standardised but not always fully integrated in bank proprietary systems, depending on the type of service that is provided.

In this chapter we have described two applications that fall under the category of bank-customer services, since the end user in both cases is the bank customer. However, they also imply the exchange of information between different organisations (in the case of mortgages, the service accesses external sites to be able to perform comparisons, while in the case of overdraft notification, the service aggregates accounts from different institutions and accesses personal accounts in consumer good companies).

SWS technology makes processes more efficient (in terms of costs and time) and simple to maintain. They optimise the manual processes currently carried out to build new added value services. For instance, they allow searching in available registries, so that the new Web services that have been deployed in the market can be discovered. They also allow composing new added value services that could not be foreseen without the use of this technology. Consequently, more services (product price comparators, broker information, deposits, etc.) can be offered by banks due to their low cost, since less human interaction is required to discover and invoke new available Semantic Web Services once the application is launched.

Though the main objective of this chapter is showing how Semantic Web Service technology can be used to create added value services for ebanking customers, these services also pose interesting technical requirements for the research being done in this area. These requirements are the following:

- **Discovery**. Discovery capabilities are needed in order to find the Semantic Web Services able to solve the goals composed by service requestors. We have shown how different types of service discovery mechanisms can be used, depending on the degree of accuracy that we want to achieve out from the service discovery process. In general, better guidelines are needed for service developers with respect to the characteristics and limitations of the service discovery process in each Semantic Web Service execution framework.
- **Mediation**. In both applications we have seen that mediation is needed, since the different information providers (e.g., the different financial entities), may use heterogeneous message syntaxes and vocabularies (ontologies). Mediation capabilities have to be provided in the different execution environments and they must be easy to configure, so that the use of external services is rather straightforward even if the vocabularies and syntaxes are too different from each other.
- **Invocation**. Once the Semantic Web Services to be used have been selected, they will be invoked by the service requestor. Invocation engines are in charge of contacting the corresponding services and executing them, receiving their responses and sending them back to the service requestor.
- **Security**. In some cases highly confidential data has to be transferred between services or used for reasoning purposes. This is the case of the overdraft notification service, which uses information about the amount of money in each account, the estimation of future invoices, etc. Security is also needed in the case of making transactions (e.g., bank transfers), to make sure that the right service is executed using the correct user account and avoiding the exposure of some pieces of data.

5 Future trends in intelligent e-services applied to personal e-banking

The evolution of the relations between banks and their customers shows that banks will no longer be proprietors of the clients, as happened in the past, but will become instruments of the clients. Current generations tend to go less to the bank, understood as the branch placed in a physical location. For their daily operations they prefer to use electronic media (such as cash dispensers, Internet, and telephone banking services). As a consequence, bank employees have been transformed from 'storekeepers' to 'consultants', due to the fact that clients only go to the physical branch to ask for some specific pieces of information.

In [2] we can find some proposals about the factors that will have to be considered in the future for their Internet channels:

- o Intermediaries between the clients and the financial institutions will appear, as it has happened in other sectors with the appearance of Internet. Banks establish direct relations with their end clients through different channels (branches, post, telephone, proprietary software, Internet, etc.). New intermediaries will gather banking data to provide new services to their customers. From the traditional banking point of view, intermediation per se is not worrying, although the loss of direct relations with the client is. Each contact of the client with the bank, whatever channel used, is a potential sales opportunity.
- o CRM systems will be integrated with banking web sites. The customization of the offer, the simplicity of use and other factors, either in a direct client to/from bank relation or through intermediaries (e.g. real estate agents who would like to complete their range of services) probably has a different meaning than the current one, which is difficult to foresee.
- o There will be a more intensive use of recognition systems for written or spoken language. Banking will tend towards simplicity and the customization of more and more complex products.

The adoption of Semantic Web Service technology will imply an easier deployment of these types of services, since they will facilitate the creation of new services, the integration with other systems and even the interactions with end users.

However, there are still potential barriers for the uptake of this technology in the core of banking service provision. These are the following:

- o Banks do not normally exchange too much information with other banks, to avoid competition. This is the reason why screen-scraping techniques are still used as the main information source for account aggregators.
- o As a result of the previous aspect, many banks do not provide yet Web services to perform operations with them, since they do not want intermediaries to be able to access easily to the data from their customers, even if their customers provide these intermediaries the access information to the bank services.

- o Finally, banks are very demanding regarding security, and current Semantic Web service technology is not too focused on this aspect, what could be seen as an important weakness for the real deployment of the technology in the future.

Acknowledgements

This work has been supported by the IST projects SWWS (IST-2001-37314) and DIP (FP6-507483).

References

1. Baader F, McGuinness D, Nardi D (2003) *The Description Logic Handbook.* Cambridge University Press.
2. Centeno C. Adoption of Internet Services in the Enlarge European Union. Lessons from the Internet Banking case. June 2003. http://fiste.jrc.es/download/eur20822en.pdf
3. Dean M, Schreiber G (2004) OWL Web Ontology Language Reference. W3C Recommendation. Latest version: http://www.w3.org/TR/owl-ref/
4. Domingue J, Cabral L, Hakimpour F, Sell D, Motta E (2004). *IRS-III: A Platform and Infrastructure for Creating WSMO-based Semantic Web Services.* Proceedings of the Workshop on WSMO Implementations (WIW 2004) Frankfurt, Germany, September 29-30, 2004, CEUR Workshop Proceedings, ISSN 1613-0073
5. Fensel D, Bussler C, Ding Y, Omelayenko B. The Web Service Modeling Framework WSMF. Electronic Commerce Research and Applications, 1(2), 2002.
6. Kifer M, Lausen G, Wu J. Logical Foundations of Object Oriented and Frame Based Languages. Journal of ACM 1995, vol. 42, p. 741-843
7. López-Cobo, J.M.; Losada, S.; Corcho, O.; Niño, M. A customer notification agent for financial overdrawn using Semantic Web Services. *14th International Conference on Knowledge Engineering and Knowledge Management (EKAW'04).* Springer-Verlag. Lecture Notes in Computer Science (LNCS) 3257:371-385. October 2004.
8. Martin D, Paolucci M, McIlraith S, Burstein M, McDermott D, McGuinness D, Parsia B, Payne T, Sabou M, Solanki M, Srinivasan N, Sycara K (2004) *Bringing Semantics to Web Services: The OWL-S Approach.* Proceedings of the First International Workshop on Semantic Web Services and Web Process Composition (SWSWPC 2004),.San Diego, California, USA
9. Martínez Montes M, Bas JL, Bellido S, Corcho O, Losada S (2004) Financial Ontology. DIP deliverable D10.3. http://dip.semanticweb.org/
10. Martínez Montes M, Bas JL, Bellido S, Corcho O, Losada S, Benjamins VR (2004) Design and Specification of ebanking application: Mortgage Comparison Service. DIP Deliverable D10.2. http://dip.semanticweb.org/
11. Martínez Montes M, Bas JL, Bellido S, López-Cobo JM, Losada S (2004) Analysis report on ebanking business needs. DIP Deliverable D10.1. http://dip.semanticweb.org/
12. Zaremba M, Moran M, Haselwanter T, Zaremba M, Oren E (2005) *DIP Revised Architecture.* DIP deliverable D6.5. http://dip.semanticweb.org/

Secure E-Transactions Protocol using Intelligent Mobile Agents with Fair Privacy

Song Han[1], Elizabeth Chang[1], and Tharam Dillon[2]

[1] School of Information Systems
Curtin Business School
Curtin University of Technology
GPO Box U1987, WA 6845, Australia
song.han, elizabeth.chang@cbs.curtin.edu.au

[2] Faculty of Information Technology
University of Technology Sydney
PO Box 123, Broadway NSW 2007, Australia
tharam@it.uts.edu.au

Abstract. Electronic commerce has pushed and benefitted from the development of intelligent mobile agents technology. One of the reasons is electronic commerce needs remote searching and negotiating between one customer and a number of E-shops. This chapter presents a new secure electronic commerce protocol. We provide the security mechanisms through using a new proxy signature scheme implied in the protocol. The digital age has enabled widespread access to and collection of data. While there are several advantages to ubiquitous access to data, there is also the potential for breaching the privacy of individuals. Therefore, preserving privacy is maintained for both the customer and the E-shops in the proposed e-transactions. In addition, fair privacy is one of the characteristics of the new protocol. The proposed e-transactions use intelligent mobile agents to help customers make buying decisions, as well as providing post-purchase service for customers, and providing post-purchase auditing for e-shops. Therefore, it will improve the successful e-commerce to satisfy customers and increase sales of e-shops.

Keywords: Auditing, Electronic commerce, Fair identifiability, Intelligent Mobile agent, Purchase plan, Privacy.

1 Introduction

With the exploration of Internet, electronic commerce has been one of the everyday life intermediation [9]. Optimal solutions to electronic commerce

S. Han et al.: *Secure E-Transactions Protocol using Intelligent Mobile Agents with Fair Privacy*, Studies in Computational Intelligence (SCI) **37**, 307–326 (2007)
www.springerlink.com

is now pursuing intelligent e-transactions. The intelligent e-transactions will enable organisations to increase online sales, as well as help customers to select and position commodity on their preference.

The mobile agent paradigm has been proposed as a promising solution to facilitate distributed computing over open and heterogeneous networks. Mobility, autonomy, and intelligence are identified as key features of mobile agent systems and enabling characteristics for the next-generation smart electronic commerce on the Internet. However, security and privacy in the mobile agent technology should be settled with caution, since the mobile agents will migrate to the heterogeneous networks for the tasks of transactions.

The reason that the intelligent mobile agent paradigm raises the new security and privacy issues is that it violates the following assumptions: (1) computer programs take actions on behalf of a person who can easily be identified, and who intends these actions to be taken; (2) computer programs are obtained from easily identifiable and generally trusted sources (thus Trojan horses, programs that attack a system by doing something else than what the user intends/expects, are rare); (3) security threats come from attackers running programs with malicious intent, therefore security measures should authenticate the user, and make sure that programs run by this user can only do those things that the user is allowed to do. However, today's computer systems already incorporate a lot of mobility features: emails can contain program code that is automatically executed when opening the email; Web pages can contain Java applets that are executed when viewing the page; Word documents can contain macrocode [1]; computer programs from unknown sources are run by users without any hesitation, and so on.

The aforementioned assumptions have thus certainly become questionable in today's computing environment, and related security problems have already resulted. It is clear that for fully featured mobile agent environments these assumptions are completely violated, thus raising new security issues. Different security aspects of mobile agents can be identified, and have already been studied in the existing works [3, 6, 12, 31]. Mobile agent systems are deployed over standard computer networks. The basic network security aspects are obviously a mobile agent security issue too. Mobile agents should be protected while they are in transit from one host to another host. The communication between agents and users, and between agents themselves, should also be protected. Agents, hosts, and users, as well as nonparticipating entities, could potentially eavesdrop or tamper with the communication, or impersonate participating entities.

Thus the typical cryptographic security services-entity authentication, data authentication (data origin authentication and data integrity), and data confidentiality should be provided. The communication channels should therefore be cryptographically secured. Standard mechanisms are available for this: SSL/TLS at the transport layer, or IPsec at the network layer. An easy solution is for agent platforms to provide this as a service to the agents [5]. Alternatively, it can be established by the agents themselves. The protection of the cryptographic keys that are needed to secure the communications is very important and is clearly dependent on the other mobile agent security issues. Malicious agents should be prevented from stealing a host's private keys. If agents carry their own keys, they should be able to protect them from malicious hosts. The authentication of both agents and hosts is not only an important aspect of the protection of the communication, it is also very relevant with respect to the malicious agents/hosts issue. Authentication is the first step in the process of determining whether an agent/host should be trusted, and thus whether, respectively, the host should execute the agent, or the agent should migrate to that host.

In an open mobile agent system, it is expected that there will be mobile agents with malicious intentions. Other agents and agent platforms should be protected from these malicious agents. As mentioned above, agent authentication allows a host to identify an agent, or in practice the signer of an agent (which can be the author, the owner, or the sender). Agents should run in a sandbox environment in which they have limited privileges, in which they are safely interpreted [24], and in which they are also protected from one another. This is exactly how Java applets are executed in a browser. Sandboxing for mobile agents can thus be provided by Java directly. Thus, to some extent, the Java security model already provides a partial solution for the malicious agents problem. Stronger protection, including resource control, can be built on top of the Java environment. In addition, resource usage control, or allocation, should prevent agents from, for example, flooding a host and denying resources to other agents. In addition to authentication, agents can carry a proof that enables a host to determine whether the agent's code is safe to install and execute. In particular, an easily checkable proof is attached to the code that the agent's execution does not violate the security policy of the receiving system. Not only the code, but also the agent's state is relevant for security. With state appraisal [6], the set of privileges an agent needs is computed depending on the state of the agent. Dangerous state modification can be detected. In some sense this protects against a malicious host that corrupts an agent's state before it migrates to another host. Another point is that groups of hosts should also be protected against malicious agents. This counters low-profile attacks (e.g., resource consuming) that are not detected by individual hosts.

Although the problem of malicious agents seems more or less easy to solve, protecting agents against malicious hosts seems a very difficult, and even impossible, task. To demonstrate the problem of malicious hosts, a model was presented in [13] in which the following attacks by malicious hosts were identified: spying out an agent's code, data, and control flow; manipulation of an agent's code, data, and control flow; masquerading of the host; denial of execution (note that there is also reexecution) of an agent; spying out interaction with other agents; returning wrong results of system calls issued by the agent. An agent is completely under the control of a host, and a malicious host can therefore do almost anything. If and how mobile agents can do secure electronic transactions in these circumstances is the question that is investigated in the following sections. It is important to notice that it is impossible to prevent the actual attacks themselves. However, the purpose of the solutions that are discussed in the remainder of the chapter is to render these attacks useless, or at least to allow for detection.

It is known that users are the root of most security problems. Besides eavesdropping on and tampering with the communication, creating and sending out malicious agents, and running malicious platforms, users can exhibit other undesirable behavior. With respect to electronic transactions, for example, users could deny having sent an agent, or could refuse to honor payments made by an agent. Nonrepudiation and auditing services should therefore be provided. Signing a mobile agent once is not enough, as this only proves who the owner is but not that this owner intended to send it to a host at that particular moment. A legal framework should determine to what extent a user remains responsible for tasks that were delegated to an agent.

We present a mobile agent based scenario for mobile electronic commerce and discuss techniques using mobile agents with trusted third party that have been implemented to provide security in this scenario.

Consider such a scenario: There is a customer **C** who decides to buy flight tickets on the Internet. We look on this decision as a *current purchase plan.* For this purchase plan, **C** defines the *purchase requirement* (e.g. travel line, maximum price of ticket, arrival time, valid period of this purchase, etc.). And then, the customer arranges some mobile agents to search over the Internet (actually these mobile agents migrate to some E-Shops, i.e. some Travel Agents Servers). These E-shops will bid for this *purchase requirement* of the *current purchase plan.* Each E-shop wishes her own bidding will be successfully selected by the Customer as an optimal one. In order to make this *bid* confirmed and accepted by the Customer, each E-Shop will put a *legal signature* on the *bid.* This will not only help the underlying E-Shop improve the possibility of the Customer accepting this bid, but also help the Customer to verify whether this bid is really from the underlying E-Shop, as well as prevent this E-Shop from denying providing this bid for the purchase plan. In addition, other E-shops could not fabricate a valid bid to impersonate the *successful bidder.* Here the successful bidder is the E-shop, whose bid is accepted and paid by the customer. An additional state of this scenario is the underlying E-shop denies either bidding any purchase plan of the customer or receiving any money to her account.

In this chapter we will propose a secure *transaction* protocol to address the above scenario [11]. The underlying techniques are based on a new proxy signature scheme[10]. Han et al have presented another new proposal for secure transactions using mobile agents with agent broker. Their method is based on the concept of undetachable signatures [8].

It should be remarked that the above scenario is essentially different from other situations addressed in the previous papers, for example [14, 15, 17, 19, 23]. In those papers, strong identifiability is maintained. That is to say, anyone can determine the identity of the corresponding E-shop by checking the verification equations. However, the *privacy of the E-shops will be compromised by anyone.* Therefore, this situation is not reasonable in virtual environment, since the rights of identifying the E-shops should be assigned to some reputable entity (e.g. the government, the trusted third party, etc) [28]. Our proposal tries to tackle this issue through introducing a TTP.

Our protocol is designed to protect *financial situations or rights* of the customer. This is maintained through the signature on the bid from the corresponding E-shop. Therefore, it can not only help the Customer to verify whether this bid is really from the underlying E-Shop, but also prevent this E-Shop from denying providing this bid for the purchase plan.

Another issue is some previous solutions have the security flaw [14, 15, 17, 20, 23]. In detail, the customer can forge valid bid on behalf of the E-shops. Consequently, the customer can blackmail the underlying E-shop by the fabricated bid [25]. Then, the financial situation of the E-shops will be spoiled by the customer. In our protocol, the customer is not able to forge any valid bid on behalf of any E-shop involved in the transactions.

However, it will be more reasonable and secure if the signing behavior is taken by the E-shop. This is because the E-shops provide the bids for the purchase plan. Also, the underlying bids will be verified by the customer.

In our protocol, a trusted third party is involved in the transaction. The trusted third party (TTP) in our proposal satisfies two conditions: (1) The trusted third party does exactly what it is expected to do. This means (a) No less than it is supposed to do, so that TTP commits no acts of commission, such as, "Oh, I forgot to lock the door." and (b) No less than it is expected to do, so that TTP commits no acts of commission, such as, "Oh, I accidently made an extra key and gave it to Eve." (2) The trusted third party always adheres to the related security law and policy. Therefore, a reputable bank (or a legal and reputable association, etc.) is able to play the role of the trusted third party [25, 28]. Based on this point, the trusted third party in our proposal is indeed a trusted authority.

The organization of the rest of this chapter is as follows: Section 2 introduces the model of secure transactions using mobile agents with a trusted third party. Section 3 presents a new protocol according to the proposed model. Section 4 provides construction analysis, security analysis, and privacy analysis. The efficiency analysis is presented in section 5. The concluding remarks are provided in section 6.

2 Model of Transactions Using MA with TTP

In this section, we will propose a model for secure transactions using mobile agents (MAs) with the trusted third party (TTP). The motivation to propose this model is the needing of a universal framework for the E-commerce protocols of secure transactions using mobile agents as a mediate. This model integrates the serviceability of a trusted third party.

Model 1 (Model of Transactions Using MA with TTP) There are at least four participants involving in the model. They include: a customer, a trusted third party, an E-shop (at least one E-shop involving), and a mobile agent (at least one mobile agent involving). Besides these participants, there are seven procedures for the proposed model. These procedures deliver the specifications for the electronic commerce protocol using mobile agents with TTP. The followings provide the details for the model.

(1) ***System Setup***: This procedure is a probabilistic polynomial time algorithm [25]. It generates *global parameters* as well as *local parameters* for the participants involving this procedure. To some extent, the *security* of the electronic commerce protocol of transactions mostly depends on the choice of the parameters.

(2) ***Interaction between E-Shop and TTP***: This procedure is a deterministic polynomial time algorithm. It generates the pseudonyms and partial private keys for the E-Shops, who plan to sell goods in the protocol. The E-shops first register themselves to the trusted third party. Thereafter, the trusted third party provides the pseudonyms and partial private keys for the *registered E-shops.*

(3) ***Preparing Mobile Agents***: This procedure is a polynomial time algorithm. It involves the interactions between the customer and its mobile agents. Firstly, the customer has a *purchase plan.* According to this purchase plan, the customer then constructs some purchase requirements. Sequentially, these *purchase requirements* are assigned to different mobile agents. In addition, the customer delegates its signing rights to the mobile agents. And then, the mobile agents deliver the signing rights to some E-shops, who accept these mobile agents. In fact, the E-shops will utilize the signing rights *to bid the purchase* initialised by the customer.

(4) ***Mobile Agents Migrating***: This procedure is a deterministic polynomial time algorithm. In this procedure, the mobile agents are equipped with a *purchase request* (It includes the *purchase requirements* and some ciphertexts of partial secrets). And then, mobile agents migrate to some E-shops. E-shops will first check whether this purchase request is legal. If it is legal, the underlying E-shop will *take part in the transaction.* That is, E-shops will *bid for the purchase requirements.*

(5) ***Processing Transactions***: This is a probabilistic polynomial time algorithm. The underlying E-shop first constructs the bidding key, by which this E-shop is able to make bidding for the purchase request. *The process of bidding for the purchase request* is, in fact, *the process of signing E-shops' bid.* After signing the bid, the underlying E-shops will arrange the mobile agents to return to the customer.

(6) ***Checking Transactions***: This procedure is a deterministic polynomial time algorithm. The customer first checks whether the *returned purchase requirement* is still the one previously delivered by the mobile agents. In addition, the time-stamp is examined whether it is still valid. If the two items are both good, the customer will verify the signature on the bid. If it is legal, and also the bid is an optimal one, the customer will accept this bid. In the end, the customer will arrange to transfer some money into the bank account of the corresponding E-shop.

(7) ***Auditing E-Shop***: This procedure is a probabilistic polynomial time algorithm. This procedure is usually off-line, except that the underlying E-shop does not take its duty in the transaction. One of the possible scenarios is the underlying E-shop withdraws the money (deposited by the customer) from her bank account,

and then denies ever taking part in the transaction. The trusted third part will be responsible for *auditing* the behaviors and financial situation of this E-shop. That is, this E-shop will be identified and then charged.

Remark 1: In the above model, the E-shops are also one type of mobile agents, that represent the real E-shops (for example, some shopping mall servers on the Internet). However, we still denote this kind of mobile agents as the E-shops. For the rest of the paper, all adhere to this symbol regulation.

3 Proposed Protocol for Transactions Using MA with TTP

In this section we will propose a new protocol for transactions using mobile agents with a TTP. The proposed protocol is specified according to the new model given in section 2. Therefore, this protocol includes the following procedures: System Setup, Interaction between E-Shops and TTP, Preparing Mobile Agents, Mobile Agents Migrating, Processing Transactions, and Checking Transactions, as well as Auditing i-th E-Shop. A new proxy signature scheme [10] is implied in the protocol. Its security is based on the security of DSS. Therefore, the proposed electronic commerce protocol has the same security level with the DSS. In addition, fair identifiability as a new security and privacy mechanism is maintained in the protocol.

3.1 System Setup

In this subsection, we will set up the system parameters for the proposed protocol. In the proposed protocol, there are at least four different participants: a Customer, a Trusted Third Party, an E-Shop, and the Mobile Agents (at least one Mobile Agent involving in the underlying transactions). The followings are the specifications of the global parameters as well as the the local parameters:

(1) **Choice of Global Parameters** There is a large prime p. Its bit-length is L, i.e. $2^{L-1} < p < 2^L$; where L is a multiple of 64, and $512 \leq L \leq 1024$. q is another large prime, where q divides $p-1$ and bit-length of q is 160. Let h be a primitive root modulo p $(1 < h < p-1)$ [9]. Set $g = h^{(p-1)/q} \bmod p$. Therefore, q is the order of g modulo p.

(2) **Choosing H()** $H()$ is a SHA hash function [25].

(3) **Private/Public Key Pair of Customer** Choosing a random number x_C, $x_C \in Z_q^*$, and computing $y_C = g^{x_C} \bmod p$. The private key of the Customer is x_C, the public key is y_C.

(4) **Identity of Customer** Denoting ID_C as the identity of the Customer. It is a bit-string that can identify the Customer.

(5) **Identity of E-Shop** Denoting ID_S as the identity of the E-Shop. It is a bit-string that can identify the E-Shop.

(6) **Private/Public Key Pair of TTP** Choosing a random number x_{TTP}, $x_{TTP} \in Z_q^*$, and computing $y_{TTP} = g^{x_{TTP}} \bmod p$. The private key of the Trusted

Third Party is x_{TTP}, the public key is y_{TTP}.

(7) **Identity of TTP** Denoting ID_{TTP} as the identity of the Trusted Third Party (TTP). It is a bit-string that is held by the Trusted Third Party.

3.2 Interaction between E-Shop and TTP

This procedure can be accomplished through "off-line" with respect to the underlying transactions. That is to say, the interaction is processed (by some E-Shops and a Trusted Third Party) earlier than the coming transactions. In this algorithm, the Trusted Third Party will issue a pseudonym and a secret key for each E-Shop (ES_1, ES_2, ..., ES_n), by which the E-Shops can take part in the underlying transactions. The details are the followings:

(1) **Registration** Each E-Shop ES_i $(1 \leq i \leq n)$ registers her/his identity $ID_S^{(i)}$ $(1 \leq i \leq n)$ and a request R_i $(1 \leq i \leq n)$ to the Trusted Third Party, respectively.

(2) **Creating Pseudonym** The Trusted Third Party chooses two different random numbers $k_{TTP_i} \in Z_q^*$ and $k_S^{(i)} \in Z_p$ for each E-Shop ES_i $(1 \leq i \leq n)$, respectively. And he/she then computes r_{TTP_i},

$$r_{TTP_i} = g^{k_{TTP_i}} \bmod p \tag{1}$$

and

$$n_S^{(i)} = H(ID_S^{(i)}, R_i, k_S^{(i)}). \tag{2}$$

The $n_S^{(i)}$ will play the role of the pseudonym for each E-Shop ES_i $(1 \leq i \leq n)$, respectively. From the computation of $n_S^{(i)}$, we know that this pseudonym is linked to the identity of the corresponding E-Shop ES_i $(1 \leq i \leq n)$, respectively.

(3) **Sending Messages** The Trusted Third Party computes $s_{TTP}^{(i)}$,

$$(s_{TTP}^{(i)} = x_{TTP} H(n_S^{(i)}, r_{TTP_i}) + k_{TTP_i}) \bmod q \tag{3}$$

and then sends the tuple $\{n_S^{(i)}, s_{TTP}^{(i)}, r_{TTP_i}\}$ to each E-Shop ES_i $(1 \leq i \leq n)$ through a secure channel, respectively. $s_{TTP}^{(i)}$ will be a partial private key of the corresponding E-Shop.

(4) **Checking Partial Private Key** After each E-Shop ES_i $(1 \leq i \leq n)$ receives the above tuple, she/he will check whether the tuple satisfies the following equation

$$y_{TTP}^{H(n_S^{(i)}, r_{TTP_i})} = g^{s_{TTP}^{(i)}}. \tag{4}$$

If it holds, the E-Shop ES_i $(1 \leq i \leq n)$ will have $s_{TTP}^{(i)}$ as a partial private key, and $n_S^{(i)}$ as the pseudonym which is linked to her/his identity $ID_S^{(i)}$. Therefore, this E-Shop keeps $s_{TTP}^{(i)}$ secret, and makes $n_S^{(i)}$ as well as r_{TTP_i} public. If the equation does not hold, this E-Shop will register to the Trusted Third Party with another request R_i'.

3.3 Preparing Mobile Agents

As soon as the Customer initializes any purchase, she/he will prepare some mobile agents MA_1, MA_2, ..., MA_n and arrange them to some E-Shops ES_i $(1 \le i \le n)$ for the purchase plan; where $n > 1$ is a positive integer. The details are the followings:

(1) **Constructing Public Parameters** The Customer chooses random numbers $k_C^{(1)} \in Z_q^*$, $k_C^{(2)} \in Z_q^*$, ..., $k_C^{(n)} \in Z_q^*$, and computes

$$r_C^{(1)} = g^{k_C^{(1)}} \bmod p, \tag{5}$$

$$r_C^{(2)} = g^{k_C^{(2)}} \bmod p, \tag{6}$$

$$\ldots,$$

$$r_C^{(n)} = g^{k_C^{(n)}} \bmod p. \tag{7}$$

These parameters will be involved in the forthcoming transactions.

(2) **Constructing Purchase Requirements** According to the current purchase plan, the Customer will construct the corresponding purchase requirements. These purchase requirements will be assigned to the corresponding mobile agents in order to seek an optimal transaction. The Customer constructs the purchase requirements as follows:

$$J_1 = Req_C^{(1)}, \tag{8}$$

$$J_2 = Req_C^{(2)}, \tag{9}$$

$$\ldots,$$

$$J_t = Req_C^{(n)}. \tag{10}$$

Since these Mobile Agents are prepared for the same purchase plan, the purchase requirements are all equal, i.e.

$$J_1 = J_2 = \ldots = J_t = J_C, \tag{11}$$

where J_C is defined as the current purchase requirement. It includes: (1) *the description of a desired product*; (2) *an expiration date and time-stamp, that implies the valid purchase period*; (3) *the maximum price that is accepted to the Customer*; (4) *a due date for the delivery of the product*; and (5) *an address for the delivery of the product.*

(3) **Constructing Partial Secrets** The Customer will construct some partial secrets, that will be used by some E-Shop in the forthcoming transactions. The details are the followings: The Customer computes

$$s_C^{(1)} = x_C H(J_1, r_C^{(1)}) + k_C^{(1)}$$

$$= (x_C H(J_C, r_C^{(1)}) + k_C^{(1)}) \bmod q; \tag{12}$$

$$s_C^{(2)} = x_C H(J_2, r_C^{(2)}) + k_C^{(2)}$$

$$= (x_C H(J_C, r_C^{(2)}) + k_C^{(2)}) \bmod q; \tag{13}$$

...

$$s_C^{(n)} = x_C H(J_n, r_C^{(n)}) + k_C^{(n)}$$
$$= (x_C H(J_C, r_C^{(n)}) + k_C^{(n)}) \bmod q. \quad (14)$$

(4) **Equipping Mobile Agents** The Customer will equip these mobile agents with the above public parameters and partial secrets. In detail, the Customer provides each Mobile Agent MA_j $(1 \leq j \leq n)$ with the corresponding tuple

$$\{J_C, E_j(s_C^{(j)}), r_C^j, E_j(ID_C)\}, \quad (15)$$

respectively. Here, $E_j(s_C^{(j)})$ is the ciphertext of the $j-th$ partial private key $s_C^{(j)}$, and $E_j()$ is a specific public key cryptosystem of an E-Shop, to whom the $j-th$ Mobile Agent MA_j will migrate for the purchase plan of the Customer.

3.4 Mobile Agents Migrating

As soon as the Mobile Agents are equipped with the corresponding tuple defined as Equation (15), the different Mobile Agent will migrate to the different E-Shop to search an optimal purchase. Without loss of generality, we may assume that the i-th Mobile Agent MA_i migrates to the i-th E-Shop ES_i, where $1 \leq i \leq n$. The followings are the details:

(1) **Migrating** The i-th Mobile Agent MA_i migrates with the tuple

$$\{J_C, E_i(s_C^{(i)}), r_C^i, E_i(ID_C)\} \quad (16)$$

to the i-th E-Shop ES_i; where $E_i()$ is the public key encryption algorithm of ES_i; and $E_i(s_C^{(i)})$ is the ciphertext of the i-th partial private key $s_C^{(i)}$ under the public key encryption algorithm $E_i()$ of the i-th E-Shop ES_i.

(2) **Checking Time-stamp** After the Mobile Agent MA_i arrives at the ES_i, the E-Shop ES_i gets the tuple

$$\{J_C, E_i(s_C^{(i)}), r_C^i, E_i(ID_C)\} \quad (17)$$

and checks whether the *purchase requirement* J_C is *legal* or not. That is, the i-th E-Shop will examine whether the time-stamp on J_C is valid. If it is not valid, this E-Shop will stop, since this *purchase request* is out of date. If it is valid, this E-Shop will go on the next step.

(3) **Obtaining Partial Secret** After the Mobile Agent MA_i arrives at the ES_i, the E-Shop ES_i gets the tuple $\{J_C, E_i(s_C^{(i)}), r_C^i, E_i(ID_C)\}$ and decrypts $E_i(s_C^{(i)})$ and $E_i(ID_C)$ by using her/his private key corresponding to the encryption algorithm $E_i()$. Consequently, the E-Shop obtains the partial secret $s_C^{(i)}$. She/he will keep $s_C^{(i)}$ secret while making $r_C^{(i)}$ public.

(4) **Checking** The E-Shop ES_i will check whether the partial secret $s_C^{(i)}$ is valid with respect to the corresponding public parameter $r_C^{(i)}$. She/he checks whether

$$y_C^{H(J_C, r_C^{(i)})} r_C^{(i)} = g^{s_C^{(i)}}. \tag{18}$$

If it is not valid, this E-Shop will stop, since the current purchase plan may be spoiled. If it is valid, this E-Shop will take part in the bidding for the purchase plan of the Customer.

3.5 Processing Transactions

In this procedure, the i-th E-Shop will first construct her/his own bidding key s_{bid}, by which this E-Shop can bid for the purchase plan initialised by the Customer. She/he will then construct the bidding of her/his goods to this purchase. And then, the i-th Mobile Agent will be equipped with this bidding and return to its owner, i.e. the Customer. Note that the bidding key is kept secret by this E-Shop. The details of this procedure is as follows:

(1) **Constructing Bidding Key** So far, the i-th E-Shop holds some parameters produced by the Trusted Third Party as well as the Customer. This E-Shop will first computes her/his bidding key as s_{bid},

$$s_{bid} = s_C^{(i)} H(s_{TTP}^{(i)}, ID_c) + s_{TTP}^{(i)} \text{mod } q. \tag{19}$$

And then, she/he computes y_{bid},

$$\begin{aligned} y_{bid} &= g^{s_{bid}} \text{mod } p \\ &= g^{s_C^{(i)} H(s_{TTP}^{(i)})} g^{s_{TTP}^{(i)}} \text{mod } p. \end{aligned} \tag{20}$$

In the end, the i-th E-Shop makes y_{bid} public while keeping s_{bid} secret.

(2) **Proposing the Bid** According to the purchase requirement J_C, the i-th E-Shop proposes the corresponding bid for J_C. This bid is defined as B_{bid}. And B_{bid} includes: (a) the description of the i-th E-Shop's goods; (b) the minimum price that will be acceptable to the i-th E-Shop; (c) a due date for the delivery of the goods; (d) a bank account number provided by the i-th E-Shop; (e) a due date for transferring money into the bank account; (f) an expiration date and time-stamp, that implies the valid period of the bid B_{bid}.

(3) **Signing the Bid** In order to make this bid confirmed and accepted by the Customer, the i-th E-Shop will put a legal signature on the bid B_{bid}. This will not only help the i-th E-Shop improve the possibility of the Customer accepting this bid, but also help the Customer to verify whether this bid is really from the i-th E-Shop, as well as prevent the i-th E-Shop from denying providing this bid for the purchase plan. The details of this procedure is as follows:

- The i-th E-Shop computes m,

$$m = H(B_{bid}, ID_C, n_S^{(i)}); \tag{21}$$

- The i-th E-Shop chooses a random number k, $k \in Z_q^*$, and sets

$$\alpha = (g^k \text{mod } p) \text{mod } q; \tag{22}$$

- The i-th E-Shop computes β,

$$\beta = k^{-1}(H(m, \alpha, n_S^i) + s_{bid}\alpha) \bmod q. \tag{23}$$

Therefore, the signature on the bid B_{bid} is $\{\alpha, \beta\}$.

(4) **Arranging MA to Return** The i-th E-Shop equips the i-th Mobile Agent MA_i with the tuple:

$$B_{bid}, r_C^{(i)}, n_S^{(i)}, ID_C, J_i, \alpha, \beta. \tag{24}$$

This tuple represents the whole transaction. The i-th Mobile Agent then returns to its owner, i.e. the Customer.

3.6 Checking Transactions

As soon as the i-th Mobile Agent returns to the Customer, the Customer first checks the transaction tuple, and then decides whether to accept this bid. The followings are the details:

(1) The Customer first checks whether $J_i = J_C$. If it holds, she/he continues the next steps. Otherwise, she/he will arrange the j-th Mobile Agent MA_j (where $1 \leq j \leq n$ and $j \neq i$) to seek an optimal bid for the current purchase plan.

(2) The Customer computes $r_1 = H(m, n_S^{(i)}, \alpha)\beta^{-1} \bmod q$.

(3) The Customer computes $r_2 = \alpha\beta^{-1} \bmod q$.

(4) The Customer verifies whether the following equation holds

$$(g^{r1} y_{bid}^{r2} \bmod p) \bmod q = \alpha.$$

If it holds, the Customer accepts this bid as valid. If it does not hold, the Customer will arrange the j-th Mobile Agent MA_j (where $1 \leq j \leq n$ and $j \neq i$) to seek an optimal bid for the current purchase plan.

3.7 Auditing i-th E-Shop

The following scenario may take place: After verifying the transaction tuple, the Customer accepts the bid B_{bid} as an optimal bid. Therefore, she transfers some money as the price listed in the bid. However, the i-th E-Shop denies ever receiving any money and sending any bid. How can we deal with this situation? Who will audit the i-th E-Shop? The details given below provides a solution to this scenario.

(1) The Customer sends the tuple $\{\alpha, n_S^{(i)}, \beta\}$ (which is from the whole transaction tuple $B_{bid}, r_C^{(i)}, n_S^{(i)}, ID_C, J_i, \alpha, \beta$.) to the Trusted Third Party.

(2) The Trusted Third Party replies the tuple $\{ID_S^{(i)}, k_S^{(i)}\}$ to the Customer.

(3) The Trusted Third Party audits the i-th E-Shop by using the following equation:

$$H(ID_S^{(i)}, R_i, k_S^{(i)}) = n_S^{(i)}.$$

Since the Trusted Third Party holds $n_S^{(i)}$ and $k_S^{(i)}$, the i-th E-Shop will be identified and audited.

4 Privacy and Security Analysis and Proofs

This chapter has presented a new electronic commerce protocol for transactions using mobile agents. And a trusted third party is involved in the proposed protocol. It is interesting to analyze how the protocol works. Most importantly, security of the protocol should be maintained, since the transactions are initiated over the Internet. And Internet is a site where there exist a number of attacks, from passive attacks to active attacks, from external attacks to internal attacks [25, 28]. Another issue is the financial situation of the participants of the transactions should be treated with caution to certain extent. That is, the privacy of the participants should be preserved, since privacy of the participants is linked to the financial situations of the corresponding participants. Motivated by the above three points, we provide three different analyses and proofs: construction analysis, security analysis, and privacy analysis.

4.1 Construction Analysis

Generally speaking, construction analysis serves as a functional deployment from the construction, operationability, and functioning points of view. This subsection will provide a deployment for the proposed transaction protocol.

In our protocol, we have introduced a customer, a trusted third party, an E-shop, and a number of mobile agents. However, in a virtual electronic commerce environment, there will be more than one customer as well as more than one E-shop. For the complex scenario, it is easy to extend the proposed protocol to a multiple level of electronic commerce transactions protocol. Therefore, in the following we only deploy the protocol from the simple and concise perspective.

(1) ***Role of the Customer***: The customer first proposes a *purchase plan*. Around the purchase plan, she constructs the purchase requirements

$$J_1 = J_2 = \ldots = J_t = J_C, \tag{25}$$

which direct the underlying E-shops to bid for the purchase plan. Here, J_C includes: (a) *the description of a desired product*; (b) *an expiration date and time-stamp, that implies the valid purchase period*; (c) *the maximum price that is accepted to the Customer*; (d) *a due date for the delivery of the product*; and (e) *an address for the delivery of the product.*

Also, the customer constructs mobile codes

$$\{J_C, E_j(s_C^{(j)}), r_C^j, E_j(ID_C)\}, \tag{26}$$

for the mobile agents. Note that a valid signature on the bid includes J_C. That is, J_C *is used to restrict the context of the bidding taken by the E-shops.* Other parts of the mobile code, i.e.

$$\{E_j(s_C^{(j)}), r_C^j, E_j(ID_C)\}$$

is used to generate the bidding key for the E-shops. Another duty of the customer is she will verify the bids

$$B_{bid}, r_C^{(i)}, n_S^{(i)}, ID_C, J_i, \alpha, \beta. \tag{27}$$

returned by the mobile agents. If it is valid, she will transfer some money to the E-shop's bank account.

(2) ***Functioning of the Mobile Agents***: The main duty of the mobile agents is to help its owner accomplish the purchase plan. They actually interact with their owner and the E-shops, respectively. (As noted in Remark 1, We know that the E-shops are also some mobile agents.) For the interaction between the mobile agents and their owner, the mobile agents are equipped with some mobile codes:

$$\{J_C, E_j(s_C^{(j)}), r_C^j, E_j(ID_C)\}, \tag{28}$$

where $1 \leq j \leq n$. For the interaction with the E-shops, the mobile agents transport some bids:

$$B_{bid}, r_C^{(i)}, n_S^{(i)}, ID_C, J_i, \alpha, \beta. \tag{29}$$

(3) ***Role of the TTP***: A trusted third party is involved in the protocol. TTP has two different roles: one is to record the registration of the E-shops, and help the E-shops generate the bidding keys . In order to fulfil this, the TTP sends the tuple

$$\{n_S^{(i)}, s_{TTP}^{(i)}, r_{TTP_i}\}$$

to each E-Shop ES_i $(1 \leq i \leq n)$ through a secure channel, respectively. Here, $s_{TTP}^{(i)}$,

$$s_{TTP}^{(i)} = x_{TTP}H(n_S^{(i)}), r_{TTP_i}) + k_{TTP_i} \bmod q. \tag{30}$$

The other role of TTP is to audit the E-shops during the course of the transactions. This service is accomplished using the following equation:

$$H(ID_S^{(i)}, R_i, k_S^{(i)}) = n_S^{(i)}.$$

(4) ***Role of the E-shops***: The E-shops take part in bidding for the purchase initiated by the customer. Therefore, The E-shops need to have bidding private key and public key: s_{bid} and y_{bid},

$$s_{bid} = s_C^{(i)} H(s_{TTP}^{(i)}, ID_c) + s_{TTP}^{(i)} \text{mod } q. \tag{31}$$

and

$$\begin{aligned} y_{bid} &= g^{s_{bid}} \text{mod } p \\ &= g^{s_C^{(i)} H(s_{TTP}^{(i)})} g^{s_{TTP}^{(i)}} \text{mod } p. \end{aligned} \tag{32}$$

4.2 Security Proofs

This subsection, we will prove that the proposed transaction protocol satisfy the following security properties: (1) strong unforgeability of any bid of the underlying E-shops. This property is valid with respect to the customer. (2) *fair identifiability* of the underlying E-shops. This property is valid with respect to the E-shops. (3) verifiability of the bid of the underlying E-shops. This property is valid with respect to any one who holds the related public parameters. (4) strong undeniability of the bid in the transactions. This property is valid with respect to the E-shops. The details are the followings:

(1) ***Strong unforgeability of any bid of the underlying E-shops.*** This property is valid with respect to the customer. This means that the customer is not able to forge valid any bid on behalf of any underlying E-shop. From Equation (3) and (31), we have

$$s_{bid} = s_C^{(i)} H(s_{TTP}^{(i)}, ID_c) + s_{TTP}^{(I)} \text{mod } q$$

$$= (s_C^{(i)} H(s_{TTP}^{(i)}, ID_c) + x_{TTP} H(n_S^{(i)}), r_{TTP_i}) + k_{TTP_i}) \text{mod } q. \tag{33}$$

It is difficult to figure out the value of s_{bid}, since k_{TTP_i} and x_{TTP} are two random and private elements of Z_q^*. If the customer tries to tackle Equation (20), she will need to solve the discrete logarithm problem [25]. On the other hand, from Equation (22) and (23), the underlying bidding signature is based on the DSS [27]. Therefore, the strong unforgeability is maintained.

(2) ***Fair Identifiability of the underlying E-shops.*** This property is valid with respect to the E-shops. Fair identifiability means that no one is able to identify the underlying E-shop whose bid is accepted by the customer. An exceptional situation is the trusted third party can identify the underlying E-shop through the pseudonym. This only takes place when the E-shop denies ever bidding and receiving money in the transactions. In fact, from the signature generated by the E-shop

$$B_{bid}, r_C^{(i)}, n_S^{(i)}, ID_C, J_i, \alpha, \beta,$$

any one except TTP cannot identify the underlying E-shop. This is because: (a) B_{bid} ={the description of the underlying E-Shop's goods; the minimum price that will be acceptable to the underlying E-Shop; a due date for the delivery of the goods; a bank account number provided by the underlyinig E-Shop; a due date for transferring money into the bank account; an expiration date and time-stamp}. It does not leak any information of the identity for the underlying E-shop. and (b) $n_S^{(i)} = H(ID_S^{(i)}, R_i, k_S^{(i)})$. Therefore, ID_S^i is mapped using a hash function.

(3) ***Verifiability of the bid of the underlying E-shops.*** This property is valid with respect to any one who holds the related public parameters. Verifiability means that any one who holds the related public parameters can check whether a bid is valid. It is easy to conclude this point from the process of Checking Transactions (see Section 3).

(4) ***Undeniability of the bid in the transactions.*** This property is valid with respect to the E-shops. Undeniability means that the underlying E-shop cannot deny she ever generated a valid signature on the bid. In fact, from Equation (29) we know that n_S^i is theoretically linked to this E-shop. More importantly, the verifying equation $(g^{r_1} y_{bid}^{r_2} \bmod p) \bmod q = \alpha$ implies this E-shop ever generated the signature on the bid. This point is derived from the procedure of Processing Transactions as well as Checking Transactions.

4.3 Privacy Analysis

We will prove that the proposed protocol that answer the questions how the privacy is preserved for both the customer and the E-shops. In a virtual community, privacy is imperative with respect to every participant. In fact, it is known that privacy is paramount particularly in respect to financial issues of the participants in the electronics transactions (known as e-transaction or e-business). Therefore, besides the security analysis, it is also necessary to analyze the privacy of the proposed protocol. We will analyze the privacy of the e-transactions protocol from the following four aspects:

Privacy of the identity of the customer

This privacy can be maintained through the encrypted communication. In fact, when the customer sends the mobile agent to some E-shops to seek "optimal purchase", she will encrypt the whole or part of the tuple if necessary for the whole content), by utilizing her private key of the underlying public key encryption algorithm. That is,

$$\{J_C, E_j(s_C^{(j)}), r_C^j, E_j(ID_C)\}.$$

Privacy of the context of the e-Transaction initiated between the customer and an E-shop

This privacy is maintained through the mutual encrypted communications between the customer and the corresponding E-shop, who will utilize the public key encryption algorithm pre-established in advance of the transactions.

$$\{B_{bid}, r_C^{(i)}, n_S^{(i)}, ID_C, J_i, \alpha, \beta\}.$$

If needed, the whole tuple can be encrypted.

Privacy of the identity of the E-shop

This privacy is maintained through the fact: when the E-shop equips the mobile agent with the bid (actually, the signature on the bid)

$$\{B_{bid}, r_C^{(i)}, n_S^{(i)}, ID_C, J_i, \alpha, \beta\}.$$

to migrate to the customer, this tuple only contains a pseudonym $n_S^{(i)}$ of the E-shop. Depending on it, no one can link it to the identity of the E-shop.

Fair Privacy of the E-shop

Fair privacy means that the privacy of the corresponding E-shop can be revealed if this E-shop is not responsible for her duty. This privacy is revealed through the fact:

(1) The Customer sends the tuple $\{\alpha, n_S^{(i)}, \beta\}$ to the Trusted Third Party.

(2) The Trusted Third Party audits the i-th E-Shop by using the following equation:

$$H(ID_S^{(i)}, R_i, k_S^{(i)}) = n_S^{(i)}.$$

Since the Trusted Third Party holds $n_S^{(i)}$ and $k_S^{(i)}$, the i-th E-Shop will be identified and audited.

5 Efficiency Analysis

The performance of the proposed electronic commerce protocol can be discussed from two aspects: off-line workloads and on-line workloads.

The off-line workloads mainly include the computation cost. The procedures of System Setup, Preparing Mobile Agents, and Processing Transactions can be all dealt with through the off-line mode. The computation cost is discussed with respect to one customer with one E-shop. The underlying computation costs are dominated by one modular exponentiation computation, one hash function computation, and two encryption computations for the procedure of System Setup; one modular multiplication, one modular exponentiation, one hash function evaluation for the procedure of Preparing Mobile Agents; one modular exponentiation, one hash function evaluation, one modular multiplication, two modular exponentiations for the procedure of Interaction between E-shops and TTP; one modular exponentiation, one modular inversion, one hash function evaluation, one modular multiplication for the procedure of Processing Transactions.

The on-line workloads mainly include the communication cost and the computation cost. The procedures of Interaction between E-shops and TTP, Mobile Agents Migrating, and Checking Transactions as well as Auditing i-th E-shop can be all dealt with through the on-line mode. We discuss the on-line workloads with respect to one-time successful transaction between the customer and the underlying E-shop. The communication cost is one round of communication between the E-shop and the TTP, one round of communication between the underlying mobile agent and the E-shop (resp. the Customer), and one round of communication between the customer and the TTP. The corresponding computation costs are dominated by two modular exponentiations for the procedure of Mobile Agents Migrating; one modular inversion evaluation, one hash function evaluation, three modular multiplications, two modular exponentiations for the procedure of Checking Transactions; one hash function evaluation for the procedure of Auditing i-th E-shop.

The proposed protocol in [14] used RSA signatures. Therefore, their setting was RSA setting, while ours is discrete logarithm (DL) setting [25]. The online workload of the protocol in [14] was dominated by two hash evaluations and six RSA modular exponentiations, while the online workload of the protocol in this chapter is only dominated by two hash evaluations, four DL modular exponentiations, and one modular inversion. Therefore, the protocol in this chapter is much more efficient. One the other hand, compared with the protocol in [14], the communication workload of the protocol in this chapter is not so efficient, since the latter involves a trusted third party. However, this is acceptable since the latter provides the auditing mechanism. Therefore, fair privacy is maintained. However, the protocol in [14] did not provide such mechanism.

6 Conclusion

This chapter has proposed a new electronic commerce protocol. The proposed protocol integrates mobile agents with the underlying transactions. The mobile agents help to accomplish the purchase plan initiated by the customer. A trusted authority is integrated and plays two important different roles: one is to help the E-shops register; the other is to help to maintain the fair privacy. We have provided proofs and analysis for construction, security and privacy, from construction, security and privacy points of view. In detail, construction analysis is presented. And, this protocol is designed with sound security cautions. That is, the following properties are maintained: strong unforgeability of any bid of the underlying E-shops, fair identifiability of the underlying E-shops, and verifiability of the bid of the underlying E-shops, as well as undeniability of the bid in the transactions. In addition, privacy proofs are provided. The limitation of our research is the trusted third party in the protocol may be compromised by some E-shop. This will decrease the fair privacy. Therefore, our future research will deploy the distributed structure for the mechanism of the trusted third party and do some tests on the new solution.

References

[1] Bryce C, Vitek J (1999) The JavaSeal mobile agent kernel, ASA/MA, 103-117.

[2] Claessens J, Preneel B, Vandewalle J (2003) (How) can mobile agents do secure electronic transactions on untrusted hosts? ACM Trans, Internet Techn; 3(1): 28-48.

[3] Chess DM (1998) Security issues in mobile code systems, In Mobile Agents and Security, G. Vigna, Ed., Lecture Notes in Computer Science, vol. 1419, Springer-Verlag, New York, 1-14.

[4] Eklund E (2006) Controlling and securing personal privacy and anonymity in the information society, http://www.niksula.cs.hut.fi/ eklund/Opinnot/netsec.html.

[5] Claessens J, Preneel B, Vandewalle J (2001) Secure communication for secure agentbased electronic commerce, In E-Commerce Agents: Marketplace Solutions, Security Issues, and Supply and Demand, J. Liu and Y. Ye, Eds., Lecture Notes in Computer Science, vol. 2033, Springer-Verlag, New York, 180-190.

[6] Farmer W, Gutmann J, Swarup V (1996) Security for mobile agents: authentication and state appraisal, Proc. of the European Symposium on Research in Computer Security (ESORICS), LNCS 1146, Springer-Verlag, 118-130.

[7] Edjlali G, Acharya A, Chaudhary V (1998) History-based access control for mobile code, ACM CCCS-98, 413-431.

[8] Han S, Chang E, Dillon T (2005) Secure e-transactions using mobile agents with agent broker, in the Proceedings of the Second IEEE ICSSSM, Jun 13-15, 2005, Chongqing, 849-855.

[9] Graff JC (2001) Cryptograhpy and e-commerce, A Wiley Tech Brief, Wiley Computer Publishing.

[10] Han S, Chang E (2005) A secure strong proxy signature scheme based on DSS, Technical Report, CBS, Curtin University of Technology.

[11] Han S, Chang E, Dillon T (2005) Secure transactions using mobile agents with TTP, in the Proceedings of the Second IEEE ICSSSM, Jun 13-15, 2005, Chongqing, 856-862.

[12] Hassler V (2000) Mobile agent security, In security fundamentals for E-Commerce, Computer Security Series. Artech House, Chapter 20, 331-351.

[13] Hohl F (1998) A model of attacks of malicious hosts against mobile agents, In Proceedings of the fourth ECOOP Workshop on Mobile Oject Systems 1998: Secure Internet Mobile Computation.

[14] Kotzanikolaous P, Burmester M, Chrissikopoulos V (2000) Secure transactions with mobile agents in hostile environments, ACISP 2000, LNCS 1841, Springer-Verlag, 289-297.

[15] Kotzanikolaous P, Katsirelos G, Chrissikopoulos V (1999) Mobile agents for secure electronic transactions, Recent Advances in Signal Processing and Communications, World Scientific and Engineering Society Press, 363-368.
[16] Kim S, Park S, Won D (1997) Proxy signatures, revisited, Proc. of ICICS'97, Y. Han et al(Eds.), LNCS 1334, Springer-Verlag, 223-232.
[17] Lee B, Kim H, Kim K (2001) Secure mobile agent using strong non-designated proxy signature, ACISP 2001, Springer-verlag, LNCS 2119, 474-486.
[18] Lee B, Kim H, Kim K (2001) Strong proxy signature and its applications, Proc. of SCIS2001, 603-608.
[19] Loureio S, Molva R (1999) Privacy for mobile code, Proc. of Distributed Object Security Workshop OOPSLA'99.
[20] Merwe J, Solms SH (1997) Electronic commerce with secure intelligent trade agents, Proc. of ICICS'97, Y. Han et al(Eds.), LNCS 1334, Springer-Verlag, 452-462.
[21] Otomura R, Soshi M, Miyaji A (2001) On digital signature schemes for mobile agents, Proc. of SCIS2001, 851-855.
[22] Petersen H, Horster P (1997) Self-certified keys - concepts and applications, Proc. Communications and Multimedia Security'97, Chapman & Hall, 102 - 116.
[23] Sander T, Tschudin CF (1997) Protecting mobile agents against malicious hosts, Mobile Agent Security, LNCS 1419, Springer-Verlag, 44-60.
[24] Singelee D, Prehneel B (2004) Secure e-commerce using mobile agents on untrusted hosts, COSIC Internal Report.
[25] Menezes A, Oorschot PCV, Vanstone SA (1997) Handbook of applied cryptography, CRC Press, Boca Raton.
[26] Rscheisen M, Winograd T (1997) A network-centric design for relationship-based security and access control, Journal of Computer Security, 5(3): 249-254.
[27] The digital signature standard, NIST, 1994.
[28] Whitman ME, Mattord HJ (2005) Principles of inforamtion security, Second Edition, Thomson Course Technology.
[29] Li CL, Li LY (2003) Integrate software agents and CORBA in computational grid, Computer Standards & Interfaces, 25(4): 357-371.
[30] Paderewski-Rodrguez P, Rodrguez-Fortiz MJ, Parets-Llorca J (2003) An architecture for dynamic and evolving cooperative software agents, Computer Standards & Interfaces, 25(3): 261-269.
[31] Tschudin CF (1999) Mobile agent security, In Intelligent Information Agents: Agent-Based Information Discovery and Management on the Internet, M. Klusch, Ed., Springer-Verlag, New York, Chapter 18, 431-446.

Trust and Reputation in E-Services: Concepts, Models and Applications

Javier Carbo[1], Jesus Garcia[2], and Jose M. Molina[3]

[1] Group of Applied Artificial Intelligence (GIAA), Computer Science Dept., Universidad Carlos III de Madrid jcarbo@inf.uc3m.es
[2] jgarcia@inf.uc3m.es
[3] molina@ia.uc3m.es

Abstract. Trust is critical for e-services, since the offline nature of relationships seems to weaken the social control of face-to-face interactions. So, more trust-related problems (such as deceptions and frauds) tend to emerge while providing e-services. The aim of the chapter is to contribute to a better understanding of trust and reputation in e-services: what these terms mean, the intelligent technology that can be applied, the different models proposed, and some useful scenarios to show the utility of trust and reputation in e-services.

1 Concept of Trust and Reputation in e-services

1.1 Introduction

Most e-services assume that a secure and reliable communications (including contracts and signatures) is enough to assure trust. But trust is more than secure communication, e.g., via public key cryptography techniques. For example, the reliability of information about the status of your trade partner has little to do with secure communication. With the growing impact of electronic societies, a broader concept of trust become more and more important.

Trust is a universal concept, it plays a very important role in social organizations, since it works as a mechanism of social control: it makes that a society produces global benefits from self-interested parts. This social order comes from the common and dynamic nature of the environment. A common environment creates interdependence at the level of payoffs, and interferences at the level of action plans and their executions. A dynamic environment provides unpredictable modifications of goals and needs of the parts.

In spite of the very different definitions [22], we can state that trust is an abstract property applied to others that helps to reduce the complexity of decisions that have to be taken in the presence of many risks. And on the other hand reputation is a concrete valuation that we use to build trust in others. So they are strongly linked, but there are (at least) some differences between our perception of both concepts.

J. Carbo et al.: *Trust and Reputation in E-Services: Concepts, Models and Applications*, Studies in Computational Intelligence (SCI) **37**, 327–345 (2007)
www.springerlink.com

Until recently, they were applicable only to human societies and therefore were a study field for sociologists, philosophers and psychologists. The emergence of electronic services add a new dimension to these old but very important concepts [2].

The scientific research in the area of trust and reputation mechanisms for virtual societies is a recent discipline oriented to increase the reliability and performance of electronic services by introducing in electronic communities these well known human social control mechanisms. Another growing trend is the use of reputation mechanisms, and in particular the interesting link between trust and reputation. Many computational and theoretical models and approaches to reputation have been developed in the last few years.

In artificial intelligence, a computational model of trust involves a cognitive approach [25]: modeling opponents to support cooperations and competitions. Therefore trust is then made up of underlying beliefs (among others, these beliefs include the reputation of the others), and trust is a vague function of the degree of these beliefs. Trust is then the result of a mental process in a cognitive sense. On the other hand a computational model of reputation involves a numerical approach, made up of utility functions, probabilities and evaluations of past interactions. The combination of both computational models intends to reproduce the reasoning mechanisms behind human decision-making.

1.2 Nature and Sources of Trust

There two opposite approaches to the problem of how to trust in others [22]: the emergent and the designed. Designed trust is inferred from explicit norms and social institutions observe the compliance of such norms. This is the view of most of the commercial online computational models of trust that consist of a central entity that certifies the satisfaction of some given evaluation criteria. Trust is then a global property shared by all the observers. This centralized nature of reputation is due to the size of these scenarios that makes repeated interactions between the same parts very improbable.

On the other hand, distributed models tackle with emergent trust where no objective evaluation criteria are universally accepted and the possibility of repeated meetings is not so low [15]. Unlike designed models of trust, with subjective evaluations, reputation emerges from a spontaneous process. This does not mean that distributed models of trust have no norms. They have them, but the are implicit, evolutive and spontaneous.

Such subjective evaluations are very common in real life and humans are used to apply them everyday. In fact, human trust is hybrid, it arise from institutional and spontaneous processes, so distributed models are not an alternative to centralized models, they are complementary. In this way, computerized human-like notion of trust requires also a distributed approach, with no certification authorities. But since different particular views of the world may coexist (as many as parts belonging to the system), and malicious intentions may also be present, a certain level of cooperation is naturally required. So, electronic providers and consumers cooperate to improve their own decisions if they believe that they share common interests and likes. And this possibility of forming cooperative clusters introduce the discussion about what are the sources of information to compute reputation, and finally build trust.

Direct experiences is supposed to be the most reliable source of information since it is the experience based on the direct interaction with the other side, although a certain level of noise can also be assumed. But we can count also on a indirect source: the witness information, also called word-of-mouth. It comes from recommendations of third parties. Since these third parties may have different subjective evaluation criteria, or they even lie intentionally, this source of reputation is surrounded of a higher uncertainty level and therefore it can not be considered as completely transitive. It would be needed overall in low-density populations, in order to reduce the level of deceptions -difference between predictions and observations- produced from the direct interaction with other agents [9].

2 Intelligent Technology Applied

2.1 Agents

The concept of an agent can be traced back to the early days of research into DAI in the 1970s. Broadly, the research on agents can be split into two main trends: the first works on intelligent agents, focused on deliberative-type agents with symbolic internal models. In this way, the classic paradigm of reasoning applied to this type of agents is based in three levels of knowledge: beliefs, desires and intentions [21].

On the other hand, there has evidently been another distinct line of research on software agents. The range of agent types being investigated is now much broader and it includes nearly every computerized entity that acts with certain autonomy.

In any case, research in agents has received an exponential growth, and most of it is strongly linked with AI foundations, since agents often had the intention to model human reasoning and to act in behalf of humans. Therefore agent systems can be studied as a society of humans (this is the so called social metaphor of agents). So, in summary, we call agents to programs with ability to represent humans, to act autonomously and to show intelligence.

Agents have been extensively applied to reputation domain due to the distributed nature of trust that we described before. Nearly all the researchers of this community assumes that their reputation models are integrated in the internal reasoning of an agent [6]. Indeed, the most important forum dedicated to trust and reputation is called *International Workshop on Trust and Reputation* takes places jointly with the *Int. Joint Conf. on Autonomous Agents and MultiAgent Systems, AAMAS*. But although researchers consider reputation models as part of agents research, most of them did not approach to agents from the deliberative strand of agents as our group of Applied Artificial Intelligence do.

2.2 Fuzzy Sets

The concepts formed in human brains for perceiving, recognizing and categorizing natural phenomena are often fuzzy concepts because the boundaries of these concepts are vague. The classifying, judging and reasoning emerging from them also are fuzzy concepts. The human brain has the incredible ability of processing fuzzy classification, fuzzy judgement and fuzzy reasoning. In fact, natural languages are full of inherent fuzziness that allow the expression of very rich information with a few words. On the other hand, classical mathematics require precise and crisp definitions to express phenomena. Overcoming such difference is the goal of fuzzy sets.

Fuzzy sets were first proposed by Zadeh in 1965. It generalizes the classical two-valued logic for reasoning under uncertainty. It appies to concepts with a not-well defined natual boundaries. So the truth of them become a matter of degree. Fuzzy sets assign a truth-value in the range [0,1] to each possible value of the domain. These values form a possibility distribution over a continuous or discrete space. This distribution represents a membership function μ: a value x of the domain will belong to the fuzzy concept in a degree defined by $\mu(x)$. Then, if we considered a continuous space of possible values [0,120], we could represent graphically the possibility distributions of terms like 'young'. And therefore we will be able to compute how much truth there is in the statement that 'a person of 35 years is young'. In order to answer this question, we will search for the value returned by the membership function μ corresponding to the possibility distribution of the fuzzy term young. This value will probably do not match with absolute logical values: 1 for true, 0 for false. The other values between them define how much young is 35-years old person. This representation of the concept is closer to human interpretation in order to allow a gradual transition from 'young' to 'not young' statements.

Fuzzy logic discipline has developed a complete mathematical theory (operations, properties, relationships, theorems) to compute and reason with such fuzzy sets.

Our research group (GIAA) is a pioneer in proposing a fuzzy definition of reputation concepts.

2.3 Adaptive filters

When agents are acting in uncertain environments, their perceptions often include some level of noise. Furthermore, such agents have to reason about that noisy environment will probably evolve. Making such time-dependent predictions is not an easy task. Adaptive filters are an useful tool to infer the next future of noisy environments. They apply a temporal statistical model to the noisy observations perceived through a linear recursive algorithm that estimate future state variable. They have been recognized as a reasoning paradigm for time-variable facts, within the Artificial Intelligence community [18]. Two of the most representative adaptive filters are Alpha Beta and Kalman:

- Alpha Beta assumes that the state variable (reputation) follows a constant velocity model, with some uncertainty characterized by a parameterized random variable (plant-noise model): starting with some initial value, the reputation's velocity evolves through time by process noise of random accelerations, constant during each sampling interval but independent. Without any noise, reputation would have constant velocity, so we are using noise to model sudden changes of behaviour (in other words, reputations with a non-constant velocity). Alpha beta also assumes that observations only are available, subject to measurement noise of constant covariance. Clearly the more is known a priori about the motion the better predictions will be, so in our application of reputation with agents this could be considered a bit of hack since noise has in fact a constant variance, but it is not a realistic assumption to know a priori the real model of the noise.

- The Kalman filter assumes certain linear stochastic models for the state dynamics and observation processes, so that it would achieve the optimum estimator (in the Bayesian sense of Minimum Squared Error) under those conditions. It has been extensively applied to different fields, outstanding the tracking systems based on sensor data [4].

Particularly, when they are applied as a reputation models, the state variable would be the reputation, while observations would be the results from direct experiences. Our research group (GIAA) is a pioneer in proposing adaptive filters as reputation models.

3 Models of reputation

3.1 Representative models from academic community

On the commercial side, the pioneer models computed reputation in simple ways, for instance: sum of ratings (eBay), and average sum of ratings (Onsale Exchange). But for some time now, academic community has generated many alternatives involving more complex computations that are extensively described in [1] and that most of them come from the Artificial Intelligence discipline and are inspired in the factors that humans are supposed to apply in real-life interactions. At least it is supposed that from the underlying model of reputation should emerge the intended intelligent behaviour of agents.

Marsh [25] proposed one of the earliest academic reputation models. It classify trust as general and situational, but it only takes into account direct interactions. On the other hand Schillo [26] proposed a model where interactions were qualified in a boolean way, so no degrees were applied to the direct experiences with other agents. Additionally Abdul-Rahman and Hailes [2] uses four possible values: very trustworthy, trustworthy, untrustworthy and very untrustworthy, but it intends only to evaluate the trust on the information given by witness and no more. Many other reputation models have been proposed in the dedicated academic forums (over all, the workshop on trust in agent societies of AAMAS Conferences). But from all the academic AI-inspired reputation models that were proposed, we will describe the foundations of four of the most representative models of reputation.

SPORAS and HISTOS

P. Maes and other researchers of the Massachussets Institute of Technology (M.I.T.) proposed two reputation algorithms: SPORAS and HISTOS [11]. SPORAS is inspired in the foundations of the chess players evaluation system called ELOS. The main point of this model is that trusted agents with very high reputation experience much smaller changes in reputation than agents with low reputation. SPORAS also computes the reliability of agents' reputation using the standard deviation of such measure.

On the other hand, HISTOS is designed to complement SPORAS including a way to deal with witness information (personal recommendations). HISTOS includes witness information as source of reputation through a recursive computation of weighted means of ratings. It computes reputation of agent i for agent j from the knowledge of all the chain of reputation beliefs corresponding to each possible path that connects agent i and agent j. It also plans to limit the length of paths that are taken into account. To make fair comparison with other proposals, that limit should be valued as 1, since most of the other views consider that agents communicate only its own beliefs (that are obviously the result from direct experiences and the recomendations), but not the beliefs of other sources that contributed to the own belief of reputation.

Based on these principles, the reputation value of a given agent at iteration i, R_i, is obtained recursively from the previous one R_{i-1} and from the subjective evaluation of the direct experience DE_i:

$$R_i = R_{i-1} + \frac{1}{\theta} \cdot \Phi(R_{i-1}) \cdot (DE_i - R_{i-1}) \tag{1}$$

Let θ be the effective number of ratings taken into account in an evaluation ($\theta > 1$). The bigger the number of considered ratings, the smaller the change in reputation is.

Furthermore, Φ stands for a damping function that slows down the changes for very reputable users:

$$\Phi(R_{i-1}) = 1 - \frac{1}{1 + e^{\frac{-(R_{i-1} - Max)}{\sigma}}} \tag{2}$$

Where dominion D is the maximum possible reputation value and σ is chosen in a way that the resulting Φ would remain above 0.9 when reputations values were below $\frac{3}{4}$ of D.

REGRET

REGRET, from the Spanish Artificial Intelligence Research Institute [8], takes into account three types of computation of indirect reputation depending on the information source: system, neighbourhood and witness reputations. From them witness reputation is the one that corresponds to the concept of reputation that we are considering. REGRET includes a measure of the social credibility of the agent and a measure of the credibility of the information in the computation of witness reputation. The first of them is computed from the social relations shared between both agents. It is computed in a similar way to neighbourhood reputation, and it uses third parties references about the recommender directly in the computation of how its recommendations are taking into account.

This view is different in the process followed to compute reputation. Often it is assumed that if recommender agent is not trusted enough, it will never be asked, but even then, the reputation of the recommender (asking whoever for references) is computed first, and then, as a different action, the influence of the recommendation in the reputation of a provider is computed. Mixing both operations in one seems to not clarify the process and it does not make any difference with the general computation of reputation with recommendations that we explained before.

On the other hand, the second measure of credibility (information credibility) is computed from the difference between the recommendation and what the agent experienced by itself. The similarity is computed matching this difference with a triangle fuzzy set centered in 0 (the value 0 stands for no difference at all). The information credibility is considered as relevant and taken into account in the experiments of this present comparison.

Both decisions are also, in some way, supported by the authors of REGRET, who also assume that the accuracy of previous pieces of information (witness) are much more reliable than the credibility based on social relations (neighbourhood), and they reduce the use of neighbourhood reputation to those situations were there is not enough information on witness reputation. The complete mathematical expression of both measures can be found in [1]. But the main point of REGRET is that it also considers the role that social relationships may play. It provides a degree of reliability for the reputation values, and it adapts them through the inclusion of a temporal dependent function in computations. The time dependent function ρ gives higher relevance to direct experiences produced at times closer to current time. The reputation held by any part at a iteration i is computed from a weighted mean of the corresponding last θ direct experiences. The general equation is of the form:

$$R_i = \sum_{j=i-\theta}^{j=i} \rho(i,j) \cdot W_j \tag{3}$$

Where $\rho(i,j)$ is a normalized value calculated from the next weight function:

$$\rho(i,j) = \frac{f(j,i)}{\sum_{k=i-\theta}^{k=i} f(k,i)} \tag{4}$$

Where $i \geq j$. Both of them represent the time or number of iteration of a direct experience. For instance, a simple example of a time dependent function f is:

$$f(j,i) = \frac{j}{i} \tag{5}$$

REGRET also computes reliability with the standard deviation of reputation values computed from:

$$STD - DVT_i = 1 - \sum_{j=i-\theta}^{j=i} \rho(i,j) \cdot \mid W_j - R_i \mid \tag{6}$$

But REGRET defines reliability as a convex combination of this deviation with a measure, $0 < NI < 1$, of whether the number of impressions, i, obtained is enough or not. REGRET establishes an intimate level of interactions, itm, to represent a minimum threshold of experiences to obtain close relationships. More interactions will not increase reliability. The next function models the level of intimate with a given agent:

$$if(i \in [0, itm]) \rightarrow NI = \sin(\frac{\pi}{2 \cdot itm} \cdot i), Otherwise \rightarrow NI = 1 \tag{7}$$

A Proposal from Singh and Yu

This trust model [10] uses Dempster-Shafer theory of evidence to aggregate recommendations from different witnesses. The main characteristic of this model is the relative importance of fails over success. It assumes that deceptions (noted as β, and valued negatively) causes stronger impressions than satisfactions (noted as α, and valued positively). So mathematically: $|\beta| > \alpha \geq 0$.

It then applies different gradients to the curves of gaining/losing reputation in order to lose easily reputation, while it is hard to acquire it. The authors of this trust model define different equations to the sign (positive/negative) of the received direct experience (satisfaction/deception) and the sign of the previous reputation corresponding to the given agent.
So in the case that both of them had the same sign, then the new reputation would take the form of the next equation:

$$R_j = \frac{DE_i + R_{i-1}}{1 - min|DE_i|, |R_{i-1}|} \tag{8}$$

In the case that those variables had different sign, then the corresponding opinion would be computed from equation 7:

$$R_j = reputation_{i-1} + DE_i \cdot (1 + R_{i-1}) \tag{9}$$

3.2 Reputation models from the Group of Applied Artificial Intelligence

A Fuzzy Reputation Agent System (AFRAS)

The agent system called AFRAS, proposed by the authors of this chapter [20], intends to integrate different features into agents with BDI architecture that implements emotive agents through the adaptive characterization of agents with sociability, susceptibility and shyness attributes (valued with fuzzy sets) [14].

The architecture adopted by these agents has three different levels of abstraction. They deal with world, social and mental models. These layers are design in such a way that each layer is bottom-up activated, and top-down executed. The former means that the higher is a layer, more complex and abstract are its competencies. The latter means that each layer uses the predicates generated by the lower layer.

The bottom level models the outside world updating reputation beliefs and it also makes binary decisions based on trust inferred from reputation rates posed by the higher layer.

The decisions that are responsibility of world model layer are the next ones: whether or not asking for references to a given agent, answering a question, and buying a certain product. These decisions are taken according to demanding levels of reputation required to ask, answer and buy. These levels represent the egoism, sociability, and susceptibility hold by agents. An agent would be acting socially, if it was avoiding isolation, and therefore, it should give more chance to newcomers and unknown agents. Egoism decides whether to execute or not the intention to answer a question posed by other agent rather than execute or not the intention to ask other agent. Susceptibility represents how much suspicious is any agent when it is taking costly decisions.

The intermediate level of abstraction, called social layer, deals with the sequence of interactions involved in a service. It controls the current different parallel courses of actions (communications) in which an agent is committed for each service with other agents. It will ask next layer to evaluate certain agents in order to ask/answer them or to buy from them. The world model level will activate this layer when the former received a expected reference or the satisfaction provided by a service, after the computation of corresponding updating of reputation rate involved is performed.

The top one, called mental level, involves the adaptation of some agent attitudes that characterise its behaviour. We have called them before shyness, egoism, susceptibility and remembrance. Their values will affect the future decisions that the two other layers will make. The social model level of abstraction would activate this control layer when the results of a service evaluation are available. These mental attitudes would change according to the evolution of the outcomes obtained by the execution of sequential services. The responsibility of this control layer also includes an artificial (and random) generation of user needs.

Additionally the other main characteristic of agent model is the extensive use of fuzzy sets, particularly to represent reputation values. The application of this formalism to reputation makes sense since human opinions about others are vague, subjective and uncertain (in other words, reputation can be a fuzzy concept, valued in fuzzy terms).

In this reputation model, direct experiences and witness information are both considered. They are aggregated through a weighted mean of fuzzy sets. Aggregation of fuzzy sets is computed with Mass Assignment assumptions based on Baldwin's theory of evidence [3] . In the case of direct experiences, weights depend on a single attribute that represents the *memory* of the agent ($0 < memory < 1$). We associated such meaning to this value because it determines the importance of past direct experiences (R_{i-1}) over a new one (DE).

$$R_i = R_{i-1} + \frac{(DE_i - R_{i-1}) \cdot (1 - memory)}{2} \tag{10}$$

It is computed as a function of the overlapping between two fuzzy sets that represents the level of success of last prediction. If the satisfaction provided by a partner was similar to the reputation estimation assigned to such partner, the relevance of past experiences (memory) would be then increased. On the other hand, if they were different, is the relevance of the last experience what would be therefore increased (the corresponding agent is 'forgetting' past experiences, 'losing' memory).

Once similarity is computed in this way, an average sum of previous memory $memory_{i-1}$ with similarity $SIM(R_{i-1}, DE_i)$ is applied to obtain $memory_i$:

$$memory_i = \frac{memory_{i-1} + SIM(R_{i-1}, DE_i)}{2} \tag{11}$$

This equation follows the next simple principles: If the prediction fitted well the rating ($SIM \approx 1$) then memory (the importance given to the past reputation over the last direct experience) would increase in $1/2 + memory/2$. On the other hand, when they were not similar at all ($SIM \approx 0$), memory would become useless, and its relevance in the next estimations of reputation would be halved.

These properties avoid *memory* being below zero and above one. The initial value of *memory* associated to any agent joining the system should be minimum (zero), although it would be soon increased when there was any success in the estimations of reputation.

Reliability of reputation values is modeled through the fuzzy sets themselves. It is implicit in them, graphically we can interpret the gradient of the sides of a trapezium representing a fuzzy reputation as its reliability. A wide fuzzy set representing a given reputation represents a high degree of uncertainty over that reputation estimation, while a narrow fuzzy set implies a reliable reputation.

Recommendations are aggregated directly with direct experiences in a similar way (weighted sum of fuzzy sets). But in this case, the weight given to each part (recommendation vs. previous opinion) is dependent on the reputation of the recommender. A recommendation would (at most) count as much as a direct experience if the recommender had the highest reputation.
Finally, to update the reputation of recommenders, an agent computes similarity (level of overlapping between the corresponding fuzzy sets) with afterwards results of the direct experience with the recommended partner. Then, reputation of recommender would be increased or decreased accordingly.

Alpha-Beta filter

Alpha Beta method [4] is an adaptive filter that tries to estimate an unknown state variable from noisy observations as we explained in the section *Intelligent Technologies.* When it is applied as a reputation model, the state variable would be the reputation, while observations would be the results from direct experiences.
So state variable (reputation) evolves following the next equation:

$$x(t+\Delta t) = F(\Delta t)x(t) + q(\Delta t) \tag{12}$$

where Δt is the time delay between last update and current observation, t_k- t_{k-1}, $F(\Delta t)$ is the transition matrix and q(t) is characterized by its covariance matrix, $Q(\Delta t)$.
Since we assume a constant velocity model of reputation, transition matrix F adopts the next value:

$$F(\Delta t) = \begin{bmatrix} 1 & \Delta t \\ 0 & 1 \end{bmatrix} \tag{13}$$

And observations result from a linear operation on state variable (reputation) corrupted by additional noise:

$$z(t) = Hx(t) + n(t) \tag{14}$$

being n(t) a random variable with covariance given by matrix R.
In our specific model for filtering reputation, we have a dynamic linear system, with vector $\widehat{x}[k|k]$ containing both the trust estimate and its time derivative for a given agent (the notation $(k|k)$ means estimation at time k, considering observations until time k, while $(k|k-1)$ is the prediction at time k from last update at time k-1).
Under this context, the equations for Alpha Beta filter to compute the best estimation for x(t) are the following:

$$\widehat{x}(k+1|k+1) = \widehat{x}(k+1|k) + \begin{bmatrix} \alpha \\ \frac{\beta}{\Delta t} \end{bmatrix} \cdot [z(k+1) - (\widehat{z}(k+1|k)] \tag{15}$$

So the state estimate is a weighted sum of a state $\widehat{x}(k+1|k)$ predicted from the last estimate to be $F(\Delta t)x(k|k)$ and innovation, computed as the difference between a predicted observation,$\widehat{z}(k+1|k)$, with the current observation, z(k+1).

We can compute the value of β from α in order to use just α as the single parameter of the estimation method:

$$\beta = 2 \cdot (2-\alpha) - 4 \cdot \sqrt{1-\alpha} \tag{16}$$

The values of α are between 0 and 1 and represent a balance between the relevance given to the history of past observations vs. the last observation. Therefore, $\alpha = 0$ would mean that the last observation has no effect in next prediction. On the other hand, $\alpha = 1$ would mean that the history of past observations were ignored in next prediction.
Estimates for covariances Q and R are 4x4 matrices. Usually, the exact models for dynamics and observation are not known, so the design for a given application is a trade-off to adjust the parameters. Matrix R is usually adjusted from observed data variability (sample variance), while matrix Q is tuned to achieve satisfactory balance between noise filtering (when the prediction model is much better than observation noise) and reactions to sudden changes (intervals while the model fails to accurately predict the state variable).

The plant noise variance, Q, has been set to 20 and observations are received evenly spaced with an uniform time interval of Δt=1. So, Q covariance matrix is computed as follows [4] :

$$Q(1) = 20 \begin{bmatrix} \frac{1}{4} & \frac{1}{2} \\ \frac{1}{2} & 1 \end{bmatrix} \tag{17}$$

The application of AlphaBeta filter as a reputation model was proposed and tested in [24]

Kalman filter

The stochastic models assumed by the algorithm can be summarized in the following equations:

- The state variable has a linear behaviour, with a certain uncertainty characterized by a parameterized random variable (plant-noise model):

$$x(t + \Delta t) = F(\Delta t)x(t) + q(\Delta t) \tag{18}$$

  where Δt is the time delay between last update and current observation, tk-tk-1, $F(\Delta t)$ is the transition matrix and q(t) is characterized by its covariance matrix, $Q(\Delta t)$.
- Observations result from a linear operation on state variable corrupted by additional noise:

$$z(t) = Hx(t) + n(t) \tag{19}$$

  being n(t) a random variable with covariance given by matrix R.

Under these models, the equations for Kalman filter to compute the best estimation for x(t) are the following:

- Prediction

$$\widehat{x}[k|k-1] = F(\Delta t) \cdot \widehat{x}[k-1] \tag{20}$$

$$P[k|k-1] = F(\Delta t) \cdot P[k-1] \cdot (F(\Delta t))^t + Q(\Delta t) \tag{21}$$

- Updating

$$K[k] = P[k] \cdot H^t \cdot (R[k] + H \cdot P[k] \cdot H^t)^{-1} \tag{22}$$

$$\widehat{x}[k|k] = \widehat{x}[k|k-1] + K[k] \cdot (z[k] - H \cdot \widehat{x}[k|k-1]) \tag{23}$$

$$P[k|k] = P[k-1] \cdot (I - H^t \cdot K[k]) \tag{24}$$

In our specific model for filtering reputation, we have a first-order dynamic linear system, with vector $\widehat{x}[k|k]$ containing both the trust estimate and its time derivative for a certain agent. So, estimates for covariances are 4x4 matrices. Usually, the exact models for dynamics and observation are not known, so the design for a certain application is a trade-off to adjust the parameters. Matrix R is usually adjusted from observed data variability (sample variance), while matrix Q is tuned to achieve satisfactory balance between noise filtering (when the prediction model is much better than observation noise) and reactions to sudden changes (intervals while the model fails to accurately predict the state variable).

This model has been extended to aggregate recommendations from third parties to our own estimation, and also to update the reputation of recommenders with the final result of direct experience. The first case is computed as a weighted combination of the recommendation and available estimation, where weights (W) are the inverse of the estimations variances in order to derive the minimum-variance combination. These values are directly computed in the Kalman filter, noted before as P[0][0].

So, assuming that these values are also provided by recommenders, together with their own trust values, we would have the following equations to integrate the recommender reputation. For instance, agent y trust (noted as $y[k|k-1]$), is combined with the own estimation (noted as $x'[k|k-1]$) as follows:

$$Wx = 1/Px[0][0] \tag{25}$$

$$Wy = 1/Py[0][0] \tag{26}$$

$$\widehat{x}[k|k] = (Wx * \widehat{x'}[k|k-1] + Wy * \widehat{y}[k|k-1])/(Wx + Wy) \tag{27}$$

However, we have to consider also that the reputation of the recommender (agent y) may diverge from our estimated reputation. This may be directly included as an increase of recommender's covariance, taking into account the differences between the two estimated vectors:

$$P'y = Py + (\widehat{x'}[k|k-1] - \widehat{y}[k|k-1])(\widehat{x'}[k|k-1] - \widehat{y}[k|k-1])^t \tag{28}$$

$$Wy = 1/P'y[0][0] \tag{29}$$

In this way, if the reputation estimated by agent y systematically diverges from the estimation of agent x, its weight is reduced in the final combination.

On the other hand, we had to update the reputation of the recommender after the interaction took place, and considering then the level of success achieved in the estimation from the recommender. Therefore we use a Kalman formulation analogous to the one of above, but the observations are no longer the direct experiences, they should be now the differences between the estimation of the recommender and the noisy observation.

Regarding the implementation details, the plant noise variance, q, has been set to 20 and observations are received evenly spaced with an uniform time interval of Δt=1. So, Q covariance matrix is computed as follows [4] :

$$Q(1) = 20 \begin{bmatrix} \frac{1}{4} & \frac{1}{2} \\ \frac{1}{2} & 1 \end{bmatrix} \tag{30}$$

Besides, the Kalman filter estimates are initialized with the first two observed interactions (it is a difference with respect to other approaches, the estimates must be always inferred from observations). Assuming that z[0], z[1] are available, the initial vector estimate is:

$$\hat{x}[1|1] = \begin{bmatrix} z[1] \\ z[1] - z[0] \end{bmatrix} \tag{31}$$

The covariance matrix for this initial estimator is computed assuming an uniform distribution of initial trust. Since the variable margins are [0, 100], we will have:

$$P[1|1] = E\left\{\hat{x}[1|1](\hat{x}[1|1])^t\right\} = \begin{bmatrix} Var(z) & Var(z) \\ Var(z) & 2 * Var(z) \end{bmatrix} = \frac{(100-0)^2}{12}\begin{bmatrix} 1 & 1 \\ 1 & 2 \end{bmatrix} \tag{32}$$

The application of Kalman filter as a reputation model was proposed and tested in [12] [23]

4 Testing reputation models

4.1 Adhoc tests from the Group of Applied Artificial Intelligence

Virtually every author runs a different set of tests to show the properties of his/her model and to establish comparisons with other models. Although there is a test-bed in progress that could fill that gap, here we analyze first a representative group of these adhoc tests that our research group applied to compare the reputation models that we proposed. They consists of the evaluation of four desirable abilities:

- Convergence of reputation
- Reactivity to sudden changes
- Contribution of cooperation
- Influence of fraudulent collusions

Agents may adopt three possible roles: consumers, providers and recommenders. Each reputation model is implanted (following the BDI paradigm) as a consumer and several recommenders. On the other hand provides are modeled as reactive agents that respond to the request of services with a value that represents the satisfaction generated by the provided service. Particularly we ran the tests with 2 recommenders for each consumer agent, and 10 provider agents along 200 iterations. Although changing that values, other slightly different scenarios can be obtained, the comparison of the results produced by the reputation algorithms show more or less equivalent relative positions.

The setting of our group of tests shares the next general properties:

- The consumer agents and the recommenders are constantly consuming services from any of the provider agents.
- Each consumer agent selects providers in the same (randomly generated) order.
- The tests measure the average square error committed by the predictions of consumer agents.
- Each of the providers behaves in a dynamic way

Convergence of reputation

The first type of tests, convergence of reputation, measures how fast the model makes reputation estimation becomes closer to the real behaviour of providers and how much accuracy has the final model of the provider. Particularly, the setting of the experiments to test the convergence of reputation estimations is as follows:

- Each of the provider agents has assigned a prefixed behaviour (uniformly distributed). And all of them use such prefixed behavior along all the services provided to generate a satisfaction value.
- The satisfaction value produced by each particular service at any time is drawn from a normal distribution. The mean of that distribution is equal to the prefixed behaviour of the provider, and standard deviation is fixed for each test. We considered one of three different standard deviation to be applied in a complete simulation for all the providers:
  - a very low value (3.33 over 100) to model a nearly constant behaviour of all providers.
  - a medium value (10 over 100) to model a variable behaviour of all providers.
  - a high value (33 over 100) to model a behaviour of all providers very difficult to predict.
- Initially the reputation of all providers may be undervalued with a very low valuation (10 over 100) or overvalued with a very high initial valuation (90 over 100) with a reliability of 1 (over 100). This allows the possibility of proving an ascendant gradient of estimations as well as a descendant one.

Then figures may show the evolution of the average error produced in estimations. They show how all the reputation models converge, but with different velocities, and with different accuracy. Both values are used to compare the reputation models. The corresponding figures of the complete comparison of reputation models using this type of tests is published in [5].

Reaction facing sudden changes

The second kind of experiments, reaction facing sudden changes, evaluates the possible advantage that any agent could obtain from past right behaviour (in other words, abuse of previously acquired good reputation). So this simulation consists of two stages. First, one provider behaves reliably until its services reach a high reputation estimation. Second, that provider begins abusing of its prior reputation providing very bad services. The rest of the providers behave in the same way as in the simulation that test the convergence of reputation. They have also assigned the same initial reputation values and standard deviation values.
In this type of tests, we change the value of an additional parameter: when such malicious provider agent would change the quality of its services:

- after the first third of its interactions.
- after the second third of its interactions.

Assuming such definition of the scenario, figures may show the evolution of the average error with a peak around the second third of interactions that corresponds to the change of behaviour of the provider. The corresponding figures of the complete comparison of reputation models using this type of tests is published in [5]

Contribution of cooperation

The third type of tests measure the contribution of cooperation. Now recommender agents are involved in the simulation, and the particular feature of this test is that the consumer agents always ask to their recommenders (those who implement the same reputation model) about the next provider to request a service. The recommendations will be weighted according to the corresponding reputation model to update the estimations about providers.
In this type of experiments we apply different values to a new setting parameter:

- The accumulated experience of recommender agents before consumer agents join the system. The recommender agents are consuming services from all the provider agents from the beginning, but consumer agents may join the system at the same time or after the 50% of the total interactions of the simulation. Then we can distinguish two scenarios where cooperation has different reliability:
  - We will note as 'sound cooperation' to the experiments where the consumer agent receives reliable referrals, since recommender agents has some previous service experience.
  - And we will also note as 'blind cooperation' to the experiments where the consumer agent has the same level of service experience than the recommenders.

Then we can define 12 different scenarios combining the type of cooperation (sound/blind), the initial reputation (overvalued/undervalued) and the variability of providers (low/medium/high).

With such kind of simulations, we can observe the evolution of the average error of estimations as we did with the previous types of tests, but now we can also measure the improvement produced by the recommendations. Percentages would show how cooperation contributes to deception avoidance in those 12 situations compared with similar situations but without cooperation. Such table of percentages and the corresponding figures of the complete comparison of reputation models using this type of tests is published in [13]

Collusion of providers and recommenders

Here the initial model of the other agents to interact with, corresponds only to a very low reputation (10 over 100). Therefore the evolution of the reputation of providers before the collusion adopts just an ascending gradient (because initial predictions about the behaviour of providers (always > 10) were undervalued).

The collusion of a recommender and a provider consists of a sudden change of behaviour of one particularly good provider to a very low value of its services (20 over 100) while the referrals about such provider from one of the recommenders are intentionally raised to a high value (80 over 100).
In this type of experiments in order to produce several testing scenarios we apply the different values to the parameter of the 'contribution of cooperation' experiments, but also to a new setting parameter:

- The moment when the provider changes its behaviour. The malicious provider starts the execution behaving just as the references from the colluding recommender say.

- We will note as 'early collusion' to situations where the provider agent starts giving bad service after the first third of the total number of services).
- On the other hand, 'late collusion' occurs after the two first thirds of the total number of services.

With such new parameter, we can produce 12 different scenarios combining the type of cooperation (sound/blind), the type of collusion (early/late) and the variability of providers (low/medium/high).

Here again we can observe the evolution of reputation, but it would be more interesting a table that shows the total amount of deception and the amount of deception caused exclusively by the false raised recommendations (computed as the percentage of increase of deception compared to a equivalent situation without recommenders). Such table and the corresponding figures of the complete comparison of reputation models using this type of tests is published in [19]

4.2 Comparison among reputation models

From these four set of simulations, we can outline a comparison of AFRAS, REGRET, SPORAS and the proposal from Singh-Yu:

- Convergence of reputation. When agents act with moderate and little variability (changes between iterations) REGRET estimations converge faster to the real behaviour of the other agents, and even in the long run produces more accurate predictions. The other models converge more or less with equal velocity and accuracy among them. But when agents are less predictable (high variability of behaviour), AFRAS estimations improve the accuracy of REGRET ones.
- Reaction facing sudden changes. When such sudden change takes place early, AFRAS suffer more deception (error in estimations) than REGRET, but less than the other ones. On the other hand, a late sudden change makes AFRAS obtain similar deception than the others.
- Contribution of cooperation. The improvement obtained with cooperation is higher with AFRAS than with the other reputation models when variability of providers is moderate and high, but in stable scenarios (variability low) all of them obtain similar results.
- Collusion of providers and recommenders. When the collusion takes place once reputation estimations are soundly based in direct experiences, the proposal of Singh-Yu obtains better results than the others, but when such collusion takes place in a early phase of the simulation, AFRAS obtains better results than the others.

So we can conclude that no reputation model is globally better than the others, since the very different scenarios faced makes different models to be the more accurate. But some of them are more suitable to certain characteristic of each scenario. Even considering the limited scope of these four simulations, there is no perfect reputation model, so considering all possible scenarios, we think that this general reputation model does not exist.

5 Future research problems

The area of reputation models in agent societies is relatively recent, and therefore there is still enough work to be done. Next we show a short list of topics that should be tackled in the next future:

- As we said in section 4, authors of reputation models run different set of tests in order to stablish comparisons with other models. Furthermore, as the number of reputation models is increasing quickly, a common method of evaluation is required. Based on that requirement, an international team of researchers has been formed in 2005 with the task of establishing a testbed for agent trust- and reputation-related technologies. Our research group has recently joined this initiative. It consists of a testbed specification in an artwork appraisal domain. It will be used to establish fair comparisons among reputation models through the international competition that will take place in AAMAS 2006 Conference [16]. This testbed was designed to:
  - to be accepted by the community members.
  - to be able to evaluate reputation models of all kind complexity.
  - to reduce the time and effort required to adapt the models to the testbed
  - to be open, in other words, that it should be easy to add new tests to the original testbed.
- An interesting aspect that has not been tacked yet is to consider the impact of the own actions in order to improve/maintain its reputation. It is also important to analyze not only the reputation of other agents, but what our own agent may do to improve/maintian its reputation.
- Another area that requires further study are cognitive models. Although most of the most recent reputation models are just a combination of mathematical elements, probably the best reputation model would be a hibrid model such as AFRAS intend to be. More effort should be driven in that direction.
- In the same line, reputation models should consider how they can integrate not only direct experiences and indirect information, but also reputation estimations computed by central entity (system reputation).
- Finally reputation models should be strongly linked to negotiation process that would take place afterwards. Researcher should consider how reputation influence the negotiation approachs to obtain satisfactory agreements. Reputation estimation should be taken into account during the negotiation, since it should be used to decide the acceptance/withdrawal point.

References

1. Sabater, J.: Trust and Reputation for Agent Societies. Monograph of the Spanish Research Institute on Artificial Intelligence (IIIA) **20**, Consell Superior d'investigacions cientifiques ISBN: 84-00-08157-9 (2003).
2. Abdul-Rahman, A.,Hailes, S.: Supporting trust in virtual communities. Procs. of 33th IEEE Int. Conf. on Systems Sciences (Maui, 2000).
3. Baldwin, J.F.: A calculus for mass assignments in evidential reasoning. Advances in Dempster-Shafer Theory of Evidence, eds. M. Fedrizzi, J. Kacprzyk and R.R. Yager (John Wiley, 1992).

4. Bar-Shalom, Y., Xiao-Rong, L.: Estimation and tracking principles, techniques and software. (Artech House, 1991).
5. Carbo, J., Molina, J.M., Davila, J.: Trust management through fuzzy reputation. Int. Journal of Cooperative Information Systems **12**(1) (2003) 135–155.
6. Castellfranchi, C., Falcone, R.: Principles of trust for multiagent systems: Cognitive anatomy, social importance and quantification. Procs. of the 3rd Int. Conf. on Multi-Agent Systems (1998) 72–79.
7. Gelb, A.: Applied Optimal Estimation. (MIT press, Cambridge Mass, 1974).
8. Sabater, J., Sierra, C.: Reputation and social network analysis in multiagent systems. Procs. of the 1st Int. Joint Conf. on Autonomous Agents and Multiagent Systems (Bologna, 2002) 475–482.
9. Shardanand, U., Maes, P.: Social information filtering: algorithms for automatic word of mouth. Procs of the ACM Conf. on Human Factors in Computing Systems (1995) 210–217.
10. Yu B., Singh, M.P.: A social mechanism for reputation management in electronic communities. Lecture Notes in Computer Science **1860** (2000) 154–165.
11. Zacharia, G., Maes, P.: Trust Management through Reputation Mechanisms, Applied Artificial Intelligence **14** (2000) 881–907.
12. Carbo, J., Garcia, J. and Molina, J.M.: Subjective Trust inferred by Kalman filtering vs. a fuzzy reputation, 23rd Lecture Notes in Computer Science 3289 (2004) 496–505.
13. Carbo, J., Molina. J.M. and Davila, J.: Fuzzy referral based cooperation in social networks of agents, Artificial Intelligence Communications **18**(1) (2005), 1–13.
14. Carbo, J., Molina. J.M. and Davila, J.: A BDI agent architecture for reasoning about reputation, Procs. of the 2001 IEEE Int. Conf. on Systems, Man and Cybernetics, (Arizona, 2001).
15. Jsang, A., Ismail, R., and Boyd, C.: A Survey of Trust and Reputation Systems for Online Service Provision (to appear). Decision Support Systems, (2005).
16. Fullam, K., Klos, T., Muller, G., Sabater, J., Schlosser, A., Topol, Z., Barber, S. Rosenschein, J., Vercouter, L. and Voss, M.: Trusting in Agent Societies Competition Game Rules (v 1.0), http://www.lips.utexas.edu/ kfullam/competition/ (2005).
17. Braubach, L., Pokahr, A., Moldt, D., Lamersdorf, W.: Goal Representation for BDI Agent Systems, 2nd Int. Workshop on Programming Multiagent Systems, 3rd Int. Joint Conference on Autonomous Agents and Multi-Agent Systems (New York, 2004) 9–20.
18. Russell, S. J., Norvig P.: Artificial intelligence: a modern approach. 2nd Edition (Prentice Hall Pearson Education International, 2003)
19. Carbo, J., Molina. J.M. and Davila, J.: Avoiding Malicious Agents in E-Commerce using Fuzzy Recommendations. Int. Journal of Organizational Computing and Electronic Commerce, in press.
20. J. Carbo, J.M. Molina and J. Davila, A fuzzy model of reputation in multi-agent systems, in: *Procs. of the 5th Int. Conf. on Autonomous Agents*, Montreal, 2001.
21. A.S. Rao and M.P. Georgeff, Modeling rational agents within a BDI architecture. In Procs. 2nd Int. Conf. on Principles of Knowledge Representation and Reasoning, R. Fikes and E. Sandewall Eds., Morgan Kauffman (1991) 473–484.
22. Conte R., Paolucci M.: Reputation in Artificial Societies. Kluwer Academic Publishers, 2002.

23. Carbo, J., Garcia J. and Molina, J.M.: Contribution of Referrals in a Reputation Model based on Kalman Filtering vs. Fuzzy Sets. Int. Workshop on Trust and Reputation, Utrecht, 2005.
24. Carbo, J., Garcia J. and Molina, J.M.: Convergence of agent reputation with Alpha-Beta filtering vs. a fuzzy system. International Conference on Computational Intelligence for Modelling Control and Automation, Wienn, 2005.
25. Marsh, S.: Trust in distributed artificial intelligence, Lecture Notes in Artificial Intelligence 830, eds. Castelfranchi and Werner, (Springer Verlag, 1994) 94–112.
26. Schillo, M., Funk, P. and Rovatsos, M.: Using trust for detecting deceitful agents in artificial societies. Applied Artificial Intelligence, Special Issue on Trust, Deception and Fraud in Agent Societies.

An Incremental Technique for Analyzing User Behaviors in an E-Business Environment

Yue-Shi Lee, Show-Jane Yen, and Min-Chi Hsieh

Dept. of Computer Science and Info. Engineering, Ming Chuan University
5 The-Ming Rd., Gwei Shan District, Taoyuan 333, Taiwan, R.O.C.
{leeys,sjyen}@mcu.edu.tw

Abstract

Web traversal pattern mining discovers most of the users' access patterns from web logs. However, the web data will grow rapidly in the short time, and some of the web data may be antiquated. The user behaviors may be changed when the new web data is inserted into and the old web data is deleted from web logs. Therefore, we must re-discover the user behaviors from the updated web logs. However, it is very time-consuming to re-find the users' access patterns. Hence, many researchers pay attention to the incremental mining in recent years. The essence of *incremental mining* is that it utilizes the previous mining results and finds new patterns just from the inserted or deleted part of the web logs such that the mining time can be reduced. In this chapter, we propose an efficient incremental web traversal pattern mining algorithm. The experimental results show that our algorithm is more efficient than the other approaches.

1 Introduction

With the trend of the information technology, huge amounts of data would be easily produced and collected from the *electronic commerce* environment every day. It causes the web data in the database to grow up at amazing speed. Consequently, how should we obtain the useful information and knowledge efficiently based on the huge amounts of web data has already been the important issue at present.

Web mining [1, 2, 3, 5, 6, 9, 10, 13, 14] refers to extracting useful information and knowledge from large amounts of web data, which can be

Y.-S. Lee et al.: *An Incremental Technique for Analyzing User Behaviors in an E-Business Environment*, Studies in Computational Intelligence (SCI) **37**, 347–364 (2007)
www.springerlink.com

used to improve the *web services*. Mining *web traversal patterns* [5, 6, 12, 13] is to discover most of users' access patterns from web logs. These patterns can not only be used to improve the website design (e.g., provide efficient access between highly correlated objects, and better authoring design for web pages, etc.), but also be able to lead to better marketing decisions (e.g., putting advertisements in proper places, better customer classification, and behavior analysis, etc.).

In the following, we describe the definitions about web traversal patterns: Let $I = \{x_1, x_2, \ldots, x_n\}$ be a set of all web pages in a website. A *traversal sequence* $S = <w_1, w_2, \ldots, w_m>$ $(w_i \in I, 1 \leq i \leq m)$ is a list of web pages which is ordered by traversal time, and each web page can repeatedly appear in a traversal sequence. The *length* of a traversal sequence S is the total number of web pages in S. A traversal sequence with length l is called an *l-traversal sequence*. Suppose that there are two traversal sequences $\alpha = <a_1, a_2, \ldots, a_m>$ and $\beta = <b_1, b_2, \ldots, b_n>$ $(m \leq n)$, if there exists $i_1 < i_2 < \ldots < i_m$, such that $b_{i1} = a_1$, $b_{i2} = a_2$, $\ldots b_{im} = a_m$, then β *contains* α, α is a *sub-sequence* of β, and β is a *super-sequence* of α. For instance, if there are two traversal sequences α = <*BEA*> and β = <*ABCEA*>, then α is a sub-sequence of β and β is a super-sequence of α.

Table 1: Traversal sequence database

TID	User sequence
1	ABCED
2	ABCD
3	CDEAD
4	CDEAB
5	CDAB
6	ABDC

A *traversal sequence database D*, as shown in Table 1, contains a set of records. Each record includes *traversal identifier (TID)* and a *user sequence*. A user sequence is a traversal sequence, which stands for a complete browsing behavior by a user. The *support* of a traversal sequence α is the ratio of user sequences which contains α to the total number of user sequences in D. It is usually denoted as *Support* (α). The *support count* of α is the number of user sequences which contain α. For a traversal sequence $<x_1, x_2, \ldots, x_l>$, if there is a link from x_i to x_{i+1} (for all i, $1 \leq i \leq l-1$) in the web site structure, then the traversal sequence is a *qualified traversal sequence*. A traversal sequence α is a *web traversal pattern* if α is a qualified traversal sequence and *Support* (α) $\geq$ *min_sup*, in which the *min_sup* is the user specified *minimum support* threshold. For instance, in Table 1, if we set *min_sup* to 80%, then *Support* (<*AB*>) = 4/5 = 80% $\geq$ *min_sup* = 80%, and there is a link from "A" to "B" in the web site structure shown in Figure 1. Hence, <*AB*> is a web traversal pattern. If the length of a web traversal pattern is *l*, then it can be called an *l-web traversal pattern*.

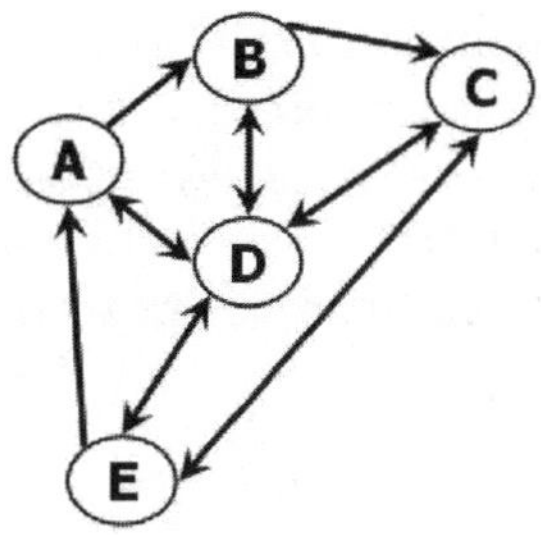

Figure 1: Web site structure

However, the user sequences will grow rapidly and some of the user sequences may be antiquated. The web traversal patterns will change from time to time. In order to keep the recent user behaviors, we need to discover the up-to-date web traversal patterns from the recent traversal sequence database. In order to keep the recent traversal sequence database, some old user sequences need to be deleted from the traversal sequence database, and the new user sequences need to be added into the traversal sequence database. For example, if a new movie "Star Wars" is coming, in a DVD movies selling web site, the users may rent or buy the new movie from the web site. Hence, the users may change their interests to the science-fiction movie. That is, the user behaviors may be changed with time. Therefore, if we do not re-discover the web traversal patterns from updated database, some of the new information (about science-fiction movie) will be lost. Hence, the mining tasks for discovering the web traversal patterns have to be redone periodically from the updated database in order to obtain the up-to-date web traversal patterns. However, the mining tasks are very time consuming since the traversal sequence database is very huge. Hence, how to keep the correct user behaviors in a short time is an important task.

In this chapter, we propose a novel *incremental web traversal pattern mining* algorithm *IncWTP* to re-discover all the web traversal patterns when the database is updated. *IncWTP* algorithm utilizes the previous mining results and finds the up-to-date web traversal patterns when the old user sequences are deleted and new user sequences are deleted from the database such that the mining time can be reduced. Therefore, how to choose a well storage structure to store previous mining results becomes very important. In this chapter, lattice structure is selected as our storage structure. Not only utilizes the previous mining results, we also use the web site structure to reduce mining time and space.

The rest of this chapter is organized as follows. Section 2 introduces the most recent researches related to this work. Section 3 describes our web traversal pattern mining algorithm *IncWTP* and storage structure. Because our approach is the first work on the maintenance of web traversal patterns, we evaluate our algorithm by comparing with web traversal pattern mining algorithm *MFTP* [13] in section 4. Finally, we conclude our work and present some future researches in section 5.

2 Related Work

Path traversal pattern mining [1, 3, 5, 6, 9, 13, 14] is the technique that find navigation behaviors for most of the users in the web environment. The web site designer can use this information to improve the web site design, and to increase the web site performance. Many researches focused on this field, e.g., *FS (Full Scan)* algorithm, *SS (Selective Scan)* algorithm [3], and *MAFTP (Maintenance of Frequent Traversal Patterns)* algorithm [14], etc. Nevertheless, these algorithms have the limitations that they can only discover the *simple path traversal pattern*, i.e., a page cannot repeat in the pattern. These algorithms just consider the forward references in the traversal sequence database. Hence, the simple path traversal patterns discovered by the above algorithms are not fit in the web environment.

MAFTP algorithm [14] is an incremental updating technique to maintain the discovered path traversal patterns when the user sequences are inserted into the database. *MAFTP* algorithm partitions the database into some segments and scans the database segment by segment. For each segment scan, the candidate traversal sequences that cannot be frequent traversal sequences can be pruned and the frequent traversal sequences can be found out earlier. However, this algorithm has the limitations that only simple path traversal patterns are discovered, i.e., the backward references do not be considered. *MAFTP* algorithm just considers the forward references in the traversal sequence database. Furthermore, *MAFTP* algorithm just considers the inserted user sequences and it cannot deal with the deleted user sequences. Our approach can discover the non-simple path traversal patterns and both database insertion and deletion are considered. Besides, only small number of the candidate traversal sequences needs to be counted from the original traversal sequence database for our algorithm.

Non-simple path traversal pattern, i.e., web traversal pattern, contains not only forward references but also backward references. This information can present user navigation behaviors completely and correctly. The related researches are *MFTP (Mining Frequent Traversal Patterns)* algorithm [13], *IPA (Integrating Path traversal patterns and Association rules)* algorithm [5, 6], and *FS-Miner* algorithm [10]. *MFTP* algorithm can discover web traversal patterns from traversal sequence database. This algorithm considers not only forward references, but also backward references. Unfortunately, *MFTP* algorithm must rediscover web traversal patterns from entire database when the database is updated. *IPA* algorithm can not only discover web traversal patterns, but also user purchase behavior. It also considers the web site structure to avoid generating un-qualified traversal sequences. Nevertheless, *IPA* algorithm does not consider incremental and interactive situations. It must rediscover web traversal patterns from entire database when the database is updated. Our approach can discover the web traversal patterns and both database insertions and deletions are also considered. Besides, our approach can use the previously discovered information to avoid re-mining entire database when the minimum support is changed.

FS-Miner algorithm can discover web traversal patterns from traversal sequence database. *FS-Miner* algorithm scans database twice to build a *FS-tree* (frequent sequences tree structure), and then it discovers web traversal patterns from the FS-tree. However, the FS-tree may be too large to fit into memory. Besides, *FS-Miner* finds the consecutive reference sequences traversed by a sufficient number of users, that is, they just consider the consecutive reference sub-sequences of the user sequences. However, there may be some noises which exist in a user sequence, that is, some pages in a user sequence may be not the pages that the user really wants to visit. If all sub-sequences for a user sequence are considered, then FS-Miner cannot work. Hence, some important web traversal patterns may be lost for *FS-Miner* algorithm.

Sequential pattern mining [4, 7, 8, 11, 15] is also similar to web traversal pattern mining, they discover sequential patterns from *customer sequence database*. The most difference between web traversal pattern and sequential pattern is that web traversal pattern considers the link between two web pages in the web structure, that is, there must be a link from each page to the next page in a web traversal pattern. Zaki et al. [11] proposed an incremental sequential pattern mining algorithm *ISL (Incremental Sequence Lattice Algorithm)*. This algorithm is based on *SPADE*

(Sequential Pattern Discovery Using Equivalence Classes) algorithm [15], *ISL* algorithm updates the lattice structure when database is updated. The lattice structure keeps all the sequential patterns, and candidate sequences and their support counts, such that just new generated candidate sequences need to be counted from the original database and the mining efficiency can be improved. The candidate sequences whose support count is 0 are also kept in the lattice. It will cause the lattice structure too huge to fit into memory. The other incremental sequential pattern mining algorithm is *IncSpan (Incremental Mining in Sequential Pattern)* which was proposed by J. Han, etc. [4]. This algorithm is based on *PrefixSpan (Prefix-Projected Sequential Pattern Mining)* algorithm [7, 8]. *IncSpan* uses the concept of *projected-database* to recursively mine the sequential patterns. However, *ISL* and *IncSpan* algorithms cannot deal with the situation that when the new user sequences are inserted into customer sequence database. They just considered inserting the transactions into the original user sequences. Because the user sequences will grow up at any time in the web environment, our work focuses on mining web traversal patterns when the user sequences are inserted into and deleted from the traversal sequence database. Besides, *ISL* and IncSpan algorithms are applied on mining sequential patterns. Our work need to consider the web site structure to avoid finding unqualified traversal sequences. For this reason, we can not apply these two algorithms on mining web traversal patterns.

3 Algorithm for Incremental Web Traversal Pattern Mining

In order to mine the web traversal patterns incrementally, we use the previous mining results to discover new patterns such that the mining time can be reduced. Therefore, how to choose a well storage structure to store previous mining results becomes very important. In this chapter, lattice structure is selected to keep the previous mining results. Figure 2 shows the simple lattice structure *O* for the database described in Table 1, when *min_sup* is set to 50%. In the lattice structure *O*, only web traversal patterns are stored in this structure. To incrementally mine the web traversal patterns and speed up the mining processes, we extend the lattice structure *O* to record more information. The extended lattice structure *E* is shown in Figure 3. In Figure 3, each node contains a traversal sequence whose support count is no less than 1. We append the support information into the upper part of each node. This information can help us to calculate and accumulate the support when the incremental mining is proceeding.

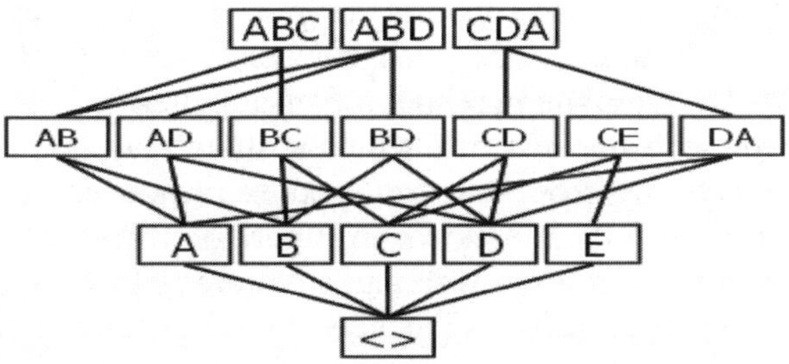

Figure 2: Simple lattice structure

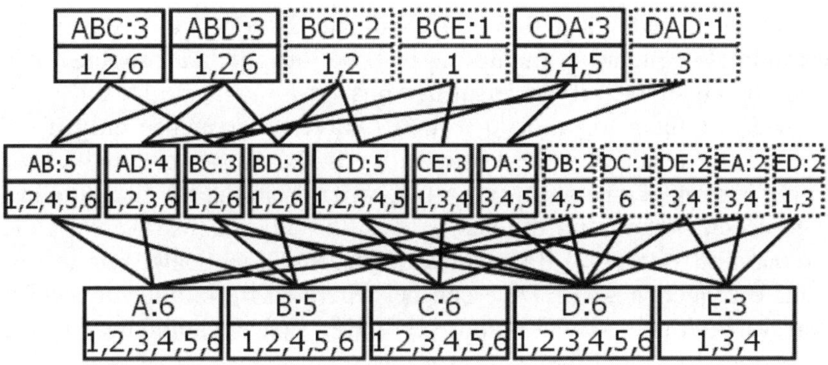

Figure 3: Extended lattice structure

Moreover, we also append the TID information, in which the traversal sequence occurs, into the lower part of each node. This information can help us to reduce the unnecessary database scans. Different from the simple lattice structure, we put all candidate traversal sequences, whose support counts are greater than or equal to one, into the lattice structure. The lattice structure is saved in hard disk level-by-level. The total size of the lattice structure (including TID information) is about 5 times larger than the original database in average. Because our method is to discover the patterns level-by-level, it will not cause the memory be broken when we just load one level of lattice structure into memory.

The lattice structure is a well storage structure. It can quickly find the relationships between patterns. For example, if we want to search for the patterns related to web page "A", we can just traverse the lattice structure from the node "A". Moreover, if we want to find the *maximal web traversal patterns* which are not sub-sequences of the other web traversal patterns, we just need to traverse the lattice structure once and output the patterns in top nodes, whose supports are greater than or equal to *min_sup*. For example, in Figure 3, the web traversal patterns <CE>, <ABC>, <ABD> and <CDA> are the maximal traversal patterns.

We utilize the web site structure which is shown in Figure 1 to mine the web traversal patterns from the traversal sequence database shown in Table 1. The final results are shown in Figure 3 when the *min_sup* set to 50%. The reason for using the web site structure is that we want to avoid the unqualified web traversal sequences to be generated in the mining process. For example, assume that our web site has 300 web pages and all of them are all 1-traversal patterns. If we do not refer to the web site structure, then 299×300=89,700 candidate 2-sequences can be generated. However, in most situations, most of them are unqualified. Assume that the average out-degree for a node is 10. If we refer to the web site structure, then just 300×10=3,000 candidate 2-sequences are generated. The candidate generation method is like the join method proposed in [3]. For any two distinct web traversal patterns , say $<s_1, \ldots, s_{k-1}>$ and $<u_1, \ldots, u_{k-1}>$, we join them together to form a k-traversal sequence only if either $<s_2, \ldots, s_{k-1}>$ exactly is the same with $<u_1, \ldots, u_{k-2}>$ or $<u_2, \ldots, u_{k-1}>$ exactly the same with $<s_1, \ldots, s_{k-2}>$ (i.e., after dropping the fist page in one web traversal pattern and the last page in the other web traversal pattern, the remaining two (k-2)-traversal sequence are identical). For example, candidate sequence <ABCDE> can be generated by joining the two web traversal patterns <ABCD> and <BCDE>. For a candidate l-traversal sequence α, if a qualified length (l-1) sub-sequence of α is not a web traversal pattern, then α must not be web traversal pattern and α can be pruned. Hence, we also check all of the qualified web traversal sub-sequences with length l-1 to reduce some unnecessary combinations. In this example, we need to check if <ABDE> and <ABCE> are the web traversal patterns. If one of them is not a web traversal pattern, <ABCDE> is also not a web traversal pattern. We do not need to check <ACDE>, because <ACDE> is an unqualified web traversal sequence (no link from A to C).

Our algorithm *IncWTP* mines the web traversal patterns from the first level to the last leveling the lattice structure. For each level k ($k \geq 1$), the k-web traversal patterns are generated. There are three main steps in each level k: In the first step, the deleted user sequences' TIDs are deleted from each node of the kth level and the support count of the node is decreased if the node contains the TID of the deleted user sequence.

In the second step, we deal with the inserted user sequences. For each inserted user sequence u, we decompose u into several traversal sequences with length k, that is, all the length k sub-sequences of the user sequence u

are generated. According to the web site structure, the unqualified traversal sequences can be pruned. For each qualified k-traversal sequence s, if s has been contained in a node of the lattice structure, then we just increase the support count of this node and add TID of user sequence u to the node. Otherwise, if all the qualified length (k-1) sub-sequences of s are web traversal patterns, then a new node n_s which contains traversal sequence s and the TID of user sequence u is generated in the kth level. The links between the nodes which contain the qualified length (k-1) sub-sequences of s in the (k-1)th level and the new node n_s are created in the lattice structure. After processing inserted and deleted user sequences, all the k-web traversal patterns can be generated. If the support count of a node is equal to 0, then the node and all the links related to the node can be deleted from the lattice structure. If the support of a node is less than *min_sup*, then all the links between the node and the nodes N in the (k+1)th level are deleted, and the nodes in N are marked. Hence, in the kth level, if a node has been marked, then this node and the links between this node and the nodes in the (k+1)th level are also deleted.

In the last step, the candidate (k+1)-traversal sequences will be generated. The new web traversal patterns in level k can be joined by themselves to generate new candidate (k+1)-traversal sequences. Besides, the original web traversal patterns in level k are also joined with the new web traversal patterns to generate new candidate (k+1)-traversal sequences. The original k-web traversal patterns need not be joined each other, because they are joined before. After generating the new candidate (k+1)-traversal sequences, the original database needs to be scanned to obtain the original support count and the TID information for each new candidate (k+1)-traversal sequence c. The new node n_c which contains c is created and inserted into the lattice structure. The links between the nodes which contain the qualified length k sub-sequences of c in the kth level and the new node n_c are created in the lattice structure. If there is no web traversal patterns generated, then the mining process terminates.

Our incremental mining algorithm *IncWTP* is shown in Algorithm 1 which is the c++ like algorithm. Algorithm 2 shows the function CandidateGen, which generates and processes the candidate traversal sequences. In Algorithm 1, D denotes the traversal sequence database, W denotes the web site structure, L denotes the lattice structure, s denotes the *min_sup*, *NewWTP* denotes new web traversal patterns, *OriWTP* denotes original web traversal patterns, *InsTID* denotes the inserted user sequences' TIDs, *DelTID* denotes the deleted user sequences' TIDs, k denotes current process level in L, and the *maximum level* of the original lattice structure is m. For instance, the maximum level of the lattice structure in Figure 3 is 3. All the web traversal patterns will be outputted as the results.

For example, in Table 1, we insert one user sequence (7, ABCEA) and delete two user sequences (1, ABCED) and (2, ABCD) as shown in Table 2. The *min_sup* also sets to 50%. At the first level of the lattice structure in Figure 3, TID 1 and TID 2 are deleted and the support count is decreased from each node which contains TID 1 or TID 2 for level 1. Then, the inserted user sequence TID 7 is decomposed into length 1 traversal sequences. The TID 7 is added and support count is increased to each node which contains one of the decomposed 1-traversal sequences. Because there is no new web traversal patterns generated in level 1, we continue to process level 2 in the lattice structure. The updated lattice structure is shown in Figure 4 when we processed the level 1. Because there is no new 1-web traversal pattern generated and no node deleted, the number of the nodes and the links between the first level and the second level are not changed. According to the deleted sequences' TIDs, the TID 1 and TID 2 are deleted and the support count is decreased in each node of level 2. Then, the inserted user sequence TID 7 is decomposed into length 2 traversal sequences. The TID 7 is added and the support count is increased to each node which contains one of the decomposed 2-traversal sequences. Finally, we can find the traversal sequence <EA> terns out to be 2-web traversal pattern and the original web traversal patterns <AD>, <BC>, and <BD> are not web traversal patterns after updating the database and <ABC>, <ABD>, <BCD>, <BCE> and <DAD> are marked. Figure 5 shows the lattice structure after processing the inserted and deleted traversal sequences in level 2. The sequence with double line is the new web traversal pattern. After generating the new 2-web traversal patterns, the two new candidate traversal sequences <CEA> and <EAB> are generated. Similarly, the last level is processed and the lattice structure about level 3 is updated. Figure 6 shows the final result in our example.

Table 2: Traversal sequence database after inserting and deleting user sequences from Table 1

TID	User sequence
3	CDEAD
4	CDEAB
5	CDAB
6	ABDC
7	ABCEA

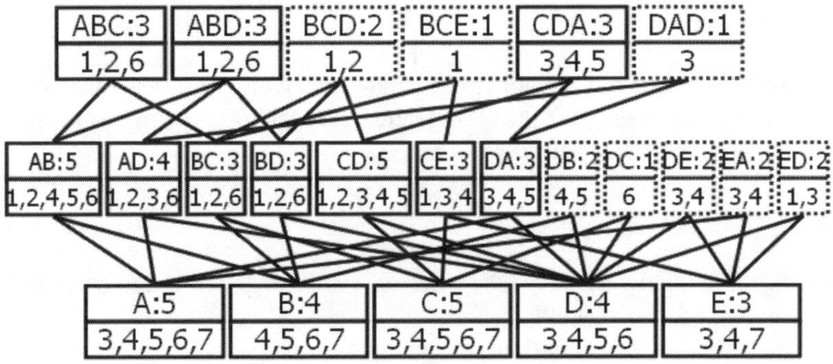

Figure 4: Updated lattice structure after processing level 1

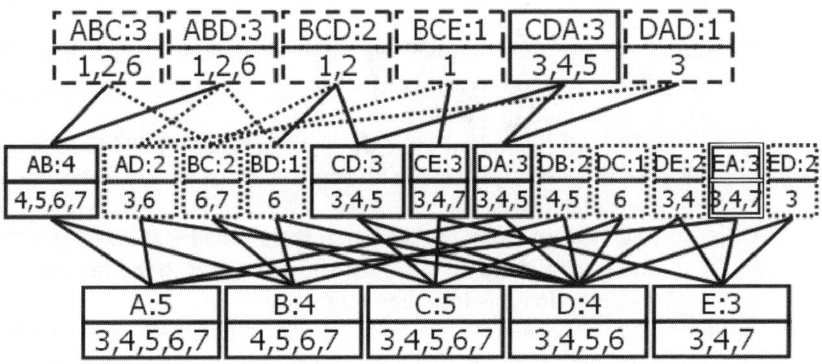

Figure 5: Updated lattice structure after processing level 2

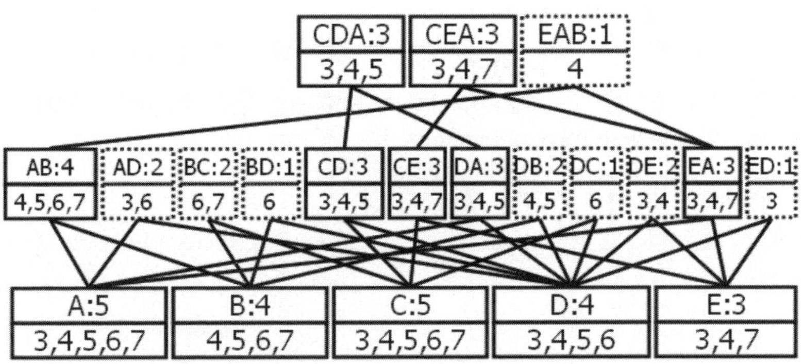

Figure 6: Updated lattice structure after processing level 3

Algorithm 1: IncWTP(*D*, *min_sup*, *W*, *L*, *InsTID*, *DelTID*, *m*)

```
Input: traversal sequence database D, min_sup, web site structure W, lattice
        structure L, insert TID InsTID, delete TID DelTID, maximum level of L m
Output: All Web traversal patterns

  k=1;

  while(k ≤ m or there are new web traversal patterns generated in level k)
     for each node n in level k
        if(the node n are marked)
           the node n and all the links related to the node can be deleted;
           the nodes in level (k+1) which have links with node n are marked;
        if(node n contains any TID in DelTID)
           delete TIDs contained in DelTID and decrease the support count
           from n;
     for each inserted user sequence u
        decompose u into several qualified traversal sequences with length k;
        for each decomposed traversal sequence s
           if(s is contained in a node n of the level k)
              add u's TID and increase the support count in the node n;
           else if(all qualified (k-1)-sub-sequences of s are web traversal
                 patterns)
              new node n_s contains s is generated in the level k;
              add u's TID and increase the support count in the node n_s;
     if(the support of a node n_m is less than min_sup)
        all the links between node n_m and the nodes in level (k+1) are deleted;
        the nodes in level (k+1) which have links with node n_m are marked;
        if(the support count of a node n_0 in level k is equal to 0)
           the node n_0 and all the links related to the node can be deleted;
     for each traversal sequence t_s in level k
        if(the support of t_s ≥ min_sup)
           WTP_k = WTP_k ∪ {t_s};
           /* WTP_k is the set of all the web traversal patterns */
     NewWTP_k = WTP_k – OriWTP_k;
     /* OriWTP_k is the set of original web traversal patterns and NewWTP_k is the
        set of new web traversal patterns */
     output all the web traversal patterns in level k;
     CandidateGen (NewWTP_k , OriWTP_k)
     k++;
```

Algorithm 2: CandidateGen (*NewWTP*$_k$, *OriWTP*$_k$)

```
for each new web traversal pattern x in NewWTP_k
        for each new web traversal pattern y in NewWTP_k
            if(x and y can be joined)
                generate a new candidate (k+1)-traversal sequence and store the
                new candidate in set C;
        for each original web traversal pattern z in OriWTP_k
            if(x and z can be joined)
                generate a new candidate (k+1)-traversal sequence and store the
                new candidate in set C;
    for each candidate (k+1)-traversal sequence c in C
        count support and record the user sequences' TIDs which contain c
        from D;
        create a new node n_c which contains c;
        for each node n_s in the kth level which contains a qualified
            k-sub-sequence of c
            create a link between n_s and n_c;
```

4 Experimental Results

Because there is no incremental mining algorithm on finding web traversal patterns currently, we use the algorithm *MFTP* [13] which is also used to find the web traversal patterns to compare with our algorithm *IncWTP*. This section presents the experimental results on the performance of our approach. The number of web pages is 300. We generate four synthetic data sets in which the numbers of user sequences are set to 30K, 50K, 70K and 100K, respectively. The four original data sets are increased by inserting 2K, 4K, 6K, 8K, 10K, 12K, 14K, 16K, 18K and 20K user sequences. In the first experiment, the *min_sup* is set to 5%. Figure 7 shows the relative execution times for *MFTP* and *IncWTP* on the four synthetic data sets. In Figure 7, we can see that our algorithm *IncWTP* outperforms *MFTP* algorithm, since our algorithm uses the lattice structure and web site structure to prune a lot of candidate sequences and keeps the previous mining results such that just inserted user sequences need to be scanned for most of the candidate sequences. The performance gap increases when the size of original database increases. This is because when the size of original database increases, *MFTP* algorithm is worse than *IncWTP* algorithm in terms of the number of candidate traversal sequences and the size of database need to be scanned, since *MFTP* algorithm must re-find all the web traversal patterns from the whole updated database, but our algorithm just need to find new web traversal patterns from the inserted user sequences. Moreover, the less the number of inserted user sequences, the less the generated new candidate sequences for our algorithm. Hence, the performance gap increases as the number of inserted user sequences decreases.

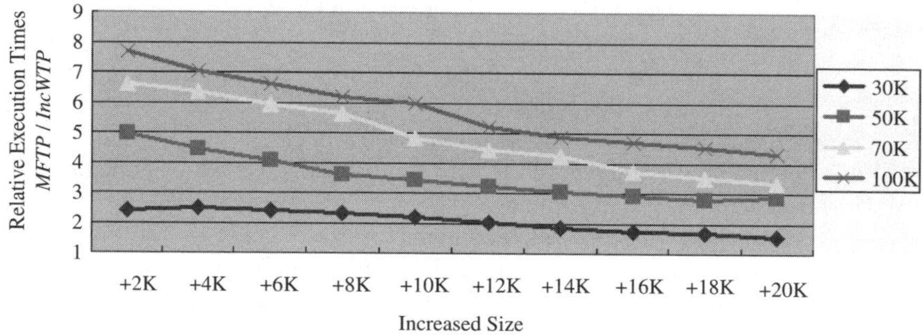

Figure 7: Relative execution times for *MFTP* and *IncWTP* (*min_sup* = 5%)

In the second experiment, we use a synthetic data set in which the numbers of user sequences is 100K, and the *min_sup* is set to 10%, 8%, 5%, 3% and 1%, respectively.

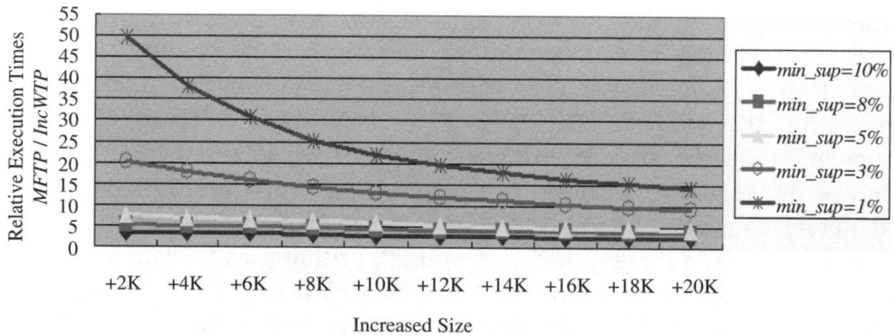

Figure 8: Relative execution times for *MFTP* and *IncWTP* (Dataset = 100K)

Figure 8 shows the relative execution times for *MFTP* and *IncWTP*, in which we can see that our algorithm *IncWTP* outperforms *MFTP* algorithm significantly. The lower the *min_sup*, the more the candidate sequences generated for *MFTP* algorithm. MFTP needs to spend a lot of time to count a large number of candidate sequences from the whole updated database. For our algorithm *IncWTP*, just few new candidate sequences generated, especially, when the number of inserted user sequences is small. Hence, if the number of inserted user sequences comes smaller, our algorithm would just pay a little time to find new web traversal patterns from the inserted user sequences.

In the third experiment, the *min_sup* is set to 5%. We also use the four synthetic data sets in the first experiment. These original data sets are decreased by deleting 2K, 4K, 6K, 8K, 10K, 12K, 14K, 16K, 18K and 20K user sequences. Figure 9 shows the relative execution times for *MFTP* and *IncWTP* on the four synthetic data sets, in which we can see that our algorithm *IncWTP* is also more efficient than *MFTP* algorithm. The more the deleted user sequences, the smaller the size of the updated database. Hence, the performance gap decreases as the number of deleted user sequences increases, since the size of the database needs to be scanned and the number of candidate sequences decrease for *MFTP* algorithm. For our algorithm, there are few or no new candidates generated when the user sequences are deleted from original database, we just need to update the lattice structure for the deleted user sequences when the number of deleted user sequences is small. Hence, *IncWTP* still outperforms *MFTP* algorithm.

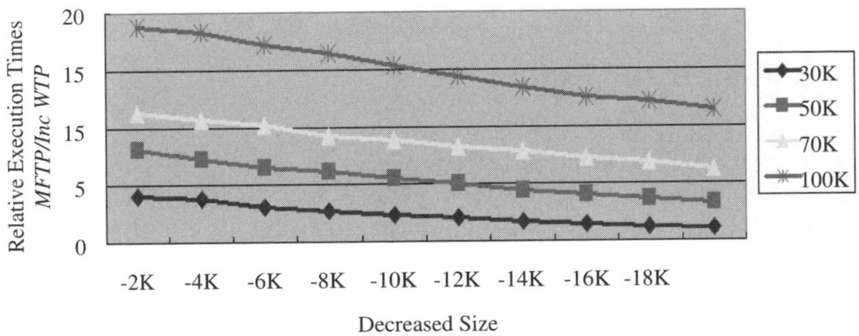

Figure 9: Relative execution times for *MFTP* and *IncWTP* (*min_sup* = 5%)

In the fourth experiment, we also use the synthetic data set in which the number of user sequences is 100K. The *min_sup* is set to 10%, 8%, 5%, 3% and 1%, respectively.

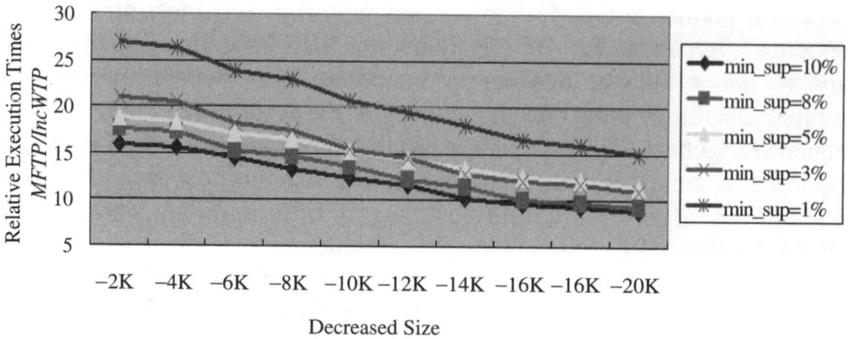

Figure 10: Relative execution times for *MFTP* and *IncWTP* (Dataset = 100K)

Figure 10 shows the relative execution times for *MFTP* and *IncWTP* on the synthetic data set. In Figure 10, we can see that our algorithm *IncWTP* outperforms *MFTP* algorithm significantly. The performance gap increases as the minimum support threshold decreases, since the number of candidate sequences increases and the whole updated database need to be scanned for the large number of candidate sequences for *MFTP* algorithm. For *IncWTP* algorithm, just the deleted user sequences need to be scanned when the number of deleted user sequences is small and the minimum support threshold is large. Hence, the performance gap increases as the number of deleted user sequences decreases.

5 Conclusion and Future Work

In this chapter, we propose an incremental data mining algorithm for discovering web traversal patterns when the user sequences are inserted into and deleted from original database. In order to avoid re-finding the original web traversal patterns and re-counting the original candidate sequences, our algorithm uses lattice structure to keep the previous mining results such that just new candidate sequences need to be computed. Hence, the web traversal patterns can be obtained rapidly when the traversal sequence database is updated. Besides, the web traversal patterns related to certain pages or maximal web traversal patterns can also be obtained easily by traversing the lattice structure.

However, the web site structure may be changed. In the future, we shall investigate how to use the lattice structure to maintain the web traversal patterns when the web site structure is changed. Besides, the number of web pages and the user sequences will grow up all the time. The lattice structure may become too large to fit into memory. Hence, we shall also investigate how to reduce the storage space and partition the lattice structure such that all the information can fit into memory for each partition.

Reference

1. M. S. Chen, X. M. Huang and I. Y. Lin, “Capturing User Access Patterns in the Web for Data Mining”, Proceedings of the IEEE International Conference on Tools with Artificial Intelligence, pp. 345-348, 1999.
2. R. Cooley, B. Mobasher, and J. Srivastava, “Web Mining: Information and Pattern Discovery on the World Wide Web”, Proceedings of the IEEE International Conference on Tools with Artificial Intelligence, 1997.
3. M. S. Chen, J. S. Park and P. S. Yu, “Efficient Data Mining for Path Traversal Patterns in a Web Environment”, IEEE Transaction on Knowledge and Data Engineering, Vol. 10, No. 2, pp. 209-221, 1998.
4. H. Cheng, X. Yan, and J. Han, “IncSpan: Incremental Mining of Sequential Patterns in Large Database”, In Proceedings of 2004 International Conference on Knowledge Discovery and Data Mining (KDD'04), Seattle, WA, Aug. 2004.
5. Yue-Shi Lee, Show-Jane Yen, Ghi-Hua Tu and Min-Chi Hsieh, “Web Usage Mining: Integrating Path Traversal Patterns and Association Rules”, Proceedings of International Conference on Informatics, Cybernetics, and Systems (ICICS'2003), pp. 1464-1469, 2003.
6. Yue-Shi Lee, Show-Jane Yen, Ghi-Hua Tu and Min-Chi Hsieh, “Mining Traveling and Purchasing Behaviors of Customers in Electronic Commerce Environment”, Proceedings of IEEE International Conference on e-Technology, e-Commerce and e-Service (EEE'2004), pp. 227-230, 2004.
7. J. Pei, J. Han, B. Mortazavi-Asl, H. Pinto, Q. Chen, U. Dayal, and M.-C. Hsu, “PrefixSpan: Mining sequential patterns efficiently by prefix-projected pattern growth” In proceeding of 2001 International Conference on Data Engineering (ICDE’01), pages 215-224, Heidelberg, Germany, April 2001.
8. J. Pei, J. Han, B. Mortazavi-Asl, J. Wang, H. Pinto, Q. Chen, U. Dayal, and M.-C. Hsu, “Mining Sequential Patterns by Pattern-Growth: The PrefixSpan Approach”, IEEE Transactions on Knowledge and Data Engineering, 16(10), 2004.
9. J. Pei, J. Han, B. Mortazavi-Asl and H.Zhu, “Mining Access Patterns Efficiently from Web Logs”, Proceedings of the Pacific-Asia Conference on Knowledge Discovery and Data Mining, pp. 396-407, 2000.

10. Maged EL-Sayed, Carolina Ruiz, and Elke A. Rundensteiner, "FS-Miner: Efficient and Incremental Mining of Frequent Sequence Patterns in Web logs", Proceedings of 6th ACM International Workshop on Web Information and Data Management (WIDM 2004), pp.128-135, 2004
11. Srinivasan Parthasarathy, Mohammed J. Zaki, Mitsunori Ogihara, Sandhya Dwarkadas, "Incremental and Interactive Sequence Mining", In Proceedings of 8th International Conference on Information and Knowledge Management , pp 251-258, Kansas City, MO, November 1999.
12. J. Srivastava, et al., "Web Usage Mining: Discovery and Applications of Usage Patterns from Web Data", SIGKDD Explorations, pp. 12-23, 2000.
13. Show-Jane Yen, "An Efficient Approach for Analyzing User Behaviors in a Web-Based Training Environment", International Journal of Distance Education Technologies, Vol. 1, No. 4, pp.55-71, 2003.
14. Show-Jane Yen, Yue-Shi Lee and Chung-Wen Cho, "Efficient Approach for the Maintenance of Path Traversal Patterns", In Proceedings of IEEE International Conference on e-Technology, e-Commerce and e-Service (EEE), pp. 207-214, 2004.
15. M. Zaki. "SPADE: An efficient algorithm for mining frequent sequences", Machine Learning, 40:31–60, 2001.

Customer Experience Management in E-Services

Zhaohao Sun and Sim Kim Lau

School of Economics and Information Systems, University of Wollongong, NSW 2522 Australia
Email: zsun@uow.edu.au, simlau@uow.edu.au

Abstract: E-services are the provision of services over the Internet. Customer experience management (CEM) is a vital part for e-services and Web services. This chapter will examine CEM in e-services by providing a unified architecture (SSES) and an intelligent system architecture (MEMES) as well as a cybernetic model for B2B e-services. The SSES unifies e-services, Web services and infrastructure services into a hierarchical framework. The MEMES is a high level system model for implementing multiagent CEM system for e-services. Both architectures tie together methodologies, techniques, and applications into a unified framework that includes both logical and intelligent embodiment of the aspects of CEM in e-services. The cybernetic model provides a new insight into B2B e-services with feedback from e-service customers. The chapter also explores case-based e-service recommendation. It is argued that the proposed approach will facilitate the development of CEM, e-business and e-service.

Keywords: Customer experience management, e-service, e-commerce, intelligent techniques, multiagent system, case-based e-service recommendation.

1 Introduction

Acquiring and retaining customers, and selling more goods and services to customers over time are essential for not only traditional commerce but also e-services [24] (p.214). E-services are the provision of service over the electronic networks such as the Internet and wireless networks [22]. With dramatic development of the Internet and the Web in the past decade, e-services have been flourishing in e-commerce, artificial intelligence (AI), information systems (IS) because they offer a number of strategic advantages such as mobility, flexibility, interactivity and interchangeability, comparing with traditional services [13].

The fundamental philosophy of e-services is to meet the needs of customers precisely and thereby increase the market share and revenue [22]. E-services have helped customers to reduce the cost of IT operations and allow them to closely focus on their own core competencies [13]. At the same time, for business marketers, e-service providers are considered very useful, as much for improving interorganizational relationships as for generating new revenue streams. Furthermore, e-services can be considered as a further development of traditional e-commerce, because it is a service-focused business paradigm that uses two-way dialogue to build customized service offerings, based on knowledge and experience about customers to build strong customer relationships [22]. It implies, however, that one of the intriguing aspects of e-services is that any e-service can not avoid similar challenges encountered in traditional services such as how to manage customer experience in order to attract more customers.

Experience management has received an increasing attention in knowledge management (KM), IS and AI [27]-[34], because for both organisations and individuals,

Z. Sun and S.K. Lau: *Customer Experience Management in E-Services*, Studies in Computational Intelligence (SCI) **37**, 365–388 (2007)
www.springerlink.com

experience is wealth, just as knowledge is power. Customer experience management (CEM) [24], as an application of experience management (EM), has found many successful applications in business and commerce. However, it is still a major issue for any development of e-services although the customer can obtain e-services without taking distance into account, but through the Web, cell phone and PDA. For example, e-services providers/marketers will need to address rational and irrational customer concerns regarding the adoption of new e-services that replace traditional processes, and improve customer support for customers who wish to customize e-service applications [13]. Furthermore, CEM is still at an empirical level and business experience level, because few attempts have been made to formalize and model CEM in e-services using intelligent techniques.

This chapter will alleviate the above mentioned issues by examining CEM, e-services and their relationships. More specifically, it proposes a unified architecture, SSES, and an intelligent system architecture, MEMES, as well as a cybernetic model for B2B e-services. The SSES unifies e-services, Web services and infrastructure services into an integrated hierarchical framework. The MEMES is a high level system model for implementing multiagent CEM system for e-services. Both architectures tie together methodologies such as multiagent system and knowledge based system, techniques such as case-based reasoning (CBR) and experience based reasoning (EBR), and applications such as in e-services and EM into a unified framework that includes both logical and intelligent embodiment of the aspects of CEM in e-services. The cybernetic model provides a new insight into B2B e-services with feedback from e-service customers. The chapter also explores case-based e-service recommendation. It argues that the proposed approach will facilitate the development of CEM, e-business and e-service.

The remainder of this chapter is organized into the following sections: Section 2 proposes a multilayer architecture for integrating e-services, Web services and infrastructure services into a hierarchical framework. Section 3 provides a cybernetic model for B2B e-services taking into account feedback information from e-service customers. Section 4 and Section 5 examine experience management (EM) and customer experience management (CEM). Section 6 proposes the MEMES, a multiagent system architecture for implementing multiagent CEM system for e-services. Section 7 explores case-based recommendation of e-services. The final section concludes the chapter with some concluding remarks.

2 A Unified Multilayer Architecture for E-Services

E-services and Web services are two different concepts with a very close relationship [1][12]. However, many people are confused by these two concepts. Therefore, this section first examines e-services, Web services and their relationships. Then it will propose a multilayer system architecture for e-services integrating Web services and infrastructure services.

2.1 E-Services and Web Services

E-services are "electronic offerings for rent" made available via the Internet that complete tasks, solve problems, or conduct transactions [13]. Song [26] demonstrates that e-services have the following features: integration, interaction, customization, self-services, flexibility and automatic response.

Interaction is an important feature of e-services. Each e-service interacts with others to fulfil customer requests [26]. Self-services are another feature of e-services. The website of the e-service allows customers to review accounts, monitor shipments, edit profiles, schedule pick-ups, adjust invoices, return merchandises and so on, as computer-mediated rather than interpersonal transactions.

Some e-services, e.g. Amazon.com, are integrated with e-commerce applications such as shipping information, package tracking and rate inquiries [26]. E-services are seamlessly integrated with e-commerce applications to make the shipping experience simple and convenient. Therefore, from a business viewpoint, e-service is a further development of traditional e-commerce [22]. In other words, traditional e-commerce is giving way to e-service. Services play a vital role in industry, especially in most developed countries, service-based businesses made up about two-thirds of the economy [24] (p. 36). Therefore, any e-business or e-commerce activity is a kind of e-service.

Web services have drawn an increasing attention in building distributed software systems across networks such as the Internet, and also in business process reengineering. Web services are defined from an e-commerce or a business logic viewpoint at one extreme, in which case, Web services are the same as e-services. At another extreme, Web services are defined from a computer science or IT viewpoint. Between these two extremes, many different definitions have been proposed based on the understanding of different authors. Even so, there are also different levels for defining Web services from a methodological viewpoint and some definitions of Web services are made from a philosophical viewpoint. For example, a Web service is a way of publishing an explicit, machine-readable, common standard description of how to use a service and access it via another program using some standard message transports [20]. Others are at an intermediary level, for example, a Web service is an operation typically addressed via a URI, declaratively described using widely accepted standards, and accessed via platform-independent XML-based messages [1] (p. 124). A more technical definition of Web services is as follows [1] (p. 125): "A standardized way of integrating Web-based applications using the XML, SOAP, WSDL, and UDDI open standards over an Internet protocol backbone. XML is used to tag the data. SOAP is used to transfer the data. WSDL is used for describing the e-service available. UDDI is used for listing what services are available."

This chapter considers Web services as simple, self contained applications which perform functions, from simple requests to complicated business processes. The "Web services" model uses protocols or open standards such as TCP/IP, HTTP, WSDL, UDDI, BPEL, and SOAP/XMLP. A WSDL description is retrieved from the UDDI directory. WSDL descriptions allow the software systems of one business to extend to be used by others directly. BPEL supports automated business processes [44]. The e-services are invoked over the Web using the SOAP/XMLP protocol. Each of the components is XML based [42]. We believe that the latter two definitions should be used together to better understand Web services.

In order to understand the difference between e-services and Web services better, we will propose a unified system architecture for e-services in the next subsection.

2.2 A Multilayer System Architecture for E-Services

There are many system architectures available in e-services and Web services literature. These architectures provide a high level system design free of implementation details in order to provide better understanding of e-services and Web services. For example, Vissers et al. [36] propose a reference model (RM) for advanced e-services, which takes into account the RM for open systems interconnection. The latter consists of seven layers: application, presentation, session, transport, network, data link and physical layer. However, they do not really discuss the interrelationship between e-services and Web services, although they provide a deep insight into e-services from a reference model viewpoint.

Kreger [16] proposes a three-layer architecture for Web services, in which the three layers are the wire layer, the description layer, and the discovery agency layer. He demonstrates that Web services technologies are being developed as the foundation of a new generation of B2B and EAI (enterprise application integration) architecture. However, he prefers the technology for Web services rather than for e-services.

Further, in many e-services literature, the discussion on related technology is either given cursory treatment or mainly discussed in relation to product specific features. The relationship between e-services and Web services has not been examined in a unified way taking into different perspectives that include business and commerce, information technology (IT) and IS, and ordinary customers. As a result, e-services and Web services are often confused in usage in the context of IS and IT. In order to resolve the above mentioned issues, we propose a unified multilayer system architecture for e-services (SSES), as shown in Fig. 1. The first layer is infrastructure services layer, the second layer is Web services layer, and the third layer is e-services layer.

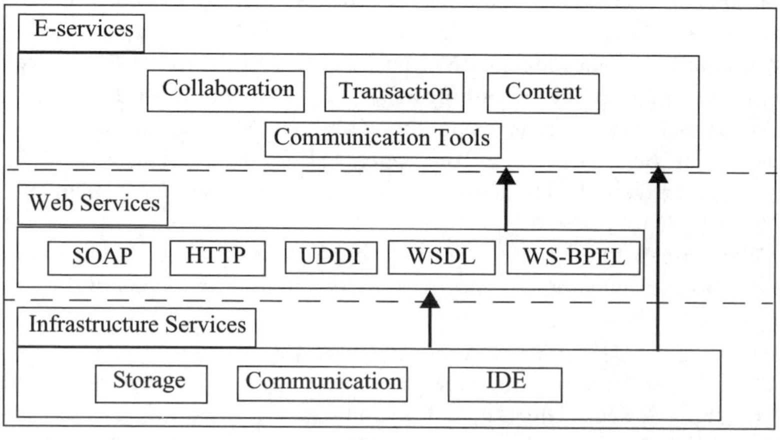

Fig.1. A unified system architecture for E-services (SSES)

The e-services layer is at the front end, which is directly interacting with the e-service customer. The Web-services layer is at the back end, which consists of software or software components that support the realization of e-services. The infrastructure services layer supports the realization of the Web services. Therefore, these three distinct services are on three different layers, and constitute an integrated systems. Based on client-server architecture [25], e-services are accessed by a client computer that is used by a customer. Web services are resided in the server computer that is normally managed by the e-service providers. The infrastructure services are also resided either in server computer or servers of internet service providers (ISP). In the following subsections, we will discuss each of these three layers in more detail.

2.2.1 First Layer: Infrastructure Services

The first layer in the SSES is the infrastructure services layer. The infrastructure services refer to the basic technology platform and features needed to implement Web services, which at least must consist of storage, communication and IDEs.

Storage, also known as a data warehouse or knowledge repository, is typically defined by its content and structure [8]. A knowledge repository could either be populated with data or documents. A data warehouse consists of many databases related to implementation of Web services. Both knowledge repository and data warehouse have been designed to capture text and graphical information such as engineering drawings and media documents.

Communication provides at least the distinct communication services such as communication between different Web services, and communication between clients and between servers as well as between themselves. ISP is one of the most important providers of such communications.

IDEs provide an integrated environment for developing Web services. IDEs include programming languages and special development tools used to implement Web services such as Java, C++, Javascript and XML.

It should be noted the infrastructure services layer in the SSES is not specific for e-services and Web services. It also provides infrastructure services to research and development of other systems and applications such as knowledge management systems [8]. Furthermore, the infrastructure services are basically studied by computer science scientists and electrical engineering or computer engineering scientists, while the IS researchers normally do not get involved in this layer.

2.2.2 Second Layer: Web Services

The second layer in the SSES is the Web Services layer. Web services refer to the software components, protocols and technology needed to implement e-services because a Web service is a software application identified by a URI, whose interfaces and bindings are capable of being defined, described and discovered as XML artifacts [1][12][11]. A Web service supports direct interactions with other software agents using XML-based messages exchanged via Internet based protocols.

Furthermore, a Web service is a server that listens for and replies with SOAP, generally via HTTP [17]. In practice, a Web service will support WSDL to describe its interfaces, and should also be listed in a UDDI registry.

Based on above discussion, Web services are not "services" in a traditional sense, because they are application enablers of e-services. However, the final goal of Web services is to realize the e-services. Therefore, we can assert that a necessary condition for successful e-services is the efficient and effective support of Web services. However, this condition is not sufficient, because we have encountered many unsuccessful e-services although they have the same or similar support of Web services. Why is this the case? One of the reasons is that they have not looked into the non-technical aspects in the customer society. Another reason is that they lack customer experience management, which will be examined in the later sections.

The second layer, Web services, is at a technical level. The goal of this service layer is to ensure different components of the Web services are operating with acceptable performances [7]. The main research and development of this layer services is on middleware infrastructure, service deployment and service management [7], which are basically studied by the computer science scientists and computer engineering scientists, while the IS researchers have been involved in the research and development of Web services to some extent, whereas the business researchers do not get involved in the research of this layer.

2.2.3 Third Layer: E-Services

The third layer in the SSES is the e-services layer, which is a business application layer that directly interacts with end users such as the e-services customer. E-services can be classified into four categories: collaborative services, transaction services, content services and communication services.

Collaborative services are designed to support groups of interacting people in their cooperative tasks [36], for example, teachers tele-teach students.

Transaction services support formal and often traceable transaction between parties by organizing the exchange of predefined messages or goods in predefined orders [36]. In practice, these services do not cover huge data per transaction. They only exchange simple messages such as orders, bills and payments.

Content services allow people to access and manipulate e-content such as accessing e-libraries [36]. These kinds of services are very popular in universities, for example, to access large amounts of data or e-journal papers.

Communication services provide at least distinct services such as communication between users, collaboration among users and workflow management [8]. The first communication services are implemented through utilities such as file sharing and e-mailing. The second communication services can be implemented through synchronous meeting and asynchronous discussion forums. The last one allows users to manage workflow processes by supporting online execution and control of workflow.

It should be noted that the objective that Vissers et al. [36] classify e-services into collaborative services, transaction services and content services is to argue their proposed reference model's applicability. They have not investigated into the business activities that are common behind these three services. In practice, e-services comprise only two major categories: free e-services and pay e-services from a business viewpoint.

Free e-services are the services provided by the e-services provider freely to potential e-service receiver. For example, a website with huge invaluable information in the fields of e-commerce, e-service, intelligent techniques for detecting fraud and deception in e-commerce is free for anybody to visit and obtain the information s/he requires. Free e-services have played an important role in promoting human culture development in the world.

Pay e-services are the services provided by the e-services provider to the e-service receiver with application and usage fees. These services are a new kind of business activity and have played a vital role for e-services development, research and development of Web services and infrastructure services. The traditional commerce and services are dramatically transferred to e-services in order to make effective use of the strategic advantages over the traditional services: mobility, flexibility, interactivity and interchangeability [13]. Collaborative services, transaction services and content services of Vissers et al. [36] all belong to pay e-services. Therefore, the classification of Vissers et al. [36] for e-services has not covered all possible e-services. Further, any pay e-service shares some of the most important business features with many traditional commerce or business activities: fraud, deception, pursuing maximum profit through bargaining and negotiation between service providers (sellers) and service receivers (buyers), etc. The service intermediary such as agent, bargainer and broker facilitates the success of these service activities [28].

The main research and development of the e-services layer is on customer satisfaction, customer relationship management and customer experience management, which are basically studied by IS and business scientists.

2.2.4 SSES: A Service-centered System Architecture

The proposed SSES is a service-centered system architecture, because a service is a behavior process of an agent (human or software agent) that can be observed and experienced by its customers [36] (p. 373). The services in the SSES have been in a hierarchical structure, shown with arrows between the layers in Fig. 1. The receiver or customer of the infrastructure services is the Web services provider and e-services provider, while the receiver or customer of the Web services is the e-services provider. The receiver or customer of e-services are ordinary customers such as the students at a university (they pay tuition fees) and the customers in a supermarket.

It should be noted that some agents play a double role in the SSES. More specifically, on one layer, they are service provider, on the other layer, they are service receiver or customer. For example, the Web services provider is the receiver of the infrastructure services, while the e-services provider is a receiver of the Web services. Therefore, the services in the SSES constitute a service chain towards e-services.

Based on the above examination, we can easily find that the competition in e-services not only occurs at the boundary between the customer market and e-services layer but also at the boundary between Web services and e-services. It can also occur at the boundary between Web services layer and infrastructure services layer in order to obtain the advance of acquiring different kinds of service customers or service receivers. Therefore, how to make customer satisfy and how to manage customer relationship and customer experience become significant issues in e-services, Web services and infrastructure services, which will be examined in more detail in the later sections.

3 A Cybernetic Model for B2B E-Services

This section will propose a cybernetic model for B2B e-services and discuss the feedback from e-service customers.

From a transaction viewpoint [28], e-services can be classified into business to business (B2B) e-services, business to customer (B2C) e-services, customer to customer (C2C) e-services. B2B and B2C e-services are dominating the e-services market and play a critical role in development of Web services and infrastructure services. However, there are few discussions on the interrelationship between them, in particular from a feedback viewpoint (based on cybernetics). In what follows, we will propose a cybernetic model with feedback to examine the interrelationship between B2B e-services and B2C e-services. The scenario for this model is as follows: The scenario consists of three kinds of actors (An actor can be either human or intelligent agent or system.) [28]: The first kind of actors are e-services providers (1, 2, ..., *n*) such as Springer, Kluwer and Elsevier, all of them are the most influential providers for e-books and e-journals to many universities in the world. The first kind of actors are e-service sellers (or agents). The second kind of actors are e-services receivers (1, 2, ..., *m*) such as librarian managers of universities in the world. They are e-services buyers (or agents). The third kind of actors are academic staff members of universities (1, 2,..., *k*). They are the e-service end consumers.

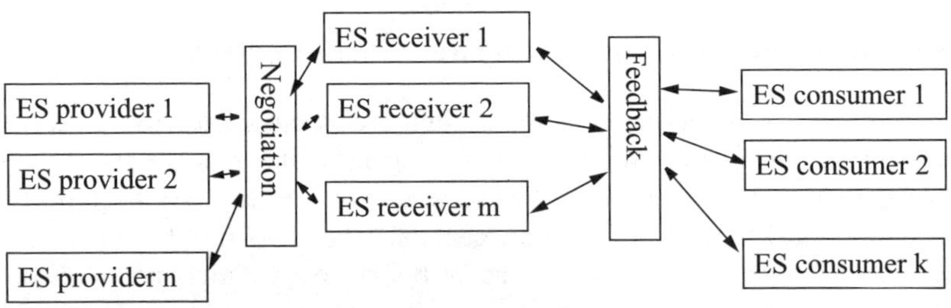

Fig.2. A cybernatic Model of E-Services (CME).

In this model, as shown in Fig. 2, e-services provider such as the agent of Springer will negotiate with the agent of the University of Bargainland (UoB) over the service scope of e-journals and e-books provided to the UoB and the price of such services. This is a B2B e-service. However, the mentioned agent of the UoB is not the consumer of the e-service provided by Springer. The academic staff members of the UoB are end consumers who need not negotiate with the e-services providers, although they participated in the B2C business activity. However, they will provide valuable feedback information to the agent of UoB and play a role in the mentioned B2B e-services indirectly, because the e-service provider and receiver require customer's feedback to improve their services. Therefore, the quality of the mentioned e-services and the negotiation between the agent of Springer and the agent of UoB will be affected by the feedback information of the end-consumer of the e-services: academic staff members.

From here we can also find that negotiation space is dwindling in this case with the development of service intermediary as an important business force, because the majority of end consumers are free of negotiation over the price of the goods or services. This is a business progress for the mankind. However, bargaining and negotiation are still an important business means for B2B e-services and B2C e-services [28]. This is because one will be very unwise if s/he tries to bargain the price of the goods sold in the supermarket such as Woolworths. However, one will also be not clever if s/he does not bargain for the selling price $660,000 in an advertisement when buying a house. Sometimes experience plays an important role in bargaining and negotiation of e-services.

4 Experience Management

Experience is wealth for any individual or organisation. Generally, experience can be taken as previous knowledge or skill that one obtained in everyday life [28] (p.13). Experience management (EM) has drawn increasing attention in IS and AI in the past few years [5][27][29][31]. In what follows, we will examine EM in some detail.

From an object-oriented viewpoint [25][31], a subclass *Y* inherits all of the attributes and methods associated with its superclass *X*; that is, all data structures and algorithms originally designed and implemented for *X* are immediately available for *Y* [21] (p. 551). This is the inheritance or reuse of attributes and operations. As we know, experience can be considered as a special case of knowledge. Methodologies, techniques and tools for knowledge management (KM) [34] can be directly reused for EM, because EM is a special kind of KM that is restricted to the management of experience. On the other hand, experience has some special features and requires special methods different from that of knowledge, just as a subclass *Y* of its superclass *X* usually possesses more special attributes and operations. Therefore, the following two issues are very important for EM:

- What features of EM are different from that of KM?
- Which special process stages does EM require?

In what follows, we will try to resolve these two issues. First of all, we define that EM is a discipline that focuses on experience processing and its corresponding management [27], as shown in Fig. 3. The experience processing mainly consists of the following process stages [5] (pp. 1-14):

- Discover experience
- Capture, gain and collect experience
- Model experience
- Store experience
- Evaluate experience
- Adapt experience
- Reuse experience
- Transform experience into knowledge

- Maintain experience.

Management has permeated each of above-mentioned process stages [27], which is distinguished from other process models of either EM or KM [34]. It is significant to separate management function from experience processing functions, and then integrate them in EM. This is also one of the characteristics of this chapter.

The management of experience processing for each of the above-mentioned process stage includes analysis, planning, organisation, support, collaboration [31], coordination and possible negotiation [27]. Generally, management issues related to each or some of the experience processing stages include:

- Organisation and control of experience
- Experience processing or management task assignment to a specific person or team.

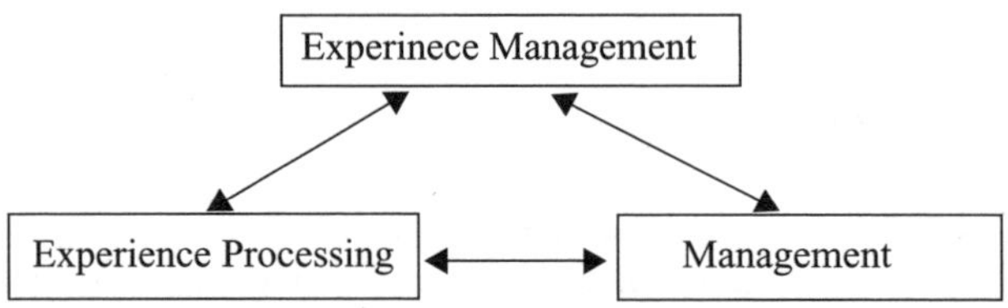

Fig.3. EM as an integration of experience processing and management [31].

In the above experience processing stages, "maintain experience" includes updating the available experience regularly, while invalid or outdated experience must be identified, removed or updated [27]. Transforming experience into knowledge is an important process stage for EM, which is the unique feature of EM that is different from those of KM. In the history of human beings, all invaluable experiences are gradually transformed into knowledge, which then is spread widely in a form of books, journals and other means such as multimedia.

It should be noted that discovery of experience from a collection of knowledge or social practice is a significant issue for EM, such as knowledge discovery from a huge database [31]. Further, the processing of experience requires an experience base, because experience processing will be conducted in experience base or through accessing experience base.

EM research is providing a new way of looking at data, knowledge, experience, in particular customer experience and its management for organizations and e-services [27]. This will include customer experience retrieval, experience similarity and experience processing. Successful solutions of these problems could provide the basis for new advances in both EM and e-services.

5 Customer Experience Management

This section will first review customer satisfaction and customer relationship management (CRM) and then examine customer experience management (CEM).

5.1 Introduction

Since the 1990's, "customer-oriented" and "customer-centred", instead of product-, technology-, or sale-focused, have become critically important for business development for any company [24] (p. 10). To make customer satisfy and improve customer relationship are a useful attempt to develop the customer-centred business development. This is an important transformation for e-services: from technology-centred e-services to customer-centred e-services, because the e-service marketers have to deal with virtual customers rather than physical customers.

CEM has drawn an increasing attention in business, commerce, and e-services, with the development of customer-centred business and service, based on the fact that the results of searched by Google.com for "customer experience management", "customer satisfaction" and "customer relationship management" are 1,720,000, 67,600,000, and 41,800,000 respectively (accessed on 5 May 2006). CEM is a new development based on customer satisfaction (CS) and CRM.

CS is vital to success of any organisation including e-services in the Internet age. CS paradigms have been widely used in the automotive industry, which has developed detailed functional measures of cars to track customer satisfaction [24] (p. 13). However, CS is only one of the most important factors in sustaining a good customer relationship for organisations such as e-services.

CRM is a strategic paradigm that calls for iterative processes designed to turn customer data into customer relationships through active use of, and learning from, the information collected in the customer database or customer data warehouse [6]. The three major success factors of CRM are: increased profit, improved customer satisfaction, and enhanced customer loyalty [6]. This is one of the reasons that CS can be considered one part of the CRM. The profit-centric success asks CRM to capture, analyze historic transaction and preference data for existing customers. The customer satisfaction-oriented success requires CRM to understand the needs and preferences of current and potential customers, and use this understanding to better service them. In order to enhance customer loyalty, CRM should also focus on managing individualized relationship with customers. This relationship helps customer retention.

It should be noted that customer satisfaction and CRM promised to help managers to better understand their customers. Although CRM is thought to be a strategy for delivering superior e-services, 60% of managers view their CRM implementations as failures [6]. Therefore, it is necessary to examine CEM. To this end, we will first examine the experiential world.

5.2 Analyzing and Understanding Experiential World

Experience is process-oriented, whereas satisfaction is outcome-oriented [24] (pp. 14-15). For example, e-services experience of a customer includes more than simply a buying process. Therefore, satisfaction can be considered as one of the results of mining the customer experience, analyzing and understanding the experiential world of the customer is the first step for a successful e-services, just as it did in traditional business activities [24] (p. 56).

Experiential user interface in the website of the e-services is a vital part for managing customer experience. Implementation of some important experience principles in e-services also provides an important means for understanding experiential world of customers. One usually hopes to experience in the e-services what s/he experienced on other social occasions such as in a traditional supermarket. For example, "similar problems have similar solutions and require similar recommendations" is one of the most important experience principles [28]. Amazon.com has effectively implemented in its e-services for buying its products such as books. The customer visiting the e-services user interface of the Amazon.com will find that when s/he searches for a book on customer experience management, the books with similar titles or "contents" are also displayed waiting for visit and access. Such a strategy for managing customer experience has been found in many other e-services such as ebay.com and baidu.com. However, how many different customer experiences for e-services should be drawn to attention in order to improve the CEM of e-services is still a big issue.

5.3 Ontology of Experiences: Type 1 and Type 2 Experiences

According to Schmidt [24] (pp. 105-106), the experience of customer can be classified into following five categories:

- The sense experience. The sense experience appeals to the five senses. Customer value is created through sight, sound, touch, taste and smell.
- Affective experience (feel), which appeals to customers' inner feelings and emotions. Customer value is created through affective experiences that range from mildly positive moods linked to a brand, to a strong emotion of joy and pride.
- Cognitive experience (think). This experience appeals to the intellect. It creates value for customers by engaging them creatively.
- Physical experiences, behaviors and lifestyles (act). The act experience appeals to behaviours and lifestyles, creating value for customers by showing them alternative lifestyles or alternative ways of doing business.
- Social identity experience (relate). The relate experience contains social experiences. It creates value for the customer by providing a social identity and sense of belonging.

We believe that only the first four experiences are useful for managing the customer experience in e-services, because the social identity experiences are basically not encountered in the context of e-services.

The above five types of experiences are all at a direct or concrete level. Therefore, they can be called direct experiences or *type 1* experiences. They are the raw material for generating and distilling experience knowledge of customer, which can be called *type 2* experience. The type 2 experience is more important in guiding further business activity of the customer, because it is the distillation and summary of numerous above-mentioned type 1 experiences through intelligent or rational processing. In the majority of cases such as in clinic practice, the type 2 experience is more significant to attract the visit of patients. In fact, we often use "experience" in the research, which is not type 1 experience but type 2 experience. For example, "experiences" in designing for the mobile devices [14] belong to the set of type 2 experience.

In practice, an e-service experience of customer usually is a combination of the above-mentioned five type 1 experiences. Different combinations of them will lead to different type 2 experiences. These combined type 2 experiences play more important roles in making customer satisfy and thus in managing customer experience in e-services.

Because there are five type 1 experiences, we have 2^5 or 32 different combinations based on these five type 1 experiences and Boolean algebra. Each of the combinations is a type 2 experience.

These combined 26 ($26 = 2^5 - 5 - 1$.) type 2 experiences (except five type 1 experiences and no combination), as shown in Fig. 4, are the direct consequence of intelligent processing with combination. In fact, more complex combination between type 1 and/or type 2 experiences as well as knowledge stored in the human mind are generated in the practice. Sometimes, these processes are called transformation of type 1 experience to type 2 experience, and type 2 experience to knowledge. It should be noted that possessing knowledge is only a necessary condition for a domain expert [27]. Experience may be more important than knowledge for a domain expert to deal with some tough problems. Accumulation of knowledge is also a necessary condition of obtaining type 2 experience for a domain expert.

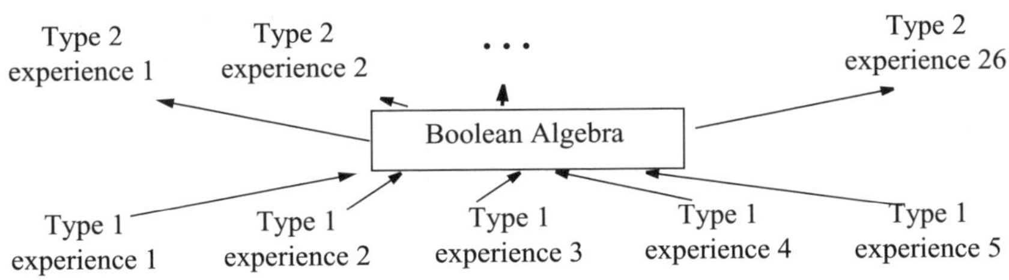

Fig.4. Type 1 experiences and type 2 experiences.

Induction or inductive reasoning is an intelligent processing of experience and generating type 2 experience. For example, let experience 1, 2,... *m* be either type 1 experience or type 2 experience respectively, and they share common facts such as "Google is an excellent search engine." Then one can inductively come to a conclusion or a new experience, a type 2 experience: "Google is an excellent search engine among all existing search engines." Therefore, inductive reasoning is an important means for human being to obtain type 2 experience (Note that inductive reasoning is also a nonmonotonic reasoning.). Whenever we believe that someone's idea is original or innovative, it means that this idea is a kind of type 2 experience.

However, induction is not a fundamental reasoning paradigm. All experience-based reasoning (EBR) activities are based on eight fundamental reasoning paradigms proposed by Sun and Finnie [28][29][30]. The different combination and network of these eight fundamental reasoning paradigms constitute a complex experience network development. We will not discuss it any more owing to space limitation.

From a viewpoint of intelligent systems, type 2 experiences can be generated and distilled into type 3 experiences, etc. Therefore, type *n* ($n = 1, 2,...$) experience can be generated by type1, type 2,..., type n-1 experience. All these experiences constitute a type hierarchy of experience. We can find that the experiential world $W_C(i)$ of a customer i consists of type 1, type 2, type 3,..., and type n experiences, ... For brevity, we only consider type 1 and type 2 experiences in the rest of the chapter, and type 2 experiences cover all other higher type experiences.

Based on the above discussion, we find that it is not enough for managing customer experience in e-services only to cover all direct experiences proposed by Schmidt [24]. This is also the reason why it is not easy to make customer satisfy and manage customer experience effectively. However, integrating intelligent approach into CEM in e-services might provide a successful strategy to overcome this difficulty.

5.4 Customer Experience Management

As an application of experience management, customer experience management (CEM) has been studied in business and commerce [24]. This section will explore CEM from a general perspective.

Generally, CEM is a process of strategically managing a customer's entire experience with a product or a company [24] (pp. 17-18). In this way, CEM changes traditional customer satisfaction from outcome-oriented to process-oriented. CEM also extends traditional CRM from recording transactions to building rich relations with customers and understanding the experiential world of customers, because it is imperative that organisation understand their customer's past experiences, current behaviors, preferences, and future needs [6].

Schmitt proposes a framework to manage customer experience [24] (p. 25), which targets the business managers or consultants, in order to manage customer experience for an organisation. This framework consists of the following five steps:

1. Analysing the experiential world of the customer
2. Building the experiential platform
3. Designing the brand experience
4. Structuring the customer interface
5. Engaging in continuous innovation.

Customer interface is one of the key implementation strategies for managing customer experience in e-services, because it affects retention through the exchanges and interactions which further determine whether the customers are satisfied with the e-services and whether they will buy the services again [24]. Most CRM solutions merely record what can be easily tracked: the history and transactions of customer-company contracts [24] (p. 141). However, this is not enough for managing customer experience in e-services because the customer in e-services believes that the interface of the e-service is the agent of the e-service, and s/he is face-to-face communicating with this agent. This is a new world, because the interaction between the customer and the agent of the e-services are different from traditional face-to-face interaction or communication in traditional business or service. However, in such an interaction, the customer will still try to obtain human-like interaction with the interface agent in the e-services.

Furthermore, humanizing and harmonizing the customer experience are important components for CEM. Humanizing the customer experience requires communicating with customers according to humanity rather than technology [24] (p. 92). This is because the customer in e-services hopes to experience a friendly human community in the virtual society such as the environment of e-services. Harmonizing the customer experience allows the customers to increase their confidence in receiving the services or products from a company.

It should be noted that Schmidt's discussion on CEM is essentially based on his business consultation experience, in particular, in the traditional business sectors, without regards to any intelligent techniques. In what follows, we will discuss this topic from an IT or AI viewpoint. The core idea behind it is that the intelligent techniques can improve management of customer experience in e-services, just as had been done in other fields such as e-commerce [28].

6 MEMES: A Multiagent System Architecture for Managing Customer Experience in E-Services

This section will propose the MEMES, a multiagent system architecture for implementing multiagent CEM system for e-services. More specifically, it will first examine the fundamentals of the MEMES. Then it will examine agents within the MEMES. It also proposes a high level system model for the EBR decision maker. Finally, it discusses the workflowing of agents within the MEMES.

6.1 Introduction

Although CEM has drawn an increasing attention in the past years, there are few studies on CEM in e-services, especially from the viewpoint of intelligent systems.

Multiagent systems (MAS) have been applied to many fields such as e-commerce [28], knowledge management [35], and Web service [20], to name a few. In fact, MAS has been used as an important design methodology [38] to develop many application systems such as experience based system [30] and e-commerce systems [28].

From the viewpoint of software engineering, systems development has experienced a human-like evolution: from module based on systems development, through object-oriented systems development, to agent-oriented systems development [33]. Therefore, the system for managing customer experience in e-services should be developed based on MAS technology. At the same time, customer interface of e-service should also be developed based on MAS technology in order to implement the human-like interaction between the customer and the intelligent agent of the e-service. Based on the above discussion, we propose a multiagent system architecture for managing customer experience in e-services (MEMES) in next subsection.

6.2 Fundamentals of MEMES

MEMES is a multilayer system architecture consisting of view layer, business logic layer, and database access layer [25]. The view layer consists of an intelligent interface agent. The business logic layer mainly comprises ESI agent, CE miner, E-service advisor, E-service negotiator, and Experience decision maker. The database access layer mainly consists of a knowledge base for customer experience (KEB), e-service information base (ESB) and a customer experience warehouse (CEW). KEB stores the generated type 2 experience and knowledge discovered from CEW. ESB consists of market information in e-services in the Web. CEW consists of three databases related to e-service customer: past experience base (PEB), current preference base (CPB) and future need base (FNB), as shown in Fig. 5.

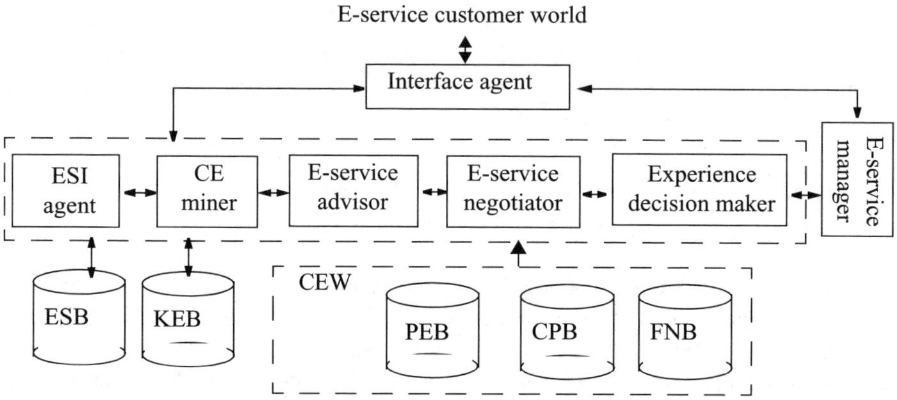

Fig.5. MEMES: A multiagent system architecture for managing customer experience in e-services.

PEB records historical data about customer's past experience or behavior which is denoted by e; CPB stores the current behavior or preference of each customer which is denoted by p; and FNB consists of data about future needs of each of customer denoted by n. Therefore, for each customer i of the e-service, we can use c_i to represent this customer's past experience, current preference and future need in the following vector form.

$$c_i = (e_i, p_i, n_i) \tag{1}$$

It should be noted that customer loyalty and satisfaction can be considered as the consequence of Eq.(1), certain inference paradigms can also be used to infer the supporting strength of customer loyalty and satisfaction. These reasoning paradigms are based on EBR and hybrid reasoning paradigms. It is obvious that one reasoning paradigm will not be appropriate to obtain real customer loyalty and satisfaction. Based on these reasoning paradigms, a hybrid multiagent system should be proposed for managing customer experience in e-services.

6.3 Agents Within MEMES

For the viewpoint of CEM, an e-service customer i creates uncertainty on three temporary levels: uncertainty about past experiences and behaviors in PEB, current behaviors and preferences in CEB, future needs in FNB. Brohman *et al.* propose the following four strategies to resolve these uncertainties [6]:

1. Transaction strategy. Data in the CEB is analysed to identify profit concerns.
2. Data strategy. Data for the PEB is captured and used to service customers by providing access to a detailed record of past transactions.
3. Inference strategy. An inference strategy or decision making is used to infer future customer needs based on data in PEB and CEB.
4. Advice strategy. A recommendation strategy is used to provide accurate and reliable advice to customers based on their future needs.

Based on these strategies, the MEMES comprise six intelligent agents: ESI agent, Customer experience (CE) miner, E-service advisor, E-service negotiator, Experience decision maker, E-services manager, and Interface agent. In what follows, we will look at each of them in some detail:

- The ESI agent is an e-service information gathering agent. It is a mobile agent that proactively roams around the main search engines in the Internet such as Google and Yahoo. It interacts and collaborates with them in order to search and analyse the required e-services information indirectly from individual websites and then puts it in the corresponding data or knowledge bases [28].
- The CE miner is an autonomous agent that discovers type 2 experiences of e-service customers based on experience discovery algorithms, which is similar to knowledge discovery algorithms [4]. It also discovers useful experience patterns, and then presents the discovered experience in a machine-readable form, in order to facilitate the e-service advisor, e-service negotiator and customer experience manager to make decisions. Induction based reasoning can be used in type 2 experience discovery as mentioned in Section 5.3. The discovered type 2 experiences of e-service customers will be stored in KEB.
- The e-service advisor is a proactive agent that makes recommendation of e-services based on the information or data available in ESB, KEB and CEW. The proposed recommendation will be normally forwarded to the e-service customer by the interface agent. There are many intelligent strategies for recommendation. Case-based recommendation is one of them, which will be discussed in some detail in Section 7.
- The e-service negotiator is a mobile and proactive agent that performs not only integrative but also distributive negotiation strategies during negotiation with the e-service buyer agent [28]. Because business negotiation is complex and difficult in some cases, the intelligence of the e-service negotiator lies in that it can change its negotiation strategies immediately according to the changing (information) resources or cases. It prepares a necessary compromise under bargaining. Thus, the e-service negotiator may use all available human inference methods such as case-based reasoning (CBR), rule-based reasoning (RBR), fuzzy reasoning, and EBR [29] in different cases, if necessary. It can thus deceive or mislead the e-service buyer agents in some cases.

- The experience decision maker that will be examined in some detail in the next subsection is an autonomous agent that performs EBR based on the available customer experience in e-services. Because of uncertainty, incompleteness and inconsistency in customer experiences, it has to use fuzzy EBR in order to make decisions in a deceptive e-service environment.
- The e-service manager is an intelligent agent that plays a leading role in the MEMES. Its main task is to decide which agent should do what and how to deal with an e-service transaction. Another task is to coordinate the tasks among all other agents.
- The transaction analysis agent, which is not shown in Fig. 5, is also an important intelligent agent in the MEMES. It analyses and computes every transaction cost of deals that the e-service negotiator suggests and then submits the related information or data to the e-service negotiator, the managing agent or interface agent, if necessary [28].
- The interface agent is an intelligent agent consisting of the dynamic interactive exchange of information and service that occurs between the customer and the e-services via the Web, online or emails or in some other ways [24], (p. 141). It proactively interacts, cooperates with the e-service buyers (agents) and gets the supply-demand information. At the same time, it gets special information about the e-service customers in the e-service market and then stores it in PEB, CPB and FNB respectively. The interface agent also interacts with the e-service manager and transfers the transaction message to the e-service buyers.
- The experience base (EB) manager, which is not shown in the Fig. 5, is responsible for administering KEB, ESB and CEW. Its main tasks are creation and maintenance of KEB, ESB and CEW, experience evaluation, experience reuse, experience revision, and experience retention. Therefore, the functions of the experience base manager are an extended form of the functions of a CBR system because case base creation, case retrieval, reuse, revision and retention are the main tasks of the CBR system [30].

It should be noted that CRM programs can be used to extract and discover type 2 experience of customers in e-services, because such a software can keep track of all customer contacts for the customer's lifetime [24] (pp.163-164). This also implies that CEM is a further development based on CRM.

6.4 EBR Decision Maker

Decision making is an important part for the MEMES based on discovered experience patterns, existing experiences and data in KEB, ESB and CEW. The EBR (experience-based reasoning) decision maker is an intelligent agent that can be considered as a high level system model of the experience decision maker of the MEMES mentioned in Section 6.3. It uses eight fundamental reasoning paradigms to perform EBR and then find an optimal decision strategy.

The system architecture of the EBR decision maker consists of experience users, U, which are either human customers or intelligent agents delegated by the human users or other systems, as shown in Fig. 6. The EBR decision maker, as a system, mainly consists of a user interface, a global experience base (GEB) and a multi-inference engine (MIE). The user interface consists of some kinds of natural language processing systems that allow the user to interact with the EBRS [19] (p. 282). The GEB consists of all the experiences that the system collects periodically and the new experiences discovered when the system is running. For example, GEB can be considered as a collection of KEB, ESB and CEW in the MEMES. The MIE is a multi-inference engine consisting of the mechanism for implementing eight reasoning paradigms based on the eight inference rules for EBR to manipulate the GEB to infer experience requested by the experience-user [30]. The remarkable difference between the mentioned EBR decision maker and the traditional KBS lies in that the latter's inference engine is based on a unique reasoning paradigm (or inference rule), while the MIE is based on many different reasoning paradigms.

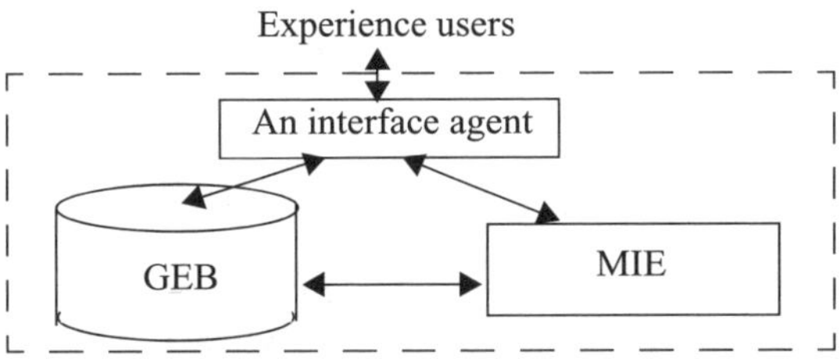

Fig.6. A system architecture of the EBR decision maker [30].

The *interface agent* is an advisor to help the experience-user to forward the problem of the experience-user for further processing.

6.5 Agents Workflowing in MEMES

Now let us have a look at how the MEMES works. The e-service customer, *C*, inquires the interface agent about the provided e-services. The interface agent asks *C* to *login* and fill in information about *C*'s past experience preferences and current preferences for e-services which will automatically be stored in PEB and CPB. Then the interface agent forwards the information from *C*, to e-service advisor or e-service negotiator. The e-service advisor can use the data of ESB and a certain recommendation strategy to recommend e-service which is then forwarded by the interface agent to *C*. Sometimes, *C* does not agree to the recommendation from the e-service advisor, s/he likes to negotiate over the price of the provided e-services. In this case, e-service negotiator has to use certain negotiation strategies [28] to negotiate with *C* over the price or related items.

The negotiation should be helped by an experience decision maker, because the e-service negotiator might not understand which negotiation strategies or reasoning paradigms that *C* has used at the negotiation. In this case, the experience decision maker will first recognize the reasoning paradigms that *C* has used and then select one of other EBR paradigms to make decisions under the deceptional environment, so does the *C*, because any negotiation usually hides some truths in order to get advantages in the interest of conflict.

If *C* accepts one of the e-services after recommendation and negotiation, then the MEMES completes this e-service transaction. In this case, the EB manager will look at whether this case is a new customer experience in the context of past experiences, current preferences and future needs. If yes, then the manager will add it to the corresponding experience bases such as PEB, CPB or FNB. Otherwise, it will keep some routine records to update all the related experience bases. If *C* does not accept the recommended e-service, the interface agent will ask *C* to adjust some attributes of her/his requirement, and then will further forward the revised requirement to the related agents within the MEMES for further processing.

Furthermore, ESI agent always searches, collects information in the e-services world and then saves the acquired information into ESB. CE miner tries to discover the experience models or templates from ESB and KEB as well as from CEW. The discovered experience models and templates will be used for e-service recommendation and negotiation. The e-service manager is overseeing and coordinating the activities of agents within the MEMES.

7 Case-based E-services Recommendation

In order to manage customer experience effectively, it is necessary to recommend different e-services to different customers based on the data in PEB, CPB and certain intelligent techniques. CBR has been successful in making recommendation of business activities such as in e-commerce [28][32]. In what follows, we will examine case-based recommendation in e-services, which is a Web-based CBR subsystem. This is a system model of the e-service advisor in the MEMES, as shown in Fig. 5, that is, the e-service advisor can be considered as a case-based e-services recommender system (CERS), as shown in Fig. 7.

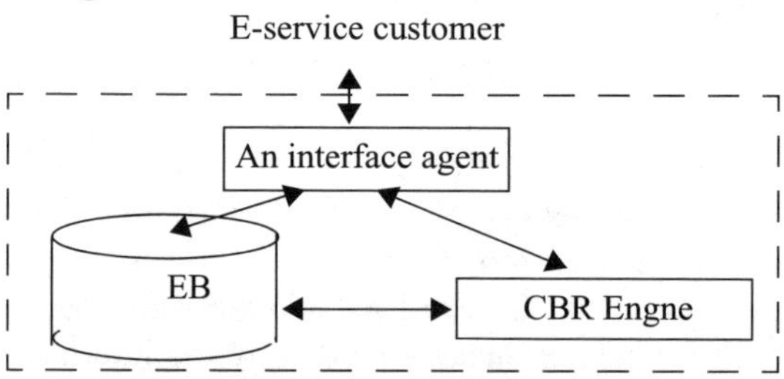

Fig.7. A general architecture of the CERS.

The e-service customer describes her/his demand p' to the CERS through its interface. This demand is normalized into a structured problem description p'. Then the CERS uses its similarity metric mechanism (the inference engine based on $\sim$) to retrieve its case base, which consists of cases, each of which is denoted as $c = (p, q)$, where p is the structured problem description and q is the solution description. In the CERS, *problem description* and *solution description* correspond to *(customer) demand description* and *e-service description* respectively. The similarity metric mechanism performs similarity-based reasoning that can be formalized as [10]:

$$\frac{P', P' \sim P, P \rightarrow Q, Q \sim Q'}{\therefore \ Q'} \tag{2}$$

where P, P', Q, and Q' represent compound propositions, $P' \sim P$ means that P and P' are similar. Q and Q' are also similar.

From a logical viewpoint, the case retrieval process is used to find the following case set from the case base EB in the CERS,

$$C(p') = \{c \mid c = (p, q), p \sim p'\} = \{c_1, c_2, \ldots, c_n\} \tag{3}$$

where n is a positive integer, c_i, $i = 1, 2, \ldots, n$ are all cases with their problem description p similar to the current problem description p'. Usually, $C(p') = \{c_1, c_2, \ldots, c_n\}$ satisfies the following property: for any integer i, $1 \le i < n$ and $c_i = (p_i, q_i)$,

$$s(p_i, p') \ge s(p_{i+1}, p') \tag{4}$$

where $s(\ \cdot\ )$ is a similarity metric, which measures the similarity between one object and another.

If n is small, then the CERS will directly recommend the e-service descriptions of $c_1, c_2, \ldots, c_n$, $q_1, q_2, \ldots, q_n$, through the interface agent. If n is very large, the CERS has to recommend the e-service descriptions of the first m cases in $c_1, c_2, \ldots, c_n$; that is, $q_1, q_2, \ldots, q_m$, to the customer, in order to meet the needs of the customer, where $1 \le m < n$. This process can be called *e-service recommendation*. More generally, this process can be considered as *case recommendation*, because the CERS usually provides the customer with not only the recommended e-services but also the customer demand description. Therefore, case recommendation is a process necessary for applying CBR in e-services.

After the customer obtains the recommended e-service descriptions from the CERS, s/he will evaluate them and then select one of the following:

1. Accept one of the recommended e-services, q_k, and order it, where $1 \le k \le m$.
2. Adjust her/his demand descriptions p' and then send them to the CERS.
3. Refuse the recommended e-services and leave the CERS.

It is obvious that among these three cases only the first two require further discussion.

For the first case, the deal was successfully done and the CERS routinely updates the successful case $c_k = (p_k, q_k)$ in the case base, EB. At the same time, the CERS has reused the case successfully.

For the second case, the demand adjustment is the process of demand adaptation that corresponds to problem adaptation. An obvious difference between e-services and traditional services really lies here, that is, it is relatively difficult for a customer in the traditional service to adjust her/his demand when s/he is in the traditional market. Usually what s/he can do is to buy what s/he sees. However, in e-services, a customer has a much broader space for selecting services. In fact, all available e-services in the Internet can be searched and selected by any customer if s/he can access the Internet. In this case, s/he frequently adjusts her/his demand and tries to get more satisfactory e-services.

After having adjusted the demand, the customer then submits it to the CERS. The CERS will do case retrieval and case recommendation again. Therefore, the problem submission, case retrieval, case recommendation, and problem (demand) adaptation constitute a cycle. This is a cyclical process mechanism of the CERS, which differs from the CBR cycle, and is illustrated in Fig. 8.

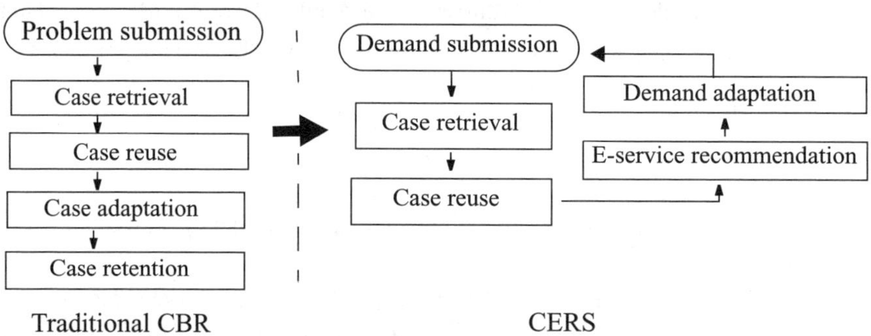

Fig.8. A cyclical process in CERS after [32].

8 Conclusions and Future Work

The chapter examined customer experience management (CEM), e-services and their relationships. More specifically, it proposed a unified architecture, SSES, and an intelligent system architecture, MEMES. The SSES unifies e-services, Web services and infrastructure services into a hierarchical framework. The MEMES is a high level system model for implementing multiagent CEM system for e-services. Both architectures tie together methodologies, techniques, and applications into a unified framework that includes both logical and intelligent embodiment of the aspects of CEM in e-services. The chapter also proposed a cybernetic model for e-services and explored case-based e-service recommendation. The cybernetic model provided a new insight into B2B e-services with feedback from the e-service customers. It is argued that the proposed approach will facilitate the development of CEM, e-business and e-services.

Applying intelligent techniques to CEM in e-services is still a new topic for experience management, AI and e-services. However, intelligent CEM will significantly alter customer processing and represent major competition for many intermediary services between e-service suppliers and customers [18]. In future work, we will develop a system prototype based on the proposed MEMES. We will also explore type

2 experience discovery algorithms and techniques which will be used for decision making in deceptional environments such as in e-services.

References

[1] Alonso G, Casati F, Kuno H, and Machiraju V. *Web Services: Concepts, Architectures and Applications*. Berlin: Springer-Verlag, 2004

[2] Avison DE and Fitzgerald G. *Information Systems Development: Methodologies, Techniques and Tools* (3rd edn), London: McGraw Hill International (UK), 2003

[3] Bartlmae K. Optimizing data-mining processes: A CBR based experience factory for data mining. In: *Proc 5th Intl Computer Science Conference* (ICSC'99), Hong Kong, China, December 13-15, 1999

[4] Becerra-Fernamdez I, Gonzalez A, and Sabherwal R. *Knowledge Management: Challenges, Solutions and Technologies*. Upper Saddle River, NJ. 2004

[5] Bergmann R. *Experience Management: Foundations, Development Methodology and Internet-Based Applications*. LNAI 2432. Berlin: Springer 2002

[6] Brohman MK, Watson RT, Piccoli G, and Parasuraman AA. Data completeness: A key to effective net-based customer service systems. *Comm of the ACM*, 46(6) 2003, 47-51

[7] Casati F, Shan E, Dayal U, and Shan M.-C. Business-oriented management of Web services. *Comm of the ACM*. 46(10) 55-59

[8] Chua A. Knowledge management systems architecture: A bridge between KM consultants and technologies. *Intl J of Information Management*, 24, 2004, 87-98

[9] Drucker P. *Post-Capitalist Society*. New York: Harper Business, 1993

[10] Finnie G and Sun Z. A logical foundation for the CBR cycle. *Intl J Intell Syst* 18(4) 2003, 367-382

[11] Ferris C and Farrell J. What are Web services, *Comm of the ACM*, 46(6) 2003, 33-35

[12] Gurugé A. *Web Services: Theory and Practice*. Amsterdam: Elsevier Inc. 2004

[13] Hoffman KD. Marketing + MIS = E-Services, *Comm of the ACM*, 46(6) 2003, 53-55

[14] Holtzblatt K. Designing for the mobile device: Experiences, challenges, and methods, *Comm of the ACM*, 48(7) 2005, 33-35

[15] Kolodner JL (ed.). Special issue on Case-Based Reasoning. *Machine Learning*, 10(3) 1993

[16] Kreger H. Fulfilling the Web services promise. *Comm of the ACM*, 46(6) 2005, 29-34

[17] Miller G. .NET vs. J2EE, *Comm of the ACM*, 48(7) 2005, 64-67

[18] Muther A. *Customer Relationship Management: Electronic Customer Care in the New Economy*.Berlin: Springer, 2002

[19] Nilsson NJ. *Artificial Intelligence. A New Synthesis*. San Francisco, California: Morgan Kaufmann Publishers Inc. 1998

[20] Petrie C, Genesereth M, et al. Adding AI to Web services. In: van Elst L, Dignum, V and Abecker A (Eds): AMKM 2003, *LNAI 2926*, 2003, 322-338

[21] Pressman RS. *Software Engineering: A Practitioner's Approach* (5th Edn), Boston: McGrawHill Higher Education, 2001

[22] Rust RT and Kannan PK. E-service: A new paradigm for business in the electronic environment. *Comm of the ACM*, 46(6) 2003, 37-42

[23] Rust RT and Kannan PK. *E-Service: New Directions in Theory and Practice*. ME Sharpe, Armonk, New York, NY, 2002

[24] Schmitt BH. *Customer Experience Management: A revolutionary approach to connecting with your customers*. John Wiley & Sons, Inc. Hoboken, NJ, 2003

[25] Satzinger, JW, Jackson, RB and Burd SD, *Systems Analysis and Design in a Changing World* (3rd edn), Boston: Thompson Learning, 2004

[26] Song H. E-services at FedEx, *Comm of the ACM*, 46(6) 2003, 45-46
[27] Sun Z. A waterfall model for knowledge management and experience management. In: *Proc. HIS 2004*, December 6-8, Kitakyushu, Japan, IEEE Press, 2004, 472-475
[28] Sun Z and Finnie G. *Intelligent Techniques in E-Commerce: A Case-based Reasoning Perspective*. Berlin, Heidelberg: Springer-Verlag, 2004
[29] Sun Z and Finnie G. Experience based reasoning for recognising fraud and deception. In: *Proc. Inter Conf on Hybrid Intelligent Systems* (HIS 2004), December 6-8, Kitakyushu, Japan, IEEE Press, 2004, 80-85
[30] Sun Z and Finnie G. MEBRS: A multiagent architecture for an experience based reasoning system. In: Khosla R, Howlett RJ and Jain LC (eds) Knowledge-Based Intelligent Information and Engineering Systems: Part I, *LNAI 3681*, Berlin: Springer, 2005, 972-978
[31] Sun Z and Finnie G. Experience management in knowledge management, In: Khosla R, Howlett RJ, Jain LC (eds) *LNAI 3681*, Berlin: Springer, 2005, 979-986
[32] Sun Z and Finnie G. A unified logical model for CBR-based e-commerce systems. *Intl J of Intelligent Systems*, 20(1), 2005, 29-26
[33] Sun Z. Human-like evolution of systems development, submitted for publication, 2006
[34] Sun Z, Hao G. HSM: A hierarchical spiral model for knowledge management, In: *Proceedings of the 2nd Intl Conf. on Information Management and Business (IMB2006)*, 13-16 February, 2006, Sydney Australia, 542-551
[35] van Elst L, Dignum V, and Abecker A. Towards agent-mediated knowledge management. In: van Elst L, Dignum V, and Abecker A (eds) *Agent-mediated Knowledge Management*. LNAI 2926, Berlin: Springer Verlag, 2004, 1-30
[36] Vissers CA, Lankhorst MM, and Slagter RJ. Reference models for advanced e-services. In: Mendes MJ, Suomi R, and Passons C (eds) *Digital Communities in A Networked Society*. Kluser Academic Publishers, Boston/ Dordrecht. 2004, 369-393
[37] Voss A. Towards a methodology for case adaptation. In: Wahlster W (ed.): *Proc 12th European Conf on Artificial Intelligence* (ECAI'96), 1996, 147-51
[38] Weiss G (ed.). *Multiagent Systems: A modern approach to Distributed Artificial Intelligence*. Cambridge, Massachusetts, USA: MIT Press, 1999
[39] Zimmermann HJ. *Fuzzy Set Theory and its Application*. Boston/Dordrecht/London: Kluwer Academic Publishers, 1991
[40] http://wm2003.aifb.uni-karlsruhe.de/workshop/ w06/GWEM2003_CfP_english.html, accessed on 21 October 2005
[41] http:// www.iese.fraunhofer.de/experience_management/, accessed on 25 October 2005
[42] http://www.google.com.au/search?hl=en&lr=&oi=defmore&defl=en&q=define:Web+Services, accessed on 22 October 2005
[43] http://en.wikipedia.org/wiki/Customer_experience_management, accessed on 26 October 2005
[44] http://www-128.ibm.com/developerworks/library/specification/ws-bpel/. Accessed on 31October 2005.

E-Service Cost Benefit Evaluation and Analysis

Jie Lu*, Chenggang Bai[#] *, and Guangquan Zhang*

*Faculty of Information Technology, University of Technology, Sydney
PO Box 123, Broadway, NSW 2007, Australia
{zhangg, jielu}@it.uts.edu.au

[#]Department of Automatic Control
Beijing University of Aeronautics and Astronautics
bcg@buaa.edu.cn

Abstract. Based on an e-service cost-benefit factor framework, an initial cost-benefit factor-relation model is proposed through analyzing a questionnaire survey results. The factor-relation model is then considered as domain knowledge, and the data collected is as evidence to the inference-based verification. This study applies Bayesian network technique to analyze and verify the relationships among cost factors and benefit factors in the development of e-services. A set of useful findings have been obtained for the costs involved in moving services online against the benefits received by adopting e-service applications. These findings have potential to improve the strategic planning of businesses by determining more effective investment items and adopting more suitable development activities in e-services development.

1 Introduction

Web-based electronic service (e-service) applications are assisting businesses in building more effective customer relationships and gaining competitive advantage through providing interactive, personalized, faster online services to fulfill customer demands (Chidambaram 2001). After several years experience of e-service development provision, businesses now urgently need to evaluate the cost involved in moving service online against the benefits received by adopting e-service for planning their further development. Importantly, businesses have obtained related e-service running data and knowledge, which makes it possible to identify in what items of investment for an e-service application effectively contribute to what benefit aspects of business objectives.

Literature review has shown increasing interest of researchers in evaluating and measuring the development of e-service applications (Wade and Nevo 2005, DeLone 2004). Many researches have been conducted to evaluate e-services from various views and using various methods. In gen-

J. Lu et al.: *E-Service Cost Benefit Evaluation and Analysis*, Studies in Computational Intelligence (SCI) **37**, 389–409 (2007)
www.springerlink.com

eral, the research in e-service evaluation can be classified under four major categories. The first one is the evaluation for the features, functions or usability of e-service systems. It is often combined with the evaluation of the use of related websites. Typical approaches used in this category of research are testing, inspection and inquiry (Hahn and Kauffman 2002). These approaches are often used together in analyzing a web search or a desk survey such as Ng et al. (1998), Smith (2001) and Lu et al. (2001). The second category is customer satisfactory evaluation. Various evaluation criteria and factors have been identified and related evaluation systems have been designed for obtaining customers' feedback and measuring the degree of their satisfaction to current e-services provided. Questionnaire-based survey and multi-criteria evaluation systems are mainly used to conduct this kind of research (Lin, 2003). The third category is e-service investment analysis which has been conducted for evaluating and justifying investment in an e-service application. Some examples can be found from Giaglis, Paul and Doukidis (1999), Drinjak, Altmann and Joyce (2001), and While Amir et al. (2000). Significant results also have been reported in the fourth category. Some evaluation models and frameworks about e-service applicationss have been established (Lee et al. 1999, Zhang and von Dran, 2000, and Hahn et al. 2002).

However, very few researches have been conducted to evaluate e-service applications from the view of e-service providers with deeply analysis in the relationships between their costs and benefits. Although some cost items on e-service is measured, there is a lack of exploration for possible factors which have internal relations to link these cost items with related business benefits. As a business, it would more like to know if its investment in e-service applications is successful by analyzing the cost caused and benefit obtained. The cost involves several aspects, such as software development, database maintains, website establishment and staff training. Similarly, the benefit obtained through developing an e-service system also includes many items, such as attracted more customers, formed more good business image, and had more competitive advantages. Therefore, a business, as an e-service provider, is more interested in and urgent to know in which aspect(s) its investment is more important and effective than other parts for achieving its business objectives, and in which aspect(s) its investment can make more obviously benefits for its business objectives. These results will directly or indirectly support better business strategy making in e-service application development.

Our research reported in (Lu and Zhang 2003) identified some inter-relationships and interactive impacts among e-service functions, e-service development attributes, the cost caused and the benefits received via providing e-services to customers. In particular, it has examined what cost items of an e-service have a more significant contribution to particular benefit items. These inter-relationships were identified by mainly using linear regression and ANOVA analysis approaches based on collected data. As some relationships are inconsistent, and some are non-linear, it is very necessary to verify and clarify these relationships by applying a more suitable intelligent technique. This paper reports how these relationships are identified and verified by applying Bayesian networks.

After the introduction, this chapter reviews our previous work including an e-service cost-benefit factor framework and hypotheses designed in Section 2. Data collection, basic data analysis and the establishment of a cost-benefit factor-relation model are shown in Section 3. Section 4 reports the process of establishing a cost benefit Bayesian network and conducting cost benefit factor inference. Section 5 shows some results on the relationships between cost factors and benefit factors in the development of e-services. Conclusions are discussed in Section 6.

2 A Cost-Benefit Factor Framework and Hypotheses Design

2.1 E-service Cost-Benefit Factor Framework

E-service benefit is concerned with the benefits gained through employing e-services, which attribute takes into account the strategies, policies and types of companies involved when developing e-service applications. Drinjak *et al.* (2001) listed a number of benefit items within three categories of web applications. These items include: providing services 24 hours a day and seven days a week (24*7); effective promotion of the company, together with the products and services it produces; enhancing the quality and speed of customer services; creating competitive advantages and subsequently avoiding competitive disadvantages; enticing shoppers and encouraging customer interaction; supporting core business functions which are integral to business strategy; providing new business opportunities, increasing market presence and facilitating online purchasing. Lu et al (2001) also listed 21 benefit factors, and identified eight as the core benefit factors, through a survey conducted in the businesses of New Zealand. The eight factors are: accessing a greater customer base; broadening market reach; lowering of entry barrier to new markets and cost of acquiring new customers; alternative communication channel to customers; increasing services to customers; enhancing perceived company image; gaining competitive advantages; and potential for increasing customer knowledge. Based on the above research results, 16 e-service benefit factors are identified and are used in this study.

E-service cost is the expenses incurred in adopting e-services. Lu et al (2001) also tested 19 cost factors and identified eight core cost factors: expense of setting up an e-service application; maintaining an e-service application; internet connection; hardware/software; security concerns; legal issues; staff training, and rapid technology changes. These eight cost items are used in this study as cost factors.

Figure 1 shows the 16 benefit factors and the 8 cost factors identified.

E-service benefits (B)

Fact.	Description
B_1	Building customer relationships
B_2	Broadening market reach
B_3	Lowering of entry barrier to new markets
B_4	Lowering the cost of acquiring new customers
B_5	Global presence
B_6	Reducing information dissemination costs
B_7	Reducing advertising media costs
B_8	Reducing operation (transaction) costs
B_9	Reducing transaction time
B_{10}	Reducing delivery time
B_{11}	Gathering information to create customer profiles
B_{12}	Customer and market research facility
B_{13}	Cooperation between companies to increase services
B_{14}	Enhancing perceived company image
B_{15}	Realizing business strategies
B_{16}	Gaining and sustaining competitive advantages

E-service costs (C)

Fact.	Description
C_1	Expense of setting up E-service
C_2	Expense of maintaining E-service
C_3	Internet connection cost
C_4	Hardware/software cost
C_5	Security concerns cost
C_6	Legal issues cost
C_7	Training cost
C_8	Rapid technology change cost

Fig. 1. E-service cost-benefit factor framework

2.2 Hypotheses design

A set of hypotheses are designed based on the proposed e-service cost benefit factor framework. The set of hypotheses are about the effects of these cost factors on these benefit factors. Hypothesis testing, being the most common method used in inferential statistics, is completed in the study. A null hypothesis is used for each hypothesis testing, to determine whether the data are strong enough to reject it. That is, the null hypothesis either will or will not be rejected as a viable possibility. This paper is restricted to presenting three interesting hypotheses proposed in the study and to discussing related test results.

Hypothesis $H_{2\text{-}16}$:

H_0**:** Cost factor 'Maintaining E-service applications' (C_2) has no effect on benefit 'Gaining and sustaining competitive advantages' (B_{16}).

$H_{2\text{-}16}$**:** Cost factor 'Maintaining E-service applications' (C_2) has an effect on benefit 'Gaining and sustaining competitive advantages' (B_{16}). There is a significant difference in the benefit 'Gaining and sustaining competitive advantages' for different groups of companies that have different levels in 'Maintaining E-service applications'.

Hypothesis $H_{3\text{-}11}$:

H_0**:** The cost factor 'Internet connection cost' (C_3) has no effect on benefit 'Gathering information to create customer profiles' (B_{11}).

$H_{3\text{-}11}$**:** The cost factor 'Internet connection cost' (C_3) has an effect on benefit factor 'Gathering information to create customer profiles' (B_{11}).

Hypothesis $H_{4\text{-}5}$:

H_0**:** The cost factor 'Hardware/software cost' (C_4) has no effect on benefit factor 'Global presence' (B_5).

$H_{4\text{-}5}$**:** The cost factor 'Hardware/software cost ' (C_4) has an effect on benefit factor 'Global presence' (B_5).

3. Establishment of Cost Benefit Factor Relation Model

3.1 Data collection

This study collected data concerning e-service development costs and benefits from a sample of Australia companies (e-service providers). In order to select the sample, this study first conducted a web search for finding companies which had adopted e-services on an appropriate level and period. A total of 100 companies were randomly selected from Yellow Pages Online (NSW, Australia) http://www.yellowpages.com.au under Tourism/Travel (including Accommodation and Entertainment) and IT/Communication categories (including Information Services). A questionnaire based survey was then conducted by sending questionnaire to these sample companies. Out of 34 questions in the questionnaires, some were related to the costs of developing an e-service application, and some were related to the benefits obtained from developing an e-service application. A total of 48 completed responses are used in this study. In the questionnaire, all cost related questions listed use a five-point Likert scales: 1--not important at all, 5--very important. For example, if a company thinks the cost of maintaining an e-service is very important it records the degree of importance of two factors as 4 or 5. A 5-point scale is also used for present benefit assessment: 1--low benefit, 5--very high benefit. For example, if a company considers that, currently, their e-services only help a little in customer relationship management, then the company would perhaps score '1' or '2' on the present benefit assessment for benefit factor B_1.

3.2 Basic data analysis

The survey data is firstly used for some basic analysis to answer the following four questions.

1) What are the main benefit factors in adopting an e-service application?

The questionnaire was designed to cover proposed benefit factors at an appropriate level, and of an appropriate form. The subjects were asked to indicate their present benefits assessment and ideal rating for each of the benefit factors. The current benefit assessment relates to the assessment of the status of each respondent's E-service application, comparing it with where they would ideally like it to be. The ideal rating for benefit factors is tested on a 5-point scale. Here '1' represents 'not important at all', and '5' 'very important'. For example, if a company considers that one of the most important benefits is enhancement of perceived company image, the company might score 5 on the ideal rating of the factor 'enhancing perceived company image'. Table 1 shows the results of the ideal rating for benefit factors. It was found that B_{14} (Enhancing perceived company image), B_{16} (Gaining and sustaining competitive advantages), B_1 (Building customer relations), B_2 (Broadening market reach), B_{15} (Realizing business strategies) and B_4 (Lowering the cost of acquiring new customers) received relatively higher rankings. This result means that the companies in our trial think these factors are more important than others for a successful business, and they have higher expectations of benefits in these areas.

Table 1. Ideal rating for benefit factors

Ideal rating	No. of companies in each benefit factor															
	B_1	B_2	B_3	B_4	B_5	B_6	B_7	B_8	B_9	B_{10}	B_{11}	B_{12}	B_{13}	B_{14}	B_{15}	B_{16}
1	0	0	3	1	5	5	3	5	5	5	4	2	4	1	1	1
2	2	4	6	2	7	3	4	3	2	4	5	4	5	2	3	2
3	8	9	9	9	12	6	9	9	6	9	4	11	6	3	8	5
4	10	12	15	16	5	16	15	13	14	6	16	12	11	9	10	8
5	27	21	12	18	17	16	15	15	19	19	15	15	18	30	23	29
NA	0	1	2	1	1	1	1	2	1	4	3	3	3	2	2	2
Average	4.3	4.1	3.6	4.0	3.5	3.8	3.8	3.6	3.9	3.7	3.8	3.8	3.8	4.4	4.1	4.4

2) What benefits have been obtained through developing an e-service application?

Companies are all interested in maximizing the business value of E-services (Giaglis *et al.,* 1999). They have adopted business strategies which address the requirements of interoperability, quality of customer service, evolution and dependability. They expect to know which factors affect which aspects of e-service benefits and how E-service can increase these business benefits, by comparing related expenses with those of associated investments. In order to complete such an analysis, this study not only explores which benefit factors are more important to business but also seeks to find in which areas companies have obtained higher benefits, and alternatively which are lower. The assessment result, shown in Table 2, indicates that companies have obtained expected benefits on B_{14} (Enhancing perceived company image), B_{16} (Gaining and sustaining competitive advantages), B_1 (Build customer relationships) and B_2 (Broadening market reach) as these factors have a relatively high average score of current benefit assessment. In another words, companies are satisfied with these areas where benefits were obtained.

Table 2. Current benefit assessment

Benefit factor (scale)	No. of companies' responses in each benefit factor															
	B_1	B_2	B_3	B_4	B_5	B_6	B_7	B_8	B_9	B_{10}	B_{11}	B_{12}	B_{13}	B_{14}	B_{15}	B_{16}
1	2	1	6	4	9	5	6	7	7	7	10	9	7	1	3	2
2	7	6	13	13	10	8	7	7	10	9	7	12	8	5	7	8
3	16	21	18	15	8	17	13	18	12	13	10	17	12	12	18	15
4	14	12	7	9	13	12	15	8	10	8	14	3	13	12	12	12
5	9	8	3	7	8	5	6	7	9	9	5	5	6	17	7	10
NA	0	0	1	0	0	1	1	1	0	2	2	2	2	1	1	1
Average	3.4	3.4	2.7	3.0	3.0	3.1	3.1	3.0	3.0	3.0	2.9	2.6	3.0	3.8	3.3	3.5

3) What cost factors are more important when adopting an E-service application?

This works the same as for the benefit factor identification. For example, if a company thinks the cost of maintaining an E-service is very important it records the degree of importance as 4 or 5. Table 3 shows C_2 (Maintaining E-service applications) as the most important factor, and C_5 (Security concerns costs) as the second most important as they received relatively high average values (3.8, 3.7) of importance (weight). This finding shows that the companies have had, or would have, a higher investment in these important cost items.

Table 3. Weights of cost factors

Cost weights	No. of companies in each cost factor							
	C_1	C_2	C_3	C_4	C_5	C_6	C_7	C_8
1	4	1	4	2	3	7	4	4
2	6	7	8	5	7	6	8	4
3	10	7	12	16	7	11	12	15
4	13	15	11	16	10	11	17	14
5	14	17	12	9	18	11	5	10
NA	0	0	0	0	2	2	2	01
Average	3.6	3.8	3.4	3.5	3.7	3.3	3.2	3.4

4) What cost items are higher than estimated when developing an E-service application?

This study also explores which items result in a higher cost than estimated, and which a lower cost. The assessment result is shown in Table 4. It was found that there was no cost factor with an average assessment value higher than 3.5. This means all costs were not much higher than estimated. However, some factors are still higher than others. For example, the average assessment values on C_1 (expense of setting up E-service) and C_4 (hardware/software) are relatively higher (3.3, 3.2) among the eight factors. Therefore, the differences between the actual cost and the estimated cost in the two areas were relatively bigger than other cost items within the companies.

Table 4. Current cost assessment

Cost assessment	No. of companies in each cost factor							
	C_1	C_2	C_3	C_4	C_5	C_6	C_7	C_8
1	7	6	15	6	7	9	9	8
2	5	10	10	7	9	11	9	7
3	10	12	15	15	11	14	17	9
4	17	17	7	13	12	8	9	19
5	9	3	1	7	7	4	2	3
NA	0	0	0	0	2	2	2	2
Average	3.3	3.0	2.4	3.2	3.1	2.7	2.7	3.0

3.3 Initial cost-benefit factor-relation model

By completing a set of ANOVA tests for data collected from the survey, a set of ‘effect’ relationships between cost and benefit factors have been obtained (Lu and Zhang 2003). These relationships reflect that certain cost factors have a significant effect on certain benefit factors. These effects are presented in a cost-benefit factor-relation model (Fig. 2). The lines in the model express the ‘effect’ relationships between related cost factors and benefit factors. For example, cost factor C_1 has significant effect on benefit factor B_5, and cost factor C_2 has significant effect on benefit factors B_1, B_2,

B_{13}, B_{14}, B_{15}. Although every cost factor makes direct or indirect contributions to all benefit factors to a certain degree, some cost factors are more important than others for improving certain benefit factors. It is therefore necessary to verify these initially identified relationships, and inference some uncertain relationships. Bayesian network approach can effectively support the analysis.

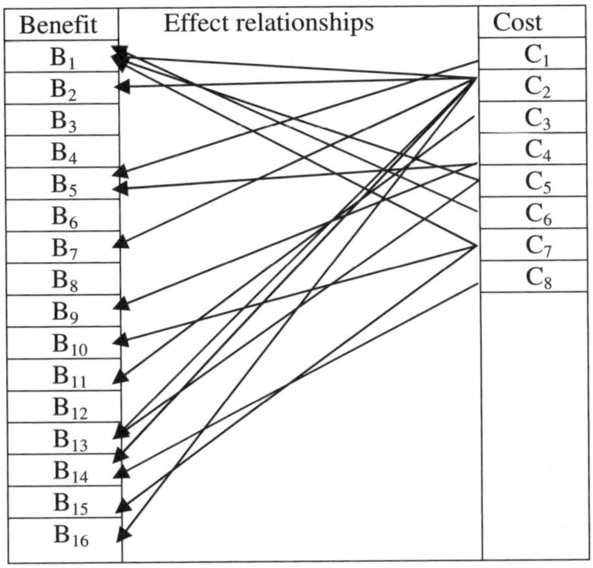

Fig. 2. Cost-benefit factor-relation model

4. Using Bayesian Networks for Cost-Benefit Factor Relationship Analysis

4.1. Bayesian Network Approach

The Bayesian network approach is a powerful knowledge representation and reasoning tool under conditions of uncertainty. A Bayesian network $B =< N, A, \Theta >$ is a directed acyclic graph (DAG) $< N, A >$ with a conditional probability distribution (CPD) for each node, collectively represented by Θ, each node $n \in N$ represents a variable, and each arc $a \in A$ between nodes represents a probabilistic dependency (Pearl 1988). In a practical application, the nodes of a Bayesian network represent uncertain factors, and the arcs are the causal or influential links between these factors. The association with each node is a set of CPDs that model the uncertain relationships between each node and its parent nodes.

Using Bayesian networks to model uncertain relationships has been well discussed in the theory by researchers such as Heckerman (1990) and Jensen (1996). Many applications have also proven that Bayesian network is an extremely powerful technique for reasoning the relationships among a number of variables under uncertainty. For example, Heckerman et al. (1995) applied Bayesian network approach successfully into lymph-node pathology diagnosis. Breese and Blake (1995) applied Bayesian network technique in the development of computer default diagnosis.

Comparing with other inference analysis approaches, Bayesian network approach has four good features of inference in its applications. Firstly, unlike neural network approach, which usually appears to users as a "black box", all the parameters in a Bayesian network have an understandable semantic interpretation (Myllymaki, 2002). This feature makes users to construct a Bayesian network directly by using domain knowledge. Secondly, Bayesian network approach has an ability to learn the relationships among its variables. This not only lets users observe the relationships among its variables easily, but also can handle some data missing issues (Heckermain 1997). Thirdly, Bayesian networks can conduct inference inversely. Many intelligent systems (such as feed-forward neural networks and fuzzy logic) are strictly one-way. That is, when a model is given, the output can be predicted from a set of inputs, but not vice versa. The Bayesian networks can conduct bi-direction inference. The fourth advanced feature is that Bayesian networks can combine prior information with current knowledge to conduct inference as it has both causal and probabilistic semantics. This is an ideal representation for users to give prior knowledge which often comes in a causal form (Heckermain 1997). These features will guaranty that using Bayesian networks is a good way to verify those initially identified relationships and inference some uncertain relationships between cost factors and benefit factors in the development of e-services.

In general, there are three main steps when applying Bayesian networks in solving a practical problem: creating a graphical Bayesian network structure for the problems, calculating related conditional probabilities to establish a Bayesian network, and finally using the established Bayesian network to conduct inference.

4.2 Creating a graphical structure for cost-benefit factor relationships

We can find from Fig. 2 that there are no connections between benefit factors B_3, B_4, B_6, B_8, and B_{12} to any of the cost factor nodes listed. A graphical Bayesian network structure of cost and benefit factors relationships can be created by deleting these unlinked factor notes, shown in Fig. 3. These lines in the graphical structure express the 'effect' relationships between these factor notes.

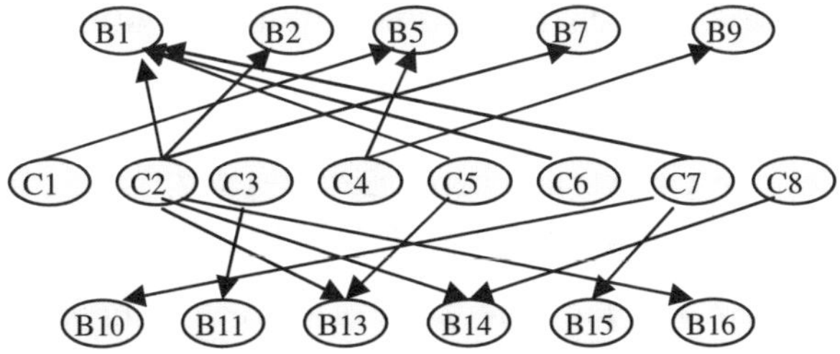

Fig. 3. Initial cost-benefit factor-relation Bayesian network

These notes and relationships shown in Fig. 3 are considered as a result obtained from domain knowledge. In order to improve the Bayesian network, structural learning is needed. Real data should be used to test these established relationships. The data collected, discussed in Section 3.1, is used to complete the structure learning of the Bayesian network.

A suitable structural learning algorithm is first selected for conducting the structural learning of the Bayesian networks. Since the number of DAGs is super-exponential in these nodes, a local search algorithm, Greedy Hill-Climbing (Heckerman 1996), is selected for the structural learning in this study. The algorithm starts at a specific point in a space, checks all nearest neighbors, and then moves to the neighbor that has the highest score. If all neighbors' scores are less than the current point, a local maximum is thus reached. The algorithm will stop and/or restart in another point of the space. By running the Greedy Hill-climbing algorithm for structure learning from collected data, an improved Bayesian network is obtained as shown in Fig 4. Comparing to Fig 3, the link between C_2 and B_1 is removed in Fig 4. Obviously, the Bayesian network shown in Fig. 4 is more consistent with the real data.

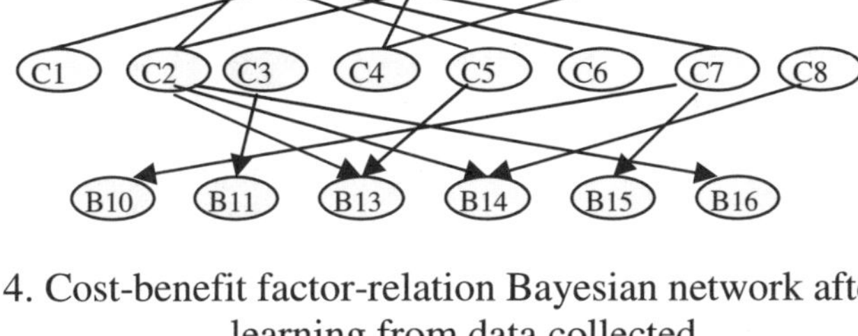

Fig. 4. Cost-benefit factor-relation Bayesian network after structure learning from data collected

This improved Bayesian network has 19 notes and 16 links, and will be used for rule based inference.

4.3 Calculating the conditional probability distributions (CPD)

Now, we let $X = (X_0, \cdots, X_m)$ be a note set, X_i (*i=0, 1,...m*) is a discrete node (variable), in a Bayesian network *B* (*m*=19) shown in Fig. 4. The CPD of the node X_i is defined as $\theta^B_{x_i|Pa_i} = P(X_i = x_i \mid Pa_i = pa_i)$ (Heckerman 1996), where Pa_i is the parent set of node X_i, pa_i is a configuration (a set of values) for the parent set Pa_i of X_i, and x_i is a value that X_i takes. Based on the data collected, the CPDs of all nodes shown in Fig. 4 are calculated.

Before using a Bayesian network to conduct inference, learning and establishing the parameters $\theta^B_{x_i|Pa_i}$ from the data collected should be completed. The easiest method to estimate the parameters $\theta^B_{x_i|Pa_i}$ is to use frequency. However, as the size of data in the study is not very large enough, using frequency methods in the study may not be very effective. This study therefore selected the Bayes method for establishing related parameters. Based on Heckerman (1996)'s suggestions, the Dirichlet distribution is choose as the prior distribution $\theta^B_{x_i|Pa_i}$ for using the Bayes method.

The Dirichlet distribution is the conjugate prior of the parameters of the multinomial distribution. The probability density of the Dirichlet distribution for variable $\theta = (\theta_1, \cdots, \theta_n)$ with parameter $\alpha = (\alpha_1, \cdots, \alpha_n)$ is defined by

$$\text{Dir}(\theta \mid \alpha) = \begin{cases} \dfrac{\Gamma(\alpha)}{\prod_{i=1}^{n} \Gamma(\alpha_i)} \prod_{i=1}^{n} \theta_i^{\alpha_i - 1} & \theta_1, \cdots, \theta_n \geq 0, \sum_{i=1}^{n} \theta_i = 1 \\ 0 & \text{others} \end{cases}$$

where $\theta_1, \cdots, \theta_n \geq 0$, $\sum_{i=1}^{n} \theta_i = 1$, and $\alpha_1, \cdots, \alpha_n > 0$. The parameter α_i can be interpreted as 'prior observation count' for events governed by θ_i.

Let $\alpha_0 = \sum_{i=1}^{n} \alpha_i$. The mean value and variance of the distribution for θ_i can be calculated by (Gelman, 1995)

$$E\,\theta_i = \frac{\alpha_i}{\alpha_0}, \text{ and Var } (\theta_i) = \frac{\alpha_i(\alpha_0 - \alpha_i)}{\alpha_0^2(\alpha_0 + 1)}.$$

When $\alpha_i \rightarrow 0$, the distribution becomes non-informative. The means of all θ_i *(i=0,1,...m)* stay the same if all α_i *(i=0,1,...m)* are scaled with the same constant. If we don't know the difference among θ_i , we should let $\alpha_1 = \cdots = \alpha_n$. The variances of the distributions will become smaller as the parameters α_i *(i=0,1,...m)* grow. As a result, if no prior information, α_i should be assigned with a small value.

After the prior distributions are determined, the Bayes method also requires to calculate the posterior distributions of $\theta^B_{x_i|Pa_i}$ and then complete the Bayes estimations of θ_i . To conduct this calculation, this study assumes that the state of each node can be one of the five values: 1 (very low), 2 (low), 3 (medium), 4 (high), and 5 (very high). Through running the approach, the CPDs of all cost and benefit factor nodes shown in Fig. 4 are obtained. The paper only presents three typical results of these relationships among the main cost factors and benefit factors corresponding to the hypotheses shown in Section 2. Table 5 shows the CPDs for node B_{16} under cost factor C_2. Table 6 shows the CPDs for node B_{11} under cost factor C_3. Table 7 shows the CPDs for node B_5 under cost factors C_1 and C_4.

Table 5. The conditional probabilities for node B_{16}

Pr(B_{16}/ C_2)	B_{16}=1	B_{16}=2	B_{16}=3	B_{16}=4	B_{16}=5
C_2=1	0.0065	0.1677	0.3290	0.3290	0.1677
C_2=2	0.2000	0.0039	0.2000	0.2000	0.3961
C_2=3	0.0033	0.3311	0.3311	0.1672	0.1672
C_2=4	0.0023	0.1186	0.3512	0.2930	0.2349
C_2=5	0.0125	0.0125	0.6375	0.3250	0.0125

Table 6. The conditional probabilities for node B_{11}

Pr(B_{11}/ C_3)	B_{11}=1	B_{11}=2	B_{11}=3	B_{11}=4	B_{11}=5
C_3=1	0.0684	0.1342	0.4632	0.2658	0.0684
C_3=2	0.2980	0.0039	0.3961	0.2000	0.1020
C_3=3	0.3316	0.2658	0.0684	0.2658	0.0684
C_3=4	0.1444	0.1444	0.1444	0.4222	0.1444
C_3=5	0.0333	0.0333	0.8667	0.0333	0.0333

Table 7. The conditional probabilities for node B_5

$Pr(B_5/ C_1, C_4)$	$B_5=1$	$B_5=2$	$B_5=3$	$B_5=4$	$B_5=5$
$C_1=1$ $C_4=1$	0.2495	0.0020	0.2495	0.0020	0.4970
$C_1=2$ $C_4=1$	0.2000	0.2000	0.2000	0.2000	0.2000
$C_1=3$ $C_4=1$	0.0039	0.4941	0.0039	0.4941	0.0039
$C_1=4$ $C_4=1$	0.2000	0.2000	0.2000	0.2000	0.2000
$C_1=5$ $C_4=1$	0.2000	0.2000	0.2000	0.2000	0.2000
$C_1=1$ $C_4=2$	0.2000	0.2000	0.2000	0.2000	0.2000
$C_1=2$ $C_4=2$	0.2000	0.0016	0.3984	0.2000	0.2000
$C_1=3$ $C_4=1$	0.0077	0.0077	0.0077	0.9692	0.0077
$C_1=4$ $C_4=1$	0.0077	0.9692	0.0077	0.0077	0.0077
$C_1=5$ $C_4=1$	0.2000	0.2000	0.2000	0.2000	0.2000
$C_1=1$ $C_4=3$	0.0077	0.0077	0.0077	0.0077	0.9692
$C_1=2$ $C_4=3$	0.0077	0.0077	0.9692	0.0077	0.0077
$C_1=3$ $C_4=3$	0.2495	0.4970	0.0020	0.0020	0.2495
$C_1=4$ $C_4=3$	0.1669	0.1669	0.3325	0.1669	0.1669
$C_1=5$ $C_4=3$	0.3316	0.0026	0.3316	0.3316	0.0026
$C_1=1$ $C_4=4$	0.2000	0.2000	0.2000	0.2000	0.2000
$C_1=3$ $C_4=4$	0.2000	0.2000	0.2000	0.2000	0.2000
$C_1=4$ $C_4=4$	0.0026	0.3316	0.0026	0.3316	0.3316
$C_1=4$ $C_4=4$	0.2852	0.1432	0.2852	0.1432	0.1432
$C_1=5$ $C_4=4$	0.0026	0.6605	0.0026	0.3316	0.0026
$C_1=1$ $C_4=5$	0.0039	0.0039	0.4941	0.4941	0.0039
$C_1=2$ $C_4=5$	0.2000	0.2000	0.2000	0.2000	0.2000
$C_1=3$ $C_4=5$	0.2000	0.2000	0.2000	0.2000	0.2000
$C_1=4$ $C_4=5$	0.6605	0.0026	0.0026	0.3316	0.0026
$C_1=5$ $C_4=5$	0.0039	0.4941	0.0039	0.4941	0.0039

Through observing these results listed in Table 5 to Table 7, we can find that the relationships among these cost and benefit factor nodes are hardly in a linear form. Therefore it is not very effective to express and test these relationships by traditional linear regression methods. On the other hand it is more convenient to use conditional probabilities to express these relationships.

4.4 Inference

The cost-benefit factor-relation Bayesian network has been now created with both its structure and all conditional probabilities defined. It can be thus used to inference the relationships between cost and benefit factors. The inference process can be handled by fixing the states of observed variables, and then propagating the beliefs around the network until all the beliefs (in the form of conditional probabilities) are consistent. Finally, the desired probability distributions can be read directly from the network.

There are a number of algorithms for conducting inference in Bayesian networks, which make different tradeoffs between speed, complexity, generality, and accuracy. The Junction-tree algorithm, developed by Lauritzen and Spiegelhalter (1988), is one of the most popular algorithms. This algorithm is based on a deep analysis of the connections between graph theory and probability theory. It uses an auxiliary data structure, called a junction tree, and is suitable for a middle and small size of samples.

The Junction-tree algorithm computes the joint distribution for each maximal clique in a decomposable graph. It contains three main steps: construction, initialization, and message passing or propagation. The construction step is to convert a Bayesian network to a junction tree. The junction tree is then initialized so that to provide a localized representation for the overall distribution. After the initialization, the junction tree can receive evidences, which consists of asserting some variables to specific states. Based on the evidences obtained for the factor nodes, we can conduct inference by using the established Bayesian network to analyse intensive and find valuable relationships between cost factors C_i $(i=1, 2, \ldots 8)$ and benefit factors B_j $(j=1,2,5,7,9,10,11,13,14,15,16)$ in e-service applications. Table 8 shows the marginal probabilities of all nodes in the Bayesion network.

Table 8. Marginal probabilities of all nodes in the cost benefit Bayesian network

Pr(node= state) state / node	1	2	3	4	5
C_1	0.1469	0.1265	0.2082	0.3510	0.1673
C_2	0.1265	0.2082	0.2490	0.3510	0.0653
C_3	0.3102	0.2082	0.3102	0.1469	0.0245
C_4	0.1265	0.1469	0.3102	0.2694	0.1469
C_5	0.1469	0.1878	0.2694	0.2490	0.1469
C_6	0.1878	0.2490	0.3102	0.1673	0.0857
C_7	0.1878	0.1878	0.3918	0.1878	0.0449
C_8	0.1673	0.1469	0.2286	0.3918	0.0653
B_1	0.1413	0.1684	0.2603	0.2273	0.2027
B_2	0.0449	0.1061	0.4327	0.2490	0.1673
B_5	0.1729	0.2388	0.1930	0.2201	0.1753
B_7	0.1265	0.1469	0.3306	0.2694	0.1265
B_9	0.1469	0.2082	0.2898	0.1878	0.1673
B_{10}	0.1469	0.1878	0.3510	0.1469	0.1673
B_{11}	0.2082	0.1469	0.2898	0.2694	0.0857
B_{13}	0.1289	0.1785	0.3468	0.2427	0.1031
B_{14}	0.0412	0.0948	0.2403	0.2670	0.3566
B_{15}	0.0653	0.1469	0.3918	0.2694	0.1265
B_{16}	0.0449	0.1469	0.3306	0.2490	0.2286

5. Result Analysis

Over all inference results obtained through running the Junction-tree algorithm, a set of significant results are obtained. Three of them are discussed in the chapter. These results are under the evidences that the factor node is 'high'. For the other situations, such as under the evidence that the node is 'low', the similar results have been obtained.

Result 1.
Assuming the cost factor 'maintaining e-service' application C_2=4 (high), we can get the probabilities of the other factor nodes under the evidence. The result is shown in Table 9.

Table 9. Probabilities of the nodes when C_2=4(high)

node \ Pr() \ state	1	2	3	4	5
C_1	0.1469	0.1265	0.2082	0.3510	0.1673
C_2				1	
C_3	0.3102	0.2082	0.3102	0.1469	0.0245
C_4	0.1265	0.1469	0.3102	0.2694	0.1469
C_5	0.1469	0.1878	0.2694	0.2490	0.1469
C_6	0.1878	0.2490	0.3102	0.1673	0.0857
C_7	0.1878	0.1878	0.3918	0.1878	0.0449
C_8	0.1673	0.1469	0.2286	0.3918	0.0653
B_1	0.1413	0.1684	0.2603	0.2273	0.2027
B_2	0.0023	0.1186	0.5256	0.1767	0.1767
B_5	0.1729	0.2388	0.1930	0.2201	0.1753
B_7	0.1767	0.0605	0.4093	0.2930	0.0605
B_9	0.1469	0.2082	0.2898	0.1878	0.1673
B_{10}	0.1469	0.1878	0.3510	0.1469	0.1673
B_{11}	0.2082	0.1469	0.2898	0.2694	0.0857
B_{13}	0.1050	0.1666	0.2741	0.3494	0.1050
B_{14}	0.0030	0.0517	0.2290	0.3667	0.3496
B_{15}	0.0653	0.1469	0.3918	0.2694	0.1265
B_{16}	0.0023	0.1186	0.3512	0.2930	0.2349

The result is also drawn in Fig. 5. We can find that when the value of C_2 (maintaining e-service) is 'high' (=4), the probability of a high B_{13} (cooperation between companies to increase services) is increased from 0.2427 to 0.3494. This indicates that C_2 and B_{13} are correlated to some extent, that is, a high C_2 tends to "cause" a high B_{13}. It is also found that the probability of a high B_{14} (enhancing perceived company image) is increased from 0.2670 to 0.3667, B_{16} (gaining and sustaining competitive advantages) is increased from 0.2490 to 0.2930. These results mean that C_2 is correlated with B_{14} and B_{16} as well. Therefore, a high investment in 'maintaining e-service applications (C_2) will also "bring" a high enhancement of company image (B_{14}) and gain competitive advantages (B_{16}).

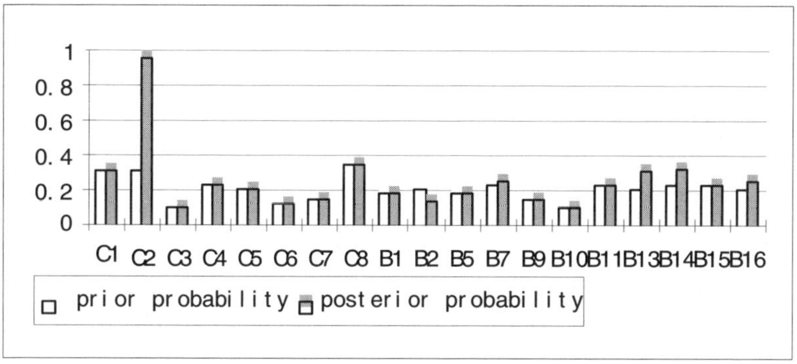

Fig. 5. Prior and posterior probability when C_2=4 (high)

Result 2.

Assuming C_3=4 (high), we can get the probabilities of the other nodes under the evidence (Table 10). Fig. 6 also shows the result. When the value of C_3 is high, the probability of a high B_{11} increases from 0.2694 to 0.4222. This fact suggests that C_3 and B_{11} are correlated to some extent, so that a high investment in 'Internet connection' (C_3) will enable businesses to gather more information to create customer profiles (B_{11}).

Table 10. Probabilities of the nodes when C_3=4(high)

Pr() state / node	1	2	3	4	5
C_1	0.1469	0.1265	0.2082	0.3510	0.1673
C_2	0.1265	0.2082	0.2490	0.3510	0.0653
C_3	0	0	0	1	0
C_4	0.1265	0.1469	0.3102	0.2694	0.1469
C_5	0.1469	0.1878	0.2694	0.2490	0.1469
C_6	0.1878	0.2490	0.3102	0.1673	0.0857
C_7	0.1878	0.1878	0.3918	0.1878	0.0449
C_8	0.1673	0.1469	0.2286	0.3918	0.0653
B_1	0.1413	0.1684	0.2603	0.2273	0.2027
B_2	0.0449	0.1061	0.4327	0.2490	0.1673
B_5	0.1729	0.2388	0.1930	0.2201	0.1753
B_7	0.1265	0.1469	0.3306	0.2694	0.1265
B_9	0.1469	0.2082	0.2898	0.1878	0.1673
B_{10}	0.1469	0.1878	0.3510	0.1469	0.1673
B_{11}	0.1444	0.1444	0.1444	0.4222	0.1444
B_{13}	0.1289	0.1785	0.3468	0.2427	0.1031
B_{14}	0.0412	0.0948	0.2403	0.2670	0.3566

B_{15}	0.0653	0.1469	0.3918	0.2694	0.1265
B_{16}	0.0449	0.1469	0.3306	0.2490	0.2286

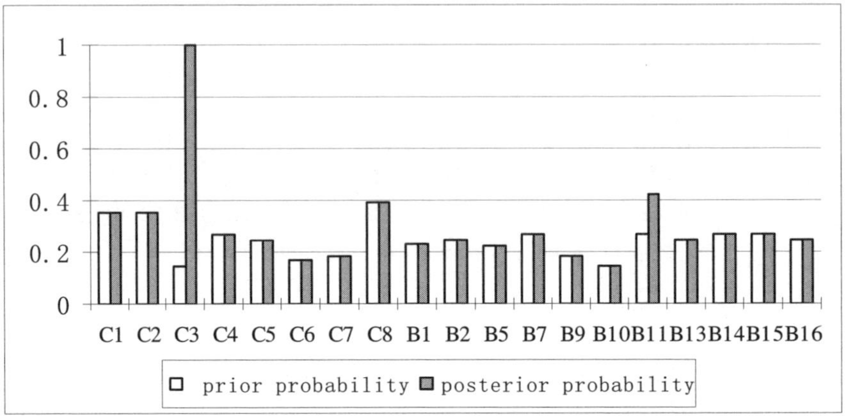

Fig. 6. Prior probability and posterior probability when C_3=4 (high)

Result 3.
Assuming C_4=4 (high), we can get the probabilities of the other nodes under the evidence (Table 11). Fig. 7 further shows the effect of observing when the value of C_4 is high. The probability of a high B_5 has increased from 0.2201 to 0.2295, which suggests that a high investment on hardware/ software (C_4) may "cause" a wonderful global presentation (B_5).

Table 11. Probabilities of the nodes when C_4=4 (high)

Pr() state / node	1	2	3	4	5
C_1	0.1469	0.1265	0.2082	0.3510	0.1673
C_2	0.1265	0.2082	0.2490	0.3510	0.0653
C_3	0.3102	0.2082	0.3102	0.1469	0.0245
C_4	0	0	0	1	0
C_5	0.1469	0.1878	0.2694	0.2490	0.1469
C_6	0.1878	0.2490	0.3102	0.1673	0.0857
C_7	0.1878	0.1878	0.3918	0.1878	0.0449
C_8	0.1673	0.1469	0.2286	0.3918	0.0653
B_1	0.1413	0.1684	0.2603	0.2273	0.2027
B_2	0.0449	0.1061	0.4327	0.2490	0.1673
B_5	0.1558	0.2845	0.1558	0.2295	0.1744
B_7	0.1265	0.1469	0.3306	0.2694	0.1265
B_9	0.3061	0.0030	0.3818	0.1545	0.1545
B_{10}	0.1469	0.1878	0.3510	0.1469	0.1673
B_{11}	0.2082	0.1469	0.2898	0.2694	0.0857

B_{13}	0.1289	0.1785	0.3468	0.2427	0.1031
B_{14}	0.0412	0.0948	0.2403	0.2670	0.3566
B_{15}	0.0653	0.1469	0.3918	0.2694	0.1265
B_{16}	0.0449	0.1469	0.3306	0.2490	0.2286

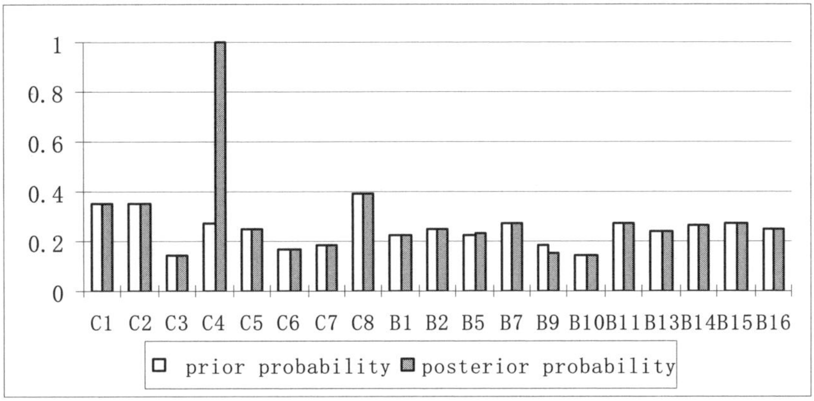

Fig. 7. Prior probability and posterior probability when C_4=4(high)

6. Conclusions

By applying Bayesian network approaches this study verifies a set of relationships between cost factors and benefit factors in the development of e-services. A cost-benefit factor-relation model proposed in our previous study is considered as domain knowledge, and the data collected through a survey is as evidence to conduct the inference-based verification. Through calculating conditional probability distributions among these cost and benefit factors, this study identified and verified some relationships where certain cost factors are more important than others to achieve certain benefit factors. One of these findings is that increased investment in maintaining e-service systems would significantly contribute to 'enhancing perceived company image' and 'gaining competitive advantages'. Therefore, in order to improve the perceived image it would be appropriate for a company to have much investment in 'maintaining e-services'. These findings will provide practical recommendations to e-service providers when they make business strategies to reduce current e-service costs, to increase benefits, or to enhance e-service functionality. These findings can also directly help e-service application developers designing new applications.

As a further study, we will address on the verification of the relationships between cost, benefit and customer satisfaction by using the Bayesian network approach. The results will help business finding more effective ways to provide personalized e-service functions.

Acknowledgment

This research is partially supported by Australian Research Council (ARC) under discovery grants DP0557154 and DP0559213.

References

1. Amir Y., Awerbuch B. and Borgstrom R. S. (2000), A Cost–Benefit framework for online management of a metacomputing system, *Decision Support Systems,* Vol 28, No 1-2, 155-164.
2. Breese J. and Blake R. (1995), Automating computer bottleneck detection with belief nets. *Proceedings of the Conference on Uncertainty in Artificial Intelligence*, Morgan Kaufmann, San Francisco, CA, 36-45.
3. Chidambaram L. (2001), The editor's column: Why e-Service Journal, *e-Service Journal*, Vol 1, No 1, 1-3.
4. DeLone H. W. and McLean R. E., (2004), Measuring e-Commerce Success: Applying Information Systems Success Model, *International Journal of Electronic Commerce*, Vol. 9, No 1, Fall 2004, 31.
5. Drinjak J., Altmann G. and Joyce P. (2001), Justifying investments in electronic commerce, *Proceedings of The Twelfth Australia conference on Information Systems,* 4-7, December 2001, Coffs Habour, Australia, 187-198.
6. Gelman, A., Carlin J., Stern, H. and Rubin, D. (1995), Bayesian Data Analysis. Chapman and Hall/CRC, Boca Raton.
7. Giaglis G. M., Paul R. J. and Doukidis G. I. (1999), Dynamic modelling to assess the business Value of electronic commerce, *International Journal of Electronic Commerce*, Vol 3, No3, 35-51.
8. Hahn J. and Kauffman R. J. (2002), Evaluating selling web site performance from a business value perspective, *Proceedings of international conference on e-Business*, May 23-26, 2002, Beijing, China, 435-443.
9. Heckerman D. (1990), An empirical comparison of three inference methods. Uncertainty in Artificial Intelligence, edited by Shachter, R., Levitt, T., Kanal, L., and Lemmer, J., North-Holland, New York, 283-302.
10. Heckerman D., Mamdani A. andWellman M. (1995), Real-world applications of Bayesian networks, *Comm ACM*, 38(3), 25-26.
11. Heckerman D. (1996), A tutorial on learning Bayesian networks. Technical Report MSRTR-95-06, Microsoft Research.

12. Heckerman D. (1997), Bayesian Networks for Data Mining, *Data Mining and Knowledge Discovery*, Vol.1, No.1, 79-119.
13. Jensen F.V. (1996), An Introduction to Bayesian Networks, UCL Press.
14. Lauritzen S. and Spiegelhalter D. (1988), Local Computations with Probabilities on Graphical Structures and their Application to Expert Systems (with discussion), *J. R. Statis. Soc. B*, Vol.50, No. 2, 157-224.
15. Lee C., Seddon P. and Corbitt B. (1999), "Evaluating business value of internet-based business-to-business electronic commerce", *Proceedings of 10th Australia Conference on Information Systems*, 508-519.
16. Lin, C. (2003) A critical appraisal of customer satisfaction and e-commerce, *Managerial Auditing Journal,* Vol. 18 No. 3, 202-212.
17. Lu, J., Tang S. and McCullough, G. (2001), An assessment for internet-based electronic commerce development in businesses of New Zealand, *Electronic Markets: International Journal of Electronic Commerce and Business Media* Vol. 11, No 2, 107-115.
18. Lu, J. and Zhang G.Q. (2003), Cost Benefit Factor Analysis in E-Services, *International Journal of Service Industry Management (IJSIM),* Vol. 14, No. 5, 570-595.
19. Myllymaki P. (2002), Advantages of Bayesian Networks in Data Mining and Knowledge Discovery, http://www.bayesit.com/docs/advantages.html.
20. Ng H., Pan Y. J. and Wilson T. D. (1998), Business use of the world wide web: A report on further investigations, *International Journal of Management*, Vol. 18, No 5, 291-314.
21. Pearl J. (1988), Probabilistic Reasoning in Intelligent Systems: Networks of Plausible Inference, Morgan Kauffman, Palo Alto, CA.
22. Smith A. G. (2001), Applying evaluation criteria to New Zealand government websites, *International Journal of Information Management,* Vol. 21, 137-149.
23. Wade R. M. and Nevo S. (2005) Development and Validation of a Perceptual Instrument to Measure E-Commerce Performance, *International Journal of Electronic Commerce*, Vol. 10, No 2, 123.
24. Zhang P. and von Dran G., (2000), Satisfiers and dissatisfiers: a two-factor model for website design and evaluation, *Journal of American Association for Information Science (JASIS),* Vol. 51, No 14, 1253-1268

Evaluation of Experience-based Support for Organizational Employees

Słota Renata[1], Majewska Marta[1], Kitowski Jacek[1,2], Lambert Simon[3], Laclavik Michal[4], Hluchy Ladislav[4], and Viano Gianni[5]

[1]Institute of Computer Science, AGH University of Science and Technology, Mickiewicza 30, 30-059 Cracow, Poland, mmajew@agh.edu.pl

[2]ACC CYFRONET-AGH, Nawojki 11, 30-950 Cracow, Poland

[3]CCLRC Rutherford Appleton Laboratory, Chilton, Didcot, OX11 0QX, UK S.C.Lambert@rl.ac.uk

[4]Institute of Informatics, SAS, Dubravska cesta 9, 845 07 Bratislava, Slovakia {laclavik.ui,hluchy.ui}@savba.sk

[5]Softeco Sismat SpA, Via De Marini 1, Torre WTC, 16149 Genova, Italy gianni.viano@softeco.it

Abstract

The purpose of this chapter is to present a study on the management of employees' experience in the e-government area and experimental results achieved in pilot sites. The conveyors of experience are Active Hints. System foundation, experience definition and modelling process using ontology approach are all described in this chapter. The quality assessment of the platform and its experience management abilities are presented. The research described herein was conducted within the EC Pellucid project.

1 Introduction

e-Government can be described as collaborative cooperation of different parts of the government, providing new, efficient and convenient ways for citizens and business to communicate with the government and receive

S. Renata et al.: *Evaluation of Experience-based Support for Organizational Employees*, Studies in Computational Intelligence (SCI) **37**, 411–434 (2007)
www.springerlink.com

services (Kolsaker and Kelley 2004). There are many ongoing initiatives in this field, such as shifting e-business and e-government policy from connectivity to adopting ICT applications, standardization and interoperability introduction of new technologies as well as job migration. These activities have resulted in the launching of the i2010 Initiative, consisting of tree pillars: information space, innovation and investment in research as well as inclusion promoting efficient and user-friendly ICT-enabled public services (Frequin 2005). Results from the case studies have determined promising investments through which to achieve quick short-term effects, consisting for example of inter-organizational cooperation, the use of proactive services and the application of private sector solutions (Driessen and Ponsioen 2005). Both, governments and business organizations must meet the rising expectations of constituents and customers by transforming the way they run their businesses. IT should be used more effectively to create better user experience, which could be supported by transforming organizational processes towards horizontal integration, collaboration and interoperability. The key aspect is business-driven development (Prister and Sage 2005), consisting of discovery of organizational processes and goals, identification of reusable assets and provision of an organizational perspective. Since the knowledge assets of any organization constitute a large part of the entire organizational assets, knowledge management has attracted a great deal of attention in the recent years. Much effort has been devoted to methods, both technological and social, for understanding and augmenting these assets and encouraging their better utilization.

Human problem solving in many fields is based on extensive experience. Experience management is a special kind of knowledge management, focusing on the dissemination of specific knowledge situated in a particular problem-solving context. Organizations with high level of staff mobility suffer significant loss if experience is regarded only as an ability that belongs to the employees. If the experience were accumulated by the organization instead, it could be reused many times.

In this chapter a detailed study of the management of experience of organizational employees as a valuable, reusable organizational asset is presented. This research has been performed within the framework of the EC IST Pellucid (6FP) project, which additionally confirms the importance of this kind of study for the currently ongoing initiatives mentioned .

In the first part of the chapter the Pellucid project is roughly described together with its technical aims, system architecture, theoretical foundations and pilot sites. In the following part the methodology of experience management adopted for the study and a model for uniform experience management for public organizations are described. The model is general in its assumptions. All, medium- and large-size organizations with semi-structured business administration processes and with knowledge-intensive activities performed by the employees are addressed. The approach to quality testing as well as experimental verification of quality of components are discussed on the basis of pilot sites. This chapter is summarized by conclusions and proposals for future work.

2 Pellucid - platform for mobile employees support in e-government

2.1 Motivation and general Pellucid aims

The fundamental objective of the Pellucid project was to develop an adaptable software platform for assisting organizationally mobile employees through experience management and to demonstrate and validate the platform at three pilot sites.

The platform was to be applicable in other organizations with experience management needs. The purpose of the platform was assisting employees of public organizations in their everyday work. This assistance was accomplished based on the accumulation, transmission and maintenance of knowledge that is acquired by individuals in the course of their work (Lambert et al 2003). Focus was on organizational mobile employees: those who move, as a normal part of their career progression, to different departments or units in a sector where they do not necessarily have direct experience. This phenomenon is becoming increasingly common in public organizations. It offers some opportunities (introducing new perspectives and ways of working), though it has its drawbacks (loss of experience). Public organizations and their employees can benefit from formalization, recording, storage and preservation of experience and knowledge. When experienced workers leave a job, a large part of their knowledge is not lost for the organization as a whole. Additionally, individual employees are supported during integration in a new department by giving access to specific knowledge and experience accumulated in the past. The time spent on gaining familiarity with new job becomes shorter.

Since knowledge management is not solely a technical problem, it is also important to take into account organizational and cultural issues in evaluation of the success or failure of knowledge management endeavours. The Pellucid project chose the knowledge management approach, which introduces enhancements to the employees' environment that would provide them with the benefits of experience management and encourage them to share their experience, but without a change of their work process.

2.2 Technical aims and architecture

Pellucid deals with organizations and processes that are fundamentally structured around the production and consumption of documentation. The core of the Pellucid platform was envisaged as an adaptable software environment. Its fundamental purpose is to extract, store and share information from the working procedures and knowledge of individual employees at different organizational levels in such a way that this information can be used in a convenient manner to assist other employees in similar positions in the future (Slota et al. 2003).

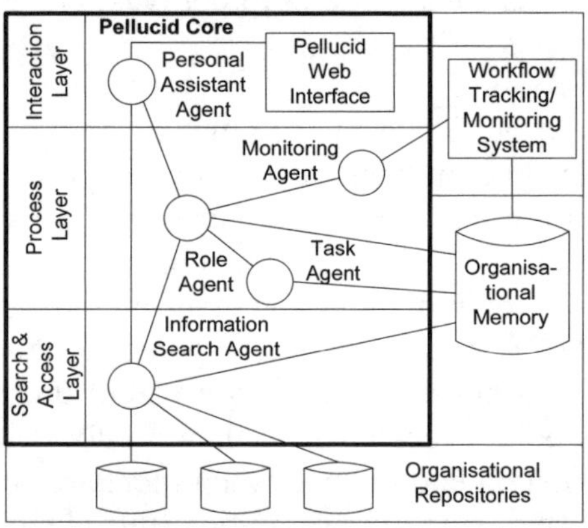

Figure 1: Pellucid platform architecture

The Pellucid system architecture is organised in three layers: the Interaction Layer, the Process Layer and the Access Layer, as shown in Fig.1. The key components of these layers, mainly software agents, fulfil the experience management requirements (Krawczyk et al.2004). They are:

(1) Personal Assistant Agents interacting with employees, providing suggestions and advice, and acquiring knowledge about good practices from the employee.

(2) Role and Task Agents flexibly configurable to correlate processes with available information in the context of the employee's task.

(3) The Information Searching Agent responsible for flexible search and retrieval of diverse kinds of documents.

(4) The Organizational Memory, a repository of metadata, organizational knowledge and experience, events that have happened as a basis for generation of advice and for analysis and enhancement of processes, etc.

2.3 Pellucid pilot sites

The pilot partners are representatives of the public organizations for which Pellucid was customised. The public organizations in the Pellucid project were diversified with respect to domain of activity, employees, work processes, computer infrastructure, software used, etc.

Comune di Genova (CdG), the Municipality of Genoa Mobility and Transport Plan Directorate, is responsible for the process of traffic lights installation, which involves numerous tasks such as technical evaluation, design and economical evaluation, installation and final assessment. Sociedad Andaluza para el Desarrollo de la Sociedad de la Información (SADESI), S.A.U. is a public organization, that was created by the Andalusian Regional Government with the objective of leading the management, technical assistance, development, implementation and exploitation of the telecommunication infrastructures and information systems in Andalusia. Mancomunidad de Municipios del Bajo Guadalquivir (MMBG) is a local corporation located in the south of Spain, in one of the fastest developing areas of Andalusia. MMBG has as its main objective contribution to the social and economic development of its area, the Southern Guadalquivir. The application chosen in Pellucid concerns the management of publicly-funded projects from their approval up to the reporting of expenses.

3 Pellucid foundations

3.1 Experience definition and context

According to (Bergmann 2002), experience is 'specific knowledge in a certain problem-solving context'. Experience management is therefore a special kind of knowledge management. It is a very wide and challenging research field, for which a general approach cannot be defined. It is necessary to select specific and realistic objectives in order to be able to deliver substantial results, providing benefits to public organizations and constituting a concrete step toward understanding of the general experience management topic.

Experience is an attribute of humans, and experience management is concerned with its dissemination between workers in an organization (across space and time). Some document-based approaches to knowledge management appear to overlap with experience management: well-known products such as Autonomy offer the ability to create access mechanisms for diverse varieties of objects (email, Web pages, office documents) and to categorise and tag these objects automatically. Users may be recommended objects based on what other users have viewed. This is a sort of experience, though of a very restricted kind, and it does not offer true support for or integration with business processes.

Pellucid can be seen as an example of an Electronic Performance Support System (EPSS) (Cole et al. 1997), which aims to support and enhance users' performance by providing them with the knowledge required by the task they are performing at the time they are actually performing this task. Other examples of EPSS include the EULE system (Reimer et al. 2000) and the VirtualOffice and KnowMore systems (Abecker et al. 2000).

The EULE system offers assistance in office work in the domain of insurance business. It includes a business-process modelling framework which is coupled with a workflow system by linking EULE office tasks to working steps of a workflow. When reaching a working step that is associated with a EULE task, the user can request EULE assistance and obtain the information missing at that specific point.

The VirtualOffice and KnowMore projects aim to support knowledge-intensive activities by providing automatic access to relevant information.

Each activity belongs to some comprehensive business process which is explicitly modelled and enacted by some workflow management system. The activities are to be supported based on an available information space, which contains information sources of various types and characteristics together with suitable access structures. An intelligent assistant bridges between the information space and the knowledge-intensive activities and performs a process identification job (similar to the context identification job carried out by Pellucid) in which the system detects the particular circumstances of a process. This work was further extended in the DECOR project (DECOR 2000).

Such approaches place the business process as central to the experience management system. In Pellucid, the two main ideas are that every employee in an organization is both a provider and a user of experience, and that employees communicate their experience to a common medium, thereafter retrieving experience from this common medium (for future use).

Experience stored in the medium has two forms: revised and unrevised. The first form applies to elements that have been analyzed and refined by knowledge administrators. Those elements can be considered as 'assessed' experience carriers. The role of the Pellucid system is to match such elements with the current working context to identify relevant ones to propose to the users by means of Active Hints (see below). Unrevised elements come from user observations or a part of user input, from the free text note tool. They also convey experience, however temporarily unverified. They are candidates for knowledge administrator assessment. The role of Pellucid again is to imitate the organization process by linking such elements with the working context. Groups of similar and analogous elements are formed. These groups are then proposed to the user.

The core of the Pellucid system is the management of Active Hints, which are conveyors of experience for employees in their normal activities. The Active Hints generated by the system from user observations have a strictly structured form, while those entered by humans are mainly in free-text form.

3.2 Experience management life cycle

The basic organizational unit of knowledge management is the 'community of practice' (Smith & Farquhar 2000) and these communities 'must have processes in place that enable them to capture, share and apply what they know in a coherent fashion across the organization.' The idea of a 'knowledge hub' linking documents, workflow, project archives, expertise directories, etc., has been introduced as a vision for this approach to knowledge management. A four-stage cycle is envisaged: (1) practitioners apply current best practices; (2) practitioners discover new practices; (3) practitioners submit new practices; (4) community validates and integrates new practices; (5) back round again.

This is analogous to the fundamental cycle of Case-Based Reasoning: search for similar cases; adapt and apply solution; store modified case. This structure has also inspired the Pellucid experience management model, consisting of three phases: Capture and Store, Analysis and Presentation and Experience Evolution (Laclavik et al. 2003).

The Capture and Store phase is concerned with transferring experience to the common medium and deals with observing and storing experience in a particular context. There are three ways of capturing experience: observing employees' actions and workflow events; analysing documents entered into the system; and direct input from workers. Capturing experience from working actions and events is particularly beneficial in repetitive tasks; they are used to create common patterns that can be retrieved in the future in order to assist other employees. Documents constitute an important asset in any organization (e.g., Johnssen and Lundblad 2005). Metadata is added to documents, enabling the system to retrieve in an automatic way the documents useful in a particular working context. Direct capture of experience from employees is carried out through free-text notes written by the employees themselves. This constitutes a good source of knowledge, particularly in dealing with exceptions and deviations from documented procedures and situations.

The Analysis and Presentation phase is concerned with providing users with just-in-time with knowledge. To do so, the concept of an Active Hint is introduced as a conveyor of experience within the organization. An Active Hint is triggered in a context and includes an action, a knowledge resource and a justification for the hint. The context is determined by the particular activity that is carried out by the employee at that time in a workflow system. An action corresponds to an atomic act on a knowledge resource, for example: 'use of a document template', 'read a document or a note', or 'consider a contact list'. The justification given to the employee explains the reason for the hint.

While an employee is performing a task, the working context is monitored by the Pellucid platform, which tries to match this context with the stored 'context prototypes'. In general, all representative contexts from the past are stored together with potential solutions in the form of hints. A perfect match is a rare occurrence, so the system has to adapt the previous context to the new one as well as to the past solution. The system suggests the past solution (possibly adapted) in the form of a hint. The user is free to follow or neglect the hint and this decision (being a kind of user evaluation) is communicated to the system, which upgrades a 'relevance measure' used to select hints to propose. With this approach obsolete knowledge can fade out progressively.

The Pellucid lifecycle has some particular characteristics in relation to other experience management cycles. First, Pellucid is continuously active. It monitors users' actions invisibly and gathers common patterns of operation (Capture and Store Phase). Second, Pellucid tries to anticipate users' problems, offering hints (even unrequested), based on the current 'working context'. Finally, user knowledge is captured by the way of free text notes - we can consider them as the 'initial' shape of hints that could be improved in the Experience Evolution Phase.

The aim of Experience Evolution is updating the available experience. Due to the rapidly changing environment, experience may only have a limited lifetime. Invalid experience must be identified and removed or updated. To this end, the final Pellucid platform includes methods and tools to allow knowledge engineers and expert users to update the experience stored in the Organizational Memory. In particular, suggestions provided can be 'evaluated' by users on the basis of a simple voting mechanism or by the revision of all kinds of Active Hints on the level of administration.

3.3 Experience model

The Pellucid system modelling follows the CommonKADS methodology for developing knowledge-based systems (Schreiber 2002). In CommonKADS, the development of a system entails constructing a set of engineering models of problem solving behaviour in its concrete organization and application contexts. A similar approach was followed in the Pellucid project and several ontology-based models were created.

In the Pellucid project a few generic and domain specific ontologies for public organizations were designed and implemented. In the generic part of the ontology schema, common to all applications, the Active Hint Ontology and Organization Ontology are included, consisting of the Document Ontology, the Contact Ontology, the Repository Ontology and the Workflow Ontology. Fragments of these ontologies are briefly discussed below. For more details on ontologies cf. (Kitowski et al. 2004; Słota et al. 2003). The Organization Ontology describes the organization in a structured, system-like way, by representing aspects such as organizational structure, processes, staff and resources. A fragment of the organization ontology is shown in Fig.2.

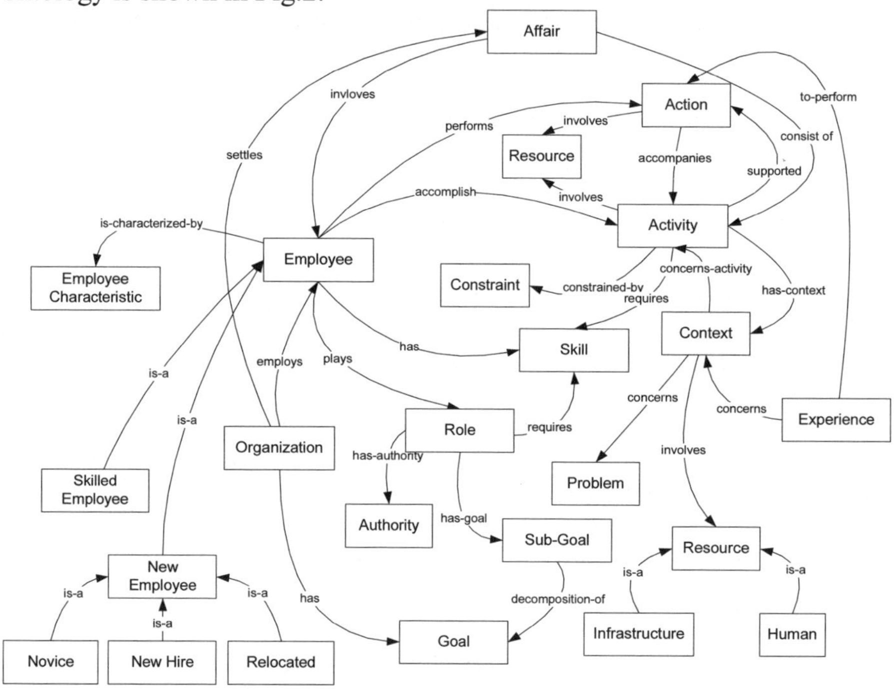

Figure 2: Public sector organization ontology

The experience model specifies the knowledge and reasoning requirements of the system. It consists of domain-specific knowledge and knowledge directed to the system tasks. In the Pellucid system, experience is understood as the ability to perform an action according to the given context. The context is composed of tasks to be performed, actors to cooperate with and resources to be exploited. It should describe all circumstances in which a particular activity is undertaken. The accomplishment of an activity requires a problem to be solved. The problem is described in terms of a domain-specific ontology. Moreover, some external factors may influence the way in which an activity is accomplished - such as resources that are involved in solving the problem (both human and infrastructural, e.g., Workflow Management System or Workflow Tracking System).

The idea of knowledge delivery through active user support, triggered according to the context in a workflow, has been developed by the DECOR project (DECOR 2000). The Pellucid project has followed the idea of Active Hints as conveyors of experience and had worked out a somewhat different approach. Active Hints are regarded as suggestions for the user to perform some actions that will assist with his/her current activity in the process instance. An Active Hint is a special element of knowledge, which consists of an action, a resource and a justification, together with a context against which the current working context is matched and which triggers the presentation of the hint (see Fig.3). This context covers both the work process ('starting or completing a task', 'opening a document') as well as domain-specific representation (relevant similarity of the case in hand to prior cases). Additional context is provided by the characteristics of the employee (whether experienced or novice) at the most basic level.

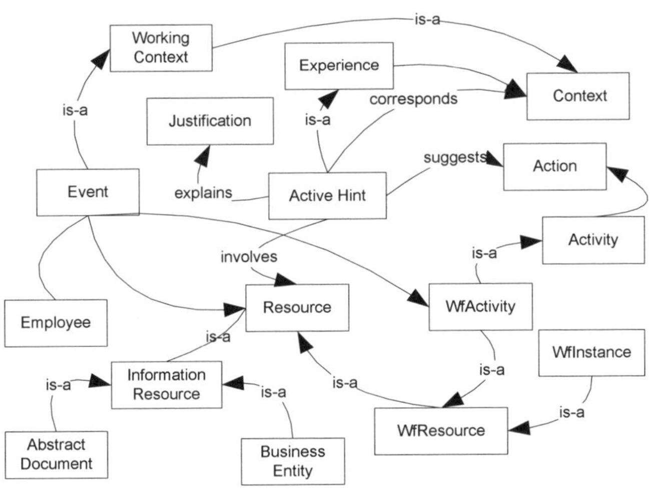

Figure 3: Active Hint ontology

The action part of the hint is what the employee is suggested to do, e.g., examine some documents, contact a person or use a template. The resource is the subject of the action. The justification presents the reason why the hint is being suggested. This model of Active Hints is very general, since it can encompass hint types built up from templates (similar to the generic query templates of DECOR) as well as retrieval of free-text notes entered by users and of documents accessible in organizational repositories. One of the important areas of work within the Pellucid project was the development of an ontology for the hints themselves.

A sample Active Hint element is presented below in Table 1.

Table 1: Sample Active Hint

	ACTIVE HINT
Context:	Creating the annual financial report concerning IST project activities.
Action:	Use
Resource:	Document A
Reason:	It is a template

During runtime, the Pellucid system tracks all actions taken by employees as well as working context and stores relevant ontological instances. These instances define the employee's experience representation.

The decision on whether an Active Hint should be shown to the employee takes into account not only the work process characteristic in a given position but also the employee profile (such as novice, newly hired employee, relocated employee or experienced employee). Active Hints are adapted to the level of employee's experience – an experienced user are not bothered with hints concerning low-level skills. Moreover, the employee's profile is defined and evolves during the employee's work period in the organization.

4 Quality assessment

Software development and project management should be performed according to a systematic approach of tracing and monitoring their objectives. Some of the main stages of software systems development include testing, integration and validation. On these stages the work is carried out on the program's unit integration into the entire application, application environment and application's conformity with the requirements. This conformity can be considered at different levels, for example taking into account functionality, correctness, efficiency and/or usefulness. This kind of validation is usually performed with reference to the requirements defined at the analysis or modelling stages of the system. The procedures related to the technical quality assessment are divided into the following aspects: documentation, conventions and metrics, software development and pilot applications development. Another very important aspect of application testing is user satisfaction, while the system works correctly from a formal viewpoint. Acceptance testing is based on specifications of the end-user as well as on the system usage by the end-users over some period of time. For the Pellucid project, an important part of quality assessment process is based on feedback from end users to developers. Since an agile methodology was adopted for the system development, keeping the approach and the realization independent of the platform (MDA, Soley 2000) enabled the users' evaluation at each stage of development to be widely respected.

4.1 Approach to quality testing in Pellucid

Since quality of software has many aspects, the methods of quality assessment are highly project-specific. The Pellucid platform is an experience management system. The purpose of this platform, i.e. to gather and reuse experience to support public employees in their everyday work, is hard to verify. It could be said that such system a fulfils its mission if its functionality actually helps employees in experience-intensive tasks. Thus, the final judgement belongs to the employees who evaluate the system. This seems to be the most suitable way to verify the quality of the experience management system.

Formal quality assessment mechanisms should be considered in addition to user assessment. Within the task of the evaluation of the final system we have elaborated formal assessment procedures concerning miscellaneous aspects of the Pellucid platform evaluated at three pilot sites. Below we present some of them, i.e. Active Hint assessment and tests of components of the Search and Access Layer, proving the capability of Pellucid to manage the employees' experience.

4.2 End-user evaluation

End-user evaluation was performed in several rounds at the three pilot sites to assess the Pellucid platform operation as an information system within the public organization structure from the users' point of view. End-users performing the evaluation of the Pellucid occupied different organizational positions and performed various roles. Evaluation by different types of users enabled us, among others, to better understand such issues as mobility scenarios best supported by the platform and the kinds of hints most sought after by novice employees.

The final tour of the user evaluation intended to verify the quality of the Pellucid platform and to specify the impact of the system at the pilot sites. For quality assessment one of important issues was the quality of Active Hints as experience conveyors.

4.2.1 Active Hints as experience conveyors

There are many aspects used to decide if an Active Hint transfers experience correctly. For instance, Active Hints, which are entered by employees could be incorrect as a result of their mistakes and the lack or excess of precision. Due to the variety of reasons that could cause an Active Hint to be incorrect or irrelevant, fully automatic quality verification is impossible. Domain experts have to occasionally revise the available sets of Active Hints. Summing up, the Active Hints are created by people and by the system as a result of users' observations, to be used by people and in a form most convenient for people. Thus, the best way to verify the quality of Active Hints is by the humans themselves.

4.2.2 User evaluation of Active Hints quality

The part of the evaluation of the quality of Active Hints described below took place at the CdG pilot site. The evaluation was focused on the accuracy, completeness and usefulness. Users were asked to assess the quality of Active Hints by filling out questionnaires. The main aspects taken into account during this evaluation are presented in Table 2.

Table 2: User assessment of Active Hint quality

Aspects of Active Hint quality	Good	Fairly good	Fairly bad	Bad	N/A
Active Hint as an experience transfer medium	7	1	-	-	-
Type of Active Hints	5	2	-	-	1
Usefulness of Active Hints concerning documents	5	2	1	-	-
Usefulness of Active Hints concerning contacts	1	5	1	-	1
Completeness of the Active Hints set	6	1	-	-	1

The users considered Active Hints as a good medium for experience transfer, both for novice and experienced employees. The available types of Active Hints were convenient for the users, who especially appreciated the free-text form of Active Hints. Users regarded as 'high' the usefulness of Active Hints concerning documents. However, CdG users mentioned that some Active Hints, namely free-text notes, are not always relevant to their current needs. "Sometimes information can be useless because each installation is very particular and it is very simple to have different dossiers even if they appear similar at first glance. This means that sometimes it is not possible to use information returned by the system related to similar dossiers due to the high specificity of the particular installation." The above is a consequence and risk taken when choosing the free-text format for Active Hints. On the one hand it is a very flexible format that allows users to formulate thoughts in their favourite way, helping convey experience. On the other hand, as the natural language is ambiguous and poses problems for automatic interpretation, the free-text notes, selected by the Pellucid and presented to the user may be of little relevance. In general, the use of the natural language may cause free-text note classification to be possible only for a human.

The usefulness of Active Hints concerning contacts was assessed as quite good. Users were appreciative of this type of Active Hints, but pointed out that they often contain an insufficient amount of information. This fact might have been caused by the short time of Pellucid operation at their pilot sites. In general, the users didn't notice the need for any additional types of hints. Only in the SADESI pilot application, users mentioned the necessity of hints related to time-critical activities.

4.3 Quality of components

A formal method of quality assessment should be applied in addition to user assessment. Within the task of the evaluation of the final system we elaborated formal assessment procedures concerning miscellaneous aspects of the Pellucid platform and implemented them at the three pilot sites. Below we present some of these procedures, to prove the Pellucid capabilities to manage the employees' experience.

Search and Access Layer

The Search and Access Layer (SAL) is at the lowest layer of the Pellucid system. The functionality of the layer is accomplished by the Information Search Agent (ISA). The key role of the ISA is to search for documents within the organizational repositories. The upper layer agents delegate ISA to answer three types of queries: similarity query, ontology-based query and full-text query. The search is based on the information retrieval techniques provided by the Lucene indexing and search engine (Hatcher and Gospodnetic 2004) empowered by semantic techniques. For ontology formalization OWL (W3C 2004) was used. The OWL ontology representation was stored in the Java-based Jena library (McBride 2001). Jena provides several storage models, methods for ontology manipulation, in particular RDF manipulation and query engine for RDQL (W3C 2003).

During the search process the ISA makes use of the Lucene full-text index and the ontological index which are derived from semantic annotation. The second responsibility of the ISA agent is monitoring various document repositories. Communication with repositories (e.g. file systems, databases, mail servers etc.) is event-driven. Events are generated by the change of the repository content, i.e. creation, modification or removal of documents. ISA handles the event received from Repository Monitors. Event handling includes updating the information gathered in the Organizational Memory and notifying other system actors.

4.3.1 Metrics

The metrics used for quality assessment of the overall document searching capabilities are Precision, Recall and F-measure (Baeza-Yates and Ribeiro-Neto 1999), adopted from information retrieval technologies. Recall and Precision are complementary metrics, while F-measure is a commonly used formula for combining them into a single metric. The definitions of Precision, Recall and F-measure are presented below.

Assume: n – number of queries, m – number of documents, Q_i - represents i-th query, $i=1,..,n$, D_j - represents j-th document, $j=1,..,m$. Let's define the *result of manual indexing/annotation* $ma(Q_i, D_j) = 1$ for $Q_i \subset_m D_j$ or 0 otherwise. The result of automatic indexing/annotation is $aa(Q_i, D_j)=1$ for $Q_i \subset_a D_j$ or 0 otherwise. For each query, Q_i, the *number of possible relevant* documents, $P(Q_i)$, and the *number of total retrieved* documents, $A(Q_i)$, are obtained. $Q_i \subset_m D_j$ and $Q_i \subset_a D_j$ mean that the document D_j contains the phrase corresponding to the query Q_i and it was found by manual or automatic indexing/annotation respectively.

$$P(Q_i) = \sum_{j=1}^{m} ma(Q_i, D_j), \qquad A(Q_i) = \sum_{j=1}^{m} aa(Q_i, D_j) \tag{1}$$

Data gathered during the manual and automatic indexing or semantic annotation are used for counting the *number of incorrect* results (retrieved but not relevant) and the *number of missing* results (relevant but not retrieved). Assume that the *wrong* result, $wrong(Q_i, D_j)$, is defined as:

$$wrong(Q_i, D_j) = \begin{cases} 1 \text{ if } aa(Q_i, D_j) = 1 \text{ and } ma(Q_i, D_j) = 0 \\ 0 \qquad\qquad \text{otherwise} \end{cases} \tag{2}$$

and the *missing* result, $miss(Q_i, D_j)$, is:

$$miss(Q_i, D_j) = \begin{cases} 1 \text{ if } aa(Q_i, D_j) = 0 \text{ and } ma(Q_i, D_j) = 1 \\ 0 \qquad\qquad \text{otherwise} \end{cases} \tag{3}$$

The *number of incorrect* documents, $W(Q_i)$, and the *number of missing* documents, $M(Q_i)$, for a given query, Q_i, are:

$$W(Q_i) = \sum_{j=1}^{m} wrong(Q_i, D_j), \qquad M(Q_i) = \sum_{j=1}^{m} miss(Q_i, D_j) \tag{4}$$

The correct results are represented by the *number of retrieved relevant* documents, $C(Q_i)$.

$$C(Q_i) = A(Q_i) - W(Q_i), \qquad \text{where } W(Q_i)=1,..,A(Q_i) \tag{5}$$

Precision, $Pr(Q_i)$ and Recall, $Re(Q_i)$ metrics are calculated as follows:

$$Pr(Q_i) = \frac{C(Q_i)}{A(Q_i)}, \qquad Re(Q_i) = \frac{C(Q_i)}{P(Q_i)} \tag{6}$$

for $A(Q_i)\neq 0$, $P(Q_i)\neq 0$. *F-measure* metric, $Fm(Q_i)$, is calculated as:

$$Fm(Q_i) = \frac{(\beta^2 + 1) \cdot Pr(Q_i) \cdot Re(Q_i)}{(\beta^2 \cdot Pr(Q_i)) + Re(Q_i)} \tag{7}$$

$Pr(Q_i \neq 0$ and $Re(Q_i)\neq 0$, β is a weight parameter ($\beta=1$ for further study).

4.3.2 Quality of Search and Access Layer

This section presents the quality tests of the overall searching abilities of the Search and Access Layer. The searching process is based on full-text indexing supported by semantic annotation. The agent uses several techniques to build an internal representation, called the summary of the document content. The summary includes both a full-text index as well as semantic annotation results. The Pellucid system uses it to effectively perform document searches requested by the end user.

Quality tests were performed for different document corpuses, both created for test purposes and originating from pilot sites. The testing procedure comprised corpus creation, ontology development, queries selection, manual annotation, automatic annotation, and result analyses.

4.3.3 Pellucid search quality

First, the corpus of test documents concerning computer hardware is discussed. The corpus for this test was created from 105 documents from the domain of computer hardware retrieved from dedicated portals. The experimental prototype of the ontology was created for semantic annotation purposes. The corpus was manually annotated using vocabulary from the ontology by a knowledge engineer. Due to quality issues, three iterations of reading and indexing were performed. The knowledge engineer also selected queries to be executed by the search engine. Next, the same corpus was automatically annotated by means of the Pellucid system. A sample of quality testing results for the Pellucid search engine (SAL) compared to the quality of Lucene's full-text search is shown in Table 3.

Table 3: Results of quality testing of the Pellucid search compared to plain full-text search

Method	Query	P(Q)	A(Q)	C(Q)	W(Q)	M(Q)	Pr(Q)[%]	Re(Q)[%]	Fm(Q)[%]
full-text	Chipset	33	25	25	0	8	100	76	86
SAL	Chipset	33	30	29	1	4	97	88	92
full-text	Intel	37	24	24	0	13	100	65	79
SAL	Intel	37	24	24	0	13	100	65	79
full-text	Producer	64	22	14	8	50	64	22	33
SAL	Producer	64	74	60	14	4	81	94	87

In general, the Pellucid search yields better results than full-text search. In some cases, the full-text method returns no results, while, for the same query, the Pellucid method gives more reasonable results. This feature follows from the ontology usage during the searching process. The Pellucid method allows understanding some semantics of the documents. Additionally, the results of the Pellucid quality testing could be further used to improve the ontology definition. Whenever the quality measures for Pellucid method are not satisfactory, improvements in the ontology are suggested.

4.3.4 Semantic annotation quality test on SADESI documents

The purpose of the second group of tests was to assess the quality of the search engine on the basis of the document corpus of the SADESI pilot site, i.e. e-mail message sets. The ISA agent was able to check if the type of each e-mail is Landline Telephony Service (LLT), Outbuilding Breakdown, Partial Breakdown, Service Cut or Service Degradation (all being specific types of the Landline Telephony Service breakdown) and to discover if the e-mail contains the affected phone number. The six different corpuses of documents were prepared and manually annotated by domain experts as presented in Table 4.

Table 4: Document corpuses used for testing semantic annotation

Document corpus	Results
1	277 documents manually annotated as containing LLT and 150 documents without LLT
2	85 document manually annotated as containing Outbuilding Breakdown and 342 document without Outbuilding Breakdown
3	72 documents manually annotated as containing Partial Breakdown and 355 documents without Partial Breakdown
4	50 documents manually annotated as containing Service Cut and 377 documents without Service Cut
5	42 documents manually annotated as containing Service Degradation and 385 documents without Service Degradation
6	76 files annotated manually as containing the phone number and 100 documents without any affected phone number (but containing other phone numbers as e.g. contacts information)

Two different tests were performed. The same procedure was applied to all tests. The first test was meant to analyze LLT discovery ability (document corpus 1,2,3,4 and 5), and the second to analyze the affected phone number discovery ability (document corpus 6). The ISA agent indexed and semantically annotated all documents and logged all operations. Table 5 presents the results of semantic annotation analysis and computed measures for each annotation concept.

Table 5: Results of semantic annotation quality assessment

Query	P(Q)	A(Q)	C(Q)	Pr(Q)[%]	Re(Q)[%]	Fm(Q)[%]
Landline Telephony Service	277	269	267	99.26	96.39	97.80
Outbuilding Breakdown	85	97	84	86.60	98.82	92.31
Partial Breakdown	72	85	71	83.53	98.61	90.45
Service Cut	50	37	36	97.30	72.00	82.76
Service Degradation	42	26	23	88.46	54.76	67.65
Affected phone number	70	65	59	90.77	84.29	87.41

The tests confirmed good quality of the ISA semantic annotation engine. For the Pellucid platform high values of Precision are especially important, since Active Hints are generated on top of the ISA semantic annotation engine. However, the Recall values are also high. The sole exception was Service Degradation and, because of its low Recall values, another iteration of ontology refinement was performed.

4.3.5 Document similarity test on the MMBG documents

The purpose of the tests described below was to assess the quality of the similar document searching functionality offered by ISA agent. The test corpus for the MMBG pilot application consisted of 80 documents.

The ISA agent indexed and semantically annotated all these documents, while the human experts categorized them manually into groups according to their similarity. The query "find documents similar to" referred to a selected document corpus. The similarity results below the threshold value were neglected. The experiments showed values of Precision between 90% and 100%, while results of two sample queries are characterized in detail in Table 6.

Table 6: Precision of document similarity searching

Threshold value	Pr(Q)[%]
0,01 (first three results)	98.33
0,01 (first six results)	91.25
0,1 (first two results)	100
0,1 (first four results)	100

Precision between 90% and 100% is a very good result. Further improvements could go in the direction of better defining the threshold, e.g., the ISA agent could first determine the type of document passed in the query and then apply the cut factor predefined for this type of document.

4.4 Quality assessment conclusions

The quality assessment consisted of evaluation of system components and the evaluation of the entire platform. For the overall quality assessment, user opinion on spontaneously-generated hints was used. Whenever possible, two kinds of data have been applied: general data selected for the experiments only and domain-specific data, typical for the pilot sites.

The tests presented in this document proved good quality of the Pellucid system components. The use of a formal methodology for defining ontology substantially increased the quality of the ontology. The long process of ontology development, carried out in strong collaboration with the end users ensured the good overall quality of Pellucid ontologies. The usage of ontology during the search process enhanced the quality of searching, as assessed with the help of typical metrics, such as Precision, Recall and F-measure. The tests verified good quality of semantic annotation. Since Active Hints are generated on top of the semantic annotation engine it is very important to have the correct information. The Lucene engine seems to be a very precise tool for searching text document sets and suitable for use in public organizations. Pellucid enabled good quality searching - the precision in the tests was nearly 100%. High precision of results is particularly important for the quality of the Active Hint activation procedure.

The quality assessment of the entire Pellucid platform was performed on the basis of the results of final user evaluation. The users assessed that the Active Hint is a proper medium for experience transfer for both novice and expert employees. The available types of Active Hints were convenient for users, particularly those expressed in free-text form. The number of Active Hints assessed was relatively small; however this was due to the short time of Pellucid operation at the pilot sites. In general, the users didn't notice the need for any additional types of hints.

5 Summary and Future Work

This chapter presents the development of the model of experience management and the evaluation of the experience management capabilities. The study was done within the framework of the EC IST Pellucid project.

This experience management model was built on the idea of Active Hints - spontaneously generated carriers of experience that are proactively presented to the user according to the current working context. The representation of the working context, capable of handling both the position in the work process as well as domain-specific characteristics of the case on hand is particularly rich. Due to this representation, the idea of Active Hints gains much strength and flexibility.

The quality assessment of the customized platform at the pilot sites was discussed. The correctness of working context modelling, the adaptability of the idea of Active Hints as transmitters of experience as well as the effectiveness of use of combined text-based and ontology-based indexing and searching were presented.

We have noted that the current version of the platform is tailored to the particular model of Active Hints, working context, etc., and has adopted and modified a fairly standard model (Capture and Store, Analysis and Presentation, and Experience Evolution). Further research could focus on incremental advances in the knowledge management lifecycle. A separate direction could constitute more advanced automatic analysis of the Organizational Memory content. Promising results have already achieved - detecting some patterns of task performance (such as sequences of activities or documents commonly consulted) and automatic construction of new hints on their basis.

We also consider wider usage of the Semantic Web achievements, since it is concerned with the integration of diverse systems and resources using technologies that allow machine reasoning about their contents. The Pellucid project has adopted a number of Semantic Web technologies, and the vision of integrating information coming from disparate sources (documents repositories, workflow systems, etc.) is consistent with the Semantic Web. Applying the Semantic Web approach to experience management in public organizations seems to be a promising direction of research.

Acknowledgments

This research has been performed in the framework of the EU IST-34519 Pellucid project. The authors would like to thank the whole project Consortium for participation in the project. Bartosz Kryza's contribution from ACC Cyfronet-AGH as well as the AGH University of Science and Technology grant are acknowledged.

References

Abecker A., Bernardi A., Maus H., Sintek M. and Wenzel C. (2000), Information supply for business processes: coupling workflow with document analysis and information retrieval, Knowledge-Based Systems, Vol. 13, No. 5, 271-284.

Baeza-Yates R. and Ribeiro-Neto B. (1999), Modern Information Retrieval, Longman, ISBN 020139829X.

Bergmann R. (2002), Experience Management. Foundations, Development Methodology, and Internet-Based Applications, LNAI, Springer,Vol.2432.

Cole K. and Fisher O., Saltzman P. (1997), Just-in-time knowledge delivery, Communications of the ACM, Vol. 40, No. 7, 49-53.

DECOR consortium (2000), DECOR Delivery of Context-Sensitive Organizational Knowledge, http://www.dfki.uni-kl.de/decor/deldec/D1-Final.pdf.

Driessen H. and Ponsioen A. (2005), Does eGovernment Pay off? in: P. Cunningham M. Cunningham (Eds.), Innovation and the Knowledge Economy. Issues, Applications, Case Studies, Part I., IOS Press, 369-374.

Frequin M. (2005), The New European ICT Agenda. in: P. Cunningham, M. Cunningham (Eds.), Innovation and the Knowledge Economy. Issues, Applications, Case Studies, Part I., IOS Press, 353-360.

Hatcher E. and Gospodnetić O. (2004), Lucene In Action, Manning Publications, ISBN 1932394281.

Johnssen G. and Lundblad N. (2005), eGovernment and the Archives of the Future. in: P. Cunningham M. Cunningham (Eds.), Innovation and the Knowledge Economy. Issues, Applications, Case Studies, Part I., IOS Press, 390-396.

Kitowski J., Krawczyk K., Majewska M., Dziewierz M., Słota R., Lambert S., Miles A., Arenas A., Hluchý L., Balogh Z., Laclavik M., Delaître S., Viano G., Stringa S. and Ferrentino P. (2004), Model of Experience for Public Organisations with Staff Mobility. Proc. of KMGov2004, Krems, Austria, LNAI, Springer, Vol.3035, 75-84.

Kolsaker A. and Kelley L. (2004), Reconceptualising Government in the New Era. Proc. of KMGov2004, May 17-19, 2004, Krems, Austria, LNAI Springer, Vol.3035, 18-26.

Krawczyk K., Majewska M., Dziewierz M., Slota R., Balogh Z., Kitowski J. and Lambert S. (2004), Reuse of Organisational Experience Harnessing Software Agents. Proc. of ICCS2004, Kraków, Poland, LNCS, Springer, Vol.3038, 583-590.

Lambert S., Stringa S., Viano G., Kitowski J., Słota R., Krawczyk K., Dziewierz M., Delaître S., Gómez AC., Hluchý L., Balogh Z., Laclavik M., Caparrós SF., Fassone M. and Contursi V. (2003), Knowledge Management for Organisationally Mobile Public Employees. Proc. of KMGov2003, Rhodes, Greece, LNAI, Springer, Vol. 2645, 203-212.

Laclavik M., Balogh A., Hluchý L., Słota R., Krawczyk K. and Dziewierz M. (2003), Distributed Knowledge Management based on Software Agents and Ontology. PPAM2003, Częstochowa, Poland, LNCS, Springer, Vol.3019, 694-699.

McBride B. (2001), Jena: Implementing the RDF Model and Syntax Specification. In S. Decker et al. eds. SemWeb2001, Hong Kong.

Pellucid Consortium (2002), Pellucid - A platform for organisationally mobile public employees. http://www.sadiel.es/Europa/pellucid/.

Prister G. and Sage J. (2005), Realizing the Potential of Government Transformation with Widespread Modernization and Innovation. in: P. Cunningham M. Cunningham (Eds.), Innovation and the Knowledge Economy. Issues, Applications, Case Studies, Part I., IOS Press, 382-390.

Reimer U., Margelisch A. and Staudt M. (2000), EULE: A knowledge-based system to support business processes, Knowledge-Based Systems, Vol.13, No. 5.

Schreiber G., Akkermans H., Anjewierden A., De Hoog R., Shadbolt N., Van de Velde W. and Wielinga B. (2000), Knowledge Engineering and Management: The CommonKADS Methodology, MIT Press.

Słota R., Majewska M., Dziewierz M., Krawczyk K., Laclavik M., Balogh Z., Hluchý L., Kitowski J. and Lambert S. (2003), Ontology Assisted Access to Document Repositories in Public Sector Organizations. Proc. of PPAM 2003, Częstochowa, Poland, LNCS, Springer, Vol. 3019, 700-705.

Smith R. and Farquhar A. (2000), The road ahead for knowledge management, AI Magazine Vol. 21, No. 4.

Soley M. (2000), Model Driven Architecture. OMG White paper, Draft 3.2, ftp://ftp.omg.org/pub/docs/omg/00-11-05.pdf.

W3C Recommendation (2003), RDQL - A Query Language for RDF. In Seaborne A. (ed), http://www.w3.org/Submission/2004/SUBM-RDQL-20040109.

W3C Recommendation (2004), OWL Web Ontology Language Guide. In Smith MK, Welty C., McGuinness DL (eds), http://www.w3.org/TR/owl-guide/.

A Web based Intelligent Sensory Evaluation System in the Textile Integrated Supply Chain

Bin Zhou[1, 2], Xianyi Zeng[1], Ludovic Koehl[1], and Yongsheng Ding[2]

[1] Laboratoire GEMTEX, ENSAIT 9, rue de l'Ermitage, 59070 Roubaix
Tél: (33)(0)3 20 25 89 67 Email:bin.zhou@ensait.fr, xianyi.zeng@ensait.fr
[2] College of Information Sciences and Technology, Donghua University, 1882 Yan-An West Road, Shanghai 200051, P. R. China

Abstract.

This chapter presents a web based intelligent sensory evaluation system of industrial products, developed according to the practical requirements of textile enterprises. This system, implemented on websites, permits to carry out normalized business transactions between distant industrial partners and describe, analyze and interpret sensory data provided by distant multiple panels. In the sector of textile/clothing/distribution, sensory data constitute a very important component in the information flow of the integrated supply chain. In our system, intelligent techniques have been used to process uncertainty and imprecision existing in sensory data and a criterion of distance between two sensory panels has been defined in order to characterize the difference between suppliers and customers. Based on this distance, we propose a procedure for converting evaluation terms between different panels, which is useful for solving business conflicts in product quality. The effectiveness of this system has been validated using a practical example of fabric hand evaluation.

1 Introduction

The European Textile/Clothing (T/C) industry is still important in the European economy, in terms of production and employment. However, there exist many barriers in this sector that are pressing textile companies to face the following competitive challenges:

B. Zhou et al.: *A Web based Intelligent Sensory Evaluation System in the Textile Integrated Supply Chain*, Studies in Computational Intelligence (SCI) **37**, 435–456 (2007)
www.springerlink.com

– Shorter product life cycles: Distributors and consumers are looking for more variety and personalization,
– Lack of flexibility in the supply chain,
– Cost reduction: Retailers wish to keep their sales margins, which leads to the competition for cheaper prices on products,
– Homogeneity need: The lack of integration, the heterogeneity and the lack of standards constitute a chronic weakness of the European T/C Industry.

Under this challenging economic pressure, there is a strong need for developing new means in order to enhance communications between all related companies in the textile/clothing/distribution supply chain and between these companies and consumers [1, 2]. In general, the structure of this supply chain should be optimized by exploiting relevant information on product quality, product design and marketing obtained from different companies in a cooperative way.

Today, industrial partners working in the textile/clothing/distribution supply chain are generally located in different parts of the world. Their professional knowledge, technical criteria, cultures and languages related to product quality and product design are quite different. In this background, a normalized sensory evaluation platform in this international supply chain can effectively help the understanding of all partners on products and decrease business conflicts between suppliers and customers. In practice, the supply chain oriented sensory evaluation work is generally done by consultation companies or evaluation centers independent of related producers. In order to optimize the cost and efficiency of the supply chain, we wish to realize the corresponding evaluation platforms on websites and perform the evaluation work on them. In this case, all partners, especially distant partners can freely carry out normalized communications and business transactions between them.

In this chapter, we present a sensory evaluation system of the textile/clothing/distribution supply chain implemented on websites. Sensory evaluation in the integrated supply chain is applied at two levels: 1) Design Oriented Sensory Evaluation (DOSE) and 2) Market Oriented Evaluation System (MOSE) [3]. In the following sections, only some basic functions at DOSE level are presented in detail because web based distant communications between business partners play a leading role in the supply chain. The functions at MOSE level, rather similar to those of DOSE, have been discussed in [4, 5]. Our chapter is organized as follows. In Section 2, we describe some basic concepts on sensory evaluation of industrial products and related application backgrounds. In Section 3, we present the general structure of the integrated supply chain in the sector of textile/clothing/distribution as well as the role of the sensory evaluation system in this supply chain. The implementation of the sensory evaluation system on websites is also described in this section. We mainly explain the functions of three subsystems (project management subsystem, the DOSE subsystem, the MOSE subsystem). In Section 4, we present three models constituting the theoretical foundation of the DOSE subsystem. These models include Data Normalization Model (DNM), Performance Analysis Model (PAD), and Conflicts Solving Model (CSM). In order to validate the effectiveness of these procedures, we apply them to fabric hand evaluation data provided by 4 sensory panels in France and China for evaluating 43 knitted cotton samples. The corresponding results and analysis are given in Section 5.

2 Sensory Evaluation and its Industrial Background

In many industrial sectors such as food, cosmetic, medical, chemical, and textile, sensory evaluation is widely used for determining the quality of end products, solving conflicts between customers and suppliers, developing new products, and exploiting new markets adapted to the consumer's preference [6-8]. In [6], sensory evaluation is defined as a scientific discipline used to evoke, measure, analyze, and interpret reactions to the characteristics of products as they are perceived by the senses of sight, smell, taste, touch, and hearing. Sensory evaluation of industrial products leads to the generation of a set of linguistic terms strongly related to consumer's preference but difficult to be normalized due to their uncertainty and imprecision. As such, this evaluation restricts the scientific understanding of product characteristics for those who wish to design high quality product by engineering means. Hence, a great number of researchers tried to develop objective evaluation systems by physical measurements in order to replace human sensory evaluation, e.g. [9, 10]. In practice, these objective evaluation systems are often expensive and lead to precise numerical data indirectly describing products but its interpretation on product quality related to consumer's preference has to be exploited. Compared with physical measures, sensory evaluation is more efficient for human related quality determination and it can not be, for a long term, completely replaced by objective evaluation.

In general, sensory evaluation can be described as follows: *under predefined conditions, a group of organized individuals evaluate some products with respect to a given target*. Consequently, there are four basic factors in sensory evaluation [3]: evaluation product, evaluation panel, and evaluation target and evaluation environment. According to different cases of these factors, we can divide sensory evaluation into two levels: (1) Design-Oriented Sensory Evaluation (DOSE); and (2) Market-Oriented Sensory Evaluation (MOSE). DOSE is done by a trained panel composed of experienced experts or consultants inside the enterprise for judging products using a number of analytical and neutral linguistic descriptors in a controlled evaluation environment, such as an evaluation laboratory. The evaluation target of DOSE is to obtain the basic sensory attributes of products for improving the quality of product design and development. MOSE is given by untrained consumer panels using analytical and hedonic descriptors according to their preference on the products to be evaluated in an uncontrolled evaluation environment, such as a supermarket. The evaluation target of market-oriented sensory evaluation is to obtain the preference degree of customers in order to forecast the market reaction to the evaluated product.

3 A Web Based Sensory Evaluation Intelligent System in the Supply Chain of Textile /Clothing / Distribution

In the modern industrial world, enterprises often perform sensory evaluation on their products in external independent consultation companies or evaluation laboratories. Compared with the creation of new sensory evaluation structures inside these enterprises, it is less expensive to deliver evaluation work to external consultants or laboratories in the aspects of human resource, time and money. However, under this structure, the evaluation quality is often fluctuating due to the difference of professional backgrounds between these consultation companies or evaluation laboratories.

In many industrial fields such as food, cosmetic and textile, there exists a strong requirement for building a normalized platform in order to solve business conflicts related to product quality and improve sensory evaluation performance of finished products in the corresponding integrated supply chain. Consequently, we have developed a web based intelligent sensory evaluation system, which offers a communication facility between remote partners in order to reduce the time, cost and error of communication and in the same time guarantee the relevance of sensory data analysis.

3.1 2-level (DOSE&MOSE) Sensory Evaluation in the Supply Chain

In the sector of textile/clothing/distribution, sensory evaluation can play an important role in the information flow of the integrated supply chain. At DOSE level, it can provide a normalized platform to the companies for designing and producing standard, flexible, customized and market oriented products and decreasing business conflicts between suppliers and customers. An example of such system in the supply chain is illustrated in Fig. 1. The symbols used in Fig.1 are explained in Table 1.

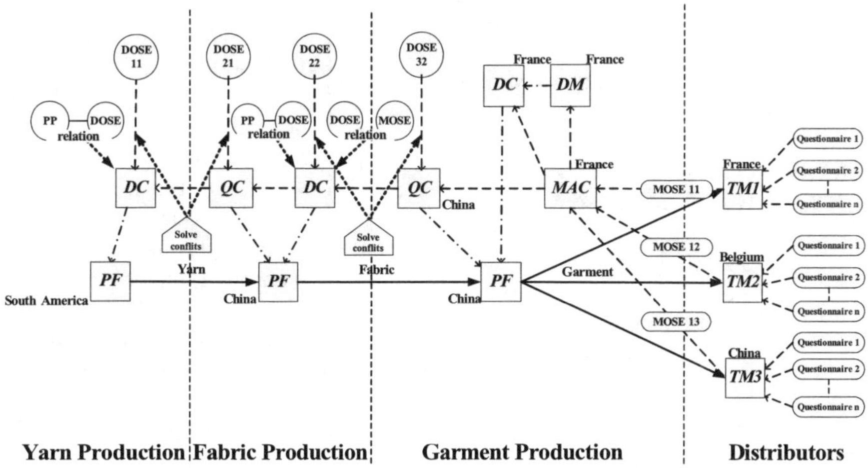

Fig. 1. Sensory evaluation in the supply chain of textile /clothing/distribution (one example)

In this example, a South-American yarn producer is the supplier of a Chinese weaving company, which provides woven fabrics to a Chinese garment making company. This garment maker only produces garments according to the design parameters and the purchasing orders given by a French company which maintains connection with a number of European design centers and delivers finished garment products to its distribution network in France and Belgium. An efficient Market Analysis Center (MAC) constitutes the kernel of this company in order to conduct all the activities of its partners (design centers, producers at different levels, distributors) according to the consumer's preference. In this multi-national industrial cooperation network, the departments of design and marketing are located in France and the production site in China. The final garment collections are sold in France, Belgium and China respectively.

Table 1. Description of the symbols used of Fig.1

Symbol	Meaning	Symbol	Meaning
⟶	Product flow	- ->	Evaluation information flow
-·-·-)	order information flow	·······►	Decision support
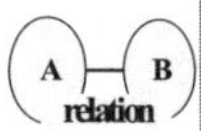	Model for relating A to B		Where the conflicts between suppliers and consumers can occur.
DOSE XY	Design-Oriented Sensory Evaluation X: No. of enterprise doing the evaluation Y: No. of product to be evaluated No. of enterprise: 1 yarn producer 2 fabric producer 3 garment maker No. of product: 1 yarn 2 fabric 3 garment	MOSE XY	Market-Oriented Sensory Evaluation X: No. of enterprise doing the evaluation Y: the target market No. of enterprise: 1 yarn producer 2 fabric producer 3 garment maker No. of target Market:1 France 2 Belgium 3 China
PF	Produce Factory	QC	Quality Center
DC	Design Center	MAC	Market Analysis Center
DMC	Decision Marking Center	TM	Target Market
PP	Produce Process Parameter		

We describe below how sensory evaluation is used for exploiting new garment collections, optimizing the relationship between related companies in the supply chain and meeting the requirements of consumers.

In this supply chain, the product flow or material flow is going from raw material suppliers to producers of higher levels, then to distributors and consumers while the evaluation information flow from distributors and consumers to the design center and the quality inspection centers associated with producers of different levels. The order information flow is going from the design centers and the quality inspection centers to their associated producers in order to produce new market oriented products and improve the quality of existing products according to the sensory evaluation results. The market oriented sensory evaluation (MOSE) is performed in MAC by filling a number of questionnaires by selected consumers in each target market. It analyzes sensory data and provides relevant information on consumer's preference related to finished products to design centers and quality centers through the evaluation information flow. The design oriented sensory evaluation (DOSE) is performed in the design centers (DCs) and the quality centers (QCs) of the supply chain. It analyzes sensory data and provides relevant normalized product quality criteria and design criteria to related partners through the evaluation information flow and the order information flow for producing satisfactory intermediate products and solving business conflicts between suppliers and customers. In the 2-level sensory evaluation of the supply chain of textile/clothing/distribution, a number of mathematical models have been developed for characterizing relations between process parameters and sensory quality criteria and sensory design criteria used by different producers as well as sensory evaluation of consumers [11, 12]. These models permit to transform quality and design criteria used by one producer into those of his partners and consumer's evaluation and preference. In this way, fabric features as well as process parameters can be adjusted using these models in order to meet the expected quality. By using these models, we can effectively enhance the relations between partners in the supply chain and conduct all elements of the chain to work in an optimized way according to the evolution of consumer's behaviors.

3.2 Realization of Web based Sensory Evaluation Intelligent System in the Supply Chain

In order to enhance communications between distant partners in the supply chain and reduce time, cost and errors of these communications, we regroup all functions of the 2-level sensory evaluation described in Section 3.1 and implement them on a web server and two other associated servers. The architecture of this web based sensory evaluation system is shown in Fig.2.

In general, different supply chains have different degrees of integration. In a supply chain where the degree of information integration is low, the link between industrial partners of the supply chain is rather loose and unstable. In order to protect their own interests, these enterprises don't wish to share more information resources with other partners. However, in a supply chain in which a core enterprise plays a leading role, the degree of information integration is rather high. In such a supply chain, the deployment of a sensory evaluation system should be more significant, more efficient and easier for manipulation [13, 14].

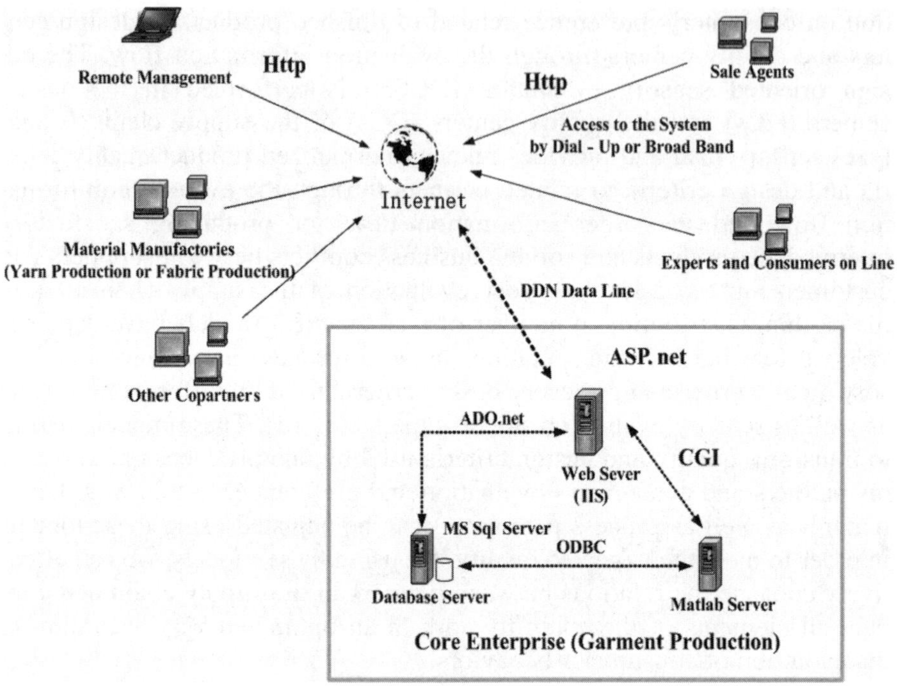

Fig. 2. Architecture of the web based sensory evaluation system

In Fig.2, we can see that this sensory evaluation system is implemented in the information system of the core enterprise of the supply chain. In the example described in Section 3.1, the Market Analysis Center (MAC) is considered as core enterprise.

In this system, all partners of the supply chain can visit the web based sensory evaluation system using Internet. The interface between Internet and the information system of the core enterprise is DDN (Digital Data Network). DDN has the advantages of high speed, low data delay and reliable secrecy. A three level Browse/Server structure is used for developing this sensory evaluation system. Its users, including material manufacturers such as yarn producer or fabric producer, sale agents, evaluation experts and consumers, can easily visit it using IE browse in any location. The B/S structure also provides convenience for remote management. Its main functions can be described as follows: (1) IIS 6.0 (Internet Information Services) is used as web server. Asp.net is used as the development technology of foreground program. (2) Microsoft SQL Server is used as database server. ADO.NET technology is used for building the connection between the web server and the database sever. ADO.NET is the next generation of ADO (ActiveX Data Objects), providing an user friendly interface for accessing, manipulating, and exchanging data in the Microsoft.NET Framework. (3) Matlab is used to process sensory data. By applying CGI (Common Gateway Interface), the Matlab Server enables to receive sensory data from WEB pages and call the corresponding Matlab programs to analyze sensory data. An ODBC (Open Database Connectivity) interface is used to transmit results of the analysis to the SQL server database.

Fig.3 presents the main modules of the web based intelligent sensory evaluation system. This system is composed of three subsystems: the project management subsystem, the DOSE subsystem, the MOSE subsystem.

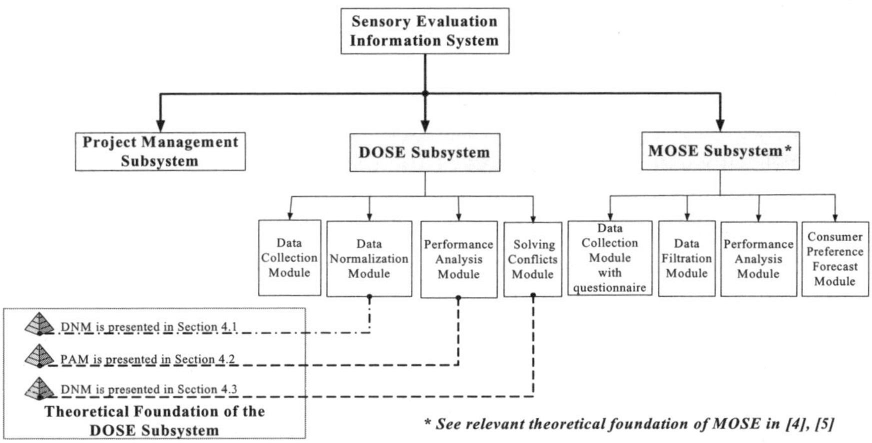

Fig. 3. Main modules of the web based intelligent sensory evaluation system

The Project Management Subsystem is used for supervising and managing projects of sensory evaluation. It permits partners of the supply chain to create, modify and remove DOSE and MOSE projects. When a DOSE project is created, we need to record the background information of the project, the related application, the corresponding evaluation terms and the types of evaluation data. When a MOSE project is created, we need to record the background information of the project and generate questionnaires according to the specific application of MOSE.

The DOSE subsystem is composed of the data collection module, the data normalization module, the performance analysis module and the conflicts solving module. In the data collection module, we design two interfaces for users. The first one permits to collect evaluation data from individual evaluators using Internet connection. These individual evaluators are often considered as leading experts of the concerned products. The second interface is designed for collecting batch sensory data provided by a panel of evaluators, i.e. a group of technicians working inside an enterprise. In the data normalization module, we realize a 2-tuple linguistic model to normalize and aggregate sensory data inside each panel using the method presented in [11]. In the performance analysis module, we compute distances between panels and between terms as well as sensitivity degrees of panels and terms. These values can be used to determine the quality of each panel and each term used as well as the quality of evaluated products. In the conflicts solving module, we calculate distances between evaluations terms used by two partners of the supply chain (a supplier and a customer), and then determine the relationship between these terms in order to convert terms of one partner into those of another one.

The MOSE subsystem is composed of the data collection module, the data filtering module, the performance analysis module and the consumer preference forecasting module [4, 5]. In the data collection module, two interfaces for users are developed. The first one is designed for collecting data from individual consumers using Internet connection. It corresponds to two cases: 1) consumers evaluate at home samples sent by evaluation organizers and then submit results using Internet; 2) selected consumers are invited to evaluate samples under controlled conditions in a laboratory and then submit results using Internet. The second interface is designed for collecting batch data. It corresponds to the case in which evaluation organizers receive at shopping centers consumers' answers to predefined questionnaires. The data filtering module is used to select the evaluation results of target consumers and latency consumers, considered as relevant response to the consumer questionnaires [4]. In the performance analysis module, we compute distances between MOSE terms and sensitivity degrees of these terms in order to determine the quality of evaluation results [4]. In the forecasting module, users can forecast the consumer's preference of a new product according to the evaluation results obtained at DOSE level, corresponding to design parameters of the product [5].

In the following section, we present the basic mathematical models (Data Normalization Model, Performance Analysis Model, and Conflicts Solving Model) used in the three main modules of the DOSE subsystem.

4 Basic Models in DOSE Level

4.1 Data Normalization Model (DNM)

In sensory evaluation, different individuals of one panel generally use unified attributes or terms but evaluation results may be on different scales for each term. This is because the sensitivity of each individual to the samples to be evaluated, strongly related to his personal experience and the corresponding experimental conditions, is often different from others. Moreover, these sensory data may be in a numerical form or a granular linguistic form. So it is necessary to develop a suitable unified scale in order to normalize and aggregate sensory data inside each panel. In our previous work, a 2-tuple fuzzy linguistic model has been used for transforming different types of sensory data (numerical, symbolic, linguistic) with different scales used by different individuals of one panel into an optimized common numerical scale [11, 15]. This transformation is denoted by $z = Tr(t, g, ug)$ in which t and z represent the evaluation scores before and after the transformation respectively and g the scale corresponding to t (number of modalities) and ug the scale corresponding to z.

For the sensory panel P_i, the evaluation results of its individuals I_{ij}'s $(j \in \{1,...,h(i)\})$ can be aggregated by transforming all the corresponding fuzzy sets to be on the unified scale ug. $h(i)$ is the number of individuals in panel P_i. The optimal scale can be calculated by taking into account the two following principles:

1) The sensory data given by the individuals I_{ij}'s should cover all the modalities of the unified scale, i.e., any modality on the scale ug is supposed to correspond to at least one datum.
2) The variation or the trend of the sensory data should not change very much with transformation of the scale.

The sensory data of I_{ij} for evaluating n samples on the term a_{il} before the transformation are $\{e_{ij}(1,l), e_{ij}(2,l),..., e_{ij}(n,l)\}$. After the transformation, these data become $\{z_{ij}(1,l), z_{ij}(2,l),..., z_{ij}(n,l)\}$, where $z_{ij}(k,l) = Tr(e_{ij}(k,l), g, ug)$ for $k = 1,2,...,n$ (k: index of the kth sample) and $l=1,2,..., m(i)$ (l: index of the term a_{il}).

According to the first principle, we first calculate the number of data for each modality q of the unified scale ug, i.e.,

$$N\_\mathrm{mod}_i(l,q) = \sum_{j=1}^{h(i)} \sum_{k=1}^{n} equ(z_{ij}(k,l), q) \text{ with } equ(p,q) = \begin{cases} 1 & \text{if } p = q \\ 0 & \text{otherwise} \end{cases}.$$

The criterion of coverage of the unified scale is then defined by

$$Cover_i(l,ug) = \min\{N\_\mathrm{mod}_i(l,q) \mid q = 0,1,2,...,ug\}.$$

According to this criterion, ug should be selected so that $Cover_i(l)$ is as big as possible. If this value is 0, it means that there exists at least one non-significant modality on the unified scale ug.

According to the second principle, the difference of the trend between two data sets $\{e_{ij}(1,l), e_{ij}(2,l),..., e_{ij}(n,l)\}$ and $\{z_{ij}(1,l), z_{ij}(2,l),..., z_{ij}(n,l)\}$ should be as small as possible. In this chapter, the corresponding criterion is defined by

$$Trend_i(l,ug) = D_t(E_i(l), Z_i(l))$$

where $D_t(E_i(l), Z_i(l))$ represents the distance criterion between two evaluation matrices $E_i(l)$ and $Z_i(l)$ defined in Section 4.2 with

$$E_i(l) = \begin{bmatrix} e_{i1}(1,l) & ... & e_{i,h(i)}(1,l) \\ ... & ... & ... \\ e_{i1}(n,l) & ... & e_{i,h(i)}(n,l) \end{bmatrix} \text{ and } Z_i(l) = \begin{bmatrix} z_{i1}(1,l) & ... & z_{i,h(i)}(1,l) \\ ... & ... & ... \\ z_{i1}(n,l) & ... & z_{i,h(i)}(n,l) \end{bmatrix}.$$

Next, we calculate $Trend_i(l,ug)$ for all possible values of the unified scale ug and then obtain its average denoted as $Trend\_avg_i(l)$. If $Trend_i(l,ug) > Trend\_avg_i(l)$, then the corresponding value of ug should not be considered as optimal scale because the information lost is too great. For the other values of ug, the optimal scale can be obtained by minimizing the linear combination of these two criteria as follows:

$$\min\{ Trend_i(l,ug) - \rho \cdot Cover_i(l,ug) \} \quad (1)$$

where ρ is a positive constant adjusting the ratio of these two criteria.

After the computation of the best value of ug for each panel P_i and each term a_{il}, the optimal unified evaluation scores can be obtained by transforming sensory data of all individuals into this scale and then aggregating all transformed data by calculating their average. Finally, for each panel P_i, we obtain a unified evaluation matrix $S_i = (s_i(k,l))_{n\times m(i)}$ with $s_i(k,l) = \frac{1}{h(i)} \cdot \sum_{j=1}^{h(i)} z_{ij}(k,l)$. This matrix will be used in the next section for the analysis and interpretation of panels and term.

4.2 Performance Analysis Model (PAM)

In the supply chain, sensory evaluation results at DOSE level are provided by different sensory panels each representing one industrial partner. In the performance analysis module of the DOSE subsystem as well as the module for resolution of business conflicts, the most important function is the calculation of dissimilarity or distance between two panels and between two evaluation terms. The other functions can be realized using these distances.

For the panel P_i, its aggregated sensory evaluation data constitute an evaluation matrix denoted by $S_i = (s_i(k,l))_{n\times m(i)}$ after applying the normalization model of section 4.1. We normalize S_i into [0, 1] and obtain $U_i = u_i(k,l)$. This transformation is as follows: $u_i(k,l) = (s_i(k,l) - s_{\min}(l))/(s_{\max}(l) - s_{\min}(l))$. $s_{\min}(l)$ and $s_{\max}(l)$ represent the minimum and the maximum of $[s_i(1,l),...,s_i(n,l)]^T$ respectively.

As the evaluation terms used by one panel is often quite different from those of another panel, the distance between two panels P_a and P_b cannot be defined using classical methods, which compute distances between vectors in the same space. So a new distance criterion between two panels P_a and P_b has been defined in [11].

In this definition, the distance criterion takes into account the degree of consistency of relative variations of two different sensory data sets. If the internal relative variations of these two data sets are close each other, then the distance between the corresponding panels is small. Otherwise, this distance is great. Formally, it is defined by:

$$D_{ab} = \frac{2}{n(n-1)} \cdot \sum_{\substack{k<k' \\ (k,k') \in \{1,\dots,n\}^2}} d_{ab}(k,k') \qquad (2)$$

It depends on the following elements:

1) The distance between P_a and P_b related to the relative variation between fabric samples t_k and $t_{k'}$: $d_{ab}(k,k') = \left| vr_a(k,k') - vr_b(k,k') \right|$.

2) The relative variations between t_k and $t_{k'}$ for P_x (x=a, b)

$$vr_x(k,k') = \frac{1}{\sqrt{m(x)}} \left\| U_{xk} - U_{xk'} \right\|, \text{ with } \begin{cases} U_{xk} = \left(u_x(k,1), u_x(k,2), \dots, u_x(k,m(x))\right)^T \\ U_{xk'} = \left(u_x(k',1), u_x(k',2), \dots, u_x(k',m(x))\right)^T \end{cases}$$

The definition of D_{ab} permits to compare between these two panels the relative variations in the set of all samples. The distance between two panels reaches its minimum only when the internal variations of their sensory data are identical.

In the same way, we also define the distance between terms used by two different panels [11]. This criterion permits to study the business conflicts between two panels related to the understanding of some specific evaluation terms. For example, the distance between two panels on the term "soft" may be very large, which means that these two panels understand this term in different ways.

It is important to physically interpret numerical values of the above criteria of distance. For this purpose, we transform these numerical values into fuzzy numbers, whose membership functions are generated according to the probability density distributions of the corresponding random matrices. The detailed procedure is given as follows.

Step 1: For fixed values n, $m(a)$ and $m(b)$, generating two random matrices S_a (dimension: $n \times m(a)$) and S_b (dimension: $n \times m(b)$) whose elements obey the uniform distribution between lower and upper bounds of evaluation scores.

Step 2: Computing the values of distance D_{ab}.

Step 3: Repeating Step 1 and Step 2 for a number of times in order to obtain the probability density distribution of D_{ab}.

Step 4: Equally dividing the area of this distribution into 5 parts. According to these divided areas, we generate 5 fuzzy subsets for D_{ab}: {*very small, small, medium, large, very large*}. The corresponding membership functions can be determined from these 5 fuzzy subsets.

In this way, each numerical value of distance criteria can be transformed into a fuzzy number whose value includes the linguistic part taken from the previous 5 terms and the corresponding membership degrees. This fuzzy number permits to interpret the distance with respect to the whole distribution of random values.

4.3 Conflicts Solving Model (CSM)

For solving business conflicts between two companies related to the understanding of evaluation terms (product quality criteria or product design criteria), there exists a strong need for interpreting evaluation terms of one panel using those of another panel. In this section, we propose a genetic algorithm based procedure to do so. The details of this procedure are given as follows.

The sensory data of two panels P_a and P_b are obtained by evaluating the same set of representative samples denoted by T. The sets of terms of P_a and P_b are denoted by $A_a = \{a_{a1}, a_{a2}, ..., a_{a,m(a)}\}$ and $A_b = \{a_{b1}, a_{b2}, ..., a_{b,m(b)}\}$ respectively. We suppose that no redundant terms exist for each panel. For each term $a_{al} (l \in 1, ...m(a)\}$ of P_a, we try to find the optimal linear combination of the terms $a_{b1}, a_{b2}, ..., a_{b,m(b)}$ to generate a new term denoted by $a(P_a, P_b, l)$ which is the closest to a_{al} in semantics, i.e. $a(P_a, P_b, l) = w_1^l \cdot a_{b1} + w_2^l \cdot a_{b2} + ... + w_{m(b)}^l \cdot a_{b,m(b)}$ with $\sum_{i=1}^{m(b)} w_i^l = 1$.

The corresponding weights $\left(w_1^l, w_2^l, ..., w_{m(b)}^l\right)^T$ are determined using a genetic algorithm with penalty strategy [16] so that the distance between a_{al} and $a(P_a, P_b, l)$ is minimal. It is an optimization problem with constraints because the sum of the weights should be 1 and each weight should be included between 0 and 1. In this case, the penalty strategy is used in the genetic algorithm. The detail of this algorithm is given in [11].

5 One Application in Fabric Hand Evaluation

At the level of DOSE, we apply our approach to sensory data on fabric hand evaluation provided by 2 sensory panels in France and 2 sensory panels in China. These panels are denoted as *FE* (French fashion experts), *FTS* (trained French students), *CE* (Chinese textile experts), *CTS* trained (Chinese students) respectively. The set T is composed of 43 knitted cotton samples produced using 3 different spinning processes. These samples can be then classified into 3 categories: Carded, Combed and Open-End, corresponding to different touch feelings. Based on the sensory data provided by these panels, we apply the three models presented in Section 4 for normalizing the sensory data, analyzing the evaluation performance and solving conflicts between different panels.

5.1 Result of sensory data normalization

In the following example, the sensory data are obtained by a panel of 6 trained Chinese students for evaluating 18 combed cotton samples on the term "soft". Table 2 presents the forms (numerical or linguistic) and values of evaluation scales used. Table 3 gives the corresponding evaluation results.

Table 2. Forms and values of evaluation scales used by 6 individuals

Individual n°	Form	Values
1	linguistic	{Very soft(*Vs*), Middle soft(*Ms*), Soft(*S*), Average(*A*), Not very soft(*Nvs*), Not soft(*Ns*), Worse(*W*)}
2	linguistic	{Very good(*Vg*), Good(*G*), Middle good(*Mg*), Average(*A*), Middle bad(*Mb*)}
3	numerical	{0,1,2,3,4,5,6,7,8,9,10}
4	numerical	{0,1,2,3,4,5,6,7,8,9,10}
5	numerical	{0,1,2,3,4,5,6,7,8,9,10}
6	linguistic	{Good(*G*), Middle good(*Mg*), A little soft(*Als*), Soft(*A*), Average(*A*), Middle bad(*Mb*)}

The two criteria for scale optimization defined in Section 4.1 are calculated using different values of ug (number of modalities in a scale). The corresponding results are shown in Table 4.

Table 3. Evaluation result

Sample n°	Individual1	Individual2	Individual3	Individual4	Individual5	Individual6
4-1	*Vs*	*G*	9	8	9	*G*
10-1	*Vs*	*G*	9	8	8	*Mg*
			… …			
34-2	*Ms*	*Mb*	2	4	6	*A*

Table 4. Calculation of the criteria for scale optimization

ug (number of modalities)	*Trend*	*Cover*	ug (number of modalities)	*Trend*	*Cover*
2	1.6660	29	7	0.3665	0
3	1.4060	5	8	0.3665	0
4	1.1460	0	9	0.2359	0
5	0.8861	0	10	0.4236	0
6	0.6261	0	11	0.6737	0

From these values, we obtain $Trend_{average} = 0.7620$. The linear combination of these two criteria is given in Fig.4. According to our experience, we select $\rho = 0.2$.

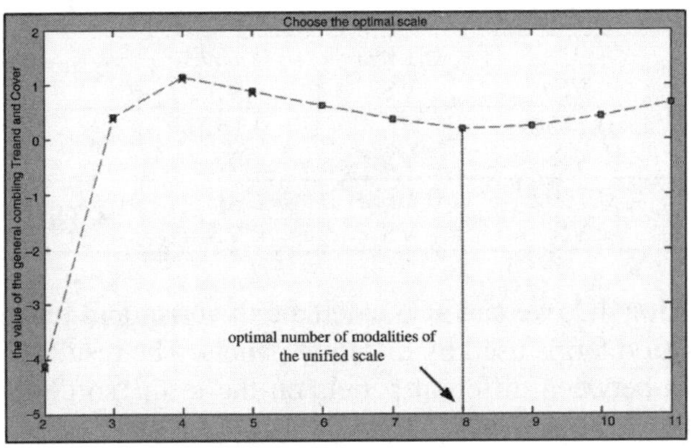

Fig. 4. Evolution of the general criterion with scale

In Fig.4, we eliminate the cases of $ug = 2,3,4,5$ whose values of the criterion *Trend* are bigger than $Trend_{average}$. The optimal scale corresponds to the case of $ug = 8$ in which the value of the general criterion combining *Trend* and *Cover* is the smallest for the remaining scales.

5.2 Result of evaluation performance analysis

We apply the model presented in section 4.2 for computing distances between panels and obtain the corresponding results in Table 5.

From Table 5, we can notice that the averaged distances between French and Chinese experts and between professional experts and students are very small. It means that the general evaluation on fabric hand is related to neither the cultural background nor the professional background. In practice, business conflicts between suppliers and customers on general fabric hand are rather limited because their evaluation results for the same collection of products are generally very similar.

Table 5. Averaged distances for different panels

	FE	FTS	CE	CTS	
Crisp#	0	0.1563	0.1284	0.1717	FE
Fuzzy set	VS	VS	VS	VS	
Crisp#	0.1563	0	0.1456	0.1692	FTS
Fuzzy set	VS	VS	VS	VS	
Crisp#	0.1284	0.1456	0	0.1622	CE
Fuzzy set	VS	VS	VS	VS	
Crisp#	0.1717	0.1692	0.1622	0	CTS
Fuzzy set	VS	VS	VS	VS	

Using the method in Section 4.2, we can also calculate the crisp and fuzzy distances between evaluation terms used by different panels. The results of crisp and fuzzy distances between different panels on the term “soft” are given in Table 6 and Table 7 respectively.

Table 6 and Table 7 show that the distances between different panels on the term “soft” are sometimes rather important (M, L, VL) although most of the fuzzy values of these distances are very small. The same phenomenon can be observed for the other terms such as “slippery” and “smooth”.

This means that one evaluation term is often semantically interpreted in different ways by different panels. In practice, this divergence in the understanding of linguistic quality criteria and design criteria constitutes a main source of business conflicts between suppliers and customers. A dictionary is then needed for the understanding of evaluation terms between different professional populations.

Table 6. Values of crisp distances between different evaluators on the common term "soft"

Process	Carded			Combed			Open-End		
Evaluators	CE	FTS	CTS	CE	FTS	CTS	CE	FTS	CTS
FE	0.13	0.21	0.17	0.23	0.23	0.27	0.10	0.21	0.18
CE		0.21	0.20		0.13	0.16		0.18	0.20
FTS			0.22			0.16			0.29

Table 7. Values of fuzzy distances between different evaluators on the common term "soft"

Process	Carded			Combed			Open-End		
Evaluators	CE	FTS	CTS	CE	FTS	CTS	CE	FTS	CTS
FE	VS	VS: 0.81 S: 0.19	VS	S: 0.4 M: 0.58	S: 0.42 M: 0.58	VL	VS	VS: 0.03 S: 0.97	VS: 0.97 S: 0.03
CE		VS: 0.81 S: 0.19	VS		VS	VS		VS: 0.97 S: 0.03	VS: 0.34 S: 0.66
FTS			VS: 0.13 S: 0.82			VS			L: 0.1 VL: 0.9

5.3 Result of solving conflicts between panels

We use the method presented in Section 4.3 to interpret each term used by the panels of FTS, CE and CTS using those of FE. For simplicity, we only discuss the case of P_a=FTS and P_b=FE in this section. Three non redundant terms of FE are chosen to interpret the terms used by FTS. By applying the genetic algorithm with penalty strategy, we obtain the optimal linear combination of the terms of FE for each term of FTS. The corresponding weights w_j^k 's of these linear combinations are shown in Table 8.

Table 8 permits to interpret the relationship between terms used by FTS and FE. Under this relationship, the terms used by FTS can be approximately expressed by linear combinations of the terms of FE. For example, the term *"Smooth"* used by FTS can be approximately expressed as a linear combination of three terms of FE: *"Soft", "Spongy" and "Intangible feeling"*, i.e.

$$\text{Smooth\_FTS} \approx 0.60 \cdot \text{soft\_FE} + 0.14 \cdot \text{spongy\_FE} + 0.26 \cdot \text{intangible feeling\_FE}$$

Table 8. Weights of optimal linear combinations of terms of FE for terms of FTS

Terms of FE \ Terms of FTS	Smooth	Slippery	Soft	Tight
Soft	0.60	0.82	0.82	0.37
Spongy	0.14	0	0	0.15
Intangible feeling	0.26	0.18	0.18	0.48

According to the results shown in Table 8, we can observe that some confusion exists between *"Smooth", "Slippery"* and *"Soft"* used by FTE and *"Tight"* is quite different from the other terms. We apply the model presented in section 4.2 to compute the crisp and fuzzy distances between these terms. The related results are given in Table 9.

Table 9. Crisp and fuzzy distances between terms of the panel FTS

Terms de FTS \ Terms de FTS		Smooth	Slippery	Soft	Tight
Smooth	Crisp#	0	-	-	-
	Fuzzy set				
Slippery	Crisp#	0.1394	0	-	-
	Fuzzy set	VS			
Soft	Crisp#	0.1008	0.1406	0	-
	Fuzzy set	VS	VS		
Tight	Crisp#	0.2494	0.2534	0.2227	0
	Fuzzy set	VS	VS	VS	

All the fuzzy distances in Table 9 are *"VS"*. This phenomenon is caused by the fact that all evaluated fabric samples have similar touch feeling. However, the distances between *"Tight"* and the other terms are evidently larger. It is conform to the results obtained in Table 8.

6 Conclusion

This chapter presents a web based intelligent sensory evaluation system of industrial products in the textile integrated supply chain. Several models at DOSE level of this system have been described in order to normalize, analyze sensory data and solve business conflicts between panels at B2B level. Fuzzy techniques and genetic algorithms have been used in these models. These models have been implemented on web sites and successfully applied in the fabric hand evaluation. In practice, this web based system can provide a normalized platform for optimizing relations between distance designers, producers at different levels and consumers.

References

[1] OECD (Organisation for Economic Co-operation and Development), A New World Map in Textiles and Clothing: Adjusting to Change, Source OECD Industry, Services & Trade, 2004(20): 1-11.
[2] J.L.W. Lo, B.Rabenasolo and A.M.Jolly-Desodt, A fashion chain which employs a speed-of-light operation to gain competitive advantage, Textile and Clothing, 2004(20): 36-38.
[3] B.Zhou, X.Zeng, L.Koehl, Y.Ding, A 2-level Model for Description of Fabric Hand Sensory Evaluation, Int. Conf. World Textile Conference - 4th AUTEX Conference, Roubaix, France, June 22-24 2004.
[4] B.Zhou, X.Zeng, L.Koehl, Y.Ding, Data Collection and Analysis Methods for Market-Oriented Sensory Evaluation, 2006 International Conference on Intelligent Systems and Knowledge Engineering, Shanghai(China), April 6-7 2006.
[5] L.Koehl, X.Zeng, B.Zhou, Y.Ding, Fabric Touch Handle Management and Assessment from Manufacturers to Customers, in Intelligent Data Mining Techniques and Applications, Eds. D.Ruan, G.Chen, EKerre, G.Wets, Springer, Berlin, 2005: 376-399.
[6] H.Stone, J. L.Sidel, Sensory Evaluation Practice, Academic Press Inc., 1993.
[7] G. B. Dijksterhuis, Multivariate Data Analysis in Sensory and Consumer Science, Food & Nutrition Press Inc., Trumbull, Connecticut, USA, 1997.
[8] J. R. Piggot, E. A. Hunter, Evaluation of assessor performance in sensory analysis, Italian Journal of Food Science, 1999,11(4): 59-64.
[9] S.Kawabata and M.Niwa, Objective measurement of fabric hand, Modern Textile Characterization Methods (Eds. M. Raheel and M. Dekker), 1996: 329-354.

[10]J.Hu, W.Vhen, and A.Newton, A psychophysical model for objective fabric hand evaluation: An application of Steven's law, J. the Textile Institute, 1993, 84(3): 354-363.
[11]B.Zhou, X.Zeng, L.Koehl, Y.Ding, An Intelligent Technology Based Method for Interpreting Sensory Evaluation Data Provided by Multiple Panels", Information Sciences, accepted in July 2004.
[12]M.Sahnoun, X.Zeng, and L.Koehl, Modeling the relationship between subjective and objective fabric hand evaluations using fuzzy techniques, IPMU 2002, Annecy, France, July 1-5, 2002.
[13]M. Hugos, Essentials of supply chain management, John Wiley & Sons Inc, 2003.
[14]J.B.Ayers, Supply Chain Project Management: A Structured Collaborative and Measurable Approach, St. Lucie Press, 2004.
[15]F.Herrera and L.Martizez, A model based on linguistic 2-tuples for dealing with multigranular hierarchical linguistic contexts in multi-expert decision-making, IEEE Tranc. Systems, Man, and Cybernetics—Part B: Cybernetics, 2001, 31(2): 227-234.
[16]F. Back, Evolutionary Algorithms in Theory and Practice, Oxford University Press, New York, 1996.

E-intelligence in Portfolio Investment Optimization

K. Stoilova, Z. Ivanova, and T. Stoilov

Institute of Computer and Communication Systems - Bulgarian Academy of Sciences, *Acad. G.Bontchev str., BL.2* , Sofia 1113, Bulgaria, tel: +(359) 2 979 27 74, fax: +(359)2 872 39 05; e-mail: todor@hsi.iccs.bas.bg

Abstract

The chapter describes the main stages of the development of a system, supporting the decision-making in e-portfolio investment optimization. The developments target the implementation of web based e-service application for financial assistance in the choice of investment strategy, according to the investor's preferences. The implementation of portfolio optimization as e-service in the web benefits the users to their inclusion in direct management and decision making in the financial investments. Here is presented a practical result from the intersection of three scientific and technological domains: portfolio theory of investments, optimization theory and information technology. A system-algorithmic model for the portfolio e-service is developed based on four-tier functional architecture. Appropriate optimization model is defined extending the investment horizon to several categories. Optimization solver is developed applying reduced sets of calculations and performing on-line evaluations for portfolio e-services. The presented results are applied in a real portfolio optimization web based system operating as financial advisor in the investment process.

1 Introduction

The electronic financial market is fast emerging. The financial products endure substantial transformation with rapid introduction of Internet technologies in the financial sector (Janal, 1998). The financial portals now allow visitors to get real-time quotes of stock-indices, to track their evolu-

K. Stoilova et al.: *E-intelligence in Portfolio Investment Optimization*, Studies in Computational Intelligence (SCI) **37**, 457–476 (2007)
www.springerlink.com

tion, to invest in mutual funds. Each of these funds has different characteristics (costs, sales charges, minimal investments) and exhibits a different performance (expected profit, risk profile). The challenge of the decision making is to find the best investment in such vast universe of possibilities. Thus, the use of sophisticated tools available through Internet sounds as a prospective solution for optimal financial investment.

Several computational issues have to be addressed in order to solve the problem. An elaboration of the information collected from the financial market has to be performed. The selection of the "best" portfolio involves numerical optimization which is a computationally intensive numerical task. The access of large databases and computational resources by the investor suggests that a Web-based architecture is the suitable choice.

The e-Portfolio is designed to give assistance in the choice of optimal strategy, according to the investor's profile. A portfolio holder may employ e-Portfolio services to access the current performance of his portfolio to make optimal decisions about a future investment strategy. However, the solution of the portfolio problem insists the implementation of optimization solver for the computational efforts. The incorporation of the solver in e-portfolio service is a prerequisite for the deployment of optimization modes on Internet based applications (Geoffrion and Krishnan, 2001).

The World Wide Web (WWW) already offers examples of remote implementation of optimization. The mode of Application Service Provision (ASP) enables to deliver the software through Internet. The early developments of distributed environments for optimization took place in the late 1990s when a number of optimization servers were designed and implemented. They have been based on client-server architecture and the user interfaces were based on protocols like ftp and smtp, which are generic Internet services. Later this generic services evolved by taking advantage from the http protocol of the WWW service. A comprehensive review of such systems is given in (Fourer and Goux, 2001).

The optimization services address modelling and solver functionalities. The solver services offered access to commercial and custom implementations of solution algorithms for linear and non-linear programming. The modelling services enabled the user to define mathematical programming models and to submit to the server the modelling definition and required data. The calculation results are sent back to the user by e-mail or by downloading files through ftp or http protocols.

In (Cohen et al., 2001) examples of application services for decision support systems implemented in Supply Chain Management, Inventory Control, and e-procurement are provided. A representative optimization tool, targeting Internet applications, is the NEOS server (Czyzyk et al., 1998). It enables optimization problems to be submitted using Web formats, e-mail or TCP/IP based client submission tool and the problem can be solved by several solvers, covering the optimization domain of linear, integer, quadratic, non-linear and stochastic programming. The system's and algorithmic architecture of the NEOS server is three-tiered. It is not based on Web server tools, which restrict its implementation in Web based services.

Another optimization system, operating in distributed environment, is AURORA Financial Management System (Pflug et al., 2000). It provides financial decision support functionalities based on optimization models for portfolio planning. The optimization relies on decomposition algorithms for the solution of linear optimization problems. The system is based on multiple GRID distributed architecture, which requires the exploitation of several powerful optimization solvers.

The Web service paradigm implemented the ASP operational mode as a business entity, which offers info services to users across a wide area network. Thus, distant access to software through Internet becomes a market service. In essence, these services provide a virtual desktop for the remote user. Many of these systems are based on proper protocols between clients and servers. With the raise of the e-commerce, a new "e-service" paradigm appears. While ASP paradigm concentrates on the provision of a solution of the optimization problem or to provide optimization tool to the user, the e-service provides the entire support and management infrastructure together with the tool addressing particular application. Thus, the e-service becomes a combination and integration of ASP solutions.

In (Rust and Kannan, 2003) the traditional e-commerce is seen as a tool for the development of a new more general paradigm, represented by e-services. For specific applications, optimization software tools are developed in order to be incorporated in e-services. The web interaction between the customer and the optimization tool allows the optimization service to be integrated in a customer application and to generate new-market product. Taking into account the e-service model, the web-based services, consisting optimization functionality, have to be developed by integrating web interface, data retrieval and integration with optimization solvers.

The optimal resource allocation for the investment process is a prospective task, solved according to the available technological support (Mital, 1992). Good examples of information services, related to the financial domain, are available at (www.schwab.com; www.etrade.com; www.ml.com; www.magnum.com; www.northinfo.com; www.fid-inv.com; www.adviceonline.co.uk; www.quicken.com). However, the services, performing optimization calculation for the investment portfolio management, are supported by locally available software suit. Good examples of financial optimization solvers are presented at (www.solver.com; www.allocationmaster.com; www.stockalpha.com; www.invest-tech.com; www.frontsys.com; www.wagner.com; www.excelbusinesstools.com). These applications do not apply client-server web based architecture and the portfolio optimization is not performed as e-service. Thus, if the portfolio optimization is implemented as e-service in the web, this will benefit the users to their inclusion in direct management and decision making in the financial investments (the aim of the presentation below).

2 System's and algorithmic models, used by the information services in Internet

The functional complexity increase of the Web services insists corresponding structural developments in the system-algorithmic models. Until now, three types of models can be defined for the design and implementation of WAN based information systems (Ivanova et al., 2005).

2.1 Information system with two-tier system model

The client-server model is the basic model widely applied for all Internet information systems (Rosenfeld, 1998), based on two-tier system model, Figure 1. The first tier is the client, who in general operates under a web browser environment. Particularly, the client does not perform data processing. The server side is implemented as Web Server (Internet Information Server – IIS, Apache, Tomcat, Java Web Server – JWS), which operates on different environments (Windows, Unix, Sun, HP). The two-tier information model does not have wide algorithmic potential for complex information treatment. It supports predefined (static) information. Such systems are applicable for user scenarios related to business and web presence in Internet, for dissemination of data, messages, and events in the global network. These information systems do not perform advanced algorithmic treatment of the data included in the services.

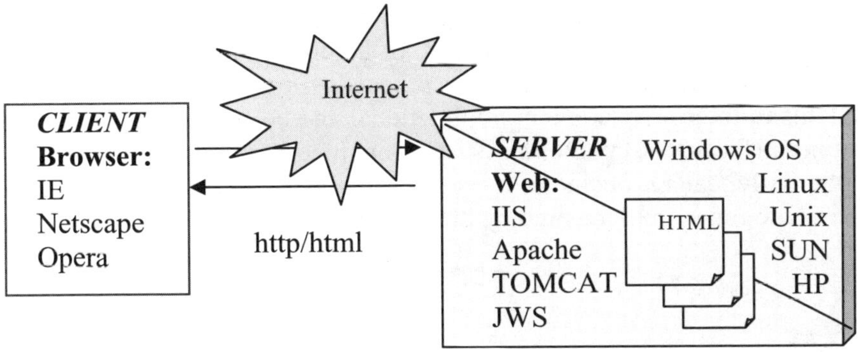

Figure 1: Two-tier client server model of WAN based information system

2.2 Information system with three-tier system model

The three-tier system model is a natural extension of the previous one, addressing the needs in complicating the functionality and the data treatment in WAN information systems. This model performs functionality, supported by a database management system (Preston, 2000), Figure 2.

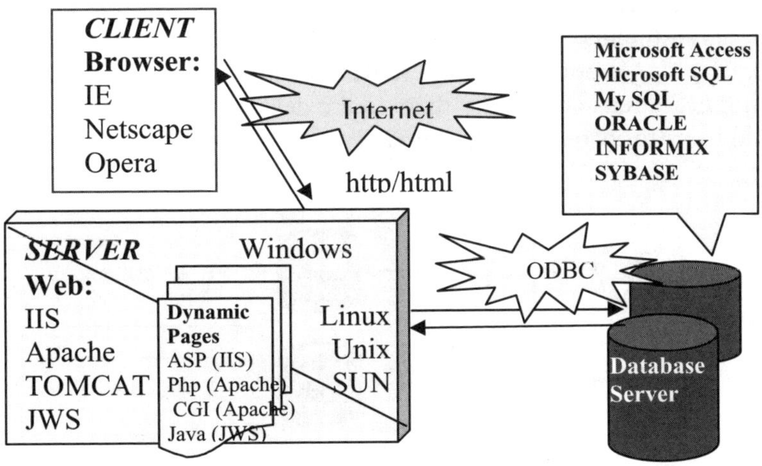

Figure 2: Three-tier client server model of WAN based information system

The third tier contains a database engine, based on different software suits like Microsoft SQL Server, Microsoft Access, Oracle, Informix, Sybase, free software solutions like MySql, etc. Additionally, the second algorithmic tier supports increased functionalities by the inclusion of the server side programming tools. These programs perform on-line communications with the databases (third tier) so that all additional functionalities in data retrieval, search, edit, and data can proceed. As a result, the information systems in Internet become more functional and cover wider area of applications in business, marketing, system engineering, culture, and science. Particularly, each e-business, e-reservation system, e-learning service, on-line catalogue is implemented by the deployment of three-tier architecture.

2.3 Information system with four- tier system model

The on-line requirements for the system management insist fast information processing. The system design and algorithmic solutions, satisfying the requirements of complex data processing and on-line system management, introduce new forth level - "Algorithmic Server" in the information architecture (Ivanova et al., 2003), Figure 3. It performs specific and complex data processing, evaluates complex mathematical problems, and supports on-line control functionality. It operates on tasks that cannot be supported by the server-side programming tools. The forth-tier information systems are implemented in real time process control, on-line market researches, investments, on-line decision-making, and resource allocation systems. The bigger potential of the fourth-tier information systems in data processing is the reason to apply them to the design of a Web based system for on-line portfolio optimization and resource allocation in financial investments.

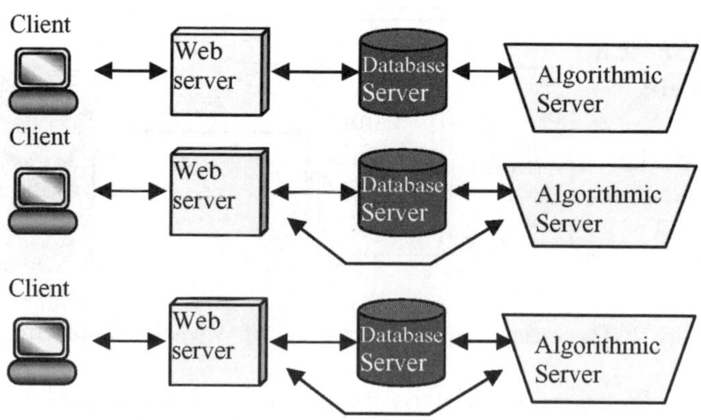

Figure 3: Fourth-tier client server model of WAN based information system

3 Portfolio theory and optimal allocation of investment resources

The Portfolio theory concerns an optimal allocation of financial resources buying and/or selling assets (securities) from the financial markets (Bodie 2000). The allocation of the financial resources is called "investment". The investor considers each asset as a prospect for future return. Thus, better combination of assets (securities) in the portfolio provides better future actual rate of return to the investor. The portfolio is a composition of securities. The portfolio theory develops models that allow the best security combinations to be found. Two general assumptions are done by the portfolio theory (Sharpe 2000):

- The actual rate of return for the securities can differ from the predicted value. The deviation between the predicted and the real value is a measure of the security risk;
- The security returns influence each other in global economy – the changes of assets' returns are not independent - they are correlated.

The problem of the optimal portfolio design is stated as follows. The financial analyst keeps records of the actual rates of the security returns for past time period, $[0,-m_i]$, m_i is the number of the periods:

$$R_{i|m_i \times 1} = \{R_i^{m_i}\}, \quad i = 1, N . \tag{1}$$

Here R_i is the vector of the actual rates of return of security i, for the time period $[0, -m_i]$; N is the number of assets that can contribute in the portfolio; m_i is the number of available actual rates of return.

The parameter R_i is used for the evaluation of the mean rate of monthly return for security i. Having the historical data of the monthly security returns, the expected (predicted) rate of return of asset i is:

$$E_i = \frac{1}{m_i} \sum_{j}^{m_i} R_i^j , \quad i = 1, N . \tag{2}$$

The portfolio expected rate of return is the weighted sum of the expected rate of returns of the assets, using the proportions of the investments x_i:

$$E_p = \sum_{i=1}^{N} x_i E_i \tag{3}$$

Thus, the investor can follow a policy for increasing the portfolio return (maximizing E_p) by changing the investment proportion x_i. The portfolio risk defines the range between the expected portfolio return from the actual one. It is assessed by the computation:

$$V_p = \tau_p^2 = \sum_i^N \sum_j^N x_i x_j c_{i,j}, \tag{4}$$

where τ_p denotes the standard deviation of the portfolio's rate of return. In general, the investor likes high portfolio return E_p, but the portfolio risk V_p has to be minimal (Sharpe 2000). The importance of the portfolio characteristics E_p and V_p are considerable for the investment process. The theory applies different methods and techniques for assessing the portfolio's behaviour in (E_p, V_p). An important theoretical conclusion is that the optimal portfolio must belong to a curve, called "efficient frontier" from (E_p, V_p) (Sharpe 2000). An inexplicit analytical description of the "efficient frontier" is given by a set of solutions of optimization problems parameterized by φ, defined on the positive half space [0,+ ∞]:

$$\min_X \{-E_p + \varphi V_p\}, \quad \sum_{i=1}^N x_i = 1, \quad \varphi = [0,+\infty], \tag{5}$$

where E_p is the expected portfolio return, (3), V_p is the portfolio risk.

For given value of the coefficient φ, the solution of (5) gives appropriate allocations $x_i(\varphi)$, $i=1,N$ of the investment per asset. The constraint $\sum_{i=1}^N x_i = 1$ denotes that the sum of the relative amount of the investment must be kept equal to relative 1. The portfolio return $E_p(x_i(\varphi))$ and risk $V_p(x_i(\varphi))$ give one point of the curve of the "efficient frontier". The different points of the "efficient frontier" are calculated for different values of φ and sequentially solving (5). The "efficient frontier" is a concave curve denoted by DOE in Figure 4.

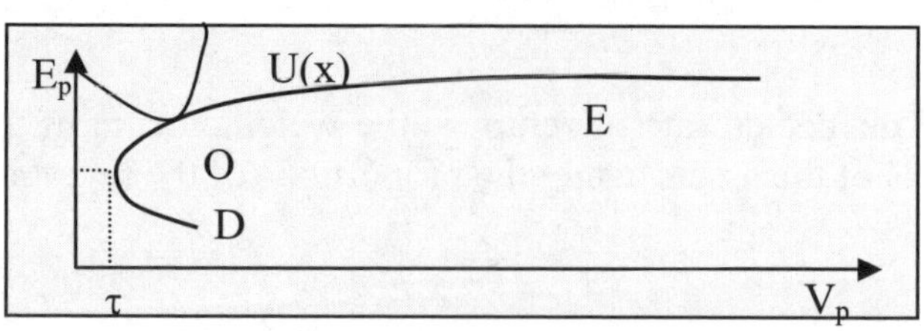

Figure 4: Portfolio efficient frontier (DOE) and optimal investor portfolio (O)

Finding the “best” point from the “efficient frontier” geometrically represents the solution of the portfolio optimization problem (Magiera 2001). The “best” point is the access point of the curve $U(E_p, V_p)$, denoted as utility function of the investor, with the “efficient frontier”. The utility function $U(E_p(x), V_p(x))$ represents the investor preferences towards the risk and return of investment. Analytically, the portfolio optimization problem concerns the evaluation of the relative amounts x_i of the investment per assets $i=1,N$ by solving the optimization problem:

$$\max_{x_i} U(E_p(x_i), V_p(x_i)), \quad x_i \in Z_i(x_i) \quad , \tag{6}$$

where

$$Z_i(x_i) \equiv \min_x \left\{-E_p(x_i) + \varphi V_p(x_i)\right\}, \quad \sum_{i=1}^{N} x_i = 1, \quad \varphi = [0, +\infty]. \tag{7}$$

The feasible set $Z_i(x_i)$ of the optimization problem is not given in an explicit analytical way. It is defined as a sequence of solutions of sub-optimization problems (5), parameterized by the value of φ, $[0, +\infty]$.

The difficulties in solving (6) originate from the fact that the “efficient frontier” can be found only via numerical recursive computations by changing φ (Sharpe 2000). The recursive evaluations will increase the computational time, which is a drawback for the real time implementation. This research develops a mixture of algorithmic and analytical sequence of calculations solving the portfolio problem (6). The calculations are based on a sequence of solutions of several optimization subproblems with smaller dimensions than the initial one (6). Additionally, the utility function $U(E_p, V_p)$ is chosen in an appropriate mathematical form, close to the financial meaning of the portfolio optimization, which targets the minimization of the investment risk and maximization of the portfolio return. This elaboration allows (6) to be expressed in explicit analytical form, speeding up the problem solution evaluation.

4 Mathematical models of the portfolio optimization problem

The classical portfolio optimization problem deals with only one time investment horizon. Here the portfolio problem has been complicated by introducing several investment horizons. Two optimization problems are derived, describing different aspects of the investment policies: invcstment with “short sales” and investment without “short sales”. The first optimization problem deals with the so-called “short sales”, i.e. the investment x_i can have positive and negative values. For $x_i>0$, the investor has to buy security i with the amount given by the relative value x_i according to the total amount of the investment. For the case $x_i < 0$, the investor has to sell security x_i. This “short sale” means that the investor has to borrow security i, to sell it and later he should restore it (Markowitz 1987).

The second optimization problem assumes non-negativity of the investments, $x_i \geq 0$, and thus the “short sales” are unfeasible. Analytically, the last problem has the mathematical description as:

$$\min_x \{\varphi x^T Vx - E^T x\}, \quad x^T\mathbf{1}=1, \quad L \leq x \leq U, \; T_r TSx \leq T_r h, \quad \varphi \geq 0 \tag{8}$$

where $x^T{}_{|Nx1} = (x_1, \ldots, x_N)$ is the vector of relative allocations of the investment per asset, $E^T{}_{|Nx1} = (E_1, \ldots, E_N)$ is the vector of the average returns of the assets, $V(.)_{|NxN}$ is the co variation matrix representing the portfolio risk, $\mathbf{1}_{|Nx1}$ is an identity vector, $L_{|Nx1}$ is the vector of lower bound constraints of the investment x, $U_{|Nx1}$ is the vector of upper bound constraints of the investment x, $h_{|3x1}$ is the vector of the relative allocation of the investment, distributed to the different time horizons,

$T_{r|3x3}$ is a triangle matrix, $T_r = \begin{vmatrix} 1 & 1 & 1 \\ 0 & 1 & 1 \\ 0 & 0 & 1 \end{vmatrix}$,

$S_{|kxN}$ is a matrix of feasible investment strategies,
$T_{|3xk}$ is a matrix of time feasible investments strategies,
φ is a scalar representing the investor risk preference.

Problem (8) can be expressed shortly in a canonical form (9)

$$\min_x \{1/2x^T Qx + R^T x\}, \quad Ax \leq C, \tag{9}$$

where the notations hold:

$$Q = 2\varphi V, \quad R = -E, \quad A = \begin{vmatrix} \mathbf{1} \\ -\mathbf{1} \\ Tr.T.S \\ -\mathbf{1} \\ \mathbf{1} \end{vmatrix}, \qquad C = \begin{vmatrix} \mathbf{1} \\ -\mathbf{1} \\ Tr.h \\ -L \\ U \end{vmatrix},$$

$C_{|mx1} = (C_i)$, $i=1,m$ is a vector of constraints, $A_{|mxN}$ ($N>m$) is a matrix of constraints, including the general investment constraint $\sum_{i=1}^{N} x_i = 1$.

The mathematical model related to the portfolio optimization problem with "short sales" has the form with equality constraints:

$$\min_x \left\{1/2x^T Qx + R^T x\right\}, \quad Ax = C. \tag{10}$$

Both optimization problems (9) and (10) are linear-quadratic ones of the mathematical programming. They can be solved applying the general methods of the non-linear and/or quadratic programming, (Murray and Prieto 1995). However, according to the specific requirements in solving (9) and (10) in real time to implement the portfolio optimization as Internet information service, this requirement insists fast computational algorithms for solving problems (9) and (10), developed below.

4.1 Optimization problem with short sales

The derivation of the analytical description of the problem solution results from the multilevel system theory, applying the concept of noniterative coordination (Stoilov and Stoilova 1998, 1999). Appropriate definitions and applications of the right and dual Lagrange problems are performed. For the initial problem (10) the right Lagrange problem is:

$$\min_x \{L(x,\lambda)\},\ L(x,\lambda) = 1/2x^T Qx + R^T x + \lambda^T (Ax - C),\ \lambda - \text{dual}. \tag{11}$$

(11) is an unconstrained optimization problem, reduced towards the solution of the linear set of equations

$$\partial L / \partial x = Qx + R + A^T \lambda = 0. \tag{12}$$

Following (12), the solution x is expressed as a vector function of λ:

$$x(\lambda) = -Q^{-1}(R + A^T \lambda). \tag{13}$$

The dual variable λ is evaluated from the dual Lagrange problem:

$$\max_\lambda H(\lambda), \quad \text{where} \quad H(\lambda) = L(x(\lambda),\lambda). \tag{14}$$

$H(\lambda)$ is described in terms of λ by substitution of vector function $x(\lambda)$ instead of the argument x in the right Lagrange function $L(x,\lambda)$ and after a substitution of (13) in (11), the explicit description of $H(\lambda)$ is:

$$H(\lambda) = -1/2\lambda^T AQ^{-1}A^T\lambda - \lambda^T(C + AQ^{-1}R). \tag{15}$$

The existence of $H(\lambda)$ in an analytical form allows the solution of the dual Lagrange problem (14) to be found as an unconstraint optimization towards λ resulting in a linear set of equations with unknown arguments λ

$$\max_{\lambda} H(\lambda) \quad \text{or} \quad dH(\lambda)/d\lambda = 0 = -AQ^{-1}A^T\lambda - (C + AQ^{-1}R). \tag{16}$$

The solution of (16) can be expressed as:

$$\lambda^{opt} = -(AQ^{-1}A^T)^{-1}(C + AQ^{-1}R) . \tag{17}$$

By substituting (17) in (13), it follows (Stoilov and Stoilova, 1999):

$$x^{opt} = -Q^{-1}\left[R - A^T\left(AQ^{-1}A^T\right)^{-1}\left(C + AQ^{-1}R\right)\right] \tag{18}$$

The explicit analytical form of (18) allows the initial problem (10) with "short sales" to be solved very fast. The information service has to ask the investor about his risk preferences φ and time horizons *h*. The information system, using the sets of risk Cov(.) and asset performances E, calculates the optimal investment allocation x_i, according to (10). Particularly, relation (18) is used, which does not insist recursive and iterative computations. The information service is able to evaluate very fast the solution of portfolio problem (10) and to derive the curve of the "efficient frontier" $E_p^T = E_p^T\left(\tau_p^T\right)$. Thus, the portfolio optimization can be implemented online as Internet information service.

4.2 Optimization problem without short sales

For problem (9) inequality constraints are presented and analytical solution cannot be found. Here a new algorithm is worked out, which does not perform recursive calculations and benefit the fast problem solution. The algorithm is founded on a sequential computational procedure, applying the right and dual Lagrange problems for the initial problem (9). Following the solution of the right Lagrange problem (12), the solution x of (9) could be expressed as linear vector function $x(\lambda)$, according to (13). The optimal value of λ is found from the dual Lagrange optimization problem of (14), adding non-negativity for λ (Stoilov and Stoilova 1998).

$$\max_{\lambda \geq 0} H(\lambda), \text{ where } L(x,\lambda) = L(x(\lambda),\lambda) . \tag{19}$$

The dual Lagrange function $H(\lambda)$ can be expressed explicitly in analytical form. The dual problem is a constrained optimization one with solution

$$\lambda^{opt} = \arg\left\{\min_{\lambda\geq 0}\left[-H(\lambda) = \frac{1}{2}\lambda^T AQ^{-1}A^T\lambda + \left(AQ^{-1}R + C\right) = \frac{1}{2}\lambda^T(G\lambda + h)\right]\right\} \quad (20)$$

where $G = AQ^{-1}A^T, \qquad h = AQ^{-1}R + C$

The initial constrained problem (9) and the simpler dual (19) can be solved by quadratic programming methods, unfortunately time consuming. It is worth to solve the simpler dual (19) and applying the analytical relation (18), the initial (9) is solved faster. The faster solution of (9) assists the implementation of the portfolio optimization as Internet on-line service.

4.3 Peculiarities of the Dual function H(λ)

The initial optimization problem is in a form (9) and the corresponding right Lagrange problem is unconstrained optimization one:

$$\min_x \mathrm{L}(\mathrm{x},\lambda) \Rightarrow \min_x L(x,\lambda) \Rightarrow dL/dx = Qx + R + A^T\lambda \quad (21)$$

$$\text{or} \qquad dL/d\lambda = Ax - C \leq 0 \; , \qquad \lambda^T(Ax - C) = 0 \; .$$

The last constrains refer to the dual Lagrange problem

$$\max_{\lambda\geq 0} H(\lambda) = -\frac{1}{2}\lambda^T G\lambda - h^T\lambda, \quad G = AQ^{-1}A^T, \quad h = R^T Q^{-1}A^T + C^T \quad (22)$$

The solution of the initial (9) has been reduced to the evaluation sequence

$$x(\lambda) = Q^{-1}(-R - A^T\lambda), \quad \lambda^{opt} \equiv \arg\left\{\max_{\lambda\geq 0}\left[-\frac{1}{2}\lambda^T G\lambda - h^T\lambda\right]\right\} \quad (23)$$

or (9) has been worked out to the simpler linear-quadratic problem(22) for the duals λ^{opt}. Here an algorithm for solving (22) is derived. If (22) is the initial optimization problem, the dual Lagrange is:

$$\min_\lambda\left\{\frac{1}{2}\lambda^T G\lambda + h^T\lambda\right\} \Rightarrow \max_\lambda\left\{L_h(\lambda,\psi) \equiv \frac{1}{2}\lambda^T G\lambda + h^T\lambda - \psi^T\lambda\right\}, \quad -I\lambda \leq 0, \quad (24)$$

$$dL_h / d\lambda = 0 = G\lambda + h - \psi = 0 \Rightarrow \lambda(\psi) = G^{-1}(-h + \psi)\,, \quad (24a)$$

$$dL_h / d\psi = -\lambda \leq 0, \quad \psi(\lambda) = G\lambda + h \quad (24b)$$

$$-\psi^T\lambda = 0 \quad . \quad (24c)$$

The substitution of (24a) in (24c) gives the nonlinear system $\lambda^T(Gx+h)=0, \lambda \geq 0$, equivalent to the linear optimization problem:

$$\min \beta = h^T\lambda, \quad G\lambda + h = 0, \quad \lambda \geq 0. \tag{25}$$

Thus, the solution of the dual Lagrange problem (22) has been reduced to the solution of the linear optimization problem (25).

Case 1: the matrix G^{-1} exists

For this case the quadratic curve $H(\lambda)$ has a central point, G^{-1} exists and the dimension of G corresponds to λ. The central point is $\lambda^* = G^{-1}h$. Hence λ^* is the unique feasible point, for the relation $G\lambda + h = 0$ but it might have negative components.

- If λ^* is a non-negative $\lambda^*\geq 0$, this is the solution of (22) and $\lambda^* = \lambda^{opt}$.
- If λ^* has negative components, $\lambda^*_i<0$, $\lambda^*_j<0$, $\lambda^*_k<0$, it is not a solution of (22). Appropriate negative component $\lambda^*_{out}<0$ must be set to zero and the optimal value is $\lambda^{opt}_{out}=0$. Because a minimum of the goal function $\beta=h^T\lambda$ is asked, the component λ^*_{out} is identified by the requirement that the product $h_{out}\,\lambda^*_{out}$ minimizes additionally β. Using these peculiarities, the choice of $\lambda^{opt}_{out}=0$ is done in a sequence:

  Rule 1: if $\lambda_i^*<0$, $\lambda_j^*<0$, $\lambda_k^*<0$ and $h_i>0$, $h_j>0$, $h_k>0$, hence λ^*_{out} is defined according to the criteria

  $$h_{out}\,\lambda^*_{out} = \min_{\lambda_i,\lambda_j\lambda_k} \{h_i\,\lambda_i^*,\ h_j\,\lambda_j^*,\ h_k\,\lambda_k^*\}.$$

  Rule 2: if $\lambda_i^*<0$, $\lambda_j^*<0$, $\lambda_k^*<0$ and $h_i<0$, $h_j<0$, $h_k<0$, hence λ^*_{out} is defined according to the criteria

  $$h_{out}\,\lambda^*_{out} = \max_{\lambda_i,\lambda_j\lambda_k} \{h_i\,\lambda_i^*,\ h_j\,\lambda_j^*,\ h_k\,\lambda_k^*\}.$$

Case II: the matrix G^{-1} does not exist

G is not in full rank, rank(G) < rank(λ) and det(G) = 0. It is worth to use only the linear independent rows of G. The lasts are found, applying QR decomposition of G (Faucett 1999) or $q\ r\ e^T = G$, where q is a orthogonal matrix, $q^T = q^{-1}$, r is an upper triangular matrix, e is an identity matrix, referring to the changes of the columns in matrix G. Including a new variable $\mu = e^T\lambda$, instead of $\lambda = e\,\mu$, the linear problem (25) becomes

$$\begin{array}{llll} \min h^T\lambda \Rightarrow & \min h^T\lambda \Rightarrow & \min h^T e\,\mu \Rightarrow & \min h_e^T\mu \\ G\lambda=h,\ \lambda\geq 0 & qre^T\lambda=h,\ \lambda\geq 0 & r\mu = q^T h,\ \mu\geq 0 & r\mu = h_q,\ \mu\geq 0 \end{array} \tag{26}$$

where the notations $h_e^T = h^T e$ and $h_q = q^T h$ are applied.

If (26) has a nonnegative solution $\mu^* \geq 0$, it corresponds to the initial variable $\lambda^* = (\lambda_+^* > 0, \lambda_0^* = 0)$. The set of the zero components will be the same for the optimal solution of the dual problem (22) and $\lambda_0^{opt} = \lambda_0^* = 0$. The nonzero components $\lambda_+^* > 0$ define the reduced linear equation system

$$G'\lambda' = h', \tag{27}$$

where G' is a square matrix, originated from G lacking the columns and rows, determined by indices of the set λ_0^*; respectively h' origins from h and it consists the components referring to the rows of G'. Because the reduced matrix G′ is defined as a full rank matrix, the evaluation of nonnegative solution follows the algorithm in Case I. At the end, the nonnegative components $\lambda^{(r)}_+ \geq 0$ of the final reduced problem (27) are found, which give the optimal solution of the dual problem (22), $\lambda^{(r)}_+ = \lambda_+^{opt} \geq 0$.

If (26) does not have a feasible solution $\mu^+ \geq 0$, respectively $\lambda^* \geq 0$, a starting point could be a solution of the rectangular linear system of equations

$$\lambda^* = \arg\{r\mu = h_q, \quad where \quad \lambda = e\mu\}, \tag{28}$$

which can consist negative components of λ^*, $\lambda_i^* < 0$, $\lambda_j^* < 0$, $\lambda_k^* < 0$. Applying Rule1 and Rule2, the solution of dual problem (22) is decomposed to a linear set of equations and several sequential logical and algebraic computational rules instead of implementation of iterative numerical calculations using the general methods of quadratic programming.

5 Computational algorithm for solving the optimization portfolio problem

The portfolio problems are (9) or (10). The optimal solution of (10) is given in an analytical form (18), which speeds up the calculations. For (9), the contact point of $H(\lambda)$ with the positive hemi space $\lambda \geq 0$ has to be found, which is reduced to the solution of linear set of equations (27). The solution of (27) depends on the rank of matrix G.

For the case det(G)=0 the algorithm is:

- QR decomposition of G is done. A linear problem (27) is defined.

- Problem (27) is solved and the feasible solution of $\lambda^* \geq 0$ is found. The zero components of λ^* match with these ones of the optimal solution of the dual problem (22), $\lambda^*_0 = \lambda^{opt}_o = 0$.
- The nonnegative components of λ^* define a reduced system (27) and the evaluations continue according to $\det(G') \neq 0$.
- If non-feasible solution origins from (27), the choice of λ^*_{out} is defined according to **rules 1** and **2**.

For the case $\det(G) \neq 0$ the algorithm is:

- Evaluation of the central point of $H(\lambda)$ according to (27) or $\lambda^*=G^{-1}\, h$.
- If $\lambda^* \geq 0$, this is the solution of dual problem (22).
- If negative components exist $\lambda^*_i<0$, $\lambda^*_j<0$, $\lambda^*_k<0$, then λ^* is not the solution of (22). Because a minimum of $\beta=h^T \lambda$ is found, $\lambda^*_{out}<0$ is defined by logical **rules1** and **2** and the optimal value is $\lambda^{opt}_{out}=0$.
- The matrix G is reduced by row=out and column=out, and h is reduced by the component $h_{out.}$ New reduced linear equation set (27) is defined with $G = G'$ and $h = h'$. The evaluation of the components of the central point λ^* is repeated, according to the case $\det(G') \neq 0$.

The optimal solution λ^{opt} of the dual (22) is found by solving a sequence of linear set of equations (27), which is simpler, than (22), Figure 5.

6 System and program architecture of the portfolio optimization information system

The computational efficiency of the algorithms for solving problems (9) and (10) was assessed in the floating point operations (flops), performed for the problem solution and compared with the Matlab Optimization toolbox (Grace 1993). The curves in Figure 6 give the relations between the number of flops, performed for the solution of (9) and (10) with dimension N for x. Considerable decrease of 40% of the computational efforts is achieved solving (9) in comparison with Matlab optimization methods. This is a benefit for the real-time solution and development of a portfolio information service. The algorithms are applied for WAN based information service for investments optimization. The information service has been implemented as a 4 - tier server side hierarchical system.

Portfolio optimization problem

$$\min \frac{1}{2} x^T Q x + R^T x , \; A x \le C.$$

⇓

$$x(\lambda) = Q^{-1} (-R + A^T \lambda)$$

⇓

$$\lambda = \arg\{ \min_{\lambda \ge 0} H(\lambda) = \frac{1}{2} \lambda^T G \lambda + h^T \lambda \},$$

$$G = A Q^{-1} A^T, \quad h^T = R^T Q^{-1} A^T + C^T$$

⇓ Dual Lagrange problem is transformed to non-linear equation system $\lambda^T (G \lambda + h) = 0, \lambda \ge 0$

⇓non-linear equation system is transformed to linear-programming problem $\min h^T \lambda , G \lambda = -h, \lambda \ge 0.$

Case I

$\det(G) \ne 0$, ellipsoid curve $H(\lambda)$

⇓

$\lambda^* = G^{-1} h$

If $\lambda_i^* < 0, \lambda_j^* < 0, \lambda_k^* < 0$,

then

Rule 1 is used

$h_{out} \lambda_{out} = \min\{h_i \lambda_i^*, h_j \lambda_j^*, h_k \lambda_k^*\}$

$h_{i,j,,k} > 0, \lambda_{i,j,,k} < 0$

else **Rule 2** is used:

$h_{out}.\lambda_{out} = \max\{h_i \lambda_i^*, h_j \lambda_j^*, h_k \lambda_k^*\}$

$h_{i,j,,k} < 0, \; \lambda_{I,,j,k} < 0$

till $\det(G') \ne 0$ and

$\lambda^*_0 = \lambda^{opt} = 0$ are found.

Case II

$\det(G) = 0$, parabolic curve $H(\lambda)$

⇓

[Q,R,E]=qr(G);

qr decomposition of G.

The linear problem is solved :

$\min h_e^T.\mu , \; r.\mu = h_q , \; \mu \ge 0$

and the solution $\lambda = e.\mu$ is found.

⇓

Continue with **Case I**.

Figure 5: Algorithm for solving the portfolio optimization problem (9) without "short"sales

The server side is operating under Linux operation system, MySQL database and Apache server. The MySQL database server contains records about the asset performances. PHP code proceeds the data from the database and evaluates the characteristics and parameters of the optimization problems: mean values of returns, risks, co-variances, the curve of the "efficient frontier". PHP server side programming and additions from the GD graphical library implement the graphical user interfaces, the communication of the results, and the input of the data. The algorithmic server performs the solution of the optimization problem. It has been developed as compiled C++ software, implementing the optimization calculations.

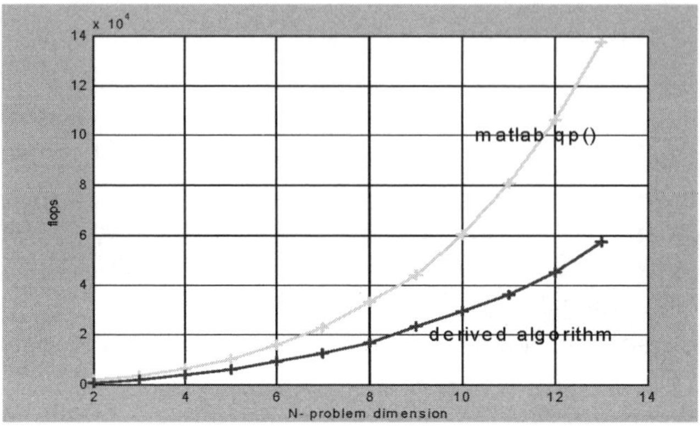

Figure 6: Comparison of the computational efficiency in "flops"

The screenshot in Figure 7 demonstrates the user interface for inputs of the investment horizons: short, medium and long terms. The "efficient frontier" is calculated and built according to the solution of 25 optimization problems. The results of the portfolio optimization are given both in numerical and graphical manner as a pie - chart, Figure 8.

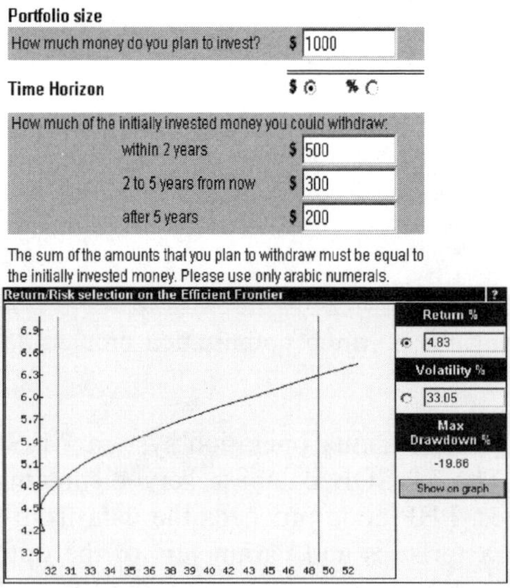

Figure 7: User interface for assessing the investor's risk preferences

Portfolio Allocation	USD	%	Strategy	Category
Fund A	250.59	25.06 %	Convertible arbitrage	Arbitrage
Fund B	214.59	21.46 %	Equity arbitrage	Arbitrage
Fund C	260.83	26.08 %	Fixed Inc. arbitrage	Arbitrage
Fund D	40.88	4.09 %	Long & short equities	Market Hedged
Fund E	178.78	17.88 %	Income arbitrage	Arbitrage
Fund F	9.33	0.93 %	Long & short equities	Sector Hedged
Fund G	27.60	2.76 %	Long & short equities	Market Hedged
Fund H	17.39	1.74 %	Risk arbitrage	Event driven
Total	1,000.00	100.00%		

Allocation to Funds Pie-chart

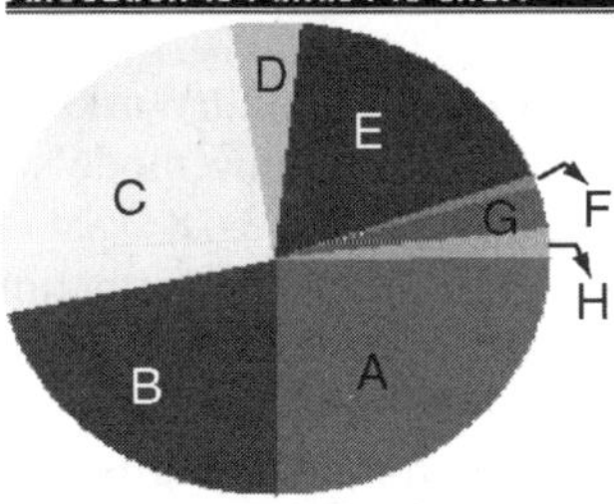

Returns	
Year-to date	10.81%
1-Year Annualized	20.73%
3-Year Annualized	18.05%
5-Year Annualized	19.25%

Risk Measures	
Maximum Drawdown	0.00%
Annualized Volatility	2.07%
Sharpe Ratio Annualized	6.77
Risk free rate - 3 m UST-Bill	5.25%

Figure 8: Screenshot of the portfolio problem's solution

7 Conclusions and achievements

The developments of information services with complex functionality and intelligence in the global network Internet is a challenge both for the developers and for users. The complexity of the functionality influences the intelligence of the server side software architecture and it is implemented as building a multi-tier software suit. Such a software engineering approach decomposes and performs the specific intelligent functionality of the e-services in appropriate algorithmic software tier. This concept has been applied for the implementation of the portfolio optimization in financial investment as Internet e-service. The portfolio optimization problems have been solved by noniterative manner, speeding up the computations. The computational efficiency and the forth - tier logical structure of the software application allow to implement a sophisticated intelligent information service. Thus the e-intelligence has been achieved by integration of three domains: portfolio theory, optimization and information technologies. The results have been implemented to a Swiss financial investment project.

8 References

Bodie Z, Kane A, Marcus A (2000) Investments. Naturela, Sofia

Cohen MD, Kelly CB, Medaglice AL (2001) Decision support with web-enabled software. Interfaces 31: 109-129

Czyzyk J, Mesnier MP, More JJ (1998) The NEOS Server. IEEE Journal of computational Science and Engineering 5:68-75

Faucett L (1999) Applied numerical analysis. Prentice Hall, NJ

Fourer R, Goux JP (2001) Optimization as an Internet Resource. Interfaces 31:130-150

Geoffrion AM, Krishnan R (2001) Prospects for Operations Research in the E-business Era. Interfaces 31:6-36

Grace L (1993) Optimization Toolbox for use with MATLAB. MathWorks Inc.

Ivanova Z, Stoilova K, Stoilov T (2003) Systems and algorithmic models of information services in Internet. J. Automatics and Informatics 3: 13-18

Ivanova Z, Stoilova K, Stoilov T (2005) Portfolio optimization – information service in Internet. Editorial house of Bulgarian Academy of Sciences “M. Drinov”, Sofia, ISBN 954-322-021-2

Janal DS (1998) On-line marketing handbook – how to promote advertise and sell your products and services on the Internet. John Willey@Sons, Inc. NY

Magiera P, Karbowski A (2001) Dynamic Portfolio Optimization with Expected Value-Variance, Proc. 9-th IFAC Conf. on LSS, Bucharest, pp 308-313

Markowitz H (1987) Portfolio Selection, Wiley, NY

Mital (1992) Knowledge Systems for Finance Advice. The Knowledge Engineering Review, vol.7, No 3, p. 281

Murray W, Prieto F (1995) A sequential quadratic programming algorithm using an incomplete solution of the subproblem. SIAM J. Optimization 5:590-640

Pflug GC, Swictanowski A, Dockner E, Moritsch H (2000) The AURORA Financial Management System: Model and Parallel Implementation Design, Annals of Operations Research 99:189-206

Preston D (2000) Basic Client Server Architecture Overview, Management Science Information System, Amir Dabirian

Rosenfeld L, Morville P (1998) Information Architecture for the World Wide Web, O'Reilly & Associates

Rust, RT, Kannan PK (2003) E-service: A new paradigm for Business in the Electronic Environment. Communications-ACM 46: 36-42

Sharpe W (2000) Portfolio Theory and Capital Markets. McGraw – Hill New York London Tokyo

Stoilov T, Stoilova K (1998) Non-iterative coordination in multilevel systems. Int J of Systems Sciences 29: 1393-1416

Stoilov T, Stoilova K (1999) Noniterative Coordination in Multilevel Systems. Kluwer Academic Publisher, Dordrecht Boston London

This work is partly supported by project 1013/2005 for the development of on-line investment portfolio model and project VU-MI-108/2005 for the design and implementation of intelligent web services of the Bulgarian Scientific Fund.

Orchestrating the Knowledge Discovery Process

Marcello Castellano[1], Giuseppe Mastronardi[1], Flaviano Fiorino[2], Giuliano Bellone de Grecis[2], Francesco Arcieri[2], and Valerio Summo [2]

1 Politecnico di Bari, Dipartimento di Elettrotecnica ed Elettronica, Via Orabona 4, 70126, Bari

2 Global Value – ACG srl – An IBM Company, Via Tridente 42/14, 70125, Bari

Abstract

Today, the analysis of huge amounts of data coming from sites distributed geographically can be problematic for both private and public organizations involved in scientific and technological research as well as in industry. Thus, the aim of the study is to produce value added knowledge in order to orient and support strategic decision-making. In this chapter we define an innovative process of knowledge discovery starting from structured and unstructured data for the realization of new e-services for knowledge. This process is designed by using the service oriented architectural pattern and implemented by using the Web Services and BPEL Technologies.

1 Introduction

Knowledge discovery is a complex and interdisciplinary field which deals with the understanding of unsuspected patterns and rules in data. The discovering of these relationships is a highly interactive process where, in order to achieve appealing results, the user must be allowed to apply various techniques or some given parameters on a permanent basis (Fayyad, Shapiro, Smith, Uthurusamy 1996). As a support to the entire process, many existing knowledge discovery systems (Matheus 1993) use legacy

M. Castellano et al.: *Orchestrating the Knowledge Discovery Process*, Studies in Computational Intelligence (SCI) **37**, 477–496 (2007)
www.springerlink.com

methodologies or, as in more recent times, the cross industry standard process for data mining (CRISP-DM).
There are several types of applications to which knowledge discovery can be applied and a variety of software tools are already in use, but they show two main problems, which are weak interoperability with one another and inability to substitute one tool with another for the same application (Cody 2002, Pena & Menasalvas 2001). As an alternative, in the present chapter we adopt a reference flexible mining architecture able to validate a process of knowledge discovery in a distributed and heterogeneous environment and take the user through its steps in a process of workflow (Castellano et al. 2005). The aim is to define and validate a sequence of steps to be followed for all mining activities, such a process to be generally applicable to different e-business sectors and able to describe how to identify interesting and new patterns by covering the whole process of the knowledge discovery. According to previous issues, many scientific works purport to provide a platform to deploy user services in a fast and efficient manner by building a software platform of components reusable across application programs. The service-oriented architecture (SOA) approach is the latest of a long series of attempts in software engineering trying to encourage the reuse of software components (Choi et al. 2004, Papazoglou & Georgakopoulos 2003). Moreover, the SOA provides a scheme to design distributed and orchestrated applications. In this chapter, we describe Knowledge Discovery process in terms of miners: that is to say reusable and elementary components designed in order to be orchestrated. In this way we present a flexible process of knowledge discovery that can be adaptively and proactively managed along with processes aiming at carrying out new e-services for knowledge. To this purpose, the Business Process Execute Language for Web Service (BPEL4WS) has been originally used to describe knowledge discovery process for Miners so as to become a means to link mining building blocks in order to compose e-services for knowledge in a business process workflow. It provides the tools to define, simulate, analyze and implement process models in terms of orchestration of elementary services (Peltz 2003). In this chapter we adopt miners orchestration as the logic to sequence, coordinate and manage the activities among miners with the purpose to compose e- services for knowledge. As a result, each e-service for knowledge represents the output of an orchestration of reusable building blocks with well defined tasks and able to interoperate among them.

2 The Knowledge Discovery Process

In this paragraph a reference model of knowledge discovery is described as a process oriented to the carrying out of e-services for Knowledge. The starting point of our work is the Knowledge Discovery in Databases (KDD) Model proposed by Piatesky-Shapiro, Matheus and Chan. They defined the KDD as a nontrivial process for identifying valid, novel, potentially useful, and ultimately understandable patterns from large collections of data. It concerns itself with extracting useful and new information from data, by combining fields of databases and data warehousing with algorithms from machine learning and methods from statistics to gain insight in hidden structures within the data.

A generic method of knowledge discovery follows these tasks (Han & Kamber 2001):

- *Data Selection*: data relevant to the analysis task are retrieved from the database.
- *Data Pre-Processing*: noise and inconsistent data are removed (data cleaning) and multiple data source are combined (data integration).
- *Data Transformation*: data are transformed or consolidated into forms appropriate for mining summary or aggregation operations.
- *Data Mining*: in this essential process intelligent methods are applied in order to extract data patterns.
- *Pattern Evaluation*: interesting patterns representing knowledge based on measures of interest are identified.
- *Knowledge Presentation*: mined knowledge is presented to the user through visualization and knowledge representation techniques.

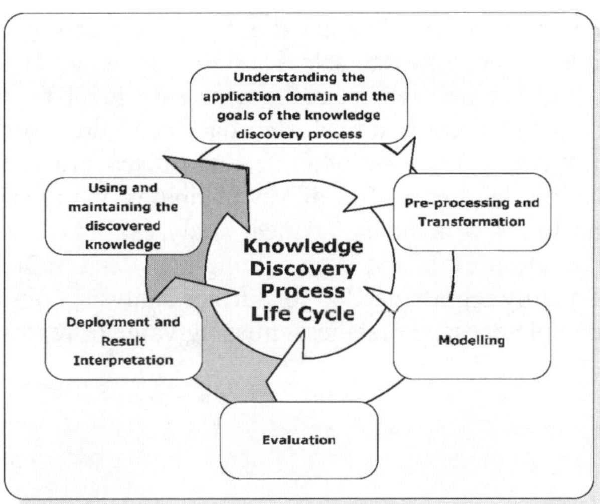

Figure 1: The Knowledge Discovery Process Life Cycle

In order to have a more concrete, industry-, tool- and application neutral approach, which is based on the practical, real-world experience of how people do data mining projects, we have considered the CRISP-DM methodology. It describes the KDD process in terms of a hierarchical process model, consisting of sets of phases, tasks and relationships, at levels of abstraction, from general to specific. As shown in Figure 1 the CRISP-DM manages the knowledge discovery life cycle for the carrying out of e-services for knowledge. In the next paragraphs, we will analyze the various steps by highlighting results and outputs.

2.1 Understanding the application domain and the goals of the knowledge discovery process

The first step in the building of a Knowledge Discovery process is to understand the domain objectives that the customer needs to accomplish. After a first analysis, whose purpose is that of getting straight to the core of the situation, a more detailed fact-finding concerning constraints, assumptions and other factors should be considered. What is necessary here is the collection of details and the identification of the problem area (Marketing, Customer Care, Business Development, etc.). This can be done by either listing the inventory of resources, the requirements, assumptions and constraints, risks and contingencies, terminology and costs and benefits, or qualifying the business object in technical terms, adding details about Data Mining and Web Mining goals. Then, a further step requires the planning of the project in order to list the stages to be executed in it, including its duration, necessary resources, inputs, outputs and dependencies. Moreover, a better understanding of the application domain implies the collection within the project of the data listed in the project resources. After collecting the data it is necessary to select and then integrate them in order to create a data set, that will represent the necessary input for the next data preparation steps. Collected and selected data need then to be described, (their format, quantity and any other features discovered), and explored (e.g. by analyzing the distribution of key attributes, the relations between numbers of attributes or simple statistical analyses). This could be useful to directly address the Data and Web Mining goals and refine the data description and quality reports needed for further analysis, considering problem such as completeness, correctness, missing value or representation.

2.2 Pre-processing and transformation

The aim of this step is the making up of the final data set starting from raw data, in a cycle that could be repeated and performed according to the precision and the necessities required. Firstly, it is necessary to understand and select those data which can be useful for the analysis, paying attention to their quality and technical constraints. In order to achieve a good quality, the data selected must be examined in order to extract clean subsets, with an eye to the possibility to insert suitable default values or estimate missing data through other techniques. Cleaned data sets are then prepared and integrated to create derived attributes, new records or values and summarization, starting from multiple tables or records that have different information about the same object.

Finally, it could be necessary to format transformed data without changing their meaning. A common situation is when dataset records are initially organized in a given order but the following modelling algorithm requires them to be arranged in a quite different order. The main outcome of this phase is the description of the dataset, including the actions necessary to address data quality issues and the process by which it was produced, table by table and field by field.

2.3 Modelling

The first step in modelling consists in selecting the most appropriate technique to achieve the business objective. When speaking about models, we refer to a specific algorithm, such as a decisional tree or a neural network, as well as a clustering or an association method; a fitting tool should already have been selected. It is extremely important to define these techniques, because many of them make specific assumptions on the data. After the selection and before the real model can be built, the latter has to be tested for its quality and validity to be determined. This could be done by separating the dataset into a train set and a test set. The model is then built on the train set and its quality estimated on the test set. In this way, the model is built by running the modelling tool in order to create one or more mining models. The models thus created are then assessed by the decision makers according to their domain knowledge and the results desired. Anyway, whereas mining engineers evaluate the successfulness of the application of modelling and discovery techniques from a technical point of view, business analysts and domain experts later discuss knowledge results in a business context. In other words, this task only takes into considerations models whereas the following evaluation step takes into account all the other results produced in the course of the project. The output of the

Modelling step should describe how models are built, tested and evaluated and the process by which they are produced. Anyway, the final outcome of model testing can be combined into one report that includes Modelling assumption, Test design, Model description and Model assessment.

2.4 Evaluation

The purpose of this phase is to evaluate the degree to which the models extracted fit the business objective and if this is not the case, determine the reasons why the model is not good or has been overlooked. An alternative way to assess the model is to test it on real applications and look at the results. After reviewing the process, the last step in the evaluation process to be taken consists in determining whether to end the project and move on to deployment, or otherwise start further iterations or set up new mining projects. What is produced at this point is a summary in which is stated whether the project meets the initial business objectives and provides hints and actions for activities that have been missed and/or should be repeated. For what concerns the summary, relevant features are:

- Assessment of data mining results, aiming at comparing results with business objectives and criteria;
- Process review, aiming at establishing a possible repetition of the project;
- List of possible actions, aiming at detailing the next steps in the project.

2.5 Deployment and Result interpretation

In this phase, evaluation results are taken into consideration and a strategy for deployment is planned by drafting a final report. This could consist either in a simple summary of the project and its experiences or in a final and comprehensive presentation of the mining results. Anyway, important results obtained are:

- Deployment plan, aiming at describing data and web mining results;
- Monitoring and maintenance plan, aiming at specifying how the results deployed are to be maintained;
- Final report, aiming at summarizing the project global results, such as process, evaluation, different plans, cost/benefit analysis and conclusions.

2.6 Using and maintaining the knowledge discovered

In order to use and maintain thc knowledge discovered, the project requires a detailed monitoring process plan. As a matter of fact, a careful preparation of a maintenance strategy helps to avoid long periods of incorrect usage of mining results.

3 A Reference Architecture for the Knowledge Discovery Process

3.1 Requirements for a Flexible Mining Architecture

The creation of a general purpose, fully automated, knowledge discovery architecture is not simply obtained. In these last years several research issues have focused on how traditional machine-learning and discovery methods can be manually applied to data stored in databases. Recently, more attention is being paid to more fully automated approaches. The Knowledge Discovery in Databases (KDD) Model proposed by Piatetsky-Shapiro represents a starting point to define the requirements for a mining architecture and provide new added value e-service for knowledge. A mining architecture must comply with a number of requirements supporting the knowledge discovery process:

- *Full control of e-services provided by the system:* each service must be a module, independent of all others.
- *Flexibility for different data and web mining techniques:* users are usually moved by different business goals when they set out to discover hidden knowledge in their data (Nasukawa 2001). This can be achieved by providing a clear separation between the process logic and the e-services for knowledge.
- *High performance for large data sets:* a mining system should incorporate optimization strategies, for large data sets in particular, in order to enable mining elaborations with acceptable response times.
- *Flexibility for changes:* factors like the inclusion of services, algorithms, changes on system topology, different user privileges or system load restrictions have a deep impact on the overall performance of the system. Important decision like task scheduling, resource management or partition schema should be adjusted nearly every time one of these factors is modified.
- *Life cycle management:* KDD can include several concurrent activities, e.g. several classification models can be created in parallel and then se-

quentially evaluated. Therefore it is necessary a workflow engine able to handle long-running complex tasks and execute them on behalf of execution plan.

3.2 The Mining Engine – A Reference Architecture

In this paragraph previous issues are resumed by taking into account the reference architecture of a Mining Engine (Castellano et al. 2005), whose purpose is to generate and make available e-services for knowledge by driving the user through the main stages of the knowledge discovery process. Figure 2 shows an UML scheme that describes at a general level how the main components of the Mining Engine work together in a four level architecture.

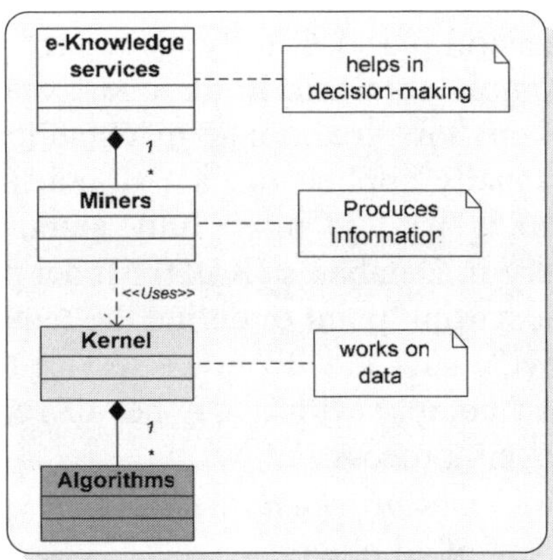

Figure 2: Logic functions of the Mining Engine architecture

The boxes on the left represent logic levels and explain the relation among the components while the boxes on the right describe the logic functions of each level. In detail, the highest level is the business one and consists of the e-Knowledge services where decisions are taken. They represent all the services that the Mining Engine has to provide. Then there are the Miners which are situated at the level where information is produced and describe the operations that produce a service. The Kernel is the core of the system. It covers the process of knowledge discovery by working on structured and unstructured data and building new mining models. Finally, one or several data and web mining algorithms can be used to provide the kernel with the techniques to operate on the data.
Figure 3 shows the logic model of the Mining Engine. It consists of a collection of components, each one of which with a well defined task; it has been designed in order to work in a distributed data and web mining envi-

ronment, where a set of services are managed and made available through a Controller. The Controller receives the request for the e-service for knowledge and then activates the business logic of the correspondent service by calling one or more Miners to provide the result. Miners are building blocks that can be used to build a complex application. In this case, the application is represented by an e-service knowledge.

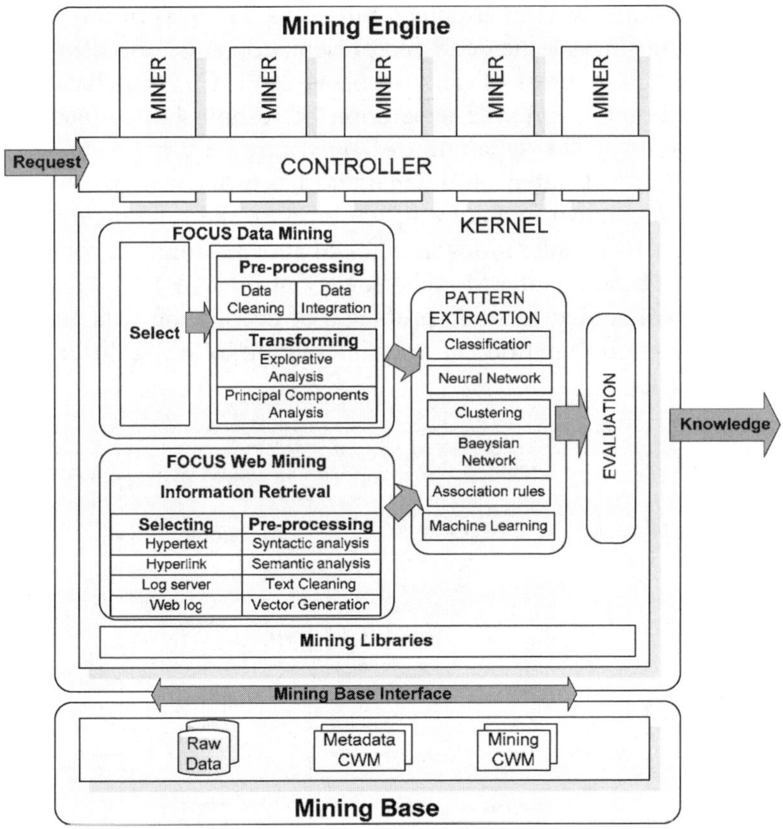

Figure 3: The logic model of the Mining Engine

Miners are situated at the level where information is produced and represent the operations that produce a service. They can either work by loading from the Mining Base the mining models associated to the required e-service for knowledge or by activating a training process of KDD or KDT in the Kernel according to the typology of the service or the mining model to be created. The Kernel follows the process of knowledge discovering starting from raw data and involving iterations of the main stages: data preparation, data and web mining, and results analysis. Finally, the Mining Base represents the knowledge repository of the whole architecture and its functions are those of repository for raw data, knowledge metadata and mining model.

3.3 The Mining Engine – A Model View Controller Implementation

For the purpose of solving architectural issues, in terms of mixture of data access code, business logic code and presentation code, the Mining Engine can be described in terms of tiers (Rantzau 2003, Chattratichat et al.1999) or layers in order to provide separation of responsibility, reuse and improved scalability. The separation of tiers may be a physical separation, where each tier is located on a separated hardware resource, or a purely logical one. As shown in Figure 4, we propose the reference architecture by following the considerations mentioned above. Moreover, our multi-tier architecture is based on the Model-View-Controller (MVC) design pattern, that can be well suited to solve problems of decoupling data access, business logic, data presentation and user interaction by using different objects for each function.

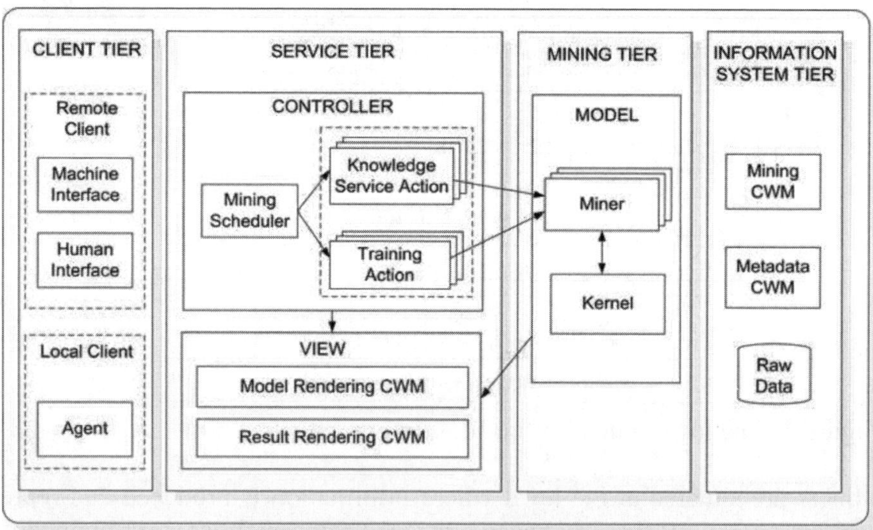

Figure 4: The multi-tier architecture of the Mining Engine

The Client Tier requires e-services for knowledge through a Remote or a Local client. The former sends a service request to the Service Tier either through a Machine Interface, that represents a software component of a remote machine, or a Human Interface, managing human interaction. The latter sends a system update request in accordance to the action of an intelligent agent. The agent has the task of updating periodically Data and Web

Mining models previously built according to a well defined Training Set. Models have to be updated because the information acquired, which must be necessarily re-analyzed in order to create new useful and more precise knowledge, is always increasing as time passes.
The Service Tier represents the front-end of the system and is composed by two of the three components of the MVC Pattern, that is the Controller and the View. The controller object translates interactions with a view into the actions to be performed by the model. The actions performed by the model include activating business processes and changing the state of the model. Basing itself on the user interactions and the outcome of the model actions, the controller responds by selecting an appropriate view. When a client requires an e-service for knowledge, the request is received by the Controller through a Mining Scheduler which analyzes and then forwards it to an Action. Each Action corresponds to an e-service for Knowledge and orchestrates different Miners to get the results. According to the type of request, there could be two kinds of Action, the Knowledge Service Actions and the Training Actions. The Knowledge Service Actions are activated by the Mining Scheduler when there is a request for an e- service for knowledge. In this case, each action is addressed to Miners that have the task to load one or more mining models, previously built by the Kernel, from the knowledge repository of the Mining Base. The Training Actions are activated by the Mining Scheduler when the system has to re-train a mining model. Each Training Action addresses the Miners that have the task to activate the process of KDD or KDT in the Kernel, according to the typology of the service. The View object renders the contents of a model. It gives access to the data through the model and specifies how those data should be presented. The view is responsible for the maintenance of consistency in its presentation when the model is changed.
The Mining Tier includes the business logic of the system, performing the Model object of the MVC design pattern which represents enterprise data and the business rules that govern access to and updates of the said data.

4 Orchestrating Miners

4.1 The Service Oriented Architecture

Nowadays it is necessary to create large structured applications through building blocks in order to use well-defined components within different business processes. The state-of-the-art solution to achieve these requirements is represented by Service Oriented Architecture (SOA). The purpose of this architecture is to address the requirements of loosely-coupled, standard-based, and protocol-independent distributed computing by mapping enterprise information systems isomorphically with respect the overall business process flow (Papazoglou 2005). The SOA is a concept which specifies how an application can be made up of a set of independent but cooperating subsystems or services. The SOA model isolates each service and exposes only those interfaces which can be useful to other services. In this way, with the changes in technology, services can be updated independently, thus limiting the impact of changes and updates to a manageable scope.

4.2 Service Oriented Miners

Today, e-services for knowledge have to be quickly adapted to decision maker needs and be able to cover the full lifecycle of the knowledge discovery process. A fundamental issue in the creation of e-services for knowledge is the support to interactions among the different components of a mining architecture and the assembling of their logic into a manageable application. The SOA provides a flexible architecture that modularizes the knowledge discovery process into Miners. For "Miners" we mean reusable components in a Knowledge Service Oriented Architecture (K-SOA) with the following characteristics:

- All operations in CRISP-DM steps are defined as Miners.
- All Miners are autonomous. The implementation and execution of the operations required are encapsulated behind the Miner Interface.

According to CRISP-DM methodology, it is possible to consider Miners in terms of a hierarchical model composed by three levels of abstraction: phase, generic task and specialized task (Figure 5).

At the first level, we consider the Knowledge Discovery process as organized into a number of phases; each phase consists then of several generic tasks, the second level, meant to be general enough to cover all possible Data and Web Mining situations and applications. Each Miner has to be as complete and stable as possible, and this particularly to make the model valid for yet unforeseen developments such as new modelling techniques or upgrading. The third level is where is described how Miners should be carried out in certain specific situations.

The example provided by figure 5 shows how a generic data cleaning task at the second level could be implemented in different situations, such as the cleaning of numeric values or categorical values.

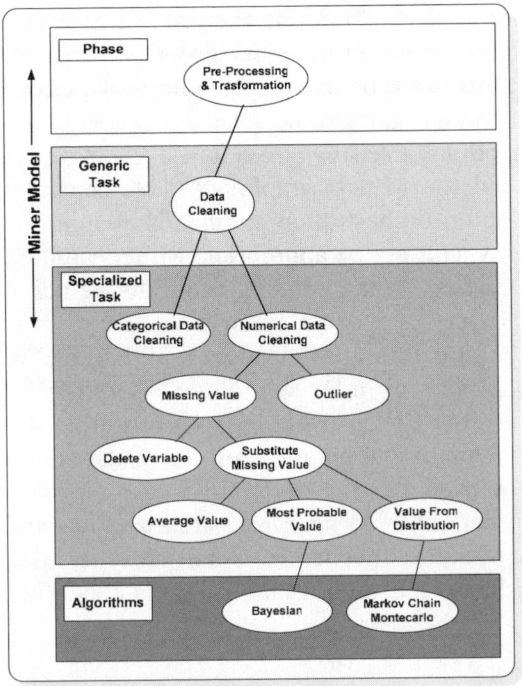

Figure 5: An example of Miner Model for the phase of Pre-Processing & Transformation.

4.3 Miner Orchestration

Adopting the SOA approach in the development of e-services for knowledge allows the definition of reusable mining components (Miners) and their orchestration in order to provide more complex functionalities in the knowledge discovery process.

Orchestration describes how services can interact with each other at the message level, including the business logic and execution order of the interactions. These interactions may span applications and/or organizations, and result in a long-lived, transactional, multi-step process model. Orchestration logic can be as simple as a single two-way conversation or as complex as a nonlinear multi-step transaction. It allows both a synchronous request-reply programming model and a conversational model based on asynchronous interactions across loosely-coupled e-services (Sherman & Doron 2002). Orchestration is necessary for the building of a Knowledge Service Oriented Architecture, and represents the layer supporting the composition of complex processes by constructing and connecting different data & web mining specialized tasks (Miners). In particular, we use the

term Miner Orchestration to define the process able to create a workflow of building blocks for the production of e-services for Knowledge. Miner Orchestration must be dynamic, flexible and adaptable so as to meet the changing needs of an organization.
Flexibility can be achieved by providing a clear separation between the process logic and the Miners employed. This separation can usually be achieved through an orchestration engine. The engine handles the overall process workflow, calling the appropriate Miners and determining the next steps to be completed. In the context of our above described reference architecture, at the very core of the knowledge discovery process, there is an Orchestration Engine aiming at managing, at the level of specialized task, a workflow of Miners in order to provide e-services for knowledge. The workflow follows the CRISP-DM steps starting from the understanding of the application domain and ending with the usage and maintenance of the discovered knowledge.
A further major benefit of Miner Orchestration is reusability as, if a new e-service for knowledge is needed, it is sufficient to insert a new Miner or simply re-coordinate the orchestration among the Miners already implemented.

4.4 Adopting Industrial Standards for Miner Orchestration

In the previous paragraphs, have been introduced SOA and Orchestration as theoretical approaches for the management of distributed applications and processes. In industrial solutions they can be implemented by using standards such as Web Services and BPEL4WS.
Web services is emerging as the third generation of integration technology able to implement the SOA. Web Services are self-contained, modular applications that can be described, published, located, and invoked over a network. These services can be new applications or be wrapped around existing legacy systems which make them distributed. They are built using XML, SOAP, WSDL, UDDI.
BPEL4WS provides a language for the formal specification of business processes and business interaction protocols, extending the Web Services interaction model. Using this emerging but already widely accepted standard, it is possible to describe the composition of stand-alone Web Services into an integrated and automated workflow (Microsoft 2000). It established a common vocabulary to describe business processes and the supporting technologies to facilitate process automation. Developing Miners like Web Services, the BPEL4WS can be used to link these mining building blocks in order to compose an e-service for knowledge like a process workflow. In this way, BPEL4WS facilitates a greater rapidity of

change in the modelling of knowledge business processes to comply with changing market conditions. This standard addresses a clear requirement for the proposed knowledge discovery framework (Mining Engine) enabling the orchestration of new, reusable business functions via Miners as the foundation of a Knowledge Service Oriented Architecture (K-SOA).

5 A Case Study: Knowledge Discovery Process for Customer Profiling

Nowadays, market strategies are constantly changing and the customer satisfaction is getting more and more important for companies and competitiveness in today's marketplace. Studies about customer satisfaction lead to adjust company targets and improve product development. Examples of this process could be suitable for e-commerce applications, dynamic contents of web pages and one-to-one advertising (Srivastava 2000). In such cases, the customer's profile has to be outlined in order to choose the right communication strategy, which is clearly focused on the client rather than on the product, as it used to be in the past. Moreover, the customer's profile provides information and knowledge that can be used to make business decisions and take appropriate actions, such as the identification of those prospects which are most likely to respond favourably to a certain type of selling campaign or offer.

Customer Profiling analysis represents the solution to the problem. It consists in the careful and complete building of profiles, aiming at letting the company have a clear description of the customer it is interacting with and his behavior.

From a technical point of view Customer Profiling is the process of discovering consistent patterns within customer or prospect data. Customer Profiling can make use of predictive data mining techniques in order to identify those prospects which are most likely to make a particular type of purchase (Krishnaswamy et al. 2000). In particular, cluster analysis can be used for the segmentation of customers and products based on different criteria. It is also possible to model individual customer behaviour through various types of conjunctive rules, including association and classification rules (Adomavicius & Tuzhilin 2001).

For the implementation of Customer Profiling in the reference architecture, our contribute is not as much as to define how the service works, as it is a well-known problem, but to demonstrate how it can be simply designed through the creation of an Orchestration of already implemented miners as building blocks.

5.1 Defining Miners for Customer Profiling

According to the CRISP model, the process of creation of Miners Orchestration can be divided into steps, i.e. Data Pre-processing and Transformation, Modeling, Evaluation and Deployment of Result, with each step made up by one or more Miners collaborating with one another.

To go into the detail, the *Data Pre-processing and Transformation* step firstly aims at understanding which data are necessary for the implementation. A complete customer profile is usually made up of two sets of data, basic and behavioural. The former contains information (name, gender, date of birth etc.) that reveals the customer's identity. A behavioural profile models the customer's actions and is usually derived from transactional data. In particular, the phase of Selecting starts from the collection of customer data from various sources. The data might include histories of customers' Web purchasing and browsing activities as well as demographic and psychographic information. After the data have been collected, they must be prepared, cleaned, and stored in a data warehouse for the preprocessing and transformation. This latter consists in the creation and cleaning of the dataset that will represent the input for the following analysis.

The *Modelling* step implements data mining techniques for the creation of the mining model of the Customer Profiling service. Customer profiles can be modeled through basic data or extracted rules describing customer behavior. In the latter case, description rules may be associative or classification rules. To discover the rules that describe the behaviour of individual customers, it is possible to use various data mining algorithms, such as Apriori for association rules and CART (Classification and Regression Trees) for classification rules (Agrawal et al. 1996, Breiman et al. 1984). Moreover, as Customer preferences represent the explicit description of a customer's habits, customer profiles can be also built by using clustering and vector space model techniques on the basis of customer behavior data (e.g. purchase history) (Li Niu et al. 2002). One more approach is the employment of neural networks. Neural networks grant high speed computing, memory, useless data studying and filtering and error tolerance, and can be thus used in several complex classification and prediction issues (Chen et al. 2004).

Finally, the *Evaluation and Deployment of Result* step importance lies in the assessment of the statistically acceptable, trivial, spurious, or just not relevant to the application generated rules. One way to validate the rules discovered is to use validation operators that let a human expert validate large numbers of rules at a time with relatively little input from the expert. Validation operators can be Similarity-based rule grouping, Template-based rule filtering, Redundant-rule elimination.

5.2 Orchestrating Miners for Customer Profiling

Figure 6 shows the orchestration among the different Miners that collaborate to accomplish the process of Customer Profiling service.

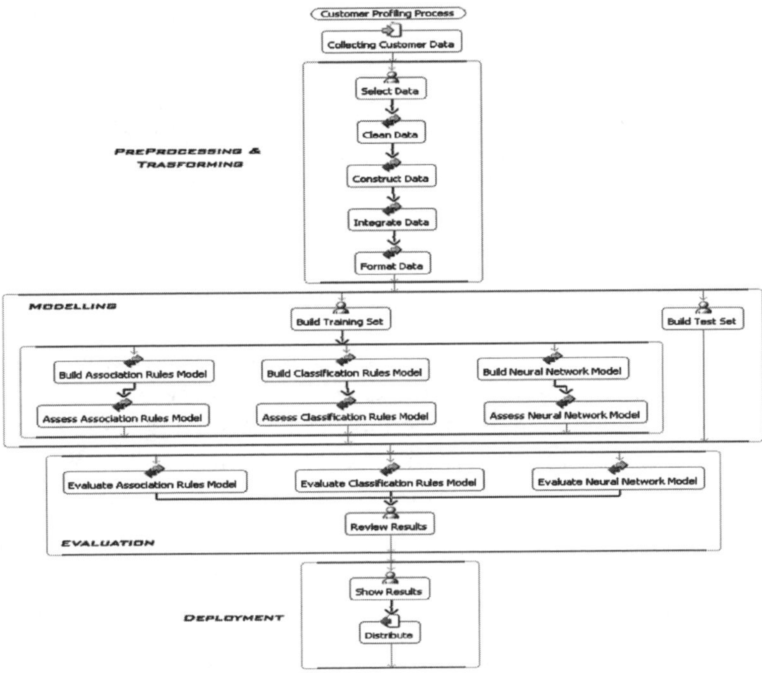

Figure 6: Customer Profiling Miners Orchestration

As shown in the previous figure, the workflow is organized in the main phases of Selecting, Data Pre-processing and Transformation, Modeling, Evaluation and Deployment of Result. Initial set of data are retrieved, selected and then pre-processed in order to build the Training Set necessary for the further step of the mining Modelling. Built models are tested and evaluated to assess their value and reliability and finally presented to the decision maker who analyzes the effectiveness of the knowledge extracted. The following rows present the essential XML code of the relative BPEL process:

```
<?xml version="1.0" encoding="UTF-8"?>
<process expressionLanguage="Java" name="CustomerProfilingProcess"
   <sequence name="MainSequence">
        <receive name="CollectingCustomerData">...</receive>
        <flow name="PreProcessingTrasformation">...
            <invoke name="SelectData">...</invoke>
            <invoke name="ConstructData">...</invoke>
            <invoke name="FormatData">...</invoke>
            <invoke name="CleanData">...</invoke>
            <invoke name="IntegrateData">...</invoke>
        </flow>
        <flow name="Modelling">...
            <invoke name="BuildTrainingSet">...</invoke>
```

```
        <flow name="Build&Assess">...
            <invoke name="BuildAssociationRulesModel">...</invoke>
            <invoke name="BuildClassificationRulesModel">...
            </invoke>
            <invoke name="BuildNeuralNetworkModel">...</invoke>
            <invoke name="AssessAssociationRulesModel">...</invoke>
            <invoke name="AssessClassificationRulesModel">...
            </invoke>
            <invoke name="AssessNeuralNetworkModel">...</invoke>
        </flow>
        <invoke name="BuildTestSet">...</invoke>
    </flow>
    <flow name="Evaluation">...
        <invoke name="EvaluateAssociationRulesModel">...</invoke>
        <invoke name="EvaluateClassificationRulesModel">...
        </invoke>
        <invoke name="EvaluateNeuralNetworkModel">...</invoke>
        <invoke name="ReviewResults">...</invoke>
    </flow>
    <flow name="DeploymentofResults">...
        <invoke name="ShowResults">...</invoke>
        <reply name="Distribute">...</reply>
    </flow>
  </sequence>
</process>
```

The sample Customer Profiling process has been created through our reference architecture; in particular, the collection of Miners has been implemented through Java Web services and Weka technology, the Controller through a BPEL4WS Engine and the Actions through BPEL4WS process definition files.

6 Conclusions

In this chapter has been presented a Knowledge discovery process architecture able to build e-services for knowledge in a modular and flexible way. To this purpose, the guide lines of the SOA and the Orchestration model have been considered as a way to manage a workflow of reusable building blocks with well defined tasks and able to interoperate with one another for the creation of new services. The main advantages offered by this architecture are the quick designing of a process according to one's own business needs and the creation of new flexible services without resorting to substantial changes. For the future, it may be possible to plan and propose a new standard based on BPEL and specific for Data Mining and knowledge discovery, where the concept of Web services can be replaced by the Data Miners one. The main advantage of the new standard could be that of enabling the exchange of knowledge information across different platforms. These interactions may span applications and organizations, and result in long-lived, transactional and distributed knowledge discovery processes.

References

1. Fayyad, U.M., Shapiro, Smith, Uthurusamy, (1996) Advances in Knowledge Discovery and Data mining, MIT Press, London.
2. Matheus, C.J. (1993) “System for Knowledge Discovery in Databases” in IEEE TKDE Special Issue on Learning & Discovery in Knowledge-Based Databases.
3. CRoss Industry Standard Process for Data Mining, [online], http://www.crisp-dm.org/
4. Cody, W.F. (2002) “The Integration of Business Intelligence and Knowledge Management”, IBM System Journal, Vol 41, No. 4.
5. Peltz C., (2003) “Web Service Orchestration: a review of emerging technologies, tools, and standards”, Hewlett Packard, Co.
6. Choi, I., Jung J., Song M. (2004) “A framework for the integration of knowledge management and business process management” in Innovation and Learning, Vol. 1, No. 4, pag 399-408.
7. IBM, BEA Systems, Microsoft, SAP AG, Siebel Systems, “Business Process Execution Language for Web Services (BPEL4WS)”.
8. Chapman, P., Clinton, J., Kerber, R., Khabaza, T., Reinartz, T., Shearer, C. and Wirth, R. (2000) “CRISP-DM 1.0 Step-by-step data mining guide”, CRISP-DM Consortium.
9. Castellano, M., Pastore, N., Arcieri, F., Summo, V., Bellone de Grecis, G., A Flexible Mining Architecture for Providing New E-Knowledge Services, Proceedings of the 38th HICSS, Hawaii Int. Conference On System Sciences, January 2005, Big Island, Hawaii, Computer Society Press.
10. Rantzau, R. (2003) “A Multi-Tier Architecture for High-Performance Data Mining” in Proceedings of the Conference Datenbanksysteme in Büro, Technik und Wissenschaft (BTW), Buchmann (Ed), A. P., Freiburg, Germany.
11. Han, J. & Kamber, (2001) M., Data Mining: Concepts and Technique, Morgan Kaufmann Publishers.
12. G. Adomavicius & A. Tuzhilin, (2001) “Using Data Mining Methods to Build Customer Profiles”, IEEE.

13. R. Agrawal et al., (1996) "Fast Discovery of Association Rules," Advances in Knowledge Discovery and Data Mining, AAAI Press, Menlo Park, Calif., chap. 12.
14. L. Breiman et al., (1984) "Classification and Regression Trees", Wadsworth, Belmont, California.
15. Li Niu, X. yan, c. Zhang, S. Zhang, (2002) "Product Hierarchy-based Customer Profiles for Electronic Commerce Recommendation", Proc. Of the First Int. Conf. on Machine Learning and Cybernetics, Beijing.
16. Q. Chen, K. Mao, Y. Zhang, L. Lv, (2004) "Catching Potential Customers: An Example of Web-mining-aided e-commerce decision making", Proc. Of the Fifth World Congress on Intelligent Control and Automation, China.
17. Nasukawa, T. (2001) "Text Analysis and Knowledge Mining System", IBM Systems Journal, Vol 40, No. 4.
18. Srivastava, J. (2000) "Web Usage Mining: Discovery and Applications of Usage Patterns from Web" in ACM SIGKDD.
19. Microsoft, (2000) "BizTalk Orchestration: A Technology for Orchestrating Business Interactions", Microsoft Corporation.
20. Sherman, Doron, (2002) "Orchestrating Asynchronous Web Services", Collaxa, [online], www.collaxa.com
21. J. Chattratichat, J. Darlington, Y. Guo, S. Hedvall, M. Kohler, and J. Syed, (1999) "An Architecture for Distributed Enterprise Data Mining", Proc. 7th International Conference on High Performance Computing and Networking Europe.
22. S. Krishnaswamy, A. Zaslavsky, S.W. Loke, (2000) "An Architecture to Support Distributed Data mining Services in E-Commerce Environments", Workshop on Advanced Issues of E-Commerce and Web/based Information Systems.
23. J.M. Pena and E. Menasalvas, (2001) "Towards Flexibility in a Distributed Mining Framework", DMKD Workshop.
24. Papazoglou, M.P. (2005): Extending the Service-Oriented Architecture. In: Business Integration Journal, pp. 18-21.
25. Papazoglou, M.P. and D. Georgakopoulos (2003): Service-Oriented Computing. In: Communications of the ACM, 46(10): 25-28.

On the Knowledge Repository Design and Management in E-Learning

Emma Kushtina, Oleg Zaikin, and Przemysław Różewski

Faculty of Computer Science and Information Technology, Szczecin University of Technology, Zolnierska 49, 71-210, Szczecin, Poland
{ekushtina, ozaikine, prozewski}@wi.ps.pl

Abstract

The authors use an ontological model and cognitive properties of human perception system to create an intelligent knowledge repository system. The concepts network creation algorithm and didactic materials compilation algorithm are proposed and integrated into the knowledge repository structure.

1 Introduction

In the present state of information society's development the task of storing and sharing knowledge is the basic research problem. Many economic initiatives (SemanticWeb, B2B), as well as social ones (Open and Distance Learning, digital libraries) are connected to the development of information systems functioning at the level of knowledge. The intelligence built into the new generation of information systems allows ensuring a personalized flow of knowledge delivered to the user on the basis of his/her preferences and characteristics [16,19]. Such approach can be found for example in modern Content Management Systems or e-learning systems [28].

The new paradigm applied in e-learning systems described by the SCORM 2004 standard [27], assumes creating a knowledge repository containing a certain domain knowledge divided into knowledge objects (called Learning Object – LO). Didactic material meant for a certain student is built through creating a sequence of LO. That creates a few significant research problems:

E. Kushtina et al.: *On the Knowledge Repository Design and Management in E-Learning*, Studies in Computational Intelligence (SCI) **37**, 497–515 (2007)
www.springerlink.com

- What should the knowledge repository structure be like to ensure a high-quality description of a given domain and at the same time to allow adapting the knowledge presented to the student depending on the education goal (e.g. the knowledge depth level) and his/her cognitive features (e.g. cognitive learning style [21]).
- Define the method of generating a personalized LO which, as an autonomous knowledge model, has to satisfy a certain education objectives and simultaneously ensure high-quality knowledge, i.e. a closed context secured in the knowledge already owned by the student.
- Build a knowledge repository management system that describes the roles of actors (such as an expert or a teacher), defines the frames of their cooperation and the tools that the actors can use to manage the knowledge repository, including their scope of operation.

The presented questions have already been analyzed in literature in the context of e-learning systems. In [20] the knowledge model is a structure of rules described in New Object-oriented Rule Model (NORM). A method establishes the content of LO in the form of a knowledge class. However, basing the knowledge repository on the rule-model makes controlling the concepts' capacity and localizing concepts from a given domain more difficult. In [33] the knowledge model has the form of a table called Knowledge Construction Model. A method of defining the order of LO in the form of a knowledge element is proposed. The structure of LO was not specified, nor were the criterions defining the content of a LO. In [31] an approach to building LO in the form of Intelligent Reusable Learning Components Object Oriented (IRLCOO) was proposed. The knowledge model uses the approach of a concept graph. For defining the sequence of LOs being delivered to the student a multi-agent approach was used. The method does not specify the way of defining the size of LO. Moreover, the method of modeling knowledge with the conceptual graphs cannot successfully model procedural and fundamental knowledge. Unfortunately, none of the propositions fulfill entirely the requirements of the SCORM standard, such as reusability, accessibility, durability and interoperability of didactical materials and environments of e-learning system.

The LO issue is intensely explored by the IT companies and main computer's vendors like IBM, Cisco System, Macromedia and many others. According to [26] the industrial effort is particularly devoted to content repositories, learning systems, and authoring tools. For the IT sector the Cisco System has made the biggest effort in terms of LO management. In [2] Cisco's Reusable Learning Object Strategy (RLOS) is presented. The

RLOS combines Reusable Learning Objects which include: overview, summary, assessment and five to nine Reusable Information Object (RIO). According to [2] each RIO is built upon a single education objective and combined content items, practice items and assessment items. The RLOS can be seen as a complete management tool for LO, which is successfully implemented in CISCO Networking Academy. However, the RLOS is missing the procedure for knowledge management and whole knowledge manipulation depends on the Subject Matter Expert and the Knowledge Engineer. Other companies focus rather on their product LO's compatibility. The LO economy is supported by the market leading learning systems like Blackboard (www.blackboard.com), WebCT (www.webct.com), TopClass (www.websystem.com), Lotus LearningSpace, IBM LMS (www.ibm.com) and free Moodle (http://moodle.org/). The reason standing behind it is that the companies follow industrial standards which continually keep missing the knowledge aspect of LOs and concentrate only on technical issue, like LO's metadata or communication.

In the given article, a method of building and managing the knowledge repository adapted to the SCORM standard and LO requirements are presented. As a base, the ontological approach was adapted, where the domain knowledge model is formulated in the form of an ontology. The material send to the student is specified on the basis of the education objectives and the student's features. The most important ones of those are his/her learning style and the cognitive features, e.g. the size of working memory. The approach to knowledge repository design is discussed. Furthermore, two algorithms facilitating the knowledge repository management are proposed. The first one, the concepts network creation algorithm, allows building a domain knowledge model in the form of an ontology and placing it in the knowledge repository in a way that allows future transformation and modeling. Additionally, a knowledge repository structure is suggested. The second algorithm, the didactic materials' compilation algorithm, allows creating a personalized stream of knowledge addressed to a student with the consideration of his/her perception abilities.

2 E-learning perspective

2.1 Learning Object

Research in the area of pedagogy, psychology and cognitive science [1] has lead to a deeper understanding of the nature of knowledge adaptation during the learning process, thus making it possible to introduce the idea of Learning Object (LO). In the asynchronous learning mode, the structure of didactic materials is module-based. Every domain can be divided into modules consistent with the e-learning standard SCORM, that can be later used in different courses, without the need for expensive recoding or re-designing operations (see fig.1).

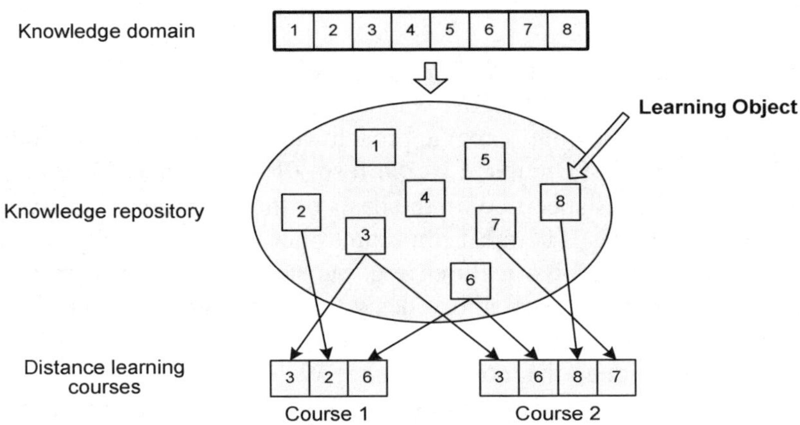

Figure 1: Concept of learning Object

The SCORM (Sharable Content Object Reference Model) standard [27], officially released in a new version at the beginning of year 2004, consists of four books. First one of them, called: The SCORM Overview, is an overview of the main idea and plays the role of a guide for the other three. The second book, called The SCORM Content Aggregation Model, contains directions for localization and creation of a structured education content, it presents the terminology and metadata model used in the standard. This part is based on IEEE LOM [17]. The third book, called The SCORM Run-Time Environment, concerns methods of activating didactic materials in the e-learning environment. It describes in detail communication between the system and the didactic material and methods of controlling and following an active learning session. The fourth book, called The SCORM Sequencing and Navigation, describes methodology of creating a sequence of LO in an interactive environment. The construction of the Learning Object is covered in detail by the SCORM standard. However, there is a lack of information about the content and structure of the Learning Object.

The international research society has been investigating the problem of the Learning Object for several years [6,18,23], however, the general solution has not yet been found. Until now, the set of guidelines and rules has been published [14].

2.2 Learning (Content) Management System

The Learning Management Systems (LMS) and Learning Content Management Systems (LCMS) can be recognized as the main software and hardware solution for e-learning [35]. The LMS is a strategic solution designed for planning, delivering and managing all the training events in the company, considering virtual classes as well as the ones taught by an instructor [12]. The LCMS [3] is what we call a system used for creating, storing, making available (sending) personal educational content in the form of LO. Usually LCMS is a part of LMS.

The LMS class systems allowed organizations to plan and track educational needs of their employees, partners, clients or students. Basic LMS functional modules can be defined as follows [7]: student management and reporting, learning event and resource, management and reporting, online course delivery infrastructure, course authoring tools, skill/competency assessment, professional development management, knowledge bases, learner-centric and organization personalization.

Contrarily to LMS, LCMS concentrate on the content of didactic materials, enabling modeling their content-related scope in the form of a sequence of LO. LCMS consists of the following modules [7]: dynamic delivery interface, administrative application, authoring application, LO repository. Some of the functions realized in LMS are presented in LCMS, but the difference in their execution concentrates on a more detailed information concerning didactic measurement, analysis of its results and using them in a dynamic adaptation of the learning path.

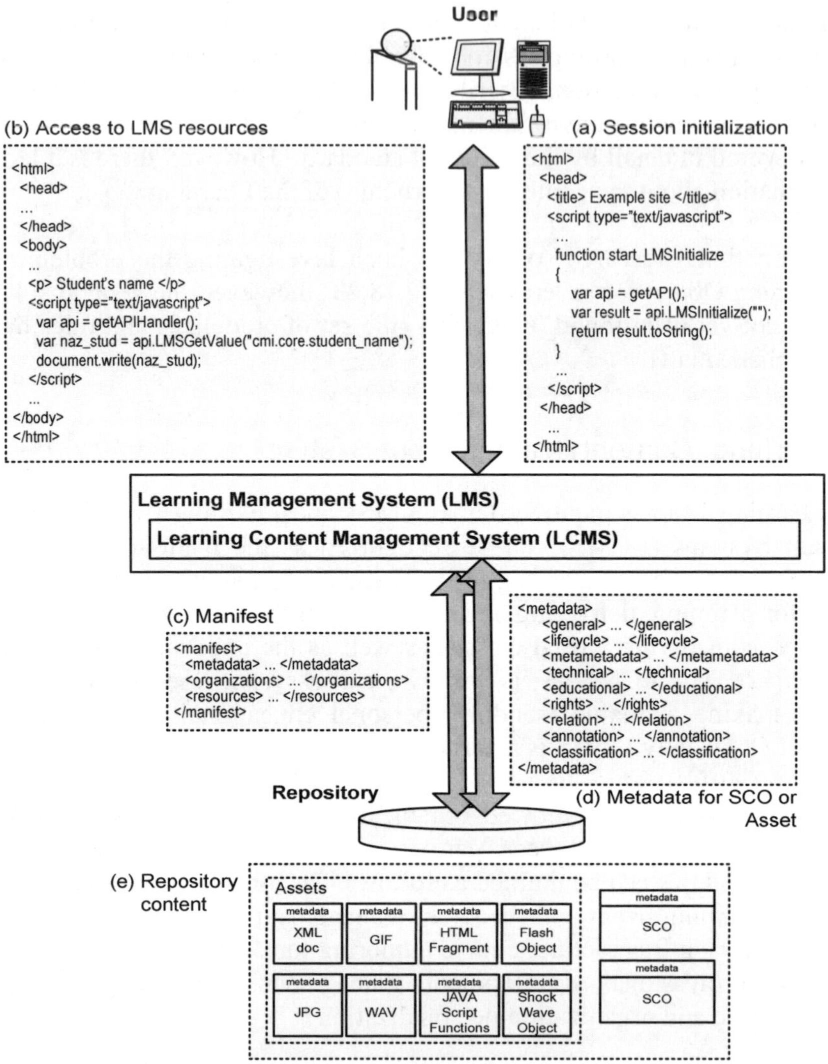

Figure 2: Student and LMS/LCMS interaction basing on SCORM standard

Figure 2 presents interaction between students, LMS/LCMS system and knowledge repository consistent to SCORM. The LO management and design take advantages of metadata approach. In SCORM nomenclature the LO is called a Sharable Content Object (SCO). The most basic form of a learning resource in SCORM is an Asset. Assets are an electronic representation of any media. One SCO consists of some number of Assets.

Let us discus the figure 2 in detail. The web browser, with some help from the user, sends requests to the LMS. In fig. 2 tasks of that kind are marked by indexes (a) and (b). Index (a) shows the session initialization. Each communication establishment requires a connection-negotiation procedure. Index (b), referring to LMS recourses, shows using the LMS base for retrieving information about the student. In SCO orders referring to LMS may be included. SCORM ensures a method that allows describing the sequence together with its components in the form of an XML file, known as a manifest. Manifest, index (c), is analyzed, and on its basis the consecutive SCOs forming the course for a student are determined.

Description based on meta-data allows an effective search through the repository. There can be many files in the repository and each of them can be a part of an SCO or an Asset. Creating a system that enables an unambiguous multi-criterion identification improves the functionality of the system. In the fig. 2, index (d) represents the structure of meta-data for an SCO or an Asset with the help of the IEEE LOM standard. The files forming computer metaphors and the meta-data files are stored in the repository (e). In the SCORM standard SCORM Content Aggregation Model is responsible for this.

3 Knowledge representation in e-learning

The analysis of existing models of knowledge representation shows that none of them satisfy the demands of e-learning. It is shown in [24] that the structure of knowledge which differs from the rules systems can be presented on the basis of semantic networks. Semantic networks can represent both abstract categories and certain objects. The big obstacle for using semantic networks is the difficulty with formal representation of semantic relations. Semantic networks are the starting point for such knowledge representation models as Mind Maps, Conceptual Maps, Conceptual Graphs and Topic Maps. The models are oriented towards a specific kind of knowledge and a certain user. Each of them is dedicated to a specific kind of tasks and a specific domain. The e-learning knowledge representation model merges the knowledge manipulation language with the corresponding pedagogical approach, which is used to learn about a subject. The adequate knowledge representation model has been discussed in detail in [25,34].

Reality is defined by an unlimited and diverse set of information and stimulus which attract human perception system. The cognitive science assumes the natural mind's ability to conceptualize [11]. The conceptual scheme exists in every domain and informs about the domain's boundaries and paradigm. The conceptual model of the knowledge domain can be created in the form of an ontology, where the concepts are the atomic, elementary semantic structures [13,29]. From the practical point of view, the ontology is a set of concepts from a specific domain. The modeling approach, considered as a cognition tool, assumes some level of the examined object's simplification.

Concept is a nomination of classes of objects, phenomena, abstract category. For each of them common features are specified in such way that there is no difficulty with distinguishing every class. The given concept's definition makes possible the modeling of the knowledge model for any domain basing on the set of basic concepts, which were specified by an expert in a verbal way.

The concept ϕ, is defined as a tuple: $\phi =< X,T >$, where X is a set of information, which provided the concept's description. The description is made basing on one of the metadata standards (e.g. DublinCore, IEEE LOM). T is a matrix of the concept's depth $T = [\hat{t}_{ij}]$ and N(ϕ) is a name of the concept ϕ. All elements of the matrix T belong to the specified domain, as a result, they are included in a ontology $\{N(\phi)$, Object i, Attribute j, $t_{ij}\} \in \Omega$, for $i = 1,\ldots, I$, $j = 1,\ldots, J$. Matrix T can be consider as concept's abstraction [30].

In the presented approach ontology is defined as a tuple: $\Omega =< S,\Pi >$, where $S = \{\phi_i\}, i = 1,\ldots, n$ is a set of concepts from a specific domain. $\Pi : S \times S \rightarrow R$ is a mapping from an ordered pair of concepts to the set of connections R= {IS_A, PART_OF, $\varnothing$}.

Developing an ontological domain model in an educational context requires analyzing a specified teaching/learning program and learning objectives, that play the role of constraints on the capacity and depth of concepts used in didactic materials. Using a matrix structure to describe concept leads to a bi-level structure of domain ontology. The first level is a network of concepts, while the second level describes the depth and capacity of knowledge contained in each concept. Rules for creating the bi-level structure can be used many times in relation to the same, pre-described ontological model. This gives the possibility to develop a multi-level ontological model. Using the proposed approach makes it possible to easily adapt the ontological domain model to specified education objectives. Ontological domain model extended in this way allows a significant automation of processing the model into a module structure of didactic materials meant to be used in the learning process.

4 Proposition for knowledge repository design

The next two subsections present the general idea of the concepts network creation algorithm and didactic materials' compilation algorithm. The formal approach to the algorithms is discussed in detail in [25,34]. The third subsection explains the proposed approach to knowledge repository design. The repository structure incorporated both algorithms in order to manage the knowledge repository.

4.1 Concepts network creation algorithm

The goal of the concepts network creation algorithm (fig. 3) is to identify the knowledge in a specific subject's domain and convert this knowledge to the form of a concepts network. Figure 3 shows an algorithm describing each stage of the concepts network creation by the domain expert (with the help of a knowledge engineer). The expert finds concepts characteristic for the given domain, joins them in a semantic way and transforms them into their media-representation. The algorithm formalizes the result into a form compatible with SCORM.

The first stage of the algorithm is performing the phase of preliminary studies of the subject domain. Performing the preliminary study phase by the expert leads to defining meta domain knowledge, which enables defining characteristics of thc given domain, such as e.g. the which of considered knowledge or context. The next step is base knowledge identification, which consists, among others, of defining knowledge sources. The expert improves the concept's set with the help of the verbalization process. Selected set of concepts is then analyzed in detail. For each concept depth analysis is performed, that means creating an abstraction of the given concept. The first phase ends with a control of conformity of the transformation of the concept with expert's ideas concerning the concept in the given context.

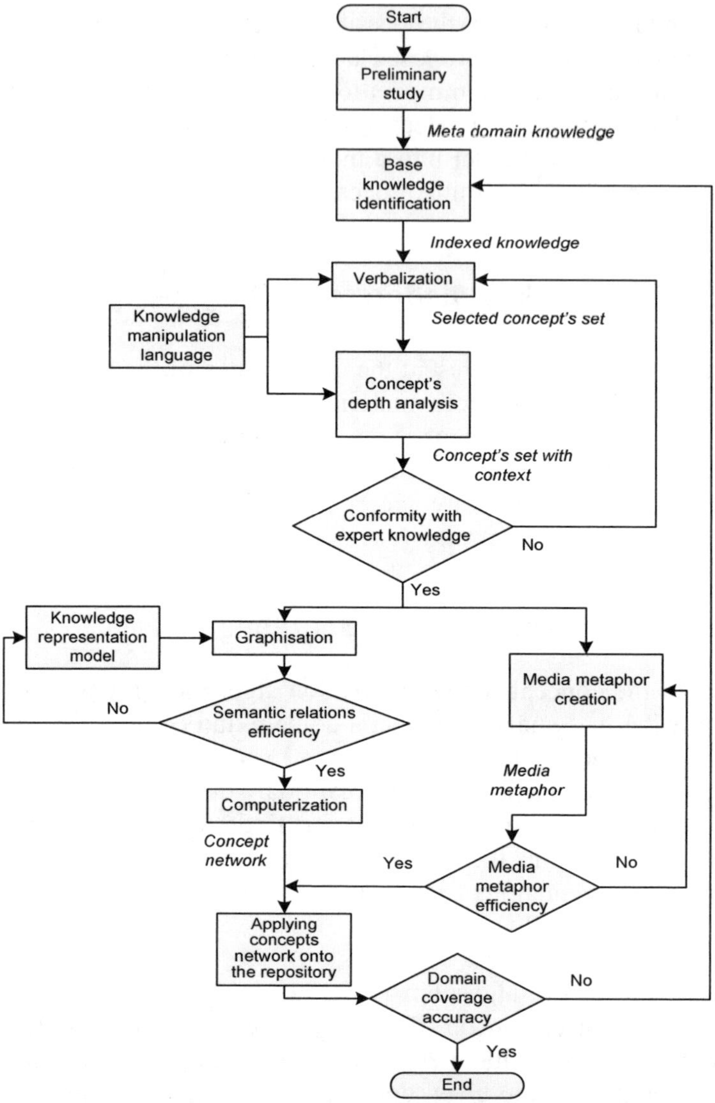

Figure 3: The algorithm of the concepts network design

The identified set of concepts is being enriched at the stage of graphisation, when relations reflecting the relationships between concepts in a given domain are added. Defining relations between concepts is made on the basis of a knowledge representation model. At the next stage, the previously prepared procedures, appropriate for the chosen knowledge representation model, allow computerization of the chosen area of knowledge into the form of a digitalized concepts network.

Parallelly, the process of representing concepts in a digital form (files) is carried out in order to create a media metaphor of all the concepts. A media metaphor – a conglomerate of data making up the concept's description – is appointed to each concept. Abstract knowledge orientation implies the possibility to base only on digital knowledge representation. The repository stores information about each concept, which includes meta-information about the concept (like: author, creation date, etc.) together with the set of files representing the media metaphor.

After a positive verification of the set of metaphors the process of applying the concepts network onto the repository takes place. The whole is then verified again, with the consideration of domain coverage accuracy. The result of completing this stage is a subject domain knowledge model consistent with the requirements of computer formalisms and placed in the repository according to the requirements of the ontological form.

The results of the algorithm are (a) a semantic structure of didactic materials, (b) a hypermedia network transforming the concepts into their media representations placed in the repository. The result of the algorithm's work is transformed to suit the SCORM standard and should be stored in a repository that includes different kinds of files compatible with different industrial data standards and described with XML. The relationships of semantic level and media representation are represented in SCORM standard in forms of adequate descriptions and meta-data structures.

The results (a) and (b) can be considered as an ontology expressed in the form of a lightweight ontology. The lightweight ontology can be described as a 3-tuple [8, 15, 32]: <CD, ID, RD>, where CD – set of classes, which defines the concepts used to the real object's description, ID – set of instances, which represents the instance of the concept, RD – set of relations. The ontology for the description of Learning Objects is developed by the presented algorithm, where result (a) represents the characteristic of a concept i.e. CD and ID. Result (b) reflects the relations between concepts i.e. RD set. Moreover, the conceptual modeling based on a lightweight ontology leads us to some simplification of the model, which results in modeling process transparency. In some sense, the lightweight ontology represents classical understanding of the knowledge phenomenon which comes from Semantic Networks [1,24].

4.2 Didactic materials' compilation algorithm

The didactic materials' compilation algorithm (fig. 4) adapts the didactic material to a certain student, with the consideration of current e-learning standards, as well as educational conditions of the process. Semantic capacity of a given concept (called in this approach: concept's depth) is determined by the level of student's competency in the given domain. The level of competency is built on the foundation of theoretical knowledge, the required basis for specialized (procedural) knowledge. Theoretical knowledge is abstract knowledge that enables effective conclusions in any technical domain. The competency of e.g. an engineer is the ability to join the theoretical knowledge with the know-how type of knowledge.

The result of the algorithm is an educational sequence (LO sequence consistent with the SCORM standard) which, through telecommunication links, will be made available to the student. The algorithm uses the following elements: concepts network, repository together with a description of each concept, student's profile. The mechanism presented in the algorithm is entirely transferable to LMS/LCMS class systems.

The first step of the algorithm assumes concepts network dimension reduction on the basis of the learning objectives, which usually reduces the target level of knowledge from the given area. Basing on the reduced concepts network and information about the level of the student's so-far-gained knowledge included in the student's profile basic concepts are selected. Basic concepts influence the structure of the reduced concepts network.

Next, an ordered multilevel graph of concepts is created, as the concepts network is being organized with consideration of the basic concepts and their relation with other concepts. The result of this organization is a hierarchical network, with basic concepts at it's top level. The concepts placed below that level are considered the learning targets (knowledge unknown to the student).

At the treelike selection step the multilevel, hierarchical network is being decomposed into a set of its sub-graphs. Concepts included in a sub-graph are in no relation with the concepts from other sub-graphs. After that, each sub-graph is being transformed into a tree, which corresponds to SCORM Activity Tree structure. The root of the tree is the final education goal – the basic concepts are placed at one of the end-levels of the hierarchical network.

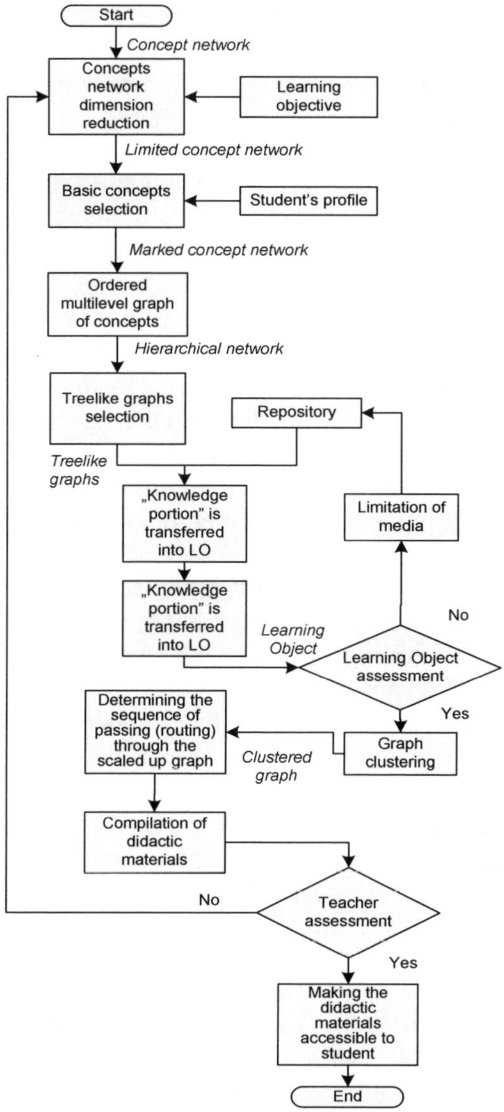

Figure 4: Didactic materials compilation algorithm

Each of the trees is analyzed from the coverage algorithm point of view in order to form a "knowledge portion". The individual knowledge portion is supposed to be a teaching/learning process element and in the form of LO will be presented to the student as an element of the learning process. Transforming the "knowledge portion" into the form of a Learning Object requires applying the concepts network onto the repository. This action is connected to knowledge portion transformation into an information structure – Learning Object, which is assessed in the aspect of technological constraints of the process caused by the telecommunication network and the digital character of the content being integrated.

As we already mentioned above, Learning Objects accumulate similar concepts into one object. The question is – how many concepts should be incorporated into one Learning Object? The answer comes from the cognitive science. Research on the mind provides a useful memory model, which consist of the following memory types: sensor memory, short-term memory and long-term memory. The information from the environment is converted into an electrical signal by the sensor memory. After that, basing on the knowledge obtained from the unlimited long-term memory, the information is analyzed in the short-term memory. Finally, some part of information is encoded in the short-term memory to the form of a cognitive structure and then sent to the long term memory as knowledge.

The most important type for our study is the short-term memory (nowadays called working memory), due to its limited capacity. The idea is to relate the "size" of the learning object to the short-term memory capacity. The short-term memory capacity is determined by the Miller Magic number (7±2) [5,22], which was updated by the Cowan (about 4) [4]. The newest research [9] brings up new estimation (about 3).

An additional significant indicator is the unique nature of the short-term capacity. The limitation is counted by chunks, not bites. A chunk [10] is a set of concepts which are strongly related to each other and, at the same time, weakly related to other concepts. In case of the previous concept's definition the concept is consider as a chunk.

The last group of activities, just before the final teacher's control, is determining the sequence of passing through the scaled up graph. Cognitive science assumes necessity of connecting the knowledge being assimilated with the already owned one. SCORM gives possibility to define the LO sequence by dedicated script language. Next, all the elements are compiled into a form consistent with SCORM. The created product is finally analyzed by the teacher, who checks the efficacy of the didactic material by referring to his educational experience in teaching the given subject. After being accepted, the course is made available to the student through LMS/LCMS mechanisms.

4.3 Knowledge repository structure

The knowledge repository conceptual scheme (fig. 5) presents cooperation between the actors of the e-learning process, based on knowledge repository. Moreover, previously described algorithms are the core knowledge repository components. One of the main projecting determinants of LMS/LCMS system is ensuring knowledge sharing and the possibility of knowledge re-usage. The task is carried out at the level of ontology layer with help of the concepts network creation algorithm. Knowledge engineer builds the model of a given domain by formalizing the concept network of the domain into the form of an ontology, basing on the concept network interface. During creating the ontology of a given domain, the knowledge engineer uses the knowledge received from an expert. Another source of knowledge could be outside knowledge bases (e.g. in the form of a global Learning Objects repository). Knowledge can also be obtained from the teacher. The ontology can also be created by the expert. Knowledge is stored in the e-learning system in the form of an ontological structure of a conceptual domain scheme. The form defined this way is useless from the point of view of modern e-learning applications based on SCORM. Therefore, there is a method implemented in the system that allows transforming the ontological form of knowledge into a module form of Learning Object. Semi-automatic didactic materials compilation algorithm based on a visual interface of conceptual network perform the task of didactic materials compilation, use the information about the student obtained from the student's profile and the participation of the teacher, who actively influences the creation of Learning Object flow considering the education goal and the given teaching methodology. All presented algorithms are examined in formal way in [34].

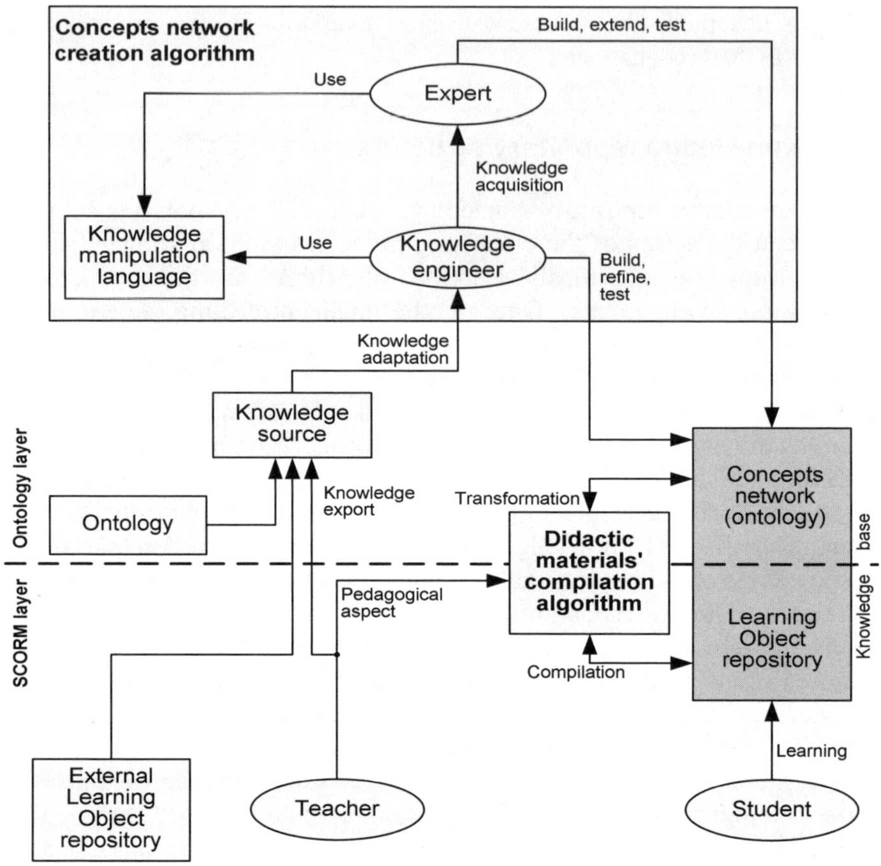

Figure 5: Knowledge repository conceptual schema

The knowledge base structure characterized by two dimensions requires a special comment. From the point of view of the task – modeling knowledge in a given ontology, knowledge base contains a structure describing domain knowledge, based on a set of concepts and their relations, presented in the form of a conceptual network. The ontological dimension allows a flexible knowledge manipulation at the lowest level of concepts and relations. In an e-learning system the possibility of communicating with the usage of a standard language and structures contained in SCORM standard is pretty important. Knowledge base, through the transformation mechanism, changes the ontology structure into a form of Learning Objects sets, which can be manipulated in the frames of standard editors and tools meant for e-learning materials construction and edition.

5 Conclusion

Creating an intelligent method of Learning Objects modeling requires a high level of personalization. Because each of the students has his/her own cognitive characteristic and his/her own style of learning, the basic structure of knowledge repository should reach a smaller level of granulation than the structure of Learning Object. Concepts network built on the basis of expert's knowledge and expressed through an ontological model seems to be a good solution for a basic form of knowledge storing.

The possibility of applying the proposed approach to existing educational systems is profitable when the method works in an automatic way on the basis of prepared algorithms and is adapted to SCORM. Describing student's knowledge, goal and methodology of the education process with a set form of knowledge formalization in a formal way enables developing methodology and algorithms that allow building Learning Objects dedicated to any given student on the basis of a concepts network.

Acknowledgements

The authors have conducted this research during EU Socrates/Minerva e-Quality project, ref. No. 110231-CP-1-2003-1-MINERVA-MP.

References

[1] J.R. Anderson, Cognitive Psychology and Its Implications, 5th edition, Worth Publishing, New York, 2000.

[2] C. Barritt, D. Lewis, Reusable Learning Object Strategy: Definition, Creation Process, and Guidelines for Building, Cisco Systems, Inc., (2000).

[3] M. Brennan, S. Funke, C. Anderson, The Learning Content Management System: A New eLearning Market Segment Emerges, IDC White Paper, Framingham, MA, 2001. (Downloadable from website http://www.idc.dk/WhitePaper/).

[4] N. Cowan, The magical number 4 in short-term memory: a reconsideration of mental storage capacity. Behavioral and Brain Sciences 24 (1): 87-114, (2001).

[5] J.L. Doumont, Magical Numbers: The Seven-Plus-or-Minus-Two Myth. IEEE Transactions of Professional Communication, 45(2): 123-127, (2001).

[6] S. Downes, Learning objects: Resource for Distance Education Worldwide. International Review of Research in Open and Distance Learning 2(1)(2001).
[7] Element K Corporate (2001), A Guide to Learning Management Systems, Rochester, New York, (Downloadable from website http://www.elementk.com/downloads/webservices_whitepaper.pdf).
[8] G. Frank, A. Farquhar, R. Fikes, Building a large knowledge base from a structured source, IEEE Intelligent Systems, 14(1): 47-54, (1999).
[9] F. Gobet, P.C.R. Lane, S. Croker, P.C-H. Cheng, G. Jones, I. Oliver, J.M. Pine, Chunking mechanisms in human learning, Trends in Cognitive Sciences, 5(6): 236-243, (2001).
[10]F. Gobet, G. Clarkson, Chunks in expert memory: Evidence for the magical number four… or is it two?. Memory, 12(6): 732-747, (2004).
[11]A. Gómez, J. Moreno, A. Pazos, A. Sierra-Alonso, Knowledge maps: An essential technique for conceptualization. Data & Knowledge Engineering, 33(2): 169-190, (2000).
[12]L. Greenberg, LMS and LSMS: What's the Difference?, Learning Circuits - ASTD's Online Magazine All About E-Learning, 2002, (Downloadable from website http://www.learningcircuits.com/ 2002/dec2002/Greenberg.htm).
[13]N. Guarino, Understanding, building and using ontologies. International Journal of Human-Computer Studies, 46(2-3): 293-310, (1997).
[14]C.J. Hamel, D. Ryan-Jones, Designing Instruction with Learning Objects. International Journal of Educational Technology 3(1)(2002).
[15]J. Heflin, J. Hendler, Dynamic Ontologies on the Web. In: Proceedings of American Association for Artificial Intelligence Conference (AAAI-2000). Menlo Park, Calif.: AAAI Press, pp. 443-449, (2000).
[16]D. Helic, H.A. Maurer, N. Scerbakov, Knowledge transfer processes in a modern WBT system. Journal of Network and Computer Applications, 27(3): 163-190, (2004).
[17]IEEE LOM - IEEE Learning Objects Metadata, The IEEE Learning Technology Standards Committee, (Downloadable from website http://ieeeltsc.org/).
[18]S. Kassanke, A. Steinacker, Learning Objects Metadata and Tools in the Area of Operations Research, In: proceedings of ED-MEDIA'01, World Conference on Educational Multimedia, Hypermedia and Telecommunications, Tampere, Finland, pp. 891-895 (2001).
[19]E. Kushtina, O. Zaikine, P. Różewski, R. Tadeusiewicz, Kusiak J, Polish experience in the didactical materials creation: the student involved in the learning/teaching process, In: proceedings of the 10th Conference of European University Information Systems (EUNIS'04), Bled, Slovenia, pp. 428-433, (2004).
[20]Y.T. Lin, S.S. Tseng, C-F Tsai, Design and implementation of new object-oriented rule base management system. Expert Systems with Applications, 25(3): 369-385, (2003).
[21]Y. Liu, D. GinTher, Cognitive Styles and Distance Education, Online Journal of Distance Learning Administration, 2(3)(1999).

[22]G. A. Miller, The magical number seven, plus or minus two: Some limits on our capacity for processing information. Psychological Review, 63(2): 81-97, (1956).

[23]P.R. Polsani, Use and Abuse of Reusable Learning Objects, Journal of Digital Information 3(4)(2003).

[24]M. Quillian, Semantic Memory, In: M. Minsky (Ed.), Semantic Information Processing, MIT Press, Cambridge, Massachusetts, 1968, pp. 227-270.

[25]P. Różewski, Method of information system design for knowledge representation and transmission in distance learning, unpublished doctoral thesis, Szczecin University of Technology, Faculty of Computer Science and Information Systems, (in Polish), (2004).

[26]D. Sampson, P. Karampiperis, Towards Next Generation Activity-Based Learning Systems, International Journal on E-Learning, 5(1): 129-150, (2006).

[27]SCORM - Sharable Content Object Reference Model, Advanced Distributed Learning Initiative, (Downloadable from website http://www.adlnet.org).

[28]L. Sheremetov, A.G. Arenas, EVA: an interactive web-based collaborative learning environment. Computers & Education, 39(2): 161-182, (2002).

[29]V. Sugumaran, V.C. Storey, Ontologies for conceptual modeling: their creation, use, and management, Data & Knowledge Engineering. 42(3): 251-271, (2002).

[30]D.C. Tsichritzis, F.H. Lochovsky, Data models, Prentice Hall – Professional Technical Reference, 1982.

[31]R.P. Valderrama, L.B. Ocaña, and L.B. Sheremetov, Development of intelligent reusable learning objects for web-based education systems. Expert Systems with Applications, 26(3): 273-283, (2005).

[32]P.R.S. Visser, D.M. Jones, T.J.M. Bench-Capon, M.J.R. Shave, An Analysis of Ontology Mismatches; Heterogeneity versus Interoperability, In; Working notes of the Spring Symposium on Ontological Engineering (AAAI'97), Stanford University, pp. 164-172, (1997).

[33]C.-H. Wu, Building knowledge structures for online instructional/learning systems via knowledge elements interrelations. Expert Systems with Applications, 26(3): 311-319, (2004).

[34]O. Zaikin, E. Kushtina, P. Różewski, Model and algorithm of the conceptual scheme formation for knowledge domain in distance learning. Accepted by European Journal of Operational Research.

[35]D. Zhang, J.F. Nunamaker, Powering E-Learning In the New Millennium: An Overview of E-Learning and Enabling Technology. Information Systems Frontiers 5(2): 207-218, (2003).

Adding Value to E-Services: a Business-Oriented Model

Matthias Fluegge[1] and Michael C. Jaeger[2]

[1] Fraunhofer FOKUS, Institute for Open Communication Systems
Kaiserin-Augusta-Allee 31, D-10589 Berlin, Germany
Matthias.Fluegge@fokus.fraunhofer.de

[2] Berlin University of Technology, Institute of Telecommunication Systems
FG FLP, SEK FR 6-10, Franklinstrasse 28/29, D-10587 Berlin, Germany
mcj@cs.tu-berlin.de

Abstract. Much of the early research in component-based development and electronic services (e-services) has been done in the area of Open Distributed Processing. Therefore, this discussion will consider the Reference Model for Open Distributed Processing (RM-ODP) published by the ISO in order to provide a generic business model that helps to clarify the actors and relationships in the service domain. Based on the RM-ODP, this discussion also gives a more practical view about applying this business model to Web services which represent the technology of choice for application integration in cross-organisational environments.

Special attention is paid to the retailer and broker roles. Service retailers add value to services offered by third-party providers. Retailers and brokers generate added value by either *a)* trading with the goal to discover the optimal services or *b)* to form composite services. This work will describe the resulting business and technical implications.

1 Introduction: E-Services Basics

The currently evolving proposal for the realisation of e-services has been introduced by the W3C as the *Web Services Architecture (WSA)* [3]. Besides this development, another work plays also an important role for defining the characteristics of distributed systems: the reference model for open distributed processing (RM-ODP) published by the ISO [13]. The RM-ODP represents a technology independent model for distributed software systems and has been adopted by many successful technologies and standardisation efforts, such as the common object request broker architecture (CORBA) [24], or the model driven architecture (MDA) proposal form the Object Management

M. Fluegge and M.C. Jaeger: *Adding Value to E-Services: a Business-Oriented Model*, Studies in Computational Intelligence (SCI) **37**, 517–534 (2007)
www.springerlink.com

Group (OMG) [25]. Basically, the WSA represents another implementation of a service oriented architecture that is covered by the RM-ODP concepts.

The Web services proposal by the W3C is considered to give a more practical discussion in addition to the theoretical foundations of an e-services architecture given in the RM-ODP. The successful evolving of the World Wide Web as a ubiquitous infrastructure for world wide available information has brought the idea of using this infrastructure also as a physical medium for application communication in the form of Web services. On the technical level Web services fulfil the need for a middleware that allows applications to collaborate over the internet irrespective of their concrete implementation. The promised success of this technology lies in the adoption of the generally accepted standards XML, SOAP [22] and the Web Service Description Language (WSDL) [5].

The mentioned standards cover the message exchange between the actors, the description of interfaces, behaviour of actors and many further specific aspects the distributed systems, such as security, quality of service or monitoring. Many of these Web service related problems are also subject to ongoing research effort. All these efforts have in common that for a detailed discussion an underlying model is necessary, i.e. a "big picture" that organises the elements and roles in the Web service domain into a coherent whole. The model given in the Web service architecture however is rather coarse-grained in this regard. The RM-ODP represents this kind of big picture.

In addition, the relevance of the RM-ODP in this respect becomes obvious when investigating the fundamental principles and architecture of Web services. Web service applications reveal a Service Oriented Architecture (SOA) in which the services are self-contained and self-describing components. Hence Web services can be regarded as distributed component-based systems with a strong emphasis on cross-organisational and inter-domain communication. Much of the early research in component-based development and distributed objects has been done to introduce the RM-ODP. The combination of the two concepts openness and distributed processing are regarded as significant for organisational flexibility. Regarding the actors in such a setup, the RM-ODP and the WSA cover the following similar roles:

- An **exporter** represents the party that provides the service. In the field of Web services an exporter is mentioned as a provider agent. An exporter submits its service description to the broker. The service description includes a description of the interface along with an information about the location where a service is available.
- A **broker** implements a trading function to trade services. Trading means to perform a matchmaking of requirements and advertisements of services. The WSA uses the concept of a discovery service for the broker.
- An **importer** is a component that imports a service. In the context of Web services the importer is named requester agent. The WSA proposes that a requester agent performs a discovery process by communicating

with a discovery service. An importer *queries* the broker for a service by submitting a description about his requirements.
- The RM ODP for trading also defines a broker administrator who *defines, manages and enforces* the trading policies. The analogy to the WSA is the owner or provider of a discovery service.

After the importer has received the interface and location of a service the importer starts the interaction with the matched service. The WSA distinguishes between a software and a organisation or single human. Accordingly a software is mentioned as a provider-/requester *agent* while when speaking about humans or organisations only requester or provider is mentioned.

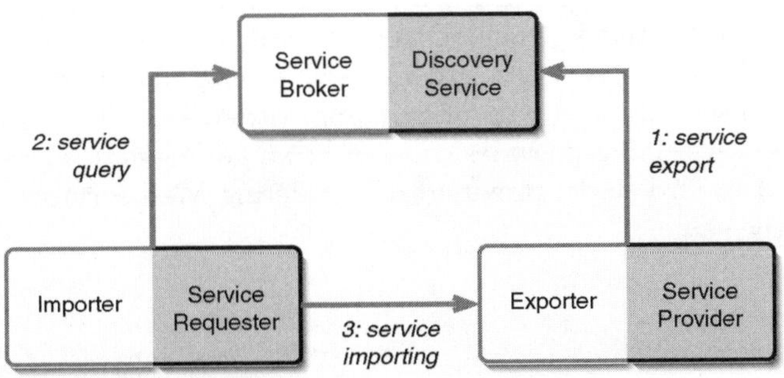

Fig. 1. Service Oriented Architecture - The RM-ODP Terminology [13] and its Web Service Counterparts [3] (shown as grey boxes)

2 A Business Model for Service Applications

Besides the basics in open distributed processing and trading of services, further key elements of RM-ODP are the different viewpoints from which to view at an abstract system structure. The RM-ODP enterprise viewpoint is concerned with the business activities (stakeholders, roles and policies) of a specified system and it is usually taken during the requirements analysis phase. Based on this discussion, we consider a concrete business model representing an instance of the RM-ODP enterprise viewpoint to clarify the key business roles present in most Web service applications.

The enterprise viewpoint introduces a specific business language for defining roles and policies in a system. Instead of starting from scratch hereafter an already existing instance of the enterprise viewpoint is used - the Telecommunications Information Networking Architecture (TINA) business model [36]. Although the roots of TINA are in the telecommunication area, the business model reflects very well key-business roles and relationships that can be found in typical cross-organisational and multi-domain environments as targeted by Web service applications. The TINA provides an existing definition about the involved roles [7] and describes this setup in a resulting business model. In this model, the main roles are:

- **Consumer.** A consumer just uses services provided by the TINA. The analogy to Web service compositions is the service requester. The consumer acts in the sense of an end costumer when referring to business relations.
- **Broker.** A broker enables stakeholders to find services or other stakeholders. In the domain of Web services, a discovery service represents the equivalent.
- **3rd Party Service Provider.** A 3rd party service provider offers his services to a retailer or other 3rd party service providers. The self-relating link from a 3rd party provider to other 3rd provider matches also to the idea of composing services: a composed service may integrate also other composed services as a part of its composition.
- **Retailer.** A retailer provides consumers an added-value access to services provided by 3rd party providers. Considering Web services, this role remains the same.

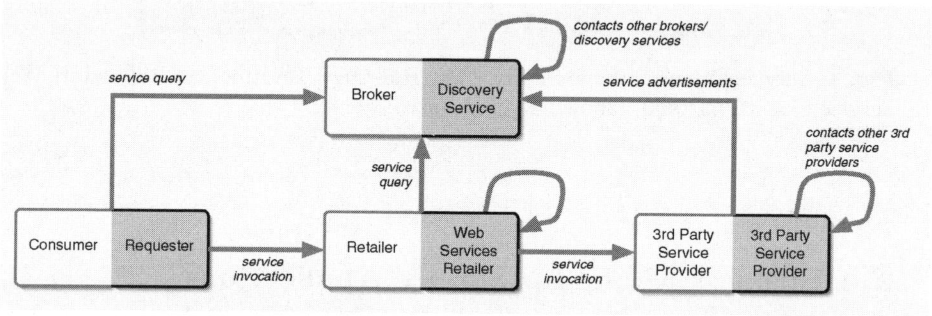

Fig. 2. The Business Roles of TINA [7] with their Representation in the Web Services Domain

In addition, the TINA also defines relations among brokers and among retailer themselves. This allows a broker to contact other brokers to enhance a discovery process, or a retailer to get access to particular services provided by other 3rd party providers. Figure 2 shows the TINA business model and points out the analogies to an environment for Web service compositions.[3]

[3] The TINA business model mentions also a *connectivity provider* In the sense of Web services, the Internet provider represents the connectivity provider. However, the use of services usually does not interfere with any aspects of the underlying network protocols in the Internet. Thus, the connectivity provider is ignored in this discussion.

3 Adding Value by Composing Services

As mentioned in the previous section the retailer adds value to services offered by third-party providers and by other retailers. Value-added services are provided to consumers and to other retailers. There are various ways of how a retailer may add value to services, e.g. by managing the accounting and billing for third-party services. A very common way however is to compose services: In order to be able to satisfy the request of a customer very often several services need to be composed. Two third-party services A and B when being linked together may be published by a retailer as service C providing a new functionality. Of course, this service in turn may be part of an even more complex service. A Service composition is a fundamental concept in a SOA. The SOA paradigm considers services as self-describing and self-contained building blocks, thus it is a common procedure to arrange these building blocks to composite ones.

A composite service can be regarded as a combination of activities (which may be other atomic or composite services), invoked in a predefined order and executed as a whole. In this way a complex service has the behaviour of a typical business process. The process of composing services involves various components and steps:

- A process model (also referred to as a plan hereinafter) specifying control and data flow among the activities has to be created.
- Concrete services to be bound to the process activities need to be discovered. The retailer usually interacts with a broker, e.g. a discovery service, in order to look up services which match with functional requirements and preference criteria.
- The composite service must be made available to potential consumers. Again, the broker is used to publish a description and the physical access point of the service.
- During invocation of a composite service a coordinating entity (e.g. a process execution engine) manages the control flow and the data flow according to the specified plan.

The retailer must arrange the determined services into the abstract composition plan in order to get an executable specification of the composition. The sequence of these steps – planning, trading, publishing and management – and the how they are performed significantly depend on the applied service composition strategy. Current service composition approaches differ with regard to two basic parameters:

- They vary in the degree of automation when creating a process model.
- They range between a early and late binding of services.

In the following section we will give a more detailed discussion about these two characteristics and resulting business and technical implications.

3.1 Creation of the Process Model

A significant characteristic of a service composition strategy is the degree of automation for the creation of a process execution plan respectively for the creation of a process model. Traditional service composition methods, as applied in EFlow [4] and similar platforms, require the user to define the data flow and the control flow of a composite service manually, either directly or by means of designer tools, e.g. in a drag-and-drop fashion. Subsequently, the process description is deployed in a process execution engine. Depending on the abstraction level provided by the tools and depending on the applied binding mechanism, the user either creates the process model based on concrete service instances or based on abstract service templates which are representatives for sets of services, i.e. for service classes. With respect to the multitude of available services and service templates it may be a time-consuming task to manually select reasonable building blocks for the composite service. Furthermore the creation of the data flow, i.e. the parameter assignments between the activities, can be complex and might require the user to have extensive knowledge about the underlying type representations.

More advanced composition strategies actively support the user with the creation of the process model, which is often referred to as semi-automated service composition. Corresponding modelling tools, such as the one presented by Sirin in [31], may interact with a broker in order to automatically look up services which match (regarding the inputs, outputs, preconditions, and effects) with the already available control and data flow, thus facilitating and accelerating the process design. The same applies for approaches in which the process model is created based on abstract functional building blocks, as it has been described in [11]. Parameter assignments between these building blocks may be automatically recommended based on an analysis of the underlying types and concepts.

Fully-automated composition approaches intend to generate a service composition plan without human interaction. Mostly AI inspired methods based on formal logic are used for that matter, such as automated reasoning as described by Berardi et al. [6], through theorem proving as discussed by Rao et al. [30]. An example for research work that resulted in a software suite is the SWORD toolkit [28] where the user can specify the initial and final state facts. By means of a planning algorithm a workflow graph containing available activities or concrete services is generated that satisfies the requirements. Besides the fully-automated generation of composition plans, such approaches can be also used to implement existing process models as discussed by Costa et al. [18]: by their approach, an existing process model is considered as a definition that covers the functional requirements of abstract individual tasks. Then, by using brokers, available and sufficiently described Web services can be either discovered in order to perform an entire task. Or, they can be automatically combined into a mini-composition to provide the required functionality that is not offered by an individual service.

If there are multiple solutions for the problem, i.e. several plans satisfy the given set of requirements, a selection is made with respect to QoS characteristics [12, 18]. This selection can either be made by the process designer or automatically through predefined weighting and ranking functions. Combining the latter with late service binding implies that the complete service composition (i.e. plan generation and service binding) can be performed at runtime. The question to which extend the composition procedure can be automated is subject to research. Fully automated service composition may work in narrow and formally well defined application domains. The more complex the context, however, the more difficult it will be to apply the automated service composition approach in real-world applications.

The applied degree of automation for generating the process model (the plan) has significant business implications for a retailer who composes services and provides them to consumers. As mentioned above, modelling the control flow and the data flow of a composite service may result in a time-consuming task. (Semi-)automated composition techniques promise to speed up this procedure, thus bringing down the costs for developing new services. Furthermore time-to-market is accelerated since the retailer may react faster and more flexible to the customer requirements. In addition, the designed composite services improves in quality as the application of "intelligent" tools helps to create more efficient processes, e.g. by proposing parallel execution of functionally independent activities.

3.2 Service Binding

The activities in a composite service may be bound at design time or they are discovered and invoked at runtime. In the former case, the bindings are static, i.e. for each instantiation the composite service will be made up of the same constituent services. In the case of late binding, the constituent services are selected at runtime, based on automatically analysable criteria, such as service functionality, signature and QoS parameters. Late binding implies the dynamic discovery and invocation of the constituent services. This presumes a sufficient level of interoperability which can be realised by either working with pre-defined interfaces or by applying more sophisticated trading involving matchmaking and mapping mechanisms. The trading will be discussed further in the next section 3.3.

Again, for a retailer the applied binding mechanism has several business implications. In a growing service market 3rd party service providers may offer the same functionality at different conditions, e.g. regarding QoS parameters like price. Applying late binding the discovery and invocation is scalable as the number of services increases. Thus the costs of a composite service offered by a retailer may decrease along with the growing competition in the associated marketplace. The cost advantage can be either handed over to the consumer or it will increase profitability at the retailer's side. Furthermore late binding may enhance fault-tolerance and thus reliability. Since the actions in a process are not hardwired to concrete services, the unavailability of a service may be compensated through the invocation of a functionally equivalent one.

On the other hand late binding may affect adversely the performance of a composite service. The interaction with a broker at runtime in order to discover suitable service candidates could be, depending on the applied matchmaking and ranking mechanisms, a time-consuming task. All in all it can be argued that composition strategies adopting late binding are more adaptable to a changing environment than strategies applying early service binding.

3.3 Trading Services

Adding value by composing services and the binding require the functionality of a broker that keeps track of available services in the Internet – the trading. The ISO has published a definition about service trading, namely the Trading Specification as a part of the RM-ODP [14]. This work partitions the process of trading into particular steps: one of them covers the matching which determines the suitability of candidate services according to their descriptions. Another main step represents a selection which performs a sort on the candidates according to preference constraints. In fact, the RM-ODP trading specification defines the concept of trading as a chain of subsequent isolation operations applied to the set of available services:

- In the beginning let the set $\mathbb{N}_1$ contain all services which the broker has available. Speaking of Web services, $\mathbb{N}_1$ would consist of all services that some discovery service contains.
- The first isolation of services is represented by the set of candidates $\mathbb{N}_2$. This set represents the result of a search for a keyword or similar search criteria among $\mathbb{N}_1$. In the Web service domain, an index of Web services could provide this functionality.
- The second isolation is obtained by applying so called matching criteria. This isolation results in the third set $\mathbb{N}_3$. Matching in this context means the comparison of descriptions from a service exporter with the requirements of an importer. This description can cover the interface as well as other metadata like organisational information.
  Some approaches consider the description of the semantics of the service. Very briefly, such approaches cover a categorisation system for services or a subsumption hierarchy of parameter types. In the literature this procedure is mentioned under the concept of semantic matchmaking [27]. With semantic matchmaking, brokers can provide additional functionality that offers more precise matching results.
- By the next step, either the importer or the broker can apply preference criteria, not to form new subset, but to give an order to $\mathbb{N}_3$. Usually, statements about the required QoS represent common preference criteria. Applying preference criteria results in a ranking of the candidates. The outcome of this step is a tuple T_4^O which is defined by the order O applied to $\mathbb{N}_3$: $T_4^O = (\mathbb{N}_3, O)$.

- The third isolation is according to return policies. Return policies can be also defined by the importer or the broker. Such policies could be to return only one candidate for a particular query. The result is a set $\mathbb{N}_5$ with an order, which is based on a subset of $\mathbb{N}_3$. This can be also seen as a tuple T_5^O with the same order O, applied to $\mathbb{N}_5$: $T_5^O = (\mathbb{N}_5, O)$.

The subsequent filtering operations put the different sets into the following relation: $|\mathbb{N}_1| \geq |\mathbb{N}_2| \geq |\mathbb{N}_3| \geq |\mathbb{N}_5|$, where the set $\mathbb{N}_5$: $T_5^O = (\mathbb{N}_5, O)$ has got an order and also represents the output of the trading process. The resulting value that a broker offers is to apply subsequent filtering operations on the set of available services to identify the optimal set $\mathbb{N}_5$ with O_5. Consequently a broker can either provide all of these operations or just parts of it while the involvement of other brokers is possible.

Trading Web Services

For trading individual Web services, a specification exists that defines a common interface to advertise and query services. This specification is named Universal Description Discovery & Integration (UDDI) [35]. Although it has reached its third revision, only very few organisations currently use a UDDI repository in order to trade Web services in the Internet as a business. Also, almost no software development product utilises the discovery of services over the Internet to consider available services when building applications. The main problem shows two aspects: the first aspects covers an organisational problem: an organisation will not likely use services of another organisations revealed by an automated service discovery. Usually, business relations need a contract or any other agreement. This problem relates to trust and contracting issues in distributed systems. It keeps developers back from using a discovery service.

The second problem lies in the description about the service: by using the current methods, a software system cannot determine matches between service offerings and requirements that cover all aspects of interoperation. Service descriptions contain at least syntactic information and metadata which a UDDI registry can support. The UDDI specification also supports this to some extent by providing a matching of syntactic interface descriptions and of predefined service categories. Different research groups have already proposed their research about how to extend the UDDI specification with the support for semantic descriptions about the service elements such as the works of Trastour et al. [34], Paolucci et al. [26], Akkiraju et al. [1], Srinivasan et al. [32], or Jaeger et al. [16]. The Web service community starts now to consider the semantic description of services as a part of future Web service infrastructures. However, this issue is still open for discussion. The main problem lies in the fact that the semantic description and their matchmaking algorithms covers mainly interface parameters but not the behaviour of the services.

In addition to the semantic description, other approaches consider the quality of the services (QoS) as criteria to describe the requirements and service capabilities. For a UDDI repository, proposals exist that process the QoS such as the work of Ran [29], Benatallah et al. [2], or Lee [17]. Also, the UDDI implementation as a part of Microsoft's server platform provides the extension of the data structure in order to support the processing of QoS information [21].

Either way, the evolving developments of trading functionality in the field of Web services show that the concepts found in the Web services domain go the way that has been defined by the RM-ODP. But the RM-ODP provides also a more detailed concept of trading compared to what has been realised with Web services so far. Since the application of preference criteria and return policies have not been adopted to the Web services domain, the currently proposed approaches for the discovery of Web services realise only parts. Thus, forthcoming commercial brokers can take advantage of more powerful trading functions that support semantic matchmaking and QoS-based selection of services.

Trading to Form Compositions

To form a composition, the trading can benefit from involving an acting party with knowledge about the whole composition to optimise the trading result. Thus, considering the TINA business model involves two roles – the retailer and the broker – for composing services, the following distinction is defined: A broker performs the trading of *individual* services. However, the trading of services to form a composition requires the knowledge about potential requirements in the composition. Thus, it is up to organisational limitations whether the retailer involves a broker to accomplish this task or covers the aspects specific to the composition on his own. Considering the case that the retailer represents the central role for forming compositions, it shows the following characteristics:

- A retailer imports services from 3rd party service providers to offer them to end costumers. When forming a composition of services, the available services are used to from a composition that offers an added value through an automated execution of individual services.
- A retailer queries a broker for discovering services. Regarding his composed service the retailer publishes his offerings of composed services to the broker as well.
- Figure 2 and also the TINA business do not mention the role that support the creation process of a composition to its full extend. For this work, the creation process is generally assigned to the retailer. The retailer represents also the *designer*, and *provider* of the composition.

This consideration brings up the question, what the possible benefits are when the trading performs with considering the whole composition rather than trading individual services. Coming back to the business model, generally all the efforts to establish the composition – planning, trading, publishing and management – represent the added value to service that is brought in by the retailer. Referring the trading, the particular tasks that generate added value are:

- A matchmaking process performs the same way as when trading individual services. Optionally, a matchmaking software could apply some matchmaking constraints to optimise the level of compatibility among the services candidates.
- The preceding matchmaking process covers the functional aspects of required and offered services. In addition to that, non-functional characteristics of services such as different QoS categories, i.e. a maximum execution time for the whole composition can represent selection criteria when selecting individual services as explained in the work of Lee [17]. Determining the selection that meets such constraints results in an added value to the composition.
- Either the retailer or – if the organisational conditions allow – the broker performs the composition-aware selection process to identify the optimal assignment. QoS categories can serve as constraint criteria as well as criteria subject to optimisation. Preceding work has shown that QoS-aware selection algorithms can substantially optimise the QoS of the composition based on evaluating individual services [38, 37, 12, 15].

4 Classification of Composition Approaches and Business Implications

The type of service binding and the degree of automation applied to the creation of a process model have been examined. In summary, service composition approaches may use early binding or late binding. And, the process model can be created manually, semi-automatically or automatically. As illustrated in Figure 3 these characteristic values can be used for a classification of existing service composition strategies in six main categories. The fact that the borders between these categories are not strict but fluent is made clear through the smooth transitions between the squares. Some categories may overlap, i.e. there are composition approaches that may be assigned to two or more categories. To give an example: Besides early and late binding there may be several variations in between, such as the specification of a restricted set of service candidates at design time from which one service is chosen and invoked at runtime.

In the preceding discussion of the business implications for an actor retailing composite services it was argued that composition approaches applying late binding mechanisms are more adaptable to a changing market with 3rd party providers frequently leaving and joining. Furthermore it was argued that a high degree of automation for the creation of the process model cuts down development costs and accelerates time-to-market, thus resulting in a higher flexibility of the retailer. Furthermore quality aspects, such as reliability, have been considered. When combining the terms adaptiveness and flexibility to the more generic term dynamics, a coarse-grained and more business-oriented classification in static, semi-dynamic and dynamic service composition strategies can be made (see Figure 3).

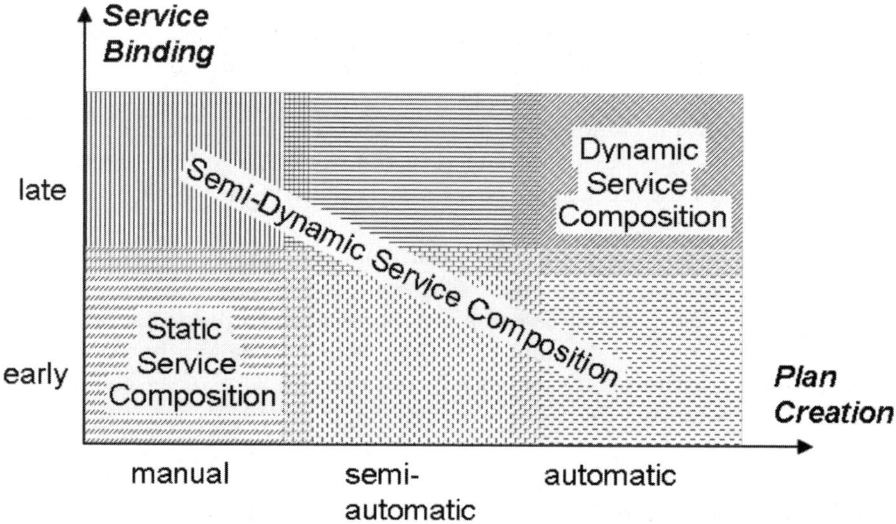

Fig. 3. Classification of Service Composition Strategies

So taking into account the above mentioned attributes cost efficiency, time-to- market and reliability it can be argued that a high degree of dynamics for service composition has positive effects on the profitability of a retailer. On the other hand this does not inherently mean the more automation the better. The degree of dynamics applicable in a real world context is limited by many more factors. Particularly this is true with regard to performance issues, since interacting with a broker for service discovery and matchmaking as well as applying sophisticated AI algorithms for automated plan generation may be time- consuming; for business critical applications probably too time-consuming to be performed at runtime.

In [8] it is stated that "functional optimisation addresses very well the technological part of a problem, but does not pay enough attention to the intentions of the users and the structure and dynamics of social groups and user communities". A retailer applying dynamic service composition usually has limited control over the way how a problem will be solved, respectively how a service will be composed. As a consequence it might become opaque with which business partners the retailer collaborates and shares data in the scope of a process. Trust is the keyword in this respect. The on-demand establishment of security and trust relationships, however, still is an unsolved problem.

5 Technological Implications

Web services are the technology of choice for application integration in cross-organisational and multi-domain environments. There are numerous process definition languages, modelling tools and workflow engines in the Web service world. Among them the Business Process Execution Language (BPEL, [23]) is the most widespread format for defining business processes consisting of simple Web services. Various orchestration tools exist that allow process designers to visually arrange and compose Web services to complex BPEL processes which are deployed and executed in corresponding workflow engines.

Most composition approaches applied in a real-world context use "traditional" Web service technologies, such as WSDL and BPEL. With respect to the above analysed criteria they can be classified as static since the process model is created manually and the services are bound at design time. Furthermore some (although very few) applications apply late binding of services based on fixed interfaces and fixed message formats, thus enabling for semi-dynamic service composition. Technically this is achieved by dividing the WSDL description in two parts, namely the service interface definition (capturing messages, port type etc.) and the service implementation definition. While the service implementation definition determines the access point of a service and differs from provider to provider, the service interface definition is a kind of contract which may be registered to a UDDI registry as a technical fingerprint. Thus within a workflow it is possible to search the registry for services that comply with a given contract.

Approaches exposing a higher degree of dynamics can hardly be found in a real world context. Apart from the obstacles discussed above, such as performance and trust, the reason is clearly related to the fact that Web services just partially kept their promise of being self-contained and self-describing software components. By using the XML-based Web Service Description Language the operations, parameters and the Internet address of a service are described in a human-readable and structured manner. However, for machines respectively for software agents the information provided about a service is barely interpretable because XML lacks a semantic background and WSDL does not define any explicit formal semantics either. As a consequence human interaction is necessary in order to understand what a service does and how it can be invoked.

As it has been discussed in [10], automated service discovery and integration that are not based on keywords or on fixed interfaces but on more flexible matchmaking algorithms are difficult to realise with traditional Web service technologies and frameworks. The functionality and the state change (i.e. preconditions and effects) caused by a service need to be formally defined in order for a software system to be able to create a process model automatically.

Having recognised the potential of the evolving Semantic Web the research community has spawned several activities in the direction of Semantic Web services. With ontology languages like the Web Ontology Language (OWL [19]) machine-understandable Web service descriptions can be created and shared. Basically this is being achieved by annotating the Web services with the concepts that have been formally defined in corresponding domain ontologies. Generic service ontologies, such as OWL Services (OWL-S [33]) and the Web Services Modelling Ontology (WSMO [9]), in combination with appropriate rule languages lay the foundations for semantically describing the functionality and the behaviour of services.

A deeper insight into these frameworks will not be given at this point. However, with respect to the service composition classification given above it can be argued that the higher the desired degree of dynamics, i.e. the closer approaching the upper left corner in Figure 3, the more comprehensive and consequently technologies need to be adopted which support the formal definition and interpretation of service semantics.

6 Conclusions

Based on the RM-ODP and the TINA a business model was presented that explains the actors and relationships in the e-services domain. In particular, the relations between brokers and the retailers of services were discussed, and how the they collaborate in order to compose and trade services. Having analysed business and technical implications of different service composition approaches we can draw conclusions regarding the strategy a retailer should adopt in order to add value to services through composition. With respect to the business model introduced in section 2 it can be said that the number of consumers and the number of 3rd party service providers participating in a business domain or in a business case are the critical factors to be taken into account. Considering the number of consumers and the number of available service, we can distinguish four main cases:

- In the case of many consumers and many service providers, a retailer would usually adopt a composition strategy that includes late binding and in which the trading of services is done considering the whole composition rather than just individual services. Furthermore the creation of composition plans should be supported in a (semi-)automated manner in order to respond to consumer needs in a flexible manner. For instance such market conditions can be found in the tourism domain. A semantic-based framework for semi-dynamic service composition in the tourism domain has been developed in the SATINE project [10].

- If there are just few consumers and many 3rd party providers, special attention needs to be paid to a flexible service binding and efficient service trading. Such combinations are characteristic for business networks that are controlled by a few dominant business partners. For instance in the automotive industry, large manufacturers have subsidiaries that are responsible for providing special product parts (e.g. cable harnesses). These subsidiaries control a great number of competing suppliers. Usually in such networks due to the limited number of consumers the process models are rather static and there is no need for the automated creation of composition plans.
- The field of eGovernmet is a typical application domain in which many service consumers but just few 3rd party providers (e.g. municipalities) can be found. A retailer may offer a one-stop eGovernment portal where consumers (e.g. citizens) may combine a set of services without having to visit several administrations separately. Thus the composition plan strongly depends on the consumers living conditions and should be created (semi-)automatically on-demand. The customised delivery of eGovernment services has been addressed in the work of Medjahed and Bouguettaya [20].
- In the case of just few consumers and few 3rd party providers there is usually no need for the (semi-)dynamic composition of services since the efforts for the implementation of an adequate infrastructure with service discovery and trading capabilities outweigh the gained flexibility benefits.

We hope that this discussion has showed that the Web services community would benefit from the concepts and reference models defined in the area of the RM-ODP before. Based on the given business model, organisations can better plan their efforts and activities when offering compositions and the trading of services.

Acknowledgments

The work of Matthias Fluegge is supported by the European Commission through the project IST-1-002104-STP SATINE (Semantic-based Interoperability Infrastructure for Integrating Web Service Platforms to Peer-to-Peer-Networks).

References

1. Rama Akkiraju, Richard Goodwin, Prashant Doshi, and Sascha Roeder. A Method for Semantically Enhancing the Service Discovery Capabilities of UDDI.

In *Proceedings of the Workshop on Information Integration on the Web*, pages 87–92, August 2003.
2. Boualem Benatallah, Marlon Dumas, Marie-Christine Fauvet, F. A. Rabhi, and Quan Z. Sheng. Overview of Some Patterns for Architecting and Managing Composite Web Services. In *ACM SIGecom Exchanges*, pages 9–16. ACM Press, August 2002.
3. David Booth, Hugo Haas, Francis McCabe, Eric Newcomer, Michael Champion, Chris Ferris, and David Orchard. Web Services Architecture. http://www.w3c.org/TR/ws-arch/, February 2004.
4. Fabio Casati, Ski Ilnicki, Li-Jie Jin, Vasudev Krishnamoorthy, and Ming-Chien Shan. eflow: A platform for developing and managing composite e-services. HP Labs Technical Report HPL-2000-36, HP Software Technology Laboratory, Palo Alto, California, USA, 2000.
5. Roberto Chinnici, Jean-Jacques Moreau, Arthur Rymanan, and Sanjiva Weerawarana. Web Services Description Language (WSDL) Version 2.0 Part 1: Core Language. Technical report, W3C, http://www.w3.org/TR/wsdl20/, 2003.
6. Daniela Berardi and Giuseppe De Giacomo and Maurizio Lenzerini and Massimo Mecella and Diego Calvanese. Synthesis of Underspecified Composite e-Services based on Automated Reasoning. In *Proceedings of the 2nd International Conference on Service Oriented Computing (ICSOC'04)*, New York City, NY, USA, November 2004. ACM.
7. Lill Kristiansen (Ed.). TINA-C Deliverable: Service Architecture. http://www.tinac.com/, June 1997.
8. Petre Dini et al. The digital ecosystem research vision: 2010 and beyond. http://www.digital-ecosystems.org/events/2005.05/de_position_paper_vf.pdf, July 2005.
9. Cristina Feier and John Domingue John Domingue and. D16.1v0.2 The Web Service Modeling Language WSML, WSML Final Draft. Wsmo final draft, DERI International, April 2005.
10. Matthias Fluegge and Diana Tourtchaninova. Middleware supporting automated inter-organisational collaboration. Position Paper, IST COCONET Roadmap Workshop, October 2002.
11. Matthias Fluegge and Diana Tourtchaninova. Ontology-derived Activity Components for Composing Travel Web Services. In *Proceedings of the International Workshop on Semantic Web Technologies in Electronic Business (SWEB'04)*, Berlin, Germany, October 2004.
12. Roy Grønmo and Michael C. Jaeger. Model-Driven Methodology for Building QoS-Optimised Web Service Compositions. In *Proceedings of the 5th IFIP International Conference on Distributed Applications and Interoperable Systems (DAIS'05)*, pages 68–82, Athens, Greece, May 2005. Springer Press.
13. ISO/IEC. ITU.TS Recommendation X.902 — ISO/IEC 10746-1: Open Distributed Processing Reference Model - Part 1: Overview, August 1996.
14. ISO/IEC. ITU.TS Recommendation X.950 — ISO/IEC 13235-1: Trading Function: Specification, August 1997.
15. Michael C. Jaeger, Gero Mühl, and Sebastian Golze. QoS-aware Composition of Web Services: An Evaluation of Selection Algorithms. In *Proceedings of the Confederated International Conferences CoopIS, DOA, and ODBASE 2005 (OTM'05)*, volume 3760 of *Lecture Notes in Computer Science (LNCS)*, pages 646–661, Agia Napa, Cyprus, November 2005. Springer Press.

16. Michael C. Jaeger, Gregor Rojec-Goldmann, Gero Mühl, Christoph Liebetruth, and Kurt Geihs. Ranked Matching for Service Descriptions using OWL-S. In Paul Müller, Reinhard Gotzhein, and Jens B. Schmitt, editors, *Kommunikation in verteilten Systemen (KiVS 2005)*, Informatik Aktuell, pages 91–102, Kaiserslautern, Germany, February 2005. Springer Press.
17. Juhnyoung Lee. Matching Algorithms for Composing Business Process Solutions with Web Services. In *Proceedings of the 4th International Conference on E-Commerce and Web Technologies (ECWEB 03)*, pages 393–402, Prague, Czechoslovakia, October 2003. Springer Verlag.
18. Luiz A. G. da Costa and Paulo F. Pires and Marta Mattoso. Automatic Composition of Web Services with Contingency Plans. In *Proceedings of the IEEE International Conference on Web Services (ICWS'04)*, San Diego, California, USA, July 2004. IEEE CS Press.
19. Deborah L. McGuinness and Frank van Harmelen. OWL Web Ontology Language Overview. Technical report, W3C, http://www.w3.org/TR/owl-features/, 2004.
20. Brahim Medjahed and Athman Bouguettaya. Customized Delivery of eGovernment Web Services. *IEEE Intelligent Systems*, 20(6):77–84, November-December 2005.
21. Microsoft. Enterprise uddi services: An introduction to evaluating, planning, deploying, and operating uddi services, February 2003.
22. Nilo Mitra. SOAP Version 1.2 Part 0: Primer. Technical report, W3C, http://www.w3.org/TR/soap12-part0/, 2003.
23. Assaf Arkin et al. OASIS WS-BPEL TC. WS-BPEL Specification Editors Draft. http://www.oasis-open.org/committees/download.php/12791/wsbpel-specification-draft-May-20-2005.html, December 2005.
24. Object Management Group (OMG). Common object request broker architecture: Core specification. Omg formal document/02-12-06, OMG, Needham, Massachusetts, USA, 2002.
25. Architecture Board ORMSC. Model Driven Architecture. Technical Report ormsc/2001-07-01, Object Management Group (OMG), Needham, Massachusetts, USA, August 2001.
26. Massimo Paolucci, Takahiro Kawamura, Terry R. Payne, and Katia Sycara. Importing the Semantic Web in UDDI. In *Revised Papers from the International Workshop on Web Services, E-Business, and the Semantic Web*, pages 225–236, Toronto, Canada, May 2002. Springer-Verlag.
27. Massimo Paolucci, Takahiro Kawamura, Terry R. Payne, and Katia Sycara. Semantic Matching of Web Service Capabilities. In *Proceedings of 1st International Semantic Web Conference. (ISWC'02)*, pages 333–347. Springer-Verlag, Berlin, 2002.
28. Shankar R. Ponnekanti and Armando Fox. SWORD: A Developer Toolkit for Web service Composition. In *Proceedings of the 11th World Wide Web Conference*, pages 83–107, Honolulu, Hawaii, USA, May 2002.
29. Shuping Ran. A model for web services discovery with QoS. *SIGecom Exch.*, 4(1):1–10, 2003.
30. Jinghai Rao, Peep Küngas, and Mihhail Matskin. Logic-based Web Services Composition: from Service Description to Process Model. In *Proceedings of the IEEE International Conference on Web Services (ICWS'04)*, San Diego, California, USA, July 2004. IEEE CS Press.

31. Evren Sirin, James Hendler, and Bijan Parsia. Semi-Automatic Composition of Web Services Using Semantic Descriptions. In *Proceedings of Web Services: Modeling, Architecture and Infrastructure workshop in conjunction with ICEIS'03*, Angers, France, April 2003.
32. Naveen Srinivasan, Massimo Paolucci, and Katia Sycara. Adding OWL-S to UDDI, Implementation and Throughput. In *Proceedings of Semantic Web Service and Web Process Composition 2004*, San Diego, California, USA, July 2004.
33. The OWL Services Coalition. OWL-S: Semantic Markup for Web Services. Technical report, http://www.daml.org/services/, 2004.
34. David Trastour, Claudio Bartolini, and Chris Preist. A Semantic Web Approach to Service Description for Matchmaking of Services. In *Proceedings of the 11th international conference on World Wide Web (WWW'02)*, pages 89–98, Honolulu, USA, May 2002. ACM Press.
35. UDDI Spec Technical Committee. UDDI Version 3.0.1. http://uddi.org/pubs/uddi-v3.0.1-20031014.pdf, 2003.
36. Martin Yates, Wataru Takita, Rickard Jansson, Laurence Demounem, and Harm Mulder. TINA-C Deliverable: TINA Business Model and Reference Points. http://www.tinac.com/, May 1997.
37. Tao Yu and Kwei-Jay Lin. Service Selection Algorithms for Web Services with End-to-End QoS Constraints. In *Proceedings of the 2005 IEEE International Conference on e-Technology, e-Commerce and e-Service (EEE'05)*, pages 129–136, Hong Kong, China, March 2005. IEEE Press.
38. Liangzhao Zeng, Boualem Benatallah, Anne H.H. Ngu, Marlon Dumas, Jayant Kalagnanam, and Henry Chang. QoS-Aware Middleware for Web Services Composition. *IEEE Transactions on Software Transactions*, 30(5):311–327, May 2004.

Developing a Knowledge-based Intelligent Services System in Sports Websites

Edmond H. Wu[1] and Michael K. Ng[2]

[1] Department of Statistics & Actuarial Science, The University of Hong Kong, Pokfulam Road, Hong Kong. hcwu@hku.hk

[2] Department of Mathematics, Hong Kong Baptist University, Kowloon Tong, Hong Kong. mng@math.hkbu.edu.hk

Abstract. Updated and relevant information retrieving is a critical factor for the success of a knowledge-based system for E-service intelligence. In this chapter, we demonstrate the methodologies as well as technical details on developing a knowledge-based intelligent services system in sports Websites to gain competitive advantages. Under the proposed framework, data integration and processing are automatically performed while effective and easy use of data across the organization is achieved. As an intelligent application, we demonstrate how the system can provide timely services through visitors' dynamic patterns during ongoing sports events. Since the Websites can spend less time responding to users' requests, management efficiency is improved and then more time can be spent on making informed decisions and developing advanced business intelligence and risk management applications based on the system.

1 Introduction

The World Wide Web has truly changed the daily life of people all over the world in the past decade. With the fast growth of the Internet and its Web users all over the world, new challenges are exposed on how to manage and discover useful patterns from tremendous and evolving Web information sources. Also, there is a great demand on designing scalable and flexible solutions for various time-critical and data-intensive Web applications.

Popular Websites attract millions of visitors surfing on the Web. It is interesting to note that the visitors leave huge amounts of Website traversal information in the Web servers, such as Web-logs. It may be transparent to the visitors, however, such data should be fully analyzed by Website administrators because it consists of valuable knowledge about the dynamic navigational behavior and preferences of visitors. The crucial issue for a Website is how to timely acquire the actionable knowledge to improve its services, especially for commercial Websites in the highly competitive environment nowadays.

E.H. Wu an M.K. Ng: *Developing a Knowledge-based Intelligent Services System in Sports Websites*, Studies in Computational Intelligence (SCI) **37**, 535–554 (2007)
www.springerlink.com

Since the development of the WWW, tremendous research has focused on Web mining. The term Web mining was first proposed by Etzioni [6] in 1996. Web mining refers to the application of data mining or other information process techniques to World Wide Web, to find useful patterns from various Web data. Web mining can be divided into three categories: Web content mining, Web structure mining, and Web usage mining.

Web content mining is an automatic process that extracts patterns from on-line information, such as the HTML files, images, or E-mails, and it already goes beyond only keyword extraction or some simple statistics of words and phrases in documents. There are two kinds of Web content mining strategies: Those that directly mine the content of documents and those that improve on the content search of other tools like search engines.

As the Web evolves, its hyperlink structure play a very important role in knowledge discovery. Web structure mining is a research field focused on using the analysis of the link structure of the Web, and one of its purposes is to identify more preferable documents.

Web usage mining is to mine user patterns from Web log records. Web servers record and accumulate data about user interactions whenever requests for resources are received. Analyzing the Web access logs of different Websites can help understand the user behavior and the web structure, thereby improving the design of the structure and content of a Website.

There are two main classification of Web Usage Mining: One is Website usage mining and the other is click-stream mining. Website usage mining methods apply data mining techniques to discover usage patterns and study the effectiveness of a Website structure from Web data collected at a single Website. Click stream mining exploits the sequential patterns of page view requests for classification of users and pages while click-stream mining exploit data collected on Web server side, client-level click-stream mining utilize data collected on the client's machine.

Notably, Web content mining, structure mining and usage mining are not independent fields in Web mining research. Techniques used in one field can also be applied in another field with satisfactory results. Moreover, a combination adoption of these Web mining techniques may achieve better understanding of the potentially useful patterns in Websites. Since our proposed intelligent system is based on Web usage analysis, we first review some of the related work in this area and then state the motivation and the problems we try to solve.

1.1 Related Work in Web Usage Mining

Understanding the user behavior is the first step to provide better Web services. Under this premise, technologies for Web applications, such as user profiling, personalization and recommendation systems are frequently used. Therefore, developing effective and efficient solutions to discover user patterns and then to optimize Web services has become an active research area in Web usage mining.

Web usage mining focuses on the adoption of data mining techniques to discover usage patterns from Web data, in order to understand and better serve the needs of Web-based applications [17]. In [17], J. Srivastva et al. proposed a three-step Web usage mining process which are called preprocessing, pattern discovery, and pattern analysis.

In the preprocessing step, different practical solutions have been proposed. In [3, 4], methods for transaction identification was proposed to find frequent traversal patterns in Websites. Researchers also presented different data mining algorithms for pattern discovery and analysis. For example, Fu et al. [8] suggested a clustering algorithm for grouping Web users from their access sequences. Mobasher et al. [11] used association rules to realize effective Web personalization. Shen et al. [16] suggested a three-step algorithm to mine the most interesting Web access associations. Garofalakis et al. [10] used page popularity to rearrange Website structure to make it more accessible and more effective. Zaiane et al [23] proposed to apply OLAP and other data mining techniques for mining access patterns based on a Web usage mining system.

The field of adaptive Websites is drawing attention from the community [13, 14, 15]. One of the new trends in Web usage mining is to develop Web usage mining system that can effectively discover users' access patterns and then intelligently optimize the Web services. Recent studies [1, 12, 18] have suggested that the structural characteristics of Websites, such as the Website topology, have a great impact on the performance or efficiency of Websites. Hence, combining with structure information of Websites, we can gain more interesting results for Web usage analysis.

Most of the existing Web analysis tools [2, 23] provide mechanisms for reporting user activities in the servers and various data management components, such data filtering and OLAP. Using such tools, for example, it is possible to evaluate the number of accesses to Web servers and the individual access to particular Web spaces. Based on these usage reports, Website masters can also know the times or time intervals of visits, and domain names and the URLs of users of the Web server. However, such reports can only provide summary statistics for analysis instead of actionable knowledge for decision making. Sometimes it is very time consuming to compare and find potential Web usage patterns. What's more, these tools are based on static databases, so they cannot deal with streaming data. Therefore, they cannot perform the tasks for further analysis of click stream data. In this context, such Web usage reporting systems have limitations in time-critical applications, such as real-time user patterns discovery.

The new generation Web usage mining system should be designed to be capable of discovering changing patterns from a data stream environment with multi-type Web sources. By analyzing server access logs and user patterns, we can find some valuable information on how to better structure a Website in order to create a more effective management and presence for the Website and the corresponding organization. Hence, it motivates us that by combining with the Website structure for Web usage analysis.

The rest of the chapter is organized as follows: In Section 2, we present the multi-level data preparation for Web usage analysis. Then, in Section 3, we introduce the patterns we focus on. In Section 4, we suggest the infrastructure of the knowledge-based intelligent E-services system. After that, we demonstrate practical applications of the system through a real case study in Section 5. Finally, we give some concluding remarks in Sections 6.

2 Data processing in Web usage analysis

In this section, we first briefly introduce different levels of data preparation in Web usage analysis and then give an overview of the data processing architecture we propose.

2.1 Web-Log Data Preparation

When people visit a Website, the Website servers automatically register the Web access log records including the URLs requested, the IP addresses of the users and the corresponding timestamps. Web-logs are the most common data used in many Web usage mining applications.

In practical cases, we need to apply some data preprocessing techniques to clean and transform the raw Web-logs into transaction itemsets (user access sessions) or a data cube [21]. Fig. 1 is a sample of Web-log records (the format of the sample Web-log is IIS 5.0, some system information is ignored). After preprocessing of these original Web-log data sets, we can use these user access sessions directly for further pattern discovery and data analysis in Websites.

```
GET /guangao/newsite/otherserver/espnchat.htm
2003-04-08 00:00:03 66.196.72.88 - 211.154.223.18 80
GET /Comment/Newscomment.asp NewsID=9632&TableName=News16
2003-04-08 00:00:08 202.108.250.198 - 221.154.223.18 80
GET /StaticNews/2000-07-25/News20a1638.htm
2003-04-08 00:00:08 61.153.18.234 - 211.154.223.18 80
GET /worldcup/worldcupnew.css
2003-04-08 00:00:09 210.22.5.36 - 211.154.223.18 80
GET /Imager/eye.swf
```

Fig. 1. A Sample of Web-log Data.

2.2 Website Topology Data Preparation

Besides Web-logs, Website structure is another data source containing potentially useful information. Website topology is the structure of a Website. The nodes in a Website topology represent the Web pages with URL addresses and the edges among the nodes represent the hyperlinks among Web pages. Mathematically, a Website topology can be regarded as a directed graph. We assume that each pair of nodes are connect to each other by at least one path, that is, all Web pages in a Website can be visited from each other through at least one path.

Fig. 2 shows an example of a Website topology. All the Web pages are assigned with unique labels. A Website topology contains linkage information among the Web pages. The hyperlinks establish an informative connection between two Web information resources. The original design of a Website topology reflects the Website administrators' expectations of user access patterns. However, it may not be consistent with the actual expectations of visitors. Hence, a Website topology combining with visitors' access tracks can help us to understand the visitors' behavior.

Table 1 is the corresponding connection matrix of the Website topology in Fig. 2. For example, the value '1' of the entry AB represents the presence of a direct hyperlink from A to B; the value '0' of the entry AE represents the absence of a direct hyperlink from A to E. However, there is at least one path from A to E, such as ACHE. Such data matrices can be regarded as Website topology data sources, which are helpful for analyzing user patterns in Websites.

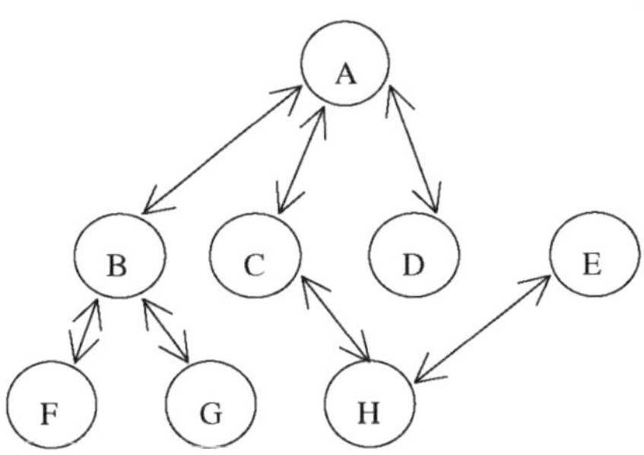

Fig. 2. Website topology

Page	A	B	C	D	E	F	G	H
A	0	1	1	1	0	0	0	0
B	1	0	0	0	0	1	1	0
C	1	0	0	0	0	0	0	1
D	1	0	0	0	0	0	0	0
E	0	0	0	0	0	0	0	1
F	0	1	0	0	0	0	0	0
G	0	1	0	0	0	0	0	0
H	0	0	1	0	1	0	0	0

Table 1. Connection Matrix

Based on the Website topology, we propose a probability model to measure the transition probabilities among the Web pages in a Website topology. Let us first consider the association probability between any two Web pages x_i and x_j in a Website. Suppose a given Website topology G contains n Web pages $X = \{x_1, x_2, \ldots, x_n\}$. We denote the number of outgoing hyperlinks from x_k by h_k for $k = 1, \ldots, n$. When a user finishes browsing the current Web page x_i, he or she may continue to visit one of the h_i Web pages connected to the current Web page or just exit the Website. Therefore, there are $h_i + 1$ choices for the user to select after visiting x_i. In our model, we assume that the probability of visiting x_j after having visited x_i is given by:

$$P(x_j|x_i) = \begin{cases} \frac{w_{i,j}}{h_i+1} & \text{if there is a link from } i \text{ to } j \\ 0 & \text{otherwise.} \end{cases}$$

Here, $w_{i,j}$ is a weighting parameter between x_i and x_j (usually we take $w_{i,j} = 1$). We also define the exit probability to be $P(\text{exit}|x_i) = 1 - \sum_j w_{i,j}/(h_i+1)$. In particular, when $w_{i,j} = 1$ for all i, j, we have $P(x_j|x_i) = P(\text{exit}|x_i) = 1/(h_i+1)$ where x_j and x_i are linked.

If a hyperlink from the pages x_i to x_j exists, then we define the distance between x_i and x_j to be $D(x_i, x_j) = \log(1/P(x_j|x_i)) = \log((h_i+1)/w_{ij})$. Otherwise, we consider a shortest path $x_1^* = x_i, x_2^*, \ldots, x_{m-1}^*, x_m^* = x_j$ from x_i to x_j. The distance between x_i and x_j is defined as $D(x_i, x_j) = \log(1/P(x_1^* x_2^* \ldots x_m^*|x_i))$ where $P(x_1^* x_2^* \ldots x_m^*|x_i)$ is the probability of the shortest path given the starting point x_i. Under the Markov assumption, $P(x_1^* x_2^* \ldots x_m^*|x_i) = \prod_{k=1}^{m-1} P(x_{k+1}^*|x_k^*)$. Thus we may also express the distance measure as $D(x_i, x_j) = \sum_{k=1}^{m-1} D(x_k^*, x_{k+1}^*)$. We remark that the Website topology is a connected graph, thus, there must be a sequence of nodes connecting x_i and x_j. We can employ the classical Floyd algorithm [7] to calculate the shortest paths between any two nodes in a Website topology.

Using the Website topology probability model and the distance measure, we can obtain information about the browsing patterns of users. On the other hand, we can also optimize the topology of a Website by minimizing a combination of the expected number of clicks and the number of hyperlinks in the Website. Therefore, Website topology provides useful information for knowledge discovery in Websites.

2.3 Website Content and Users Data Preparation

In some applications, we are interested in what kinds of users would like to browse which types of Web pages at particular time periods. In such cases, Website content or user information can help us to analyze. For this purpose, we propose a data cube model called PUT-Cube which integrates Website topology, content, user and session information for multiple usage mining tasks. The PUT-Cube is defined as follows:

Definition 1. *A PUT-Cube model is a four-tuple* $\langle P, U, T, \mathcal{V} \rangle$ *where* P, U, T *are the sets of indices for the three main dimensions (Page, User, Time) where*

1. P *is the set of all page-related attributes* $P = P_1, P_2, ..., P_n$*.*
2. U *is the set of all user-related attributes* $U = U_1, U_2, ..., U_m$ *which identifies groups of users or individuals.*
3. T *is the set of all temporal-related attributes,* $T = T_1, T_2, ..., T_r$*, where each* T_i *describes the occurrence time or duration of user accesses.*
4. $\mathcal{V}$ *is a bag of values of all attribute combinations.*

The PUT-Cube model focuses on the most important factors in Web usage mining. If we also consider the Website topology, the page factor not only provides information about which pages have been accessed, but also provides the information about their access order and relative positions in Websites. The user and time factors suggest who and when is involved in the Web access events. Therefore, based on the 'who', 'when' and 'which' user access information from the PUT-Cube, we can discover more useful user patterns, and then try to explain 'why'. This mining process and results are desired by domain experts because the cube model concentrates on the most important factors for analyzing user behavior. The PUT-Cube model also provides the flexibility of selecting relevant dimensions or attributes for analysis. Using the PUT-Cube, we can perform data integration over multiple Web data resources in a compact data model.

2.4 Levels of Web Data Preparation

In order to meet different data requirements of various data mining tasks, we need to prepare different levels of data resources from the massive amount of Web data available. Fig. 3 shows a four-level data processing architecture we propose for effective Web usage mining.

The first level consists of the raw Web data, including Web-logs, Website content and structural information etc. These data are usually stored in Web servers. In the second level of data preparation, we adopt some data cleaning and filtering techniques [21] to convert the original data into Website topology and access session datasets. Data in the first two levels contain the most complete information of historical data. However, due to the massive amount of the data in the elemental levels, the data preparation will usually be proceeded off-line.

In the third level of data preparation, we will integrate different kinds of Web data in the first two levels. In order to effectively reduce the data sizes, we will use a discretization model to transform different types of data values and employ the PUT-cube model to aggregate the access information. In the fourth level, we can direct use the data prepared in the third level to perform higher level data clustering for pattern detection or reduction. The aggregate data can be used to support intelligent Web applications, such as dynamic Web usage monitoring, Web prefetching, and Website optimization.

Since the modest sizes of the data prepared in the last two levels, the meta data may be stored in memory. It greatly accelerates the mining process for some on-line data analysis and knowledge discovery applications, which is also our motivation of this chapter. Because in some particular Websites with dynamic nature, such as sports, entertainment, or news Websites, a large proportion of the Website contents will be updated everyday, even intraday, how to timely detect and forecast visitors' changing interests is the key factor to succeed for such Websites. Therefore, we try to explore some practical solutions to improve the users interaction to Websites by adopting the proposed multilevel Web data processing scheme.

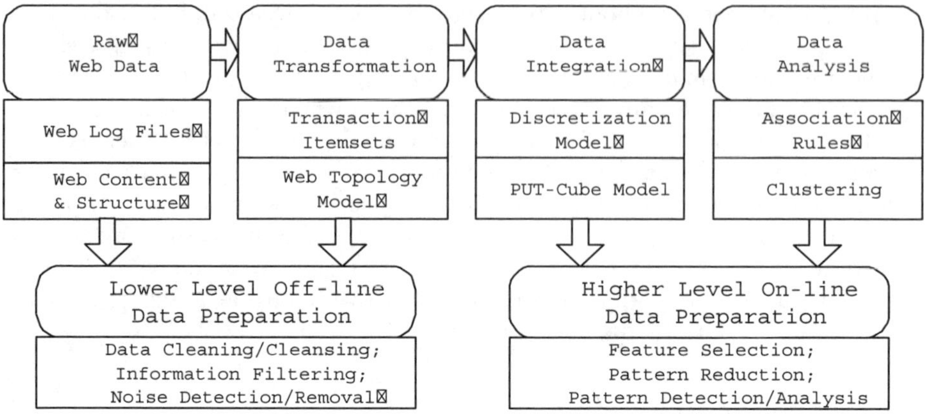

Fig. 3. Levels of Data Preparation for Web Usage Analysis

3 Dynamic Web Usage Monitoring and Pattern Discovery

In this section, we present the methods we develop and give some technical details on how to monitor the evolving Web usage and discover useful user patterns based on the ready-prepared Web data.

3.1 User Access Aggregation

From the access sessions, we first need to aggregate the user accesses to quantitatively measure the frequencies of Website visiting.

A data structure for aggregating access sessions is as follows: given a Website topology $G = \{V, E\}$, where $V = \{V_1, V_2, ..., V_n\}$ represents the page set and $E = \{E_1, E_2, ..., E_m\}$ represents the link set. Also, given a user access session dataset $D = \{S_1, S_2, ..., S_n\}$, for each access session $S \in D$, $S = \{P_1, P_2, ..., P_r\}$, where $P_i \in V$, $i = 1, ..., r$. We set $C_{ij} = Count(P_j|P_i)$ to indicate the number of accesses from P_i to P_j in the access session dataset D.

Therefore, there are two cases of such counting, one is that P_i and P_j are adjacent in session S which also means the link of $P_i, P_j \in E$. The other case is that P_i and P_j are not adjacent which means there exists at least a page between P_i and P_j. For the first case, we will count it whenever it appears. As to the second case, we will only count once in the access session.

For example, given access session $S = \{A, B, C, B\}$. As to the first case, we will count AB, BC, and CB once since they are adjacent in the session. As to the second case, we count AC once. But we do not count AB again since it has been counted once in the first case. So, $Count(B|A) = 1$, $Count(C|B) = 1$, $Count(B|C) = 1$, $Count(C|A) = 1$.

Using the page frequency counting model, we can add another page attribute $C_{ij} = Count(V_j|V_i)$ to indicate the access frequency from page V_i to V_j.

3.2 Indicators for Access Patterns Discovery

Then, we further propose two numerical indicators which measure the unusualness of a pattern.

Given access matrix C and topology probability matrix P, the Access Distinctness (AD) is defined as below:

$$AD(i,j) = \ln\left(\frac{C_{ij}}{P_{ij}} + 1\right)$$

where C_{ij} is the number of accessing from Web page V_i to V_j, and P_{ij} is the probability from V_i to V_j.

If the AD value is high, it means that the pattern is a frequent user access pattern, but the access pattern is relatively hard to access in original Website topology. A pattern may raise our concerns if its AD value is higher than most other patterns.

Usually, visitors will spend more time on the Web pages they interested in. Hence, the temporal factor can help to analyze user patterns. Thus, we can use access matrix C, topology probability matrix P, and page staying time T as the major factors in a new index to identify really interesting access patterns. So, an interestingness index named Access Interest (AI) as below:

$$AI(i,j) = \ln\left(\frac{C_{ij}T_{ij}}{P_{ij}} + 1\right)$$

where C_{ij} is the number of accesses from Web page V_i to V_j, P_{ij} is the topological probability from V_i to V_j and T_{ij} is the total staying time of visiting V_i and V_j. Here, we define $T_{ij} = T_i + T_j$, where T_i and T_j are the average staying time on page V_i and V_j, respectively.

3.3 Automatic Pattern Detection and Summation

Based on the indicators as well as other statistical measures, we suggest how to find the informative patterns. There are three levels (discretization, dense regions, dense clusters) we concern in the pattern discovery. We briefly introduce them as follows:

Definition of Discretization

As a key step to ensure data and information quality, discretization is defined as a process that divides continuous numeric values into a set of intervals that can be regarded as discrete categorical values [9]. In [5], Dougherty *et. al.* suggested three different axes to classify discretization methods: supervised vs. unsupervised, global vs. local and static vs. dynamic. Supervised methods use the information of class labels while unsupervised methods do not. Global methods are applied before the learning process while local methods produce partitions that are applied to localized regions of the instance space. The difference between static and dynamic methods is that in static methods, attributes are discretized independently of each other, while dynamic methods take into account the interdependencies among the attributes.

Definition of Dense Regions

We then introduce the concept of dense region discovery. We now fix some notations and give the definition of dense regions. Let R and C be an index set of a subset of rows and columns of X respectively. Since we do not distinguish between a matrix and its permuted versions, we assume that R and C are sorted in the ascending manner. A submatrix of X formed by its rows R and columns C is denoted by $X(R,C)$. We also identify $X(R,C)$ by the index set $D = R \times C$. For example, let $R = \{3\}$ and $C = \{1,2\}$, then $R \times C = X(R,C) = (x_{31}x_{32})$.

Definition 2 (Dense regions (DR)). *A submatrix $X(R,C)$ is called a maximal dense region with respect to v, or simply a dense region with respect to v, if*

- *$X(R,C)$ is a constant matrix whose entries are v (density), and,*
- *Any proper superset of $X(R,C)$ is a non-constant matrix is non-constant (maximality).*

For example, let X be a data matrix given by the first matrix below. Then, the dense regions of X with respect to 1 are given by the four matrices in brace.

$$\begin{pmatrix} 1\,0\,0\,1 \\ 1\,1\,0\,1 \\ 1\,1\,0\,1 \\ 1\,1\,2\,0 \end{pmatrix}; \quad \left\{ \begin{pmatrix} 1 * * * \\ 1 * * * \\ 1 * * * \\ 1 * * * \end{pmatrix}, \begin{pmatrix} 1 * * 1 \\ 1 * * 1 \\ 1 * * 1 \\ * * * * \end{pmatrix}, \begin{pmatrix} * * * * \\ 1\,1 * 1 \\ 1\,1 * 1 \\ * * * * \end{pmatrix}, \begin{pmatrix} * * * * \\ 1\,1 * * \\ 1\,1 * * \\ 1\,1 * * \end{pmatrix} \right\}.$$

Alternatively, we may denote the above dense regions by $\{1,2,3,4\} \times \{1\}$, $\{1,2,3\} \times \{1,4\}$, $\{2,3\} \times \{1,2,4\}$ and $\{2,3,4\} \times \{1,2\}$ respectively.

From the definition of dense regions, we can see that finding dense regions has a practical meaning in knowledge discovery. For example, if the rows present visitor IDs and the columns present Web pages IDs, then a dense region is a group of visitors who browse the same set of Web pages they are interested in. Such patterns can be used in analyzing user behavior.

Note that the notion of 'dense regions' does not represent the sets of data points that are close to each other in distance. It denotes the sub-matrices with common values. In [22], we present an algorithm for mining dense regions in large data matrices. In this chapter, we just employ the algorithm and omit a detailed introduction of it.

A limitation of existing dense regions discovery algorithms is that they assume the discrete values of entries in matrices. Therefore, in order to handle data with continuous values, we need to perform data discretization in the preprocessing stage before mining dense regions.

Definition of Dense Clusters

Based on the dense regions, we present a clustering model for dense regions reduction. Here, we use $|D|$ to represent the total number of entries in a dense region D or a set of dense regions. We first give the definitions of two measures of dense regions.

Definition 3 (Dense Region Pairwise Overlap Rate). *Given two dense regions D_i and D_j, the Dense Region Pairwise Overlap Rate (DPOR) of D_i is defined as the ratio:*

$$DPOR(D_i, D_j) = \frac{|D_i \bigcap D_j|}{|D_i|} \tag{1}$$

Definition 4 (Dense Region Union Overlap Rate). *Given a set of DRs $\mathcal{D} = \{D_1, ..., D_n\}$, the Dense Region Union Overlap Rate (DUOR) is defined as the ratio:*

$$DUOR(\mathcal{D}) = \frac{|\bigcap_{i=1}^{n} D_i|}{|\bigcup_{i=1}^{n} D_i|} \tag{2}$$

Here, we use $DPOR$ and $DUOR$ to measure the extent of association (overlap) among different dense regions. Based on them, we give the definition of dense clusters as follows:

Definition 5 (Dense Clusters (DC)). *Given a set of dense regions $\mathcal{D} = \{D_1, ..., D_n\}$, a Dense Cluster $\mathcal{DC} = \bigcup_{i=1}^{k} D_i$ is defined as a subset of $\mathcal{D}$ with k DRs such that:*

- *For any $D_i \in \mathcal{DC}$, $DPOR(D_i, \mathcal{DC}) \geq MinDPOR$ and for any $D_j \notin \mathcal{DC}$ but $D_j \in \mathcal{D}$, $DPOR(D_j, \mathcal{DC}) < MinDPOR$, where $MinDPOR$ is the minimal threshold of $DPOR$.*
- *For $\mathcal{DC}$, $DUOR(\mathcal{DC}) \geq MinDUOR$, where $MinDUOR$ is the minimal threshold of $DUOR$.*

Besides MinDPOR and MinDUOR, we also set a threshold MinDC (the minimal size of a dense cluster) to restrain the size of the dense clusters found by a clustering algorithm. It means that for any dense cluster, the total number of entries $|\mathcal{DC}| = |\bigcup_{i=1}^{n} D_i| \geq MinDC$. The benefit of setting MinDC is that we can filter out trivial clusters which are not so useful to analyze data patterns. In [19], we propose a clustering algorithm to find the dense clusters by grouping overlapping dense regions together. Therefore, this clustering method combining with data discretization can be used as a pattern detection and reduction solution for Web usage mining. For example, track the changes of user access patterns in Websites over time by comparing the clusters found from different time periods.

From the above description, we see that discretization is the preliminary data processing stage for informative dense regions or dense clusters discovery. The original data matrices and discretization results are the lower level information for data collection whereas dense regions and dense clusters are higher level information for knowledge discovery. In the case study, we try to find interesting patterns from user browsing behavior based on dense clusters. However, the intelligent system can also adopt other pattern recognition or data mining algorithms to meet the needs of specified applications.

4 The Infrastructure of the Web Intelligent System

The Infrastructure of the Knowledge-based Website Intelligent System we propose is shown in Fig 4. It consists of five interactive components: The requirements of Website users, the requirements of Website staffs, the Web data preparation unit, the Web usage monitoring unit, and the knowledge-based core system integrating the functions of other components consistently with a knowledge discovery process.

The primary aspect we respect is the human factor, which also meets the mission of human-orientation for most Websites. For this purpose, we need to have clear definitions of the requirements and then aim to develop an intelligent system that can satisfy the requirements of Website visitors as well as the Website staffs on a regular basis. For visitors, they hope to experience interactive, personalized, and recommended Website services in a timely manner. Therefore, the major function of the system is how to accurately and effectively deliver their requests to the services providers.

On the other hand, Website staffs can develop a new knowledge-based working environment through the updated information and knowledge acquired from the intelligent system. In the traditional organizational structures of commercial Websites, senior managers or editors allocate resources and tasks to junior staffs. However, this top-down and one-way working style lacks of flexibility and relies on subjective experience. With the assistance of the intelligent system, senior managers can identify the popular services and decide the new services to explore. Meanwhile, with the information and knowledge sharing, other staffs can improve the quality of their work by better targeting the needs of potential customers.

We have introduced the Web data preparation unit in Section 2. The major function of this unit is to provide quality data and information to the intelligent system. The Web usage monitoring unit includes the indicators as well as other system profile we need to monitor. This unit provides real-time analysis of user patterns and Website performance. It can raise warning signals when connectivity problems occur and then prevent system failure.

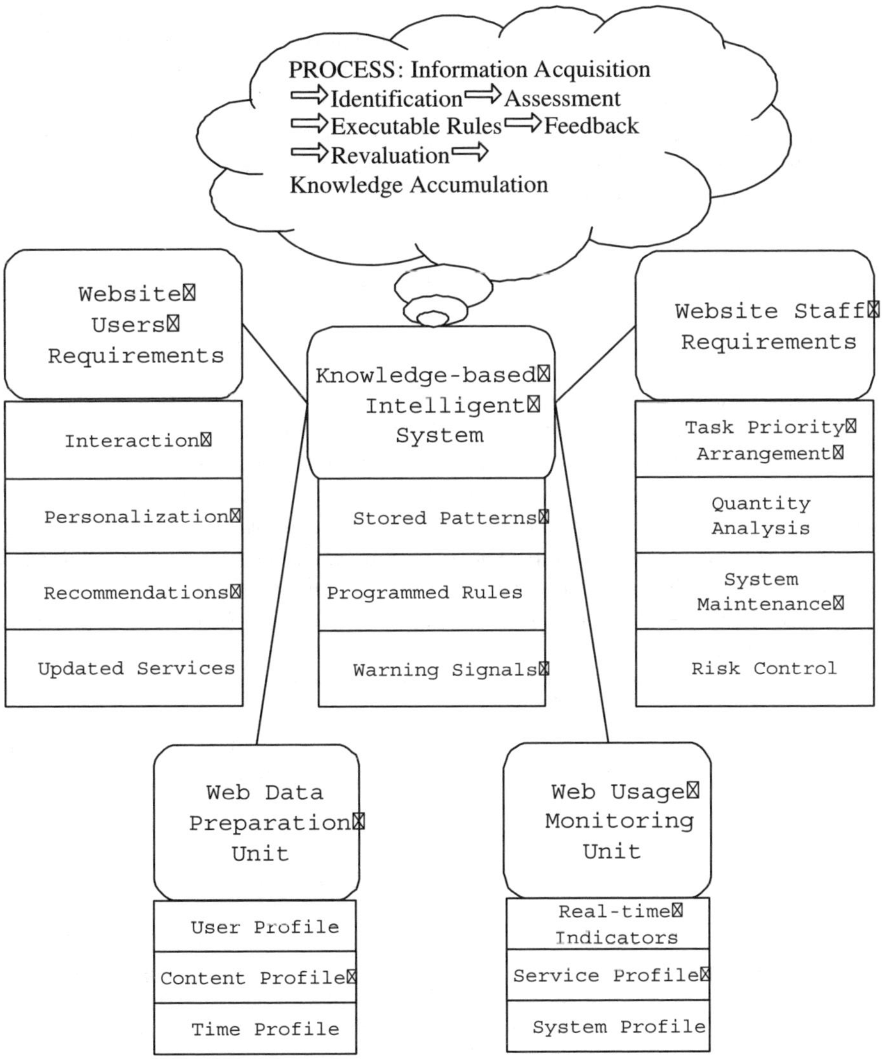

Fig. 4. The Infrastructure of the Knowledge-based Website Intelligent System

The core of the Knowledge-based Intelligent System plays a key role to acquire information from other components and produce actionable knowledge to improve Website services. It stores the current and historical user patterns. Based on the patterns, some programmed rules derived from the Website expertises can be executed if conditions satisfy. Thus, it can improve the efficiency of services when regular patterns occur. Notably, the intelligent system cautiously follows a knowledge discovery process. First, identify and assess the emerging patterns. Then, if feasible, attempt to solve it by existing rules. Otherwise, submit the pattern and related information to experts. The feedback of the solution will be revaluated before it is stored into the knowledge base. Also, the system allows human interaction and manual operations if needed. Therefore, the intelligent system can be self-evolving and it permits further extensions under the infrastructure. In the next section, a real case study will show how the system can intelligently discover patterns to support decision making of E-services enhancement.

5 Capturing Dynamic User Patterns in a Sports Website

5.1 Problem Statement

Nowadays, with the rapid growth of multimedia channels, such as TV broadcast and Internet, more and more people can watch various sporting activities and access related information in a timely manner. For some world-class sporting events (e.g., Olympic games, World Cup), billions of audiences all over the world pay close attention to the matches during the whole time period, which suggests the huge market potential for Websites to explore their business by providing online interactive sporting information services.

Since there are different kinds of sporting events from time to time, in order to attract visitors, a sports Website needs to update their content very fast. Also, some unexpected affairs or results often occur in some sporting events which can cause the user access patterns change drastically. Therefore, the dynamic nature of sports events and visitors' interests determines that how to accurately understand and fast response their needs is a critical issue for a sports Website.

5.2 The Activities and Services Dimensions

In Fig 5, we demonstrate the User-Website interaction process of a portal sports Website with the design of the knowledge-based intelligent E-services system. Based on visitors' interests and domain experts' suggestions, we classify the sports Website's content and services as well as visitors' main browsing activities into five dimensions.

When users enter the sports Website, most of them will choose the topics they prefer. Some users have particular interests in certain kind of sports, e.g., football fans who are persistent visitors will be keen on football related information, not only news, but also match statistics, fans' discussion etc. However, some users will only be active when some important sporting events happen, e.g., Yao ming's Fans will be attractive by the news about the ongoing progress of the NBA season-off. Therefore, they can be regard as event-driven visitors.

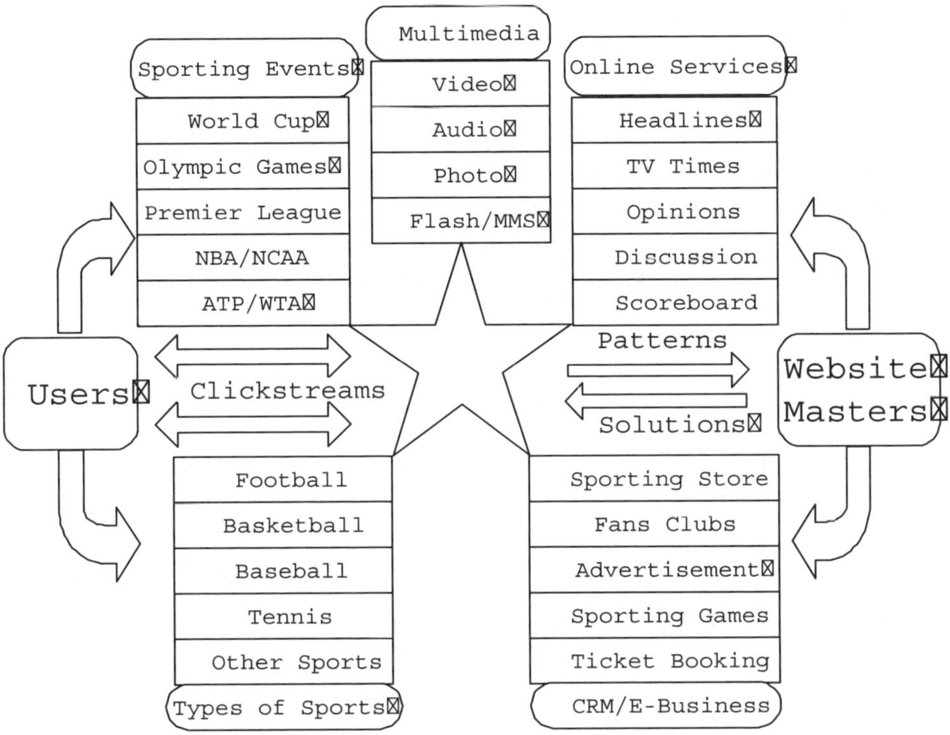

Fig. 5. Patterns and Services Mining in Sporting Websites

The sources of Multimedia data are also of great interests to analyze potentially meaningful patterns. For instance, Zhu *et. al.* [24] proposed some intelligent solutions to detect events and mine associations from sports video data. In sports Websites, some visitors prefer to read text news, some people prefer to watch match video review (e.g., the highlight video chips available in NBA.com) while some people would like to hear the audio content from sports commentators (e.g., the on-line or off-line audio chips available in ESPNSTAR.com.cn). Therefore, the visitors can also be classified by the type of multimedia content they access.

Since TV broadcast cannot provide interactive exploration for audience, more and more people would like to give comments in Websites' discussion forums. With the same reason, many people have great interests in reading the opinions and analysis by professional sports commentators. Hence, we also suggest the interactive online services dimension for visitors' preference analysis.

Some sports Websites, such as NBA.com, also provide E-Business services to promote vendors' sports related products or services. Moreover, advertisement is also a major source of revenue for a sports Website. Therefore, building the E-Business dimension is also quite necessary for a new-generation sports

Website. The star in Fig 5 suggests that the five analytical dimensions mentioned above for dynamic Web usage analysis are highly correlated.

Then, we can employ our intelligent system to detect the changing user access patterns based on these dimensions. By know this information, we can immediately provide the services that can best match visitor's current browsing activities. Improving the real-time User-Website interaction and communication is also the primary requirement of quality E-services.

5.3 Monitoring the User Patterns during Sporting Events

We are interested in detecting the user patterns during different time periods of a sporting event. There are two reasons for this: first, we have observed that during important sporting events, the number of visitors and accesses to a sports Website is significantly larger than normal time periods. The much more frequent requests to Web servers will make the systems unstable, which is a negative influence to the Website. Second, we also find that during sporting events, user access patterns are fast changing, if the Website cannot provide timely information or services that the visitors want to have, the consequence is the lose of visitors. Therefore, our objective is trying to detect user patterns as well as improve the system performance during sporting events.

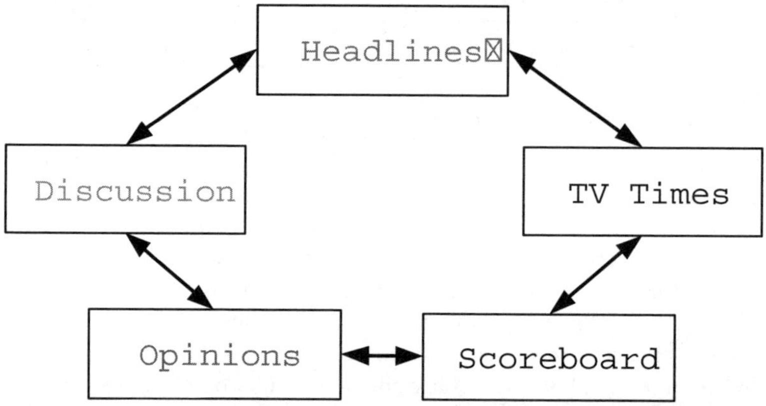

Fig. 6. Analysis of the Time-dependent and Event-driven Services

More specifically, we intent to detect the changing user patterns before, during and after a sporting event. We employ the recent data from ESPN-STAR.com.cn which contains the user access data from March to April, 2005. We select 10 English Premier League matches in this period as the sporting events we investigate. Normally, a football match will last 2 hours, including halftime. In this case, we want to study the changes of user patterns to the Website's five services (Headlines, TV Times, Scoreboard, Opinion, and Discussion which are shown in Fig 6) in the three time intervals (2 hours before the matches, 2 hours during the matches, and 2 hours after the matches).

The intelligent system automatically identifies the cluster patterns for further analysis.

Before we see the results, we first present the basic statistics of Website visiting during the periods. Fig 7 shows the average number of visitors and access requests during the 6 hours periods of the matches. We can see that there are more visitors before the matches than during the matches, it can be explained that some visitors leave to watch live TV broadcast of the matches. For the same reason, after the matches, increasing visitors come to the Website to access latest information.

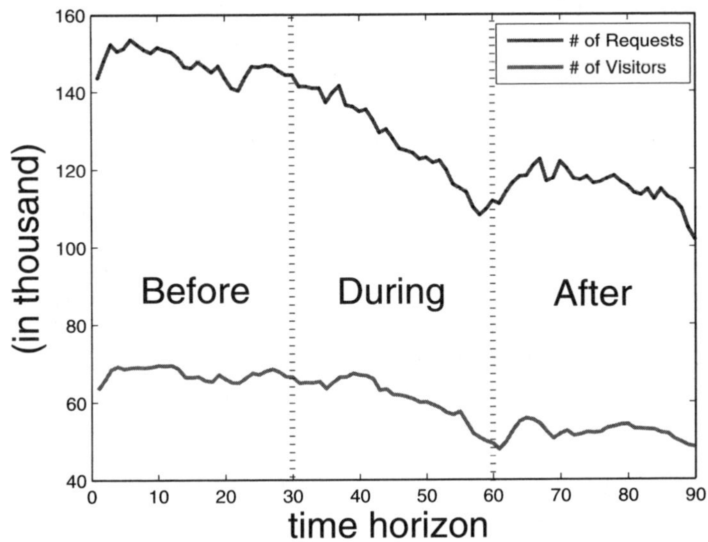

Fig. 7. The Changing Visitors & Requests

Then, we employ the system to detect the active user access patterns. We define an active access pattern as the corresponding Web pages which have more user requests than 90% of total pages. In Fig 8, we can see the clustering results in the three intervals, respectively. Here, we set the minimal size of the DRs to be found is 5×5. The thresholds for DCs are MinDUOR=0.6, MinDPOR=0.6, MinDC=50. The clusters represent the active access patterns in the five services, the more clusters to be found, the more active the service is.

For Headlines, there are not significant differences. However, we detect that there is much more visitors going to browse the TV Times section before a match starts. It can be explained that people would like to know the exact time-schedule and TV channel they can watch the match. It is somewhat surprising that much more people would like to browse the scoreboard after the matches. Maybe many people are eager to know how the match results affect the ranking of their favorable football teams. Moreover, we detect that more visitors prefer to read commentator's opinion before the matches and discuss in the public forums during and after the matches.

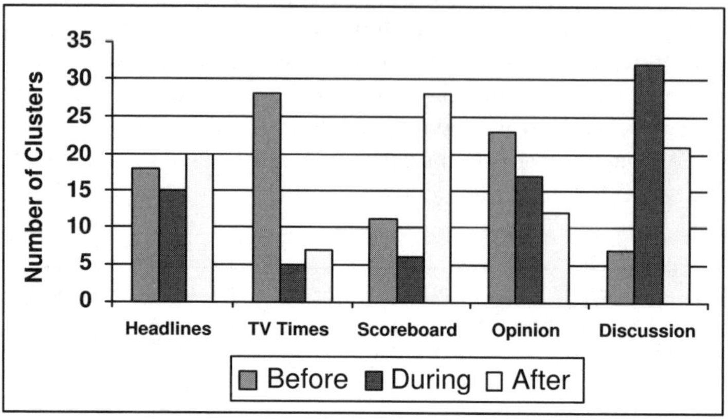

Fig. 8. The Changing User Patterns

Based on these user patterns, we can reorganize the Website's content and services for the easy accesses by visitors. For example, we can put a match's TV Times in an attractive homepage position before a match and update the scoreboard quickly after a match. Also, to promote the opinion and discussion community services and attract more visitors, the Website administrators can invite the commentators to discuss with the fans online during a match to raise some interesting discussion, especially during the halftime periods. Such improvement of services arrangement can greatly increase visitors' satisfaction and loyalty to the sports Website.

Moreover, such expertise knowledge can be coded into the intelligent system so that whenever similar patterns occur, the system can smartly provide relevant and interesting Web services to target customers. The feedbacks from Website staffs also indicate that the system can push up their working efficiency by providing them accurate and timely guidelines of customer requests.

In summary, the integration of the intelligent system into daily operations of the Website can upgrade the E-services offered by a sports Website.

6 Conclusion

In this chapter, we have developed a knowledge-based intelligent system for improving the E-services of sports Websites. It can detect potential patterns of users in timely manner. The case study shows that systems based on the proposed infrastructure will be adaptive, effective and efficient to support the online intelligent services, decision making as well as daily operations. It can be employed in various Web applications, such as personalized recommendations, intelligent search of interesting contents and E-commerce promotions etc.

7 Acknowledgment

The authors would like to thank Joshua Huang and Jian Li for their valuable suggestions and supports in this chapter. The research is supported in part by RGC Grant Nos. HKU 7046/03P, 7035/04P, 7035/05P, and HKBU FRGs.

References

1. Bettina Berendt, Bamshad Mobasher, Miki Nakagawa, and Myra Spiliopoulou, The Impact of Site Structure and User Environment on Session Reconstruction in Web Usage Analysis, *Proceeding of the WEBKDD 2002 Workshop*, Canada, 2002.
2. Robert Cooley, Pang-Ning Tan, and Jaideep Srivastava. Websift: The web site information filter system.*In Proceedings of the Web Usage Analysis and User Profiling Workshop*, 1999.
3. M. S. Chen, J. S. Park and P. S. Yu, Efficient Data Mining for Path Traversal Patterns in Distributed Systems, Proc. of the 16th IEEE Intern'l Conf. on Distributed Computing Systems, May 27-30, 1996, pp. 385-392.
4. M. S. Chen, J. S. Park and P. S. Yu, Efficient Data Mining for Path Traversal Patterns, IEEE Trans. on Knowledge and Data Engineering, Vol. 10, No. 2, pp. 209-221, Arpil 1998.
5. J. Dougherty, R. Kohavi and M. Sahami, *Supervised and Unsupervised Discretization of Continuous Features*, Proceedings of International Conference on Machine Learning, Tahoe City, CA, 1995, pp. 194-202.
6. Oren Etzioni, The World Wide Web: quagmire or gold mine?, Communications of the ACM, vol.39, no. 11, Nov, 1996, pp.65-68.
7. Robert W. Floyd, Algorithm 97: Shortest path, Communications of the ACM, v.5 n.6, p.345, June 1962.
8. Y. Fu, K. Sandhu, and M. Shih, Clustering of Web Users Based on Access Patterns, International Workshop on Web Usage Analysis and User Profiling (WEBKDD'99), San Diego, CA, 1999.
9. J. Gama, L. Torgo and C. Soares, *Dynamic discretization of continuous attributes.* In Proceedings of the Sixth Ibero-American Conference on AI (1998), pp. 160–169.
10. John D. Garofalakis, Panagiotis Kappos, Dimitris Mourloukos: Website Optimization Using Page Popularity. *IEEE Internet Computing* **3**(4): 22-29, 1999.
11. Bamshad Mobasher, Honghua Dai, Tao Luo, Miki Nakagawa, Yuqing Sun, Jim Wiltshire (2000). *Discovery of aggregate usage profiles for Web personalization.*
12. Miki Nakagawa, Bamshad Mobasher, A Hybrid Web Personalization Model Based on Site Connectivity, *WEBKDD*, 2003.
13. M. Perkowitz and O. Etzioni, Adaptive Websites: an AI Challenge, *In Proceedings of the Fifteenth International Joint Conference on Artificial Intelligence*, 1997.
14. M. Perkowitz and O. Etzioni, Adaptive Websites: Automatically Synthesizing Web Pages, *In Proceedings of the Fifteenth National Conference on Artificial Intelligence*, 1998.

15. M. Perkowitz and O. Etzioni, Adaptive Websites: Conceptual cluster mining, *In Proc. 16th Joint Int. Conf. on Artificial Intelligence (IJCAI99)*, pages 264-269, Stockholm, Sweden, 1999.
16. L. Shen, L. Cheng, J.Ford, F.Makedon, V. Megalooi-konomou, T. Steinberg,*Mining the most interesting web access associations*, Proc. the 5th International Conference on Knowledge Discovery and Data Mining (KDD'99) (1999) pp.145-154
17. J. Srivastava, R. Cooley, M. Deshpande, and P. N. Tan. Web Usage Mining: Discovery and applications of usage patterns from web data.*SIGKDD Explorations,* 1:12-23, 2000.
18. Edmond H. Wu, Michael K. Ng, *A graph-based optimization algorithm for Website topology using interesting association rules*, Proc. of the Seventh Pacific-Asia Conference on Knowledge Discovery and Data Mining (PAKDD 2003), Korea, 2003.
19. Edmond H. Wu, Michael K. Ng, Andy M. Yip, Tony F. Chan: A Clustering Model for Mining Evolving Web User Patterns in Data Stream Environment. IDEAL 2004: 565-571.
20. Edmond H. Wu, Michael K. Ng, and Joshua Z. Huang, *On improving website connectivity by using web-log data streams*, Proc. of the 9th International Conference on Database Systems for Advanced Applications (DASFAA 2004), Korea, 2004.
21. Q. Yang, J. Z. Huang and M. K. Ng, *A data cube model for prediction-based Web prefetching*, Journal of Intelligent Information Systems, 20:11-30, 2003.
22. Andy M. Yip, Edmond H. Wu, Michael K. Ng, Tony F. Chan, *An efficient algorithm for dense regions discovery from large-scale data stream*, Proc. of the 8th Pacific-Asia Conference on Knowledge Discovery and Data Mining (PAKDD2004), 2004.
23. Osmar R. Zaiane, Man Xin, Jiawei Han, *Discovering Web Access Patterns and Trends by Applying OLAP and Data Mining Technology on Web Logs*, in Proc. ADL'98 (Advances in Digital Libraries), Santa Barbara, April 1998.
24. X. Zhu, X. Wu, Ahmed K. Elmagarmid, Z. Feng, and L. Wu, Video Data Mining: Semantic Indexing and Event Detection from the Association Perspective, IEEE Transactions on Knowledge and Data Engineering, 17(2005), 5: 665-677.

Developing a Model Agent-based E-Commerce System

Costin Bădică[1], Maria Ganzha[2,4], and Marcin Paprzycki[3,4]

[1] University of Craiova, Software Engineering Department, Bvd.Decebal 107, Craiova, RO-200440, Romania, badica_costin@software.ucv.ro
[2] Elbląg University of Humanities and Economy, ul. Lotnicza 2, 82-300 Elbląg, Poland,
[3] Computer Science Institute, SWPS, ul. Chodakowska 19/31, 03-815 Warsaw, Poland,
[4] System Research Institute Polish Academy of Science, ul. Newelska 6, 01-447 Warsaw, Poland, {Maria.Ganzha, Marcin.Paprzycki}@ibspan.waw.pl

Abstract

It is easy to realize that goals set behind a large class of *agent systems* match these put forward for systems defined as *e-service intelligence*. In this chapter we describe a model agent-based e-commerce system that utilizes rule-based approach for price negotiations. Furthermore, the proposed system attempts at mediating the apparent contradiction between agent mobility and intelligence.

1 Introduction and Overview

Recently an increasing interest in combining Internet-based electronic services (e-services) with "intelligent" functions can be observed (these new e-services are often called *e-service intelligence*). While this particular trend is relatively new, creation of intelligent distributed systems in form of software agents can be traced back a least to the seminal paper of P. Maes [29]. While her main concern was development of an infrastructure dealing with information overload, further research concerned applications of software agents in a number of areas including e-government, e-learning, e-shopping, e-marketing, e-banking, e-logistics etc. There, software agents are to facilitate much higher quality information, personalized recommendation, decision support, quasi-direct user participation in organizational planning, knowledge discovery etc. When developed and implemented, agent systems are to be adaptive, personalized, proactive and accessible from a broad variety of devices [40]. It is therefore easy to see how software agents, and agent systems in general, can be viewed as an incarnation of e-service intelligence.

While there exist a large number of attempts at developing agent-based systems, they are mostly small-scale demonstrator systems—later described in academic publications. Separately, some applications utilize the agent metaphor, but not existing agent tools and environments. Finally it is almost impossible to find out if actual

C. Bădică et al.: *Developing a Model Agent-based E-Commerce System*, Studies in Computational Intelligence (SCI) **37**, 555–578 (2007)
www.springerlink.com

agent systems exist in the industry; e.g. the true role of the Concordia agent system within the Mitsubishi Corp, or the extent to which software agents are used within Daimler-Chrysler. While a number of possible reasons for this situation have been suggested (for instance see [31, 32]), one of them has been recently dispelled. It was shown that modern agent environments (e.g. JADE [21]), even when running on an antiquated hardware, can easily scale to 2000 agents and 300,000 messages [14, 15]. Thus it was experimentally established that *it is possible to build, and experiment with, large-scale agent systems*. Therefore, it is extremely important to follow the positive program put forward by Nwana and Ndumu [31] and focus on developing and implementing such systems.

One of the well-known applications where software agents are to play an important role is e-commerce. Modern agent environments (such as JADE) can support implementation of quasi-realistic model e-commerce scenarios. Moreover, advances in auction theory have produced a general methodology for describing price negotiations. Combination of these factors gave new impetus to research on automating e-commerce. In this context, autonomous, and sometimes mobile, software agents are cited as a potentially fruitful way of approaching e-commerce automation [25].

Since *autonomy* is a broad concept that can be defined in many ways, we would like to narrow it down and focus on *adaptability* viewed as ability to update the negotiation "mechanism" to engage in unknown in advance forms of price negotiations. Obviously, another aspect of autonomy is *decision autonomy* that can be understood as capability to reason over past experiences and domain knowledge in order to maximize "utility" (making it very closely related to "intelligence").

Finally, the notion of agent *mobility* refers to its capacity to migrate from one computer to another. While the goal of such a migration is typically related to acting on behalf of some software or human entity, it does not depend on the intelligence that agents are possibly equipped with. However, to be able to facilitate e-service intelligence, we have to be able to combine the two—as mobile agents have to be able to dynamically adapt to situations found within visited sites. Therefore, agent mobility requires transfer of code, data, process and authority between machines. This makes mobile intelligent agents very heavy [40] and later in this chapter we discuss a partial solution of this problem.

In our work we have been developing a skeleton system in which autonomous agents interact in a way that models realistic scenarios arising in an e-marketplace (for a summary of our early results see [19] and references collected there). Here, we have two long-term goals in mind. The first one is to broaden understanding of technical aspects of developing agent systems, such as agent functionalities, their interactions and communication, agent mobility etc. We are also concerned with the fact that without agents systems being actually implemented using tools that are apparently designed to do this, agent research will never be able to reach beyond academia. Success in achieving the first goal will allow utilization of our systems as a tool-box for modeling processes occurring in an e-marketplace. For instance, it will be possible to apply it to study: effects of pricing strategies, of negotiation protocols and strategies, flow of commodities etc. Due to agent flexibility it will be relatively easy to experiment with various e-commerce scenarios.

In this chapter we proceed as follows. In the next section we provide background information and follow with the description of our system formalized through a complete set of UML diagrams. We then discuss in some detail (including implementation specifics) how rule based engine can be used to facilitate autonomous price negotiations.

2 Background

2.1 Agent Systems in E-Commerce

While there exist many definitions of agents, for the purpose of this chapter we will conceptualize them as: encapsulated computer programs, situated in an environment, and capable of flexible, autonomous actions focused on meeting their design objectives [40]. For such agents, e-commerce is considered to be one of the paradigmatic application areas [25].

Proliferation of e-commerce is strongly related to the explosive growth of the Internet. For example, the total number of Internet hosts with domain names was estimated at 150 millions in 2003, while in the same year, Web content was estimated at 8000 millions of Web pages ([26]). At the same time, e-commerce revenue projections were estimated to reach in 2006 up to $0.3 trillions for B2C e-commerce and up to $5.4 trillions for B2B e-commerce ([26]).

E-commerce utilizes (to various degrees) digital technologies to mediate commercial transactions. As a part of our research we have modified slightly Laudons approach ([26]) and conceptualized a commercial transaction as consisting of four phases:

- *pre-contractual phase* including activities like need identification, product brokering, merchant brokering, and matchmaking;
- *negotiation* where participants negotiate according to the rules of the market mechanism and using their private negotiation strategies;
- *contract execution* including activities like order submission, logistics, and payment;
- *post-contractual phase* that includes activities like collecting managerial information and product or service evaluation.

While there exist many scenarios of applying agents in e-commerce, automated trading is one of the more promising ones. In particular, we are interested in using agents to support all four, outlined above, phases of a commercial transaction, by addressing questions like: how is an e-shop to negotiate price with e-buyers, what happens before negotiations start and after they are finished, which e-store is the purchase actually made from etc, thus going beyond the phase of negotiation itself.

Unfortunately, our research indicates that most existing automated trading systems are not yet ready to become the foundation of the next generation of e-commerce. For example, the Kasbah Trading System ([12]) supports buying and selling but does not include auctions; SILKROAD ([30]), FENAs ([23]) and InterMarket ([24]) exist as "frameworks" but lack an actual implementation (which is typical for most agent systems in general [31]).

2.2 Automated and Agent-based Negotiations

In the context of this chapter we understand negotiations as a process by which agents come to a mutually acceptable agreement on a price ([28]). When designing systems for automated negotiations, we distinguish between *negotiation mechanisms* (*protocols*) and *negotiation strategies*. Protocol defines "rules of encounter" between negotiation participants by specifying requirements that enable their interaction. The strategy defines the behavior of participants aiming at achieving a desired outcome. This behavior must be consistent with the negotiation protocol, and usually is specified to maximize "gains" of each individual participant.

Auctions are one of the most popular and well-understood forms of automated negotiations ([41]). An increased interest has been manifested recently in attempts to parameterize the auction design space with the goal of facilitating more flexible automated negotiations in multi-agent systems ([41, 28]). One of the first attempts for standardizing negotiation protocols was introduced by the Foundation for Intelligent Physical Agents—FIPA ([17]). FIPA defined a set of standard specifications of agent negotiation protocols including English and Dutch auctions.

Authors of [9, 10] analyzed the existing approaches to formalizing negotiations (including FIPA protocols) and argued that they do not provide enough structure for the development of truly portable systems. Consequently, they outlined a complete framework comprising: (1) negotiation infrastructure, (2) a generic negotiation protocol and (3) taxonomy of declarative rules. The *negotiation infrastructure* defines roles of negotiation participants and of a host. Participants negotiate by exchanging proposals and, depending on the negotiations type, the host can also become a participant. The *generic negotiation protocol* defines the three phases of a negotiation: admission, exchange of proposals and formation of an agreement, in terms of how, when and what types of messages should be exchanged between the host and participants. *Negotiation rules* are used for enforcing the negotiation mechanism. Rules are organized into a taxonomy: rules for admission of participants to negotiations, rules for checking the validity of negotiation proposals, rules for protocol enforcement, rules for updating the negotiation status and informing participants, rules for agreement formation and rules for controlling the negotiation termination. Finally, they introduce a *negotiation template* that contains parameters that distinguish one form of negotiations from another, as well as specific values characterizing given negotiation. In this context it should be noted that rule-based approaches have been indicated as a very promising technique for introducing "intelligence" into negotiating agents ([9, 11, 16, 27, 37, 41, 42, 20]). Furthermore, proposals have been put forward to use rules for describing both negotiation mechanisms ([9, 38]) and strategies ([16, 37]).

With so much work already done in the area of agents and agent systems emerging in the context of autonomous price negotiations, let us underline what makes our approach unique.

- In most, if not all, papers only a "single price negotiation" is considered. Specifically, negotiations of a single item or a single collection of items is contemplated. Once such a negotiation is over, a group of agents (agent system) that participated in it completes its work. We are interested in a different (and a considerably more realistic) scenario when a number of products of a given type are placed for sale one after another. While this situation closely resembles what happens in any Internet store, it is practically omitted from research considerations. In this chapter, for clarity of enclosed UML diagrams, we depict situation where an almost unlimited number of items is to be sold. However, this assumption has only aesthetical reasons.
- Fact that multiple items are to be sold has also an important consequence for the way that price negotiations are organized. In the literature it is very often assumed that agents join an ongoing negotiation process as soon as they are ready (see for instance [9]), while agent-actions that take place after price negotiation is completed are disregarded. Since we sell multiple items one after another, we have decided to treat price negotiations as a "discrete process." Here, except of a specific case of fixed price mechanism, buyer agents are "collected" and released in a group to participate in a given price negotiation. While the negotiation takes place, buyer agents communicate only with the seller agent—e.g. the host (they can be envisioned as being placed in a closed negotiation room). At the same time the next group of buyer agents is collected (as they arrive) and will participate in the next negotiation.
- Fact that multiple subsequent auctions (involving the same product) take place allows us to go beyond one more popular "limitation" of known to us agent systems. While sometimes they involve rather complicated price negotiations, e.g. mixed auctions (see for instance [35, 36]), since only a single item or a single collection of items are sold, it is only that given price negotiation mechanism that is taken into account. In our case, since multiple negotiations are used to sell items of the same product we conceptualize situation in which price negotiation mechanism changes. For instance, first 25 items may be sold using English Auction, while the next 37 using fixed price with a deep discount.
- Furthermore, we consider the complete e-commerce system, which means that after negotiation is completed we conceptualize subsequent actions that may, or may not result in an actual purchase. In the case when purchase does not take place, we specify what should happen to all involved agents.
- While agent mobility is often considered to be important in the context of e-commerce systems, above described conflict between agent mobility and intelligence is rarely recognized. In our work we address this question by designing modular agents and clearly delineating which modules have to be sent, when, by whom and where.
- Finally, the complete system is being implemented using JADE; an actual agent environment.

3 Code Mobility in an Agent-Based E-Commerce System

Code mobility has been recognized as one of key enablers of large scale distributed applications, while its specific technologies, design patterns and applications have been systematically analyzed ([18]). Furthermore, recent research results suggest that blending mobility and intelligence can have important benefits, especially in advanced e-commerce; by providing application components with automated decision-making capabilities and ubiquity as required in networked environments ([25]). At the same time it has been argued that, as a general feature, agent mobility is unnecessary. Therefore, we asked a basic question: why, in the case of e-commerce, should one use mobile agents instead of messaging? To answer it, let us consider someone who, behind a slow Internet connection (which is not an uncommon situation), tries to participate in an eBay auction. In this case it is almost impossible to assure that this persons bid (1) reaches eBay server in time, (2) is sufficiently large to outbid opponents that have been bidding simultaneously (information about auction progress as well as user responses may not be able to reach their destinations sufficiently fast). As a result, network-caused delays may prevent purchase of the desired product. Obviously, this would not be the case if an autonomous agent representing that user was co-located with the negotiation host. In this context, one can obviously ask about the price of moving buyer agents across the network. Naturally, it may happen that an agent may not be able to participate in an auction because it does not reach the host in time. In response let us observe that: (1) if it is a particular single auction that the user is interested in, then agent not reaching the host has exactly the same effect as not being able to win because of bid(s) being late and/or too small; (2) therefore, it is only an agent that reaches the host in time that gives its user any chance to effectively participate in price negotiations; (3) furthermore, if an agent reaches its destination, it will be able to effectively participate in all subsequent negotiations within that host (and we assume across this paper that multiple negotiations involving items of the same product take place), while delays caused by network traffic may permanently prevent user from effective participation in any of them. For an extended discussion of the need for agent mobility in e-commerce see [7].

Let us now sketch proposed resolution of an above mentioned obvious contradiction between agent mobility and adaptivity. In our work, we utilize the negotiation framework introduced in [9, 10], where the *negotiation protocol* is a generic set of rules that describes all negotiations, while the *negotiation template* is a set of parameters that establishes the form of negotiation and its details. Finally, there is the *negotiation strategy* defining outcome optimizing actions of individual negotiation participants. It should be obvious that the *negotiation protocol* is generic and public—all agents participating in all negotiations have to use it. Therefore buyer agent can receive it upon its arrival at the host; similarly to the negotiation template which has to be "local" as it describes currently used form of negotiations (and which can change over time). It is only the strategy that is "private" and has to be obtained from the client agent (we name *client agent* agents representing User-Clients). At the same time, it has to be assumed that depending on the form of negotiation, different strategies will be used, and thus strategy is not known in advance. Therefore, since the protocol and the template can be obtained within the e-store, carrying them across the network is unnecessary. Unfortunately, it is not possible to establish the negotiation form in advance and send buyers with the negotiation strategy pre-loaded. Recall that in our system we assume that e-stores respond to the flow of commodities by actively changing forms of price negotiations. This being the case, by the time the buyer agent reaches its destination its strategy module may be useless, as the form of negotiations has already changed. We thus propose two network-traffic minimizing approaches to agent mobility. In the first case (named thereafter *agent mobility*) only an agent skeleton is sent across the network and upon arrival it obtains the negotiation protocol and the template and then requests the strategy module from the client agent. In the second case (named thereafter *code mobility*) buyer agents are created by the host (on the basis of a request from the client cgent) and assembled including (1) protocol, (2) actual template, and (3) information who they represent. Then, again, they request an appropriate strategy module from their designated client agent. Observe, that since only the strategy module is "secret," while the remaining parts of the buyer are public and open to any scrutiny at any time, this latter solution should not directly result in an increased security risk.

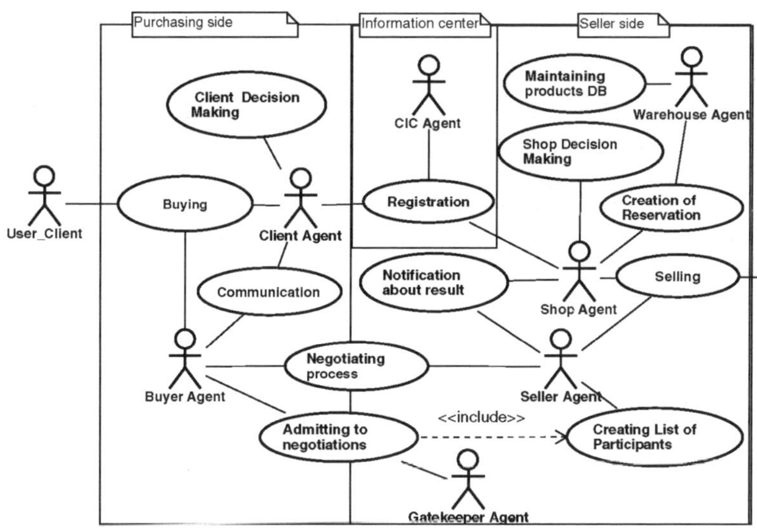

Fig. 1. Use case diagram

4 Description of the System

4.1 Conceptual Architecture

In our description of the system we utilize its UML-based formalization. Due to lack of space we have decided to present a set of UML diagrams of the system, rather than lengthy descriptions of its features and underlying assumptions. Interested readers should consult ([6, 7, 19]) for more details. In Figure 1 we present the use case diagram of our system that depicts all of its agents and their interactions. We can distinguish three major parts of the system: (1) the *information center* where white-page and yellow-page type data is stored—this is our current solution of the matchmaking problem [38], (2) the *purchasing side* where agents and activities representing User-Client reside, and (3) the *seller side* where the same is depicted for the User-Seller. Let us now describe in detail each of the agents (except the CIC agent that plays only an auxiliary role; see [19]) found in Figure 1.

4.2 UML Models of Agents in the System

Client agent

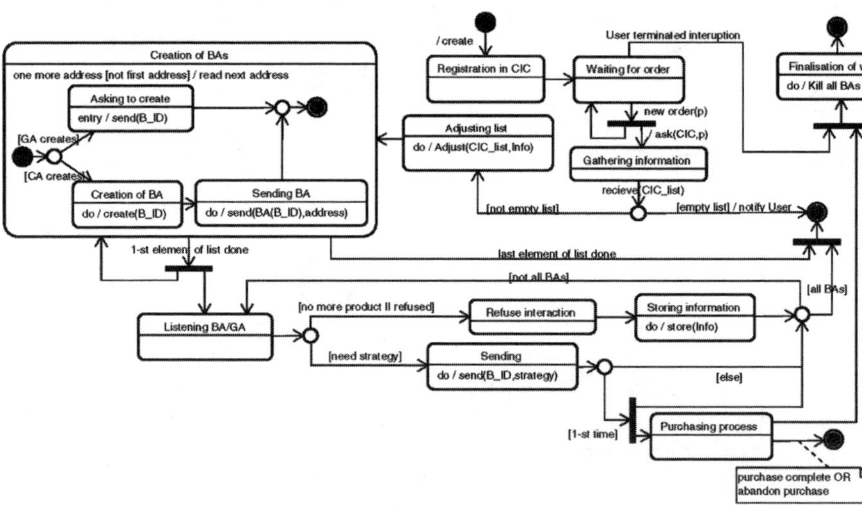

Fig. 2. Client Agent Statechart diagram

On the purchasing side, we have two agents. The *Client* agent, represented in Figures 2 and 3 exists in a complex state. On the one hand it listens for orders from the User-Client and, to fulfill them: (1) queries the *CIC* agent which has access to information which stores sell the requested product and if they create *Buyer* agents locally (or if such agent has to be sent to them), (2) then it dispatches or requests creation of *Buyer* agents to / by each such e-store (identified by its *Gatekeeper* agent).

At the same time, it directly manages the process of making purchases on behalf of the User-Client (Figure 3), on the basis of *Buyer* agent messages informing about results of price negotiations (let us note that in the case of multiple orders separate groups of *Buyer* agents—corresponding to separate products—will be managed in the same fashion). For a certain amount of time the *Client* agent collects reports sent by *Buyer* agents. When the wait-time is over (or when all *Buyer* agents have reported back), *Client* agent enters a complex state. On the one hand it continues listening for messages from *Buyer* agents (obviously, if all have reported already then none will be coming). On the other hand it goes through a multi-criteria decision making procedure (the "MCDM" box) that has one of three possible outcomes: (i) to attempt at completing a selected purchase, (ii) to await better opportunity, or (iii) to declare the purchase impossible and notify the User-Client accordingly. Note that, in a realistic system, the *MCDM* analysis should be truly multi-criteria and include factors such as: price, history of dealing with a given e-shop, delivery conditions etc.

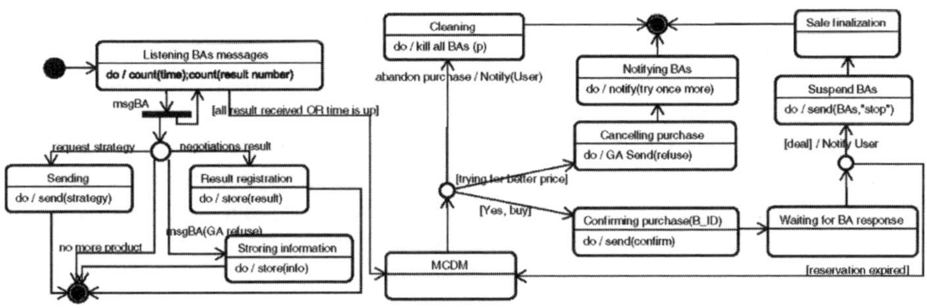

Fig. 3. Client Agent Statechart diagram

When attempt at completing a purchase is successful, the *Client* agent sends messages to all *Buyer* agents to cease to exist. The situation is slightly more complicated when the attempt was unsuccessful. Note that it is quite possible that the first *MCDM* analysis was undertaken before all *Buyer* agents have complete their "first round" of price negotiations. They could have contacted the *Client* while it was "thinking" which of the existing offers to choose. Therefore, when the *Client* agent analyses available reservations, they include not only reservations that have been already considered, but also possibly new ones that have arrived in the meantime. As a result of the MCDM procedure another attempt at making a purchase can be made. If none of available offers is acceptable, but purchase was not declared impossible, the *Client* agent undertakes the following actions: (1) informs all *Buyer* agents that have already reported to cancel current reservations and return to price negotiations (or just to return to price negotiations if they previously failed) and (2) resets timer establishing how long it will wait before the next round of *MCDM* analysis. Observe that in this way, in the proposed system, it is possible that some agents make their second attempt at negotiating prices, while some agents have just finished the first. As this process continues in an asynchronous fashion *Buyer* agents will make different number of attempts at negotiating price that is acceptable to the *Client* agent. This process will terminate when all orders submitted by the customer have been either honored or abandoned. For the time being we assume that the "Sale finalization" process seen in Figure 3 is always successful. In the future we plan to remove this somewhat artificial restriction.

Let us note that it is possible that since it is the *Client* agent that makes the final determination which offer to accept, and that it has to communicate with one of its remotely located *Buyer* agents to actually complete a purchase, the request to attempt at making that purchase could be network-delayed resulting in an expired reservation and inability to complete the task. Unfortunately, this problem does not seem to have a simple solution, since price comparison requires communication between agents participating in price negotiations. In our system we have selected a central point—*Client* agent—that will collect all offers, instead of all-to-all communication. Since not all sites will conduct their price negotiations at the same time, and with the same urgency, it is impossible to assure that the best offer will still be available, when the "remaining" agents complete their negotiations. Therefore, our solution remains optimal only in terms of reducing total network congestion by sending only minimal-size agents and minimizing the total number of messages sent over the network.

Finally, let us note that a complete information about all events taking place during servicing User-Client request (such as: refusal to admit to negotiations, results of price negotiations, length of reservation, etc.) is stored for further information extraction. For instance, as a result of data analysis, store that is constantly selling good(s) at very high prices may be simply avoided.

Buyer agent

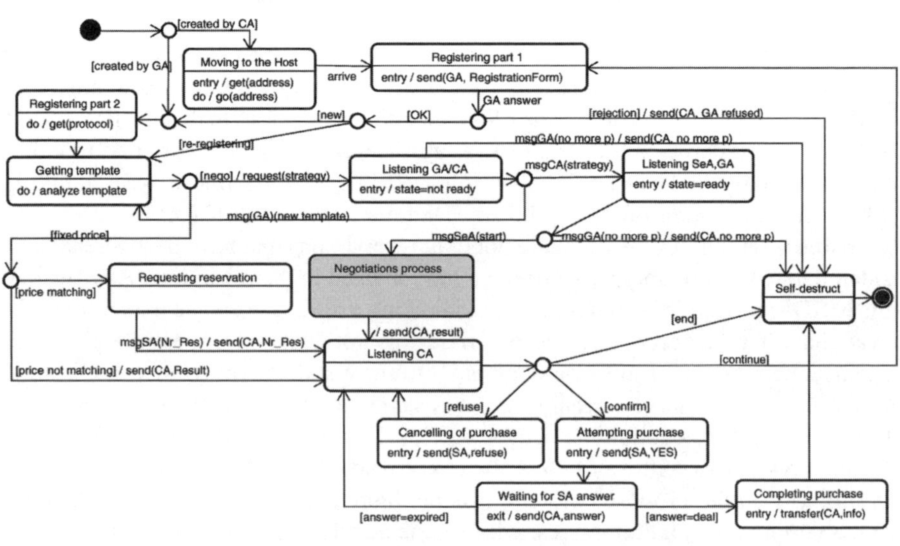

Fig. 4. Buyer Agent Statechart diagram

Buyer agent (see Figure 4) is the only agent in the system that involves mobility. It is either dispatched by the *Client* agent or created by the *Gatekeeper* agent on the request of the *Client*. If it is dispatched, then upon arrival at the store it communicates with the *Gatekeeper* (see Figurc 6,7) to obtain entry to negotiations (in case when entry is not granted it informs its *Client* agent and is killed). In the case of *Client* requesting creation of the *Buyer*, the *Gatkeeper* may deny such request. In both cases, information about refusal to cooperate is stored by the *Client* agent (e.g. e-stores that do not want to cooperate may be removed from the list obtained from the *CIC*; see box "Adjusting list" in Figure 2). Next, the preregistered *Buyer* obtains from the *Gatekeeper* the negotiation protocol and the current negotiation template and requests (and obtains) an appropriate strategy module from the *Client* agent (see Figure 2). When all these modules are installed *Buyer* informs the *Gatekeeper* that it is ready to negotiate (it is then registered as such). Price negotiations start when the *Buyer* receives a start message from the *Seller* agent (see Figure 9; note that the "Negotiations box" appearing there is "the same" for both the *Seller* and the *Buyer* agents) that later participate in price negotiations; note also special treatment of fixed-price negotiations by both the *Buyer* and the *Gatekeeper* agents. Upon completion of negotiations, *Buyer* informs the *Client* about their result and, if necessary (when an attempt at completing purchase is made), acts as an intermediary between the *Client* and the *Shop* agents. In the case when purchase was not attempted or was not successful, *Buyer* agent awaits the decision of the *Client* and if requested proceeds back to participate in price negotiations (before doing so it requests permission to re-enter that may be granted or denied; updates its negotiation template and, possibly, the strategy module—if the template has changed). This process continues until the *Buyer* agent self-destructs on the request of the *Client* agent.

Shop agent

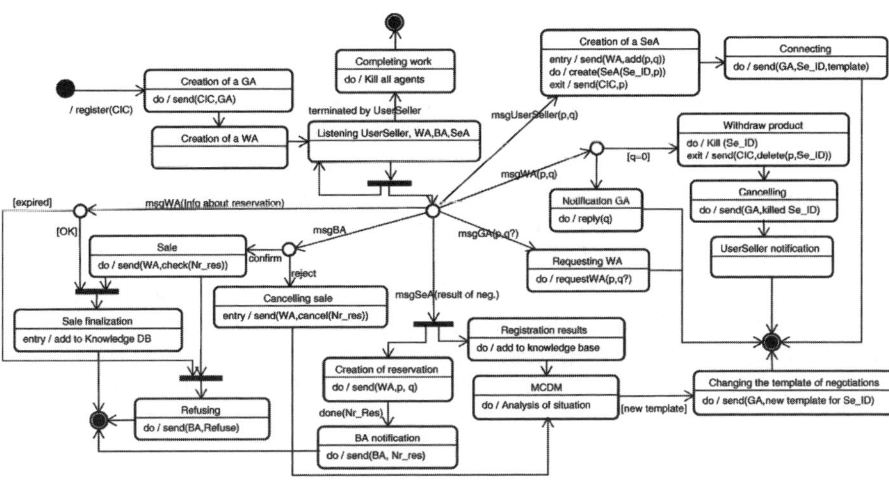

Fig. 5. Shop Agent Statechart diagram

On the “selling side” of the system, the *Shop* agent acts as the representative of the User-Seller. We assume that after it is created, it persistently exists in the system until the User-Seller decides that it is no longer needed. The UML diagram representing the *Shop* agent is presented in Figure 5. Upon its instantiation, the *Shop* agent creates and initializes its co-workers: a *Gatekeeper* agent, a *Warehouse* agent and *Seller* agents (one for each product sold). Initialization of the *Warehouse* agent involves passing information about goods that are initially available for sale (see Figure 8), while initialization of the *Gatekeeper* agent involves providing it with templates that are to be used initially in price negotiations of each product sold. Furthermore, the *Gatekeeper* agent and the list of products available in the store are registered with the *CIC* agent.

After the initialization stage, the *Shop* agent enters a complex state where it supervises negotiations and product flow. First, it waits for finish of any price negotiation. If the negotiation was successful, a given *Seller* informs the *Shop* agent, which is asking the *Warehouse* agent to reserve a given quantity of a particular product for a specific amount of time. (Currently we assume that a single item of a given product is sold each time, but this, somewhat limiting, assumption will be removed in the future.) Events can then proceed according to following scenarios.

1. If the winning *Buyer* confirms purchase then the *Shop* asks the *Warehouse* agent to check the reservation.
   - If the reservation did not expire then the *Shop* informs the *Buyer* agent about acceptance of transaction. This event starts the final stage, named “Sale finalization” which includes such actions as payment and delivery.
   - In the opposite case, the *Shop* agent sends rejection to the *Buyer* agent
2. If the *Client* agent rejects purchase (and informs the *Shop* agent about it through the *Buyer* agent) then the *Shop* agent asks the *Warehouse* agent to cancel the reservation.

Completing one of these scenarios “closes” this branch of *Shop* agent execution. Separately, the *Shop* agent keeps track of all negotiations and transactions and periodically performs multi-criteria analysis (the *MCDM* module) that may result in changes in the negotiation template for one or more products (e.g. minimal price, type of price negotiation mechanism, etc.). For instance, when only a few items are left they may be deeply-discounted, or put on sale through an auction. In this case a new template is generated and sent to the *Gatekeeper* agent that switches it in an appropriate moment (see below, Figures 6, 7).

Let us also note that, similarly to the *Client* agent, the *Shop* agent stores complete information about all events taking place in the e-store (such as: results of price negotiation, information about agents that actually purchased reserved product, information of agents that canceled reservations, etc.). This information, when analyzed, may result for instance in a given *Client* agent being barred from entering negotiations.

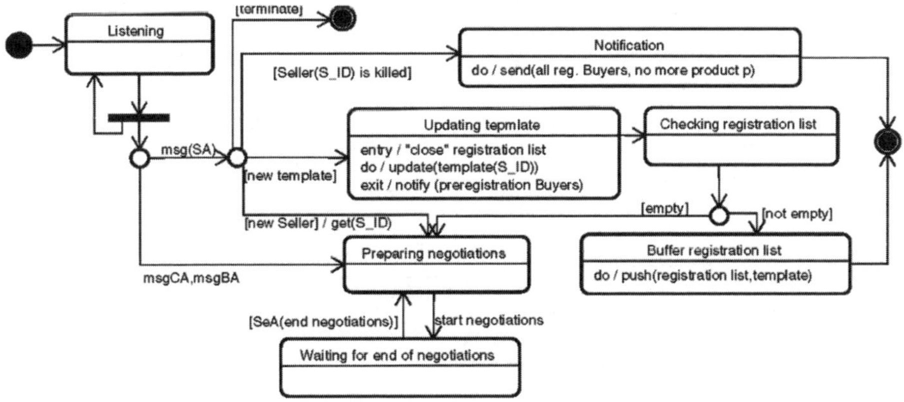

Fig. 6. Gatekeeper Agent Statechart diagram

Gatekeeper agent

Shop agents cooperate directly with their *Gatekeeper* agents that (1) either interact with incoming *Buyer* agents, and admit them to the negotiations (or reject their attempt at entering the host), or interact with *Client* agents and, on their request, create *Buyer* agents (or reject such a request), and provide admitted / created *Buyer* agents with the protocol and the current negotiation template (2) in appropriate moments release *Buyer* agents to appropriate *Seller*s and (3) manage updates of template modules. The statechart diagram of the *Gatekeeper* agent is presented in Figure 6 (the top level description of *Gatekeeper* functionality) and continued in Figure 7 (depicting negotiation related activities). Each created or allowed to enter *Buyer* agent is put on a list of preregistered agents and provided with protocol and current template. *Buyer* agents remain on that list until they receive their strategy module and complete self-assembling. Assembled *Buyer* agents are put on a list of registered agents that await start of price negotiations. When a minimum number of *Buyer* agents have registered (minimum for a given form of negotiations) and the wait-time has passed, the *Gatekeeper* passes their identifiers and the current negotiation template to the *Seller* agent. Then it cleans the current list of registered *Buyer* agents and the admission/monitoring process is restarted (assuming that the *Seller* agent is still alive). As stated above, our system allows *Buyer* agents that lost negotiations or that decided not to make a purchase to stay at the host and try to re-enter negotiations. They have to ask permission to be re-admitted and if allowed back they receive an updated template ("old Buyer" path). When a new template module is delivered by the *Shop* agent, a list of currently registered *Buyer* agents is put into a buffer ("Buffer registration list" box). These agents have to be serviced first, using the current template that they have been provided with upon entering the e-store. At the same time the new incoming agents will then be given the new template. Finally, in a special case, when a given product has been sold-off and the *Shop* agent terminates the *Seller* responsible for selling it, the *Gatekeeper* informs awaiting *Buyer* agents about this fact.

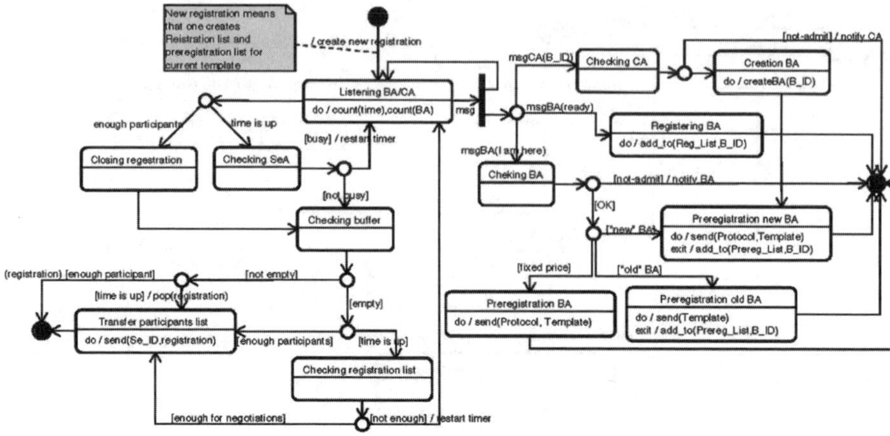

Fig. 7. Statechart diagram for Preparing Negotiations State

Warehouse agent

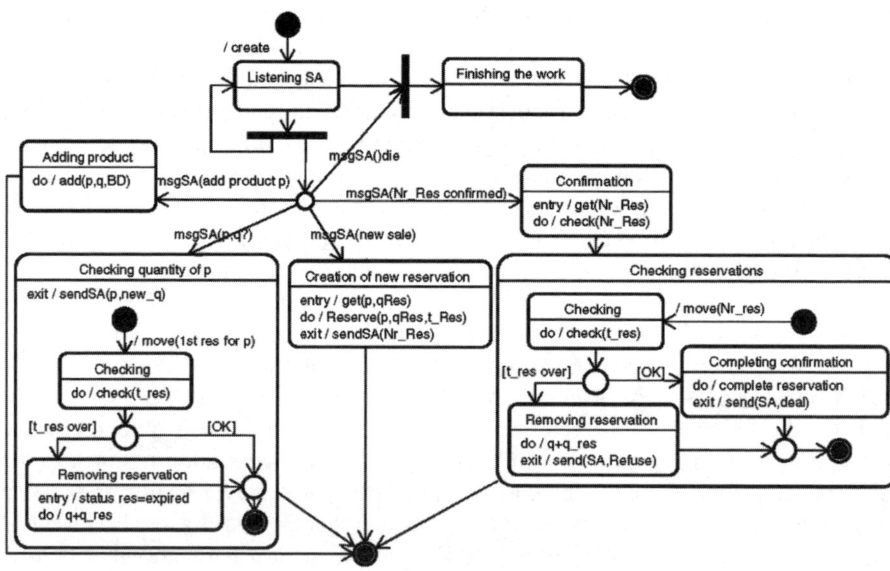

Fig. 8. Warehouse Agent Statechart diagram

Shop agent interacts also directly with the *Warehouse* agent (presented in Figure 8). In the early stages of its functioning the *Warehouse* agent is supplied (by the *Shop* agent) with information about products and their quantities (to be saved in a database). Then it enters a complex state where it (a) awaits notifications from the *Shop* agent and (b) acts on them. The *Shop* agent notifies the *Warehouse* agent about: (i) registration of new products for sale, (ii) product reservations, (iii) purchase confirmations, and (iv) purchase terminations. Each of these notifications is followed by an appropriate response: (i) product registration, (ii) product reservation, (iii) checking status of a reservation, (iv) cancellation of a reservation. Finally, if quantity of some product becomes 0, the *Warehouse* agent informs about it the *Shop* agent, which (in the current state of our system) terminates the corresponding *Seller* agent, and informs about it both the *CIC* and the *Gatekeeper* agents.

Seller agent

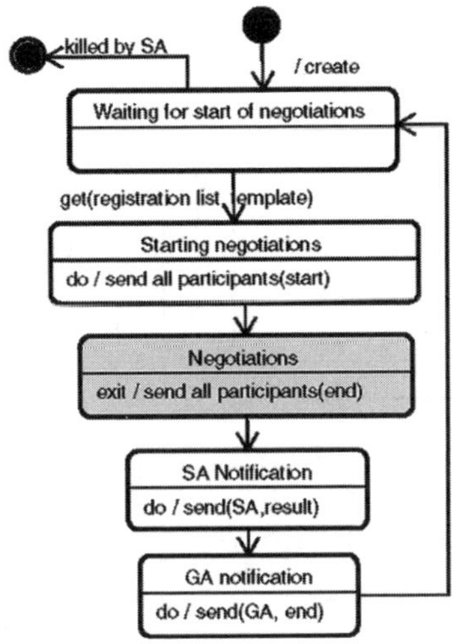

Fig. 9. Seller Agent Statechart diagram

Finally, the last agent working on the "selling side" of the system is the *Seller* agent. It is characterized by a rather simple statechart diagram (see Figure 9). The simplicity comes form the fact that, in the "Negotiations box," the *complete* negotiation framework proposed in [9, 10] in enclosed. Observe that not all negotiations have to end in finding a winner and our system is able to handle such an event. At the same time, all data about negotiations is collected and analyzed by the *Shop* agent and, for instance, a sequence of failures could result in a change of the negotiation template.

System activity diagram

Let us now combine activitics of all agents in the system into one diagram (see Figure 10, 11). This diagram represents flow of actions presented from the perspective of the two main agents in the system: the *Shop* and the *Client*. Obviously, to keep that diagram readable, we had to omit large number of details that have been represented within statechart diagrams of individual agents that should be "co-viewed" with the activity diagram.

4.3 Rule-Based Mechanism Representation

Let us now describe how we have implemented in our system rule-based mechanisms. We start by summarizing the framework for automated negotiation introduced in [9, 10] which is based on an abstract negotiation process that comprises: a negotiation infrastructure, a generic negotiation protocol and a taxonomy of declarative

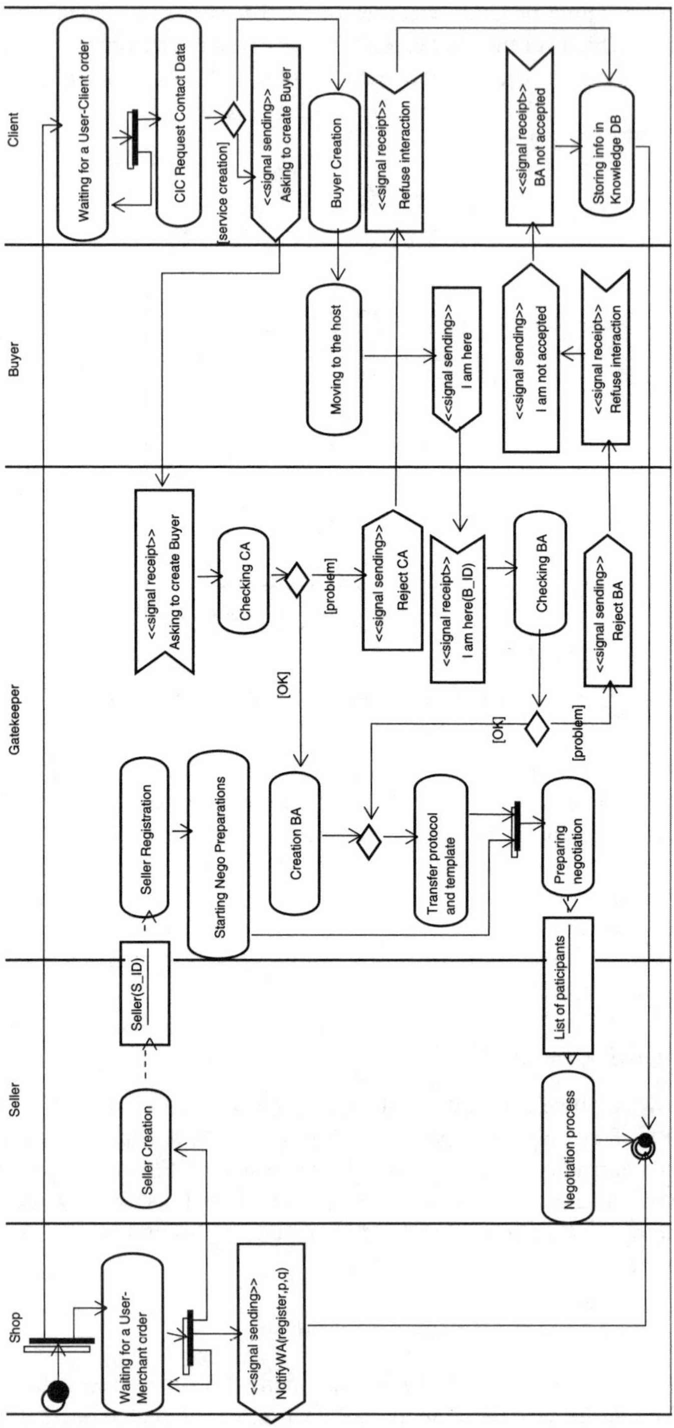

Fig. 10. Activity Diagram—before negotiation process

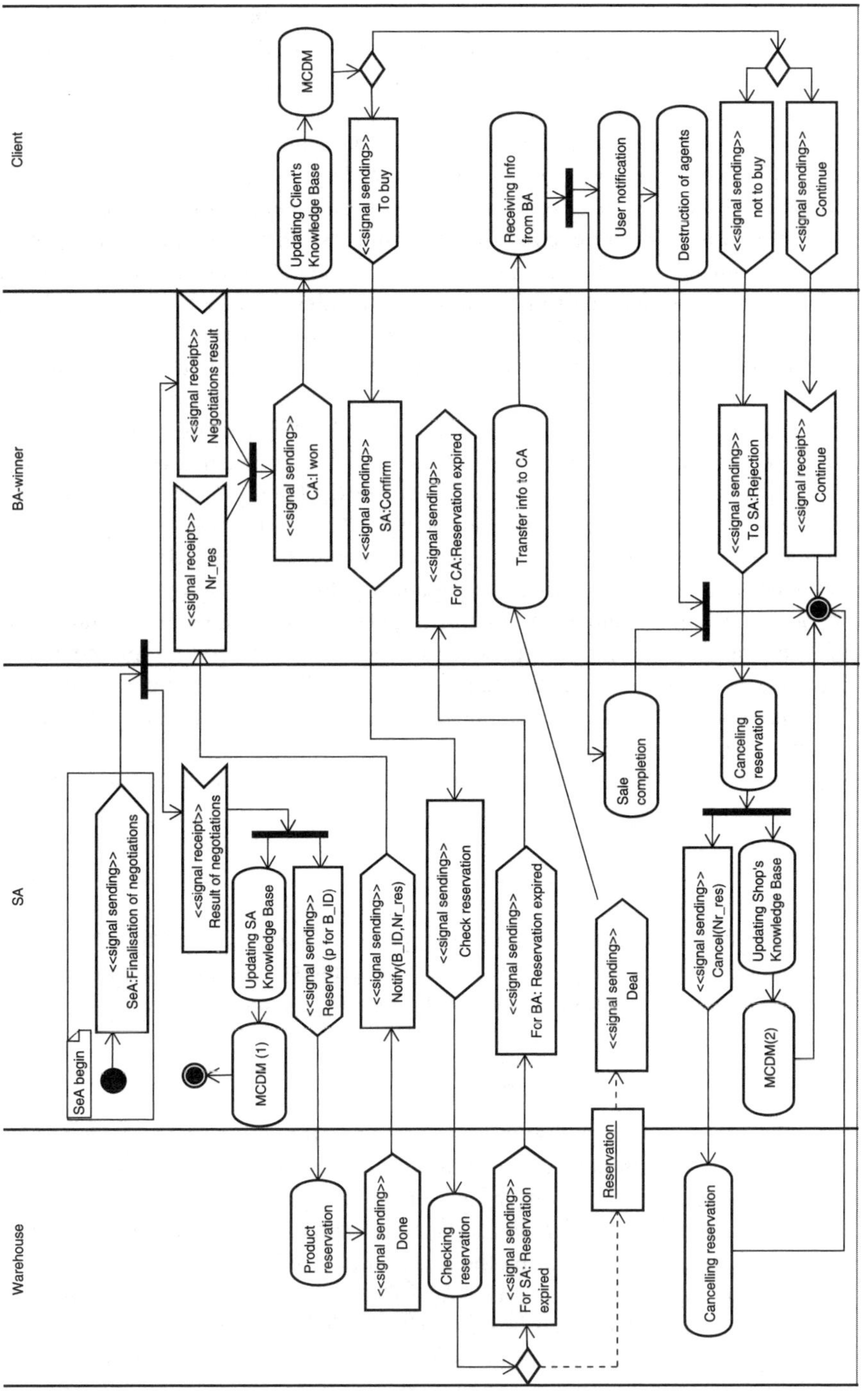

Fig. 11. Activity Diagram—after negotiation process

rules. Here, the *negotiation infrastructure* defines roles involved in the negotiation process: participants (in our system *Buyer* agents) and a host (*Seller* agent). Participants negotiate by exchanging proposals within a "negotiation locale" that is managed by the negotiation host. Depending on the type of negotiations, the host can also play the role of a participant (for example in an iterative bargaining scenario). The *generic negotiation protocol* defines, in terms of how and when messages should be exchanged between the host and negotiation participants, the three main phases of negotiations: (1) admission, (2) exchange of proposals and (3) formation of an agreement. *Negotiation rules* are needed for enforcing a specific negotiation mechanism. Rules are organized into a taxonomy that contains the following categories: (a) rules for admission of participants to negotiations, (b) rules for checking the validity of negotiation proposals, (c) rules for protocol enforcement, (d) rules for updating the negotiation status and informing participants, (e) rules for agreement formation and (f) rules for controlling the negotiation termination. Based on the categories of rules identified as necessary to facilitate negotiations, in [9, 10] it is suggested to partition the negotiation host into a number of corresponding components: *Gatekeeper*, *Proposal Validator*, *Protocol Enforcer*, *Information Updater*, *Negotiation Terminator* and *Agreement Maker* (that are called sub-agents). Each component is responsible for enforcing a specific category of rules. Host components interact with each-other via a blackboard and with negotiation participants by direct messaging. Note that these components are conceptualized as a part of the host (sub-agents), not as stand-alone agents. This fact will have consequences as to how they have been implemented.

Before proceeding let us recall that we have modified the proposed framework and upgraded the *Gatekeeper* to become a full-fledged agent [19]. In its new role, the *Gatekeeper* agent has also an increased scope of responsibilities (described above). This also means that admission rules are no longer part of the negotiation process itself.

Let us now show: (i) how the negotiation host agent (*Seller*) is structured into components (sub-agents); (ii) how rules are executed by the negotiation host in response to various messages received from negotiation participants and how rule firing control is switched between various components of the negotiation host, and (iii) how the generic negotiation protocol was implemented using JADE agent behaviors and ACL message exchanges between host and participants.

The Negotiation Host — *Seller* agent

It should be obvious by now that what was defined in [9, 10] as a *negotiation host* became a *Seller* agent in our system. We will thus use these two terms interchangingly. Host and *Buyer* agents are implemented as ordinary JADE agents and thus they extend the *jade.core.Agent* class. The *Seller* agent encapsulates the negotiation controlling sub-agents that are implemented as ordinary Java classes (see Figure 12): *Proposal Validator*, *Protocol Enforcer*, *Information Updater*, *Negotiation Terminator* and *Agreement Maker*. Each host component defines a *handle()* method that is activated to check the category of rules that are to be dealt with and delegates the call to the responsible component. Each component activates the rule engine via the

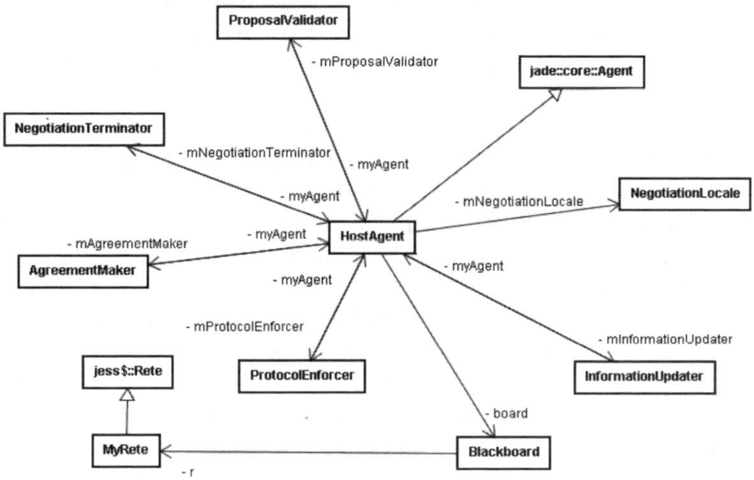

Fig. 12. The class diagram showing the structure of the *Seller* agent

myAgent member object that points to the parent host agent (see Figure 12). Note, again, that these components are not full-blown JADE agents, but ordinary member objects within the *Seller* agent.

In addition to sub-agents responsible for protocol enforcement, the host encapsulates two member objects representing the negotiation locale and the rule engine (see Figure 12): *Negotiation Locale* and *Blackboard* "boxes". The *Negotiation Locale* object stores the *negotiation template* (a structure that defines negotiation parameters; see [9]) and the list of participants that were admitted to a given negotiation (obtained from the *Gatekeeper* agent—see above). The *Negotiation Locale* is operated on directly as a Java object. The *Blackboard* object is a Java wrapper for a JESS rule engine (class *jess.Rete*) that is initialized with negotiation rules and JESS templates for storing JESS facts within the blackboard. Whenever the category of negotiation rules is checked, the rule engine is activated and rules are fired to update facts within that "JESS blackboard." Note that there is a clear distinction between the Java object called *Blackboard* that encapsulates the JESS rule engine and the actual blackboard which is a set of JESS facts that are updated by firing rules via the JESS engine.

Controlling Rule Execution

Rather then implementing each component of the negotiation host as a separate rule engine, we are using a single JESS rule engine that is shared by *all* host components. This rule engine is implemented using class *jess.Rete*. The advantage is that we now have a single rule engine per negotiation host rather than 6 engines as suggested in [9]. Furthermore, this means that in the case of *m* products sold, we will utilize *m* instances of the JESS rule engine, instead of 6*m* instances necessary in [9, 10].

Rules and facts managed by the rule engine are partitioned into JESS modules. Currently we are using one JESS module for storing the blackboard facts and a separate JESS module for storing rules used by each component. Facts within the blackboard are instances of JESS templates (*deftemplate* statements) and they can represent: (1) the negotiation template; (2) the active proposal that was validated by the *Proposal Validator* and the *Proposal Enforcer* components; (3) a withdrawn proposal; (4) seller reservation price (not visible to participants); (5) negotiation participants; (6) the negotiation agreement that may eventually be generated at the end of a negotiation; (7) the information digest that is visible to the negotiation participants; (8) the maximum time interval for submitting a new bid before the negotiation is declared complete; or (9) the value of the current highest bid. Note that these facts have been currently adapted to represent English auctions (and will be appropriately modified to represent other price negotiation mechanisms).

Each category of rules for mechanism enforcement is stored in a separate JESS module. This module is controlled by the corresponding component of the *Seller* agent. Whenever the component handles a message it activates the rules for enforcing the negotiation mechanism. Taking into account that all pertinent rules pertinent are stored internally in a single JESS rule-base (attached to a single JESS rule engine), the JESS *focus* statement is used to control the firing of rules located only in the focus module. This way, the JESS facility for partitioning the rule-base into disjoint JESS modules proves very useful to efficiently control the separate activation of each category of rules. Note also that JADE behaviors are scheduled for execution in a non-preemptive way and this implies that firings of rule categories are correctly serialized and thus they do not cause any synchronization problems. This fact also supports our decision to utilize a single rule engine for each host.

Generic Negotiation Protocol and Agent Behaviors

The *generic negotiation protocol* specifies a minimal set of constraints on sequences of messages exchanged between the host and participants. As specified in [9], the negotiation process has three phases: (1) admission, (2) proposal submission and (3) agreement formation. The admission phase has been removed from the *negotiation process* described in [9], but it was implemented in exactly the same way as suggested there. For instance, in the case of *agent mobility* it starts when a new participant (*Buyer* agent) requires admission to the negotiation, by sending an ACL PROPOSE message to the *Gatekeeper* agent. The *Gatekeeper* grants (or not) the admission of the participant to the negotiation and responds accordingly with either an ACL ACCEPT-PROPOSAL or an ACL REJECT-PROPOSAL message (currently admission is granted by default). In the way that the system is currently implemented, the PROPOSE message is sent by the participant agent immediately after its initialization stage, just before its *setup()* method returns. The task of receiving the admission proposal and issuing an appropriate response is implemented as a separate behavior of the *Gatekeeper* agent.

When a *Buyer* agent is accepted to the negotiation, it also receives (from the *Gatekeeper* agent) the negotiation protocol and template (representing parameters of auctions: auction type, auctioned product, minimum bid increment, termination time window, currently highest bid). *Buyer* agent will enter the phase of submitting proposals after it was dispatched to the negotiation (here, a number of *Buyer* agents that were granted admission is "simultaneously" released by the *Seller* (that sends them a start message) and they—possibly immediately—start submitting bids according to their strategies [19]). The generic negotiation protocol states also that a participant will be notified by the negotiation host if its proposal was either accepted (with an ACL ACCEPT-PROPOSAL) or rejected (with an ACL REJECT-PROPOSAL). In the case when a proposal was accepted, the protocol requires that the remaining participants will be notified accordingly with ACL INFORM messages.

Strategies of participant agents must be defined in accordance with the constraints stated by the *generic negotiation protocol*. Basically, the strategy defines when a negotiation participant will submit a proposal and what are the values of the proposal parameters. In our system (where the English auction has been implemented), for the time being, we opted for an extremely simple solution: the participant will submit a first bid immediately after it was released to the negotiation and subsequently, whenever it gets a notification that another participant issued a proposal that was accepted by the host. The value of the bid is equal to the sum of the currently highest bid and an increment value that is private to the participant. Additionally, each participant has its own valuation of the negotiated product in terms of a reservation price. If the value of the new bid exceeds this reservation price then the proposal submission is canceled. The implementation of the participant agent defines two JADE agent behaviors for dealing with situations stated above. Obviously, as the system matures, we plan to add more complicated price negotiation mechanisms. For these price negotiations we will develop, implement and experiment with a number of negotiation strategies that can be found in the literature (e.g. see [20]).

Finally, the agreement formation phase can be triggered at any time. When the agreement formation rules signal that an agreement was reached, the protocol states that all participants involved in the agreement will be notified by the host with ACL INFORM messages. The agreement formation check is implemented as a timer task (class *java.util.TimerTask*) that is executed in the background thread of a *java.util.Timer* object.

5 Concluding Remarks

In this chapter we have described an agent-based model e-commerce system that is currently being developed and implemented in our group. This system, as it is being extended, is slowly converging toward the main ideas underlying e-service intelligence systems. After presenting background information about software agents and automatic negotiations we have provided a description of the system, illustrated by its complete formal UML-based definition. We have also argued that the proposed solution is able to mediate the existing contradiction between agent mobility and intelligence, by precisely delineating which components, and when, have to be pushed across the network. Furthermore, we have discussed in detail how the negotiation framework, utilizing a rule-based engine is implemented in the system.

Currently, the proposed system is systematically being implemented and extended. We have experimented with its earlier versions and were able to see that it scales well (on a network consisting of 22 computers). Furthermore, we were able to successfully run it in a heterogeneous environment consisting of Windows and Linux workstations. The results have been reported in [4]. More recently we have implemented and successfully experimented with the above described rule-based engine approach, applied to the English auction mechanism, in which more than 140 agents negotiated prices within a single shop. Additional information can be found in [1, 2].

As the next steps we envision, among others: (1) completion of integration of the original system skeleton with the rule-based engine, (2) addition of rules for a number of additional price negotiation protocols (e.g. Vickery auction, Dutch auction etc.), (3) implementation of an initial set of non-trivial negotiation strategies for both buyers and sellers, (4) conceptualization of MCDM processes, starting from the ways in which data concerning results of price negotiations has to be stored so that it can be effectively utilized in support of decision making in the system etc. We will report on the results in subsequent publications.

Acknowledgement

Work of Maria Ganzha and Marcin Paprzycki has been partially sponsored by the Maria Curie IRG grant (project E–CAP).

References

1. Bădică, C., Ganzha, M., Paprzycki M.: Rule-Based Automated Price Negotiation: an Overview and an Experiment. In: *Proceedings of the International Conference on Artificial Intelligence and Soft Computing, ICAISC'2006*, Zakopane, Poland. Springer LNAI, 2006 (in press)
2. Bădică, C., Bădiţă, A., Ganzha, M., Iordache, A., Paprzycki M.: Implementing Rule-based Mechanisms for Agent-based Price Negotiations. In: L. Liebrock (ed.): *Proceedings of the ACM Symposyum on Applied Computing, SAC'2006*, Dijon, France. ACM Press, New York, NY, pp.96-100, 2006.
3. Bădică, C., Ganzha, M., Paprzycki, M., Pîrvănescu, A.: Combining Rule-Based and Plug-in Components in Agents for Flexible Dynamic Negotiations. In: M. Pěchouček, P. Petta, and L.Z. Varga (Eds.): *Proceedings of CEEMAS'05*, Budapest, Hungary. LNAI 3690, Springer-Verlag, pp.555-558, 2005.
4. Bădică, C., Ganzha, M., Paprzycki, M., Pîrvănescu, A.: Experimenting With a Multi-Agent E-Commerce Environment. In: V. Malyshkin (Ed.): *Proceedings of PaCT'2005*, Krasnoyarsk, Russia. LNCS 3606, Springer-Verlag, pp.393-402, 2005.
5. Bădică, C., Ganzha, M., Paprzycki, M.: Mobile Agents in a Multi-Agent E-Commerce System. In: *Proceedings 7th International Symposium on Symbolic and Numeric Algorithms for Scientific Computing, SYNASC'05*, Timişoara, Romania. IEEE Computer Society Press, Los Alamitos, CA, pp.207-214, 2005.

6. Bădică, C., Ganzha, M., Paprzycki, M.: UML Models of Agents in a Multi-Agent E-Commerce System. In: *Proceedings of the IEEE Conference of E-Business Engineering, ICEBE 2005*, Beijing, China. IEEE Computer Society Press, Los Alamitos, CA, pp.56-61, 2005.
7. Bădică, C., Ganzha, M., Paprzycki, M.: Two Approaches to Code Mobility in an Agent-based E-commerce System. In: C. Ardil (ed.): *Enformatika*, Volume 7, pp.101-107, 2005.
8. Bădică, C., Bădiţă, A., Ganzha, M., Iordache, A., Parzycki, M.: Rule-Based Framework for Automated Negotiation: Initial Implementation. In: *Proceedings 1st Conference on Rules and Rule Markup Languages for the Semantic Web, RuleML'2005*, Galway, Ireland. Lecture Notes in Computer Science 3791, Springer-Verlag, pp.193-198, 2005.
9. Bartolini, C., Preist, C., Jennings, N.R.: Architecting for Reuse: A Software Framework for Automated Negotiation. In: *Proceedings of AOSE'2002: Int. Workshop on Agent-Oriented Software Engineering*, Bologna, Italy, LNCS 2585, Springer Verlag, pp.88-100, 2002.
10. Bartolini, C., Preist, C., Jennings, N.R.: A Software Framework for Automated Negotiation. In: *Proceedings of SELMAS'2004*. LNCS 3390, Springer-Verlag, pp.213-235, 2005.
11. Benyoucef, M., Alj, H., Levy, K., Keller, R.K.: A Rule-Driven Approach for Defining the Behaviour of Negotiating Software Agents. In: J.Plaice et al. (eds.): *Proceedings of DCW'2002*, LNCS 2468. Springer-Verlag, pp.165-181, 2002.
12. Chavez, V., Maes, P.: Kasbah: An Agent Marketplace for Buying and Selling Goods. In: *Proc. of the First Int. Conf. on the Practical Application of Intelligent Agents and Multi-Agent Technology*. London, UK, 1996.
13. Chmiel, K., Czech, D., Paprzycki, M.: Agent Technology in Modelling E-commerce Process; Sample Implementation. In: C. Danilowicz (ed.): *Multimedia and Network Information Systems*, Volume 2, Wroclaw University of Technology Press, pp.13-22, 2004.
14. Chmiel, K., Tomiak, D., Gawinecki, M., Karczmarek, P., Szymczak, Paprzycki, M.: Testing the Efficiency of JADE Agent Platform. In: *Proceedings of the 3rd International Symposium on Parallel and Distributed Computing*, Cork, Ireland. IEEE Computer Society Press, Los Alamitos, CA, USA, pp.49-57, 2004.
15. Chmiel, K., Gawinecki, M., Kaczmarek, P., Szymczak, M., Marcin Paprzycki: Efficiency of JADE Agent Platform. in: *Scientific Programming*, vol.13, no.2, pp.159-172, 2005.
16. Dumas, M., Governatori, G., ter Hofstede, A.H.M., Oaks, P.: A Formal Approach to Negotiating Agents Development. In: *Electronic Commerce Research and Applications*, Vol.1, Issue 2 Summer, Elsevier Science, pp.193-207, 2002.
17. FIPA: Foundation for Physical Agents. See `http://www.fipa.org`.
18. Fuggetta, A., Picco, G.P., Vigna, G.: Understanding Code Mobility. In: *IEEE Transactions on Software Engineering*, vol.24, no.5, IEEE Computer Science Press, pp.342-361, 1998.
19. Ganzha, M., Paprzycki, M., Pîrvănescu, A., Bădică, C, Abraham, A.: JADE-based Multi-Agent E-commerce Environment: Initial Implementation, In: *Analele Universităţii din Timişoara, Seria Matematică-Informatică*, Vol. XLII, Fasc. special, pp.79-100. 2004.
20. Governatori, G., Dumas, M., ter Hofstede, A.H.M., and Oaks, P.: A formal approach to protocols and strategies for (legal) negotiation. In: Henry Prakken (ed.): *Procedings of the 8th Int. Conference on Artificial Intelligence and Law*, IAAIL, ACM Press, pp.168-177, 2001.
21. JADE: Java Agent Development Framework. See `http://jade.cselt.it`.
22. JESS: Java Expert System Shell. See `http://herzberg.ca.sandia.gov/jess/`.
23. Kowalczyk, R.: On Fuzzy e-Negotiation Agents: Autonomous negotiation with incomplete and imprecise information, In: *Proc.DEXA'2000*, London, UK, pp.1034-1038, 2000.
24. Kowalczyk, R., Franczyk, B., Speck, A.: Inter-Market, towards intelligent mobile agent E-Market places. In: *Proc. 9th Annual IEEE Internmational Conference and Workshop on the Engineering of Computer-Based Systems, ECBS'2002*, Lund, Sweden, pp.268–276, 2002.

25. Kowalczyk, R., Ulieru, M., Unland, R.: Integrating Mobile and Intelligent Agents in Advanced E-commerce: A Survey. In: *Agent Technologies, Infrastructures, Tools, and Applications for E-Services, Proceedings NODe'2002 Agent-Related Workshops*, Erfurt, Germany. LNAI 2592, Springer-Verlag, pp.295-313, 2002.
26. Laudon, K.C., Traver, C.G.: *E-commerce. business. technology. society* (2nd ed.). Pearson Addison-Wesley, 2004.
27. Lochner, K.M., Wellman, M.P.: Rule-Based Specification of Auction Mechanisms. In: *Proc. AAMAS'04*, ACM Press, New York, USA, 2004.
28. Lomuscio, A.R., Wooldridge, M., Jennings, N.R.: A classification scheme for negotiation in electronic commerce. In: F. Dignum, C. Sierra (Eds.): *Agent Mediated Electronic Commerce: The European AgentLink Perspective*. LNCS 1991, Springer-Verlag, 19-33, 2002.
29. Maes, P., Guttman, R.H., Moukas, A.G.: Agents that Buy and Sell: Transforming Commerce as we Know It. In *Communications of the ACM*, Vol.42, No.3, pp.81-91, 1999.
30. Michael, S.: Design of Roles and Protocols for Electronic Negotiations. In: *Electronic Commerce Research Journal*, Vol.1 No.3, pp.335-353, 2001.
31. Nwana, H., Ndumu, D.: A Perspective on Software Agents Research. In: *The Knowledge Engineering Review*, 14(2), pp.1-18, 1999.
32. Paprzycki, M., Abraham, A.: Agent Systems Today; Methodological Considerations. In: *Proceedings of 2003 International Conference on Management of e-Commerce and e-Government*, Nanchang, China. Jangxi Science and Technology Press, China, pp.416-421, 2003.
33. Pîrvănescu, A., Bădică, C., Ghanza, M., Paprzycki, M.: Developing a JADE-based Multi-Agent E-Commerce Environment. In: Nuno Guimares and Pedro Isaias (eds.): *Proceedings IADIS International Conference on Applied Computing, AC'05*, Algarve, Portugal. IADIS Press, Lisbon, pp.425-432, 2005.
34. Pîrvănescu, A., Bădică, C., Ghanza, M., Paprzycki, M.: Conceptual Architecture and Sample Implementation of a Multi-Agent E-Commerce System. In: Ion Dumitrache, Catalin Buiu, (Eds.): *Proceedings of the* 15th *International Conference on Control Systems and Computer Science CSCS'15*). "Politehnica Press" Publishing House, Bucharest, 2005, Vol.2, pp.620-625
35. Rolli, D., Eberhart, A.: An Auction Reference Model for Describing and Running Auctions, Wirtschaftsinformatik 2005, Physica-Verlag, pp.289-308.
36. Rolli, D., Luckner, S., Gimpel, H., Weinhardt, C.: A Descriptive Auction Language. In: *International Journal of Electronic Markets*, 2006, 16(1), pp. 51-62.
37. Skylogiannis, T., Antoniou, G., Bassiliades, N.: A System for Automated Agent Negotiation with Defeasible Logic-Based Strategies – Preliminary Report. In: Boley, H., Antoniou, G. (eds): *Proceedings RuleML'04*, Hiroshima, Japan. LNCS 3323, Springer-Verlag, pp.205-213, 2004.
38. Tamma, V., Wooldridge, M., Dickinson, I: An Ontology Based Approach to Automated Negotiation. In: *Proceedings Agent Mediated Electronic Commerce, AMEC'02*. LNAI 2531, Springer-Verlag, pp.219-237, 2002.
39. Trastour, D., Bartolini, C., Preist, C.: Semantic Web Support for the Business-to-Business E-Commerce Lifecycle. In: *Proceedings of the WWW'02: International World Wide Web Conference*, Hawaii, USA. ACM Press, New York, USA, pp.89-98, 2002.
40. Wooldridge, M.: *An Introduction to MultiAgent Systems*, John Wiley & Sons, 2002.
41. Wurman, P.R., Wellman, M.P., Walsh, W.E.: A Parameterization of the Auction Design Space. In: *Games and Economic Behavior*, 35, Vol.1/2, pp.271-303, 2001.
42. Wurman, P.R., Wellman, M.P., Walsh, W.E.: Specifying Rules for Electronic Auctions. In: *AI Magazine* 23(3), pp.15-23, 2002.

Creating Visual Browsers for Large-Scale Online Auctions

Mao Lin Huang[1], Quang Vinh Nguyen[1], and Wei Lai[2]

[1]Faculty of Information Technology, University of Technology, Sydney, PO Box 123 Broadway, NSW 2007, Australia

[2]School of Information Technology, Swinburne University of Technology, P.O. Box 218, Hawthorn, VIC 3126, Australia

Abstract

This chapter discusses the requirements raised for running online auctions as well as the technical issues on the design of graphical user interfaces and how we could use these graphical interfaces to help users navigate through the large on-line auction sites. We will introduce a very efficient visualization technique called *EncCon* as well as the design of graphic attributes that can be used to present the domain specific attributes of the auction items and the relational structures among these items, and these graphic presentations will provide users with a clear map showing the possible paths to the target items. We will demonstrate the effectiveness of our techniques by illustrating an online auction prototype that simulates the ordinary auction activities with the assistance of visualization.

1 Introduction

Over the past few years, electronic commerce (or e-commerce) has emerged as a dramatic new model of business (Bakos, 1998). One of the greatest potentials of e-commerce is its ability to bring the effective-ness and unprecedented massive scale of buyers and sellers from all over the world. This property benefits both sides so that the buyers might have greater product diversity with potentially lower prices, and the sellers are able to reach a greater numbers of potential customers (Hahn, 2001). At any time, through the online shopping stores (or auction websites), cus-

M.L. Huang et al.: *Creating Visual Browsers for Large-Scale Online Auctions*, Studies in Computational Intelligence (SCI) **37**, 579–600 (2007)
www.springerlink.com

tomers can learn more about the products, buy goods with electronic cash, and even have information goods delivered over the network. On the other hands, suppliers can reduce the overhead costs by investigating less in physical stores and distribution channels (Kim, 1999).

An important precondition to the success of e-commerce systems, or specifically online auctions, is the construction of appropriate customer interfaces, from which online product catalogues can be retrieved, is one of the key elements. Many extensive research projects have been done on both components of the online product catalogue including the content management and the catalogue interface. For content management, a number of products have been developed and used at commercial website such as *CardoNet*, *Interwoven*, *OnDisplay*, *Poet Software*, *Vignette*, etc (Neumann, 2000). Various methods that support product search and navigation have been developed for catalogue interface such as those systems in (Callahan & Koenemann, 2000), and (Huang & Zhang, 2002).

Currently, a majority of commercial auction websites, including ebay.com, ubid.com, bidz.com, best-bidz.com and others, provide users with both the basic click-through navigation scheme, which is based on HTML pages, tables and lists, and add-on navigation aids. These auction websites typically categorize products as hierarchies so that users can retrieve auctioned items by hierarchically clicking on a number of HTML pages. The add-on navigation aids aim to provide navigation functions customizable to each user's need, such as search engines and personalized recommendations. In addition, multiple views of lists are usually used for the ease of seeking interesting items. These views include 'Current' (i.e. the de-fault view of all items), 'New Today' (i.e. the new items posted today), 'Ending Today' (i.e. the items ended today), and others.

Although the available navigation techniques can effectively assist sellers/buyers in searching and accessing product information over the *World Wide Web*, they mostly use the text-based interface that allows users to navigate by clicking-through several pages via URL links. Therefore, it could be difficult for the users to perceive the overall view of structure of the product hierarchy by reading these textural lists.
Figure 1 shows an example of a text-based interface that is used on eBay's online auction website for browsing the product catalogue. The add-on navigation aids from those dominant commercial auction websites could help users to quickly locate interesting items. However, these aids, including search engines and personalized recommendations, might not be comprehensible enough for users who would like to do further research and

analysis on several thousands items. Consequently, graphical visualization and navigation techniques are necessary for commercial auction websites so that they can effectively provide an overview of several thousands of items as well as quickly navigate to particular items.

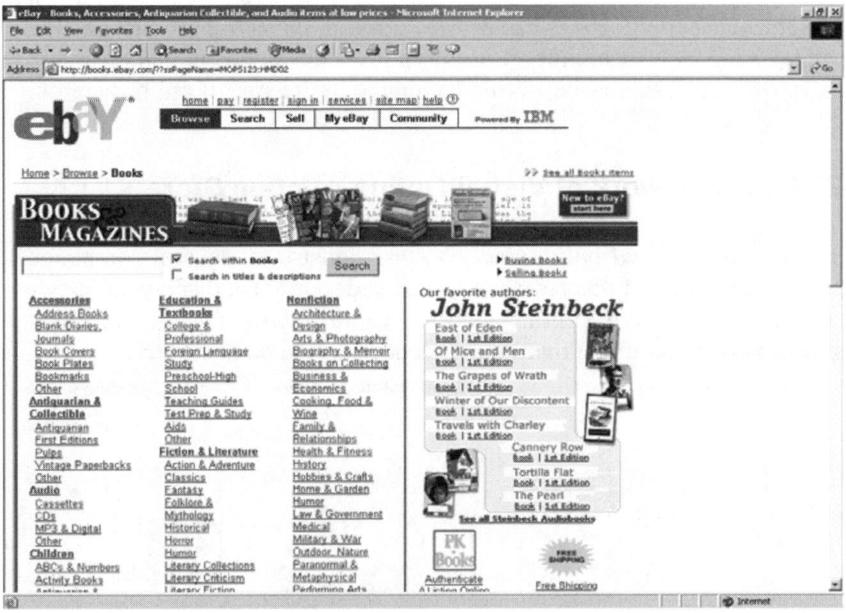

Figure 1: An example of the traditional text-based interface for online auction. (Source from: http://www.ebay.com, accessed 20/03/2004)

Some newly developed visualization approaches have been proposed and implemented to enhance the presentation of product hierarchies for navigation. They aim to improve the readability, understandability, and comprehension of underlying hierarchical structure and to reduce the cognitive overhead for understanding thc structure. These techniques are primarily for two-dimensional graph/tree visualization techniques to display and navigate the product catalogues. The technical detail of these visualization techniques can be found at (Huang & Zhang, 2002), (Lee, Lee & Wang, 2001), and (Inxight). However, there is still little research on the visual navigation of online auctions.

This chapter describes a new visualization approach for navigating large-scale online product catalogues of online auction stores. The visualization technique uses *EncCon tree* layout algorithm (Nguyen & Huang, 2005) that displays the entire product hierarchy as well as a small portion of focused sub-hierarchy. We also introduce a new attributed visualization model that can be used to visualize the relational data with many associated attributes. We split the visualization model into two mappings between the abstract data and physical pictures, Geometric Mapping and Graphical Mapping. Users can browse through the entire product catalogue via *layering view* technique. A prototype was developed to demonstrate the effectiveness of this visualization technique in the area of online auction.

2 The Framework of Visual Online Auction Store

The proposed *visual online auction store* consists of several components. Within the scope of this research, we consider only the display and navigation components of the online auction. Figure 2 shows the components and interconnections among them in the context of online auction.

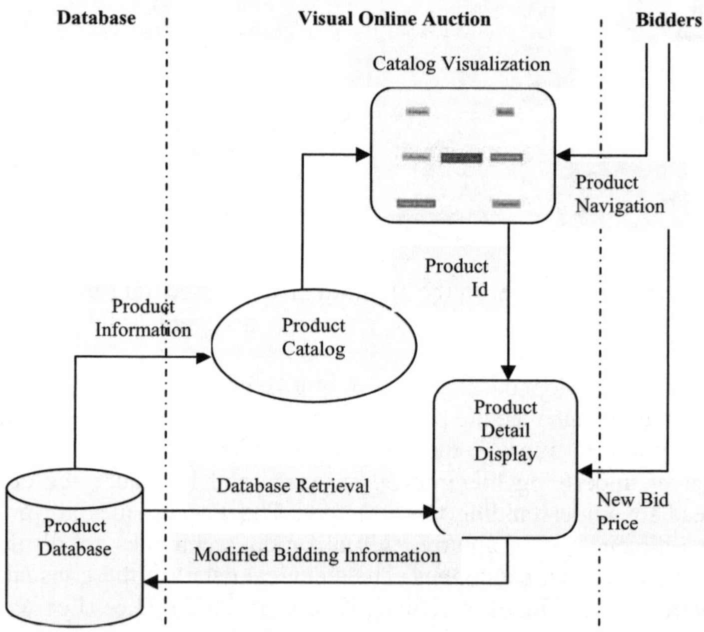

Figure 2: The framework of a visual online auction store.

- **Product database:** is a relational database used to store product information, including all data fields, attributes, and bidding information associated with a particular product that is available for auctioning. We used a MySQL database in our implementation.

- **Product Catalogue:** is a content management system that assembles, indexes, aggregates and normalizes product information from the product database, and quickly distributes the product catalogue information.
- **Catalogue Visualization:** is a visual navigational interface that automatically displays the entire product catalogue's hierarchy, including categories, subcategories, and products. This component employs the *focus+context* visual layout and navigation mechanism (Nguyen & Huang, 2004) that allows users not only to view the entire product hierarchy, but also to interactively browse down to a particular auctioned item.
- **Product Detail Display:** is a web page generated on the server side by a particular scripting language, PHP, in our implementation, to show all the appropriate information of the selected product. This page also displays the product's bidding information, and it allows the authenticated bidder to input the bidding price for the product.

2.1 The Procedure of Online Auction

- **The Role of Costumer:** our costumer is defined as an Internet User who accessed to our auction site for either obtaining information or conducting business as a seller or a buyer of a particular item for auction. For those costumers who want only to obtain some information about the auction we classify them as non-members. In contrast, we classify others who want to participate in the auction business process as members of the website.
- **Navigation/Bidding Session:** this procedure allows users, including members & non-members to freely browse through a multi-level hierarchical structure of items for viewing and bidding at the site. This session allows registered members to browse, search and bid for items. Non-member visitors are only allowed to browse and search the items. In the item list page, a summarized description of found item will be displayed, including: Item Name, Current Bid (A$), Starting Date, Closing Date and Number of Bids. Each item has a link to another page showing its complete information.
- **Sell Session:** this session is for registered members only. It is used to place items for online auction. This page includes an online form allowing sellers to enter the details of their items, including Item Name, Item Image File, Starting Price, Reserved Price, Ending Date and Description of the Item, and add them under a particular Category and Subcategory.

The information will then be saved into the database. The image file(s) will be stored somewhere in the server.

- **Registration Session**: this session allows a visitor to register on the system by filling and submitting an online form. The Registration Form includes Username, Password, Confirm Password, First Name, Last Name, Email Address, Phone No, Street Address, Suburb, State, Postcode, Country and Credit Card Information. A confirmation email will be send to the user if her/his application is accepted. Once the registration is completed successfully, the user then becomes a formal member of the site.
- **Member Session:** this session is for registered members only. In this system, a member can play both roles: the seller and the buyer, i.e. places items for selling or places bids for buying. The member session includes several functions allowing users to browse and bid on items, place books for selling, update his/her personal information, and change their password.
- **Administration Session:** this session includes all functions for managing the system. The administration includes the following functions: Managing the members, including viewing the information of all members, and removing particular members, a function that checks the end-date for all items. If an item reaches its close date, an email will be send to the seller with information telling her/him whether the highest bidding price reaches the reserve price. If the highest bidding price is higher than the reserve price, then it will give the details about the potential buyers and the highest bidding price, and the money that she/he has to pay for the commission (This amount is 10% of the value of the sold item). An email will also be sent to the highest-price bidder to congratulate them for winning the item. It will also contain information about the seller. However if the highest bidding price does not reach the reserve price, then it will ask the seller to reset a new reserve price for the next period of auction.

3 Dynamic Visualization of Online Auction's Product Catalogue

The visualization of the product catalogue for online auction is implemented using the Java programming language. This visual browsing window does not replace entirely the traditional text-based interface, but provides extra assistance to users. In addition, the size of this applet window can be adjusted to suit each user's preference.

EncCon tree algorithm was used to lay out the structure of the product catalogues. In this visualization, nodes are used to represent the objects (such as categories, subcategories and auction items), while edges are used to present relationships among the objects or the relations among auction items and categories.

There are several alternative approaches in the design of a navigational structure for online auction sites. The navigational structure can be either *breadth-oriented* or *depth-oriented.* The *breadth-oriented* structure has the advantage of guiding users to their target item with the minimized number of mouse clicks, while *depth-oriented* structure enables the user to browse through more specific sub-category of interesting items effectively. However, the *depth-oriented* navigational structure requires more intermediate levels of retrieval (Hahn, 2001). Although the use of appropriate navigational structure purely depends on the nature of applications, our auction prototype system uses *breadth-orientated* structure in its implementation. On the other hand, the navigational structure can be either single-only or multiple hierarchies. The use of multiple hierarchies may increase the chance of locating a target item of interest, but it often confuses the user because of its inconsistency through the site. The navigation structure used in our implementation is a single-only hierarchy.

It is desired that the chosen layout algorithm takes its advantage of geometric space efficiency, speed, and aesthetics. The above features and advantages of our layout technique ensure the capability of handling large or very large scale visualization with several levels of hierarchical views, i.e. a complex online auction's product catalogue with thousands of auction items. In other words, this could improve the scalability of the traditional interface. The layout also provides an overview of the entire category. This helps users have a better understanding of the overall structure of the product catalogue. Figure 3 shows an example of the visual navigational window (that displays all categories, subcategories and auction items) and the main window for browsing online auction's product catalogue.

We use our new *layering view* technique for navigating product catalogues. In the visualization, the layout of the overall context is overlapped with the layout of focused sub-hierarchy. Specifically, a semi-transparent technique is employed to display these dual views in the same geometric area, in which the focused sub-hierarchy is called the detail-view and the context of the hierarchy is called the global-view. This allows users to explore the entire hierarchy quickly by moving around the sub-hierarchies.

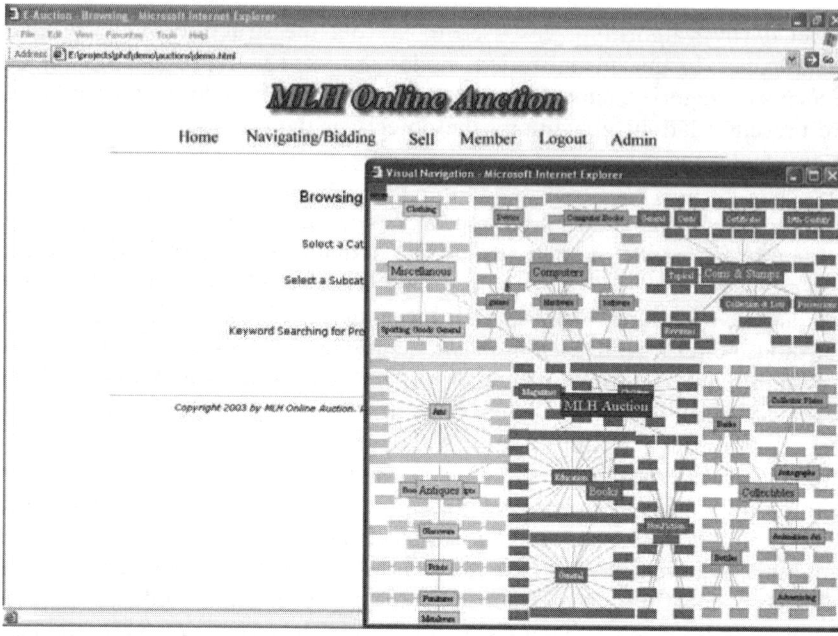

Figure 3: An example of the visual navigation window and the main window of the online auction's prototype.

Each visual interaction is accommodated by an animation in order to preserve the mental-map of the user during the navigation. In more detail, there are two states of the visualization: *normal* and *context*. Normally, the users' attention is on a particular sub-catalogue from the *detail-view*, and when the *context* state turns on, users' attention moves to the content of the *global-view*. At the *normal* state, the selected sub-hierarchy is displayed with no transparency, while the context is partly transparent and is displayed in brighter color (see Figure 5). At the *context* state, the context is brought from the back to the front and displayed with no transparency, while the detail information is sent from the front to the back and displayed with brighter and partly transparent color. These two views can be shifted interactively by using a left mouse-click on the background of each layer.

The visualization uses different colors to present items and subcategories of different categories. The categories and subcategories are also presented with bold boundary to identify auction items within the domain. These displays aim to improve the clarity of the visualization. The system also provides a mechanism to highlight the new products, ending-today products, and others. This aim to improve the overall display where the users can easily find the special items through the product catalogue (see Figure 4). We also provide an interactive menu allowing users to adjust the display to their preferred styles. When the mouse is moving over a node, the sub-hierarchy of the focused node is highlighted to emphasize the selection. In addition, if the focused node is an auctioned item, brief information of this product will be displayed. This property reduces the navigation time since the users can quickly view information of the item from the visual navigation window. In our prototype, the brief information includes current bid price, starting date and closing date (see Figure 6). Finally, from the focused item, the bidders can also double-click on a particular product node in order to display all of the information associated with that auction item in the main window (see Figure 7).

Figure 4 shows a global view of a product catalogue of the prototype system, MLH Online Auction. From this figure, we can quickly identify a new product at the 'Computers-Games' categories. This item is highlighted by being painted with darker color at their front-end. The figure also indicates that the user is focusing on the category 'Computers'. Figure 5 shows the next display when the node 'Computers' is selected. You can see from the display of this figure that the subcategories and product of the category 'Computers' are enlarged and occupied the entire screen, while the previous context view in Figure 4 is reduced and sent to the background with semi-transparency.. One can see that the display is reversed and that the context is brought from the background to the front and displayed with full colors, and the detail-view is sent from the front to the back and displayed with semi-transparent colors.

4 Attributed Visualization

Traditional techniques of relational information visualization (Battista, Eades, Tamassia, and Tollis, 1999, Eades, 1984) concern only the viewing of the abstract data and relations amount data items. They use *graphs* to model relational structures: the entities are *nodes*, and the relationships are *edges* (sometimes called *links*). For example, the structure of an Online Auction site can be modeled as a graph: the nodes are auction items, and a catalogue relationship is represented as a directed edge. These graphs are typically drawn as diagrams with text at the nodes and line segments joining the nodes as edges. These drawings provide a readable and comprehensive graphic format of the relational structure for easy understanding.

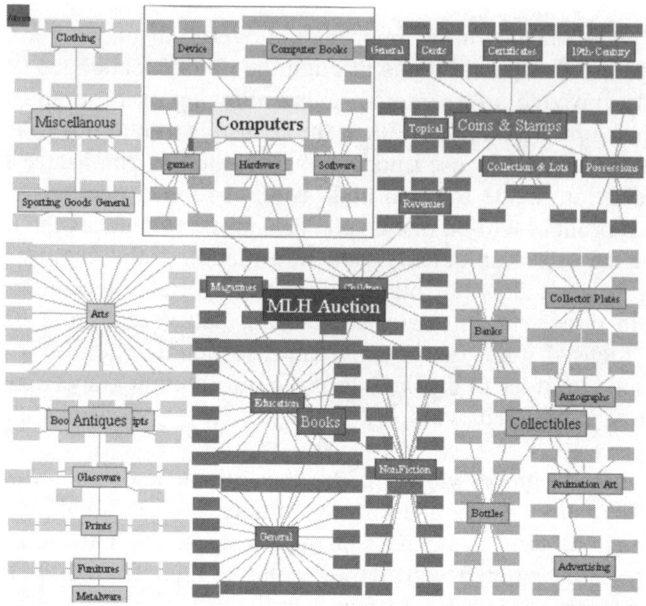

Figure 4: The global view of the entire product catalogue.

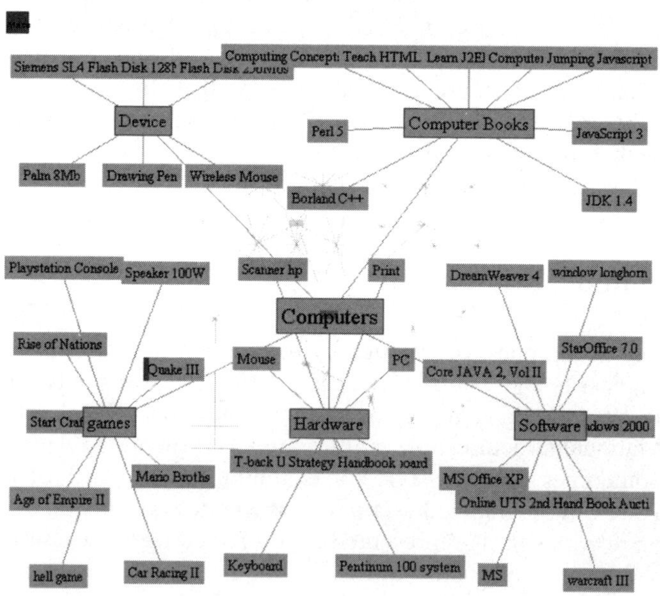

Figure 5: The display of all subcategories and auctioned items belonging to the category "Computer".

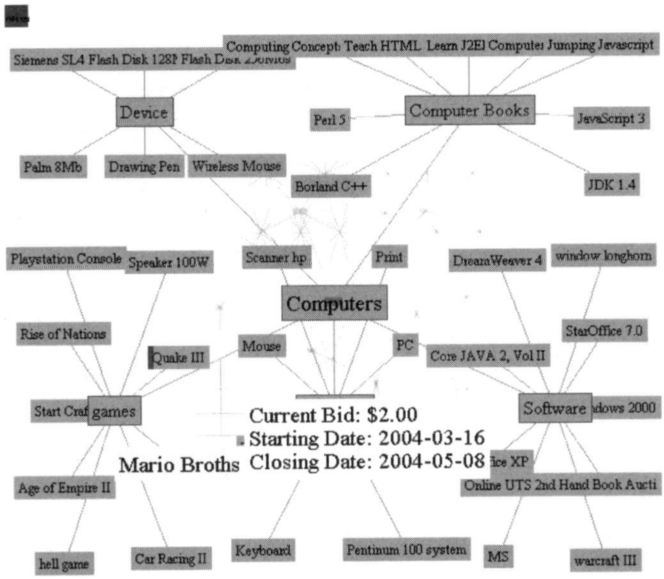

Figure 6: The display when the mouse is over a product. The system pop-ups a layer to show more detail of the auctioned item.

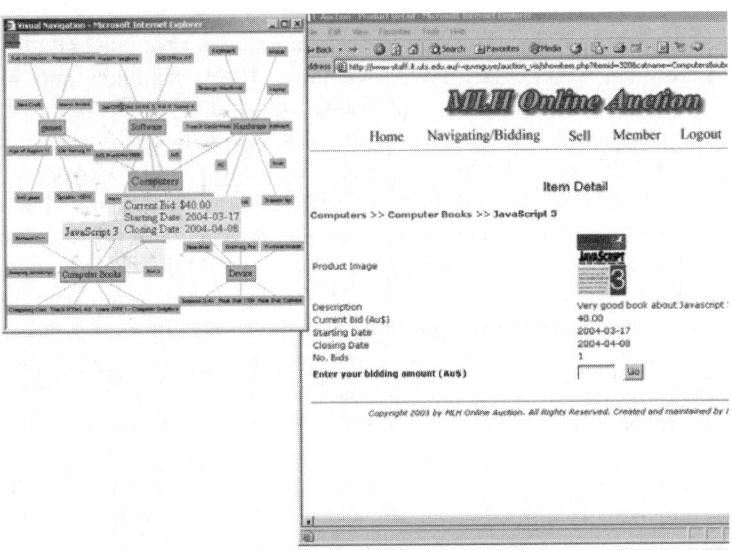

Figure 7: An example of the display of both visual navigation and the main window when the bidder double-clicks on the item “JavaScript 3” from categories “Computers-Computer Books”.

The goal of the traditional *graph drawing* is to convert the abstract relational structure into a 2D/3D geometric space for visualization. This geometrical mapping involves the appending of two geometric attributes (x_a and y_a coordinates) to each *node a*, and four geometric attributes ((x_a, y_a), (x_b, y_b)) to each *edge* (*a, b*) in a graph. There are many applications in which the traditional graph drawing methods can be used to convey information, such as *family trees, software design diagrams* and *web site-maps*.

However, the data we want to visualize in the real world is much more complex than those can be simply presented in a 2D/3D geometric plane. For example, in the drawing of a web graph, an *edge* can only be used to present a hyperlink with four geometrical attributes ((x_a, y_a), (x_b, y_b)), and it cannot present the communication protocol (such as *http, ftp*, *mailto* or *telnet*) associated with the hyperlink. The pure geometric presentation of a web graph does not show any further detail about a particular HTML document, such as the *access frequency* and *the connectivity* of the document. Thus the traditional information visualization techniques and graph drawing methods tend to be inadequate to represent the relational data which has more attributes associated with it. We call this type of the data *Attributed Relational Data.*

Some alternative graph drawing methods have been proposed (Eades, Lai, and Mendonca, 1994, Lai, 1993, Kamada, 1989, Kamada and Kawai, 1988). They used a weighted graph model to address the problem of visualizing attributed relational data, such as the data describing the *E-mail Traffic* on the Internet (Eades, Lai, and Mendonca, 1994). However, these methods still restrict the solutions of visualization within the domain of graph drawing, which concerns only *the geometric mapping* between the data and target pictorial representation. They do not take the advantages of rich graphics available on most of the PC/workstations to achieve the second mapping: *the graphical mapping* in which a set of rich graphical attributes is used to represent the domain-specific attributes of the data. Other systems, such as described by (Becker, Eick and Wilks, 1995), concentrate on specific applications.

We are proposing techniques to visualize attributed relational data. We introduce our new visualization model that converts the attributed relational data into a target pictorial representation through two mappings: *the geometric mapping* and *the graphical mapping.* Our aim is to provide techniques that help human to distinguish the variety of attributes associated with the relational data through the use of graphical attributes. This technique is implemented as a system for visualizing web graphs. It uses a force-directed graph drawing method (Eades, 1984) to achieve the *geometric mapping*, and uses variety of graphical attributes to represent the metrics (attributes) associated with each *HTML document* and each *hyperlink*.

4.1 The Model of Attributed Visualization

A graph G consists of a finite set N of nodes and a finite set E of edges, where each edge is an unordered pair of nodes. A node μ is said to be *adjacent* to a node v if (μ, v) is an edge of G; in this case, the edge (μ, v) is said to be incident with μ and v. A *drawing* of a graph $G = (N, E,)$ consists of two functions, D_n: $N \rightarrow R^2$ that associates a location $D_n(v)$ to each node $v \in N$, and a function D_e: $E \rightarrow C$ that assigns a curve $D_e(u,v)$ to each edge (u,v) in E such that the endpoints of $D_e(u,v)$ are $D_n(u)$ and $D_n(v)$. In practice, the nodes and edges have domain-specific attributes. For example, a hyperlink in a web graph has a protocol attribute (*http, ftp, mailto, etc*), and a node in a network may have a numerical attribute representing the amount of traffic passing through the node.

An attributed graph $A(G) = (A(N), A(E))$ consists of a finite set $A(N)$ of attributed nodes and a finite set $A(E)$ of attributed edges. Each attributed node $a(u) \in A(N)$ consists of $(u, DA(u))$ where $DA(u)$ is a set of domain-specific attributes associated with node u in the graph. Respectively for each attributed edge $a(u,v) \in A(E)$ we have that $a(u,v) = ((u,v), DA(u,v))$ where $DA(u,v)$ is a set of domain-specific attributes associated with edge (u,v) in the graph. In practice, the underlying graphics system has a set of available *glyphs* and a set of *linetypes*. Using this terminology, a drawing of a graph G maps each node u to a glyph g at a location $D_n(u)$ on the screen and each edge (u,v) to a linetype l with endpoints attached to the glyphs at $D_n(u)$ and $D_n(v)$.

Each glyph g (respectively each linetype l) has a set A_g (respectively A_l) of *attributes*. The attributes of a glyph include *shape, size*, and *colour*. The attributes of a linetype include *shape* (*Bezier, B-spline, etc*.), *thickness, linestyle* (*solid, dashed, dotted etc*.) and *colour*. The attributes $\{a_g^1, a_g^2, a_g^3, \ldots a_g^x\}$ of a glyph g is a set of specific values of different types (*boolean, Integer, character, etc*) that associated with g. For example, suppose that a_g^i is a *shape* attribute associated with glyph g, we then can have a set of possible values, *"Rectangle", "Oval", "Curve", "Polygon"*, etc of the *character* type assign to this attribute. If a_g^i is a *size* attribute, then we may assign a non-negative integer value (from 0 to 1024) to it as the actual number of pixels.

An *Attributed Visualization* (the underlying graphics) $AV(G) = (GLYPH(N), LINETYPE(E))$ of a graph G consists of a finite set $GLYPH(N)$ of *glyphs* (graphical entities) and a finite set $LINETYPE(E)$ of *linetypes* (graphical entities). Each glyph $g(u) \in GLYPH(N)$ consists of $(A_g, D_n(u))$ where A_g is a set of graphical attributes associated with g in the visualization, and the drawing $D_n(u)$ is a geometric location of u in a 2D plane with x, y coordinates associated with u. Respectively for each linetype $l(u,v) \in LINETYPE(E)$ we have $l(u,v) = (A_l, D_e(u,v))$, where A_l is a set of graphical attributes associated with $l(u,v)$ and the drawing $D_e(u,v)$ is a geometric curve representing the relation (u,v) in a 2D plane, which includes two endpoints for the curve (u,v). Under this schema, the visualization problem becomes the problem of mapping between attributed graphs $A(G)$'s and attributed visualizations $AV(G)$'s.

5 Translation of Data into Pictures

In this section, we describe a framework for visualizing attributed relational data. In this framework, the translation of data into pictures includes two steps: *the geometric mapping* and *the graphical mapping*. We present the conceptual model of the data translation, and implement this model by using Java.

5.1 The Transition Diagram of Attributed Visualization

Figure 8 illustrates a transition diagram for attributed visualizing. This diagram is made up of several components. These components are described below:

- **Attributed Relation Data:** The real world relational data that we want to visualize. It includes a set of relationships among the data objects, and a set of domain-specific attributes associated with each data object and relationship (relation object).
- **Attributes:** A set of domain-specific attributes (properties) that are associated with a particular data object or relation object in a specific domain of the real world.
- **Relations:** Relationships among the data objects. As a certain type of the objects, a relation can associate a set of domain-specific attributes

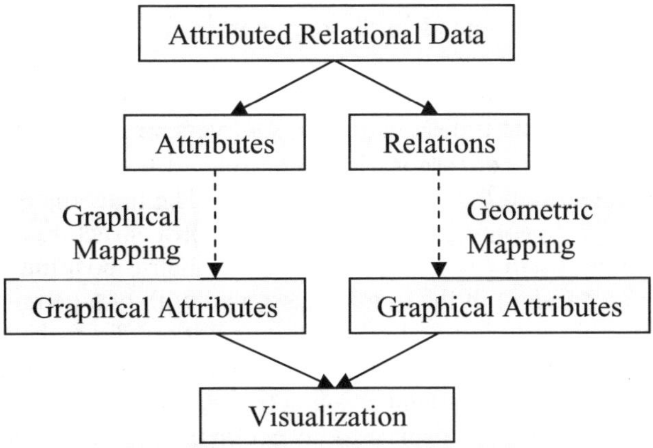

Proposed Visualization Model

Figure 8: The Framework of Attributed Visualization.

with it. These attributes determine the type of the relationship between two objects.

- **Graphical Attributes:** A set of graphical properties that are associated with a particular graphical object *glyph* or *linetype*, which is the graphical shadow of a particular data object or relation object.
- **Geometric Attributes:** A set of geometric properties that are associated with a data object /or a relation object. For example, in a 2D plane, a data object usually associates two geometric properties, *x* and *y* coordinates in a plane.
- **Visualization:** The final pictorial representation of attributed relational data. This includes representations for data, relational structure and domain-specific attributes that are associated with the data/relation objects. The basic elements of the visualization are *glyphs* and *linetypes*. Each glyph *g* (respectively each linetype *l*) represents a data object $v \in N$ (respectively a relation object $e \in E$). The attributes $A_g = \{a_g^1, a_g^2, a_g^3, \ldots a_g^x\}$ of a glyph g is used to represent the domain-specific attributes of a data object $v \in N$ that is graphically mirrored on *g*. Respectivcly the attributes $A_l = \{a_l^1, a_l^2, a_l^3, \ldots a_l^y\}$ of a linetype *l* is used to represent the domain-specific attributes of a relation object $e \in E$ where *l* is a graphical shadow of *e* in the visualization.

During the transition, the original data could have three different representations in each stage of the translation process.

- **The abstract graph representation:** This level of the representation describes the abstract relational structure of the data objects and relations among the objects in the real world. These data objects are represented as a set of *nodes*, and these relations are represented as a set of *edges* in a graph model. Different applications of the relational date in the real world are translated into this universal representation.

- **The geometric representation:** The second level of the representation is generated after the geometric mapping. The outcome of the geometric mapping is a geometric structure. A geometric structure means that each data object (node) is assigned with a geometric position and each relation object (edge) in the graph is assigned with two geometric positions, a *start-point* position and a *end-point* position. To make this geometric structure visible and allow users to view this geometric structure in a 2D plane, some simple graphical objects, such as *rectangles* and *lines*, are used to graphically display these geometric objects. This level of the representation is sufficient for the visualization, as the target representation, of *non-attributed relational data.*
- **The geometric + graphical representation:** This level of the representation is the target representation - the actual pictorial representation for the visualization of *attributed relational data.* It is generated after the graphical mapping. The outcome of the graphical mapping is a picture. A picture consists of many graphical objects, and each graphical object can have many graphical properties (including geometric and graphical properties). These properties, such as *position, shape, size, color, brightness* and *z-coordinate*, can be used to represent the main features of the data/relation objects as well as the attributes associated with the data and relations in the real world.

5.2 The Geometric Mapping

The geometric mapping is the process of creating geometric representation of data. It converts the abstract data structure into a network with geometric *points* and *links* on a 2D plane. It assigns a geometric position (x, y) to each graphical glyph $g(v)$ that is the shadow of a data object v in a visualization. It also defines the *start-point* and the *end-point* for each graphical linetype $l(e)$ that is the shadow of a relation e in the visualization. Suppose that we have a finite set N of data objects and a finite set E of relation objects; then the actual geometric mapping from data set N to the drawing $D_n(N)$ can be illustrated below:

$$N = \{ n_1, n_2, n_3, \ldots n_x \}$$

$$D_n = \{D_n(n_1), D_n(n_2), D_n(n_3), \ldots. D_n(n_x)\}$$

The actual geometric mapping from relation set E to the drawing $D_e(E)$ can be illustrated below:

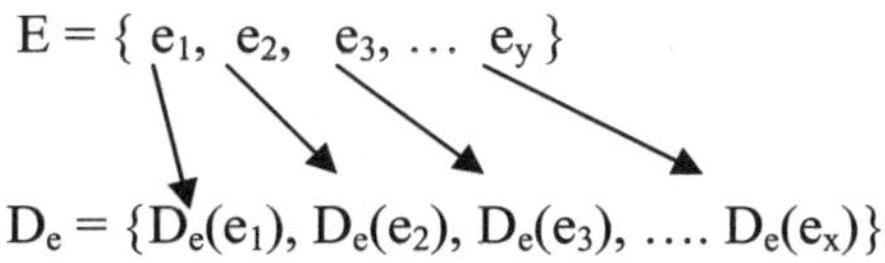

These mappings are one-to-one which means that there is only one geometric shadow created for each data object v and each relation object e after the mapping process. It assigns a location $D_n(v)$ to each data object v and assigns a curve $D_e(e)$ to each relation object e.

5.3 The Graphical Mapping

The graphical mapping is the process of creating pictorial representation of the data. It converts a set of domain-specific attributes DA_v of a data object v (respectively a relation object e) into a set of graphical attributes A_g (respectively A_l) of a graphical object g (respectively l) for visualization. Note that g (or l) is a graphical shadow (representation) of data v (or relation e). Suppose that v is a data object associated with a set DA_v of domain-specific attributes. There is a graphical object g that can have a set A_g of graphical attributes associated with it. The actual graphical mapping from v to its graphical shadow g can be done as illustrated below:

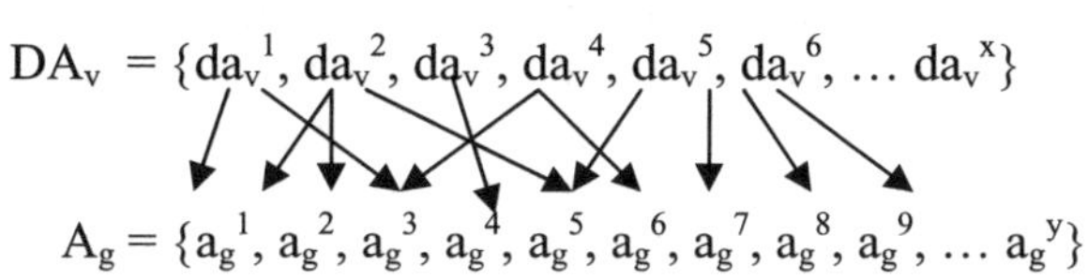

Respectively the graphical mapping from a relation object e to its graphical shadow l can be done in the same way.

These mappings are many-to-many mapping which means that there are more than one graphical attributes can be used to represent a domain-specific attribute da_v^i (respectively da_e^i) that is associated with a data object v (respectively a relation object e). This also means that a graphical attribute a_g^i can be used to represent more than one domain-specific attribute in the visualization.

5.4 Implementation on Auction Web Graphs

The incredible size of the online auction sites accompanied with its large access frequency, introduces great challenges for information discovery. The web has no navigation structure or any sort of complete index of the content available. The problem of navigation across a huge auction site within a minimal time for finding particular items hardly perfectly solved.

One of the ways in which web site designers are trying to address this problem is by providing what is commonly called "site-maps". The idea of a web site-map is based on the geographical metaphor of map. It is used to provide the user with an overall visual picture of the contents of a web site so that the user can easily navigate and obtain the interested information.

Since web site mapping is essentially a process of visualization of the information content of a web site, various approaches are adopted based on human visualization and perception capabilities. Each approach or technique for web site mapping has adopted one or a combination of these capabilities hoping to exploit them to help the user in navigation and comprehending the contents.

In web site-maps, a HTML document can be presented as a *node* in the graph, and a hyperlink can be presented as an *edge* in the graph. However, most of these approaches are only focusing on the pure geometric representations, rather than the graphical representations, of web graphs and they usually just assign some very simple graphical objects *glyphs* (perhaps some simple rectangles of the same size) to the *nodes* with the same graphical properties (such as *size, color* and *shape*) for visualization. These simple graphics are unable to represent the domain-specific attributes associated with the auction item, such as *the access frequency, bidding frequency* and *the connectivity* of an item page. Therefore the user gains no knowledge about the domain-specific attributes of nodes (auction items) and edges (relationships), which are very important to the user where she/he is surfing through the visual structure of a auction catalogue graph.

We apply our new attribute visualization to create auction site-maps. We want to use a set of graphical attributes A_g (respectively A_l) associated with a glyph g (respectively l) to represent a set of domain-specific attributes DA_v (respectively DA_e) of an auction document v (or a hyperlink e) , where g (or l) is the graphical shadow (representation) of a HTML document v (or a hyperlink e).

5.5 Graphical Attributes Associated with Glyphs and Linetypes

In our implementation, each glyph *g* in a web site-map has a set

$A_g = \{a_g^1, a_g^2, a_g^3, a_g^4, a_g^5, a_g^6\}$ of six graphical attributes. They are:

a_g^1 → size of the graphic entity
a_g^2 → background color
a_g^3 → shape of the nodes
a_g^4 → brightness
a_g^5 → highlight/shadows
a_g^6 → Z-coordinate at overlaps

In a web site-map, each linetype *l* has a set $A_l = \{a_l^1, a_l^2, a_l^3, a_l^4\}$ of four graphical attributes. They are:

a_l^1 → length
a_l^2 → thickness
a_l^3 →brightness
a_l^4 → Z-coordinate at crossings

5.6 Creating Pictorial Representation of Auction Graphs

The actual pictorial representation of a web site-map is generated after the *graphical mapping*, in which a set of domain-specific attributes associated with each web object *v* (or *e*) in the graph is mapped to a set of graphical attributes associated with graphical objects *glyphs* (or *linetypes*) in the visualization.

Figure 9 shows the auction web graph. This graph is drawn using an *Enc-Con* algorithm (Nguyen and Huang, 2005) (a kind of space-efficient drawing), without applying Attributed Visualization technique. In the visualization, we can only see the relationships (links) among the nodes (items), and cannot see any domain-specific attributes of auction objects in this visual site-map. Therefore, the user gains no knowledge about the properties (attributes) associated with these objects (items & links) in an auction site-map while she/he is browsing through the map.

Figure 10 shows the similar layout of an auction web graph as shown in Figure 9. However, it uses graphical attributes to represent the domain-specific attributes associated with web objects. This gives the user some ideas of what nodes are more important in the auction localities and worthwhile to have a look.

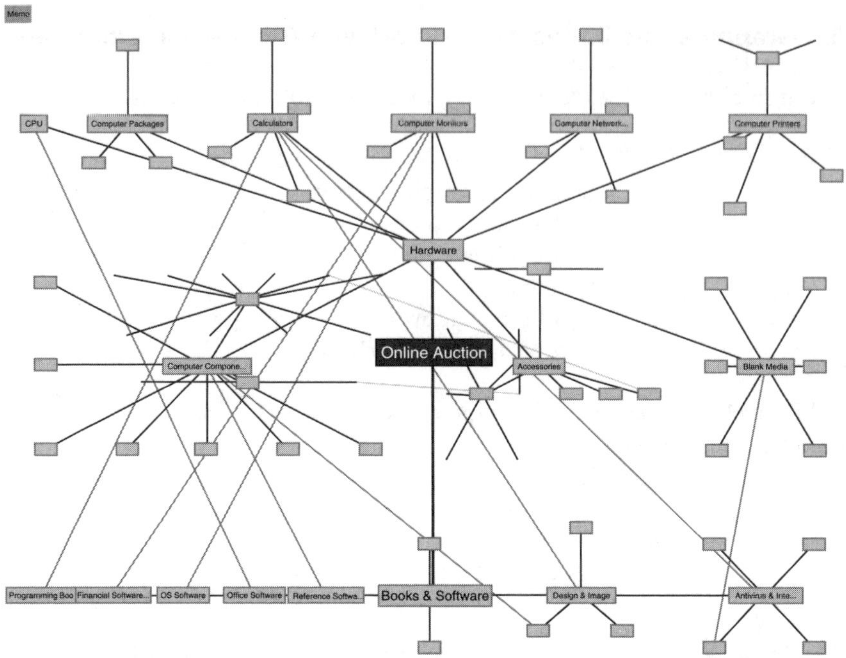

Figure 9: A screen of the original graph visualization of an auction website.

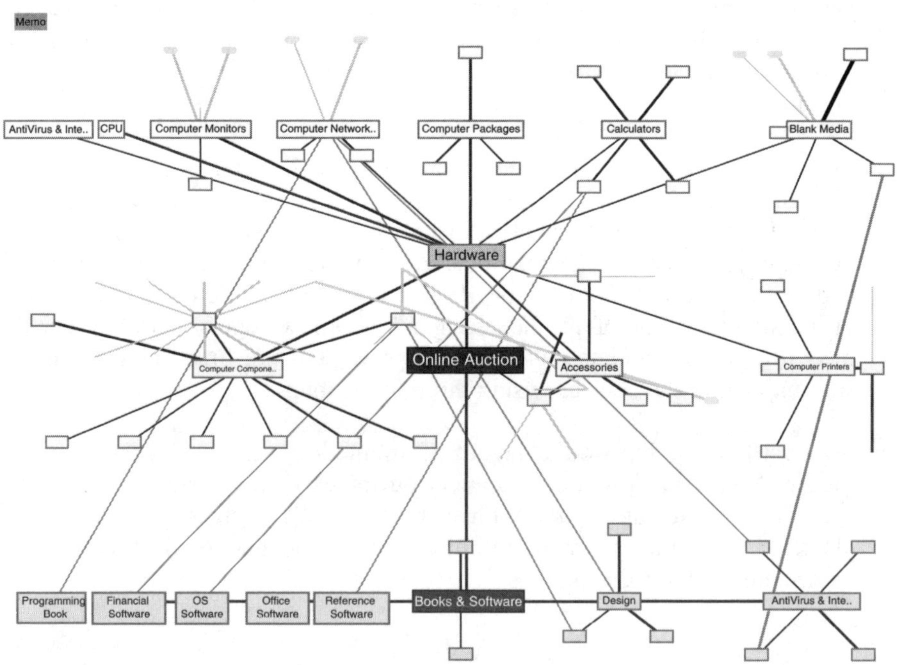

Figure 10: A screen of an attributed visualization of the same auction site as shown in Figure 9.

6 Conclusions

This chapter has presented a new dynamic visual user interface that appears as an additional window in the auction site for assisting the online auction process. This visual interface enables users to view and browse the auction catalogue with a large number of the auction items. A new *focus+context* viewing technique called *layering view* was employed to handle the overlapped display of both the *context view* and the *focus view*. This visualization combines the *EncCon tree layout* and the *layering view* method to provide users with a visual aid for the fast and effective product navigation.

In addition to the *EncCon* layout method, we also use a new attributed visualization to visualize not only the relational structure of the data, but also the domain-specific attributes that are associated with the data. The visual representations of these attributes are very important for users, who are browsing through the visual structure, in understanding of the data objects and the relationship objects that are represented as a set of *glyphs* and *linetypes* in a picture. Pictures produced by attributed visualization are much meaningful than those produced by traditional information visualization techniques.

The use of attributed visualization for auction website mapping can improve the quality of visual auction catalogue browsing. Under this scheme, the user can gain the knowledge about each node (item) in the auction web locality from the visual map. This will directly help the user in making up her/his decision of where to go next for finding particular items, while he/she is interactively browsing through the website via the visualization.

References

Bakos, Y. (1998): The Emerging Role of Electronic Marketplaces on the Internet, *Communications of the ACM*, 41(8): 35-42.

Battista, G., Eades, P., Tamassia R. & Tollis, I. (1999): *Drawing Graphs: Algorithms for the Visualization of Graphs*. USA, Prentice Hall.

Becker, R., Eick, S. and Wilks, A. (1995): Visualizing Network Data. *IEEE Transactions on Visualization and Graphic* 1(1): 16 - 28.

Callahan, E. and Koenemann, J. (2000): A Comparative Usability Evaluation of User Interfaces for Online Product Catalog, *In 2nd ACM Conference on Electronic Commerce (EC-00)*, ACM Press, pages 197-206.

Eades, P. (1984): A Heuristic for Graph Drawing. *Congressus Numerantium* 42:149–160.

Eades, P., Lai, W. and Mendonca, X. (1994): A Visualizer for E-mail Traffic. *In 4th Int. Conf. Proc. Pacific Graphics'94 / CADDM'94,* pages 64-67.

Hahn, J. (2001): The Dynamics of Mass Online Marketplaces: A Case Study of an Online Auction, *In SIG-CHI on Human factors in Computing Systems*, ACM Press, pages 317-24.

Huang, M.L. and Zhang, K. (2002): Navigating Product Catalogs Through OFDAV Graph Visualization, *In International Conference on Distributed Multimedia Systems (DMS'2002)*, Knowledge Systems Institute, pages 555-561.

Inxight. (2004): Hyperbolic Browser (Start Tree) for Online Shopping Stores, http://www.inxight.com (accessed 20/03/2004).

Lai, W. (1993): Building Interactive Diagram Application. Ph.D. thesis. University of Newcastle, Australia.

Kamada, T. (1989): Visualizing Abstract Objects and Relations. *World Scientific Series in Computer Science*, vol 5.

Lee, J., Wang, p. and Lee, H.S. (2001): A Visual One-Page Catalog Interface for Analytical Product Selection. *In Electronic Commerce and Web Technologies, Second International Conference, EC-Web 2001*, Springer, pages 240-249.

Kamada, T. and Kawai, S. (1988): Automatic Display of Network Structures for Human Understanding. Technical Report 88:007. Department of Information Science, Faculty of Science, University of Tokyo.

Kim, J. (1999): An Empirical Study of Navigation Aids in Customer Interfaces, *Behaviour & Information Technology*, 18(3): 213-224.

Mackinlay, J. (1986): Automating the Design of Graphical Presentations of Relational Data. *ACM Transactions on Graphics* 5(2): 110 - 141.

Neumann, A. (2000): A Better Mousetrap Catalog, *Business 2.0*, pages 117-118.

Nguyen, Q.V. and Huang, M.L. (2005): EncCon: An Approach to Constructing Interactive Visualization of Large Hierarchical Data, *Information Visualization Journal*, 4(1): 1-21.

Nguyen, Q.V. and Huang, M.L. (2004): A Focus+Context Visualization Technique Using Semi-transparency, *In the 4th International Conference on Computer and Information Technology (CIT2004)*, IEEE, pages 101-108.

Design and Implementation of Multi-Agents for Learner-oriented Course Scheduling on the Internet

Dong Chun Lee[1] and Keun Wang Lee[2]

[1]Dept. of Computer Science Howon Univ., Korea
ldch@sunny.howon.ac.kr
[2]Dept. of Multimedia Science Chungwoon Univ., Korea
kwlee@chungwoon.ac.kr

Abstract

Recently e-Learning model which is based on Web has been proposed in the part of the new activity model of teaching-learning due to the role of the application of multimedia technology, computer communication technology and multimedia application contents. The demand for the customized courseware which is required from the learners is to be increased, and then the needs of the efficient and automated education agents in the e-Learning are recognized. But many education systems recently did not service fluently the courses which learners had been wanting and could not provide the way for the learners to study the learning weakness which is observed in the continuous feedback of the course. In this paper we propose design of multi-agent system for learner-oriented course scheduling using weakness analysis algorithm on the Internet. The proposed system monitors learner's behavior constantly, and evaluates them and calculates his accomplishment.

1. Introduction

The e-Learning usually realized in the form of Web-Based Instruction (WBI) is a fast-emerging field in education according to the rapid growth of the Internet and information technology. WBI takes advantage of the interaction of hypermedia and a convenient medium to deliver the instruc-

D.C. Lee and K.W. Lee: *Design and Implementation of Multi-Agents for Learner-oriented Course Scheduling on the Internet*, Studies in Computational Intelligence (SCI) **37**, 601–626 (2007)
www.springerlink.com

tional materials on the Internet. WBI can realize the instructional strategies within a constructivist and the collaborative learning environment with its interactive characteristics. WBI also supports learner-centered individual instruction and distance learning with asynchronous or synchronous interactions.

Though WBI offers a very rich and flexible environment and some amount of adaptation and intelligence for education systems, there still remain some inherent problems that hinder the development of such systems. To activate e-Learning paradigm, the e-learning standardization process is an active, continuously evolving process that will last for years to come, until a clear, precise, and generally accepted set of standards for educational-related systems is developed. Among the main contributors to this effort are the IEEE's Learning Technology Standardization Committee (LTSC) (Hamalainen, M.19961), the IMS Global Learning Consortium, the Aviation Industry CBT Committee (AICC), the U.S. Department of Defense's Advanced Distributed Learning (ADL) initiative and the reference model known as Sharable Content Object Reference Model or SCORM. The accomplishments of these standardization efforts can be identified into two specifications for Learning Management System (LMS) and Learning Content Management System (LCMS): one is the specification of the information models and the other is Specifications of the architectures, software components and interfaces. As this e-Learning system is spread widely to the public, the users demand more diverse education service, and that results facilitating study on applied education service being very active (Moore, M.G and Kearsley, G. 1998). Since the agent and broker for the domestic and foreign education software are organized to meet the demands of the average public more rather than customized service for individual learner, it is very difficult to accommodate the various needs for knowledge and evaluated level for each and every individual.

Three types of learning, that is the self-study, the lecturing and the discussions, can be taken into consideration when we transform classroom-based environment into the e-Learning system (Moore, M.G and M, Whinston and Hamalainen, M, Whinston 1996). The self-study can be done by utilizing learning source data from the instruction for complementing the weak point of learner. As similar to the traditional classroom environment, the lecturing, the learner progress his study with the study material and lecture schedule that are presented by instructor or tutor, however, in contrast to the traditional lecturing, the learner can learn by his own schedule in the place where he set randomly (Kirshner, D. and Whitson, J. 1997). Also, like traditional classroom system, questioning and answering can be possible to solve the problem. The discussions can be made by using the bulletin board or chat room that registered in the corresponding course and with the help of those, students can communicate each other to progress lecturing activity. Similar to that fact that classroom education is not done by only one type of method, those Web sites that provide learning also use more than one method for their education service. Therefore, for this Web-based education system, assigning and making-up an appropriate course for each individual is very important information for improving effectiveness of learning for individual (Badrul H. Khan 1997).

Tools used for supporting the interaction between the instructor and the learner occurred while transmitting knowledge, are the email and electronic bulletin board that use asynchronous mode, and the text, the voice chatting and the video conference system that use synchronous mode (Agogino, A. 1994). Although tools to help interaction between learners had been supported in many ways, in instructor's perspective, it is very hard to provide the right course schedule and combinations by analyzing each learner status after facing all registered learners. Hence, agent who can deliver feedback such as effective way of learning, course formation and course schedule to the learner is needed in this Web-based education system (Whinston, A. 1994).

This paper propose multi-agent system, which can provides the appropriate active course scheduling and feed back to learner after evaluating the learner's education level and method. By developing agent which provides the fast and suitable feedback to learner's learning status, we are going to recompose course that is suitable for each learner to increase the effectiveness in the learning.

2. Related Work

The most typical and popular WBI system is a Customized on Demand Education (CODE) system in the University of Texas (Thomas, R.1992). This system is built with electronic commerce concept, and it defines conceptual model for designing education and provide value added service such as production on demand for course. Also as a broker between the learner and the supplier of education, it design the model for education delivery, appoint potential supplier to produce new material based on the standard that is already defined in advance, and suggested the methodology for production of course and its delivery based on the intermediation utilizing course data storage. Besides, by including the method and tools for learning and evaluating learning, it proposed the Web-based model for course delivery and description in the learning environment. However many problem occurred in this theoretical Web based education system when it was practically applied and implemented as an application, and one of the biggest problem among those is the customization and rating a degree of satisfaction. The CODE system provide course required by learner based on the electronic commerce, but it does not present a methodology to improve a degree of accomplishment of learning and learning effect. Moreover this system does not have a proper feedback function necessary for evaluating a degree of accomplishment of learning for each learner.

Playades project that is in progressed in Carnegie Mellon University, applied multi-agent structure Reusable Task Structure-based Intelligent Network Agents (RESTINA) for integrating independent agent to decision-making domain in distributed environment (Katia S. and Dajun Z. 1996). The visitor management system which was implemented in this project arranges the meeting by changing the schedules of the visitors and corresponding researchers in the area of visitors concern. Agents of RETSINA will be classified as following three agents; interface agent, work agent, and information agent. For communication between agents, it has a communication module and Knowledge Query and Manipulation Language is used. In addition to that it has a plan module for planning and a scheduling module for individual sub operation. Agents manage schedule within a certain frame and use planning library to produce working tree.

To solve the complicated work, Playades project focused on dividing the work, then distributing that to proper agents and making notification of results. This can be called agent based structural prototype that leads cooperation by direct communication between agents in distributed computing environment.

Unlike other meeting scheduler, the distributed meeting scheduler developed by Tulsa University has independent scheduler for each user (Sandip Sen. 1994). If you want to have a meeting with other users, request the meeting proposal and negotiate with the agents of other users and create a meeting. The experimented result of this system shows that it requires around 2 to 3 rounds for the meeting, under the assumption that it has three to five attendees, require two or three hours and difference between agents user calendar is about 70 to 80 percent. But for several users, many massages passing is expected and, since agents answer for all massages, so chances are, meeting can be arranged apart from users intention and it might be the revealed limitation for the system. Recently, ontologies are regarded as a core technology and fundamental data model for knowledge systems. Semantic technologies initiated by ontologies, show great promise for the next generation of more capable information technology solutions because they can solve some problems much more simply than before and make it possible to provide certain capabilities that have otherwise been very difficult to support. According to the expansive future vision for semantic technologies by ontologies, numerous researches have been carried out for ontology-based e-Learning modeling [Ronald Denaux1 and Lora Aroyo 2004].

As discussed above, the problem for the previous work is that it does not provide the interaction between the instructor and the learner, face-to-face education system which is a necessary and sufficient condition, and absence of proper support system for feedback between the learner and the instructor in online. The problem from the previous work is summarizing as follows; first, it maintains static course formation. In other words, course should be appropriately reorganized by the performance of learners, but in traditional study initially scheduled course will be applied to all learners without any difference. Second, it does not provide the course that reflecting the level of each learner. Course proper to the learner's level should be provided with mutual interaction between learners and agents to fully understand the level of learners and assign the course that is appropriate, however, the simple evaluation method and test could not provide such courses. Third, function for accurate course measurement of agents is not present. It does not have the course organization agent function, which has been applied by the algorithm designed for getting accurate measurement on learning progress and the objective and optimized course. Therefore, in conjunction with web based the agent skill and education system infrastructure, we are planning to develop course scheduling agent system that is practical and appropriate for learning.

In this paper we propose the learner focused course scheduling agent providing course that can satisfy the learner's tendency by evaluating their level and learning method. The scheduling agent will continue learning about the information regarding understanding individual course and learning result, then after matching new learners profile with proper course source acquired by agent if the other learners require some or similar course, schedule the most suitable course for the learner using the course scheduling algorithm. To maximize effective in the learning for learners and to schedule proper course to each learner, we make various information of learner into learner's profile to help agent learning it.

3. Course Scheduling Multi-agent (CSMA) System

Learning Technology System Architecture (LTSA) is learning system specifications that are written for establishing international standard for virtual education by IEEE 1484 Learning Technology Standards Committee (LTSC), with the consideration of user's perspective in the aspect of information technology when implementing mutually interactive system. The basic structure of learning system using the CSMA is designed based on LTSA standard model.

3.1 Learning System Structure

In this learning system, the learner and the CSMA are connected via Web Interface (WI), and through WI the request and transfer for course scheduling occurs between the learner and the CSMA. The learner's study of the course provides by the CSMA in this system.

All the information created by the CSMA will be stored into the database and if required, it will be loaded by the CSMA and used for reorganizing course. The learner's profile and information obtained by their learning activity as well will be stored into database via WI, then it will be regenerated and stored again as necessary information to the learner such as learning achievement level; course, scheduling, evaluation data, feedback and etc. The learner can ask questions about information needed for their learning activity not only through the WI but also by email, and the CSMA provides many kind of information or feedback to the learner by email that enables learners to get information in addition to the logged learning hour to the Web server. Fig. 1 shows the structure for CSMA system.

Investigated learning behavior of the learner that is appeared thorough the WI is summarized as following; 1) studying about the learning material, 2) questioning and answering for studied contents and 3) evaluation for the learning material. For the assigned learning material in corresponding time interval, a learner can ask questions and expect for an answer, and after finishing studying on that learning material, the studied contents will be tested. Like this way, the CSMA can be course scheduling according to the test result and the learner will study again with learning activity with reorganized course provided by the CSMA.

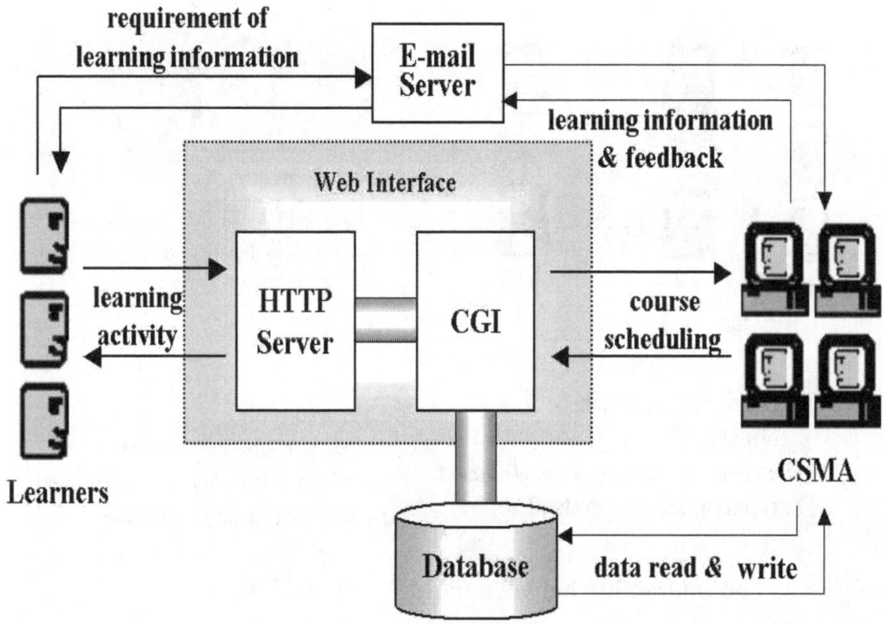

Figure 1: Structure for CSMA System

3.2 Multi-agents of CSMA System

The key component of CSMA consists of four agents; Course Recomposition Agent (CRA), Learning Accomplishment Agent (LAA), Learning Evaluation Agent (LEA) and Feedback Agent (FA). The CRA is delivered the information on the degree of accomplishment of learning from the LAA, and creates it and provides a new and most customized learner-oriented course. The LAA estimates the degree of learning accomplishment based on the test results from the LEA and tracks the effectiveness of learning. The LEA is carrying out learning evaluation at every stage. The FA provides relevant feedback to learners in accordance with the learner's profile and calculated degree of accomplishment of learning. Fig. 2 demonstrates the interaction between multi-agent and database.

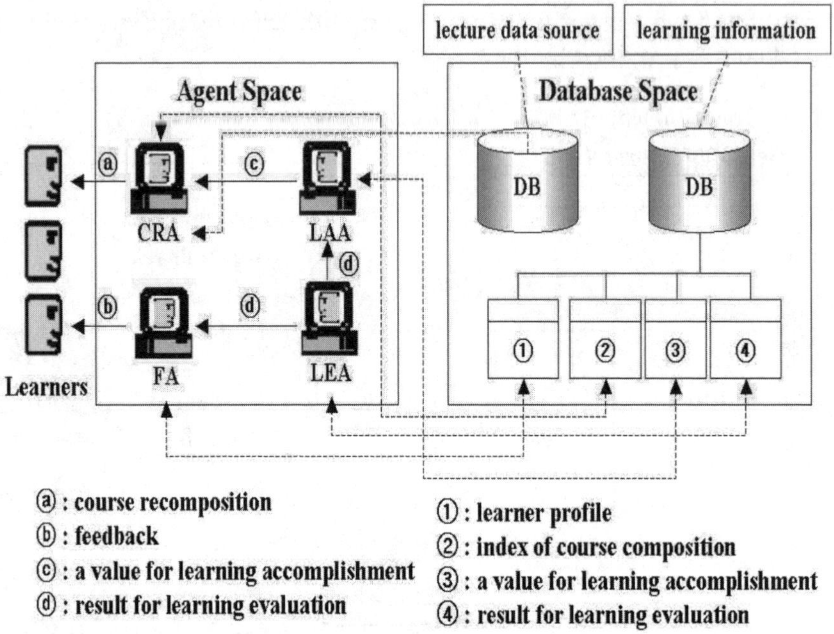

Figure 2: The Interaction between multi-agent and database.

3.3 Course Scheduling Scheme

Although functioning independently of the course re-composition agent, course scheduling cannot make a good performance unless each agent is properly assigned its role and smoothly accomplishes. The term "section" means learning source that is represented by N and "subsection" represented by n in Fig. 3. Each course has a series of stage with a section and a subsection, and learner should step up from the first stage to the next as their learning level. Figure 3 shows the course state diagram to explain the course scheduling scheme.

The course consists of sections from 1 to N and each section is composed of subsection from 1 to n. After completing a subsection, learners can decide whether they study it over again or go back to other subsection they have already studied through the subsection test. If they study the prerequisite subjects by the learning flow focused on section by tutor, they can choose and learn the section.

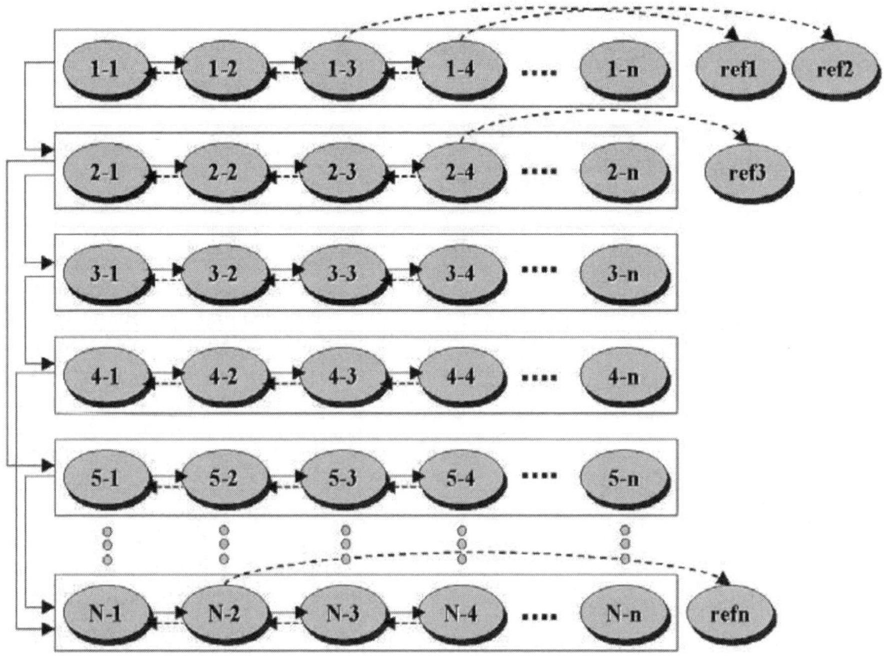

Figure 3: State diagram for course scheduling

Accordingly, the section learning pattern proceeds flexibly as learner's choice not with sequential pace. Furthermore tutors can insert necessary reference materials at their discretion into the subsection for the learner to refer it whenever they want and to continue studying effectively. Table 1 shows the definition of variable for CSMA to enhance the understanding of course scheduling and standardize it.

According to the course scheduling utilizing thc above definition of variable, we get to understand that a learner who studied firstly subsection 1 can proceed to subsection 2 only after he or she gets a certain point on the subsection test. This is the principle of general course learning. According to the learner level set by each test, the course is recomposed by the help of course scheduling agent. Table 2 shows the evaluation rule of subsection and course procedure from stage S (I, 1) to S (I, n-1). If i is from 1to n and i-1is 0, i-1 is resulting in 1. If i+1 is n+1, it is proceed to the subsection test. The learning time is given by the equations (9)-(13).

Table 1. Definition of variable for CSMA

Symbol	Description	Symbol	Description
$S(I, i)$	Learning content of subsection	$W_r(I)$	Weakness of section about repeated learning
I	Index of section	$Q_c(I, i)$	The number of subsection question item in section test
i	Index of subsection	$t_r(I, i)$	Required time for solving subsection question Item in section test
N	The number of sections	$t_d(I, i)$	Needed time for solving subsection question item in section test
n	The number of subsections in a section	$t_a(I, i)$	Average required time for solving one subsection question item in section test
$T(I, i)$	Test of subsection	$t_{ar}(I, i)$	Rate of required time for solving one subsection question item in subsection test
$T(I)$	Test of section	$R(I, i)$	Rate of correct answer for subsection question item in a section test
T_T	Test of final course	$A(I, i)$	Achievement degree for each subsection
T_s	Score of course test	$W(I, i)$	Weakness for each subsection
$W(I)$	Weakness of section	$W_t(I, i)$	Weakness of solving time for each subsection
$P(I, i)$	Evaluation score of subsection	$W_{tR}(I, i)$	Weakness of solving time and correct answer item for each subsection
$G(I, i)$	Evaluation grade of subsection	T_s	Evaluated point of subsection
$t_{ls}(I, i)$	Required standard time for learning subsection	α_1	Coefficient applied for subsection learning time
$t_{lr}(I, i)$	Required time for subsection learning	α_2	Rate of average achievement
$L_c(I, i)$	Count of repeat for subsection learning	β	Maximum reflection rate in the system
$R_a(I, i)$	Average rate of correct answer for subsection	P_b	Initial points applied for a question item
P	Rate of correct answer for question item	P_m	Maximum applied Rate for correct answer item
E_p	Expected rate of correct answer item	W_{ri}	Points of correct answer item
W	Point of each question item	W_j	Points of all question items
W_{tl}	Maximum point of each question item	$T_s(I, i)$	Score of subsection test
W_{bl}	Minimum point of each question item	$T_s(I)$	Score of section test

Table 2. The Evaluation rule of subsection and course procedure

Evaluation score T_s *(I, I)*	*0～60*	*60～69*	*70～79*	*80～89*	*90～100*
Evaluation grade *(G)*	F	D	C	B	A
Moving stage *(S)*	*S (I, i-1)*	*S (I, i-1)*	*S (I, i)*	*S (I, i+1)*	*S (I, i+1)*
Learning time (t_{lr})	t_{lsF} *(I, i)*	t_{lsD} *(I, i)*	t_{lsC} *(I, i)*	t_{lsB} *(I, i)*	t_{lsA} *(I, i)*

With the degree resulted from the subsection test, the learners learning styles become varied by re-composition of moving state and the learning time. The algorithm of moving state and learning time is as follows:

1) Score and Evaluation of Questions:

The degree of difficulty of every question in every subsection can be automatically measured by weighed mean of correct answer item the learner choose, based on the initial basic score; higher marks allotted to the lower rate of correct answer item and lower marks to the higher rate of correct answer item. And the tutors allot the maximum and minimum score initially to keep the balance among the question item. Also, the degree of difficulty can escape being distorted due to the learner groups under the standard level. The score and evaluation of questions in subsection is given by the following equation.

$$W = P_b \tag{1}$$

$$\text{if } (P < P_m) \text{ then } P = P_m; \tag{2}$$

$$W = W * (1 - (P - E_p)) \tag{3}$$

$$\text{if } (W < Wbl) \text{ then } W = W_{bl} \tag{4}$$

$$\text{if } (W > Wtl) \text{ then } W = W_{tl} \tag{5}$$

From the Equ. (1) to (5), the score of test for subsection is given by the following equation.

$$T_s\,(I, i) = \Sigma W_{ri} / \Sigma W_j * 100 \tag{6}$$

2) The Flexible Learning Time with the Degree of Difficulty:

The learning time is varied according to the accomplishment of learners and the degree of difficulty. Of course at the first stage, professional tutors give the basic initial value. If the average score of learners gets higher, the basic learning time can be reduced. At this time, the maximum reflection ratio of the system to avoid the incorrect distortion by distributing the time as the small unit learning time reflecting coefficient is imported as well. The required time for subsection learning is given by the Equ. (7) and (8).

$$\text{if } ((R_a(I, i) < \beta) \text{ then } R_a(I, i) = \beta \quad (7)$$
$$t_{lr}(I, i) = t_{lr}(I, i) * (1 - (R_a(I, i) - \alpha_2)) \quad (8)$$

The learning time is decided by the score earned by the following equation at each subsection selected by the test score. The learning time is given by the Equ. (9) to (13).

$$t_{lrA}(I, i) = t_{ls}(I, i) * (1 - \alpha_1 * (T_s(I, i) - 90) / 20), \text{ when } \mathbf{A} \text{ grade} \quad (9)$$
$$t_{lrB}(I, i) = t_{ls}(I, i) * (1 - \alpha_1 * (T_s(I, i) - 80) / 20), \text{ when } \mathbf{B} \text{ grade} \quad (10)$$
$$t_{lrC}(I, i) = t_{ls}(I, i) * (1 - \alpha_1 * (T_s(I, i) - 70) / 10), \text{ when } \mathbf{C} \text{ grade} \quad (11)$$
$$t_{lrD}(I, i) = t_{ls}(I, i) * (1 - \alpha_1 * (T_s(I, i) - 60) / 10), \text{ when } \mathbf{D} \text{ grade} \quad (12)$$
$$t_{lrF}(I, i) = t_{ls}(I, i), \quad \text{when } \mathbf{F} \text{ grade} \quad (13)$$

Learners can step up to the next learning level only if they get beyond **B** grade, a basic score in subsection test. And the next learning time is depending on the required marks at the basic learning time t_{is} *(I, i)*. If the learners get **D** grade and **F** grade, they can choose whether to go back to the former subsection or to remain the current subsection.

The length of learning time decided by the previous subsection test has a flexibility to reflect the learner's accomplishment. With this system, the advanced learner can have less learning time while the underdeveloped the learner can secure enough review hours. The evaluation rule of subsections is shown in Fig. 4.

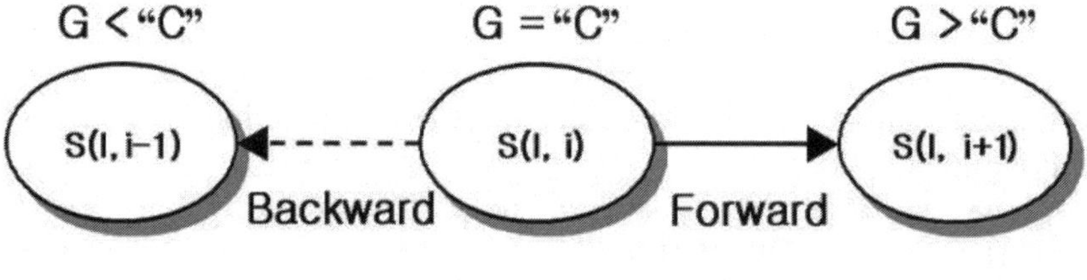

Figure 4: The evaluation rule of subsection

The learner completing the subsection evaluation proceeds the previous, the current and the next level according to the result and the learning time and schedule are also recomposed. There is a case that learner proceed to stage S (1, 0). For example, a student get under **D** grade under the basic **B** grade after he learned S (1, 1), the learner should go back to stage S (1, 0) by the subsection test rule. But the stage S (1, 0) does not actually exist so that he remained at S (1, 1). After completing the S (1, n), the final subsection of section 1, the learner takes the subsection test T (1, n), and the result will decide whether to proceed S (2, 1). If they don't get beyond **B** grade in this test, they can't step up to the next stage S (2, 1). If they get under **C** grade, they should remain at S (1, n) or go back to S (1, n-1).

The learning time for the first subsection of the next section is decided by the same test rule previously referred. The evaluation rule of final subsection to proceed to the next section represented in Table 3. If we are from 1 to N and I+1 are N+1, the learner should learn the same stage.

Table 3. Evaluation rule

Evaluation score *Ts (I, i)*	*0 ~60*	*61 ~70*	*71 ~80*	*81 ~90*	*91 ~100*
Evaluation grade *(G)*	*F*	*D*	*C*	*B*	*A*
Moving stage *(S)*	*S (I, n-1)*	*S (I, n-1)*	*S (I, n)*	*S (I+1, 1)*	*S (I+1, 1)*
Learning time *(tlr)*	$t_{lsF}(I, i)$	$t_{lsD}(I, i)$	$t_{lsC}(I, i)$	$t_{lsB}(I, i)$	$t_{lsA}(I, i)$

3.4 The Calculation of the Learning Accomplishment Degree

In the course scheduling scheme, the moving stage after the subsection test represent only the moving stage within the subsection test not the section test. That is the weakness of this scheme in that the learner can repeat studying with the result of the subsection test but they cannot recognize which is the weak subsection with the total section test. Therefore a new schedule scheme is necessary so that the learner can review the weak subsection based on the analysis of the section test result. For this scheme, exact calculation of the degree of learning accomplishment to tract the weak subsection is required. The degree of learning accomplishment can be calculated by the comparison between the current test result and the previous test result, and the analysis of the learning effectiveness growth.

Let the maximum degree of learning accomplishment be 1 and we can give certain amount as a degree of weakness. Therefore, 1 minus the amount is the degree of learning accomplishment. This can be defined by the following Equ. (14).

$$A\ (I, i) = 1\text{-}W\ (I, i) \qquad (14)$$

The reason the degree of learner's weakness is under 1 is to represent it with percentage. It has observation from 0 to 1. The result of section test T (I) is used to calculate the degree of learning accomplishment by evaluation agent and serves as a key information for course re-composition with the T_t, the final total test of relative course. Accordingly, the section test T (I) is not related with the proceeding to the first subsection of the next section and the result is utilized to analyze the degree of learning accomplishment.

Evaluation agent decides the degree of learning accomplishment by test. Evaluation agent also estimates the weighted value of the weak problems by calculating the marking time of individual question. Fig. 5 is a diagram in which section test is inserted into the subsection test.

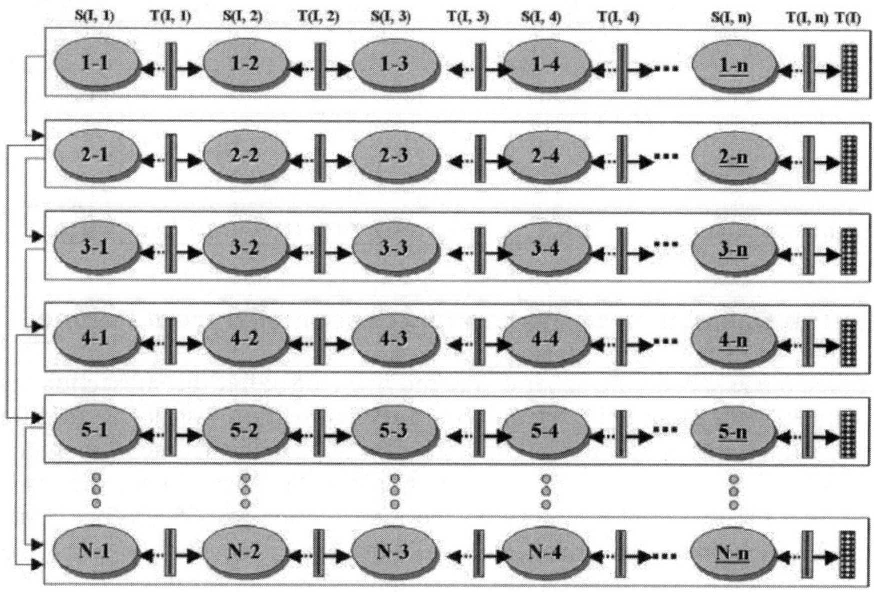

Figure 5: Diagram of the subsection and the section test

For example, if the number of questions item Q_c (I) in the section test is 20 and the marking time is represented with percentage, the average marking time ratio of every question item becomes 5%. Each marking time of individual question compared with the average marking time of all questions is used as a weighted value in course scheduling. This weighted value serves as a significant parameter in calculating the degree of weakness in every subsection. The average standard marking time is made as the following Equ. (15) and (16).

$$t_{ar}(I, i) = 1 / \ Q_c(I, i) \tag{15}$$

$$t_a(I, i) = t_r(I, i) \ * \ t_{ar}(I, i) \tag{16}$$

For the example, we assume that the learner **A** has learned stage S (2, 5), the final subsection of second section and took the section test. By calculating 50 minutes (t_r (I, i)) * 0.05(t_a (I, i), we can earn the average marking time t_a (I, i) of average 2.5 minutes at every question item. Accordingly the average marking time of learner **A** is 2.5 minutes. The learner **A** is given a total of 20 questions and the subsection incorporated in each question shown in Table 4.

Table 4. The Question number of subsection

Question Number	1~4	5~8	9~12	13~16	17~20
Subsection	*S (2, 1)*	*S (2, 2)*	*S (2, 3)*	*S (2, 4)*	*S (2, 5)*

All the questions are created in proportion to the subsection in the section test T (2). From the question 1 to 4 are presented from stage S (2, 1) and four questions are presented each from four subsections. The learner **A** solved all question from two sections for 50 minutes and the each marking time at each question item is shown in Fig. 6.

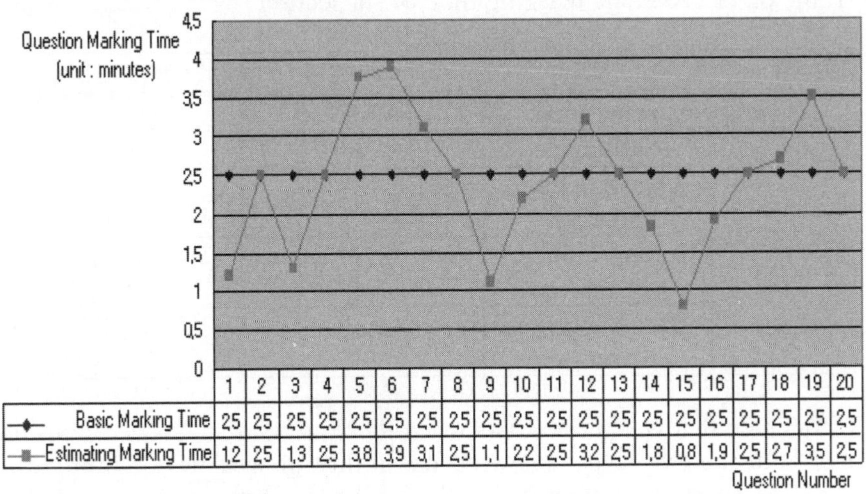

Figure 6: Marking time analysis for each question

We can hypothesize that the learner **A** is weak in the questions with over 2.5 minutes of marking time. And he is possibly good at the questions with marking time of under 2.5 minutes. But it is impossible to decide about the degree of weakness according to the marking time far over 2.5 minutes.

Because we cannot say the learner **A** exactly is far weaker in question 6 with 3.9 minute than question 5 with 3.8 minutes of marking time. Accordingly it is better to calculate the degree of weakness by estimating the marking time at each stage considering which stage the questions includes. The average marking time of every subsection is calculated as in Table 5.

Table 5. The Average marking time of subsection

Subsection	*S (2, 1)*	*S (2, 2)*	*S (2, 3)*	*S (2, 4)*	*S (2, 5)*	Total average time
Average marking time	*1.9*	*3.3*	*2.3*	*1.8*	*2.8*	*2.39*
Difference	*- 0.6*	*+ 0.8*	*- 0.2*	*- 0.7*	*+ 0.3*	*- 0.11*

The average time in stage S (2, 2) and stage S (2, 5) is higher than the basic marking time as 3.3 and 2.8 respectively. Accordingly the learner **A** shows higher weakness in S (2, 2) and S (2, 5) than other subsections. The rate of correct answer items to the questions of S (2, 2) and S (2, 5) is represented in Table 6.

Table 6. The Average marking time of subsection

Weakness subsection		*S (2, 2)*				*S (2, 5)*			
Average marking time		3.3				2.8			
Degree of weakness	Difference	+ 0.8				+ 0.3			
	Question item number	5	6	7	8	17	18	19	20
	Correct or incorrect answers	O	X	O	O	X	X	X	O
	Rate of correct answer item	75%				25%			

We conclude that weakness exists when there are correct answer items under 60% at the two stages showing possibility of weakness. Through the section test result, we can calculate the weakness of the relative subsection by the time lag of marking and the rate of correct answer. The degree of weakness of subsections $W_{tr}(I, i)$ can be represented based on the marking time and ratio of correct answers as following Equ. (17) and (18).

$$W_t(I, i) = \begin{array}{ll} 0 & , t_d(I, i) < t_a(I, i) \\ 1 & , t_d(I, i) \geq (4 * t_a(I, i)) \end{array}$$

$$(t_d (I, i) - t_a (I, i)) / (3 * t_a (I, i)) \quad , t_d (I, i) < (4 * t_a (I, i)) \tag{17}$$

$$W_{tR} (I, i) = W_t (I, i) * \alpha_t + (1 - R (I, i)) * (1 - \alpha_t) \tag{18}$$

The degree of learning accomplishment is figured out from the section test. In calculating learner weakness, we don't consider weakness showing at the previous subsection test, only using the time lag of marking and ratio of correct answer. The index of weakness is figured out by the times the learner review a sector. The evaluation agent remembers the reviewing number of the same subsection and stores at database. For example, the frequency that learner **A** studied in accordance with the stage test result is in Table 7. The equations to calculate the weakness of subsection analyzing repeated learning is defined by:

$$W_r (I, i) = (L_c (I, i) - 1) * 0.3 \tag{19}$$

Accordingly the degree of learning weakness according to the course test can be calculated as follows Equ. (20).

$$W(I, i) = W_{tR}(I, i) * (1 - \alpha_r) + W_r(I, i) * \alpha_r \tag{20}$$

Table 7. The Learning number of subsection

Subsection	S (1, 1)	S (1, 2)	S (1, 3)	S (1, 4)	S (1, 5)
Learning number	1	2	4	1	1
Subsection	S (2, 1)	S (2, 2)	S (2, 3)	S (2, 4)	S (2, 5)
Learning number	1	3	1	1	4
Subsection	S (3, 1)	S (3, 2)	S (3, 3)	S (3, 4)	S (3, 5)
Learning number	2	1	4	3	1
Subsection	S (4, 1)	S (4, 2)	S (4, 3)	S (4, 4)	S (4, 5)
Learning number	1	1	1	3	1

The degree of learning weakness from the analysis of learning repetition represents the weakness of total subsection along with the marking time. Therefore the degree of learning weakness at each subsection is calculated by the weighted value at between the subsection weakness analyzed by the marking time and the rate of correct answer. With this learning weakness, we can estimate the degree of learning accomplishment. And by the degree of learning accomplishment, we track the subsection showing weakness and recompose the course. The result of subsection-specific weakness against the result of CSMA of learner **A** is represented in Table 8.

Table 8. The Result of subsection-specific weakness

Subsection	S (1, 1)	S (1, 2)	S (1, 3)	S (1, 4)	S (1, 5)
Learning weakness degree	0.345	0.231	0.789	0.122	0.342
Subsection	S (2, 1)	S (2, 2)	S (2, 3)	S (2, 4)	S (2, 5)
Learning weakness degree	0.345	0.894	0.232	0.824	0.689
Subsection	S (3, 1)	S (3, 2)	S (3, 3)	S (3, 4)	S (3, 5)
Learning weakness degree	0.231	0.278	0.782	0.351	0.325
Subsection	S (4, 1)	S (4, 2)	S (4, 3)	S (4, 4)	S (4, 5)
Learning weakness degree	0.129	0.258	0.174	0.824	0.323

The subsections representing weakness over 0.4 degrees are stages S (1,3), S (2,2), S (2,4), S (2,5), S (3,3) and S (4,4). The course is recomposed by the CSMA about the subsection with 0.4 of weakness degree. Figure 7 shows the subsection scheduling.

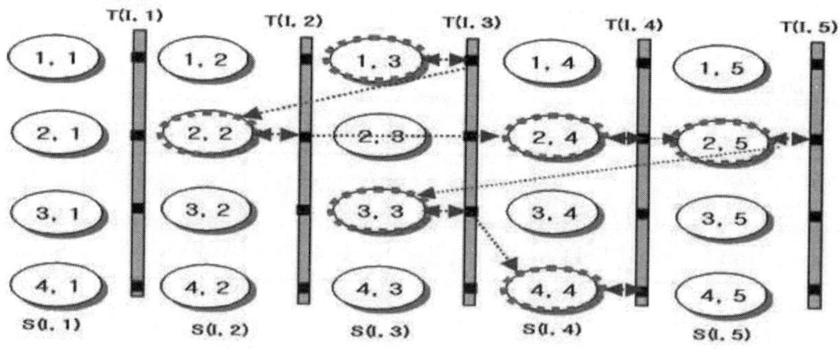

Figure 7: The Course re-composition by scheduling

The learning schedule agent which requires learners to study over again recomposes each subsection showing learners weakness. The learner can enhance their learning effectiveness in accordance with this course schedule. The learner should review S (1,3), S (2,2), S (2,4), S (2,5), S (3,3) and S (4,4) in sequence by the schedule of the CSMA and get the final course test and finish the learning.

4. Implementation

After studying the section and completing the section test, the learning evaluation agent of the CSMA estimates the learning result and reports it to the learner. It shows the marking number and the number of correct answers which enable the learner to self-analyze the test result. Fig. 8 represents the test result. After the learner finishes the final test, the learning accomplishment estimating agent of the CSMA begins to analyze the degree of learning accomplishment and offers the final information on test and weakness of learner and the recomposed course. The section-specific weakness is demonstrated with the graph and figures and the final test degree so that the learner can compare it with their target score. The learner under the target degree can begin the repetition program by the course schedule provided by the CSMA system. Figure 9 offers the information on the degree of learning accomplishment.

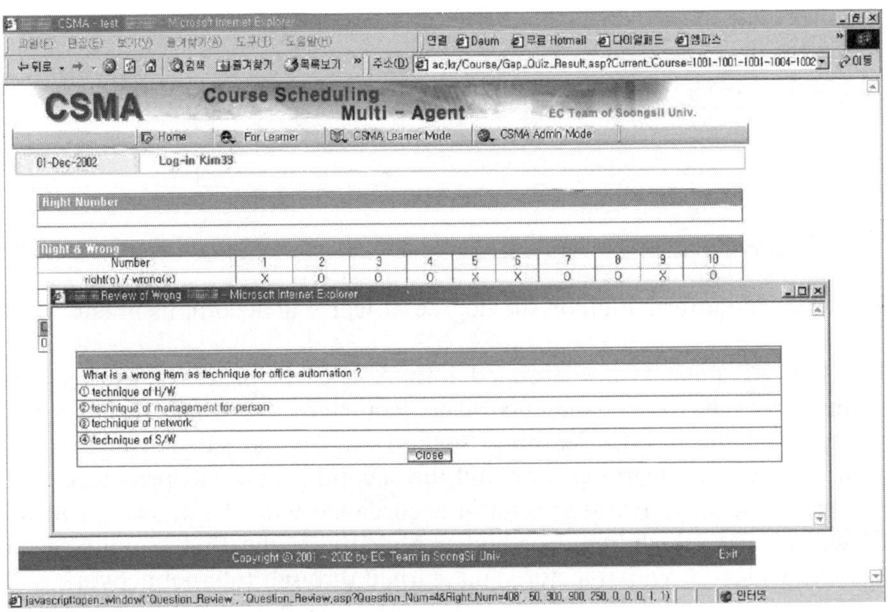

Figure 8: The result of learning test

5. Experiments

5.1 Experiments environment

For experiment the course scheduling multi-agents, 42 persons who studied ordinary learning method were extracted randomly from unspecified persons and the same courseware was offered. And another 44 persons were extracted with the Web-based learning system using the CSMA. The summary of experimental environment is represented in Table 9.

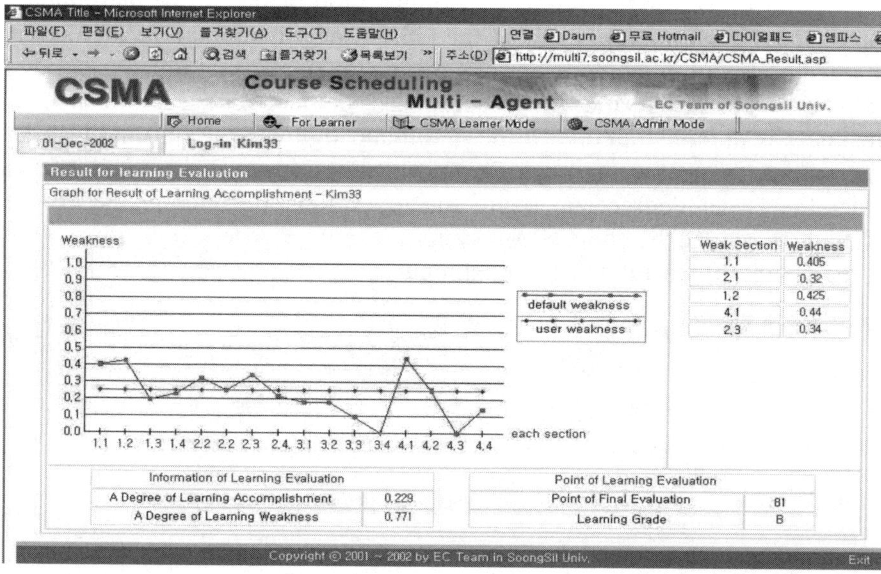

Figure 9: The information on the degree of learning accomplishment

The first 42 persons are grouped into **A** and the next 44 learners are grouped into **B** for convenience. The factors of the traditional learning method and the CSMA learning method are all the same. But the first group allotted the learning time and the second group was presented with the learner-specific learning time in accordance with the algorithm of the CSMA course scheduling. For higher objectivity, the first group was not delivered specific verbal lecture and learned through Internet network with the same materials with the second group. The target degree was **A** for both the two groups.

5.2 Experiments Analysis

CHI-SUARE tests model was used as a way to identify the different results of the experiment. This model makes frequency table for two groups by class value, squares the difference of each class frequency and standard distribution frequency, divides standard distribution frequency from above value and sums the values for each class. CHI-SUARE TESTS model was used as a way to identify the different results of the experiment. Then this model verifies statistically whether there exists any deference for two groups. The simple example is shown in Table 10.

Table 9. A Comparative table about tested two evaluation method

Learning method Object	Traditional learning method	Proposed CSMA method
Learner	42 college students not majored for computer science	44 colleges student not majored for computer science
Subject	How to use computer in the office	
Content to learn	Number of chapters: 2 Number of sections in a chapter: 4	
How to learn	Learning contents in HTML document	CSMA course learning
Where to learn	PC laboratory	PC laboratory
Evaluation	Web environment (An objective test)	Web environment (An objective test)
Learning time for section	Self- study by the learner	Self- study guided by the CSMA
Time for evaluation	Last test: 15 minutes	
Question for evaluation	20 questions	
Repeated learning for weak section	Selected weak section by the learner	Selected weak section by the CSMA

The formula of the pseudo CHI-SQUARE distribution using the previous table is as following Equ. (21).

$$X = SUM_{i=1..C}\ (SUM_{j=1..2}((f_{ij} - (f_{is} * f_{cj} / n))^{2} / (f_{is} * f_{cj} / n))) \qquad (21)$$

The comparison between the calculated value and the value of statistical distribution table can verify the actual reason of difference by probability. During the experiment, the two groups were encouraged to learn and were presented the learning method through exercises with the help of 10-minute guidance and has studied on the Internet network for 200 minutes and then tested at respective laboratory room at the same time.

Table 10. Model for frequency distribution table

Frequency distribution table			
Grade	**A** group	**B** group	Total
Grade1	*f11*	*f12*	*f1s*
Grade1	*f21*	*f22*	*f2s*
....	....	....	....
Total	*fc1*	*fc2*	*n*

Table 11. Frequency distribution table of two experiment groups

Frequency distribution table of experiment groups			
Score grade	**A** group	**B** group	Total
70	5	2	7
75	*13*	*4*	*17*
80	*6*	*3*	*9*
85	*9*	*8*	*17*
90	*3*	*11*	*14*
95	*5*	*13*	*18*
100	*1*	*3*	*4*
Total	*42*	*44*	*86*
Average	81.3	88.3	

The first group carried out the self-study with the Internet-based learning materials and exchanged questions and answers, while the second group accomplished CSMA course scheduling learning, repeating 15 minutes section learning and then five-minute test on 10 questions. Finally the two groups took the 15 minutes online test on 20 questions. The results are shown in Table 11.

Figure 10 represents the frequency distribution table in graph. The **B** group shows higher frequency with the higher marks and **A** group shows relatively lower marks. The average score of **B** group is much higher than that of the **A** group.

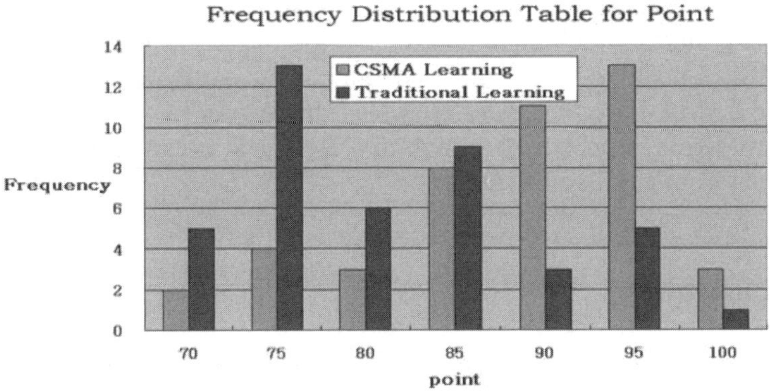

Figure 10: The frequency distribution table

The nature of score gab between the two groups can be recognized by statistics with the help of value of CHI-SQUARE distribution. Table 12 represents the calculated value of CHI-SQUARE.

Table 12. CHI-SQUARE vale of two experiment groups

CHI-SQUARE value of two experiment groups		
Score grade	**A** group	**B** group
70	0.73	0.70
75	2.66	2.54
80	0.59	0.56
85	0.06	0.06
90	2.15	2.06
95	1.63	1.56
100	0.47	0.44
Total	16.20	

According to the table above, the classes are seven, so the probability that can occur the difference on the free rate 7 and significance level 95 would be X=12.59. Accordingly the value 16.20 of the approximate CHI-SQUARE distribution is enough large figures. Therefore the proposed learning method is proved to be excellent within over the 95% of reliability.

Most Web-learning systems are lacking for agents to provide convenience for the learner and analyze the degree of learning accomplishment. And the traditional learning systems using agent just keeps information on general learning states which cannot calculate the degree of learning accomplishment to encourage learners to continuously participate in learning. While the CSMA-based learning propelled the learner to consistently participate in learning and keep reaction and feedback, creating an engaging learning environment like the game.

6. Conclusions

In this paper we proposed multi-agent system for the learner courseware scheduling creating the courseware customized for the individual learner by evaluating the learning. These agents continuously learn the individual learning course and the feedback on the learning, offering customized scheduling course and giving the maximum learning effectiveness. Accordingly the course ordered by the learner can be fittest to the learner with the help of course scheduling agent. The learner can continuously interact with agents until completing the course. If the agent judges the course schedule for learners not to be effective, they recompose the course schedule and offer the new course to the learner.

In future work, the CSMA system will be implemented independently of web-based learning system and manage all Web-learning environment, even in the heterogeneous protocol environment. Also the multi-agent system proposed in this paper can be enhanced in the form of intelligent agent with the assistance of the shared, reusable ontologies for learning context and student's profile.

References

Hamalainen, M.(1996), A Model for Delivering Customized on-Demand Education on the Internet, The Road to the Information Society. New Technologies for Education and Training. European Commission, DGXIII, Luxembourg, pp. 104-125.

Learning Technologies Standardization Committee (LTSC).
Web site at http://www.ltsc.ieee.org/.

IMS Global Learning Consortium. Web site at http://www.imsproject.org/.

Hamalainen, M, Whinston, A, and Vishik, S.(1996), Electronic Markets for Learning: Education Brokerages on the Internet, Communications of the ACM, vol. 39 no 6 (June), 51-58.

Badrul H. Khan (1997), Web-Based Instruction (WBI): What Is It and Why Is It?, Education Technology Publications, Inc..

Agogino, A (1994), The Synthesis Coalition: Information Technologies Enabling a Paradigm Shift in Engineering Education, Proc. of Hypermedia in Vaasa '94, Vaasa Institute of Technology, 3-10.
Thomas, R (1997). Implications of Electronic Communication for the Open University, in Mindweave, Communication, Computers, and Distance Education, R. Mason and A. Kaye (eds.), Pergamon Press, 166-177.

http://grouper.ieee.org/p1484 IEEE Learning Technology Standards Committee (LTSC)

Whinston, A.(1994), Re-engineering MIS Education, Journal of Information Science Education, Fall 1994, 126-133.

Sandip Sen., Edmund H. Durfee (1994), On the Design of an Adaptive Meeting Scheduler, Proc. of the Tenth IEEE AI Application.

Katia Sycara, Dajun Zeng (1996), Coordination of Multiple intelligent Software Agent, International Journal of Cooperative Information System.

Agogino, A(1994), The Synthesis Coalition: Information Technologies Enabling a Paradigm Shift in Engineering Education, Proc. of Hypermedia in Vaasa '94, Vaasa Institute of Technology, 3-10.

US Department of Defense, Advanced Distributed Learning (ADL) Initiative. Web site at http://www.adlnet.org/.

The Alliance of Remote Instructional Authoring and Distribution Networks for Europe (ARIADNE). Web site at http://www.ariadne.unil.ch/.

Akiko Inaba, Mitsuru Ikeda, Riichiro Mizoguchi, and Jun'ichi Toyoda (2001), Design and Analysis of Learners', Interaction based on Collaborative Learning Ontology, Proc. of Euro-CSCL2001, pp.308-315.

Judy Kay, Andrew Lum (2004), Ontologies for Scrutable Student Modeling in Adaptive E-Learning, Adaptive Hypermedia and Adaptive Web-Based Systems 2004 Workshop.

Gilbert Paquette and Ioan Rosca (2004), "An Ontology-based Referencing of Actors, Operations and Resources in e-Learning Systems", Proc. of SWEL-04.

Ronald Denaux1, Vania Dimitrova, Lora Aroyo (2004), "Interactive Ontology-Based User Modeling for Personalized Learning Content Management", Workshop on Applications of Semantic Web Technologies for e-Learning (SW-EL).

AGrIP – Agent Grid Intelligence Platform

Zhongzhi Shi[1], He Huang[1,2], Yuncheng Jiang[3], Jiewen Luo[1,2], Zheng Zheng[1,2], and Fen Lin[1,2]

[1]Institute of Computing Technology, Chinese Academy of Sciences, Beijing 100080, China {shizz,huangh,luojw,zhengz,linf}@ics.ict.ac.cn
[2]Graduate University of the Chinese Academy of Sciences, Beijing, China
[3]College of Computer Science and Information Engineering, Guangxi Normal University, Guilin 541004, China ycjiang@mailbox.gxnu.edu.cn

Abstract

The exponential growth of information and proliferation of service offers on the Web require a robust platform which can not merely offer networked computing capabilities but also support intelligent data and service handling. Agent Grid Intelligence Platform (AGrIP) is such an Internet intelligent platform that enables greater access to both content and services. Based on agents, AGrIP re-organizes the global information into a well-ordered and semantic-rich space by using ontology techniques, structures dynamic aspects of Web services in a formal way by using Dynamic Description Logic, and allows users to query global information and invoke a global Web service without being aware of the site, structure, query language, execution details and service holder. AGrIP provides a collaborative e-Commerce scenarios for service offerings on the Semantic Web.

1. Introduction

Nowadays, we are witnessing an exponential growth of information and the proliferation of service offers accumulated within universities, corporations, and government organizations. **A**gent **Gr**id **I**ntelligence **P**latform (AGrIP) is an Internet intelligent platform enabling greater access not only to content but also to services on the Web. By using agent technology AGrIP is aimed at the realization of collaborative e-Commerce scenarios for service offerings on the Semantic Web [1].

According to functionalities, Web services can roughly be categorized as information-providing services, such as weather information providers,

Z. Shi et al.: *AGrIP – Agent Grid Intelligence Platform*, Studies in Computational Intelligence (SCI) **37**, 627–646 (2007)
www.springerlink.com

and world-altering services, such as book-selling programs [2]. Usually, the specification of services presumes a (sophisticated) representation of parts of the world. In addition, realizing online service offerings requires suppliers to structure and store information and knowledge about their service offerings in a machine-readable way, so that agents can discover, select, employ, reason about services [3].

AGrIP allows users to query global information and invoke a global Web service without being aware of the site, structure, query language, execution details and service holder. Three main aspects of AGrIP are considered in this chapter: first, how to re-organize the global information into a well-ordered and semantic-rich space in a well-defined way [4,5,6]; second, how to structure dynamic aspects of Web services in a machine-readable way so that agent can automatically find proper services and reason about them [7,8]; and the agent-based architecture of the intelligent platform [9].

The rest of this chapter is organized as follows. Section 2 describes the re-organization of global information based on ontologies. Section 3 describes the modeling of dynamic aspects of Web services. Section 4 shows architecture of AGrIP and how agents construct the intelligent platform. A conclusion and future research are presented in section 5.

2. DB Resource management

The specification of the effects of services presumes a (sophisticated) representation of parts of the (physical) world. Those parts consist of the environment of the services and the objects in the world are affected by services. Databases are often used for containing some part of the real world and they are a kind of very important information resources in the current Web. Re-organizing database resources into a well-ordered space provides a basis for Internet intelligence. First of all, AGrIP should provide an infrastructure enabling re-organization of DB resources into a semantic context and domain environment for services.

Currently, many research and development activities have focused on building infrastructures for data and information handling on the Web. In [10], two additional layers are defined on top of grid toolkits [11] to support data caches, data transfer, etc. The SDSC SRB [12] provides a uniform interface for connecting to heterogeneous data resources and for accessing replicated datasets. In these methods, naming mechanisms enable datasets accessed based on their attributes and/or logical names rather than their physical locations. However, these methods at least leaves two problems unresolved [4,5]: firstly, although it is possible to uniquely name a billion objects, it will be very difficult to discover a certain object without a context associated with it; secondly, only the naming mechanism is not sufficient to support intelligent data and information handling. The deficiency is mainly caused by the lack of representation of data semantics.

Ontologies can bring structure to the meaningful content of a domain and define a common vocabulary for people and computer programs to reuse and share domain knowledge across heterogeneous platforms. In this section, we introduce our work on service-oriented and ontology-driven knowledge management in AGrIP. In AGrIP, ontologies are used to capture domain knowledge, and two kinds of mappings (or links) are built: mappings between ontologies and their underlying databases, and mappings between terms of different ontologies. Then ontologies construct a distributed knowledge space on top of database resources. Discovering data objects via the knowledge space can work much more effectively and efficiently than via the huge namespace, since the knowledge space has meaningful content and is far smaller in size than the namespace. And semantic interoperability can be presented to agent applications.

2.1 General Architecture

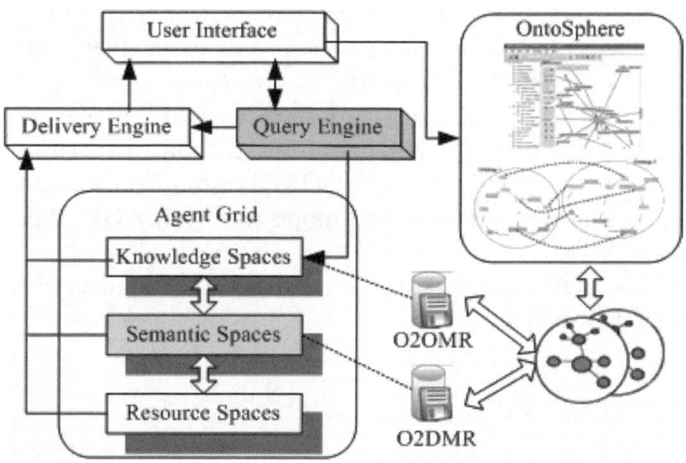

Fig. 1. The General Architecture for DB Resource Management

The general architecture of knowledge management is illustrated in figure 1. The agent grid ties together distributed data, information, knowledge, handling tools and platforms into globally accessible resources. In AGrIP, data is viewed as uninterpreted bit streams; information corresponds to tagged data elements denoting the meaning of data objects, e.g. relational DB; and knowledge corresponds to relationships among pieces of information facilitating the use of information to solve problems, e.g. domain ontologies. The OntoSphere provides supports for ontology visualization and editing, and for building two kinds of mappings. The query engine is responsible for processing the query and checking results. The delivering engine is responsible for result delivering to users or other computer programs.

Two services are defined based on a multi-agent system --MAGE [9], which enables the use of agent facilities to dynamically discover the necessary resources. *AGrid Ontology Service* (AGOS) is to support ontology editing, link ontologies with underlying databases, reconcile heterogeneities between ontologies, and process ontology queries, etc. *AGrid Ontology Service Registry* (AGOSR) is a facility for the publication of AGOSs. In addition, following interfaces are defined in AGOS:

- The *Ontology Extension* (OE) interface is used to build mappings between ontologies and their underlying data. The primary operation in the OE interface is **extend**.
- The *Ontology Mediation* (OM) interface is to reconcile heterogeneities between ontologies by building ontology mappings. The primary operation in the OM port type is **mediate**.
- The *Ontology Query* (OQ) interface enables queries on ontologies. This port type accepts query requests and processes these queries. The primary operation in the OQ interface is **perform**.
- The *Ontology Transport* (OT) interface enables the transport of ontologies or sub-ontologies, with or without extension, between AGOSs or between client and AGOS. The implementation of OT is based on transportation facilities supported by MAGE, thus we don't discuss OT here.

The design and functionalities of OE, OM and OQ are described in detail in the rest of this section.

2.2 Design of AGOS

Today a rapid growth of information is accumulated in various autonomous databases on the Web. This makes it infeasible to build from scratch the instance base of an ontology which includes voluminous information. In fact, we choose to build mappings between the intensional part of ontologies and the underlying databases. These mappings are called ontology-to-database mappings (O2D mappings) and stored in ontology-to-database mapping repositories (O2DMR). OE provides an interface for building O2D mappings. The operation **extend** receives a mapping request between an ontology and data repositories, and then initiates the O2DMapping Builder. An O2D mapping is composed of two parts: a term defined in ontologies and a mapping expression. The O2D mappings are encoded into XML document and stored in O2DMR. An example of an O2D mapping for a relational DB is shown in follows:

```
<O2DMappings srcOntology="Onto1">
 <O2DMapping>
  <concept name="InstructionBook"/>
   <mappingExpression>
      <queryStatement name="s1"  destSource="DataSource1">
         <expression>
           select * from Table1 where type="Instruction"
         </expression>
         <objectKey>
           <keyAttribute name="ISBN" type="string"/>
         </objectKey>
      </queryStatement>
   </mappingExpression>
 </O2DMapping >
 ...
</O2DMappings>
```

Although ontologies aim to capture consensual knowledge of a given domain, the decentralized nature of the Web makes it difficult to achieving consensus across communities. Heterogeneities exist between ontologies that model the same kind of knowledge or domain. Mappings between terms of two ontologies are needed to bridge these heterogeneities. These mappings are called ontology-to-ontology mappings (O2O mappings) and stored in ontology-to-ontology mapping repositories (O2OMR). We focus on two kinds of O2O mappings, i.e. synonym and hyponym. OM is an interface to reconcile heterogeneities between ontologies by discovering and specifying ontology mappings. The **mediate** operation accepts a request for building O2O mappings between two ontologies. This port type allows the AGOS instance to import ontologies required in the document and visit their O2DMRs when needed. Then O2OMapping builder is initiated to implement two important phrases: finding similarities and specifying mappings. The O2O mappings are encoded into XML document stored in O2OMR. An example of an O2O mapping is shown in follows:

```
<O2OMappings srcOntology="Onto1" destOntology="Onto2">
<O2OMapping>
  <srcConcept name="Print-Media"/>
  <subsumedBy>
   <destConcept name="Document"/>
  </subsumedBy>
 </O2OMapping>
 <O2OMapping>
  <srcConcept name="Periodical"/>
   <sameAs>
    <union>
     <destConcept name="Periodical"/>
     <destConcept name="Press"/>
    </union>
   </sameAs>
 </O2OMapping>
 …
</O2OMappings>
```

OQ is an interface for queries on ontologies. The operation **perform** takes a request document and returns a response document. The request document contains details of the queries to be performed and thus instructs the AGOS instance to interact with ontologies. For example, to find emails of all the Chinese researchers who have published at least one paper about Semantic Web, the XML fragment of the query that involves two ontologies is shown in follows:

```
<ontologyServiceRequest>
 <Header> … </Header>
 <Body>
  <rdqlQueryStatement name="s1" >
   SELECT ?email
     WHERE (O1:Researcher, <O1:name>, ?name)
           (O2:Paper, <O2:authors>, ?name)
           (O1:Researcher, <O1:nationaility>, "China")
           (O2:Paper, <O2:subjects>, "Semantic Web")
           (O1:Researcher, <O1:hasEmail>, ?email)
     USING O1 FOR <http://somewhere/reseacherInfo#>
           O2 FOR <http://somewhere/publicationInfo#>
   </rdqlQueryStatement>
  <Delivery name="delivery" from="s1" to="consumer">
   <Mechanism type="bulk"/>
   <Mode type="full"/>
  </Delivery>
 </Body>
```

</ontologyServiceRequest>

The processing of an ontology query includes three functionalities: query rewriting, decomposition and local evaluation. The query rewriting is to rewrite a query or part of it in terms of ontologies into new queries in terms of other related ontologies. The decomposition is to decompose it into two groups of sub-queries: sub-queries on the intension of ontologies and sub-queries on the instance bases (data queries). In the first group, there is no need to access the underlying data resources. While in the second, it is needed to translate each sub-query, by referring to the O2DMR, into a query for the underlying data repositories. The local evaluation is to execute each decomposed query.

3. DDL – Modeling of Dynamic Aspects of Services

From the view of implementation, the execution of a Web service can be viewed as an action that means a change of the world. Corresponding to information providing or world altering services, actions can be distinguished into two kinds, too. One kind of actions means a change of the physical world, another kind means mental changes involved in social or mental activities like acts of communicating.

Many web-based ontology languages have been created for implementing ontologies, e.g. OWL [13]. These languages describe domain knowledge in a static sense. Therefore, they are unable to describe the temporal and behavioral aspects of dynamic domains. Artale et al [14] presented a kind of semantic models based on interval to temporally extend (nontemporal) ontology models. But actions' effects to the world were not defined. Wolter et al [15] tried to introduce a dynamic dimension to description logic by making trade-off between expressivity and decidability. Actions were interpreted as modal operators and their effects were not explicitly defined.

In this section, we introduce our work on dynamic description logics (DDL)[7,8], which supports representation and reasoning of static knowledge and dynamic knowledge.

3.1 Description Logics

Description Logics (DLs) [16] are a family of knowledge representation languages that are able to represent structural knowledge in a formal and well-understood way. Based on DL, a knowledge base $\mathcal{K} = \langle \mathcal{T}, \mathcal{A} \rangle$ consists of two parts: TBox $\mathcal{T}$ and ABox $\mathcal{A}$. TBox consists of subsumption assertions in forms of $C \sqsubseteq D$, where C and D are concepts. ABox consists of instance assertions in forms of $C(a)$ or $P(a, b)$, where a and b are individual names, C is a concept name and p a role.

An interpretation $I = (\Delta^{I}, \cdot^{I})$ consists of domain Δ^{I} and interpretation function $\cdot^{I}$, where interpretation function $\cdot^{I}$ maps each primitive concept A to subset of domain Δ^{I}, and maps each primitive role to subset of domain $\Delta^{I} \times \Delta^{I}$.

A simple kind of DL is called ***ALC***. Table 1 is the syntax and semantics of description logic ***ALC***.

Table 1. Syntax and Semantics of ***ALC***

constructor	Syntax	Semantics	example
primitive concept	A	$A^I \subseteq \Delta^I$	Human
primitive concept	P	$P^I \subseteq \Delta^I \times \Delta^I$	has-child
top	$\top$	Δ^I	True
bottom	$\bot$	Φ	False
intersection	$C \sqcap D$	$C^I \cap D^I$	Human ⊓ Male
union	$C \sqcup D$	$C^I \cup D^I$	Doctor ⊔ Lawyer
negation	$\neg C$	Δ^I / C^I	¬Male
existential quantification	$\exists R.C$	$\{x \mid \exists y, (x, y) \in R^I \wedge y \in C^I\}$	∃has-child.Male
value restriction	$\forall R.C$	$\{x \mid \forall y, (x, y) \in R^I \Rightarrow y \in C^I\}$	∀has-child.Male

The reasoning problems in description logic include concept satisfiability, concept subsumption, instance checking, and consistency checking and so on, among which the concept satisfiability is the basic one and others can be reduced to it. Given two concepts C and D, there exist the following proposition:

(i) $C \sqsubseteq D \Leftrightarrow C \sqcap \neg D$ is unsatisfiable;

(ii) $C \doteq D$ (concept C and D are equivalent) $\Leftrightarrow$ $(C \sqcap \neg D)$ and $(D \sqcap \neg C)$ all are unsatisfiable;

(iii) C and D are disjoint $\Leftrightarrow$ $C \sqcap D$ is unsatisfiable.

3.2 Syntax of DDL

Dynamic description logic (DDL) is defined by extending traditional description logic ***ALC***. But the framework for reasoning about services (actions) proposed in this paper is not restricted to particular description logics, and it can be instantiated with any description logic that seems appropriate for the application domain at hand.

Actions without variables are called ground actions. Actions can be parameterized by introducing variables in place of object names. Ground actions can be viewed as actions where parameters have already been instantiated by object names, while parametric actions should be viewed as a compact representation of all its ground instances. For simplicity, we concentrate on ground actions.

Definition 1 Let $\mathcal{T}$ be an acyclic TBox, a primitive action description on $\mathcal{T}$ is in the form of $\alpha \equiv (P_\alpha, E_\alpha)$, where:

(1) α is the action name;

(2) the *pre-condition*, P_α, is a finite set of ABox assertions;

(3) the *effects* of action, E_α, is a finite set of pair *head*/*body*, where *head* is a finite set of formulae and *body* is a literal of ABox assertion.

Intuitively, a primitive action is specified by first stating the pre-condition under which the action is applicable. Secondly, one must specify how the action affects the state of the world with effects: *head / body*, called conditional post-conditions, describes under the condition *body* doing the action leads in the successor state the addition of *head*. According to the law of inertia, only those facts that are forced to change by the post-conditions should be changed by the performance of the action.

Definition 2 Actions in DDL are defined as the following:

(1) Primitive action α is action.

(2) If α and β are actions, then

  a) $\alpha;\beta$ is an action, which means the sequential execution of α and β, i.e., performing α and then β;

  b) $\alpha \cup \beta$ is an action, which means non-deterministic choice of actions, i.e., do either α or β;

  c) α^* is an action, which means iteration of action, i.e., iteratively performing α finitely many (including zero) times.

(3) If φ is an assertion formula, then $\varphi?$ is an action, which means the test of current truth value of φ.

In this paper, we concentrate on acyclic actions, which is an action can not be composed of itself or can not appear on the right side of its own definition. Given an acyclic action, it can be always expanded into a sequence that is composed with only primitive actions. Actually, an action defines the transition relation of state, i.e. an action α transit a state u to a state v. The transition relation is denoted as $uT_{\alpha}v$.

Definition 3 Let C is concept, R is role, a, b are individuals, formulas in DDL are defined as following:

(1) $C(a)$ and $R(a, b)$ are called assertion formulas.

(2) If φ and ψ are formulas, then $\neg\varphi$, $\varphi\wedge\psi$, $\varphi\rightarrow\psi$ are formulae.

(3) $[\alpha]C$, $[\alpha]R$ are formulae.

3.3 Semantics of DDL

Now we will explain the semantics of DDL in detail. Firstly, for a state u in DDL, an explanation $I(u) = (\Delta, \bullet^{I(u)})$ in u is composed of two components, written as $I(u) = (\Delta, \bullet^{I(u)})$, where explanation function $\bullet^{I(u)}$ maps each concept into a subset of Δ, maps each role into a binary relations on $\Delta\times\Delta$, and maps:

- $([\alpha]C)^{I(u)} = \{a \mid \exists v, uT_{\alpha}v, a\in C^{I(v)}\}$
- $([\alpha]R)^{I(u)} = \{(a,b) \mid \exists v, uT_{\alpha}v, (a,b)\in R^{I(v)}\}$.

In each state u, assertion formulas connect individual constants to concepts and roles. Concept assertions show the instantiation relations of individuals and concepts. Role assertions show the relations of two individual objects. Semantics of concept assertion and role assertion can be interpreted as following:

- $u\models C(a)$ iff $a\in C^{I(u)}$;
- $u\models\neg C(a)$ iff $a\notin C^{I(u)}$;
- $u\models R(a_1, a_2)$ iff $(a_1, a_2)\in R^{I(u)}$;
- $u\models\neg R(a_1, a_2)$ iff $(a_1, a_2)\notin R^{I(u)}$.

Similarly, formulas that composed of assertion formulas can be interpreted in u as the following, where φ and ψ are assertion formulas:

- $u\models\neg\varphi$ iff $u\not\models\varphi$;
- $u\models\varphi\wedge\psi$ iff $u\models\varphi$ 且 $u\models\psi$;
- $u\models \varphi\rightarrow\psi$ iff $u\models\varphi \Rightarrow u\models\psi$.

Definition 4 Given two interpretation $I(u) = (\Delta, \bullet^{I(u)})$ and $I(v) = (\Delta, \bullet^{I(v)})$ under state u and v, an action $\alpha = (P_{\alpha}, E_{\alpha})$ can generate state v when applied to state u (written $u \rightarrow_\alpha v$), if $I(u)$ and $I(v)$ satisfy the following conditions:

- $I(u)$ satisfies each formula of P_α;
- For each pair *head/body* of E_α, if $I(u)$ satisfies *body*, then $I(v)$ satisfies *head*.

Given a set of state $\mathcal{W}$, the semantics of primitive and complex actions are shown as following:

- $\alpha = \{<u, v> \mid u, v \in \mathcal{W},\ u \rightarrow_\alpha v\}$;
- $\alpha ; \beta = \{<u, v> \mid u, v, w \in \mathcal{W},\ u \rightarrow_\alpha w \wedge w \rightarrow_\alpha v\}$;
- $\alpha \cup \beta = \{<u, v> \mid u, v \in \mathcal{W},\ u \rightarrow_\alpha v \wedge u \rightarrow_\alpha v\}$;
- $\alpha^* = \{<u, v> \mid u, v \in \mathcal{W},\ u \rightarrow_\alpha v \vee u \rightarrow_\alpha v \vee u \rightarrow_\alpha v \vee \ldots\}$;
- $\varphi? = \{<u, u> \mid u \in \mathcal{W},\ u \models \varphi\}$.

Definition 5 Given action α, if there exist two interpretations $I(u) = (\Delta, \bullet^{I(u)})$ and $I(v) = (\Delta, \bullet^{I(v)})$ satisfy $u \rightarrow_\alpha v$, then action α is executable.

3.4 Reasoning in DDL

Consistency problem of DDL formulae is the basic reasoning problem of DDL, and other reasoning problems may be reduced to consistency problem of assertion formulas.

An algorithm is given to determine if a set of formulae $\mathcal{F}$ is consistent. The algorithm utilizes inference rule to expand $\mathcal{F}$, then checks if there is a conflict. Let $\mathcal{F}$ be a set of formulas, the algorithm is described as follows.

1. Use the following rules to expand the assertion formulas $\mathcal{F}$, until there is not rule which may expand the assertion formulas $\mathcal{F}$:
   a) $\sqcap$ rule, if $C_1 \sqcap C_2(x) \in \mathcal{F}$, and $C_1(x) \notin \mathcal{F}$, $C_2(x) \notin \mathcal{F}$, then add $\{C_1(x), C_2(x)\}$ to the assertion formulas $\mathcal{F}$.
   b) $\sqcup$ rule, if $C_1 \sqcup C_2(x) \in \mathcal{F}$, and $C_1(x) \notin \mathcal{F}$, $C_2(x) \notin \mathcal{F}$, then add $\{D(x)\}$ to the assertion formulas $\mathcal{F}$, where $D=C_1$ or $D=C_2$.
   c) $\exists$ rule, if $\exists R.\ C(x) \in \mathcal{F}$, and there exists y, $R(x, y) \in \mathcal{F}$ and $C(y) \in \mathcal{F}$, then add $\{C(y), R(x, y)\}$ to the assertion formulas $\mathcal{F}$.
   d) $\forall$ rule, if $\forall R.\ C(x) \in \mathcal{F}$, $R(x, y) \in \mathcal{F}$ and $C(y) \notin \mathcal{F}$, then add $\{C(y)\}$ to the assertion formulas $\mathcal{F}$.
   e) action α rule, if $[\alpha]C \in \mathcal{F}$, where $\alpha = (P_\alpha, E_\alpha)$, then all *pre-conditions* of P_α will be deleted from $\mathcal{F}$, and all *post-conditions* of E_α will be added to $\mathcal{F}$, and concept C will be added to $\mathcal{F}$ too.

2. Check if there exists conflict in the assertion formulas $\mathcal{F}$, if has conflict, then the assertion formulas $\mathcal{F}$ is consistent; else the assertion formulas $\mathcal{F}$ is inconsistent.

In above steps, $\sqcup$ rule is uncertain, i.e. it may generate two branches $\mathcal{F}_1$ and $\mathcal{F}_2$. If there exists a formulas set which is consistent in $\mathcal{F}_1$ or $\mathcal{F}_2$, then $\mathcal{F}$ is consistent. If both branches contain conflict, then $\mathcal{F}$ is inconsistent. In real application, if one branch has conflict, then it is deleted. Other rules (i.e., $\sqcap$ rule, $\sqcup$ rule, $\exists$ rule, $\forall$ rule) are same as the rules of description logic Tableaux algorithm. But action α rule is given based on action description and reasoning mechanism, and it modifies assertion formulas $\mathcal{F}$. All these rules insure algorithm correctness.

Theorem 1 The consistent problem of a set of formulae is decidable.

Definition 6 A action description $\alpha \equiv (P_\alpha, E_\alpha)$ is consistent, if and only if both P_α and P_b are consistent.

Theorem 2 The consistency problem of action description of DDL is decidable.

The DDL has clear and formally defined semantics. It provides decidable reasoning services, and it can support effective representation and reasoning of the static knowledge, dynamic process and running mechanism.

4. Architecture of Agent Grid Intelligence Platform

AGrIP is a highly open software environment whose structure is capable of dynamical changes. From the implementation point of view, AGrIP is a four-layer model illustrated in figure 2, where:

- data resources consist of various resources distributed in Internet, such as web pages, databases, knowledge bases etc.;
- multi-agent environment is the kernel of AGrIP, which is responsible for resources location and allocation, authentication, unified information access, communication, task assignment, agent library and others;
- middle wares provide developing environment, containing agent creation, information retrieval, data mining, case base reasoning, expert system etc, to let users effectively use grid resources;

- application services automatically organize agents for specific purpose application, such as power supply, oil supply e-business, distance education, e-government.

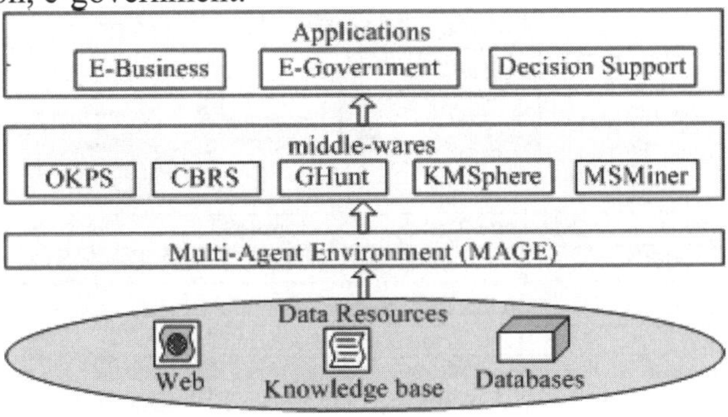

Fig. 2. Implementation-oriented Model for AGrIP

In this section, we mainly introduce agent environment – MAGE, and the services environment architecture built on MAGE.

4.1 Multi-AGent Environment -- MAGE

MAGE[9] is designed to develop multi-agent systems and . It is to create a relatively general purpose and customizable toolkit that could be used by software users with only basic competence in agent technology to analyze, design, implement and deploy multi-agent systems.

MAGE is designed to be compliant with FIPA[18]. Figure 3 illustrates the architecture of MAGE. It mainly consists of four subsystems: Agent Management System, Directory Facilitator, Agent, and Message Transport System. **Agent Management System** is a mandatory component of MAGE. It maintains a directory of AIDs (Agent Identifiers), which contain transport addresses for agents registered in MAGE and offer white pages services to other agents. **Directory Facilitator (DF)** is an indispensable component of MAGE. It provides yellow page services to other agents. Agents may register their services with the DF or query the DF to find out which services are offered by other agents. **Message Transport Service (MTS)** is the default communication approach between agents on different FIPA-Compliant agent platforms. It uses FIPA ACL as the standard communication language. **Agent** is the fundamental actor in MAGE, which combines one or more service capabilities into a unified and integrated execution model that may include access to external software, human users and communications facilities. **Software** represents all non-agent, external components accessible to an agent.

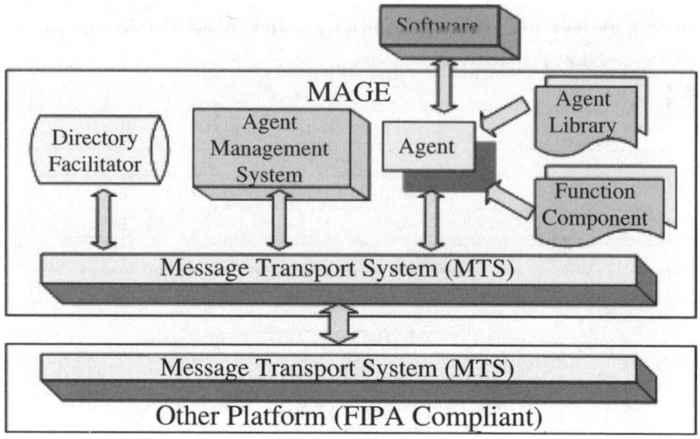

Fig. 3. Architecture of MAGE

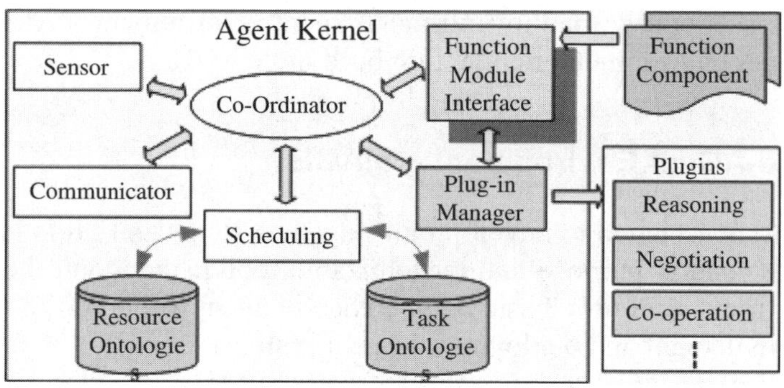

Fig. 4. Architecture of the generic MAGE agent

Figure 4 is the extensible architecture of a generic MAGE agent that shows our philosophy. Agent kernel consists of the following parts. **Sensor** perceives the outside world. **Function Module Interface** makes an effect to the outside world. **Communicator** handles communications between the agent and other agents. **Co-Ordinator** makes decisions concerning the agent's goals, and it is also responsible for co-coordinating the agent interactions with other agents using given co-ordination protocols and strategies, e.g. the contract net protocol or different kinds of auction protocol. **Scheduler** plans the agent's tasks based on decisions taken by the Co-ordination Engine and the resources and task specifications available to the agent. **Resource Ontologies** maintains a list of ontologies describing the semantic content of resources that are owned by and available to the agent. **Task Ontologies** provide logical descriptions of tasks known to the agent. **Plug-In Manager** manages the components provided by MAGE or by users that can be plugged into agent kernel.

4.2 Middle Wares

Middle ware layer provides developing environment, containing information retrieval, data mining, case base reasoning, expert system etc, to let users effectively use agent grid resources. In this section, we only expound the information retrieval and data mining because of page limits. Refer to http://www.intsci.ac.cn/ for more information.

Information Retrieval Toolkit Ghunt[19], from the function aspect, is an all-sided solution for information retrieval on the Internet. When it runs on the internet, a parallel, distributed and configurable Spider is used for information gather; a multi-hierarchy document classification approach combining the information gain initially processes gathered web documents; a swarm intelligence based document clustering method is used for information organization; a concept-based retrieval interface is applied for user interactive retrieval. It was integrated as a module of the AGrIP platform, which provides a powerful information retrieval function.

Data Mining Toolkit MSMiner[17] is a generic multi-strategy data-mining tool, which include database access, data modeling, data preprocessing, data mining, and data visualization. MSMiner not only provides convenient tools to develop new data mining algorithms, but also includes many build-in algorithms such as SOM and C4.5. MSMiner has an open interface for adding data preprocessing function and can access a variety of databases such as SQL Server, Oracle, and Informix. It can collect information from web, text, database and multimedia database, clean these data, and store them in the data warehouse. Machine learning, rough set, CBR and Statistics techniques are integrated in the algorithms library which provides a strong data mining for decision support.

4.3 Agent Service Description Language -- SDLSIN

Handling dynamic services is an important function of AGrIP. Agents are the core components of AGrIP and they are real entities that realize all kinds of activities related to services. In this part, we present SDLSIN – an Agent **S**ervice **D**escription **L**anguage with **S**emantics and **I**nheritance and Supporting **N**egotiation) based on DDL. The core of SDLSIN is DDL, and we implement SDLSIN in Java. This language considered not only semantic service description of agent, but also the inheritance and negotiation mechanism of agent service description, agent state language, and data types. SDLSIN is a kind of framework language with slots, and its formal criterion is the following:

<asdl-descr>::=(**ctype**
:**service-name** name
:**context** context-name+
:**types** (type-name = <modifier> type)+
:**isa** name
:**inputs** (variable: <modifier> put-type-name)+
:**outputs** (variable: <modifier> put-type-name)+
:**precondition** (DL formulae)
:**effects** (head/body)*
:**concept-description** (ontology-name = ontology-body)+
:**attributes** (attributes-name : attributes-value)+
:**text-description** name
)
ctype ::= capability | task
context-name ::= name<* ontology-name >
type-name ::= name
modifier ::=listof | setof
type ::= (name : name<* ontology-name >)+
put-type-name ::= (type-name | name<* ontology-name >)+
variable ::= name<* ontology-name >
DL formulae::=null | (DL formulae)(,DL assertion)*
head::= null | (DL assertion)
body::= DL formulae
ontology-body ::= (expression in concept-language)
attributes-name ::= name <* ontology-name >
attributes-value ::= name
name ::= String
ontology-name ::= name

In the formal criterion of SDLSIN, the meaning of each component is the following: **ctype** has two values, i.e. capability and task, where capability is the identifier which service provider (SP) registers its capability to the middle agent, while task is the identifier which service requester (SR) requests services from the middle agent; **service-name** denotes the name of service (i.e. identifier); **context** uses several keywords (from domain ontologies) to description the main characteristics of service and it may be used in (syntax based or semantics based) service matchmaking; **types** denotes the definition of the data types (i.e. Integer, Real, and String) used in the service description; **isa** allows the name of a service from which this service will inherit the description; **inputs** denote input variable declarations for the service; **outputs** denote output variable declarations for the service i.e., the outcome of invoking service; **precondition** denotes the executability of service; **effects** denote the effects of performing the service; **concept-description** denotes the meaning of words used in the service description i.e., some terminologies defined in a given local domain ontology; **attributes** mainly support for service negotiation and its values include cost, quality of service, style of service, and performance of service etc; **text-description** is mainly used to descript the service in natural language; Except for attribute service-name, all of these attributes are optional.

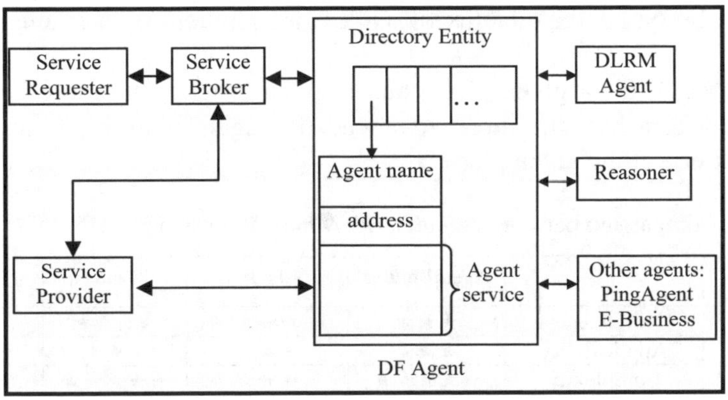

Fig. 5. MAGE services environment architecture

In multi-agent environment MAGE, we have realized the service description language SDLSIN and DF (Directory Facilitator) agents and SBroker (Service Broker) agents, where DF agent stores agent's capability (service) description and finishes service matchmaking, and SBroker agent finishes service invoke, service negotiation, service composition, service cooperation and service control, etc. DF and SBroker make up of the middle agent of our MAGE. Service providers register their service to DF agent in SDLSIN. Service requester puts in its request to SBroker in SDLSIN, then SBroker acts as this requester to finish all service reasoning work (such as service matchmaking, service negotiation, service composition), at last SBroker returns the answer to this requester. The services environment architecture is illustrated in Figure 5.

4.4 Comparison between multi-agent platforms

We show comparison between MAGE and some multi-agent platforms. Some multi-agent platforms are selected to compare with MAGE: AgentBuilder®[20], Jack™ [21] and Zeus[22]. The comparison between

MAGE, AgentBuilder, Jack and Zeus is listed in Table 4.1, according to the following four criteria for each stage of the systematic analysis, design, construction and deployment of agent-oriented applications. The four criteria are:

- completeness, i.e. the degree of coverage the platform provides for this stage;
- applicability, i.e. the range of possibilities offered, the restrictions imposed by the proposed stage;
- complexity, i.e. the difficulty to complete the stage;
- reusability, i.e. the quantity of work gained by reusing previous works.

In Table 2, for a given phase and a given criteria, the more "★" one agent platform has, the more advanced this agent platform is about the given criteria at the given phase.

Table 2. Comparison between AgentBuilder, Jack, Zeus and MAGE

		AgentBuilder	**Jack**	**Zeus**	**MAGE**
Analysis	Completeness	★★★	★★★	★★★★	★★★★★
	Applicability	★★★	★★★	★★★	★★★★
	Complexity	★★★★★	★★★	★★★★★	★★★★
	Reusability	★★★	★★★	★★★	★★★★★
Design	Completeness	★★★	★	★★★★	★★★★★
	Applicability	★★★	★	★★★	★★★★
	Complexity	★★★★	★	★★★	★★
	Reusability	★★	★★★	★★★	★★★★
Develop ment	Completeness	★★★★★	★★★★	★★★★★	★★★★★
	Applicability	★★★	★★★★★	★★★	★★★★
	Complexity	★★★★	★★	★★★★	★★★
	Reusability	★★	★★★★★★	★★	★★★
Deploym ent	Completeness	★★★	★★★	★★★★	★★★★★
	Applicability	★★★	★★★★	★★★	★★★
	Complexity	★★★★	★★★	★★★★	★★★★★
	Reusability				★

5. Conclusion

In an increasing number of scientific disciplines and business applications, an exponential growth of information and the proliferation of service offers are important community resources. Geographically distributed users often need to access information and invoke the services in collaborated ways. This requires a robust infrastructure which can not only support networked intelligent data handling but also offer synergetic environment for applications on Web services.

Agent **Gr**id **I**ntelligent **P**latform (AGrIP) is an Internet intelligent platform built on top of MAGE. In order to enable greater access to content, AGrIP manages DB resources by using ontology techniques and provides an interoperability mechanism for turning remote, heterogeneous resources into a globally well-ordered knowledge space. In order to enable greater access to services, DDL is proposed as a kind of formally logical framework to process static knowledge and dynamic knowledge. Based on DDL, an agent service description language SDLSIN is defined. And the language is implemented in AGrIP to build collaborative service environment on MAGE.

Acknowledgment

This work has been funded by the National Basic Research Priorities Programme (No.2003CB317004) and the National Science Foundation of China (No.90604017).

References

[1] T. Berners-Lee, J. Hendler, and O. Lassila. The Semantic Web. Scientific American, 284(5): 34-43, 2001.

[2] S. McIlraith, T.C. Son and H. Zeng. Semantic Web Services. IEEE Intelligent Systems, Special Issue on the Semantic Web, 16(2): 46~53, 2001.

[3] The OWL-S Coalition. OWL-S: Semantic Markup for Web Services. http://www.daml.org/services/.

[4] He Huang, Zhongzhi Shi, Lirong Qiu and Yong Cheng. Ontology-Driven Knowledge Management on the Grid. In: Proceedings of The 2005 IEEE/WIC/ACM International Conference on Web Intelligence (WI'05), pp. 475-478, 2005.

[5] He Huang, Zhongzhi Shi, Yong Cheng, and Lirong Qiu. Service-Oriented Knowledge Management on Virtual Organizations. In: Proceedings of The Fifth International Conference on Computer and Information Technology (CIT'05), pp. 1050-1054, 2005.

[6] He Huang, Zhongzhi Shi, Yong Cheng, Lirong Qiu and Xiaoxiao He. Semantic-based data access services on the grid. In: PROCEEDINGS OF THE 8TH JOINT CONFERENCE ON INFORMATION SCIENCES, vols 1-3, pp. 1554-1557, 2005.

[7] Zhongzhi Shi, Mingkai Dong, Yuncheng Jiang, Haijun Zhang. A logical foundation for the semantic Web. SCIENCE IN CHINA SERIES F-INFORMATION SCIENCES, 48(2): 161-178, 2005.
[8] Mingkai Dong, Yuncheng Jiang, Zhongzhi Shi. A description logic with default reasoning. Journal of Computer, 26(6): 729-736, 2003.
[9] Zhongzhi Shi, Haijun Zhang, Yong Cheng, Yuncheng Jiang, Qiujian Sheng, Zhikung Zhao. MAGE: An Agent-Oriented Programming Environment. IEEE ICCI 2004: 250-257.
[10] A. Chervenak, I. Foster, C, Kesselman, C. Salisbury, and S. Tuecke, "The Data Grid: Towards an Architecture for the Distributed Management and Analysis of Large Scientific Datasets", Journal of Network and Computer Applications, 23:187-200, 2001.
[11] I. Foster, C. Kesselman, and S. Tuecke, "The Anatomy of the Grid: Enabling Scalable Virtual Organizations", International J. Supercomputer Applications, 15(3), 2001.
[12] A. Rajasekar, M. Wan, R. Moore, and W. Schroeder, "Storage Resource Broker - Managing Distributed Data in a Grid", Computer Society of India Journal, Special Issue on SAN, 33(4): 42-54, 2003.
[13] P.F. Patel-Schneider, P.H. and I. Horrocks, OWL Web Ontology Language Semantics and Abstract Syntax, W3C Recommendation 10 February 2004.
[14] A. Artale and E. Franconi. Temporal Description Logics. Annals of Mathematics and Artificial Intelligence, 30(1-4), 2000.
[15] F. Wolter and M. Zakharyaschev. Dynamic description logic. In: Segerberg K, et al, editors, Advances in Modal Logic, 2: 449-463, 2000.
[16] F. Baader, et al. The Description Logic Handbook: Theory, Implementation and Applications. Cambridge: Cambridge University Press, 2002.
[17] Zhongzhi Shi. MSMiner: A platform for data mining. In: Proceedings of 2002 International Conference on Machine Learning and Cybernetics, v1, 2002.
[18] http://www.fipa.org/
[19] http://www.intsci.ac.cn/GHuntWeb/
[20] http://www.agentbuilder.com/
[21] http://www.agent-software.com.au/
[22] http://labs.bt.com/projects/agents/zeus/

Web-based Service Information Systems based on Fuzzy Linguistic Techniques and Semantic Web Technologies

Enrique Herrera-Viedma[1], Eduardo Peis[2], José M. Morales-del-Castillo[2], and Karina Anaya[1]

[1]Dpt. Computer Science and AI. University of Granada. viedma@decsai.ugr.es, karina@ugr.es

[2] Dpt. Library and Information Science. University of Granada. epeis@ugr.es, josemdc@ugr.es

Abstract. The aim of this paper is to present a model of a web multi-agent system which combines the use of Semantic Web technologies together with the application of user profiles to provide an enhanced Web retrieval service. This system uses fuzzy linguistic techniques to deal with qualitative information in a user-friendly way. The system activity is developed in two phases: retrieval phase to gather the documents from the Web, and feedback phase to update the user profiles and the recommendation values of resources. In this paper we focus on the analysis of the retrieval phase. With this multi-agent system model the retrieval capabilities on the Web can be considerably increased.

1 Introduction

As it is known, one of the main problems of the Web today is to efficiently manage the overwhelming quantity of resources available to Internet users. To avoid arriving to a situation of collapse, it is becoming necessary to develop tools capable to offer solutions to this problem, since the available instruments have shown little efficiency in easing users' time consuming tasks such as gathering and selecting relevant documents. So, it could be useful to develop Web-based service information systems that permit to improve the access to information in a more efficient way. Most of the solutions proposed to face this problem involve different technologies as intelligent software agents [14, 27], information filtering techniques [31], and Semantic Web technologies [4, 5].

H.-V. Enrique et al.: *Web-based Service Infromation Systems based on Fuzzy Linguistic Techniques and Semantic Web Technologies*, Studies in Computational Intelligence (SCI) **37**, 647–666 (2007)
www.springerlink.com

Software agents applied to a Web-based framework are usually organized in distributed architectures [7, 13, 14, 25] to mainly perform tasks of intermediation between users and the Web. In other words, we could say that agents play the role of *infomediators* that assist users in the information retrieval process [7, 14, 25, 31]. These agents are entities capable to act in an autonomous way, processing and exchanging results with other agents [19]. Nevertheless, to develop these tasks agents need a knowledge base that can be supported on Web ontologies [18, 19] and/or on implicit or explicit information about the users (obtained by direct observation and imitation of users' behaviour, or by registering users' feedback, respectively [27])

However, the main problem of using agents is to find a flexible and agile communication protocol for exchanging information among agents, and between users and agents because of the great variety of forms the information is represented in the Web. One possibility to facilitate the communication processes consists in the application of the fuzzy linguistic approach [39], that provides a flexible representation model of information by means of linguistic labels. The application of fuzzy linguistic techniques enables us to handle information with several degrees of truth and solving the problem of quantifying qualitative concepts. Some examples of the use of fuzzy linguistic techniques in the design of multi-agent systems can be found in [9, 11, 21, 24].

On the other hand, information filtering techniques ease users the task of sorting out relevant documents from those that are not, thanks to the previous selection (carried out by system) of the resources that better fit users' needs, requirements and preferences. These needs, requirements and preferences are mostly defined in the form of user profiles [26] that can contribute to improve the performance of information systems.

Another possibility to improve the activity of a multi-agent system could be the use of some of the technologies of the Semantic Web project [4, 5]. These semantic technologies allow developing ontology-based infrastructures [29, 35] where agents can operate at semantic level with resources described using RDF (Resource Description Framework)[3] in a manner both interpretable by humans and machines. This common syntactic framework allows us to define a unique communication vocabulary among agents that could also be used to characterize the knowledge base of the system and even the semantics of resources and user profiles.

The aim of this paper is to present a new model of fuzzy linguistic multi-agent system that involves the use of the Semantic Web technologies and user profiles dynamically updated to improve the information access on the Web. The Semantic Web technologies are used to endow the agents with a common communication language, to develop the ontologies that describe the elements of the system and their interrelation, and to characterize resources and user profiles in a standardized way using RDF. As in [24], the system activity presents two phases, retrieval and feedback. In this paper we focus on the first one and show the structure of the multi-agent system that allows develop it.

The paper is structured as follows. Section 2 reviews the technologies employed in this research: the fuzzy linguistic approach, filtering tools and user profiles, and the Semantic Web. Section 3 presents the new multi-agent model and describes the retrieval phase. Section 4 shows an example of the system functionality, and finally, some concluding remarks are pointed out in section 5.

2 Preliminaries

2.1 Fuzzy linguistic approach

The *fuzzy linguistic approach* [39] and in particular, the *ordinal fuzzy linguistic approach* [20, 22, 23] are approximate techniques appropriate to deal with qualitative aspects of problems. An ordinal fuzzy linguistic approach is defined by considering a finite and totally ordered label set in the usual sense

$$S = \{s_i, i \in H = \{0, \ldots, T\}\}$$

and with odd cardinality (7 or 9 labels). The mid term representing an assessment of "approximately 0.5" and the rest of the terms being placed symmetrically around it. The semantics of the linguistic term set is established from the ordered structure of the term set by considering that each linguistic term for the pair (s_i, s_{T-i}) is equally informative. Furthermore, for each label s_i could be given a fuzzy number defined on the [0,1] interval, which is described by a linear trapezoidal membership function represented by the 4-tuple (a_i, b_i, α_i β_i) (the first two parameters indicate the interval in which the membership value is 1.0; the third and fourth parameters indicate the left and right widths of the distribution). Additionally, we require the following properties:

$$1.-\textit{The set is ordered}: s_i \geq s_j \quad \textit{if} \quad i \geq j.$$
$$2.-\textit{There is the negation operator}: Neg(s_i) = s_j, \textit{ with } j = T - i.$$
$$3.-\textit{Maximization operator}: MAX(s_i, s_j) = s_i \textit{ if } s_i \geq s_j.$$
$$4.-\textit{Minimization operator}: MIN(s_i, s_j) = s_i \textit{ if } s_i \leq s_j.$$

To combine the linguistic information we need to define an aggregation operator such as the Linguistic Ordered Weighted Averaging (LOWA) operator [21]. It is used to aggregate non-weighted linguistic information (i.e., linguistic information values with equal importance) and it has been satisfactorily applied in different fields [20, 23, 24]. It is based on the OWA operator [38] and the convex combination of linguistic labels [10]. The LOWA operator is an *"or-and"* operator [21], i.e., its result is located between the maximum and minimum of the set of aggregated linguistic values. Its main advantage is that it allows to aggregate automatically linguistic information without to use linguistic approximation processes [39].

2.2 Filtering techniques and user profiles

Information filtering techniques deal with a variety of processes involving the delivery of information to people who need it. Operating in textual domains, *filtering systems* or *recommender systems* evaluate and filter the resources available on the Web (usually in HTML or XML documents) to assist people in their search processes [32], in most cases through the use of filtering agents [33]. Traditionally, these systems have fallen into two main categories [31]. *Collaborative filtering systems* use the information provided by many users to filter and recommend documents to a given user, ignoring the representation of documents. *Content-based filtering systems* filter and recommend the information by matching user query terms with the index terms used in the representation of documents, ignoring data from other users. These may be a drawback when little is known about user needs, so it becomes a necessity to apply user profiles for providing a fast and efficient filtering [37].

User profiles are the basis for the performance of information filtering systems. User profiles represent the user's long-term information needs or interests. There are two main distinct types of user profiles [26]: i) *collaborative profile*, which is based on the rating patterns of similar users, and hence, it can be represented by a community of similar users; and ii) content-based profile, which is represented by a vector of interest areas.

User profiling is a critical task in information filtering systems. As it is pointed out in [26] "... an improper user profile causes poor filtering performance (for example, the user may be overloaded with irrelevant information, or not get relevant information that has been erroneously filtered out)". Two desire properties that any user profiling should support are the following:

- User profiles should be adaptable or dynamic since user's interests are changing continuously and rapidly over time. This implies the necessity to include a learning module in the information filtering system to adapt the user profile according to feedback from user reaction to information provided by the information filtering system.
- The generating and updating of user profiles should be carried out with a minimal explicit involvement of the users, i.e. by minimizing the degree of the user intervention to reduce user effort and facilitate the system-user interaction.

2.3 Semantic Web technologies

The Semantic Web is an extension of the present Web, in which the information is gifted of a well defined meaning, permitting a better cooperation between humans and machines [4, 5]. It is based on two main ideas: i) semantic mark up of resources, and ii) development of "intelligent" software agents capable to understand and to operate with these resources at semantic level [4, 18].

The semantic backbone of the model is RDF/XML [3], a vocabulary that provides the necessary infrastructure to codify, exchange, link, merge and reuse structured metadata in order to make them directly interpretable by machines. RDF/XML structures the information in assertions (resource-property-value triples), and uniquely identifies resources by means of URI's (Universal Resource Identifier), allowing intelligent software agents the knowledge extraction from and inference reasoning over resources (such as documents, user profiles or even queries) using web ontologies. Ontologies, in the Semantic Web context, represent exhaustively specific knowledge shared by the members of a specific domain structured as a hierarchy of concepts, the relations between these concepts and the axioms defined upon these concepts and relations [6, 15, 16]. Therefore, Web ontologies provide an "invisible" semantic lattice where complex systems can be built over, defining and interrelating the different elements that form their structure. There exist several ontology languages that can be used for designing web-based ontologies, but the recommendation proposed by the World Wide Web Consortium (W3C) is the Ontology Web Language (OWL) [28], a language with a great expressive capacity that allow defining ontologies maintaining the RDF/XML syntactic convention.

This feature allows querying both resources and ontologies using a common semantic query language (for example [1, 30, 34]), that allows agents extracting information and inferring knowledge from RDF graphs.

3 A model of fuzzy linguistic multi-agent system based on Semantic Web and user profiles

In [24] we define a model of fuzzy linguistic multi-agent system to gather information on the Web with a hierarchical architecture of seven action levels: *internet users, interface agent, collaborative filtering agent, task agent, content-based filtering agent, information agent* and *information sources*. Its activity is developed in two phases: i) *Retrieval phase:* This first phase coincide with the information gathering process developed by the multi-agent model itself; and ii) *Feedback phase:* The second phase correspond with the updating process of recommendations on desired documents existing in a collaborative recommender system.

The main drawback of this model is that it does not utilize user profiles to characterize users' preferences and consequently its retrieval capabilities are clearly handicapped.

To overcome the limitations of the model presented in [24] we define a new and enhanced model of a fuzzy linguistic multi-agent system that improves information retrieval by means of the application of user profiles to enrich the filtering activity, and Semantic Web technologies to set a base for the operation of software agents. This model has been specifically designed for its use in academic environments, although it can be easily adapted for its implementation on different domain-dependant environments where a very specialized information retrieval and filtering is required (such as bio-medical or enterprise information systems). It presents a hierarchical structure with six action levels (*internet users, interface agent, filtering agent, task agent, information access* and *information bases*) and also two main activity phases (see Fig. 1):

1. **Retrieval phase**: This phase involves three processes. The first one is the "semantic retrieval" process and coincides with the information gathering process developed by the multi-agent model itself, although

the query language used is not a Boolean one as in [24], but a semantic query language capable of comparing both literal and semantics structures [17, 36]. In such a way it is possible to obtain more accurate and contextualized answers to queries. The second process is a "filtering" process, which consists on the selection of those resources that better fit both the explicit and the implicit preferences of the users. The third process is the "display" process the filtered resources to users.

2. **Feedback phase**: This phase involves two processes: "recommendation" process and "profile updating" process. In both processes the system needs users to qualitatively appraise the selected documents and the global answer provided by the system to a specific query, respectively. The "recommendation" process is similar to that defined in [24]: users express their opinion about any retrieved documents and with the appraisal provided the system can recalculate and update the recommendations of the desired documents. The second process consists on the dynamic updating of user profiles on the basis of the satisfaction degree the user expresses regard to the global results provided by the system.

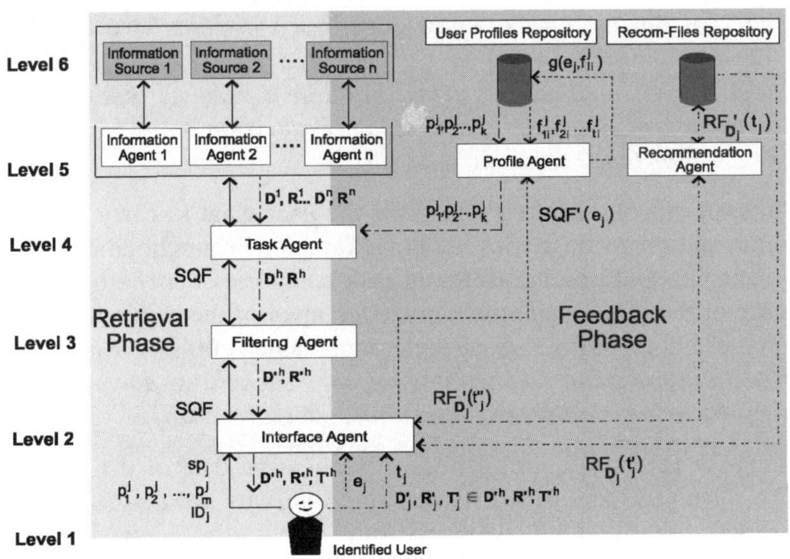

Fig. 1. Model architecture: "Semantic" retrieval and feedback phases

In the following sections, we analyze in detail the retrieval phase together with its action processes.

3.1 Information Retrieval Phase: "Semantic" retrieval process

This process begins when an identified user defines a query and ends when the information agents retrieve sets of relevant resources from the different document repositories. Users define their queries using basically both *preferences* and *search parameters*. *Preferences* refer to the search options that users can select to scope queries, defining constraints upon the characteristics of the documents to be retrieved (such as its semantic context or typology). For example, the user could provide his/her preferences about any of these four categories of basic preferences:

- *Document Type*: This preference establishes the type of document that user prefers to retrieve. For example, we could consider the following F_1= *{SciArticle, Proceedings, BookChapter, all}*.
- *Search Context:* It consists of general topic categories that represent the main areas of the system domain. For example, if the domain of work of our system is "knowledge-based systems" we could define the following set of values [2]: F_2= *{case-based reasoning, knowledge-based intelligent systems, intelligent systems, multi-agent systems, neural networks, fuzzy systems, decision support systems, genetic algorithms, semantic web, all}*.
- *Search aim:* It defines those tasks the user want to carry out with the information to be retrieved. For example, in a scholarship environment we could define different task categories depending on the nature of these tasks and the knowledge level of the different kind of users. Then, a possible set of preference values could be F_3= *{research, teaching_bachelor, teaching_master, teaching_doctorate, studies_bachelor, studies_master, studies_doctorate, all}*.
- *Date:* It refers to the updating or publication date of the resources to be retrieved. A set of different time intervals are defined to cover a wide range of values that vary from few months to several years. For example, F_4= *{3months, 6months, 1year, 3years, 5years,+5years}*.

On the other hand, *search parameters* correspond with both natural language *keywords* that better define user's own information needs, and *structural elements* of the document where the search must be performed (e.g., in the whole document or just in the abstract). The *structural elements* define the logical structure of each document type (which is, in turn, defined and validated through its corresponding XML Schema [12]). Therefore, the set of structural elements available to define a query depends on the selected value for the *document type* preference. Therefore, the set of structural elements vary from a document type to another. For example, suppose a user choosing the *Scientific article* type, then he/she could select any structural element from the following set that has been exclusively defined for this document type, e.g., E ={*title, authors, abstract, introduction, body, conclusions, bibliography, whole_document*}.

Once the user formulates a query, it is assigned a unique "semantic query file" (SQF) in RDF format that contains the user's idenfier ID_j, the search parameters (sp_j) and the set of selected values $\{d_1^j, d_2^j, ..., d_k^j\}$ for the preferences $\{p_1^j, p_2^j, ..., p_k^j\}$ (see Fig. 2 below).

```
...
<Query rdf:ID="query384">
  <user_ID>http://www.ugr.es/~user/U022005</user_ID>
  <preferences_e>
    <preferences rdf:ID="pref_384">
      <docType>sciArticle</docType>
      <context>user_modeling</context>
      <aim>research_article</aim>
      <date>3months</date>
      <value>NULL</value>
    </preferences>
  </preferences_e>
  <search_parameters_e>
   <keyW>ontologies</keyW>
   <keyW>machine learning</keyW>
   <struct_e>abstract</struct_e>
  </search_parameters_e>
</Query>
...
```

Fig. 2. Semantic Query File (SQF)

The *semantic retrieval process* is developed in the following steps:

- **Step 1:** To define a query any registered user j must specify the search parameters sp_j (keywords and an optional structural element) and a set of k $(0 \leq k \leq m)$ preferences $\{p_1^j, p_2^j, \ldots p_k^j\}$, being m the number of properties used to define a user profile and $p_i^j \in F_i$, being F_i the expression domain associated to the property i. From this query the interface agent generates the associated SQF, storing in it the inputs given by the user (i.e., search parameters and preferences) and his/her ID. Those preferences not explicitly given will appear with NULL value.
- **Step 2:** The SQF is transferred from the interface agent to the task agent.

- **Step 3:** The task agent proceeds to complete the SQF replacing every NULL preference with the value of its expression domain (stored in the user's profile) with a highest associated frequency, obtaining as a result a SQF with no NULL values.
- **Step 4:** Using the keywords and the structural element stored in the SQF, the task agent composes a semantic query using a pre-agreed semantic query language. This semantic query is sent to the different information agents. The query, as in ordinary search engines, is defined using clauses and operators set by default to combine the keywords. In our model, the chosen semantic query language is SeRQL [1] due to its simplicity and flexibility. With this language we can define simple queries using a structure similar to the following:

```
SELECT SciArt
FROM {SciArt} doc:hasAbstract {} doc:abstract {LIT}
WHERE Lit LIKE "keyword1" or Lit LIKE "keyword2"
   IGNORE CASE
USING NAMESPACE
   doc = <http://www.ugr.es/local/kishimaru/SciOnt#>
```

Fig. 3. Semantic query sample

In this case, query indicates that the different information agents should retrieve those scientific articles whose abstracts contain the terms *keyword1* or *keyword2*.

- **Step 5:** The information agents apply these common "semantic" searches in their associated document repository (one per agent), retrieving sets of pertinent documents $\{D^1, D^2, \ldots, D^n\}$, supposing n information agents. For each resource D^i_j in a set of retrieved documents D^i, being $1 \leq i \leq n$, the information agents calculate a relevance degree $R^i_j \in R^i$ (being R^i the set of relevance degrees for the set of documents D^i). These relevance degrees must be interpreted as the relative importance of the selected keywords in a specific structural element for a particular document. In other words, the retrieved resources will be, for example, documents with relevant abstracts if the chosen structural element was *abstract* or with relevant conclusions if it was the *conclusions* element, and so on (it is obvious that the relevance degree is relative to the whole document when the search is performed using the *whole_document* element). Each information agent sends the resulting sets of relevant documents to the task agent.
- **Step 6:** The task agent aggregates the different sets of resources (obtained from distributed sources) into a single set (D^h, R^h) to make the information more tractable for its filtering.

3.2 Information Retrieval Phase: Filtering process

This process basically consists in performing different “semantic” searches (one per preference) over the set of resources retrieved by the information agents to match users’ preferences with the characteristics of the documents. Afterwards, a ranked list of filtered resources is generated.

```
...
<User rdf:ID="user02555">
  <updated>01/06/2005</updated>
  <personalInfo_e>
     <PersonalInfo rdf:ID="inf-pr001">
       <photo>pht0445.jpg</photo>
       <name>Juan</name>
       <surname1>Doe</surname1>
       ...
     </PersonalInfo>
   </personalInfo_e>
   <preferences_e>
     <DocType rdf:ID="docType-pr001">
       <type_e>
         <Type rdf:ID="type1-pr001">
           <type>SciArticle</type>
           <freq>AlmostAlways</freq>
         </Type>
       </type_e>
     ...
     </DocType>
     ...
   </preferences_e>
</User>
...
```

Fig. 4. Representation of a user profile

This process needs three basic inputs:

***The user profile*:** We assume a repository storing a set of user profiles, defined as a structured representation of an individual in RDF format, which is determined by users’ ID and characterized by the particular values that each user has assigned to the categories of basic preferences. Each preference has an associated linguistic frequency property (tagged as *<freq>*) representing how often a specific value is used in queries assessed by users as “satisfactory”. Thus, being $F_i = \{g_{1i}, g_{2i}, \ldots, g_{ti}\}$ the set of basic preferences, then we define $f^j_i = \{f^j_{1i}, f^j_{2i}, \ldots, f^j_{ti}\}$ as the set of frequency values associated with each possible value $l \in \{1, \ldots, t\}$ of the property i in the profile of the user *j*. The range of possible values for the frequencies is defined in a set of seven linguistic labels, *S={always, almostAlways, mostTimes, sometimes, aFewTimes, almostNever, never}*, i.e., $f^{j}_{li} \in S$. An example is given in Fig. 4. The utilization of user profiles makes necessary a registration process previous to the retrieval phase. When a user logs into the system for the first time, he/she must fill in a simple form with personal information and interests, professional aims, etc., in order to get an approximate structured representation of the individual. On the base of these data, the system is able to assign to each user a basic stereotypic profile (by means of clustering algorithms), that serves as a basis where the user profile can be developed. Each one of these stereotypic profiles describes a specific user type through a set of characteristics, constraints and preferences set by default. For example, in a scholarship environment we could define three different stereotypes: researchers, teachers and students. Although stereotypes may not exactly reflect the characteristics of each individual, they are valid approximations that avoid the problem of "cold start". Another characteristic of these basic stereotypic profiles is that they may evolve over time and be modified when a significant change is detected in the behavior of users pertaining to a specific stereotype. In this process the user is also automatically assigned an ID (a URI) that will uniquely identify him/her in the system, so therefore it will be possible to relate users with other elements and actions they perform (as for example when formulating a query, or giving a recommendation about a resource).

***The Semantic Query Files (SQFs)*:** As it was explained above, this element contains the search parameters and preferences of a particular user for a specific query. To filter the set of retrieved documents is necessary to match concrete documents' characteristics with users' preferences.

***The Description Files of the Documents*:** We assume that each document in a resource repositories has associated a content file (see Fig. 5) and two auxiliary description files, *a classification file (CF)* and a *recommendation file (RF),* in RDF format. This description allows a more flexible and complete characterization of documents. In such a way, we can easily update the recommendation values or classification terms of a given document. Both auxiliary metadata files can be directly referenced from the content file. While RF stores a historical log of the recommendations assigned to the document since it was accessed, the CF (see Fig. 6) contains additional data about the resource, such as its level of complexity, a *document type* property, and a set of content classification categories. The *level of complexity* is an attribute defined in a bid to match the search aim of users with the complexity of the content of the resources. Therefore, the rank of possible values for the level of *complexity* element must be the same defined for the *search aim* preference.

```
...
<SciArt rdf:ID="S442">
  <hasTitle>
    <Title_e rdf:ID="Title_SA-001">
      <title>Introduction to RDF</title>
    </Title_e>
  </hasTitle>
  <hasAuthor>
    <Author_e rdf:ID="Author_SA-001">
      <author>Martina Branshaw</author>
    </Author_e>
  </hasAuthor>
   ...
</SciArt>
...
```

Fig. 5. Sample of the content file of a document

```
...
<Class_File rdf:ID="CS442">
  <resource>
   http://www.ugr.es/local/kishimaru/SciOnt#S442
  </resource>
  <doctype>Scientific article</doctype>
  <class_e>
    <Class rdf:ID="ce_CS442">
      <catg>web usage mining</catg>
      <dex:mat>minería de datos</dex:mat>
    </Class>
  </class_e>
  <complex>Researcher</complex>
</Class_File>
...
```

Fig. 6. Representation of a classification file (CF)

Then, the filtering process is developed in the following steps:

- **Step 1:** The task agent sends the ranked set of aggregated resources (D^h, R^h) to the filtering agent.
- **Step 2:** The filtering agent matches the characteristics of the resources (stored both in the content file and in its associated CF) with the preferences stored in the SQF generated for the current query, discarding

those documents that don't fit user's requirements. To do so, defines a set of queries (one per preference) to filter the resulting ranked set of resources. For example, according to the *search context* preference, the filtering agent could define a query with a structure similar to the following one:

```
SELECT Class_file
FROM {Class_file} cf:class_e {} cf:catg {"data mining"}
USING NAMESPACE
   cf = <http://www.ugr.es/~CF/class#>
```

Fig 7. Filtering query sample

In this example, the search is carried out in the CF of each retrieved resource, matching the classification topics with the "*Search Context*" preference in the SQF. As a result, those resources that doesn't include in their classification categories the term "data mining" are discarded.

– **Step 3:** The filtering agent seeks the URIs of the documents that have matched all the preferences defined in the SQF, therefore generating a ranked list with the set of already filtered resources (D'^h, R'^h), that is consequently sent to the interface agent.

3.3 Information Retrieval Phase: Display process

Once the retrieved documents are filtered, the system proceeds to display the results to the users. They can choose those particular documents of their interest and their display format.

This process requires the following input:

-Recommendation Files of the Documents: As aforementioned, we assume that each document has associated a RF in RDF format (see Fig. 8) where is contained information about all the appraisals made by users that have read it previously. The RF contains data as its URI, the last recommendation value displayed and a set of log items containing previous appraisals about that document. Each log item is defined by an user's ID, his/her corresponding appraisal and the search context used in the query formulated by the user to retrieve that document. This representation enables the adoption of different recommendation policies, allowing us, for example, to recalculate recommendation values based on the opinion of all the users, or just of some of them (e.g., using the appraisals given by those users who looked for information by the same topic).

```
...
<RecomFile rdf:ID="recomf001">
  <resource>
   http://www.ugr.es/local/kishimaru/SciOnt#S442
  </resource>
  <accesed>2005-08-17T13:25:42Z</accesed>
  <modified>2005-03-08T16:32:11Z</modified>
  <recom_value>High</recom_value>
  <recom_history>
    <R_history rdf:ID="histf001">
      <item>
        <RecomItem rdf:ID="form01-pr001">
          <appraissal>VeryHigh</appraissal>
          <topic>Data mining</topic>
          <user_ID>user-pr022005</user_ID>
        </RecomItem>
      </item>
      ....
    </R_history>
  </recom_history>
</RecomFile>
...
```

Fig. 8. Representation of a Recommendation File (RF)

This process is developed in four steps:

- **Step 1:** The interface agent asks the recommendation agent for the recommendation values corresponding to each one of the documents to be displayed.
- **Step 2:** For each particular document, the recommendation agent checks out if its corresponding recommendation file $RF_{D'j}$ has been modified since the last time it was accessed. If not, the interface agent receives the last recommendation value displayed t'_j that was stored in the *<recom_value>* tag of the $RF_{D'j}$. If in the $RF_{D'j}$ was added a new log item, the recommendation agent proceeds to recalculate a new global recommendation value by means of the LOWA operator [21] from the set of historical values, generating a new recommendation value (t''_j) that replaces the old one t'_j.
- **Step 3:** The documents are displayed coupled with their respective recommendation value.

- **Step 4:** Once the user selects a document of his/her interest, it is displayed by the interface agent in the preferred file format defined by the user (e.g., txt, xml, html, pdf, etc.) by means of XSLT stylesheets [8].

4 Application example

Suppose the following framework. John Quest, a researcher member of RECSEM (an academic institution specialized in the study of "information systems"), has to write a paper about "*semantic information systems*" for a prestigious scientific journal. John decides to search in RECSEM's documents repositories for recent resources about this topic and, in such a way, to build a solid knowledge background for his work. He proceeds to log into RECSEM's site and reaches the search interface page. Then, he writes a pair of keywords ("*information systems*" and "*ontologies*") in the search box, and explicitly specifies that these keywords must be searched in the abstract of each document. To scope the search, John selects *"knowledge representation"* as preferred search context, *"research"* as search aim and *"3months"* as preferred date of publication, but doesn't specify any value for the *"Document type"* preference. The system checks his profile and works out that the most satisfying value for the *"Document type"* preference is *"Scientific Article"*.

The task agent composes a semantic query that is sent to the different information agents. Each information agent retrieves a set of relevant documents and calculates their corresponding relevance degrees. The task agent aggregates all the retrieved documents into a set of relevant resources. This relevant set is filtered, being discarded those documents that don't match the preferences that appear in the SQF.

Before displaying the filtered documents to John, the task agent proceeds to check the RF of each resource and dynamically calculates their recommendation value. For example, let *H={high, medium, veryHigh, medium, low, high}* the set of historical recommendations values of a specific resource. If this set has been modified since the last time the document was accessed then the recommendation agent aggregates the different historical recommendation values using the LOWA operator and calculates a new recommendation value for the resource.

As a result of this search, on John's laptop screen are displayed a list of ranked documents that fit his explicit requirements (see Table 1).

Table 1. Query answer sample

D_j^h	R_j^h	T_j^h
http://www.ugr.es/~Arep/N0021	0.95	high
http://www.ugr.es/~Arep/L57641	0.93	NULL
http://www.ugr.es/~Erep/P70435	0.87	medium

Each document appears with an associated relevance degree and a recommendation value. This extra information eases John's task of deciding which resources can be more useful for him.

After he has used a document, he is asked to voluntary provide an opinion about its quality that will be stored in the list of historical recommendation values of that document. Furthermore, and before John can leave the current search session, he must provide his satisfaction degree relative to the answer provided by the system to his query, thus triggering the user profile updating mechanism.

After the search process, John finishes his session with the feeling that he has saved a lot of time, and with the certainty that he has obtained relevant and useful resources for his purposes.

5 Concluding remarks

We have described the architecture and elements of a fuzzy linguistic multi-agent system specially designed to perform information retrieval and filtering tasks in domain dependant environments (specifically in academic environments).

The key aspect of the system is the joint application of Semantic Web technologies and the use of user profiles. Semantic Web technologies provide the system with the necessary semantic infrastructure to improve inference and communication capacities of agents and with means to represent the information (resources and user profiles) in a common vocabulary both human and machine interpretable. The use of user profiles allows a better users' characterization and a better performance of the system can be achieved.

In the future, we shall study the development of an enhanced user profile updating process based on web usage analysis and rule discovery techniques, and the adaptation of this system to different work contexts.

Acknowledgements

Authors wish to acknowledge support from the Spanish Ministry of Education and Culture (project ref. TIC-2003-07977).

References

1. B.V. Aduna, A. Sirma, The SERQL query language (revision 1.2). In *User guide for Sesame* (2005). Available at http://www.openrdf.org/doc/sesame/users/ch06.html. Accessed 10/05/2005.
2. G. Avram, Empirical study on knowledge based systems, *The electronic journal of Information Systems Evaluation*. 8 (1) (2005), 11-20.
3. D. Beckett (ed.), RDF/XML Syntax Specification (Revised). W3C Recommendation. (2004). Available at *http://www.w3.org/TR/rdf-syntax-grammar/*. Accessed 02/19/2005.
4. T. Berners-Lee, Semantic Web Road map (1998). Available at *http://www.w3.org/DesignIssues/Semantic.html*. Accessed 02/15/2005.
5. T. Berners-Lee, J. Hendler, O. Lassila, The Semantic Web: A new form of Web content that is meaningful to computers will unleash a revolution of new possibilities, *Scientific American* May (2001).
6. Y.A. Bishr, H. Pundt, W. Kuhn, M. Radwan, Probing the concept of information communities: a first step toward semantic interoperability. In M. Goodchild, M. Egen-Hofer, R. Fegeas, C. Kottman (eds), *Interoperating Geographic Information Systems*, (Kluwer Academic, 1999), 55–69.
7. W. Brenner, R. Zarnekow, H. Witting, Intelligent Software Agent, Foundations and Applications, (Springer-Verlag. Berlin Heidelberg, 1998).
8. J. Clark (ed.), XSL Transformations (XSLT), Version 1.2 (1999). Available at http://www.w3.org/TR/xslt. Accessed 07/16/2005.
9. M. Delgado, F. Herrera, E. Herrera-Viedma, M.J. Martín-Bautista, M.A Vila, Combining linguistic information in a distributed intelligent agent model for information gathering on the Internet. In P.P. Wang (ed), *Computing with Word*, (John Wiley & Son, 2001), 251-276.
10. M. Delgado, J.L. Verdegay, and M.A. Vila, On aggregation operations of linguistic labels, *International Journal of Intelligent Systems*, 8 (1993) 351-370.
11. M. Delgado, F. Herrera, E. Herrera-Viedma, M.J. Martín-Bautista, L. Martínez, M.A. Vila, A communication model based on the 2-tuple fuzzy linguistic representation for a distributed intelligent agent system on Internet, *Soft Computing* 6 (2002), 320-328.
12. D. C. Fallside, P. Walmsley (eds.), XML Schema Part 0: Primer Second Edition, (2004). Available at http://www.w3.org/TR/xmlschema-0/. Accessed 09/10/2005.

13. B. Fazlollahi, R.M. Vahidov, R.A. Aliev, Multi-agent distributed intelligent system based on fuzzy decision making, *Int. J. of Intelligent Systems* 15 (2000), 849-858.
14. J. Ferber, Multi-Agent Systems: An Introduction to Distributed Artificial Intelligence (Addison-Wesley Longman, New York, 1999).
15. T.R. Gruber, Toward principles for the design of ontologies used for knowledge sharing, *Int. J. of Human-Computer Studies* 43 (5-6) (1995), 907-928.
16. N. Guarino, Formal ontology and information systems. In N. Guarino (ed) *Formal Ontology in Information Systems, Proceedings of FOIS'98*. Trento (Italy), (IOS Press, Amsterdam, 1998), 3–17.
17. R. Guha, R. McCool, E. Miller, Semantic search. *12th Int. World Wide Web Conference 2003 (WWW2003)*, Budapest (Hungary), (2003) 700 – 709
18. J. Hendler, Is there an intelligent agent in your future? (1999). Available at http://www.nature.com/nature/webmatters/agents/agents.html. Accessed 02/20/2005.
19. J. Hendler, Agents and the Semantic Web, *IEEE Intelligent Systems*, March, April (2001), 30-37.
20. F. Herrera, E. Herrera-Viedma, J.L. Verdegay, A model of Consensus in Group Decision Problems under Linguistic Assessments, *Fuzzy Sets and Systems* 78 (1996), 73-87.
21. F. Herrera, E. Herrera-Viedma, J.L. Verdegay, Direct Approach Processes in Group Decision Making using Linguistic OWA operators, *Fuzzy Sets and Systems* 79 (1996), 175-190.
22. E. Herrera-Viedma, Modelling the Retrieval Process of an Information Retrieval System Using an Ordinal Fuzzy Linguistic Approach, *J. of the American Society for Information Science and Technology (JASIST)* 52 (6) (2001), 460-475.
23. E. Herrera-Viedma, E. Peis, Evaluating the informative quality of documents in SGML format using fuzzy linguistic techniques based on computing with words, *Information Processing & Management* 39 (2) (2003), 195-213.
24. E. Herrera-Viedma, F. Herrera, L. Martínez, J.C. Herrera, A.G. López, Incorporating Filtering Techniques in a Fuzzy Multi-Agent Model for Gathering of Information on the Web, *Fuzzy Sets and Systems* 148 (1) (2004), 61-83.
25. N. Jennings, K. Sycara, M. Wooldridge, A roadmap of agent research and development, *Autonomous Agents and Multi-Agents Systems* 1 (1998), 7-38.
26. T. Kuflik, P. Shoval, Generation of user profiles for information filtering-research agenda. *Proc. of the 23rd Annual Int. ACM SIGIR Conf. on Research and Development Information Retrieval*, Athens (Greece) (2000), 313-315.
27. P. Maes, Agents that reduce the work and information overload, *Communications of the ACM*, 37 (7) (1994), 30-40.
28. D.L. McGuinness, F. van Harmelen (eds.), OWL Web Ontology Language Overview. W3C Recommendation. 10 February 2004, (2004). Available at http://www.w3.org/TR/2004/REC-owl-features-20040210/. Accessed 02/16/2005.
29. Ontoknowledge Project. Available at http://www.ontoknowledge.org/. Accessed 02/18/2005.

30. E. Prud'hommeaux, A. Seaborne (eds.), SPARQL query language for RDF (2004). Available at www.w3.org/TR/2004/WD-rdf-sparql-query-20041012/. Accessed 03/07/2005.
31. A. Popescul, L.H. Ungar, D.M. Pennock, S. Lawrence, Probabilistic models for unified-collaborative and content-based recommendation in sparse-data environments, *Proc. of the Seventeenth Conf. on Uncertainty in Artificial Intelligence (UAI)*, San Francisco (2001), 437-444.
32. P. Reisnick, H.R. Varian, Recommender Systems. *Special issue of Comm. of the ACM* 40 (3) (1997).
33. J.B. Schafer, J.A. Konstan, J. Riedl, E-Commerce recommendation applications. *Data Mining and Knowledge Discovery* 5 (1/2) (2001), 115-153.
34. A. Seaborne (eds.), RDQL: A query language for RDF, (2004). Available at http://www.w3.org/Submission/2004/SUBM-RDQL-20040109/. Accessed 02/02/2005.
35. Semantic Web Advanced Development for Europe (SWAD-Europe). Available at http://www.w3.org/2001/sw/Europe/. Accessed 02/17/2005.
36. U. Shah, T. Finin, Y. Peng, J. Mayfield, Information Retrieval on the Semantic Web, *Proc. of the 10th In.l Conf. on Information and Knowledge Management* (2002), 461-468.
37. B. Shapira, U. Hanani, A. Raveh, P. Shoval, Information filtering: A new two-phase model using stereotypic user profiling, *J. of Intelligent Information Systems* 8 (1997), 155–165.
38. R.R. Yager, On ordered weighted averaging aggregation operators in multicriteria decision making, *IEEE Transactions on Systems, Man, and Cybernetics* 18 (1988) 183-190.
39. L.A. Zadeh, The concept of a linguistic variable and its applications to approximate reasoning. Part I, In *Information Sciences* 8 (1975), 199-249. Part II, *Information Sciences* 8 (1975), 301-357. Part III, *Information Sciences* 9 (1975), 43-80.

Application of Chaos-based Pseudo-Random-Bit Generators in Internet-based Online Payments

Ping Li, Zhong Li, Siegfried Fettinger, Yaobing Mao, and Wolfgang A. Halang

Faculty of Electrical and Computer Engineering, FernUniversitaet in Hagen, UniversitaetStr. 27, 58084 Hagen, Germany ping.li@fernuni-hagen.de

Abstract

Soft computing (SC) has been well applied in web mining, search engine, E-service, and some other areas. As a constituent of SC, chaos theory has matured as a science (although is still evolving) and has wide applications. In this chapter, a pseudo-random-bit generator (PRBG) based on a spatiotemporal coupled map lattice (CML) is proposed. Synchronizations in the CML should be avoided in order to make the PRBG efficient and analyzed via the Lyapunov exponent spectrum and cross-correlation among the sites. The cryptographic properties, such as period, probability distribution, auto-correlation and cross-correlation, of the PRBG with various parameters are investigated numerically to determine the ranges of the parameters, within which the PRBG has good cryptographic properties. As a practical application, the PRBG has been employed in Internet-based online payments application, where the PRBG is used as a transaction-number (TAN) generator. Investigation shows that the TAN generator has satisfactory statistical properties and high efficiency.

1 Introduction

Soft computing (SC), initiated by Lotfi A. Zadeh, the founder of fuzzy set theory, consists of fuzzy logic (FL), neural network theory (NN) and probabilistic reasoning (PR), with the latter subsuming parts of belief networks, genetic algorithms, chaos theory and learning theory. It has been well applied in web mining, search engine, E-service, and some other ar-

P. Li et al.: *Application of Chaos-based Pseudo-Random-Bit Generators in Internet-based Online Payments*, Studies in Computational Intelligence (SCI) **37**, 667–685 (2007)
www.springerlink.com

eas. Among the constituents of SC, chaos theory has independently developed and has matured as a science (although is still evolving). In particular, chaos is now considered together with relativity and quantum mechanics as one of the three monumental scientific discoveries of the twentieth century.

Over the past two decades, applying chaos theory in cryptography has attracted much interest due to the fundamental features of chaotic systems, such as the sensitive dependence on initial conditions, ergodicity and mixing, which are quite advantageous to cryptography [Kocarev98, Avarez99, Kocarev01, Daselt01]. One of the applications of chaotic systems in cryptography is to design pseudo-random-number generators (PRNGs), which are turned out to play a central role in the construction of ciphers. Till now, lots of chaos-based PRNGs have been proposed [Li03a]. However, there exists dynamical degradation of chaotic systems in their realization with digital computers [Wheeler89, Li03b], this damages the cryptographic properties of the chaos-based PRNGs [Forre91, Zhou97]. In addition, since most of them adopt simple chaotic systems, the randomness of the pseudo-random numbers generated from the simple orbits may not meet cryptographic requirements. To overcome these drawbacks, higher finite precision [Wheeler91], multiple chaotic systems [Li01] and perturbation-based algorithms [Sang98a,b] have been proposed to improve chaos-based PRNGs. Especially, using spatiotemporal chaotic systems [Tang03, Lu04] may be a significant advance in this aspect because of the following special inherent features of spatiotemporal chaos. The orbit of a spatiotemporal chaotic system has long period even with dynamical degradation of digital chaos [Wang04]. Moreover, the randomness of the orbit of a spatiotemporal system is guaranteed by the complexity of system dynamics with large number of positive Lyapunov exponents. Finally, there are multiple sites in a spatiotemporal chaotic system, which can generate independent pseudo-random-bit sequences (PRBS) simultaneously.

In this paper, a pseudo-random-bit generator (PRBG) based on a spatiotemporal chaotic system is proposed, where a one-way coupled map lattice (CML) consisting of logistic maps is used as a spatiotemporal chaotic system with low computation expense. Synchronizations in the CML should be avoided in order to make the PRBG efficient and analyzed via the Lyapunov exponent spectrum and cross-correlation among the sites. The cryptographic properties, such as period, probability distribution, autocorrelation and cross-correlation, of the PRBG with various parameters are investigated numerically to determine the ranges of the parameters within which the PRBG has good cryptographic properties. As a practical application, the PRBG has been employed in Internet-based online payments application, where the PRBG is used as a transaction-number (TAN) generator. Investigation shows that the TAN generator has satisfactory statistical properties and high efficiency.

The rest of the paper is organized as follows. In Section 2, the PRBG based on a spatiotemporal chaotic system is proposed, the problem of how to avoid synchronizations in the PRBG is discussed, and the cryptographic properties of the PRBG with various parameters are investigated. The application of the PRBG as a TAN generator in Internet-based online payments is described in Section 3. Finally, conclusions are drawn in Section 4.

2 PRBG based on spatiotemporal chaotic system

2.1 Spatiotemporal chaotic system

A spatiotemporal chaotic system is a spatially extended system, which can exhibit spatiotemporal chaos, namely, nonlinear dynamics in both space and time. It is often modeled by a partial differential equation (PDE), a coupled ordinary differential equation (CODE), or a coupled map lattice (CML). A CML is a dynamical system with discrete-time, discrete-space and continuous states [Kaneko93]. It consists of nonlinear maps located on the lattice sites, named as local maps. Each local map is coupled with other local maps in terms of certain coupling rules. Because of the intrinsic nonlinear dynamics of each local map and the diffusion due to the spatial coupling among the local maps, a CML can exhibit spatiotemporal chaos. Since a CML captures the essential features of a spatiotemporal chaos and can be easily handled both analytically and numerically, a CML is used as a spatiotemporal chaotic system in this paper. In addition, various CMLs can be constructed by adopting various local maps, such as logistic map, tent map and skew tent map, and coupling methods, such as nearest-neighbor coupling and one-way coupling.

2.2 Algorithm of PRBG

To make CML-based PRBG efficient in computer realization, simple local maps and coupling methods with low computation expense are preferred. Therefore, the logistic map and one-way coupling are adopted to construct a CML for designing a PRBG. The CML is described as

$$x_{n+1}^{i} = (1-\varepsilon)f(x_n^i) + \varepsilon f(x_n^{i-1}), \qquad (1)$$
$$f(x) = rx(1-x),$$

where x_n^i represents the state variable for the site i (i=1,2,...,L, L is the number of the sites) at time n (n=1,2,...), $\varepsilon \in (0,1)$ is a coupling constant, , f is the logistic map and $r \in (0,4)$ is the parameter of the logistic map. The nearest-neighbor coupling with a periodic boundary condition, i.e., $x_n^0 = x_n^L$, $x_n^{L+1} = x_n^1$, is used in the CML.

When the CML described in Eq. (1) behaves chaotically, x_n^i can be regarded as a pseudo-random number, which means that $\{ x_n^i \}(n = 1, 2,\ldots)$ can be used as a pseudo-random-number sequence (PRNS), denoted by PRNS$_i$. Therefore, L PRNSs can be generated simultaneously from the CML with L sites. By digitizing the PRNSs, i.e., transforming the real number sequences to binary sequences, PRBSs can be obtained. Multiple PRBSs can be generated from the PRBG in the following two steps: 1) one CML with multiple sites generates multiple PRNSs; 2) each PRNS generates multiple PRBSs. To illustrate the mechanism clearly, a diagram is shown in Fig. 1, where LM is the abbreviation of the logistic map.

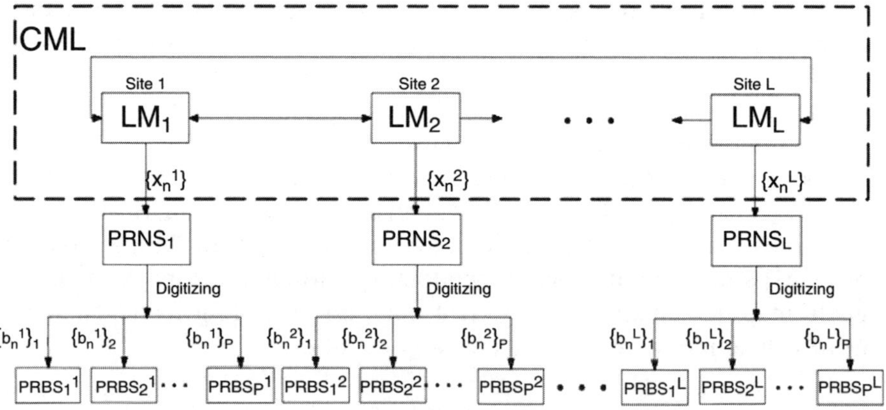

Fig. 1. Multiple PRBSs generated from the CML

There are several ways to digitize a real number to a binary number. One of them is approved suitable for cryptography [Li05a] and used in this PRBG. The digitization method is described in detail as follows.

x_n^i can be represented as a binary sequence [Kohda93]

$$x_n^i = 0.b_{n_1}^i, b_{n_2}^i, \ldots, b_{n_P}^i,$$

where P stands for a certain precision. The maximum P is equal to 52 in computer realization. Based on the binary representation, the digitization method can be shown in the Fig. 2.

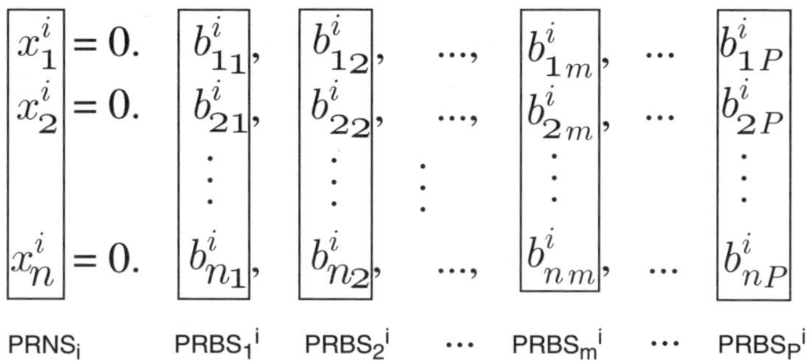

Fig. 2. Digitization method

It is dedicated that the bits at the ith positions of the binary representations of PRNSi comprise a binary sequence, i.e., $\{b_{n_m}^i\}(m = 1,2,\ldots,52)$, denoted by $\mathrm{PRBS}_m{}^i$. Thus, 52 PRBSs are generated from one PRNS by using this digitization method. Totally, $52L$ PRBSs can be generated at one time from the CML.

2.3 Avoidance of synchronizations in CML

If the multiple PRBSs are independent, the PRBG can be quite effective. Therefore, the PRNSs, from which the multiple PRBSs are generated, are desired to be independent. That is, the synchronization among the sites of the CML should be avoided. The worst case is that a complete synchronization of all the sites happens, i.e., the PRBG outputs only one PRBS actually. On the other hand, the ideal case is that there is no synchronization between arbitrary two sites, i.e., no partial synchronization. In this section, how to avoid the synchronizations is discussed.

Avoiding complete synchronization

A Lyapunov exponent spectrum of the synchronous chaos of the CML is used here to investigated the complete synchronization of the CML. Denote λ_1 as the Lyapunov exponent of the logistic map, which is equal to $\ln r$, the Lyapunov exponent spectrum is listed as the following Lyapunov exponents in a descendent order [Li05b]

$$\lambda_j = \begin{cases} \lambda_1 + \frac{1}{2}\ln\left[1 - 2\varepsilon(1-\varepsilon)(1-\cos\frac{\pi j}{L})\right], & j \text{ even}, \\ \lambda_1 + \frac{1}{2}\ln\left\{1 - 2\varepsilon(1-\varepsilon)\left[1-\cos\frac{\pi(j-1)}{L}\right]\right\}, & j \text{ odd}. \end{cases}$$

Since the synchronous chaos of the CML can be observed if $\lambda_2 < 0$ [Ding97], in order to avoid complete synchronization in the CML, the parameters of the CML, r, ε and L, should meet the following inequality

$$r\sqrt{1 - 2\varepsilon(1-\varepsilon)(1-\cos\tfrac{2\pi}{L})} > 0.$$

Avoiding partial synchronization

The cross-covariance is employed here to investigate the synchronization between arbitrary two sites of L sites. The smallest-region partial synchronization in the CML can be guaranteed by a close-to-zero cross-covariance between arbitrary two PRNSs of L PRNSs. The cross-covariance between PRNS_i and PRNS_j of length N is defined as [Xiao96]

$$C_{ij}(\tau) = \hat{C}_{ij}(\tau) / \hat{C}_{ij}(0),$$

$$\hat{C}_{ij}(\tau) = \frac{1}{N}\sum_{n=1}^{N}(x_n^i - \overline{x}_n^i)(x_n^j - \overline{x}_n^j),$$

$$\overline{x}_n^i = \frac{1}{N}\sum_{n=1}^{N} x_n^i, \overline{x}_n^j = \frac{1}{N}\sum_{n=1}^{N} x_n^j, |\tau| = 0,1,\ldots,N-1$$

The parameters of the PRBG, r, ε and L, may have effects on the cross-covariance. For the symmetric configuration of the CML (1), it is reasonable to analyze only the cross-covariance between arbitrary two PRNSs generated from the CML with various parameters. To analyze the effect of each parameter, the maximum of the cross-covariance, i.e., the maximum of $C_{ij}(\tau) \mid \tau \mid = 0,1,\ldots,N-1$, among two arbitrary PRNSs of the PRBG with one various parameter and other fixed ones is investigated. The length of the PRNSs is set as 10^4. The parameters are assumed here and therethrough in the following way: firstly, increase r from 3.56 to 4 by 0.01 each time to let the logistic map be chaotic, while fix ε as 0.95 and L as 8; then increase ε from 0.01 to 0.99 by 0.02 each time, while fix r as 4 and L as 8; finally, increase L from 8 to 64 by 1 each time with fixing r as 4 and ε as 0.95. Thus, the maximum of the cross-correlation among PRNSs is plot against various parameters of the PRBG from which the PRNSs generated in Fig. 3.

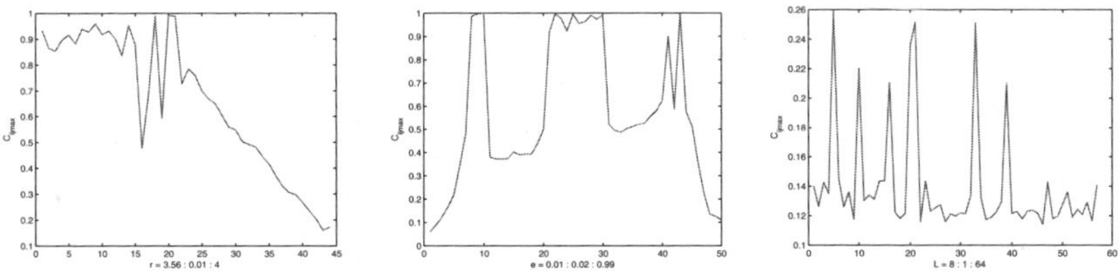

(a) (b) (c)

Fig. 3. The maximum of the cross-correlation among PRNSs of the PRBG with various parameters

It is shown that the cross-correlation is close to zero if r is close to 4 and ε is close to 0 or 1. While, L has no evident influence on the cross-correlation.

To avoid complete and partial synchronization, r should be close to 4 and ε close to 1.

2.4 Cryptographic properties of PRBG

As for a PRBG useful for cryptography, the following cryptographic properties should be satisfied: 1) long period; 2) uniform probability distribution; 3) δ-like auto-correlation; 4) close-to-zero cross-correlation. Since the parameters of the CML may influence the statistical properties of the CML-based PRBG, these properties of the PRBG with various parameters are investigated in this section.

Period

Generally, if chaotic maps are realized in the computers with a finite precision, there exist short periods of the chaotic orbits. However, this problem can be avoided in spatiotemporal chaotic systems. The period of the CML with L sites is derived numerically as around 10^{7L} [Wang04]. Since the CML has a symmetric configuration, the periods of all PRNSs generated from the CML are the same as that of the CML. Therefore, 52 PRBSs generated from the PRNS have the period $O(10^{7L})$, which is long enough for cryptographic applications.

Probability distribution

Probability distributions of 52 PRBSs generated from arbitrary one PRNS, which stand for those of the rest 52(L-1) PRBSs, from the PRBG with various r, ε and L are analyzed by computing a scaled difference ΔP between $P\{(b_n)=0\}$ and $P\{(b_n)=1\}$ of each PRBS, which is actually described as $\Delta P= |N_1-N_0|/(N/2)$ (N_1 and N_0, N, are the number of "*1*" and "*0*", the length of the PRBS, respectively).

ΔP of these 52 PRBSs output from the PRBG with various r is computed and shown in Fig. 4(a), where the x-axis is the index of 52 PRBS, "i", the y-axis denotes the various r and the z-axis stands for ΔP. The lengths of the 52 PRBSs are assumed as 10^4 here and thereafter. It is shown that ΔP of the first four PRBSs are much bigger than zero. Additionally, we set a threshold of ΔP as *0.07* and get the Figs. 4(d) in the following way. If ΔP of the PRBS is smaller than the threshold, the point corresponding to the index of the PRBS and r of the PRBG from which the PRBS output is drawn black, otherwise, the point is drawn white. In the same way, ΔP of 52 PRBSs from the PRBG with various ε and various L are plot in Figs. 4(b), 4(c), 4(e) and 4(f), respectively. It is shown that the 5th-52th PRBSs have uniform probability distribution whatever the parameters are.

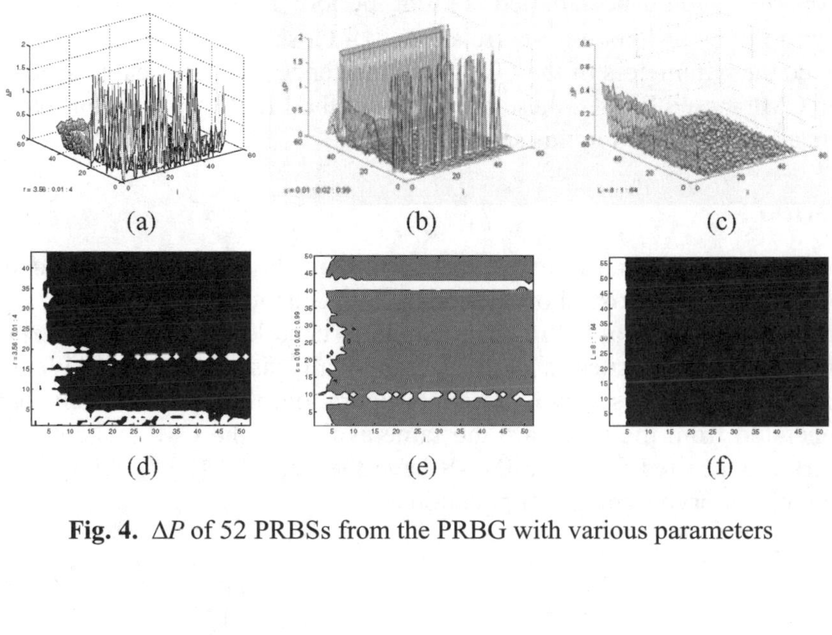

Fig. 4. ΔP of 52 PRBSs from the PRBG with various parameters

Auto-correlation

The auto-covariance, which equals mean-removed auto-correlation, is used here to investigate the auto-correlation, which is a special case of the cross-correlation. The auto-covariance of a PRBS, $\{b_n\}(n=0,\ldots,N-1)$, is described as

$$C_{ii}(\tau) = \hat{C}_{ii}(\tau)/\hat{C}_{ii}(0),$$

$$\hat{C}_{ii}(\tau) = \tfrac{1}{N}\sum_{n=1}^{N}(b_n - \overline{b}_n)(b_{n+|\tau|} - \overline{b}_n),$$

$$\overline{b}_n = \tfrac{1}{N}\sum_{n=1}^{N} b_n, |\tau| = 0,1,\ldots,N-1$$

δ-like auto-correlation means $C_{ii}(0)=1$ and $C_{ii}(\tau)(|\tau|=1,\ldots,N-1)$ or the maximum of $C_{ii}(\tau)(|\tau|=1,\ldots,N-1)$ is close to zero. The maximum auto-correlation of 52 PRBSs generated from arbitrary one PRNS of the PRBG with various parameters are computed and shown in Fig. 5(a)(b)(c). It is indicated that the maximum auto-correlations of the first 4 PRBSs generated from the PRBG with arbitrary parameters and the last 48 PRBSs generated are the PRBGs with some parameters are much far from zero, and the rest are close to zero. Additionally, by setting the threshold of the maximums as 0.055, Fig. 5(d), 5(e) and 5(f) are got in the same way as that of the previous section. It is shown that the maximum auto-correlations of 5th-52th PRBSs output from the PRBG with certain parameter values are larger than 0.055.

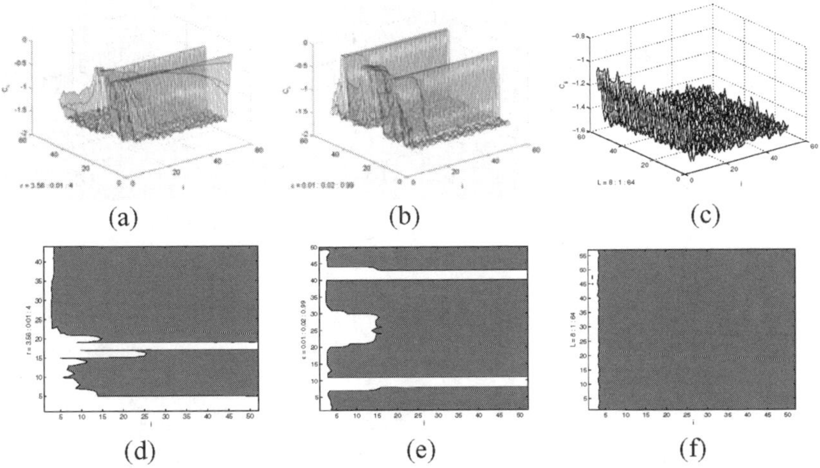

Fig. 5. The maximum auto-correlation of 52 PRBSs from the PRBG with various parameters

Cross-correlation

In order that the multiple PRBSs output from the PRBG can be applied in parallel, the cross-covariance between arbitrary two of them should be close-to-zero. The cross-covariance between a PRBS, $\{a_n\}(n=0,\ldots,N-1)$, and another PRBS, $\{b_n\}(n=0,\ldots,N-1)$, is described as

$$C_{ij}(\tau) = \hat{C}_{ij}(\tau) / \hat{C}_{ij}(0),$$

$$\hat{C}_{ij}(\tau) = \tfrac{1}{N}\sum_{n=1}^{N}(a_n - \overline{a}_n)(b_{n+|\tau|} - \overline{b}_n),$$

$$\overline{a}_n = \tfrac{1}{N}\sum_{n=1}^{N} a_n, \overline{b}_n = \tfrac{1}{N}\sum_{n=1}^{N} b_n, |\tau| = 0,1,\ldots,N-1$$

A maximum cross-covariance, denoted by C_{ij}, between arbitrary two PRBSs output from the PRBG with various parameters are computed, which is shown in Fig. 6(a), 6(b) and 6(c) where z-axis stands for $\mathrm{Lg}(C_{ij})$. By setting the threshold of the maximum cross-covariance as 0.055, Fig. 6(d), 6(e) and 6(f) is obtained. It is indicated that the maximum cross-covariance between some pairs of PRBSs output from the PRBG with certain parameter values are larger than 0.055. It is argued, additionally, that the unsatisfactory cross-correlation among PRNSs, which is demonstrated above, cannot influence the close-to-zero cross-correlation among PRBSs generated from them by the adopted digitization method.

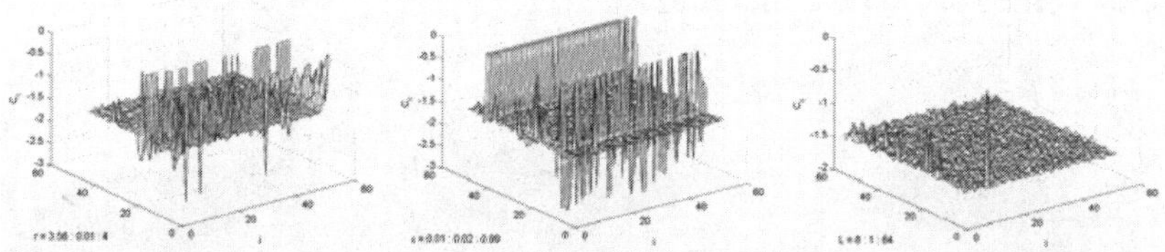

(a) (b) (c)

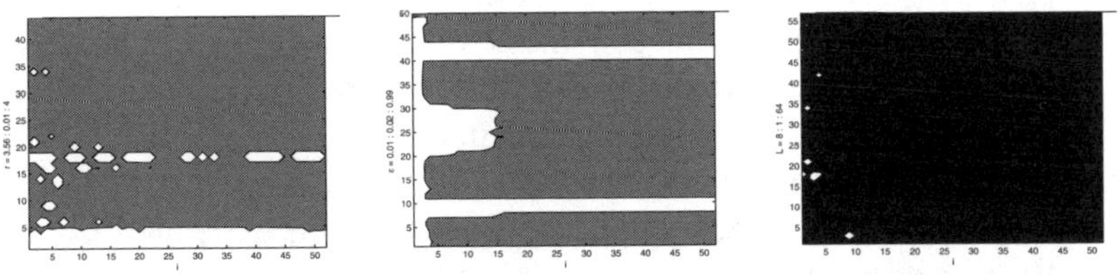

(d) (e) (f)

Fig. 6. The maximum cross-correlation of 52 PRBSs from the PRBG with various parameters

According to the investigation of the statistical properties of the 52 PRBSs generated from the PRBGs with various parameters, the following remarks are drawn,

1. The first 4 PRBSs should be discarded from the PRBG because the statistical properties of them do not meet the cryptographic requirements whatever the parameters of the PRBGs are.
2. The parameter of the local map and the coupling strength of the CML have evident influence on the statistical properties of the CML-based PRBG; whereas, the size of the CML not.
3. The values of r close to 4 and the values of ε close to 1 are preferred because the CMLs with such parameters possess the best cryptographic properties.

3 Internet-based online payments with PRBG as TAN Generator

Since the PRBG with certain parameters has satisfactory cryptographic properties, it can be applied in practice, such as Internet-based online payments, where the PRBG is used as a TAN generator. The TAN generator has been implemented in the system of paysafecard.com in Vienna, which issues electronic cash codes for payments via Internet-based online payments, embedded in a project to automate the logistics process for distribution of so-called cash codes via internet communication. In this section, the application of the PRBG in the internet online payment is described.

3.1 Environment of Internet-based online payments

The cash codes for Internet-based online payments are delivered by the issuer to the distributors who sell them to the customers. The customer uses the cash codes for Internet-based online transactions. The delivery of the cash codes, which are generally printed on a scratch card, has so far meant a partially costly manual process. With the implementation of the proposed PRBG this process shall be automated and expedited in order to meet various requirements, e.g., short reaction times, reduction of manual processes, less production costs, etc.. Changing from physical to electronic delivery of the cash codes brings up more possibilities of distribution channels and models (e.g., PIN-on-demand).

The interface for the distributors to connect to the distribution system is web-based. The accessibility of these processes via public internet necessitates a high level of security to protect transactions representing very high economic values, i.e., electronic money transport. The security concept provides for the following three-step authorization and authentication processes: at connection layer, the link to the system is controlled by digital certificates with client authentication via https protocol; at application layer, the use of the functions is controlled by access data, i.e., username and password; for sending an order, which is an actual high-risk operation, single-use crypto codes (TANs) are utilised to prevent any misuse by unauthorized persons, to authorize the process by means of a digital signature and to prevent malfunctions, such as unintended multiple orders.

Once connected to the interface the distributor can order sets of TANs, download and activate them, which is shown in Fig. 7.

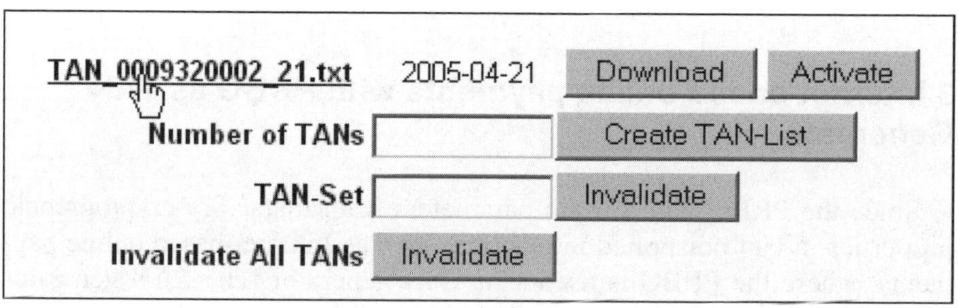

Figure 7: Menu of administrator interface.

The system behind the interface consists of a web server, an application server running a Java application and a database, connected with internal interfaces to an enterprise resource planning (ERP) system and a business warehouse (BW) system. Each distributor is identified in the system by a so-called Distributor-ID (DID). Upon the request of the administrator a set of TANs will be created and stored in the database related to the respective DID where they are available for single-use after activation.

The format of the TANs is 8 characters in HEX format which means that one TAN is a 32 bit word. The administrator gets the TAN set via his web interface as a text file containing some common information, such as DID, set number, date and the TANs listed per line, which are shown in Fig. 8.

```
TAN_0009320002_21.txt
```

```
TAN-file, version 1.0, host TEST1
-----HEADER-----
SETNUMBER=21
QUANTITY=100
DISTRIBUTORID=0009320002
GENERATIONDATE=2005-04-21
-----DATA-----
83354A1F
4A3F2C63
D1F0E25B
[...]
-----END-----
```

Figure 8: A TAN file

The TANs are handed over to the distributor via digital medium where they are saved, e.g., USB stick, or a paper where they are printed on. The distributor has to store them in a safe environment. After acknowledgement of receipt, i.e., handover, legibility, integrity, by the distributor the TANs are activated and then useable. When sending an order for cash codes the distributor has to use a new or unused TAN, which is checked by the system for validity (assigned to the distributor, activated, not invalidated, not used) and then marked as used. Once used, they become invalid.

3.2 TAN generator based on the PRBG

The proposed PRBG with multiple output can be used to generate TANs with 32-bit. The modified PRBG generating TAN, named as a TAN generator, is described as

$$x_{n+1}^i = (1-\varepsilon) f(x_n^i) + \varepsilon f(x_n^{i-1}),$$
$$f(x) = rx(1-x),$$
$$K_n^i = \text{int}\left[x_n^i \times 2^u\right] \text{mod}\, 2^v,$$

where K_n^i, keystream, is used as TANs. Since TANs are desired to be random-like, the parameters of the TAN generator are chosen by considering the following points

1. According to Fig. 4-6, ε and r are assumed as 0.95 and 4, respectively, to obtain good statistical properties of the multiple output of the PRBG. This benefits the statistical properties of the TAN generator.
2. u, v are fixed as 52 and 32, respectively, to exploit the last 32 bits of the binary representation of the state variables of the CML for generating TANs with length of 32-bit. Thus, the first 4 bits which comprise PRBS with bad statistical properties are discarded.
3. L is set as 4 to make the period of the keystream equal to about 2^{100}, which meets the requirement from cryptography to a random-like sequence.

3.3 Properties of TAN generator

To assess the randomness of the TAN generator, its statistical properties, such as order-1 and order-2 probability distributions, probability of runs, auto-correlation, are investigated.

The order-1 probability distribution [Lu04] of the keystream, $\rho(k_n^i)$ $(= \rho(K_n^i)/2^{32})$ is plotted in Fig. 9 (a). The order-2 probability distribution of the keystream, $\rho(k_n^i, k_{n-1}^i)(= \rho(K_n^i, K_{n-1}^i)/2^{32})$ is plotted in Fig. 9 (b). The length of the keystream is 10^6. It is shown that the probability distributions are uniform.

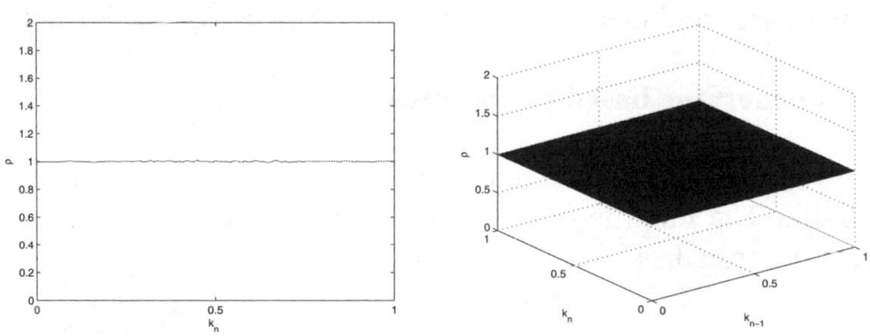

(a) order-1 probability (c) order-2 probability

Fig. 9. Probability of the TAN generator

A run of a binary sequence S is another postulate of randomness, and defined as a subsequence of S consisting consecutive 0's or consecutive 1's which is neither preceded nor succeeded by the same symbol [Menezes97]. The probabilities of 0/1 runs of length $n(n=1,2,\ldots,N)$, denoted by $p_0(n)$ / $p_1(n)$ or $p_{0/1}(n)$ of K^j, are investigated, where $p_{0/1}(n) = R_{0/1}(n)/R_{0/1}$, $R_{0/1} = \sum_{n=1}^{N} R_{0/1}(n)$ with $R_{0/1}(n)$ being the number of 0/1 runs of length n. $p_{0/1}(n)$ vs n is plotted in Fig. 10.

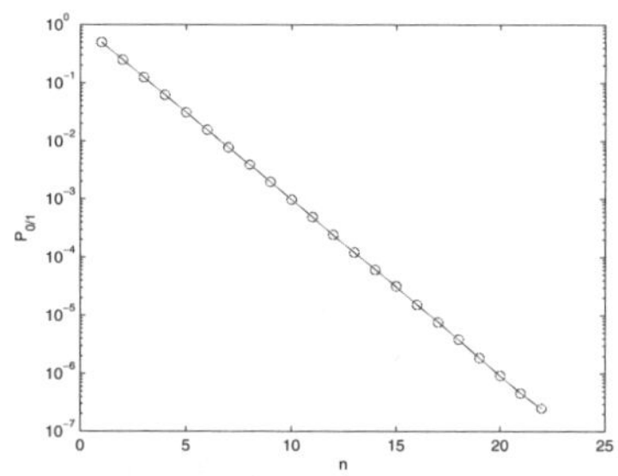

Fig. 10. Probability of the run of the TAN sequence

It is shown that $p_{0/1}(n)$ is in direct proportion to n, which is a characteristic of a truly random binary sequence of an infinite length [Lu04].

Auto-correlation of the TAN sequence is analyzed via auto-covariance of the TAN sequence with length 10^5, which is plotted in Fig. 11. It is shown that the TAN sequence has δ-like auto-correlation.

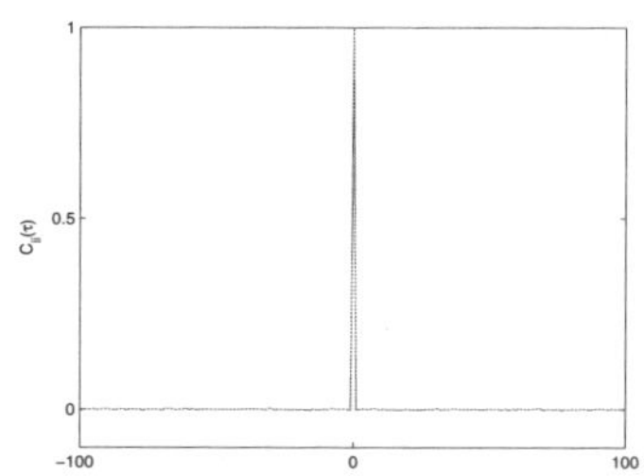

Fig. 11. Auto-correlation of the TAN sequence

According to the analysis above, the TAN generaror has satisfactory random-like statistic properties.

4 keystream are generated at one time, in order to make the parallel operation efficient, they should be independent, which is investigated via the cross-correlation among them. The cross-correlation among any two of them is shown in Fig. 12.

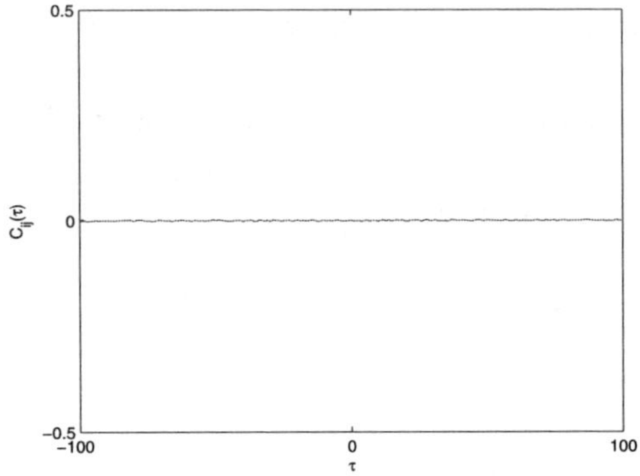

Fig. 12. Cross-correlation of the

It is indicated that the cross-correlations among them are close to zero. Therefore, the generation speed of TANs is up to 450M bits per second in the computer with 1.8GHz CPU and 1.5GB RAM.

The DID, system times, the number of runs of the PRBG and the number of the distributor are used as the initial conditions. To get different TANs for different DID, system times, the number of runs of the PRBG and the number of the distributor, the sensitivity of the TANs to the initial conditions is desirable. The difference function between two different TAN sequences generated from two different initial conditions is used here to investigate the sensitivity. The difference function is described as

$$d(j,\Delta x_0^t)=\frac{1}{T}\sum_{n=1}^{T}|\,k_n'^j-k_n^j\,|,$$

$$k_n'^j=\frac{K_n'^j}{2^{32}},k_n^j=\frac{K_n^j}{2^{32}},$$

$$t,j=1,2,3,4$$

where T is the run times of the PRBG. The difference function vs Δx_0^1 with T=10^5 is plot in Fig. 13.

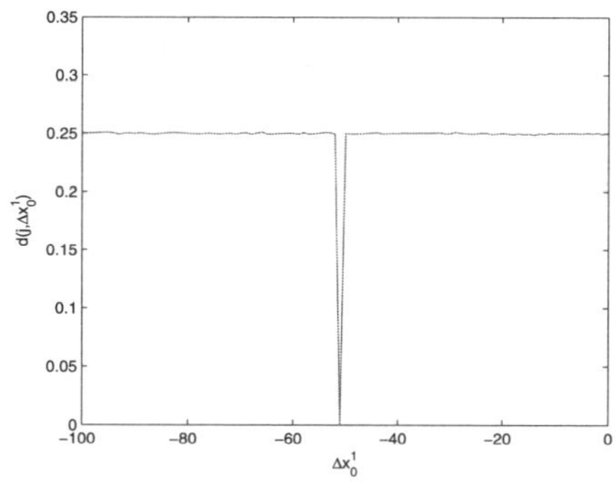

Fig. 13. Sensitivity of the TAN generator to initial conditions

It is shown that the difference function does not equal zero but 0.25 even if Δx_0^1 is an extremely small value 2^{-48}. In other word, the TAN sequence is sensitive to any differences of the initial condition, x_0^1, equal to or larger than 2^{-48}. Similarly, the difference function of $\Delta x_0^t (t = 2,3,4)$ are computed, and it is shown that the TAN sequence is also sensitive to any differences of the initial conditions, $x_0^t (t = 2,3,4)$, equal to or larger than 2^{-48}. Since the initial conditions of the TAN generator, i.e., the DID, system times, the number of runs of the PRBG and the number of the distributor, are 8-bit words. Therefore, the TAN generator with different DID, system times, the number of runs of the generator or the number of the distributor with deviation equal to or larger than 2^{-48} can generate different TANs.

4 Conclustions

A PRBG based on a spatiotemporal chaotic system is proposed. A CML consisting of logistic maps in one-way coupling is adopted as the spatiotemporal chaotic system to construct a PRBG with low computation expense. To make the PRBG efficient, synchronizations in the CML should be avoided. By analyzing the Lypunov exponent spectrum and the cross-correlation among the sites of the CML, it is argued that the parameters of the CML should be in certain ranges in order that synchronizations do not occur. The cryptographic properties, such as period, probability distribution, auto-correlation and cross-correlation, of the PRBG with various parameters are investigated. As a result, the ranges of the parameters within which the PRBG has satisfactory cryptographic properties are determined. Moreover, a TAN generator based on the PRBG is designed for Internet-based online payments. The statistical properties of the TAN generator are investigated and approved as satisfactory. The high sensitivity of the TAN generator to the initial conditions makes it efficient. In addition, its close-to-zero cross-correlation guarantees the high speed of generating TANs. Therefore, the TAN generator based on a spatiotemporal chaotic system provides a low-cost, efficient and secure means for Internet-online payments.

References

L. Kocarev, G. Jakimoski, T. Stojanovski, and Ulrich Parlitz. From chaotic maps to encryption schemes. *In Proc. IEEE Int. Symposium Circuits and Systems* 98: 4(514–517). IEEE, 1998.

G. Alvarez, G. Pastor F. Monotoya, and M. Romera. Chaotic cryptosystems. *In Proc. IEEE Int. Carnahan Conf. Security Technology*, pages 332–338. IEEE, 1999.

L. Kocarev. Chaos-based cryptography: A brief overview. *IEEE Circuits and Systems Maganize*, 1(3):6–21, 2001.

F. Dachselt and W. Schwarz. Chaos and cryptography. *IEEE Trans. Circuits and Systems–I*, 48(12):1498–1509, 2001.

S.J. Li, *Analyses and new designs of digital chaotic ciphers*, Ph.D. thesis, School of Electronics and Information Engineering, Xi'an Jiaotong University, Xi'an, China, 2003, available online at http://www.hooklee.com/pub.html.

D.D. Wheeler. Problems with chaotic cryptosystems. *Cryptologia*, XIII(3):243-250, 1989.

S.J Li, X.Q. Mou, B.L. Yang, Z. Ji and J.H. Zhang, Problems with a Probabilistic Encryption Scheme based on Chaotic Systems, *International Journal of Bifurcation and Chaos*, vol. 13, no. 10, pp. 3063-3077, 2003

R. Forr´e. The H´enon attractor as a keystream generator. *In Advances in Cryptology– EuroCrypt'91, Lecture Notes in Computer Science* 0547, pages 76–81, Berlin, 1991. Spinger-Verlag.

H. Zhou and X.T. Ling. Generating chaotic secure sequences with desired statistical properties and high security. *Int. J. Bifurcation and Chaos*, 7(1):205–213, 1997.

D. D. Wheeler and R. Matthews. Supercomputer investigations of a chaotic encryption algorithm. *Cryptologia*, XV(2):140–151, 1991.

S.J. Li, X.Q. Mou, and Y.L. Cai. Pseudo-random bit generator based on couple chaotic systems and its application in stream-ciphers cryptography. *Progress in Cryptology – INDOCRYPT 2001, Lecture Notes in Computer Science* 2247:316–329. Springer-Verlag, Berlin, 2001.

T. Sang, R.L. Wang and Y.X. Yan. Perturbance-based algorithm to expand cycle length of chaotic key stream. *Electronics Letters*, 34(9):873–874, 1998.

T. Sang, R.L. Wang and Y.X. Yan. Clock-controlled chaotic keystream generators. *Electronics Letters*, 34(20):1932–1934, 1998.

H.P. Lu and S.H. Wang and X.W. Li and G.N. Tang and J.Y. Kuang and W.P. Ye and G. Hu, A new spatiotemporally chaotic cryptosystem and its security and performance analyses, *Chaos*, 14(3):617-629, 2004.

G. Tang, S. Wang, H. LÄu, and G. Hu. Chaos-based cryptograph incorporated with S-box algebraic operation. *Physics Letters A*, 318:388-398, 2003.

K. Kaneko(ed.). *Theory and Application of Coupled Map Lattices*, chapter 1. John Wiley and Sons, 1993.

A. Lasota and M.C. Mackey. *Chaos, Fractals, and Noise: stochastic aspects of dynamics*. Spinger-Verlag, 1997.

A.M. Batista and S.E.S. Pinto and R.L. Viana and S.R. Lopes. Lyapunov spectrum and synchronization of piecewise linear map lattiecs with power-law coupling. *Physical Review E*, 65:056209(9), 2002.

T. Habutsu, Y. Nishio, I. Sasase, and S. Mori. A secret key cryptosystem by iterating a chaotic map. In D.W. Davies, editor, *Lecture Notes in Computer Science*, volume 547, pages127-140, Brighton, UK, April 1991. Advances in Cryptology - EUROCRYPT '91: Workshop on the Theory and Application of Cryptographic Techniques, Proceeding, Springer-Verlag, Heidelberg.

N. Masuda and K. Aihara. Cryptosystems with discretized chaotic maps. *IEEE Trans.Circuits and Systems-I*, 49(1):28-40, 2002.

H. Zhou and X.T. Ling. Problems with the chaotic inverse system encryption approach. *IEEE Trans. Circuits and Systems-I*, 44(3):268-271, 1997.

T. Kohda and A. Tsuneda. Pseudonoise sequence by chaotic nonlinear maps and their correlation properties. *IEICE Trans. Commun.*, E76-B:855-862, 1993.

S.H. Wang, W.R. Liu, H.P. Lu, J.Y. Kuang, and G. Hu. Periodicity of chaotic trajectories in realizations of finite computer precisions and its implication in chaos communications. *International Journal of Modern Physics. B*, 18(17-19):2617-2622, 2004.

M.Z. Ding and W.M. Yang. Stability of synchronous chaos and on-off intermittency in coupled map lattices. *Physical Review E*, 6(4):4009-4016, 1997.

J. Xiao, G. Hu, and Z. Qu. Synchronization of spatiotemporal chaos and its application to multichannel spread spectrum communication. *Phys. Rev. Lett.*, 77:4162-4165, 1996.

P. Li, Z. Li, W.A. Halang and G.R. Chen, A Multiple Pseudorandom-Bit Generator Based on a Spatiotemporal Chaotic Map, *Phys. Lett. A*, 349:.467-473, 2006.

P. Li, Z. Li, W.A. Halang and G.R. Chen, Li-Yorke chaos in a spatiotemporal chaotic system, *Chaos, Solitons, and Fractals*, in press.

A. Menezes, P.V. Oorschot, and S. Vanstone. *Handbook of Applied Cryptography*. CRC Press, 1997.

Computer Hardware Devices in Efficient E-Servicing: Case Study of Disk Scheduling by Soft Computing

A.B. Patki[1], Tapasya Patki[2], Swati Khurana[2], Revati Patki[3], and Aditi Kapoor[2]

[1]Department of Information Technology (DIT), Government of India, 6 CGO Complex, Lodhi Road, New Delhi - 110003, India, apatki@mit.gov.in,

[2]Research Trainees at DIT, pursuing B. Tech at MSIT, GGSIP Univ. New Delhi, India,

[3]Research Trainee at DIT, pursuing B.E. at HVPMCOE, Amravati Univ., Amravati, India

Abstract

With the growing use of networked solutions and services in the Information Society, many citizen-oriented services are planned not only in utility or e-governance sectors, but also in several other walks of life. The formulation of next generation e-Service infrastructure needs to be addressed as it forms the basis for defining the successful functioning of the e-Service Sector, which is likely to proliferate surpassing all global boundaries. The roles, responsibilities and governance of the organizations are bond to rely greatly on the ancillary modules of this sector. Furthermore, companies may find that well-defined products and product portfolios are a single most important determinant. Thus, large scale computing vis-à-vis large scope computing workload definitions for delivering efficient e-Services, are gaining significance in the e-Service Sector. The infrastructure functions' success depends on mature computer hardware, which is adaptive enough to cater to the needs of the future e-Service product technologies. This paper discusses the characteristic requirements of computer hardware in the context of e-Service scenario and brings out the satisfiable solutions for increasing the interactive throughput. Application and usage of Fuzzy Logic to e-Services encompassing disk scheduling has been presented.

A.B. Patki et al.: *Computer Hardware Devices in Efficient E-Servicing: Case Study of Disk Scheduling by Soft Computing*, Studies in Computational Intelligence (SCI) **37**, 687–704 (2007)
www.springerlink.com

1 Introduction

With the availability of e-governance and similar e-Services, in the current decade of the 21st century, a larger cross section of end-user expectations is gaining momentum for 'acceptable' level of response time. Thus, it is not only the small scientific community using supercomputers (say for weather forecasting or identical compute bound numerical processing) but also a citizen remotely working in a registrar of marriages office who needs to verify marriage registration certificates using digital watermarks through nationwide grid computing infrastructure. While connected world has already delivered a revolution in terms of human-computer interface, web services are poised to take us into another realm entirely.

E-Service intelligence is incorporated using Web services technology for deploying automated interactions amongst heterogeneous distributed applications. Web services are becoming more and more popular in both the industry and academic research.

The recent advent of powerful personal computers and workstations has created a demand for higher capability in magnetic rigid disk drives and other peripheral storage devices in the context of capacity as well as performance enhancement. Further, with the introduction of grid computing, capacity and performance of disks are increasingly drawing attention of manufacturers of the mass-storage media. In the past, the emphasis was on providing high-performance CPUs on almost every computing platform to support new applications ranging from multimedia to consumer credit card and point of sale interfaces. A survey of literature in the past two decades indicates that the trend of high performance CPUs through architecture as well as silicon process improvement has been the focus of hardware manufacturers. Thus, for CPUs we have Reduced Instruction Set Computing (RISC) architecture and CISC (Complex Instruction Set Computer) architecture and sub micron VLSI fabrication which gave a face-lift to information and communication technology (ICT) for intelligent e-Services on mass scale. However, similar developments have not been witnessed in mass-storage device generation for delivering efficient e-Services in networked community; even though enhanced capacity Hard Disk Drives (HDDs) have been produced to meet data warehousing requirements and support grid computing. Further, in the present era, where e-Services are gaining momentum, data storage as well as database management in the real-time web environment pose a serious challenge for developers and researchers

(Neema et al. 1999). Not only the context sensitive web search engines will be integrated into ICT specifications but also cognitive off loading and Machine Intelligence Quotient (MIQ) will be the routine parameters for selecting ICT applications for e-Service sector.

In the present era of Soft Computing, a strong need to shift towards a paradigm of 'thinking' devices is felt. The devices are the real interfaces between the user and the processing unit of the computer, and thus are the user's 'eyes' to the networked world. The conceptual distinction of workload definition for large-scale computing scenario like supercomputer applications vis-à-vis large-scope computing catering primarily the cognitive offloading in a web-based networked society, is to be reckoned for building successful next generation e-Services infrastructure. It is here that the computer hardware devices come into the picture and need to be addressed for their deployment. In this context, we can provide the following taxonomy of computer hardware devices:

a. Input-Output (I/O) Devices

b. Storage Devices

The I/O devices encompass a large number of devices that are the interface between the user and the computer system. The peripheral and interactive devices come under this category. Mass-storage devices, on the other hand handle the passive secondary memory of a computer system. The disk is the most fundamental device in this regard.

In creating e-Services infrastructure, considerable importance needs to be given for hardware support systems and disks draw significant attention especially in *intelligent setup institutions*, housing e-Service establishments, providing its end-users with solutions from entry to the utility points through smart cards to secure exits through smart escort systems. Thus, the era of supercomputers that was for the scientific community exploring space applications reappears in a new incarnation for providing e-Services in the Information Society. Scientific community focused on high-speed processors and parallel processing to enhance CPU throughput (Deitel 1984, Henzinger et al 2003). The present e-Service scenario needs to focus on increasing the I/O throughput with the usage of intelligent devices. In this paper, we consider secondary mass-storage devices as a base for application of soft computing techniques for providing intelligent e-services. This approach can be extended to other I/O devices as well.

2 Emerging Trends in E-Services

E-service intelligence encompasses a range of issues regarding managing the next-generation IT infrastructure. A decade into the challenging transition to distributed computing indicates that the infrastructure groups are managing client-server and web-centered architectures with growing authority. E-Services are assuming new meaning in the evolving information society. A web service is not merely software like application that can be published, located and invoked over the web via browsers or other client software. Now-a- days, high performance processors, storage units, and the networks ensure that infrastructure elements rarely need hand tuning to meet the requirements of applications. Web services are beginning to fundamentally change the way in which business is conducted. In a world increasingly connected by networks, potential to integrate conventional value chain modules, are assuming new dimensions. Large IT organizations support thousands of applications, hundreds of physical sites, and millions of end users. All three of these elements, are critical drivers of infrastructural demands. These can be categorized as follows:

a. Applications require servers and storage
b. Websites need network connectivity
c. Users want access to desktops, notebooks, PDAs etc.

Computation intensive applications like pricing and risk management and transaction processing applications like program trading emerge as new issues in the e-Service sector as compared to stand-alone in-house applications.

In the context of providing modern e-services, situations arise where a user's service request cannot be met with a single available web site or a portal and calls for a composite solution in the form of combination of available services. This aspect relies heavily on the context in which the composition and execution of web services occur. Context can be perceived as interaction amongst various resources supporting database clustering. The usefulness and usability of these databases for providing e-services intelligence depend how quickly data can be retrieved. A serious concern is about the number of I/Os required in response to a query. The time to access a randomly chosen page on hard disk requires about 10 ms (Silberschatz et al 2003), which is several orders of magnitude slower as compared to semiconductor memory. With the introduction of 'thin-client' architecture the disk capacity enhancement and up gradation at server side are critical performance characteristics. Ubiquitous computing poses additional demands on the secondary storage benchmarking parameters. E-Service is in its transition phase emerging from the shadow of Information Technology and is poised for independent identity. E-Service industry sector is a strong force driving and shaping future of information society. However, for the success of intelligent e-Service environment, the relevant problems including the modeling, composition, verification and monitoring of the e-Services need to be attended. Prior to the research on these problems, we have to know what kind of services actually exist and on the other hand, from the academic research point of view, we have to figure out the shortcomings and limitations of the current web-service models (Motwani et al 2003). Database servers are occupying considerable importance in providing web services. In this section, we briefly discuss the data base aspects of relevance as a preamble to the disk storage topic of the paper.

Efficient and good physical clustering of data on disk is necessary and the issues of relevance are whether the entire database is clustered or only partial clustering be introduced. This is an important consideration in providing intelligent e-service as dynamic clustering demands are generated and re-clustering the entire database is costlier. Buffering, clustering, indexing, parallelism techniques have been resorted in the past to reduce the performance penalty between primary (main) and secondary (disc) memory system Conventional database clustering is based on grouping together objects in the database using similarity criteria along attribute clustering and record clustering. While attribute clustering is of immense importance for e-service intelligence, the record clustering is of less consequence. Most record clustering techniques deploy statistical analysis (Silberschatz et al 2003, Ullman et al 2002). Attribute clustering incorporates attributes of relation that are divided into groups based on their affinity. Clusters consisting of smaller records, dealing with a subset of attributes from the relation result in better performance as fewer pages from disc is accessed to process transactions. This calls for a methodology of query analysis frequency for providing e-service intelligently. In order to facilitate such provisions, symbolic objects that are extension of classical data types are considered. In conventional data sets, the objects are "individualized" where as in symbolic data sets, they are more "unified' by means of fuzzy relationships. Each symbolic object will be of type mixed features consisting of quantitative and qualitative values. A fuzzy logic frequency analysis to cluster the data base attributes on the lines of FUZOS (Patki et al 1996, 1997) is helpful for dynamic clustering. It uses the notion of clustering; and has become inadequate due to modifications in query patterns.

3 Disk Hardware Technology

In order to get a feel and appreciate the need of introducing soft computing techniques for improvising the HDD technology, this section provides a brief review of secondary storage devices.

A hard disk drive provides bulk of secondary storage for modern computer systems. For rigid and Winchester disk technology, an aluminum base coated with magnetic material has been provided. It is used to record data in the form of zeros and ones as tiny spots of magnetization. Fig. 1 depicts a disk assembly.

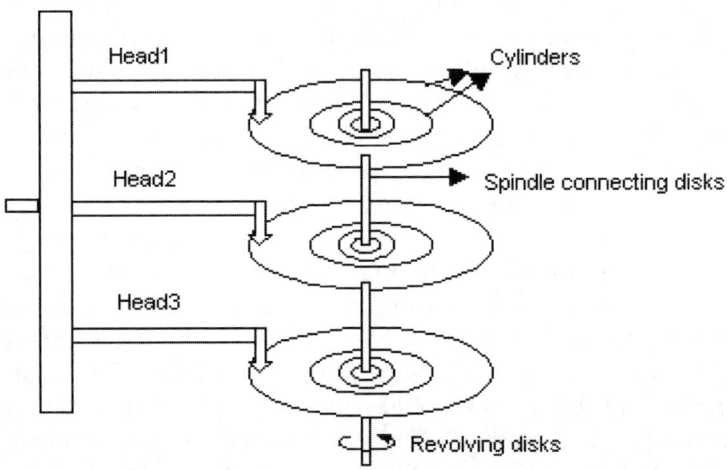

Fig. 1. A typical disk assembly

Normally both sides of *platter* are used for recording and there is a read/write head on each side, which gives the current position of the disk arm that can move radially in towards the axle or out towards the edges of the platters. When in use, the disk arm spins rapidly and the read/write head slides along the surface. All the heads are attached to same arm and are equidistant from the center of their platters.

Modern disk drives are addressed as large one-dimensional array of logical blocks, where block is the smallest unit of transfer. These are then mapped onto sectors of a disk sequentially. This mapping is used to convert logical address to physical address that consists of a cylinder number, a track number within that cylinder, and a sector number within that track. A drive is made ready by loading the heads on the surface by a linear motor or a rotary stepper motor. Once loaded, the heads fly over the disk surface with a few micro-inch air-gap. In a Direct Access Storage Device (DASD), the access to a certain data block is implemented by straight access to a cylinder followed by a serial search for the data within a track in the selected cylinder. The there basic commands executed by a disk drive controller are Seek, Read and Write. *Disk bandwidth* is the total number of bytes transferred, divided by the total time taken from the first request for service till the last request.

To read or write a bit of data on the disk, a head has to be right over the spot where the data is stored. This may require three operations, giving rise to four kinds of delay:

- The head has to be moved to the correct distance from the center of the disk. This movement is called *seeking* and involves physically moving the arm in or out. Because the arm has mass (inertia), it must be accelerated and decelerated.
- The disk has to rotate until the correct spot is under the selected disk. Since the disk is constantly spinning, all that the drive has to do is wait for the correct spot to come around.
- Finally, the actual data has to be transferred. On a read operation, the data is usually transferred to a RAM buffer in the device and then copied, by DMA, to the computer's main memory. Similarly, on write, the data is transferred by DMA to a buffer in the disk, and then copied onto the surface of a platter.

When a process requests two cases can occur:

- If the disk is idle, the request is processed immediately.
- If the disk is performing some other task, then the request has to be added to the input queue and wait till the disk is free. This total time delay for the process is called **access time**. It has four components:
    - The *overhead* of getting into and out of the OS, and the time the OS spends fiddling with queues, etc.

- The *queuing time* spent waiting for the disk to become available.
- The *latency* spent waiting for the disk to get the right track and sector.
- The *transfer time* spent actually reading or writing the data.

Redundant Array of Independent Disks (RAID) is an industry accepted standardized scheme for multiple disk database design. RAID scheme consists of seven levels that describe different design architectures sharing following three common characteristics:

(i) RAID is as set of physical drives viewed by the OS as a single drive

(ii) Data are distributed across the physical drives of an array

(iii) Redundant disk capacity is used to store parity information, which generates data recoverability in case of disk failure

RAID methodology replaces large capacity disk drives with multiple smaller capacity drives and distinguishes data in such a way as to enable simultaneous access to data from multiple drives, thereby improving I/O performance & allowing easier incremental increases in the capacity.

In RAID strategy data are striped across the available disks. All of the user and system data are viewed as being stored on a logical disk. The disk is divided into strips; these strips may be physical blocks or sectors. The strips are mapped round robin to consecutive array members. A set of logically consecutive strips that maps exactly one strip to each array member is referred to as stripe. In an n-disk array, the first n logical strips are physically stored as the first strip on each of the n disks, forming the first stripe and likewise the entire mapping is carried out by array management software between logical and physical disk place.. This results in an advantage in that if a single I/O request consists of multiple logically contiguous strips, then up to n strips for that request can be handled in parallel, greatly reducing the I/O transfer. Out of the seven layer of RAID architecture, in RAID 0, the user and system data are distributed across all of the disks in the array. This is advantageous over a single large disk. If two different I/O requests are pending for two different blocks of data, then there is a fair chance that the requested blocks are on different disks. Thus, two requests can be issued in parallel, reducing the I/O queuing time. The performance is influenced by the strip size. If the strip size is relatively large, so that a single I/O request only involves a single disk access, then multiple waiting I/O requests can be handled in parallel, reducing the queuing time for each request. Disk striping can improve data transfer rate when the strip size is small compared to the I/O request size.

In RAID levels 2 through 6, some form of parity calculation is used to introduce redundancy, whereas in RAID 1, redundancy is achieved by duplicating all the data (i.e. Mirroring). For multiple mirrored web sites providing reliable e-Service, at the outset, RAID 1 level may appear to be a simple solution but it is very expensive, as it requires twice the disk space of the logical disk that it supports (Deitel, 1984, Friedman 1996, Pal Chaudhuri 2003). In e-service supporting heavy transaction oriented application environment, RAID 1 configuration also provides high I/O request rates if the bulk of the requests are reads (and very few are write request) e.g. Banking passbook balance enquiring Servers, Credit card verification Servers. This is however, still recommended for disaster management and recovery systems supporting e-Services in the case of natural disasters as well as for critical infrastructure protection providing e-services in the corporate sector. A classification methodology using fuzzy clustering, without a prior knowledge about data has potential to analyze large volume of remotely sensed data. Usage of fuzzy alpha-cut technique helps in minimizing memory requirements and computation time. But the disk input/output is still a constraint and in the next section, we discuss disk-scheduling methodology. Interpretation of reduced data set as symbolic object helps in the disaster management situations in quickly providing continuity of e-services fuzzy partition coefficient is typically used to find the number of clusters present in the data. A fuzzy partition coefficient indicates the average relative amount of membership sharing done between pairs of fuzzy subsets.

In the subsequent sections, we briefly introduce disk-scheduling practices currently in vogue and then describe fuzzy logic based soft computing trends for disk scheduling.

4 Disk Scheduling

Operating Systems need to provide a wide range of functionality to applications to allow them to control various aspects of I/O devices, while optimizing I/O for maximum concurrency. A serial port device has a simple device controller in the form of a simplc chip but for complex I/O devices like the hard disk, built-in controllers are provided by disk manufacturers and OS developers have very little scope to modify the behavior of such proprietary controllers. The main processor communicates with device controller by reading and writing bit patterns in the controller registers. Disk scheduling is an important issue for providing a quick response time in e-Service query processing. Operating System, being the interface between the user and the hardware, has the responsibility to use hardware efficiently. One crucial branch of this function is to decrease the access time and allow for better bandwidth through disk scheduling. Disk scheduling is used when processes running on the machine have multiple requests for data from the disks and a particular order is needed to access them most efficiently.

There are six existing algorithms for disk scheduling at present (Deitel 1984, Silberschatz 2003). Table 1 summarizes these algorithms.

No.	Algorithm details	Advantages	Disadvantages
1	**FCFS**: This is the first come first serve algorithm in which requests are processed in the order in which they are brought in the input queue.	No reordering. No starvation. Good for priority queues. Easy to program.	Poor performance.
2	**SSTF**: This is shortest seek time first algorithm. It chooses the job with minimum seek time, by selecting the pending request closest to current head position.	Better than FCFS in terms of seek time.	Starvation possible as a request from a remote side of a disk may never be handled. Switching directions increase seek time. It is not optimal.

3	**SCAN**: In this method, the read / write head goes from one end of disk to the other until it reaches the end of that side, servicing all the requests it encounters. The direction is reversed in the next iteration.	It overcomes the problem of starvation. It has fewer fluctuations.	The head goes to the end of the disk even if there is no request.
4	**CSCAN**: In this method also the head travels in one direction first till it reaches the end, servicing requests. After that, it immediately returns to the beginning of the track and treats cylinders as circular lists that traverse from final cylinder to first.	It provides more uniform wait time.	The head still traverses to the end of the disk unnecessarily. If a request is added later on in the path that the arm has traversed then it has to wait till the end.
5	**LOOK**: It is somewhat similar to SCAN but in this method, the read/write head moves from one end of queue to the other instead of that of disk. The arm goes as far as the final request before reversing direction.	It overcomes the drawback of the head going to other end when there is no request there. So it provides uniform wait time.	If a request is added at a later stage which beyond the final element in the current queue then it has more waiting time.
6	**CLOOK**: It is an optimization over LOOK method. It also treats the input queue as a circular list, thus inheriting features of CSCAN.	Considered as advantageous over LOOK and SCAN	

Table 1. Comparisons of Various Disk Scheduling Algorithms

We can thus infer that for the conventional disk schedulers:

- FCFS is best option in case of priority queues and also when the list is sorted. Thus, the factor to be taken into consideration is priority.
- SSTF depends upon the difference between the various arm requests. It has the potential to provide minimal seek time.
- CLOOK is based on the distribution on the requests in comparison to the initial request.
- Out of SCAN, CSAN, LOOK and CLOOK, all except CLOOK can be ignored, LOOK being an optimization over SCAN and CLOOK over LOOK.

5 Applying Fuzzy Logic to e-Services

The theory of binary logic is based on the assumption of crisp membership of an element to a certain set. An element x, thus, either *belongs to* (i.e. has a membership value of 1) or *doesn't belong to* (i.e. has a membership value of 0) a particular set X. Conventional logic systems can be extended to encompass normalized values in the range of (0,1), thus introducing the notion of *partial membership* of an element to a particular set. Such a logic system that allows us to represent variables in a natural form with infinite degrees of membership is referred to as Fuzzy Logic System (Klir 1995, Klir et al 1999, Zadeh 1968). The variable in a fuzzy system is generally described linguistically prior to its mathematical description, as it is more important to visualize a problem in totality to devise a practical solution.

A fuzzy set F, on a collection of objects, X, is a mapping

$$\mu_F(x): X \rightarrow [0,a]$$

Here, $\mu_F(x)$ indicates the extent to which $x \in X$ has the attribute F, thus it is the *membership function*. In general, we use a normalized fuzzy domain set, for which

$$a = \sup \mu_F(x) = 1$$

The membership function can be generated with the help of mathematical equations. Typically, it can be in trapezoidal form, triangular form or in the form of S or π- curve.

The *support* of a fuzzy set, F, S(F) is the crisp set of all $x \in X$ such that $\mu(x) > 0$.

The three basic logical operations of intersection, union and complementation can be performed on fuzzy sets as well.

i). The membership $\mu_C(x)$ of the *intersection* $C = A \cap B$ satisfies for each $x \in X$,

$$\mu_C(x) = \min\{\mu_A(x), \mu_B(x)\}$$

ii). The membership $\mu_C(x)$ of the *union* $C = A \cup B$ satisfies for each $x \in X$,

$$\mu_C(x) = \max\{\mu_A(x), \mu_B(x)\}$$

iii). The membership $\mu_C(x)$ of the *complementation* $C = \bar{A}$ satisfies for each $x \in X$,

$$\mu_C(x) = 1 - \mu_A(x)$$

At the highest level of e-Service abstraction, we need to analyse the following things. The terms in the brackets are fuzzy terms and the membership functions for these needs to be generated a-priori. This is useful for Segmenting User demand.

- Frequency of the query (Frequent, Occasional)
- Socio-economic User application Class (Socially active, Socially passive, Socially deprived)
- Urgency of the query (Normal, Immediate)

Dynamically we generate a fuzzy priority curve by overlapping the previously generated fuzzy membership functions (Type-I) of the above three requirements by examining the degree of correlation amongst these. This is not analogous to Type II fuzzy set (Zadeh 1968). This curve is the core basis of the operation of the fuzzy controller in response to a particular demand.

A simple cognitive-level algorithm has been presented to implement such a system in this paper. The authors are at the nascent phase of software development based on this algorithm, and thus there may be many modifications in it in future.

5.1 Algorithm:

The steps are summarized as follows:

1. The shape of the current demand for infrastructure services as well as how that demand will most likely evolve needs to be identified using fuzzy hedges.
2. Categorize demands into segments (e.g. Uptime, throughput, scalability)
3. Clusterize the requirements

   Ex 1: A pharmaceutical manufacturer may find that applications can be in the following clusters

   - Round the clock support-
   - Off-line availability
   - Applications that must scale up to thousand of users and handle transactions efficiently

   Ex 2: A wholesale Banking Environment can be in the following clusters

   - Computation Intensive analysis
   - Applications with little or no down time: Funds Transfer
4. Generation of Fuzzy membership function for Fuzzy variables
5. Defining Operational Fuzzy hedges- e.g. Very, more-or-less using concentration and dilation operation
6. Fuzzy Rule Generation and Fuzzy Inference (including defuzzification as applicable)

The RAID mapping software discussed in section 3, will use the fuzzy inferences to define the strip size as small, medium and large instead of a standard fixed size currently adopted. The membership function for small, medium and large is user programmable. This is required to provide dynamic facility to improve e-Service performance depending on the user feedback. The periodicity could be monthly. Thus using the fuzzy logic based algorithm for RAID strip mapping software, we have introduced fuzzy control at RAID level.

6 Impediments in Deploying Soft-Computing for Computer Hardware Devices

The present storage device market does not permit the OS to completely control the disk scheduling operations. The major reasons, as stated earlier in this chapter, include the 'fixing' of a particular algorithm to a HDD for it's scheduling and not providing alternatives for the OS to access or modify it. Traditionally, all manufacturers have used proprietary interfaces and techniques to monitor and control the hardware. Soft computing techniques are not popular among the manufacturers. Also, research laboratories, which have the capability of practically implementing the fuzzy disk scheduling algorithms, need extra financial support, thus not many hardware manufacturers find it lucrative to establish such laboratories in their production units. However, the grid-computing and e-service sector is likely to influence this scenario. We describe disk scheduling using fuzzy logic controller. Typically a fuzzy controller receives inputs in the form of an error signal e(t) and change in error c(t) analogous to a parameter-time varying Proportional Derivative (PD) controller. A typical fuzzy logic controller (Fig. 2) has three building blocks, viz. a fuzzifier, a fuzzy inference/decision reasoner and a defuzzifier.

Usually a designer formulates the fuzzy control rules using linguistic hedges. Since we have to accommodate error as well as change in error, the normalized universe of discourse [-1, +1] is used to describe negative and positive terms along with big, medium and small to give membership functions, as shown in Table 2.

Negative	Positive
Negative Big (NB)	Positive Big (PB)
Negative Medium (NM)	Positive Medium (PM)
Negative Small (NS)	Positive Small (PS)

Table 2. Membership Functions

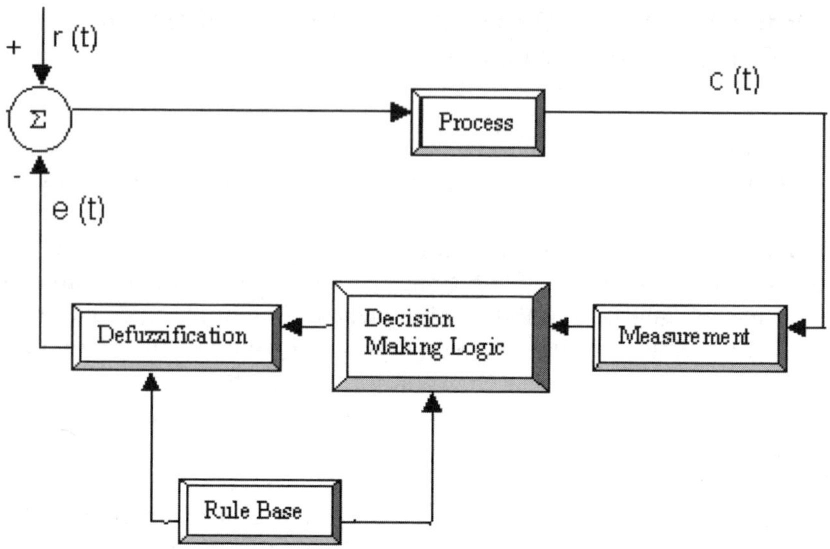

Fig. 2. Block Diagram of a Fuzzy Controller

In addition, a zero-membership function is also defined. It has been seen that triangular membership function representation is more appropriate and popularly used based on empirical knowledge and engineering experience in the design in fuzzy controllers. Unlike the conventional PD controller, a fuzzy controller uses a linguistic context that leads to better performance and throughput (Chen 1990, Mendel 1995).

The aim of various methods described in previous sections is to minimize the seek time. Seek time is the time taken by the head reading the data to move from head data cylinder to target cylinder. It is limited by the performance of actuator moving and also the control method adopted. This can be further improved by implementing fuzzy logic controller along with a method for correcting changes in actuator coil resistance. Limited amount of research work has been reported in this regard in the open literature. One of these is the *bang-bang seek method* (Yoshida et al 1992). In this method, the relation between target distance and the bang-bang acceleration time has to be determined in advance, and the result stored in a table with the unevenness in the driving force being omitted. The interpolation of this table gives the bang-bang acceleration time corresponding to any given target distance. The following relation gives the relation:

$$X = KT^2$$

Fuzzy logic is used to determine the amount of correction corresponding to the extent to which the region of acceleration contains ranges of force unevenness. For this, time management is implemented to switch between acceleration and deceleration, and fuzzy logic is employed both for estimation of the bang-bang acceleration time and for correction of unevenness of the actuator force at different positions. This method decreases the seek time by up to 30%. Amount of time lost due to rotational delay can be minimized. Rotational ordering in which requests are ordered in each track so that first sector of second track is next request to be served does this.

7 Conclusions

E-Service intelligence encompasses a range of issues regarding managing the next-generation IT infrastructure. A crucial contribution of this paper is to deploy fuzzy logic techniques for disk scheduling where the algorithm can be implemented in hardware electronics. It is expected that the Soft computing methodologies suggested in this paper would ensure increased reliability, availability and serviceability that are essential for e-Service sector.

References

Deitel H. M. (1984), An Introduction to Operating Systems, Addison-Wesley Reading (Mass.)

Friedman M. (1996), RAID keeps going and going and.., IEEE Spectrum, Vo. 33., No. 4

Henzinger M. R., Motwani R., Silverstein C. (2003), Challenges in Web Search Engines, Proceedings of the 18th Int'l Joint conf on AI, pp 1573-1579

Klir G. J. (1995), Fuzzy Logic, IEEE Potentials, October- November, pp. 10-15

Klir C.J., Folger T. (1999), Fuzzy Sets, Uncertainty and Information, Prentice Hall India

Lee Chuen Chen (1990), Fuzzy Logic in Control Systems: Fuzzy Logic Controller, Part II, IEEE Tran. On Systems, Man and Cybernetics, Vol. 20, pp. 419-435

Mendel J. M. (1995), Fuzzy Logic Systems for Engineering: A Tutorial, Proceedings of the IEEE, Vol. 83, No. 3, 345-376

Motwani R. et al. (2003), Query Processing, Resource Management, and approximation in a Data Stream Management System, Proceedings of the CDIR Conference

Neema F., Waid D. (1999), Data Storage Trends, UNIX Review, 17 (7),

Pal Chaudhuri P. (2003), Computer Organization and Design, Prentice Hall of India

Patki A.B (1996).-Fuzzy Logic Based Hardware: Some Experiences, Proceedings of First International Discourse on Fuzzy Logic And The Management of Complexity (FLAMOC '96), January, 15-18, Sydney, Australia, Vol.3, pp 247-251

Patki A.B., Raghunathan G.V., Khurshid A.(1997), -FUZOS –Fuzzy Operating System support for Information Technology, Proceedings of Second On-line World Conference On Soft Computing In Engineering, Design And Manufacturing, June 23-27, Cranfield University, UK.

Silberschatz A., Korth H.F., Sudarshan S. (1998), Database System Concepts, Mc-Graw Hill, 1998

Silberschatz, Galvin, and Gagne (2003), Operating System Concepts, 6th edition, John Wiley & sons Inc. Sections 13.1,13.2

Ullman J. D., Widom J. (2002), A first course in database systems Second Edition, Prentice Hall, NJ

Yoshida S., Wakabayashi N. (1992), Fuzzy Logic Controller for Rigid Disk Drive, *IEEE Control Systems*,

Zadeh L.A. (1968), Fuzzy Algorithm, *Information & Control*, Vol. 12, pp 94-102

SUBJECT INDEX

Printing: Krips bv, Meppel
Binding: Stürtz, Würzburg